Vitamin D
Oxidative Stress, Immunity, and Aging

OXIDATIVE STRESS AND DISEASE

Series Editors

LESTER PACKER, PhD
ENRIQUE CADENAS, MD, PhD

UNIVERSITY OF SOUTHERN CALIFORNIA SCHOOL OF PHARMACY
LOS ANGELES, CALIFORNIA

20. Reversionaries ... ...
    Shelley's Trustee.

27. Creative Rules and ... Mutter ... Subordinate ... ... and
    Lesser Estate.

22. Valuable Easements ... Unresolved Abbot O'Keefe, edited by Major Vernon.
    ... ...

23. Common Ghosts in Births ... ... by Polar ... ... Problems ...
    Revised ... ...

23 ... ...
    Major ... ...

... ...

24 ... ...
    Human Psychoanalysis in a Single Organ.

25. Adjustment and Information ... ... and ... Revised
26. Recent Medical Biomedical and Clinical Aspects. Second Edition, edited by
    ... F. Barnes and S... Warren, Ltd.

29 ... ... and Reversed Systems by ... Dr. Philip ... and ... , edited by
    Al ... F. B. Major and ... Crozier.

... ...
    ... ...

Vera's Disease ... ... and Accumulated by ... and Gerrard.

Vitamin D

Oxidative Stress, Immunity, and Aging

Edited by
Adrian F. Gombart

CRC Press
Taylor & Francis Group
Boca Raton London New York

CRC Press is an imprint of the
Taylor & Francis Group, an **informa** business

CRC Press
Taylor & Francis Group
6000 Broken Sound Parkway NW, Suite 300
Boca Raton, FL 33487-2742

First issued in paperback 2016

Version Date: 20120627

ISBN 13: 978-1-138-19944-6 (pbk)
ISBN 13: 978-1-4398-5020-6 (hbk)

Library of Congress Cataloging-in-Publication Data

Vitamin D : oxidative stress, immunity, and aging / editor, Adrian F. Gombart.
    p. cm. -- (Oxidative stress and disease ; 31)
   "A CRC title."
   Includes bibliographical references and index.
   ISBN 978-1-4398-5020-6 (alk. paper)
   1. Vitamin D in human nutrition. 2. Vitamin D deficiency. I. Gombart, Adrian F.

QP772.V53.V596 2013
615.3'28--dc23
                                                2012019020

Visit the Taylor & Francis Web site at
http://www.taylorandfrancis.com

and the CRC Press Web site at
http://www.crcpress.com

Contents

SECTION III   Immunity and Disease

SECTION IV   Aging

Series Preface

Through evolution, oxygen—itself a free radical—was chosen as the terminal electron acceptor for respiration; hence, the formation of oxygen-derived free radicals is a consequence of aerobic metabolism. These oxygen-derived radicals are involved in oxidative damage to cell components inherent in several pathophysiological situations. Conversely, cells convene antioxidant mechanisms to counteract the effects of oxidants by either a highly specific manner (e.g., superoxide dismutases) or in a less specific manner (e.g., through small molecules such as glutathione, vitamin E, vitamin C, etc.). Oxidative stress—as classically defined—entails an imbalance between oxidants and antioxidants. However, the same free radicals that are generated during oxidative stress are produced during normal metabolism and, as a corollary, are involved in both human health and disease by virtue of their involvement in the regulation of signal transduction and gene expression, activation of receptors and nuclear transcription factors, antimicrobial and cytotoxic actions of immune system cells, as well as aging and age-related degenerative diseases.

In recent years, the research disciplines interested in oxidative stress have increased our knowledge of the importance of the cell redox status and the recognition of oxidative stress as a process with implications for many pathophysiological states. From this multidisciplinary and interdisciplinary interest in oxidative stress emerges a concept that attests to the vast consequences of the complex and dynamic interplay of oxidants and antioxidants in cellular and tissue settings. Consequently, our view of oxidative stress is growing in scope and new future directions. Likewise, the term *reactive oxygen species*, adopted at some stage in order to highlight nonradical/radical oxidants, now fails to reflect the rich variety of other species in free radical biology and medicine, encompassing nitrogen-, sulfur-, oxygen-, and carbon-centered radicals. These reactive species are involved in the redox regulation of cell functions, and, as a corollary, oxidative stress is increasingly viewed as a major upstream component in cell signaling cascades involved in inflammatory responses, stimulation of cell adhesion molecules, and chemoattractant production and as an early component in age-related neurodegenerative disorders such as Alzheimer's, Parkinson's, and Huntington's diseases and amyotrophic lateral sclerosis. Hydrogen peroxide is probably the most important redox signaling molecule that, among others, can activate NFκB, Nrf2, and other universal transcription factors and is involved in the redox regulation of insulin and MAPK signaling. These pleiotropic effects of hydrogen peroxide are largely accounted for by changes in the thiol/disulfide status of the cell, an important determinant of the cell's redox status with clear involvement in adaptation, proliferation, differentiation, apoptosis, and necrosis.

The identification of oxidants in regulation of redox cell signaling and gene expression was a significant breakthrough in the field of oxidative stress: the classical definition of oxidative stress as an imbalance between the production of oxidants and the occurrence of antioxidant defenses now seems to provide a limited depiction of oxidative stress, but it emphasizes the significance of cell redox status. Because individual signaling and control events occur through discrete redox pathways rather than through global balances, a new definition of oxidative stress was advanced by Dean P. Jones as a disruption of redox signaling and control that recognizes the occurrence of compartmentalized cellular redox circuits. These concepts are anticipated to serve as platforms for the development of tissue-specific therapeutics tailored to discrete, compartmentalized redox circuits. This, in essence, dictates principles of drug development-guided knowledge of mechanisms of oxidative stress. Hence, successful interventions will take advantage of new knowledge of compartmentalized redox control and free radical scavenging.

Virtually all diseases thus far examined involve free radicals. In most cases, free radicals are secondary to the disease process, but in some instances, causality is established by free radicals.

Thus, there is a delicate balance between oxidant and antioxidants in health and diseases. Their proper balance is essential for ensuring healthy aging. Compelling support for the involvement of free radicals in disease development originates from epidemiological studies showing that enhanced antioxidant status is associated with reduced risk of several diseases. Of great significance is the role that micronutrients play in modulation of cell signaling: this establishes a strong linking of diet and health and disease centered on the abilities of micronutrients to regulate redox cell signaling and modify gene expression.

Oxidative stress is an underlying factor in health and disease. In this series of books, the importance of oxidative stress and diseases associated with organ systems is highlighted by exploring the scientific evidence and clinical applications of this knowledge. The series is intended for researchers in the basic biomedical sciences and clinicians. The potential of such knowledge for healthy aging and disease prevention warrants further knowledge about how oxidants and antioxidants modulate cell and tissue function.

This series volume provides an update on the essential role of vitamin D in optimal human nutritional requirements and in health and disease. Further insights are provided into the protective role of vitamin D in autoimmune, infectious, and inflammatory diseases, oxygen metabolism, and healthy aging. In recent years, many new tissue-specific actions of vitamin D have been recognized. This area of research is rapidly expanding. We congratulate Adrian Gombart, the editor-in-chief, for providing this valuable addition to the science of vitamin D.

Enrique Cadenas
Lester Packer

Preface

Vitamin D insufficiency/deficiency is a worldwide public health problem in both developed and developing countries. Vitamin D promotes and maintains healthy bones and teeth, but with the near eradication of rickets in the early part of the twentieth century by the fortification of foods, chronic insufficiency has gone largely unrecognized. However, with (1) the current reemergence of nutritional rickets among infants, (2) recent evidence that low levels of circulating vitamin D are associated with increased risk and mortality from cancer, and (3) evidence of the potential beneficial effects of vitamin D on multiple sclerosis, rheumatoid arthritis, diabetes, cardiovascular disease, aging, and microbial infections, there has been renewed interest in this vitamin. In 2007, *Time* magazine cited the benefits of vitamin D in its list of Top 10 Medical Breakthroughs, and vitamin D continues to receive extensive coverage in both the scientific and lay press. Although extensive research has been done on vitamin D, the molecular and cellular mechanisms responsible for its many benefits have not been fully elucidated. In November 2010, the Institute of Medicine issued new Dietary Reference Intakes for vitamin D. The committee provided an exhaustive review of studies on potential health outcomes and found that the evidence supported only a role for vitamin D in bone health but not in other health conditions. However, it is clear that there is a preponderance of studies demonstrating a biologically plausible role for vitamin D in more than bone health. The focus of this book is on the role of vitamin D in oxidation, immunity, and aging. These topics are receiving increased attention in the research community but have not been covered extensively in past books.

This book is a state-of-the-art compilation of recent information and is divided into four sections that cover studies of vitamin D in regulating numerous aspects of health, disease, and aging. Section I, Vitamin D: An Overview, reviews literature regarding vitamin D and its genomic and nongenomic effects, the role of therapeutic analogs in treating disease, and the production of vitamin D by the body. The areas reviewed in this section provide a background for the subsequent sections. The chapters in Section II, Oxidative Stress, cover the role that vitamin D plays in modulating oxidative stress, with areas of focus including cancer, stress-mediated diseases, photoprotection of the skin, and energy metabolism. The chapters in Section III, Immunity and Disease, review evidence for vitamin D in regulating the immune response and the importance that it plays in protecting against autoimmune, infectious, and inflammatory diseases. The final section, Aging, focuses on the role that the vitamin D pathway plays in the regulation of the aging process, including aspects of oxidative stress, senescence, and mortality. Furthermore, its role in protection against cardiovascular disease and nervous system disorders is discussed.

The editor greatly appreciates the time and effort given by all of the authors, who have shared their knowledge in writing outstanding, timely, and scholarly chapters on the biological actions of vitamin D. The chapters in this book represent important contributions toward understanding the mechanisms by which vitamin D promotes health. In addition, the information presented greatly increases awareness of the importance that vitamin D plays during development, at birth, and throughout the aging process. It will serve as a valuable reference to researchers in academia, nutrition, medicine, and industry.

Editor

Adrian F. Gombart earned his PhD in microbiology from the University of Washington. For many years, he was an assistant and associate professor at Cedars-Sinai Medical Center and the David Geffen School of Medicine at the University of California, Los Angeles, in the Department of Biomedical Sciences and Division of Hematology and Oncology. He is currently a principal investigator in the Linus Pauling Institute and an associate professor in the Department of Biochemistry and Biophysics, Oregon State University. Dr. Gombart is a member of the American Society for Hematology and the Society for Leukocyte Biology.

Dr. Gombart's research interests focus on the role of vitamin D in the innate immune response against infection. His laboratory studies the regulation of antimicrobial peptide gene expression by vitamin D and other nutritional compounds. Dr. Gombart was recognized for his contributions to the field of vitamin D and immunity with a Young Investigator Award from the Vitamin D Workshop in 2005.

Contributors

Nasimul Ahsan
Department of Medicine
Fayetteville VA Medical Center
Fayetteville, North Carolina

Cédric Annweiler
Department of Neuroscience
Angers University Hospital
University of Angers
Angers, France

and

Division of Geriatric Medicine
Department of Medicine
The University of Western Ontario
London, Ontario, Canada

Jim Bartley
Department of Surgery
University of Auckland
Auckland, New Zealand

Shane Batie
Division of Mathematical and Natural Sciences
Arizona State University
Phoenix, Arizona

Antje Bruckbauer
NuSirt Sciences, Inc.
Knoxville, Tennessee

Carlos A. Camargo, Jr.
Department of Emergency Medicine
Massachusetts General Hospital
Harvard Medical School
Boston, Massachusetts

Moray J. Campbell
Department of Pharmacology and Therapeutics
Roswell Park Cancer Institute
Buffalo, New York

Inpyo Choi
Cell Therapy Research Center
Korea Research Institute of Bioscience and
    Biotechnology
and
Department of Functional Genomics
University of Science and Technology
Yuseong, Republic of Korea

Ryan Forster
Basic Medical Sciences
University of Arizona College of Medicine
Phoenix, Arizona

Clare Gordon-Thomson
School of Medical Science (Physiology) and
    Bosch Institute
Sydney Medical School
University of Sydney
Sydney, Australia

Mark R. Haussler
Basic Medical Sciences
University of Arizona College of Medicine
Phoenix, Arizona

Colleen E. Hayes
Department of Biochemistry
College of Agricultural and Life Sciences
University of Wisconsin–Madison
Madison, Wisconsin

Martin Hewison
Department of Orthopaedic Surgery and
    Molecular Biology Institute
David Geffen School of Medicine
University of California, Los Angeles
Los Angeles, California

Eun-Kyeong Jo
Department of Microbiology
Infection Signaling Network Research Center
College of Medicine
    Chungnam National University
Daejeon, South Korea

Glenville Jones
Department of Biomedical and Molecular
    Sciences
and
Department of Medicine
Queen's University
Kingston, Ontario, Canada

Haiyoung Jung
Cell Therapy Research Center
Korea Research Institute of Bioscience and
    Biotechnology
Yuseong, Republic of Korea

Peter W. Jurutka
Division of Mathematical and Natural Sciences
Arizona State University
and
Basic Medical Sciences
University of Arizona College of Medicine
Phoenix, Arizona

Ramesh C. Khanal
Department of Nutrition and Food Sciences
and
Center for Integrated BioSystems
Utah State University
Logan, Utah

Dong Oh Kim
Cell Therapy Research Center
Korea Research Institute of Bioscience and
    Biotechnology
and
Department of Functional Genomics
University of Science and Technology
Yuseong, Republic of Korea

Beate Lanske
Department of Developmental Biology
Harvard School of Dental Medicine
Boston, Massachusetts

Jamie Lee
Division of Mathematical and Natural Sciences
Arizona State University
Phoenix, Arizona

Yi-Fen Lee
Department of Urology
University of Rochester Medical Center
Rochester, New York

Pamela L. Lutsey
Division of Epidemiology and Community
    Health
School of Public Health
University of Minnesota
Minneapolis, Minnesota

Rebecca S. Mason
School of Medical Science (Physiology) and
    Bosch Institute
Sydney Medical School
University of Sydney
Sydney, Australia

Robert L. Modlin
Division of Dermatology
Department of Microbiology, Immunology, and
    Molecular Genetics
University of California, Los Angeles
Los Angeles, California

Corwin D. Nelson
Department of Biochemistry
College of Agricultural and Life Sciences
University of Wisconsin–Madison
Madison, Wisconsin

Ilka Nemere
Department of Nutrition and Food Sciences
and
Center for Integrated BioSystems
Utah State University
Logan, Utah

Young Jun Park
Cell Therapy Research Center
Korea Research Institute of Bioscience and
    Biotechnology
Yuseong, Republic of Korea

Syed K. Rafi
Department of Anatomy and Cell Biology
University of Kansas Medical Center
Kansas City, Kansas

Mohammed S. Razzaque
Department of Oral Medicine, Infection and
    Immunity
Harvard School of Dental Medicine
Boston, Massachusetts

Jared P. Reis
Division of Cardiovascular Sciences
National Heart, Lung, and Blood Institute
National Institutes of Health
Bethesda, Maryland

Dong-Min Shin
Department of Microbiology
Infection Signaling Network Research Center
College of Medicine
    Chungnam National University
Daejeon, South Korea

Prashant K. Singh
Department of Pharmacology and Therapeutics
Roswell Park Cancer Institute
Buffalo, New York

Justin A. Spanier
Department of Biochemistry
College of Agricultural and Life Sciences
University of Wisconsin–Madison
Madison, Wisconsin

Hyun-Woo Suh
Cell Therapy Research Center
Korea Research Institute of Bioscience and
    Biotechnology
Yuseong, Republic of Korea

Jun Sun
Department of Biochemistry
Rush University
Chicago, Illinois

Huei-Ju Ting
Department of Urology
University of Rochester Medical Center
Rochester, New York

Wannit Tongkao-on
School of Medical Science (Physiology) and
    Bosch Institute
Sydney Medical School
University of Sydney
Sydney, Australia

Pentti Tuohimaa
Medical School
University of Tampere
and
Centre for Laboratory Medicine
Tampere University Hospital
Tampere, Finland

G. Kerr Whitfield
Basic Medical Sciences
University of Arizona College of Medicine
Phoenix, Arizona

Michael B. Zemel
Department of Nutrition
The University of Tennessee
Knoxville, Tennessee

Section I

Vitamin D: An Overview

1 Vitamin D: A Fountain of Youth in Gene Regulation

Peter W. Jurutka, G. Kerr Whitfield, Ryan Forster,
Shane Batie, Jamie Lee, and Mark R. Haussler

CONTENTS

1.1 VITAMIN D BIOACTIVATION AND ITS ENDOCRINE/MINERAL FEEDBACK CONTROL

The hormonal precursor and parent compound, vitamin D_3, either can be obtained in the diet or formed from 7-dehydrocholesterol in skin (epidermis) via a nonenzymatic, UV light-dependent reaction (Figure 1.1). Vitamin D_3 is then transported to the liver, where it is hydroxylated at the C-25 position of the side chain to produce 25-hydroxyvitamin D_3 (25D), which is the major circulating form of vitamin D_3. The final step in the production of the hormonal form occurs mainly, but not exclusively, in the kidney via a tightly regulated 1α-hydroxylation reaction (Figure 1.1). The cytochrome P450-containing (CYP) enzymes that catalyze 25- and 1α-hydroxylations are microsomal CYP2R1 (Cheng et al. 2003) and mitochondrial CYP27B1, respectively. As depicted in Figure 1.1, 1,25-dihydroxyvitamin D_3 (1,25D) circulates, bound to plasma vitamin D binding protein, to various target tissues to exert its endocrine actions, which are mediated by the vitamin D receptor (VDR). Many of the long-recognized functions of 1,25D involve the regulation of calcium and phosphate metabolism, raising the blood levels of these ions to facilitate bone mineralization, as well as activating bone resorption as part of the remodeling cycle (Haussler et al. 2010).

In addition to affecting bone mineral homeostasis by functioning at the small intestine and bone, 1,25D also acts through its VDR mediator to influence a number of other cell types. These extraosseous actions of 1,25D-VDR include differentiation of certain cells in skin (Bikle and Pillai 1993) and in the immune system (Mora et al. 2008; Figure 1.1). Interestingly, the skin and the immune system are now recognized as extrarenal sites of CYP27B1 action to produce 1,25D locally for autocrine and paracrine effects (Adams et al. 1985; Omdahl et al. 2002), creating intracrine systems (Figure 1.1) for extraosseous 1,25D-VDR functions distinct from the renal endocrine actions of 1,25D-VDR on the small intestine and skeleton. Apparently, higher circulating 25D levels are required for optimal intracrine actions of 1,25D (Figure 1.1). This insight stems from the importance of attaining

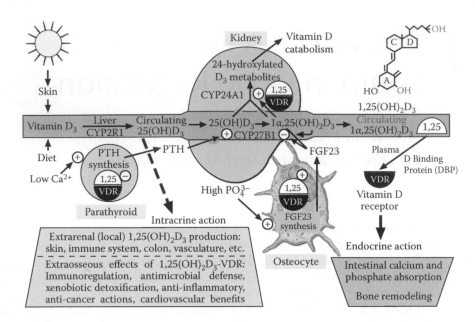

FIGURE 1.1   Vitamin D acquisition, regulation of metabolic activation/catabolism, and receptor-mediated endocrine and intracrine actions of the 1,25D hormone.

adequate levels of circulating 25D revealed in a multitude of epidemiologic associations between low 25D levels and chronic disease, coupled with statistically significant protection against a host of pathologies by much higher circulating 25D (Bikle 2009). Thus, as depicted schematically in Figure 1.1, locally produced 1,25D appears to be capable of benefitting the vasculature to reduce the risk of heart attack and stroke, controlling the adaptive immune system to lower the incidence of autoimmune disease while boosting the innate immune system to fight infection, effecting xenobiotic detoxification, and exerting antiinflammatory and anticancer pressure on epithelial cells prone to fatal malignancies.

The parathyroid gland also expresses VDR (Brumbaugh et al. 1975; Wecksler et al. 1977), and when the receptor is liganded with 1,25D, parathyroid hormone (PTH) synthesis is suppressed by a direct action on gene transcription (De May et al. 1992). This negative feedback loop, which curtails the stimulation of CYP27B1 by PTH under low calcium conditions (Figure 1.1), serves to limit the bone-resorbing effects of PTH in anticipation of 1,25D-mediated increases in both intestinal calcium absorption and bone resorption, thus preventing hypercalcemia. More recent understanding of the homeostatic control of phosphate has emerged, emanating originally from characterization of unsolved familial hypo- or hyperphosphatemic disorders, which we now know are caused by deranged levels of bone-derived FGF23 (Bergwitz and Juppner 2010). In short, FGF23 has materialized as a dramatic new phosphate regulator and a second phosphaturic hormone after PTH. We (Kolek et al. 2005) and others (Quarles 2008) proved that 1,25D induces the release of FGF23 from bone, specifically from osteocytes of the osteoblastic lineage (Figure 1.1), which is a process that is independently stimulated by high circulating phosphate levels (Figure 1.1). Thus, in a striking and elegant example of biological symmetry, PTH is repressed by 1,25D and calcium, whereas FGF23 is induced by 1,25D and phosphate, protecting mammals against hypercalcemia and hyperphosphatemia, respectively, either of which can elicit ectopic calcification.

As illustrated in Figure 1.1, using the kidney as an example, an important mechanism by which the 1,25D-VDR-mediated endocrine or intracrine signal is terminated in all target cells is the catalytic action of CYP24A1, which is an enzyme that initiates the process of 1,25D catabolism (St-Arnaud 2010). The CYP24A1 gene is transcriptionally activated by 1,25D (Ohyama et al. 1994a;

Zierold et al. 1994b), as well as by FGF23 (Figure 1.1). In addition, the 1α-hydroxylase (1α-OHase) CYP27B1 gene is repressed by FGF23 and 1,25D, with the latter regulation affected by epigenetic demethylation (Kim et al. 2009) in a short negative feedback loop to limit the production of 1,25D (Murayama et al. 1999). Therefore, the vitamin D endocrine system is elegantly governed by feedback controls of vitamin D bioactivation, which interpret bone mineral ion status, and via feedforward induction of 1,25D catabolism to prevent the pathologies of hypervitaminosis D. The vitamin D intracrine system, in contrast, appears to be dependent more on the availability of ample 25D substrate to generate local 1,25D to lower the risk of chronic diseases of the epithelial (e.g., skin and colon), immune, cardiovascular, and possibly nervous systems.

1.2 BIOLOGICAL RESPONSES TO THE 1,25D HORMONE ARE WIDESPREAD

We are the first group to propose that 1,25D, either alone or in combination with bona fide antiaging gene products like klotho, is a mediator of healthful aging (Haussler et al. 2010). This hypothesis could explain epidemiologic/association studies that suggest that 25D, in the newly recognized optimal high-normal range in blood, confers a lower risk of virtually all of the fatal diseases of aging such as heart attack, stroke, and cancers. Thus, as depicted in Figure 1.2, the endocrine/intracrine actions of vitamin D/klotho protect the vascular system, as well as epithelial cells subject to fatal cancers (Mordan-McCombs et al. 2007; breast, prostate, colon, and skin), the immune system (Liu et al. 2006a; Mora et al. 2008; Raghuwanshi et al. 2008; Figure 1.1), and possibly the central nervous system (Keisala et al. 2009). With respect to the chronic diseases of aging, it is now becoming clear that the kidney represents the nexus of control, and we contend that klotho is a third renal hormone after 1,25D and erythropoietin. Therefore, in this chapter, we emphasize the importance of renal health during aging and unveil the kidney as a focal point for the prevention of chronic diseases (Figure 1.2).

Figure 1.2 summarizes and integrates the endocrine regulation and actions of 1,25D-VDR in exerting the bone mineral homeostatic, immune, cardiovascular, and anticancer effects. While the predominant action of 1,25D-VDR is promoting intestinal calcium and phosphate absorption to prevent osteopenia, the signal for this function is PTH reacting to low calcium, whereas the hormonal agent that feedback controls these events to preclude ectopic calcification is FGF23. In this fashion, bone resorption and mineralization remain coupled to protect the integrity of the mineralized skeleton. FGF23 functions acutely in concert with PTH and chronically when PTH is suppressed by calcium and 1,25D (Figure 1.2). In fact, FGF23 directly represses PTH (Ben-Dov et al. 2007; Figure 1.2) to abolish the activation of CYP27B1 by PTH while, at the same time, appropriating from PTH the role of phosphate elimination. Like PTH, FGF23 inhibits renal Npt2a and Npt2c to elicit phosphaturia (Shimada et al. 2004a; Figure 1.2). In contrast to PTH, which is downregulated by 1,25D in parathyroid glands, FGF23 is upregulated by 1,25D in osteocytes (Bergwitz and Juppner 2010; Kolek et al. 2005; Liu et al. 2006b), which is a major source of FGF23 endocrine production by bone. As illustrated in Figure 1.2 (lower right), hyperphosphatemia enhances osteocytic FGF23 production independently of 1,25D, rendering FGF23 the perfect phosphaturic counter-1,25D hormone because it inhibits renal phosphate reabsorption and 1,25D biosynthesis via inhibition of CYP27B1 while enhancing 1,25D degradation by inducing CYP24A1 in all tissues (Figure 1.2). In this fashion, FGF23 allows osteocytes to communicate with the kidney to govern circulating 1,25D, as well as phosphate levels, thereby preventing excess 1,25D function and hyperphosphatemia. FGF23 signals via renal FGFR/klotho coreceptors to promulgate phosphaturia (Razzaque 2009), repress CYP27B1 (Perwad et al. 2007), and induce CYP24A1 (Razzaque 2009; Shimada et al. 2004a; Figure 1.2).

FGF23 regulation is complex and multifactorial, including the suppressive proteins PHEX and dentin matrix acidic phosphoprotein 1 (DMP-1; Figure 1.2). 1,25D represses PHEX expression in UMR-106 osteocyte-like cells, which is in accordance with the induction of FGF23 in that the PHEX suppressor is attenuated to permit maximal induction of FGF23 by 1,25D (Hines et al. 2004). It is conceivable that the mechanism of FGF23 induction by 1,25D is, in part or entirely, a consequence of PHEX repression, yet the PHEX substrate which ultimately regulates FGF23

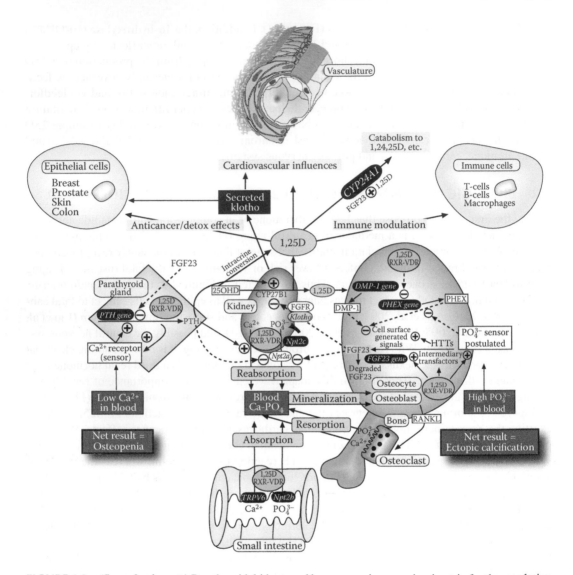

FIGURE 1.2 (See color insert.) Parathyroid, kidney, and bone comprise an endocrine trio for the regulation of phosphate and calcium metabolism to prevent osteopenia/osteoporosis and ectopic calcification (shaded in light blue). Renal hormones 1,25D (shaded in light blue) and klotho (shaded in dark blue) reach beyond bone mineral homeostasis to delay other chronic disorders of aging besides osteoporosis, such as cardiovascular disease, epithelial cell cancers, and autoimmune disease.

transcription is not known. Although DMP-1 is apparently not a PHEX substrate, loss of function mutations in DMP1 cause a phenotype identical to XLH, with excess FGF23 producing hypophosphatemia (Quarles 2008). This suggests that DMP1, like PHEX, normally represses FGF23 expression in osteocytes, although this is inconsistent with the observation (Farrow et al. 2009) that 1,25D induces DMP1 in UMR-106 cells. Fascinatingly, it has been shown recently (Martin et al. 2011) that PHEX and DMP1 regulate FGF23 expression in osteocytes through a common pathway involving FGF receptor (FGFR) signaling, intimating that PHEX and DMP-1 regulate FGF23 expression by impacting an autocrine loop in the osteocyte whereby FGF23 governs its own synthesis (not shown in Figure 1.2). FGF23 synthesis is also governed by high phosphate (Ito et al. 2005), possibly via an undiscovered transcription factor (analogous to signaling through Gq by the calcium sensing receptor in the parathyroid and other tissues) to induce the FGF23 gene. Such factors that play a specific role in transduction of the phosphate signal are herein termed "hyperphosphatemia transducing transfactors" (Figure 1.2). The targeting of these factors will be of great interest to those attempting to modulate FGF23 in patients such as those in renal failure who may benefit from reduced FGF23 secretion (Fukumoto 2010; Juppner et al. 2010). Finally, identification of these factors will also increase our comprehension of the control of FGF23, and we may, for the first time, be able to integrate the 1,25D and phosphate arms of FGF23 regulation with other known osteocyte players such as PHEX and DMP1 (Figure 1.2), as well as renal phosphate transporters, Npt2a/c. Indeed, Demay and colleagues (Miedlich et al. 2010) have recently shown, by ablation of renal phosphate transporter Npt2a, that phosphate is the central regulator of the FGF23 gene and is capable of prevailing over vitamin D because 1,25D fails to induce FGF23 when hypophosphatemia and elevated 1,25D occur concurrently.

Intestinal calcium absorption is mediated by 1,25D-VDR induction of TRPV6 (Barthel et al. 2007; Meyer et al. 2006), which supplies dietary calcium via transport to build the mineralized skeleton (Figure 1.2). Indeed, TRPV6 null mice have 60% decreased intestinal calcium absorption, decreased bone mineral density (BMD), and, strikingly, 20% of animals exhibit alopecia and dermatitis (Bianco et al. 2007) similar to VDR knockout mice (Li et al. 1997). Since the skin phenotype in VDR null mice is not ameliorated by the high-calcium rescue diet (Amling et al. 1999), we speculate that TRPV6 may mediate calcium entry into keratinocytes to elicit differentiation and hair cycling. Because calcium is protective against colon cancer (Garland et al. 1985), while hair plus a full-stratum corneum reduce UV-induced skin damage and cancer, VDR-induced TRPV6 could also function in colon and skin to lower the risk of neoplasia in these two epithelial cell types (Figure 1.2).

Although 1,25D also enhances intestinal phosphate absorption via the induction of Npt2b (Katai et al. 1999), because phosphate is abundant in the diet and constitutively absorbed by the small intestine, the phosphate absorption effect of 1,25D may not be as physiologically important as the profound effect of 1,25D to trigger calcium transport.

In the osteoblast, RANKL constitutes one of the most dramatically 1,25D-upregulated bone genes, the product of which affects 1,25D-VDR-mediated bone resorption through osteoclastogenesis (Figure 1.2). We have shown that RANKL is induced more than 5000-fold by 1,25D in mouse ST-2 stromal cells in culture (Haussler et al. 2010). OPG, which is the soluble decoy receptor for RANKL that tempers its activity, is simultaneously repressed by 86% (Haussler et al. 2010) to amplify the bioeffect of displayed (or secreted) RANKL. Thus, like PTH, 1,25D is a potent bone-resorbing, hypercalcemic hormone, and although chronic excess of either hormone elicits severe osteopenic pathology, physiologic bone remodeling can be argued to strengthen the skeleton. In other words, like a well-mineralized bone, an appropriately remodeled bone is a healthy bone and is less susceptible to fractures and the eventual ravages of senile osteoporosis. Beyond bone, renal 1,25D and, especially, locally generated extrarenal 1,25D benefit the cardiovascular system in which VDR is expressed in endothelial cells, smooth muscle cells, and cardiac myocytes. Finally, kidney- and locally derived 1,25D also influences many cells in the immune system to modulate its functions, as well as to exert anticancer actions in virtually all epithelial cells (Figure 1.2).

1.3 STRUCTURE–FUNCTION OF VDR AND MECHANISMS OF GENE REGULATION

Various domains of the 427 amino acid human VDR are highlighted on a linear schematic of the protein (Figure 1.3a), with the two major functional units being the N-terminal zinc finger DNA binding domain (DBD), and the C-terminal ligand binding (LBD)/heterodimerization domain. To date, the Protein Data Bank (PDB) database contains over 50 x-ray crystal structures for the VDR LBD and four of the DBD bound as a homo- or heterodimer on VDRE DNA sequences. The

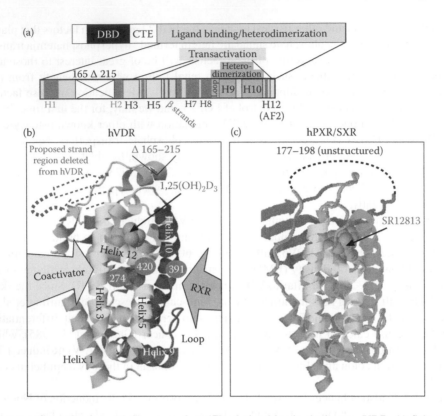

FIGURE 1.3 (See color insert.) Structure–function relationships in the human VDR. (a) Schematic view of human VDR, in which the following subdomains are highlighted: the 3rd, 5th, and 12th helices (H3, H5 and H12, yellow), which have been implicated in binding of coactivators and transactivation; the 9th and 10th helices (H9 and H10) and the loop between helices 8 and 9, which comprise an interface for interaction with the RXR heterodimeric partner; and finally, three beta strands in the VDR crystal structure (Rochel et al. 2000). A section of the VDR LBD/heterodimerization domain from position 165–215 was deleted (a box with crossed lines) by Rochel et al. (2000) in order to facilitate purification and crystallization. (b) X-ray crystal structure of human VDR (residues 118–164 spliced to residues 216–425; Rochel et al. 2000) bound to its natural 1,25D ligand, as viewed in iMol (Martz 2002). Three examples of natural mutations that cause HVDRR are indicated with hVDR residue numbers (274, 391, and 420) and are discussed in the text. (c) Human PXR (or SXR, steroid and xenobiotic receptor) LBD/heterodimerization domain bound to the synthetic ligand, SR12813 (Watkins et al. 2001), created in Protein Explorer to approximate the same view of hVDR in panel B. A contiguous fragment of PXR was used for crystallization (residues 142–431).

original x-ray crystallographic structure of the VDR LBD consisting of 12 α-helices (Rochel et al. 2000) has been updated (PDB 3A78) and now contains 14 α-helices, including two new short segments of three to five amino acids each between residues 149 and 292 of the human receptor. To be consistent with the majority of the literature, in this chapter, we retain the original nomenclature based on 12 α-helices. As shown in Figure 1.3b, the VDR LBD is a sandwichlike structure presenting VDR surfaces for heterodimerization with RXR (predominantly helices (H) 9 and 10 and the loop between helices 8 and 9), as well as for transactivation via interaction with coactivators. Coactivator interfaces in VDR, as shown in Figure 1.3b, consist of portions of helices H3, H5, and H12 (with the last constituting the AF-2 or activation function-2 domain). The LBD domain of human VDR has been cocrystallized with 1,25D (Rochel et al. 2000), as well as with many vitamin D analogs occupying the hydrophobic pocket. One human vitamin D receptor (hVDR) mutant detected in a patient with hereditary hypocalcemic vitamin D–resistant rickets (HVDRR) and highlighted in Figure 1.3b is an arginine 274 to leucine alteration, which leads to a loss in binding of the 1,25D ligand (Whitfield et al. 1996). Arginine 274 directly contacts the 1,25D ligand via hydrogen bonding of the 1α-hydroxyl moiety (Rochel et al. 2000). Moreover, there is a human HVDRR patient harboring a VDR in which arginine 391 is mutated to cysteine (Figure 1.3b), and in vitro experiments have confirmed that RXR heterodimerization by the R391C mutant hVDR is severely impaired (Whitfield et al. 1996). The R391C HVDRR patient, like all those with DNA-binding mutations in hVDR, displayed an alopecic phenotype, indicating that RXR heterodimerization is functionally linked to VDRE binding and that both of these actions of VDR are required for hair cycling. Finally, testing has demonstrated that HVDRR mutant E420K hVDR (Malloy et al. 2002), as well as its synthetic counterpart, E420A (Jurutka et al. 1997), has abrogated transactivation capacity caused by their inability to interact with the comodulator steroid receptor coactivator-1 (SRC-1) and VDR interacting protein 205 (DRIP205). Thus, the three natural point mutations in the hVDR LBD highlighted in Figure 1.3b represent loss of function alterations for each of the three major molecular actions of this domain, namely LBD, heterodimerization, and transactivation.

The availability of a pregnane X receptor (PXR) structure (Watkins et al. 2001) permits interesting comparisons between the LBD/heterodimerization domain of VDR and that of PXR, which is its closest relative in the nuclear receptor superfamily. The overall structures of the domains of these two receptors are nearly superimposable (Rochel et al. 2000, Watkins et al. 2001; compare Figure 1.3b and c). The PXR LBD has a particularly large LBD pocket (1150 Å^3; Watkins et al. 2003); the LBD pocket of VDR, even with the deletion of residues 165–215, is also large (approx. 700 Å^3) compared with other nuclear receptors that have been crystallized [approx. 400 Å^3 (Rochel et al. 2000) but ranging as high as 800 Å^3 for LXRβ; Hoerer et al. 2003]. This suggests that, like PXR and LXR, VDR may be able to accommodate a variety of lipophilic ligands beyond 1,25D. Of additional interest is the fact that the PXR crystal, which was created from a continuous stretch of sequence from positions 142–431 of overexpressed human PXR (Watkins et al. 2001), contains residues that correspond to most of the 165–215 region (or its equivalent) that is absent in the VDR LBD crystals. Thus, the PXR x-ray crystallographic solution can serve as a model for the VDR structure in this region. Of particular note are the additional two β-strands in PXR, for which the corresponding hypothetical positions in VDR are suggested by the dotted outline in Figure 1.3b. Indeed, inclusion of these two β-strands, plus associated turns, in this speculative hVDR LBD structure suggests that VDR may indeed have a very large binding pocket for diverse ligands.

Very recently, the structure of the hVDR DBD and LBD together in the same protein, heterodimerized with full-length RXRα, docked on a VDRE, and occupied with 1,25D plus a single coactivator, has been determined in solution via Small Angle X-ray Scattering and Fluorescence Resonance Energy Transfer techniques (Rochel et al. 2011) and the allosteric communication between the interaction surfaces of the VDR-RXR complex determined by hydrogen-deuterium exchange (Zhang et al. 2011). These new advances render it possible to visualize how the DBD and the LBD/heterodimerization domains are arranged relative to one another and how their binding to ligand, DNA, and coactivators influence one another. The process of gene modulation is best understood for VDR mediation of 1,25D-stimulated transcription, where RXR heterodimerization constitutes an obligatory initial step in the VDR activation pathway. Figure 1.4 illustrates in schematic fashion how the hormonal ligand could be influencing VDR to interact more efficaciously with its heterodimeric partner, with a VDRE, and with coactivators. From experiments with VDR (Bettoun et al. 2003; Jurutka et al. 2001; Thompson et al. 2001), as well as insight from the mode of action of other nuclear receptors (McKenna and O'Malley 2002; Thompson and Kumar 2003; Warnmark et al. 2003), the key event in the allosteric model presented in Figure 1.4 is the binding of a ligand, namely the 1,25D ligand for VDR.

There are several steps that apparently are set in motion by the LBD event. The presence of the 1,25D ligand in the VDR binding pocket results in a dramatic conformational change in the position

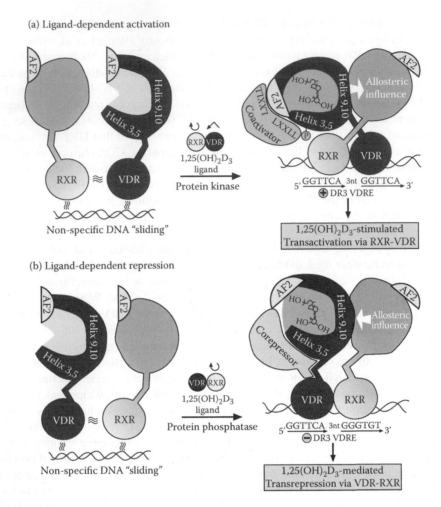

FIGURE 1.4 Proposed mechanisms of gene induction and repression by VDR. (a) Allosteric model of RXR-VDR activation after binding 1,25D and coactivator, phosphorylation, and docking on a high-affinity positive VDRE (mouse osteopontin). See text for explanation. (b) Allosteric model for VDR-RXR inactivation after binding 1,25D and corepressor, dephosphorylation, and docking in reverse polarity on a high-affinity negative VDRE (chicken PTH). See text for explanation.

of helix 12 at the C-terminus of VDR, bringing it to the "closed" position to serve in its AF2 role as part of a platform for coactivator binding (Jurutka et al. 1997; Masuyama et al. 1997; Rachez et al. 2000). The attraction of a coactivator to the helix-3, -5, and -12 platform of liganded VDR likely allosterically stabilizes the VDR-RXR heterodimer on the VDRE and may even assist in triggering strong heterodimerization by inducing the VDR LBD to migrate to the 5′ side of the RXR LBD and, in so doing, rotate the RXR LBD 180° employing the driving force of the ionic and hydrophobic interactions between helices 9 and 10 in hVDR and the corresponding helices in RXR (Figure 1.4a). Therefore, ligand-intensified heterodimerization, VDRE docking, and coactivator recruitment by VDR appear to be functionally inseparable, yet experimentally dissociable, events that occur in concert to effect 1,25D-elicited gene transcription. There is also evidence that DNA binding influences the stability of helix 12 for coactivator binding (Zhang et al. 2011).

The conformational changes allosterically elicited in VDR by the previously described interactions with ligand, RXR, and DNA have the added effect of converting VDR into a more efficient substrate for one or more serine protein kinases (Jurutka et al. 2002). The most prominent phosphorylation is catalyzed by casein kinase II (CK2) on hVDR serine 208 (Hilliard et al. 1994; Jurutka et al. 1993), which is an event that has been shown to potentiate the transcriptional activity of the VDR-RXR heterodimer (Jurutka et al. 1996), likely by enhancing interactions with coactivators such as DRIP205 (Figure 1.4a; Arriagada et al. 2007; Barletta et al. 2002).

Finally, as depicted in Figure 1.4a and supported experimentally (Bettoun et al. 2003; Pathrose et al. 2002; Thompson et al. 2001), the liganding of VDR conformationally influences its RXR heteropartner and appears to cause the AF2 region of RXR to pivot into the "closed" or active position. The RXR member of the heterodimer may now be endowed with the potential to bind an additional coactivator (not shown in Figure 1.4), and the allosteric repositioning of the RXR AF2 appears to greatly reduce the affinity of RXR for its 9-*cis* retinoic acid (9-*cis* RA) ligand. With RXR serving as a subordinate partner, VDR is referred to as a nonpermissive primary receptor within the heterodimer because RXR may not be able to bind 9-*cis* RA when heterodimerized to liganded VDR (Bettoun et al. 2003; Pathrose et al. 2002; Thompson et al. 2001). Alternatively, as depicted in Figure 1.4, we leave open the possibility that, in specific cell contexts with certain promoters containing unique VDREs, the LBD pocket of RXR in the heterodimer is available to occupation by 9-*cis* RA (Zhang et al. 2011) or other RXR ligands such as docosahexaenoic acid (DHA; de Urquiza et al. 2000), rendering the RXR heteropartner capable of ligand occupation and coactivator recruitment for the purpose of synergistic activation by 1,25D and retinoids/fatty acids of genes such as CYP24A1 (Zou et al. 1997).

Ligand-dependent repression of gene transcription by VDR-RXR likely shares some molecular features with induction, but no doubt is more complex mechanistically because it appears to occur via multiple routes. One theme of repression is probably the recruitment of nuclear receptor repressor(s) to alter the architecture of chromatin in the vicinity of the target gene to that of heterochromatin. This restructuring of chromatin would be catalyzed by histone deacetylases and demethylases attracted to the receptor-tethered corepressor. The initial targeting of the repressed gene, as illustrated in Figure 1.4b, is hypothesized to be docking of liganded VDR-RXR on a negative VDRE. In the case of repression, liganded VDR is apparently conformed such that it binds corepressor rather than coactivator. We postulate that the information driving this allosteric transformation of VDR is intrinsic to the negative VDRE DNA sequence (Whitfield et al. 2005). Furthermore, because nonconsensus nucleotides in negative VDREs appear to occur in either or both half-elements, we contend that such base-pair changes may be sufficient to drive RXR-VDR into reverse polarity on the negative VDRE (Figure 1.4b), which is an event that is documented to switch liganded RAR-RXR into a repressor (Kurokawa et al. 1995), albeit on a DR1, instead of its normal DR5 enhancer. Our assumption is that docking in reverse polarity on the negative VDRE transforms the conformation of liganded VDR such that it favors the recruitment of corepressor over coactivator to their overlapping docking sites in helices 3–6. VDR may also be prone to protein phosphatase rather than protein kinase activity in this altered conformation, again favoring corepressor attraction. Evidence for this model is provided by the observation (Jurutka et al. 2002) that okadaic acid, which is a protein phosphatase inhibitor, potentiates transactivation by VDR. A key question is the role of the RXR heteropartner in gene repression by VDR. One possibility is that RXR is simply a "silent" partner in VDRE binding, with negative nucleotides alone in the 5′ half-site allosterically conforming VDR to attract corepressor. However, because the negative cPTH VDR (Figure 1.4b) does not possess nonconsensus variations in its 5′ half-site and it can be converted to a positive VDRE by altering the 3′ terminal bases from GT to CA (Koszewski et al. 1999), we favor a role for the RXR LBD in allosterically locking VDR into a corepressor docking motif (Figure 1.4b). Thus, since nonconsensus nucleotides occur only in the 3′ half-element of the negative chicken PTH VDRE, it is conceivable that the RXR partner receives information from the 3′ half-site nucleotides on which it is docked and transmits a signal to its VDR heteropartner that allosterically alters VDR to attract corepressors. This concept is consistent with the data of Zhang and colleagues obtained using hydrogen-deuterium exchange (Zhang et al. 2011), but more direct experiments will be required to verify the hypothetical model presented in Figure 1.4b.

1.4   VDR BINDS NONVITAMIN D LIGANDS

As implied by Figure 1.3b and c, VDR and PXR share a similar tertiary structure and may thus regulate some of the same genes. Like VDR, PXR utilizes direct repeat-3 (DR3) and everted repeat-6 (ER6) responsive elements in DNA for transcriptional activation (Goodwin et al. 2002). The differential transcriptional effects of VDR and PXR may reside not in the genes that are regulated but rather in the overlapping, yet distinct, tissue expression and ligand profile; for example, PXR does not respond to 1,25D (Kliewer et al. 1998). It should be noted that CAR, which is a closely related detoxification nuclear receptor, is also implicated in CYP regulation (Guo et al. 2003, Honkakoski et al. 2003). While CAR normally binds DR4 elements, this relatively uncharacterized receptor may exhibit some cross-over binding to DR3 elements; conversely, PXR may cross over to DR4 elements (Maglich et al. 2002). There is some preliminary evidence that VDR may also cross over to DR4 elements (Drocourt et al. 2002), as it was reported that VDR can activate transcription from at least three elements that strongly resemble degenerate DR4s (Gill and Christakos 1993; Kitazawa and Kitazawa 2002; Saeki et al. 2008).

Thus, VDR, PXR, and CAR are three nuclear receptors that heterodimerize with RXR to signal detoxification of xenobiotics and overlap somewhat in their target gene repertoires, which are laden with CYPs.

One notable feature of the subfamily of nuclear receptors involved in detoxification is the ability of these receptors to recognize multiple ligands (Moore et al. 2002). The diversity of ligands for PXR is especially broad and includes not only endogenous steroids but also an array of other lipophilic compounds such as the secondary bile acid LCA, the antibiotic rifampicin, and xenobiotics such as hyperforin, which is the active ingredient of St. John's Wort (Moore et al. 2002). We have identified several additional nutritional lipids as candidate low-affinity VDR ligands, which may function locally in high concentrations. Figure 1.5 reveals that these novel putative VDR ligands

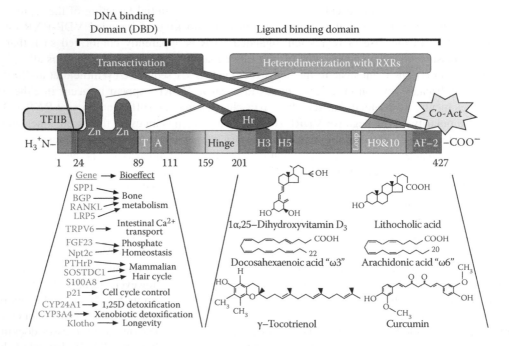

FIGURE 1.5 Functional domains in human VDR. Highlighted at the left is the human VDR zinc finger DBD, which, in cooperation with the corresponding domain in the RXR heteropartner, mediates direct association with the target genes listed at the lower left, leading to the indicated physiological effects. The official gene symbol for BGP is BGLAP, for RANKL is TNFSF11, for Npt2c is SLC34A3, for PTHrP is PTHLH, and for klotho is KL. Below the LBD domain (at the right) are illustrated selected VDR ligands, including several novel ligands discussed in the text.

include ω3- and ω6-essential polyunsaturated fatty acids (PUFAs), DHA and arachidonic acid, respectively; the vitamin E derivative γ-tocotrienol; and curcumin (Bartik et al. 2010), which is a turmeric-derived polyphenol found in curry. Thus, it is now recognized that VDR binds several ligands beyond the 1,25D hormone. The first nonvitamin D-related VDR ligands discovered were LCA and its derivative, 3-ketolithocholic acid (3-ketoLCA; Makishima et al. 2002). This finding is of considerable interest to human medicine since it is known that LCA is a secondary bile acid with significant carcinogenic potential (Hamada et al. 1994; Makishima et al. 2002; Nagengast et al. 1995), and its binding to VDR could trigger the detoxification of this ligand via the induction of CYP3A4 mediated by the VDREs in this gene (Table 1.1). Considering the structures for prototypical PXR, FXR, and LXR ligands and comparing these compounds with the LBD profile of VDR, expanded to include LCA and its 3-keto derivative, it is evident that VDR exhibits a ligand profile resembling that of the closest VDR relatives in the nuclear receptor superfamily, especially when it is noted that both PXR (Staudinger et al. 2001; Xie et al. 2001) and FXR (Makishima et al. 2002; Parks et al. 1999) are also activated by LCA to some extent.

The general conclusion based on examination of the evolutionary position of VDR among the 48 human nuclear receptors is that it is extremely closely related to PXR, both structurally and functionally. In one respect, VDR appears to have evolved as a "specialty" regulator of intestinal calcium absorption and hair growth in terrestrial animals, providing both a mineralized skeleton for locomotion in a calcium-scarce environment and physical protection against the harmful UV radiation of the sun. However, VDR also has retained its PXR-like ability to affect xenobiotic detoxification via CYP induction. VDR could complement PXR by serving as a guardian of epithelial cell integrity, especially at environmentally or xenobiotically exposed sites such as skin, intestine, and kidney.

1.5 VDR-MEDIATED CONTROL OF NETWORKS OF VITAL GENES

The binding of ligand by VDR triggers tight association between VDR and its heterodimeric partner, RXR, and only this liganded VDR-RXR heterodimer is able to penetrate the deep groove of DNA and recognize VDREs in the DNA sequence of vitamin D–regulated genes (Hsieh et al. 2003; Whitfield et al. 2005). Table 1.1 provides a list of VDR-RXR target genes recognized by the combined DBD zinc fingers of the two receptors, and their C-terminal extensions (Figure 1.3). These VDR-RXR-controlled genes encode proteins which determine bone growth and remodeling, bone mineral homeostasis, detoxification, the mammalian hair cycle, cell proliferation/differentiation, apoptosis, immune function, lipid metabolism, and likely longevity. In general, VDREs possess either a direct repeat of two hexanucleotide half-elements with a spacer of three nucleotides (DR3) or an everted repeat of two half-elements with a spacer of six nucleotides (ER6) motif, with DR3s being the most common. In positive DR3 VDREs, VDR has been shown to occupy the 3′ half-element, with RXR residing on the 5′ half-site (Jin et al. 1996). The "optimal" VDRE, which was experimentally determined via binding of randomized oligonucleotides to a VDR-RXR heterodimer (Colnot et al. 1995; Nishikawa et al. 1994), is in general agreement with the repertoire of natural VDREs (Table 1.1) and defines the optimal VDRE as a direct repeat of two six-base half-elements that resemble estrogen responsive element (ERE) half-sites, i.e., AGGTCA, separated by a spacer of three nucleotides. The highest affinity 3′ (VDR) half-site is PGTTCA, where P is a purine base, and the highest affinity 5′ (RXR) half-site is PGGTCA. In contrast to the human CYP3A4 ER6 VDRE, in which both half-elements coincide with the randomly selected sequences, the half-sites that exist in natural DR3 elements usually contain one to three bases that do not match the optimal VDRE. The fact that functional VDREs generally conform to this sequence, but with numerous minor variations, has recently been verified by genome-wide studies of the VDR/RXR "cistrome" (Meyer et al. 2012). The multiple sequence variations in natural VDREs (Table 1.1) may provide a spectrum of affinities for the VDR-RXR heterodimer, thus enabling these elements to respond to differing concentrations of the receptors (or their ligands; van Den Bemd et al. 2002).

TABLE 1.1
VDREs in Genes Directly Modulated in Their Expression by 1,25D and Possibly Other VDR Ligands

Gene	Bioeffect	Type	Location	5'-Half	Spacer	3'-Half	Ref	Group
rBGP	Bone metabolism	Positive	−456	GGGTGA	atg	AGGACA	(Terpening et al. 1991)	Bone
mBGP	Bone metabolism	Negative	−444	GGGCAA	atg	AGGACA	(Lian et al. 1997)	Bone
hBGP	Bone metabolism	Positive	−485	GGGTGA	acg	GGGGCA	(Kerner et al. 1989)	Bone
mSPP1	Bone metabolism	Positive	−757	GGTTCA	cga	GGTTCA	(Noda et al. 1990)	Bone
mSPP1	Bone metabolism	Positive	−2000	GGGTCA	tat	GGTTCA	(Pike et al. 2007)	Bone
mLRP5	Bone anabolism	Positive	+656	GGGTCA	ctg	GGGTCA	(Barthel et al. 2007)	Bone
mLRP5	Bone anabolism	Positive	+19 kb	GGGTCA	tgc	AGGTTC	(Fretz et al. 2006)	Bone
rRUNX2	Bone anabolism	Negative	−78	AGTACT	gtg	AGGTCA	(Drissi et al. 2002)	Bone
mRANKL	Bone resorption	Positive	−22.7 kb	TGACCT	cctttg	GGGTCA	(Haussler et al. 2008)	Bone
mRANKL	Bone resorption	Positive	−76 kb	GAGTCA	ccg	AGTTGT	(Kim et al. 2006)	Bone
mRANKL	Bone resorption	Positive	−76 kb	GGTTGC	ctg	AGTTCA	(Kim et al. 2006)	Bone
cIntegrin-beta3	Bone resorption, platelet aggregation	Positive	−756	GAGGCA	gaa	GGGAGA	(Cao et al. 1993)	Bone
cCarbonic anhydrase II	Bone resorption, brain function	Positive	−39	AGGGCA	tgg	AGTTCG	(Quelo et al. 1998)	Bone
cPTH	Mineral homeostasis	Negative	−60	GGGTCA	gga	GGGTGT	(Liu et al. 1996b)	Mineral
mVDR	Autoregulation of VDR	Positive	+8467	GGGTTA	gag	AGGACA	(Zella et al. 2007)	Mineral
hTRPV6	Intestinal Ca^{2+} transport	Positive	−1270	AGGTCA	ttt	AGTTCA	(Meyer et al. 2006)	Mineral
hTRPV6	Intestinal Ca^{2+} transport	Positive	−2100	GGGTCA	gtg	GGTTCG	(Meyer et al. 2006)	Mineral
hTRPV6	Intestinal Ca^{2+} transport	Positive	−2155	AGGTCT	tgg	GGTTCA	(Meyer et al. 2006)	Mineral
hTRPV6	Intestinal Ca^{2+} transport	Positive	−4287	GGGGTA	gtg	AGGTCA	(Meyer et al. 2006)	Mineral
hTRPV6	Intestinal Ca^{2+} transport	Positive	−4337	CAGTCA	ctg	GGTTCA	(Meyer et al. 2006)	Mineral
hNpt2a	Renal phosphate reabsorption	Positive	−1963	GGGGCA	gca	AGGGCA	(Taketani et al. 1998)	Mineral
hNpt2c	Renal phosphate reabsorption	Positive	−556	AGGTCA	gag	GGTTCA	(Barthel et al. 2007)	Mineral
hFGF23	Renal phosphate reabsorption	Positive	−32.9 kb	TGAACT	caaggg	AGGGCA	(Haussler et al. 2011)	Mineral
hklotho	Renal phosphate reabsorption	Positive	−31 kb	AGTTCA	aga	AGTTCA	(Forster et al. 2011)	Mineral
hklotho	Renal phosphate reabsorption	Positive	−46 kb	GGTTCG	tag	AGTTCA	(Forster et al. 2011)	Mineral
mklotho	Renal phosphate reabsorption	Positive	−35 kb	AGGTCA	gag	AGTTCA	(Forster et al. 2011)	Mineral
rCYP24A1	1,25D detoxification	Positive	−151	AGGTGA	gtg	AGGGCG	(Ohyama et al. 1994b)	Detox

Gene	Function	Regulation	Position	Sequence	Spacer	Sequence	Reference	Category
rCYP24A1	1,25D detoxification	Positive	−238	GGTTCA	gcg	GGTGCG	(Zierold et al. 1994a)	Detox
hCYP24A1	1,25D detoxification	Positive	−164	AGGTGA	gcg	AGGGCG	(Zou et al. 1997)	Detox
hCYP24A1	1,25D detoxification	Positive	−285	AGTTCA	ccg	GGTGTG	(Zou et al. 1997)	Detox
hCYP3A4	Xenobiotic detoxification	Positive	−169	TGAACT	caaagg	AGGTCA	(Thompson et al. 2002a, Thummel et al. 2001)	Detox
hCYP3A4	Xenobiotic detoxification	Positive	−7.7 kb	GGGTCA	gca	AGTTCA	(Makishima et al. 2002)	Detox
rCYP3A23	Xenobiotic detoxification	Positive	−120	AGTTCA	tga	AGTTCA	(Barwick et al. 1996, Thompson et al. 2002a)	Detox
hMDR1	P-glycoprotein, drug resistance	Positive	−7863	AGTTCA	atg	AGGTAA	(Saeki et al. 2008)	Detox
hMDR1	P-glycoprotein, drug resistance	Positive	−7853	AGGTCA	agtt	AGTTCA	(Saeki et al. 2008)	Detox
hp21	Cell cycle control	Positive	−765	AGGGAG	att	GGTTCA	(Liu et al. 1996a)	Cell life
hFOXO1	Cell cycle control	Positive	−2856	GGGTCA	cca	AGGTGA	(Wang et al. 2005)	Cell life
hIGFBP-3	Cell proliferation/apoptosis	Positive	−3282	GGTTCA	ccg	GGTGCA	(Peng et al. 2004)	Cell life
hInvolucrin	Skin barrier function	Positive	−2083	GGCAGA	tct	GGCAGA	(Bikle et al. 2002)	Cell life
hPLD1	Keratinocyte differentiation	Positive	−246	GGGTGA	tgc	GGTCGA	(Kikuchi et al. 2007)	Cell life
hCCR10	Homing of T-cells to skin	Positive	−110	GGGTCT	acg	GGGTCA	(Shirakawa et al. 2008)	Cell life
rPTHrP	Mammalian hair cycle	Negative	−805	AGGTTA	ctc	AGTGAA	(Falzon 1996)	Cell life
hSOSTDC1	Mammalian hair cycle	Negative	−6215	AGGACA	gca	GGGACA	(Haussler et al. 2008)	Cell life
rVEGF	Angiogenesis	Positive	−2730	AGGTGA	ctc	AGGGCA	(Cardus et al. 2009)	Cell life
hMIS	Müllerian-inhibiting substance	Positive	−381	GGGTGA	gca	GGGACA	(Malloy et al. 2009)	Cell life
hHLA-DRB1	Major histocompatibility complex	Positive	−1	GGGTGG	agg	GGTTCA	(Ramagopalan et al. 2009)	Immune
hCAMP	Antimicrobial peptide	Positive	−615	GGTTCA	atg	GGTTCA	(Gombart et al. 2005)	Immune
hKSR-1	Monocytic differentiation	Positive	−8156	GGTGCA	tat	AGGTCA	(Wang et al. 2006)	Immune
hKSR-2	Monocytic differentiation	Positive	−2501	AGTTCA	gca	TGGTCA	(Wang et al. 2007)	Immune
hKSR-2	Monocytic differentiation	Positive	+3185	GGTTCA	aac	AGTTCT	(Wang et al. 2007)	Immune
mInsig-2	Regulation of lipid synthesis	Positive	−2470	AGGGTA	acg	AGGGCA	(Lee et al. 2005)	Metabolism
hPCFT	Intestinal folate transporter	Positive	−1680	AGGTTA	ttc	AGTTCA	(Eloranta et al. 2009)	Metabolism
hCystathionine β synthase	Homocysteine clearance	Positive	+2563	AGGGCA	gtg	AGGACA	(Kriebitzsch et al. 2009)	Metabolism
hCystathionine β synthase	Homocysteine clearance	Positive	+7849	GGGACA	gat	AGTTCA	(Kriebitzsch et al. 2009)	Metabolism

Furthermore, as depicted conceptually in Figure 1.4, evidence is accumulating (Zhang et al. 2011) that variant VDRE sequences induce unique conformations in the VDR-RXR complex, thereby promoting association of the heterodimer with distinct subsets of comodulators (Staal et al. 1996) or permitting differential actions in the context of diverse tissues (van Den Bemd et al. 2002).

Many VDREs occur as a single copy in the proximal promoter of vitamin D–regulated genes, for instance those in rat osteocalcin (BGP), rat RUNX2 (rRUNX2), chicken PTH (cPTH), human sodium phosphate cotransporter 2c (hNpt2c), human p21 (hp21), and rat PTHrP (rPTHrP). However, CYP24A1 VDREs are at least bipartite, as is the human CYP3A4 VDRE, with the 5′ DR3 located some 7.5 kb upstream of the proximal ER6 VDRE in the latter case. Studies of these vitamin D–controlled CYP genes introduced the concepts of multiplicity and remoteness to VDREs, as confirmed more recently by ChIP or ChIP scanning (Barthel et al. 2007, Fretz et al. 2006, Kim et al. 2007, Meyer et al. 2006) of genomic DNA surrounding the transient receptor potential vanilloid type 6 (TRPV6), LRP5, and receptor activator of nuclear factor κB ligand (RANKL) genes, uncovering novel VDREs at some distance from the transcription start site (Table 1.1). Indeed, the recent genome-wide study of the VDR/RXR cistrome previously cited found that the vast majority (98%) of VDR/RXR binding sites in LS180 cells were located >500 bp upstream or downstream from the transcriptional start site of the nearest gene (Meyer et al. 2012). Genes possessing multiple VDREs require all VDR-RXR docking sites for maximal induction by $1,25(OH)_2D_3$ and the individual VDREs appear to function synergistically in attracting coactivators and basal factors for transactivation. The most attractive model is that remote VDREs are juxtaposited with more proximal VDREs via DNA looping in chromatin, creating a single platform that supports the transcription machine, and this conclusion has been verified by chromatin conformation capture studies from two independent groups (Saramaki et al. 2009; Wang et al. 2010).

1.5.1 Detoxification of Endobiotics and Xenobiotics

A common theme for VDR, PXR, and CAR is the induction of CYP enzymes that participate in xenobiotic detoxification. A major target for VDR and PXR in humans is CYP3A4 (Makishima et al. 2002; Thompson et al. 2002a; Thummel et al. 2001), for which the detoxification substrates include LCA (Araya and Wikvall 1999), as illustrated in Figure 1.6. Initial studies focused on VDR liganded to 1,25D as a regulator of CYP3A4 (Thompson et al. 2002a; Thummel et al. 2001), but later experiments revealed that LCA is also capable of binding VDR to upregulate expression of human CYP3A4 or its equivalent in rats (CYP3A23) or mice (CYP3A11) (Makishima et al. 2002). There is also evidence that other CYP enzymes may be VDR targets (Drocourt et al. 2002; Meyer et al. 2012). Additionally, 1,25D induces SULT2A (Figure 1.6), which is an enzyme that detoxifies sterols via 3α-sulfation (Echchgadda et al. 2004).

A hypothetical model for the pathophysiologic significance in humans of LCA as a VDR ligand is depicted in Figure 1.6. The precursor of LCA, chenodeoxycholic acid is produced in the liver via a pathway that is controlled in a positive fashion by LXR and in a negative feedback loop by FXR (Chawla et al. 2001; Lu et al. 2001). (Both of these receptors form heterodimers with RXR, which are not shown.) LCA, formed through 7-dehydroxylation by gut bacteria, is not a good substrate for the enterohepatic bile acid reuptake system and thus remains in the enteric tract and passes to the colon, where it can exert carcinogenic effects (Kozoni et al. 2000). VDR in the colonocyte is proposed to bind LCA or its 3-keto derivative and activate CYP3A4 (Makishima et al. 2002). CYP3A4 then catalyzes the 6α-hydroxylation of LCA (Araya and Wikvall 1999), thus converting it into a substrate for the ABC efflux transporter (Chawla et al. 2001). We (Thompson et al. 2002b) and others (Thummel et al. 2001) have shown that 1,25D, which is formed in the kidney or locally in colon via the action of CYP27B1, can also induce CYP3A4. Thus, natural ligands for VDR, including the high-affinity 1,25D hormonal metabolite and the lower affinity, nutritionally-modulated bile acids, seem to possess the important potential to serve as agents for promoting detoxification of LCA and

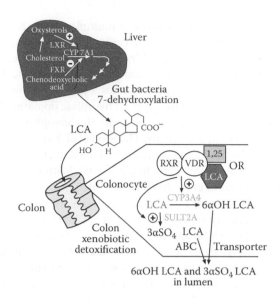

FIGURE 1.6 Physiologic roles of two VDR ligands. LCA (center) is produced from liver-derived chenode-oxycholic acid (CDCA) by the action of gut bacteria. LCA, which is not recycled in the terminal ileum due to the loss of the 7α-hydroxyl, instead travels to the colon, where it exerts tumorigenic actions on the colonocyte (see text). However, the ability of VDR to bind LCA and induce the CYP3A4 and SULT2A genes, catalyzing the detoxification of LCA by 6α-hydroxylation and 3α-sulfation, respectively, followed by export of both polar metabolites via the ABC efflux transporter, lessens the risk of colon cancer.

possibly other intestinal endobiotics or xenobiotics, with the end result likely being a reduction in colon cancer incidence. It is intriguing to consider that additional, naturally-occurring VDR ligands with similar, or even complementary, activities may remain to be discovered.

1.5.2 PHOSPHATE HOMEOSTASIS ATTENUATES SENESCENCE

Phosphate is abundant in a normal diet and is a fundamental biologic component of not only min-eralized bone but also essential biomolecules such as DNA, RNA, phospholipids, phosphoproteins, ATP, and metabolic intermediates. However, phosphate excess may act as a pro-senescence factor independently of hypervitaminosis D. For example, an excess of phosphate in the blood can lead to ectopic calcification and arteriosclerosis, COPD, chronic kidney disease, and loss of hearing. Fortunately, FGF23 and klotho are designed to signal the elimination of circulating phosphate and therefore promote healthspan. Klotho appears to have systemic antiaging properties independent of its phosphaturic actions, perhaps through its glycosyl hydrolase enzymatic activity (Cha et al. 2009). By analogy, the kidney also releases renin, which is a hormone with proteolytic enzyme activ-ity. Conversely, although FGF23 is antiaging at the kidney by eliciting phosphate elimination and detoxifying 1,25D, its "off-target" actions could actually be proaging in terms of coronary artery disease, and it is possible that these off-target FGF23 pathologies are opposed by secreted klotho (Bergwitz and Juppner 2010).

Klotho is the only reported single gene mutation that leads to a premature aging phenotype in the mouse (Kuro-o et al. 1997), and a recessive inactivating mutation in the human klotho gene elicits a phenotype of severe tumoral calcinosis (Ichikawa et al. 2007). Klotho- and FGF23-null mice have identical hyperphosphatemic phenotypes of short life-span/premature aging, ectopic cal-cification, arteriosclerosis, osteoporosis, muscle atrophy, skin atrophy, and hearing loss (Kuro-o et al. 1997; Shimada et al. 2004b). Klotho is a known coreceptor for FGF23, whereas β-klotho is a coreceptor for other FGF hormones. Klotho exists in multiple forms (Kuro-o 2010): a full-length (130 kDa) transmembrane coreceptor with a minimal cytoplasmic region of 11 amino acids and two extracellular domains with homology to glycosyl hydrolases, at least three proteolyzed forms that are shed into the circulation (Chen et al. 2007), and finally, a hypothetical 80 kDa secreted form produced by alternative splicing in exon 3 that generates a protein species possessing a portion of the extracellular domain containing one of the glycosyl hydrolase domains (Matsumura et al. 1998). Because of the potential significance of these forms in klotho biology, much remains to be learned about how their expression is regulated. Recently, Thurston et al. have demonstrated that TNFα and γ-interferon are suppressors of renal klotho expression (Thurston et al. 2010); the FGF23 ligand is also a putative repressor of klotho expression (Quarles 2008). With the exception of a preliminary report by Tsujikawa et al. (Tsujikawa et al. 2003) who treated mice with various dietary and pharmacologic regimens, including vitamin D, inducers of klotho are poorly characterized. However, as detailed here, it has been recently reported (Forster et al. 2011) that 1,25D significantly induces klotho mRNA expression at the cellular and molecular level in human and mouse renal cell lines.

Analysis by Forster et al. (Forster et al. 2011) of RNA isolated from mouse distal convoluted tubule (mpkDCT) cells, the primary expression site for klotho in kidney, with primers designed to capture both alternatively spliced mRNAs for the membrane and secreted forms of klotho, demonstrated that 1,25D treatment (100 nM for 24 h) induces klotho mRNA expression (Figure 1.7a). This induction is evident for both the membrane and secreted splice forms of klotho mRNA, suggesting that 1,25D may be capable of both amplifying FGF responsiveness and eliciting secretion of circulating klotho hormone. These data complement the previous demonstration of klotho mRNA induction by 1,25D in human proximal kidney (HK-2) cells (Haussler et al. 2010). Interestingly, curcumin (Figure 1.5), which is an alternative VDR ligand (Bartik et al. 2010), selectively upregulates membrane klotho mRNA in mpkDCT cells (Figure 1.7b), indicating that distinct VDR ligands can differentially modulate the membrane and secreted forms of klotho. These data lead to the hypothesis that designer vitamin D analogs could promote the healthful aging benefits of systemic klotho without accentuating FGF23 action to perhaps elicit hypophosphatemia.

Bioinformatic analysis (Forster et al. 2011) of both the human and mouse klotho genes revealed 11 candidate VDREs in the human gene and 17 putative VDREs in the mouse gene (Figure 1.7c). Electrophoretic mobility shift assays of 11 candidate VDREs in the human klotho gene revealed that three of these (−46 kb, −31 kb, and +3.2 kb) displayed an ability to bind VDR/RXR that was abrogated by the 9A7 anti-VDR mAb. Two of these VDREs (−46 and −31 kb; Table 1.1) were more potent in this assay than the established rat osteocalcin VDRE (Figure 1.7d). Interestingly, the remote location of these two VDREs in relation to the transcriptional start site of klotho, yet residing between two insulators, is consistent with the new paradigm of VDR/RXR action in which the chromatin conformation loops out DNA to bring distant enhancers together adjacent to the RNA Pol II docking site (Bishop et al. 2009). Forster et al. (Forster et al. 2011) analyzed the functional activity of candidate VDREs at −46 kb, −31 kb, and +3.2 kb in transfected HK-2 renal cells and observed striking (>10-fold) 1,25D responsiveness of VDREs corresponding to sequences at −46 kb and −31 kb, but not +3.2 kb, in the context of synthetic VDRE-luciferase reporter constructs (Figure 1.7e). A similar analysis of 17 candidate mouse VDREs (Forster et al. 2011), of which only one shows a degree of positional (but not sequence) conservation with a human VDRE (Figure 1.7c, red outline), revealed only two mouse VDREs at −35 kb, and +9 kb displayed gel shift activity comparable to that of rat osteocalcin VDRE. However, only the mouse klotho VDRE located at −35 kb (Table 1.1) displays transactivation ability (Forster et al. 2011). Thus, it appears that 1,25D-liganded VDR-RXR induces klotho expression by binding to functional VDREs in the range of 31–46 kb 5′ of the transcriptional start site of both the human and mouse klotho genes. In combination with the data of Tsujikawa et al. (Tsujikawa et al. 2003) that 1,25D increases steady-state klotho mRNA levels in mouse kidney, *in vivo*, the results of Forster et al. (Forster et al. 2011) indicate that 1,25D is the first discovered inducer of the longevity gene, klotho.

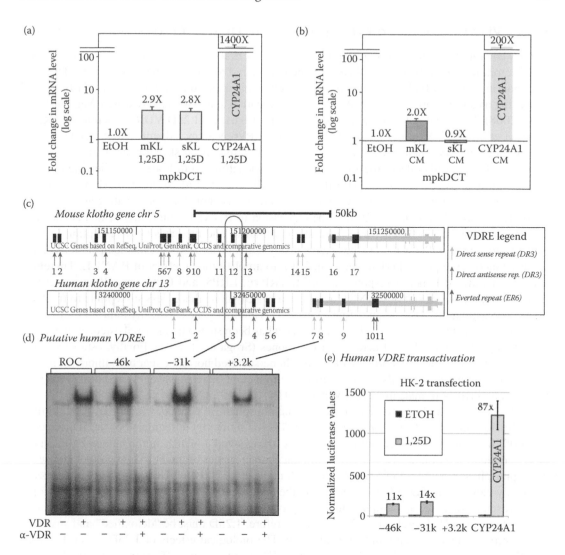

FIGURE 1.7 Upregulation of klotho by liganded VDR. (a) VDR-mediated regulation of mouse klotho mRNA as assessed by qrtPCR. The mouse klotho gene is induced by 1,25D in the distal convoluted tubule cell line mpkDCT. (b) Curcumin (50 µM, 24 h), which is an alternative VDR ligand, also induces klotho in mpkDCT cells, but the effect is selective for membrane klotho (mKL); on the other hand, 1,25D significantly upregulates both klotho mRNA splice forms in mpkDCT cells. (c) Candidate VDREs in the human and mouse klotho loci. Seventeen mouse and 11 human elements located bioinformatically are shown as solid boxes numbered starting upstream and distal to the transcription start site. (d) Double-stranded oligonucleotides for each element (including four flanking bases on either side) were ^{32}P-labeled and subjected to electrophoretic mobility shift analysis. A rat osteocalcin (ROC) element served as a positive control. Each candidate VDRE was tested without cell lysate (first lane), with lysate containing VDR and RXRα (second lane), and lysate plus a specific anti-VDR monoclonal antibody (α-VDR; third lane). (e) Candidate VDREs were cloned into a pLUC-MCS reporter vector, cotransfected into the HK-2 human kidney cells along with a pSG5-VDR cDNA expression plasmid and treated with 1,25D (10^{-8} M) for 24 h. Firefly luciferase values were normalized to expression of Renilla luciferase. Data are depicted as a fold effect of 1,25D.

1.5.3 VDR Ligands Promote Health Span via the Delay of Chronic Diseases of Aging

125D-VDR regulates the expression of at least 11 genes that encode bone and mineral homeostasis effectors for which their control can also be considered to facilitate healthful aging. The first, osteopontin or SPP1 (Table 1.1), triggers ossification and is induced by 1,25D in osteoblasts, as well as serving as an inducible inhibitor of vascular calcification and associated disease (Weissen-Plenz et al. 2008). Intestinal calcium uptake is mediated, in part, by 1,25D-VDR induction of TRPV6 (Barthel et al. 2007; Meyer et al. 2006; Table 1.1). TRPV6 is a key calcium channel gene product that supplies dietary calcium via transport to build the mineralized skeleton and thereby not only prevent rickets but also delay the inevitable calcium leaching from bone in senile osteoporosis. 1,25D significantly induces LRP5 (Barthel et al. 2007; Fretz et al. 2006; Table 1.1), which is a gene product that promotes osteoblastogenesis via enhanced canonical Wnt signaling and is thereby anabolic to bone (Milat and Ng 2009). The expression of RANKL (Table 1.1), which is catabolic to bone, is enhanced by 1,25D-VDR (Haussler et al. 2010) and mediates bone resorption through osteoclastogenesis. OPG, which is the soluble decoy receptor for RANKL that tempers its activity, is simultaneously repressed (Haussler et al. 2010) to amplify the bioeffect of RANKL. Therefore, a well-mineralized bone, in response to SPP1, TRPV6, LRP5, and OPG, as well as an actively remodeled bone as a result of RANKL action, is a healthy bone that is less susceptible to fractures associated with aging. Osteocalcin (BGP, Table 1.1) is another gene classically induced by 1,25D in osteoblasts. Recently, utilizing BGP null animals, it has been shown that normal osteocalcin expression is important for robust, fracture-resistant bones (Sroga et al. 2011). Finally, osteocalcin has been identified by Karsenty and coworkers as a bone-secreted hormone that both improves insulin release from pancreatic β-cells and increases insulin metabolic responsiveness and is also required for optimal fertility in male mice (Oury et al. 2011).

FGF23 is a second bone-secreted hormone and phosphaturic peptide, which, like PTH, inhibits renal Npt2a and Npt2c to elicit phosphaturia. FGF23 is markedly upregulated by 1,25D (Kolek et al. 2005) in osteocyte-like cells of the osteoblast lineage, and a VDRE in the human FGF23 gene has been identified (Table 1.1). The two dominant characteristics of the FGF23 knockout mouse are hyperphosphatemia and ectopic calcification (Shimada et al. 2004b). FGF23 null mice also possess markedly elevated 1,25D in blood, generating the additional phenotypes of skin atrophy, osteoporosis, vascular disease, and emphysema. Many of these pathologies are also the consequence of hypervitaminosis D (Keisala et al. 2009), and therefore, 1,25D must be "detoxified" and sustained in an optimal range to maintain healthful aging. The biological effects of 1,25D are curtailed by CYP24A1-catalyzed catabolism (Table 1.1) of 1,25D, providing an "off" signal once the hormone has executed its physiologic modulation of gene expression. Mice with ablation of the CYP24A1 gene die early because of 1,25D toxicity (Masuda et al. 2005).

FGF23 liganding signals via renal FGFR and klotho coreceptors to elicit phosphaturia and repress CYP27B1 while inducing CYP24A1 (Razzaque 2009). 1,25D induces klotho mRNA in cultured kidney cells and VDREs have been identified in both the human and mouse klotho genes (Forster et al. 2011). Upregulation of klotho by 1,25D is consistent with potentiation of FGF23 signaling in kidney and perhaps protection of other cell types (e.g., vascular), where a secreted form of klotho is considered a potential beneficial renal hormone (Wang and Sun 2009).

The final network of 1,25D VDR–regulated genes is composed of those encoding factors impacting cell survival/cancer, the immune system, and metabolism. The VDR null mouse is supersensitive to DMBA-induced skin cancer (Zinser et al. 2002b), as well as UV-light-induced skin malignancy (Ellison et al. 2008). Moreover, VDR likely reduces risk for many cancers by inducing the p53 (Audo et al. 2003) and p21 (Audo et al. 2003; Table 1.1) tumor suppressors, as well as DNA mismatch repair enzymes in colon (Sidelnikov et al. 2010). VDR knockout mice exhibit enhanced colonic proliferation (Kallay et al. 2001) plus amplified mammary gland ductal extension, end buds, and density (Zinser et al. 2002a), indicating that the fundamental actions of VDR to promote cell differentiation and apoptosis (Egan et al. 2010) play an important role in reducing the risk of age-related epithelial cell cancers such as those of the breast and colon. Interestingly, curcumin, which is found in curry and known to be antiinflammatory to the degree that it reduces inflammatory bowel disease (Egan et al. 2004), was compared with LCA as a potential VDR ligand. Strikingly, we observed that curcumin is slightly more active than LCA in driving VDR-mediated transcription and that it also binds to VDR with approximately the same affinity as LCA based on ligand competition assays (Bartik et al. 2010). The mechanism of action of curcumin is not known, and we suggest that at least part of its beneficial functions are mediated by the nuclear VDR, perhaps even its ability to lower the risk of colon cancer (Johnson and Mukhtar 2007).

Another class of nutritionally available lipids, which is critical in maintaining cell membrane fluidity and serves as precursors of the prostanoids and leukotrienes, is the essential PUFAs. DHA, for example, is an ω-3 PUFA responsible for infant brain development, which is also a known ligand for RXR (de Urquiza et al. 2000), which, in turn, is the heterodimeric partner of VDR. ω-3 PUFAs are also ligands for PPARα, through which they lower VLDL and, eventually, LDL cholesterol to lessen coronary artery disease, as well as reduce the incidence of metabolic syndrome (Dussault and Forman 2000). It has been reported that ω-3 PUFAs such as DHA and eicosapentaenoic acid, as well as ω-6 PUFAs such as linoleic acid and arachidonic acid, compete with tritiated 1,25D for binding to VDR with affinities for the receptor some four orders of magnitude lower than that of the 1,25D hormonal ligand (Haussler et al. 2008). Nevertheless, it can be concluded that high local concentrations of PUFAs could occur in select cells or tissues and exert VDR-mediated antiproliferation/prodifferentiation effects that may partially explain the chemoprotective nature of diets rich in PUFAs, plus their cardioprotective and antiinflammatory influences. The lack of specificity among the PUFAs for VDR binding and activation (Jurutka et al. 2007) is unusual for ligand receptor interactions. However, not all lipophilic compounds bind VDR with low affinity, as dexamethasone, the synthetic glucocorticoid, and α-tocopherol (the antioxidant vitamin E) do not compete with 1,25D for occupation of VDR. Surprisingly, the vitamin E metabolite, γ-tocotrienol (Figure 1.5), is a low-affinity VDR ligand that is capable of activating the receptor (Bartik et al. 2007). This finding reveals that, similar to the concept that vitamin D and even 25D are ineffective ligands for VDR, whereas metabolism to 1,25D generates a high-affinity hormone, the basic PUFA "core" structure could be metabolically activated to yield an array of higher affinity ligands that function as cell-specific activators of VDR.

In addition to PUFAs such as DHA serving as nutritionally derived lipids that affect VDR signaling to potentiate health span, resveratrol, which is a potent antioxidant found in the skin of red grapes and as a component in red wine, may serve to augment 1,25D-VDR antiaging actions (Hayes 2011). There is an overall structural symmetry and parallel configuration of resveratrol and known VDR ligands, which may indicate that resveratrol is yet another low-affinity VDR ligand with the ability to activate VDR. In fact, as shown by Batie et al. (Batie et al. 2011), resveratrol activates VDR-mediated transcription in human embryonic kidney cells employing a VDRE from the human xenobiotic detoxification gene, CYP3A4 (Figure 1.8a; sixfold over vehicle control). Moreover, in renal cells, the combined presence of 1,25D and resveratrol results in synergism (300%) in VDR transactivation, compared to 1,25D alone (Figure 1.8a). Indeed, in the context of human colonic cells, the synergism between 1,25D and resveratrol is amplified (Figure 1.8a), unlike that observed with alternative VDR ligands such as LCA (Egan et al. 2010). This synergism between resveratrol and 1,25D suggests that resveratrol may not constitute an alternative VDR ligand like LCA and curcumin, with the antioxidant from red wine instead functioning as a potentiator of 1,25D-VDR signaling. Competitive displacement binding assays of resveratrol with tritiated 1,25D-bound VDR (Figure 1.8b) illuminate the molecular mechanism of the cooperation between resveratrol and 1,25D in activating VDR. Surprisingly, unlike DHA, which competes with 1,25D for VDR occupancy, resveratrol significantly enhances tritiated 1,25D binding to VDR in a COS-7 cell extract (Figure 1.8b). Thus, resveratrol confers VDR with an increased capacity for 1,25D binding, perhaps by activating SIRT1 to deacetylate VDR or one of its comodulators. Since 1,25D and curcumin also induce klotho (as discussed above), while resveratrol activates SIRT1 (Baur 2010), we propose the

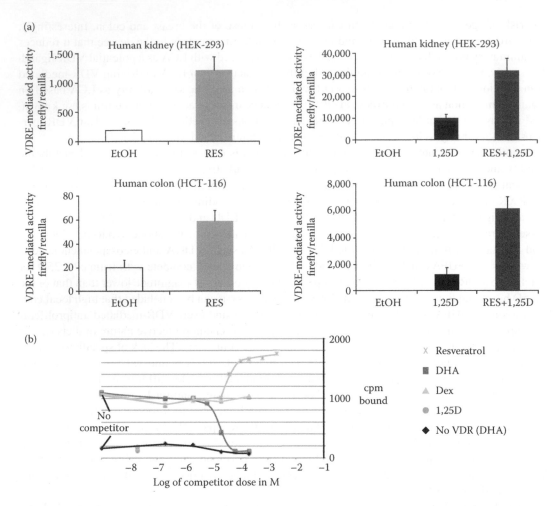

FIGURE 1.8 **(See color insert.)** Resveratrol cooperates with 1,25D to enhance VDR signaling. (a) A transcriptional assay was employed to evaluate resveratrol (RES) for modulation of VDR signaling. A vector expressing human VDR and a reporter construct containing the proximal VDRE from the human CYP3A4 gene linked to a luciferase reporter were used to transfect either human embryonic kidney (HEK-293) or colon cancer (HCT-116) cells that were then treated with ethanol vehicle, 10^{-8} M 1,25D, 3×10^{-4} M RES, or a combination of both lipophilic compounds. Firefly luciferase values were normalized to expression of Renilla luciferase. (b) Ability of resveratrol (blue X symbols) to enhance 1,25D binding by VDR. Competition binding curves were generated essentially as described previously (Bartik et al. 2010) using lysates from COS-7 cells transfected with expression plasmids for both human VDR and human RXRα. Five μL of lysate were combined with 4 μL of [³H]1,25D (approximately 4×10^{-10} M) and the indicated concentrations of competitor in a final incubation volume of 210 μL. After overnight incubation at 4 C, unbound [³H]1,25D was removed with dextran-coated charcoal, and a 200 μL aliquot of supernatant containing bound hormone was removed and counted as described (Bartik et al. 2010) to obtain bound cpm (out of a total of 4050 cpm added to each reaction). Controls included unlabeled 1,25D (orange circles), which shows complete competition at a concentration of 1.9×10^{-8} M; another known competitor, docosahexaenoic acid (DHA; red squares); dexamethasone (dex; green triangles), which has no effect on 1,25D binding to VDR; and, finally, lysates not receiving VDR/RXR (black triangles) treated with DHA. Binding in the absence of competitor is shown at the far left of the figure.

(c)

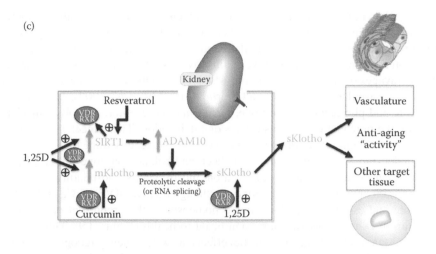

FIGURE 1.8 (Continued) (c) Hypothetical model for resveratrol activation of VDR via stimulation of SIRT1 (Baur 2010). SIRT1 catalyzes deacetylation of VDR, RXR, or comodulators to increase the capacity of 1,25D binding to VDR. SIRT1 activation also leads to ADAM10 stimulation (Donmez et al. 2010) to produce soluble (s)-klotho via ADAM10-mediated cleavage of membrane (m)-klotho. Curcumin induces m-klotho (Figure 1.7), while 1,25D stimulates SIRT1 activity (An et al. 2010) and expression of both m- and s-klotho (Figure 1.7). The integration of these regulatory circuits, which are controlled by the levels of nutritionally-derived bioactive "healthy" lipids (1,25D, curcumin, and resveratrol), culminates in the elaboration of s-klotho from the kidney to exert proposed endocrine anti-aging effects in the vasculature and other target tissues.

existence of a molecular mechanism (Figure 1.8c) that orchestrates collaborative crosstalk in 1,25D/klotho/resveratrol/SIRT signaling to achieve the antiaging activity of klotho, *in vivo*, that is thought to occur in the vasculature and other tissues (Wang and Sun 2009).

Liganded VDR also functions as a detoxification nuclear receptor by inducing CYP3A4 (Makishima et al. 2002; Table 1.1) and SULT2A (Echchgadda et al. 2004) to eliminate toxic xenobiotics such as LCA that might affect the gastrointestinal tract (Figure 1.6). In the case of immune function, 1,25D-VDR induces cathelicidin (Liu et al. 2006a) to activate the innate immune system to fight infection (Table 1.1) and represses IL-17 (Mora et al. 2008) to temper the adaptive immune system and lower the risk of autoimmune disorders such as type-I diabetes mellitus, multiple sclerosis, lupus, and rheumatoid arthritis. 1,25D-VDR is antiinflammatory by blunting NFκB (Cohen-Lahav et al. 2006) and COX2 (Moreno et al. 2005), and inflammation is considered a common denominator in maladies such as heart disease and stroke, as well as cancer. In the realm of cardiovascular disease (Guilliams 2004) and neurodegenerative disorders of aging such as Alzheimer's disease (Seshadri et al. 2002), excess circulating homocysteine is considered a negative risk factor. Indeed, 1,25D-VDR has recently been shown by the group of Bouillon (Kriebitzsch et al. 2009) to induce cystathionine β-synthase (Table 1.1), which is a major enzyme catalyzing the metabolic elimination of homocysteine. In addition, 1,25D-VDR induces FOXO3 (Eelen et al. 2009), which is a significant molecular player in preventing oxidative damage, which is the leading candidate for the cause of aging (Lin and Beal 2003). Clearly, through control of vital genes, VDR allows one to age well by delaying fractures, ectopic calcification, malignancy, oxidative damage, infections, autoimmunity, inflammation/pain, cardiovascular and neurodegenerative diseases.

1.6 CONCLUSION AND PERSPECTIVES

The healthful aging facets of vitamin D and novel VDR ligands described in this chapter reveal new roles for vitamin D and its nutritional surrogates that go far beyond vitamin D as a simple promoter of dietary calcium and phosphate absorption to ensure adequate bone mineralization. We now understand that vitamin D hormonal ligands and their nuclear receptor also mediate both the sculptured delimiting and remodeling of the skeleton, and the prevention of ectopic calcification through novel peptide mediators, such as FGF23 and klotho. These latter roles of 1,25D-VDR can be considered, on the one hand, as protective against osteoporotic fractures and, on the other hand, in reducing the ravages of ectopic calcification, which occur with aging, especially with respect to diminishing cardiovascular calcification and mortality (Stubbs et al. 2007). It is striking indeed that the calcemic and phosphatemic hormone, 1,25D, and its receptor, coevolved mechanisms for countermanding the potential deleterious effects of calcification. These mechanisms include feedback repression of calcemic PTH in one endocrine loop and feedforward induction of FGF23 in yet a second endocrine loop. FGF23 acts as a check and balance on bone mineral by delimiting skeletal calcification, retarding ectopic calcification, mostly via its phosphaturic action, and feedback repressing 1,25D production by the kidney. Although both 1,25D/PTH/Ca and 1,25D/FGF23/PO$_4$ are intricate and essential axes for mineral homeostasis, they represent only "the tip of the iceberg" in vitamin D and VDR functions significant to health (Figure 1.9). Thus, the traditional bone and mineral (antirachitic/antiosteoporotic) effects, as well as newly recognized bone anabolic and counterectopic calcification functions, comprise the fraction of the iceberg above the water line.

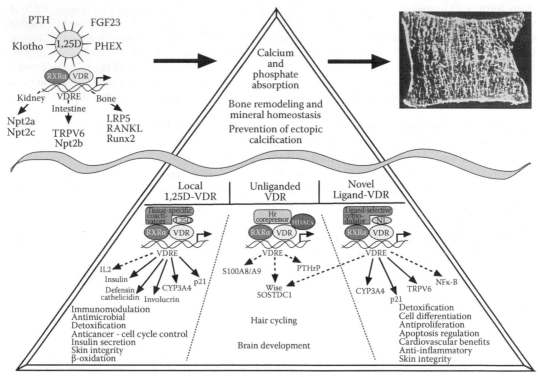

FIGURE 1.9 Osseous and extraosseous effects mediated by VDR. Shown at the upper left are the calcemic and phosphaturic hormones that participate in feedback loops to maintain bone mineral homeostasis as discussed in the text; a normally mineralized human vertebral body with its trabeculations is illustrated at the upper right. The upper portion depicts actions of 1,25D liganded VDR to maintain bone health, including interactions with other hormones (PTH, CT, FGF23). The lower portion summarizes the many extraosseous effects of VDR in the 1,25D-bound, unliganded, and novel ligand-bound states, as discussed in the text. Repressive actions of VDR are depicted as dashed arrows. We hypothesize that VDR occupied by locally generated 1,25(OH)$_2$D$_3$ uses cell-context-specific coactivators, and that VDR occupied by a novel ligand may utilize ligand-selective comodulators.

As illustrated, the bulk of VDR's actions are novel extraosseous effects that are diagrammed in the submerged portion of the iceberg.

Many of the extraosseous effects of VDR appear to be triggered by locally produced 1,25D. Numerous tissues besides the kidney express the 1α-OHase enzyme, including cells of the immune system (e.g., T-cells), the pancreas, skin, etc. This locally produced 1,25D does not contribute significantly to circulating 1,25D, but it retains the capacity to be active in a cell- and tissue-specific manner. Examples of local 1,25D-VDR actions include repression of IL-2 in T-cells (Haussler et al. 1998), induction of defensin and cathelicidin as local antimicrobial effectors (Liu et al. 2006a), stimulation of involucrin synthesis in skin (Bikle et al. 2004), CYP3A4 and p21 induction in epithelial cells—especially in the colon (Thompson et al. 2002a), and promotion of insulin secretion from the β-cells of the pancreas (Norman et al. 1980). By locally stimulating the aforementioned genes, the vitamin D/VDR system emerges, likely redundantly with other regulators, as an immunomodulator that stimulates the innate and suppresses the adaptive immune system to effect both antimicrobial and anti-autoimmune actions, detoxifies xenobiotics to be chemoprotective, controls cell proliferation and regulates apoptosis to reduce cancer, and moderates type-II diabetes by promoting insulin release, as well as possibly enhancing fatty acid β-oxidation via induction of FOXO1 (Table 1.1).

Another possibility obviating the need to locally generate 1,25(OH)$_2$D$_3$ would be for VDR to function unliganded. VDR, but not vitamin D, is required to sustain the mammalian hair cycle (Sakai et al. 2001). Thus, as depicted in Figure 1.9 (lower center), the Hr corepressor could function as a surrogate VDR "ligand" to suppress SOSTDC1 (Wise), S100A8/A9 (Haussler et al. 2010), or other genes that normally keep the hair cycle in check. Also, unlike the case of intestine, kidney, and bone, calbindin induction by VDR does not require vitamin D in the brain (Clemens et al. 1985). VDR is widely expressed in the central nervous system, as is Hr, raising the possibility that unliganded VDR, along with Hr, acts in select neurons. Notably, it has been reported that VDR-null mice exhibit behavioral abnormalities including anxiety (Kalueff et al. 2006).

The ability of VDR to function unliganded is difficult to justify physicochemically because the tertiary structure of VDR and its functionally interactive surfaces cannot be stabilized unless the hydrophobic binding pocket is occupied by a lipophilic ligand. We therefore suggest that VDR binds one or more naturally occurring nonvitamin D ligands to affect many of its extraosseous actions. As previously discussed, we have identified several potential examples of nonvitamin D related VDR ligands, including LCA, curcumin, γ-tocotrienol, and PUFAs, as well as a naturally occurring compound, resveratrol, that can potentiate the activity of liganded VDR. Because VDR is capable of binding alternative lipid ligands, albeit with low affinity, the receptor may have retained its promiscuity for ligand binding that presumably originated with its primitive detoxification function. Also, the LBD pocket of VDR is second only to PXR in volume among the crystallized nuclear receptor LBD domains, suggesting (but not proving) that it can accommodate a broad array of lipids. The question remains whether, in the course of its evolution, VDR coevolved higher affinity local ligands that would explain the broad health benefits of vitamin D and other lipid nutrients beyond bone. For example, 1,25D-VDR is anti-inflammatory and suppresses NFκB (Yu et al. 1995). This action would be desirable for instance in preventing atherosclerosis. Is there perhaps a local novel VDR ligand in endothelial cells that could trigger the antiinflammatory influence of VDR? Combined with the anticalcification effect of FGF23/klotho, VDR would then be able to exert a two-pronged attack in preventing arteriosclerosis. Only the future will reveal the actual mechanisms for the apparent cardiovascular benefits of vitamin D/VDR. Clearly, VDR will emerge as a versatile therapeutic and preventative target once we understand fully the pleiotropic extraosseous effects of vitamin D/VDR. As depicted in schematic form in Figure 1.10, through control of key genes, we profess that VDR feeds the "Fountain of Youth" and allows one to age well by delaying fractures, ectopic calcification, oxidative damage, infections, autoimmunity, inflammation, pain, cardiovascular disease, and malignancy.

FIGURE 1.10 **(See color insert.)** Stylistic representation of VDR regulation of vital gene expression by 1,25D and resveratrol to "feed the Fountain of Youth." Genes listed on the upper portion of the fountain (green) are induced by liganded VDR, whereas those appearing in *italics* in the lower tier of the fountain (red) are repressed.

REFERENCES

Adams, J.S., F.R. Singer, M.A. Gacad, O.P. Sharma, M.J. Hayes, P. Vouros, and M.F. Holick. 1985. Isolation and structural identification of 1,25-dihydroxyvitamin D_3 produced by cultured alveolar macrophages in sarcoidosis. *J. Clin. Endocrinol. Metab.* 60: 960–966.

Amling, M., M. Priemel, T. Holzmann, K. Chapin, J.M. Rueger, R. Baron, and M.B. Demay. 1999. Rescue of the skeletal phenotype of vitamin D receptor-ablated mice in the setting of normal mineral ion homeostasis: Formal histomorphometric and biomechanical analyses. *Endocrinology* 140: 4982–4987.

An, B.S., L.E. Tavera-Mendoza, V. Dimitrov, X. Wang, M.R. Calderon, H.J. Wang, and J.H. White. 2010. Stimulation of Sirt1-regulated FoxO protein function by the ligand-bound vitamin D receptor. *Mol. Cell. Biol.* 30: 4890–4900.

Araya, Z. and K. Wikvall. 1999. 6alpha-hydroxylation of taurochenodeoxycholic acid and lithocholic acid by CYP3A4 in human liver microsomes. *Biochem. Biophys. Acta.* 1438: 47–54.

Arriagada, G., R. Paredes, J. Olate, A. van Wijnen, J.B. Lian, G.S. Stein, J.L. Stein, S. Onate, and M. Montecino. 2007. Phosphorylation at serine 208 of the 1alpha,25-dihydroxy Vitamin D3 receptor modulates the interaction with transcriptional coactivators. *J. Steroid Biochem. Mol. Biol.* 103: 425–429.

Audo, I., S.R. Darjatmoko, C.L. Schlamp, J.M. Lokken, M.J. Lindstrom, D.M. Albert, and R.W. Nickells. 2003. Vitamin D analogues increase p53, p21, and apoptosis in a xenograft model of human retinoblastoma. *Invest Ophthalmol Vis. Sci.* 44: 4192–4199.

Barletta, F., L.P. Freedman, and S. Christakos. 2002. Enhancement of VDR-mediated transcription by phosphorylation: Correlation with increased interaction between the VDR and DRIP205, a subunit of the VDR-interacting protein coactivator complex. *Mol. Endocrinol.* 16: 301–314.

Barthel, T.K., D.R. Mathern, G.K. Whitfield, C.A. Haussler, H.A.T. Hopper, J.C. Hsieh, S.A. Slater, G. Hsieh, M. Kaczmarska, P.W. Jurutka et al. 2007. 1,25-Dihydroxyvitamin D3/VDR-mediated induction of FGF23 as well as transcriptional control of other bone anabolic and catabolic genes that orchestrate the regulation of phosphate and calcium mineral metabolism. *J. Steroid Biochem. Mol. Biol.* 103: 381–388.

Bartik, L., G.K. Whitfield, M. Kaczmarska, C.L. Lowmiller, E.W. Moffet, J.K. Furmick, Z. Hernandez, C.A. Haussler, M.R. Haussler, and P.W. Jurutka. 2010. Curcumin: A novel nutritionally derived ligand of the vitamin D receptor with implications for colon cancer chemoprevention. *J. Nutr. Biochem.* 21: 1153–1161.

Bartik, L., G.K. Whitfield, M.J. Kaczmarska, T.L. Archuleta, C.A. Haussler, M.R. Haussler, and P.W. Jurutka. 2007. Discovery of nutritionally-derived novel ligands of the vitamin D receptor: Curcumin and tocotrienols. Paper presented at The Endocrine Society (Toronto, ON, Canada).

Barwick, J.L., L.C. Quattrochi, A.S. Mills, C. Potenza, R.H. Tukey, and P.S. Guzelian. 1996. Trans-species gene transfer for analysis of glucocorticoid-inducible transcriptional activation of transiently expressed human CYP3A4 and rabbit CYP3A6 in primary cultures of adult rat and rabbit hepatocytes. *Mol. Pharmacol.* 50: 10–16.

Batie, S., J.H. Lee, M.R. Haussler, and P.W. Jurutka. 2011. Nutritional crosstalk between resveratrol and vitamin D signaling is mediated by the vitamin D receptor. Proceedings of the Arizona-Nevada Academy of Science 46: 32.

Baur, J.A. 2010. Resveratrol, sirtuins, and the promise of a DR mimetic. *Mech. Ageing Dev.* 131: 261–269.

Ben-Dov, I.Z., H. Galitzer, V. Lavi-Moshayoff, R. Goetz, M. Kuro-o, M. Mohammadi, R. Sirkis, T. Naveh-Many, and J. Silver. 2007. The parathyroid is a target organ for FGF23 in rats. *J. Clin. Invest.* 117: 4003–4008.

Bergwitz, C., and H. Juppner. 2010. Regulation of phosphate homeostasis by PTH, vitamin D, and FGF23. *Annu. Rev. Med.* 61: 91–104.

Bettoun, D.J., T.P. Burris, K.A. Houck, D.W. Buck, II, K.R. Stayrook, B. Khalifa, J. Lu, W.W. Chin, and S. Nagpal. 2003. Retinoid X Receptor Is a Nonsilent Major Contributor to Vitamin D Receptor-Mediated Transcriptional Activation. *Mol. Endocrinol.* 17: 2320–2328.

Bianco, S.D., J.B. Peng, H. Takanaga, Y. Suzuki, A. Crescenzi, C.H. Kos, L. Zhuang, M.R. Freeman, C.H. Gouveia, J. Wu et al. 2007. Marked disturbance of calcium homeostasis in mice with targeted disruption of the Trpv6 calcium channel gene. *J. Bone Miner. Res.* 22: 274–285.

Bikle, D. 2009. Extrarenal synthesis of 1,25-dihydroxyvitamin D and its health implications. *Clin. Rev. Bone Miner. Metab.* 7: 114–125.

Bikle, D.D., D. Ng, Y. Oda, K. Hanley, K. Feingold, and Z. Xie. 2002. The vitamin D response element of the involucrin gene mediates its regulation by 1,25-dihydroxyvitamin D_3. *J. Invest. Dermatol.* 119: 1109–1113.

Bikle, D.D., Y. Oda, and Z. Xie. 2004. Calcium and 1,25(OH)2D: Interacting drivers of epidermal differentiation. *J. Steroid Biochem. Mol. Biol.* 89–90: 355–360.

Bikle, D.D., and S. Pillai. 1993. Vitamin D, calcium and epidermal differentiation. *Endocr. Rev.* 14: 3–19.

Bishop, K.A., M.B. Meyer, and J.W. Pike. 2009. A novel distal enhancer mediates cytokine induction of mouse RANKl gene expression. *Mol. Endocrinol.* 23: 2095–2110.

Brumbaugh, P.F., M.R. Hughes, and M.R. Haussler. 1975. Cytoplasmic and nuclear binding components for 1a,25-dihydroxyvitamin D_3 in chick parathyroid glands. *Proc. Natl. Acad. Sci. USA* 72: 4871–4875.

Cao, X., F.P. Ross, L. Zhang, P.N. MacDonald, J. Chappel, and S.L. Teitelbaum. 1993. Cloning of the promoter for the avian integrin β_3 subunit gene and its regulation by 1,25-dihydroxyvitamin D_3. *J. Biol. Chem.* 268: 27371–27380.

Cardus, A., S. Panizo, M. Encinas, X. Dolcet, C. Gallego, M. Aldea, E. Fernandez, and J.M. Valdivielso. 2009. 1,25-dihydroxyvitamin D3 regulates VEGF production through a vitamin D response element in the VEGF promoter. *Atherosclerosis* 204: 85–89.

Cha, S.K., M.C. Hu, H. Kurosu, M. Kuro-o, O. Moe, and C.L. Huang. 2009. Regulation of renal outer medullary potassium channel and renal K(+) excretion by Klotho. *Mol. Pharmacol.* 76: 38–46.

Chawla, A., J.J. Repa, R.M. Evans, and D.J. Mangelsdorf. 2001. Nuclear receptors and lipid physiology: Opening the X-files. *Science* 294: 1866–1870.

Chen, C.D., S. Podvin, E. Gillespie, S.E. Leeman, and C.R. Abraham. 2007. Insulin stimulates the cleavage and release of the extracellular domain of Klotho by ADAM10 and ADAM17. *Proc. Natl. Acad. Sci. USA* 104: 19796–19801.

Cheng, J.B., D.L. Motola, D.J. Mangelsdorf, and D.W. Russell. 2003. De-orphanization of cytochrome P450 2R1: A microsomal vitamin D 25-hydroxylase. *J. Biol. Chem.* 278: 38084–38093.

Clemens, T.L., X.Y. Zhou, J.W. Pike, M.R. Haussler, and R.S. Sloviter. 1985. 1,25-Dihydroxyvitamin D receptor and vitamin D-dependent calcium binding protein in rat brain: Comparative immunocytochemical localization. In *Vitamin D: Chemical, Biochemiscal and Clinical Update*, ed. A.W. Norman, K. Schaefer, H.-G. Grigoleit, and D.V. Herrath, 95–96. Berlin, Germany: Walter de Gruyter.

Cohen-Lahav, M., S. Shany, D. Tobvin, C. Chaimovitz, and A. Douvdevani. 2006. Vitamin D decreases NFkappaB activity by increasing IkappaBalpha levels. *Nephrol. Dial. Transplant* 21: 889–897.

Colnot, S., M. Lambert, C. Blin, M. Thomasset, and C. Perret. 1995. Identification of DNA sequences that bind retinoid X receptor-1,25$(OH)_2D_3$-receptor heterodimers with high affinity. *Mol. Cell. Endocrinol.* 113: 89–98.

de Urquiza, A.M., S. Liu, M. Sjoberg, R.H. Zetterstrom, W. Griffiths, J. Sjovall, and T. Perlmann. 2000. Docosahexaenoic acid, a ligand for the retinoid X receptor in mouse brain. *Science* 290: 2140–2144.

DeMay, M.B., M.S. Kiernan, H.F. DeLuca, and H.M. Kronenberg. 1992. Sequences in the human parathyroid hormone gene that bind the 1,25-dihydroxyvitamin D_3 receptor and mediate transcriptional repression in response to 1,25-dihydroxyvitamin D_3. *Proc. Natl. Acad. Sci. USA* 89: 8097–8101.

Donmez, G., D. Wang, D.E. Cohen, and L. Guarente. 2010. SIRT1 suppresses beta-amyloid production by activating the alpha-secretase gene ADAM10. *Cell.* 142: 320–332.

Drissi, H., A. Pouliot, C. Koolloos, J.L. Stein, J.B. Lian, G.S. Stein, and A.J. van Wijnen. 2002. 1,25-$(OH)_2$-vitamin D_3 suppresses the bone-related Runx2/Cbfa1 gene promoter. *Exp. Cell. Res.* 274: 323–333.

Drocourt, L., J.C. Ourlin, J.M. Pascussi, P. Maurel, and M.J. Vilarem. 2002. Expression of CYP3A4, CYP2B6, and CYP2C9 is regulated by the vitamin D receptor pathway in primary human hepatocytes. *J. Biol. Chem.* 277: 25125–25132.

Dussault, I., and B.M. Forman. 2000. Prostaglandins and fatty acids regulate transcriptional signaling via the peroxisome proliferator activated receptor nuclear receptors. *Prostaglandins Other Lipid Mediat.* 62: 1–13.

Echchgadda, I., C.S. Song, A.K. Roy, and B. Chatterjee. 2004. Dehydroepiandrosterone sulfotransferase is a target for transcriptional induction by the vitamin D receptor. *Mol. Pharmacol.* 65: 720–729.

Eelen, G., C. Gysemans, L. Verlinden, R. Gijsbers, I. Beullens, M. Van Camp, J. Paik, R. DePinho, R. Bouillon, and A. Verstuyf. 2009. Induction of FOXO3a by 1,25D in MC3T3E1 cells mediates resistance to oxidative stress. Abstracts from the 14th Workshop on Vitamin D; Brugge, Belgium, October 4–8, 2009, 60.

Egan, J.B., P.A. Thompson, M.V. Vitanov, L. Bartik, E.T. Jacobs, M.R. Haussler, E.W. Gerner, and P.W. Jurutka. 2010. Vitamin D receptor ligands, adenomatous polyposis coli, and the vitamin D receptor FokI polymorphism collectively modulate beta-catenin activity in colon cancer cells. *Mol. Carcinog.* 49: 337–352.

Egan, M.E., M. Pearson, S.A. Weiner, V. Rajendran, D. Rubin, J. Glockner-Pagel, S. Canny, K. Du, G.L., Lukacs, and M.J. Caplan. 2004. Curcumin, a major constituent of turmeric, corrects cystic fibrosis defects. *Science* 304: 600–602.

Ellison, T.I., M.K. Smith, A.C. Gilliam, and P.N. MacDonald. 2008. Inactivation of the vitamin D receptor enhances susceptibility of murine skin to UV-induced tumorigenesis. *J. Invest. Dermatol.* 128: 2508–2517.

Eloranta, J.J., Z.M. Zair, C. Hiller, S. Hausler, B. Stieger, and G.A. Kullak-Ublick. 2009. Vitamin D3 and its nuclear receptor increase the expression and activity of the human proton-coupled folate transporter. *Mol. Pharmacol.* 76: 1062–1071.

Falzon, M. 1996. DNA sequences in the rat parathyroid hormone-related peptide gene responsible for 1,25-dihydroxyvitamin D_3-mediated transcriptional repression. *Mol. Endocrinol.* 10: 672–681.

Farrow, E.G., S.I. Davis, L.M. Ward, L.J. Summers, J.S. Bubbear, R. Keen, T.C. Stamp, L.R. Baker, L.F. Bonewald, and K.E. White. 2009. Molecular analysis of DMP1 mutants causing autosomal recessive hypophosphatemic rickets. *Bone* 44: 287–294.

Forster, R.E., P.W. Jurutka, J.C. Hsieh, C.A. Haussler, C.L. Lowmiller, I. Kaneko, M.R. Haussler, and G. Kerr Whitfield. 2011. Vitamin D receptor controls expression of the anti-aging klotho gene in mouse and human renal cells. *Biochem. Biophys. Res. Commun.* 414: 557–562.

Fretz, J.A., L.A. Zella, S. Kim, N.K. Shevde, and J.W. Pike. 2006. 1,25-Dihydroxyvitamin D3 regulates the expression of low-density lipoprotein receptor-related protein 5 via deoxyribonucleic acid sequence elements located downstream of the start site of transcription. *Mol. Endocrinol.* 20: 2215–2230.

Fukumoto, S. 2010. FGF23: Phosphate Metabolism and Beyond. *IBMS BoneKEy* 7: 268–278.

Garland, C., R.B. Shekelle, E. Barrett-Connor, M.H. Criqui, A.H. Rossof, and O. Paul. 1985. Dietary vitamin D and calcium and risk of colorectal cancer: A 19-year prospective study in men. *Lancet* 1: 307–309.

Gill, R.K., and S. Christakos. 1993. Identification of sequence elements in mouse calbindin-D28K gene that confer 1,25-dihydroxyvitamin D3- and butyrate-inducible responses. *Proc. Natl. Acad. Sci. USA* 90: 2984–2988.

Gombart, A.F., N. Borregaard, and H.P. Koeffler. 2005. Human cathelicidin antimicrobial peptide (CAMP) gene is a direct target of the vitamin D receptor and is strongly upregulated in myeloid cells by 1,25-dihydroxyvitamin D_3. *FASEB J* 19: 1067–1077.

Goodwin, B., M.R. Redinbo, and S.A. Kliewer. 2002. Regulation of *CYP3A* gene transcription by the pregnane X receptor. *Annu. Rev. Pharmacol. Toxicol.* 42: 1–23.

Guilliams, T.G. 2004. Homocysteine—A risk factor for vascular diseases: Guidelines for the clinical practice. *JANA* 7: 11–24.

Guo, G.L., G. Lambert, M. Negishi, J.M. Ward, H.B. Brewer, Jr., Kliewer, F.J. Gonzalez, and C.J. Sinal. 2003. Complementary roles of farnesoid X receptor, pregnane X receptor, and constitutive androstane receptor in protection against bile acid toxicity. *J. Biol. Chem.* 278: 45062–45071.

Hamada, K., A. Umemoto, A. Kajikawa, M.J. Seraj, and Y. Monden. 1994. In vitro formation of DNA adducts with bile acids. *Carcinogenesis* 15: 1911–1915.

Haussler, M.R., C.A. Haussler, L. Bartik, G.K. Whitfield, J.C. Hsieh, S. Slater, and P.W. Jurutka. 2008. Vitamin D receptor: Molecular signaling and actions of nutritional ligands in disease prevention. *Nutr. Rev.* 66: S98–112.

Haussler, M.R., C.A. Haussler, G.K. Whitfield, J.C. Hsieh, P.D. Thompson, T.K. Barthel, L. Bartik, J.B., Egan, Y. Wu, J.L. Kubicek et al. 2010. The nuclear vitamin D receptor controls the expression of genes encoding factors which feed the "Fountain of Youth" to mediate healthful aging. *J. Steroid Biochem. Mol. Biol.* 121: 88–97.

Haussler, M.R., P.W. Jurutka, M. Mizwicki, and A.W. Norman. 2011. Vitamin D receptor (VDR)-mediated actions of 1alpha,25(OH)vitamin D: Genomic and nongenomic mechanisms. *Best Pract. Res. Clin. En.* 25: 543–559.

Haussler, M.R., G.K. Whitfield, C.A. Haussler, J.C. Hsieh, P.D. Thompson, S.H. Selznick, C.E. Dominguez, and P.W. Jurutka. 1998. The nuclear vitamin D receptor: Biological and molecular regulatory properties revealed. *J. Bone Miner. Res.* 13: 325–349.

Hayes, D.P. 2011. Resveratrol and vitamin D: Significant potential interpretative problems arising from their mutual processes, interactions and effects. *Med. Hypotheses* 77: 765–772.

Hilliard, G.M.T., R.G. Cook, N.L. Weigel, and J.W. Pike. 1994. 1,25-Dihydroxyvitamin D_3 modulates phosphorylation of serine 205 in the human vitamin D receptor: Site-directed mutagenesis of this residue promotes alternative phosphorylation. *Biochemistry (Mosc)* 33: 4300–4311.

Hines, E.R., O.I. Kolek, M.D. Jones, S.H. Serey, N.B. Sirjani, P.R. Kiela, P.W. Jurutka, M.R. Haussler, J.F. Collins, and F.K. Ghishan. 2004. 1,25-dihydroxyvitamin D_3 downregulation of PHEX gene expression is mediated by apparent repression of a 110 kDa transfactor that binds to a polyadenine element in the promoter. *J. Biol. Chem.* 279: 46406–46414.

Hoerer, S., A. Schmid, A. Heckel, R.M. Budzinski, and H. Nar. 2003. Crystal Structure of the Human Liver X Receptor β Ligand-binding Domain in Complex with a Synthetic Agonist. *J. Mol. Biol.* 334: 853–861.

Honkakoski, P., T. Sueyoshi, and M. Negishi. 2003. Drug-activated nuclear receptors CAR and PXR. *Ann. Med.* 35: 172–182.

Hsieh, J.-C., G.K. Whitfield, P.W. Jurutka, C.A. Haussler, M.L. Thatcher, P.D. Thompson, H.T.L. Dang, M.A. Galligan, A.K. Oza, and M.R. Haussler. 2003. Two basic amino acids C-terminal of the P-box specify functional binding of the vitamin D receptor to its rat osteocalcin DNA responsive element. *Endocrinology* 144: 5065–5080.

Ichikawa, S., E.A. Imel, M.L. Kreiter, X. Yu, D.S. Mackenzie, A.H. Sorenson, R. Goetz, M. Mohammadi, K.E. White, and M.J. Econs. 2007. A homozygous missense mutation in human KLOTHO causes severe tumoral calcinosis. *J. Clin. Invest.* 117: 2684–2691.

Ito, M., Y. Sakai, M. Furumoto, H. Segawa, S. Haito, S. Yamanaka, R. Nakamura, M. Kuwahata, and K. Miyamoto. 2005. Vitamin D and phosphate regulate fibroblast growth factor-23 in K-562 cells. *Am. J. Physiol. Endocrinol. Metab.* 288: E1101–1109.

Jin, C.H., S.A. Kerner, M.H. Hong, and J.W. Pike. 1996. Transcriptional activation and dimerization functions in the human vitamin D receptor. *Mol. Endocrinol.* 10: 945–957.

Johnson, J.J., and H. Mukhtar. 2007. Curcumin for chemoprevention of colon cancer. *Cancer Lett.* 255: 170–181.

Juppner, H., M. Wolf, and I.B. Salusky. 2010. FGF-23: More than a regulator of renal phosphate handling? *J. Bone Miner. Res.* 25: 2091–2097.

Jurutka, P.W., L. Bartik, G.K. Whitfield, D.R. Mathern, T.K. Barthel, M. Gurevich, J.C. Hsieh, M. Kaczmarska, C.A. Haussler, and M.R. Haussler. 2007. Vitamin D receptor: Key roles in bone mineral pathophysiology, molecular mechanism of action, and novel nutritional ligands. *J. Bone Miner. Res. 22 Suppl 2*, V2-10.

Jurutka, P.W., J.-C. Hsieh, P.N. MacDonald, C.M. Terpening, C.A. Haussler, M.R. Haussler, and G.K. Whitfield 1993. Phosphorylation of serine 208 in the human vitamin D receptor: The predominant amino acid phosphorylated by casein kinase II, *in vitro*, and identification as a significant phosphorylation site in intact cells. *J. Biol. Chem.* 268: 6791–6799.

Jurutka, P.W., J.-C. Hsieh, S. Nakajima, C.A. Haussler, G.K. Whitfield, and M.R. Haussler. 1996. Human vitamin D receptor phosphorylation by casein kinase II at ser-208 potentiates transcriptional activation. *Proc. Natl. Acad. Sci. USA* 93: 3519–3524.

Jurutka, P.W., J.-C. Hsieh, L.S. Remus, G.K. Whitfield, P.D. Thompson, C.A. Haussler, J.C.G. Blanco, K. Ozato, and M.R. Haussler. 1997. Mutations in the 1,25-dihydroxyvitamin D_3 receptor identifying C-terminal amino acids required for transcriptional activation that are functionally dissociated from hormone binding, heterodimeric DNA binding and interaction with basal transcription factor IIB, *in vitro*. *J. Biol. Chem.* 272: 14592–14599.

Jurutka, P.W., P.N. MacDonald, S. Nakajima, J.-C. Hsieh, P.D. Thompson, G.K. Whitfield, M.A. Galligan, C.A. Haussler, and M.R. Haussler. 2002. Isolation of baculovirus-expressed human vitamin D receptor: DNA responsive element interactions and phosphorylation of the purified receptor. *J. Cell. Biochem.* 85: 435–457.

Jurutka, P.W., G.K. Whitfield, J.-C. Hsieh, P.D. Thompson, C.A. Haussler, and M.R. Haussler. 2001. Molecular nature of the vitamin D receptor and its role in regulation of gene expression. *Rev. Endocr. Metab. Disord.* 2: 203–216.

Kallay, E., P. Pietschmann, S. Toyokuni, E. Bajna, P. Hahn, K. Mazzucco, C. Bieglmayer, S. Kato, and H.S. Cross. 2001. Characterization of a vitamin D receptor knockout mouse as a model of colorectal hyperproliferation and DNA damage. *Carcinogenesis* 22: 1429–1435.

Kalueff, A.V., T. Keisala, A. Minasyan, M. Kuuslahti, S. Miettinen, and P. Tuohimaa. 2006. Behavioural anomalies in mice evoked by "Tokyo" disruption of the Vitamin D receptor gene. *Neurosci. Res.* 54: 254–260.

Katai, K., K. Miyamoto, S. Kishida, H. Segawa, T. Nii, H. Tanaka, Y. Tani, H. Arai, S. Tatsumi, K. Morita et al. 1999. Regulation of intestinal Na+-dependent phosphate cotransporters by a low-phosphate diet and 1,25-dihydroxyvitamin D_3. *Biochem. J.* (Pt 3) 343: 705–712.

Keisala, T., A. Minasyan, Y.R. Lou, J. Zou, A.V. Kalueff, I. Pyykko, and P. Tuohimaa. 2009. Premature aging in vitamin D receptor mutant mice. *J. Steroid Biochem. Mol. Biol.* 115: 91–97.

Kerner, S.A., R.A. Scott, and J.W. Pike. 1989. Sequence elements in the human osteocalcin gene confer basal activation and inducible response to hormonal vitamin D_3. *Proc. Natl. Acad. Sci. USA* 86: 4455–4459.

Kikuchi, R., S. Sobue, M. Murakami, H. Ito, A. Kimura, T. Iwasaki, S. Shibayama, A. Takagi, T. Kojima, M. Suzuki et al. 2007. Mechanism of vitamin D3-induced transcription of phospholipase D1 in HaCat human keratinocytes. *FEBS Lett.* 581: 1800–1804.

Kim, M.S., T. Kondo, I. Takada, M.Y. Youn, Y. Yamamoto, S. Takahashi, T. Matsumoto, S. Fujiyama, Y. Shirode, I. Yamaoka et al. 2009. DNA demethylation in hormone-induced transcriptional derepression. *Nature* 461: 1007–1012.

Kim, S., M. Yamazaki, N.K. Shevde, and J.W. Pike. 2007. Transcriptional control of receptor activator of nuclear factor-kappaB ligand by the protein kinase A activator forskolin and the transmembrane glycoprotein 130-activating cytokine, oncostatin M, is exerted through multiple distal enhancers. *Mol. Endocrinol.* 21: 197–214.

Kim, S., M. Yamazaki, L.A. Zella, N.K. Shevde, and J.W. Pike. 2006. Activation of receptor activator of NF-kappaB ligand gene expression by 1,25-dihydroxyvitamin D3 is mediated through multiple long-range enhancers. *Mol. Cell. Biol.* 26: 6469–6486.

Kitazawa, R., and S. Kitazawa. 2002. Vitamin D_3 augments osteoclastogenesis via vitamin D-responsive element of mouse RANKL gene promoter. *Biochem. Biophys. Res. Commun.* 290: 650–655.

Kliewer, S.A., J.T. Moore, L. Wade, J.L. Staudinger, M.A. Watson, S.A. Jones, D.D. McKee, B.B. Oliver, T.M. Willson, R.H. Zetterstrom et al. 1998. An orphan nuclear receptor activated by pregnanes defines a novel steroid signaling pathway. *Cell.* 92: 73–82.

Kolek, O.I., E.R. Hines, M.D. Jones, L.K. Lesueur, M.A. Lipko, P.R. Kiela, J.F. Collins, M.R. Haussler, and F.K. Ghishan. 2005. 1{alpha},25-Dihydroxyvitamin D_3 upregulates FGF23 gene expression in bone: The final link in a renal-gastrointestinal-skeletal axis that controls phosphate transport. *Am. J. Physiol. Gastrointest Liver Physiol.* 289: G1036–G1042.

Koszewski, N.J., S. Ashok, and J. Russell. 1999. Turning a negative into a positive: Vitamin D receptor interactions with the avian parathyroid hormone response element. *Mol. Endocrinol.* 13: 455–465.

Kozoni, V., G. Tsioulias, S. Shiff, and B. Rigas. 2000. The effect of lithocholic acid on proliferation and apoptosis during the early stages of colon carcinogenesis: Differential effect on apoptosis in the presence of a colon carcinogen. *Carcinogenesis* 21: 999–1005.

Kriebitzsch, C., L. Verlinden, G. Eelen, K. Marchal, B. De Moor, I. Beullens, M. Van Camp, S. Marcelis, R. Bouillon, and A. Versruyf. 2009. Regulation of cystathionine beta synthase by 1alpha,25-dihydroxyvitamin D_3, 10. Abstracts from the 14th Workshop on Vitamin D. Brugge, Belgium, October 4–8, 2009.

Kuro-o, M. 2010. Klotho. *Pflugers Arch.* 459: 333–343.

Kuro-o, M., Y. Matsumura, H. Aizawa, H. Kawaguchi, T. Suga, T. Utsugi, Y. Ohyama, M. Kurabayashi, T. Kaname, E. Kume et al. 1997. Mutation of the mouse klotho gene leads to a syndrome resembling ageing. *Nature* 390: 45–51.

Kurokawa, R., M. Söderström, A. Hörlein, S. Halachmi, M. Brown, M.G. Rosenfeld, and C.K. Glass. 1995. Polarity-specific activities of retinoic acid receptors determined by a corepressor. *Nature* 377: 451–454.

Lee, S., D.K. Lee, E. Choi, and J.W. Lee. 2005. Identification of a functional vitamin D response element in the murine Insig-2 promoter and its potential role in the differentiation of 3T3-L1 preadipocytes. *Mol. Endocrinol.* 19: 399–408.

Li, Y.C., A.E. Pirro, M. Amling, G. Delling, R. Baron, R. Bronson, and M. B. Demay. 1997. Targeted ablation of the vitamin D receptor: An animal model of vitamin D-dependent rickets type II with alopecia. *Proc. Natl. Acad. Sci. USA* 94: 9831–9835.

Lian, J.B., V. Shalhoub, F. Aslam, B. Frenkel, J. Green, M. Hamrah, G.S. Stein, and J.L. Stein. 1997. Species-specific glucocorticoid and 1,25-dihydroxyvitamin D responsiveness in mouse MC3T3-E1 osteoblasts: Dexamethasone inhibits osteoblast differentiation and vitamin D downregulates osteocalcin gene expression. *Endocrinology* 138: 2117–2127.

Lin, M., and M. Beal. 2003. The oxidative damage theory of aging. *Clin. Neurosci. Res.* 2: 305–315.

Liu, M., M.H. Lee, M. Cohen, M. Bommakanti, and L.P. Freedman. 1996a. Transcriptional activation of the Cdk inhibitor p21 by vitamin D_3 leads to the induced differentiation of the myelomonocytic cell line U937. *Genes Dev.* 10: 142–153.

Liu, P.T., S. Stenger, H. Li, L. Wenzel, B.H. Tan, S.R. Krutzik, M.T. Ochoa, J. Schauber, K. Wu, C. Meinken et al. 2006). Toll-like receptor triggering of a vitamin D-mediated human antimicrobial response. *Science* 311: 1770–1773.

Liu, S., W. Tang, J. Zhou, J.R. Stubbs, Q, Luo, M. Pi, and L.D. Quarles. 2006b. Fibroblast growth factor 23 is a counterregulatory phosphaturic hormone for vitamin D. *J. Am. Soc. Nephrol.* 17: 1305–1315.

Liu, S.M., N. Koszewski, M. Lupez, H.H. Malluche, A. Olivera, and J. Russell. 1996b. Characterization of a response element in the 5′-flanking region of the avian (chicken) PTH gene that mediates negative regulation of gene transcription by 1,25-dihydroxyvitamin D_3 and binds the vitamin D_3 receptor. *Mol. Endocrinol.* 10: 206–215.

Lu, T.T., J.J. Repa, and D.J. Mangelsdorf. 2001. Orphan nuclear receptors as eLiXiRs and FiXeRs of sterol metabolism. *J. Biol. Chem.* 276: 37735–37738.

Maglich, J.M., C.M. Stoltz, B. Goodwin, D. Hawkins-Brown, J.T. Moore, and S.A. Kliewer. 2002. Nuclear pregnane X receptor and constitutive androstane receptor regulate overlapping but distinct sets of genes involved in xenobiotic detoxification. *Mol. Pharmacol.* 62: 638–646.

Makishima, M., T.T. Lu, W. Xie, G.K. Whitfield, H. Domoto, R.M. Evans, M.R. Haussler, and D.J. Mangelsdorf. 2002. Vitamin D receptor as an intestinal bile acid sensor. *Science* 296: 1313–1316.

Malloy, P.J., L. Peng, J. Wang, and D. Feldman. 2009. Interaction of the vitamin D receptor with a vitamin D response element in the Mullerian-inhibiting substance (MIS) promoter: Regulation of MIS expression by calcitriol in prostate cancer cells. *Endocrinology* 150: 1580–1587.

Malloy, P.J., R. Xu, L. Peng, P.A. Clark, and D. Feldman. 2002. A novel mutation in helix 12 of the vitamin D receptor impairs coactivator interaction and causes hereditary 1,25-dihydroxyvitamin D-resistant rickets without alopecia. *Mol. Endocrinol.* 16: 2538–2546.

Martin, A., S. Liu, V. David, H. Li, A. Karydis, J.Q. Feng, and L.D. Quarles. 2011. Bone proteins PHEX and DMP1 regulate fibroblastic growth factor Fgf23 expression in osteocytes through a common pathway involving FGF receptor (FGFR) signaling. *FASEB J* 25: 2551–2562.

Martz, E. 2002. Protein explorer: Easy yet powerful macromolecular visualization. *Trends Biochem. Sci.* 27: 107–109.

Masuda, S., V. Byford, A. Arabian, Y. Sakai, M.B. Demay, R. St-Arnaud, and G. Jones. 2005. Altered pharmacokinetics of 1alpha,25-dihydroxyvitamin D3 and 25-hydroxyvitamin D3 in the blood and tissues of the 25-hydroxyvitamin D-24-hydroxylase (Cyp24a1) null mouse. *Endocrinology* 146: 825–834.

Masuyama, H., C.M. Brownfield, R. St-Arnaud, and P.N. MacDonald. 1997. Evidence for ligand-dependent intramolecular folding of the AF-2 domain in vitamin D receptor-activated transcription and coactivator interaction. *Mol. Endocrinol.* 11: 1507–1517.

Matsumura, Y., H. Aizawa, T. Shiraki-Iida, R. Nagai, M. Kuro-o, and Y. Nabeshima. 1998. Identification of the human klotho gene and its two transcripts encoding membrane and secreted klotho protein. *Biochem. Biophys. Res. Commun.* 242: 626–630.

McKenna, N.J., and B.W. O'Malley. 2002. Combinatorial control of gene expression by nuclear receptors and coregulators. *Cell* 108: 465–474.

Meyer, M.B., P.D. Goetsch, and J.W. Pike. 2012. VDR/RXR and TCF4/beta-Catenin Cistromes in Colonic Cells of Colorectal Tumor Origin: Impact on c-FOS and c-MYC Gene Expression. *Mol. Endocrinol.* 26: 37–51.

Meyer, M.B., M. Watanuki, S. Kim, N.K. Shevde, and J.W. Pike. 2006. The human transient receptor potential vanilloid type 6 distal promoter contains multiple vitamin D receptor binding sites that mediate activation by 1,25-dihydroxyvitamin D3 in intestinal cells. *Mol. Endocrinol.* 20: 1447–1461.

Miedlich, S.U., E.D. Zhu, Y. Sabbagh, and M.B. Demay. 2010. The receptor-dependent actions of 1,25-dihydroxyvitamin D are required for normal growth plate maturation in NPt2a knockout mice. *Endocrinology* 151: 4607–4612.

Milat, F., and K.W. Ng. 2009. Is Wnt signalling the final common pathway leading to bone formation? *Mol. Cell. Endocrinol.* 310: 52–62.

Moore, L.B., J.M. Maglich, D.D. McKee, B. Wisely, T.M. Willson, S.A. Kliewer, M.H. Lambert, and J.T. Moore. 2002. Pregnane X receptor (PXR), constitutive androstane receptor (CAR), and benzoate X receptor (BXR) define three pharmacologically distinct classes of nuclear receptors. *Mol. Endocrinol.* 16: 977–986.

Mora, J.R., M. Iwata, and U.H. von Andrian. 2008. Vitamin effects on the immune system: Vitamins A and D take centre stage. *Nat. Rev. Immunol.* 685–698.

Mordan-McCombs, S., M. Valrance, G. Zinser, M., Tenniswood, and J. Welsh. 2007. Calcium, vitamin D and the vitamin D receptor: Impact on prostate and breast cancer in preclinical models. *Nutr. Rev.* 65: S131–133.

Moreno, J., A.V. Krishnan, S. Swami, L. Nonn, D.M. Peehl, and D. Feldman. 2005. Regulation of prostaglandin metabolism by calcitriol attenuates growth stimulation in prostate cancer cells. *Cancer Res.* 65: 7917–7925.

Murayama, A., K. Takeyama, S. Kitanaka, Y. Kodera, Y. Kawaguchi, T. Hosoya, and S. Kato. 1999. Positive and negative regulations of the renal 25-hydroxyvitamin D_3 1alpha-hydroxylase gene by parathyroid hormone, calcitonin, and 1alpha,25(OH)$_2$D$_3$ in intact animals. *Endocrinology* 140: 2224–2231.

Nagengast, F.M., M.J. Grubben, and I.P. van Munster. 1995. Role of bile acids in colorectal carcinogenesis. *Eur. J. Cancer* 31A: 1067–1070.

Nishikawa, J., M. Kitaura, M. Matsumoto, M. Imagawa, and T. Nishihara. 1994. Difference and similarity of DNA sequence recognized by VDR homodimer and VDR/RXR heterodimer. *Nucleic Acids Res.* 22: 2902–2907.

Noda, M., R.L. Vogel, A.M. Craig, J. Prahl, H.F. DeLuca, and D.T. Denhardt. 1990. Identification of a DNA sequence responsible for binding of the 1,25-dihydroxyvitamin D_3 receptor and 1,25-dihydroxyvitamin D_3 enhancement of mouse secreted phosphoprotein 1 (Spp-1 or osteopontin) gene expression. *Proc. Natl. Acad. Sci. USA* 87: 9995–9999.

Norman, A.W., J.B. Frankel, A.M. Heldt, and G.M. Grodsky. 1980. Vitamin D deficiency inhibits pancreatic secretion of insulin. *Science* 209: 823–825.

Ohyama, Y., K. Ozono, M. Uchida, T. Shinki, S. Kato, T. Suda, O. Yamamoto, M. Noshiro, and Y. Kato. 1994a. Identification of a vitamin D-responsive element in the 5′ flanking region of the rat 25-hydroxyvitamin D_3 24-hydroxylase gene. *J. Biol. Chem.* 269: 10545–10550.

Ohyama, Y., K. Ozono, M. Uchida, T. Shinki, S. Kato, T. Suda, O. Yamamoto, M. Noshiro, and Y. Kato. 1994b. Identification of a vitamin D-responsive element in the 5′-flanking region of the rat 25-hydroxyvitamin D3 24-hydroxylase gene. *J. Biol. Chem.* 269: 10545–10550.

Omdahl, J.L., H.A. Morris, and B.K. May. 2002. Hydroxylase enzymes of the vitamin D pathway: Expression, function, and regulation. *Annu. Rev. Nutr.* 22: 139–166.

Oury, F., G. Sumara, O. Sumara, M. Ferron, H. Chang, C.E. Smith, L. Hermo, S. Suarez, B.L. Roth, P. Ducy et al. 2011. Endocrine regulation of male fertility by the skeleton. *Cell* 144: 796–809.

Parks, D.J., S.G. Blanchard, R.K. Bledsoe, G. Chandra, T.G. Consler, S.A. Kliewer, J.B. Stimmel, T.M. Willson, A.M. Zavacki, D.D. Moore et al. 1999. Bile acids: Natural ligands for an orphan nuclear receptor. *Science* 284: 1365–1368.

Pathrose, P., O. Barmina, C.Y. Chang, D.P. McDonnell, N.K. Shevde, and J.W. Pike. 2002. Inhibition of 1,25-dihydroxyvitamin D_3-dependent transcription by synthetic LXXLL peptide antagonists that target the activation domains of the vitamin D and retinoid X receptors. *J. Bone Miner. Res.* 17: 2196–2205.

Peng, L., P.J. Malloy, and D. Feldman. 2004. Identification of a functional vitamin D response element in the human insulin-like growth factor binding protein-3 promoter. *Mol. Endocrinol.* 18: 1109–1119.

Perwad, F., M.Y. Zhang, H.S. Tenenhouse, and A.A. Portale. 2007. Fibroblast growth factor 23 impairs phosphorus and vitamin D metabolism in vivo and suppresses 25-hydroxyvitamin D-1alpha-hydroxylase expression in vitro. *Am. J. Physiol. Renal Physiol.* 293: F1577–1583.

Pike, J.W., M.B. Meyer, M. Watanuki, S. Kim, L.A. Zella, J.A. Fretz, M. Yamazaki, and N.K. Shevde. 2007. Perspectives on mechanisms of gene regulation by 1,25-dihydroxyvitamin D_3 and its receptor. *J. Steroid Biochem. Mol. Biol.* 103: 389–395.

Quarles, L.D. 2008. Endocrine functions of bone in mineral metabolism regulation. *J. Clin. Invest.* 118: 3820–3828.

Quelo, I., I. Machuca, and P. Jurdic. 1998. Identification of a vitamin D response element in the proximal promoter of the chicken carbonic anhydrase II gene. *J. Biol. Chem.* 273: 10638–10646.

Rachez, C., M. Gamble, C.P. Chang, G.B. Atkins, M.A. Lazar, and L.P. Freedman. 2000. The DRIP complex and SRC-1/p160 coactivators share similar nuclear receptor binding determinants but constitute functionally distinct complexes. *Mol. Cell. Biol.* 20: 2718–2726.

Raghuwanshi, A., S.S. Joshi, and S. Christakos 2008. Vitamin D and multiple sclerosis. *J. Cell. Biochem.* 105: 338–343.

Ramagopalan, S.V., N.J. Maugeri, L. Handunnetthi, M.R. Lincoln, S.M. Orton, D.A. Dyment, G.C. Deluca, B.M. Herrera, M.J. Chao, A.D. Sadovnick et al. 2009. Expression of the multiple sclerosis-associated MHC class II Allele HLA-DRB1*1501 is regulated by vitamin D. *PLoS Genetics* 5: e1000369.

Razzaque, M.S. 2009. The FGF23-Klotho axis: Endocrine regulation of phosphate homeostasis. *Nat. Rev. Endocrinol.* 5: 611–619.

Rochel, N., F. Ciesielski, J. Godet, E. Moman, M. Roessle, C. Peluso-Iltis, M. Moulin, M. Haertlein, P. Callow, Y. Mely et al. 2011. Common architecture of nuclear receptor heterodimers on DNA direct repeat elements with different spacings. *Nat. Struct. Mol. Biol.* 18: 564–570.

Rochel, N., J.M. Wurtz, A. Mitschler, B. Klaholz, and D. Moras. 2000. The crystal structure of the nuclear receptor for vitamin D bound to its natural ligand. *Mol. Cell.* 5: 173–179.

Saeki, M., K. Kurose, M., Tohkin, and R. Hasegawa 2008. Identification of the functional vitamin D response elements in the human MDR1 gene. *Biochem. Pharmacol.* 76: 531–542.

Sakai, Y., J. Kishimoto, and M.B. Demay 2001. Metabolic and cellular analysis of alopecia in vitamin D receptor knockout mice. *J. Clin. Invest.* 107: 961–966.

Saramaki, A., S. Diermeier, R. Kellner, H. Laitinen, S. Vaisanen, and C. Carlberg. 2009. Cyclical chromatin looping and transcription factor association on the regulatory regions of the p21 (CDKN1A) gene in response to 1alpha,25-dihydroxyvitamin D_3. *J. Biol. Chem.* 284: 8073–8082.

Seshadri, S., A. Beiser, J. Selhub, P.F. Jacques, I.H. Rosenberg, R.B. D'Agostino, P.W. Wilson, and P.A. Wolf. 2002. Plasma homocysteine as a risk factor for dementia and Alzheimer's disease. *N. Engl. J. Med.* 346: 476–483.

Shimada, T., H. Hasegawa, Y. Yamazaki, T. Muto, R. Hino, Y. Takeuchi, T. Fujita, K. Nakahara, S. Fukumoto, and T. Yamashita. 2004a. FGF-23 is a potent regulator of vitamin D metabolism and phosphate homeostasis. *J. Bone Miner. Res.* 19: 429–435.

Shimada, T., M. Kakitani, Y. Yamazaki, H. Hasegawa, Y. Takeuchi, T. Fujita, S. Fukumoto, K. Tomizuka, and T. Yamashita. 2004b. Targeted ablation of Fgf23 demonstrates an essential physiological role of FGF23 in phosphate and vitamin D metabolism. *J. Clin. Invest.* 113: 561–568.

Shirakawa, A.K., D. Nagakubo, K. Hieshima, T. Nakayama, Z. Jin, and O. Yoshie. 2008. 1,25-dihydroxyvitamin D_3 induces CCR10 expression in terminally differentiating human B cells. *J. Immunol.* 180: 2786–2795.

Sidelnikov, E., R.M. Bostick, W.D. Flanders, Q. Long, V. Fedirko, A. Shaukat, C.R. Daniel, and R.E. Rutherford. 2010. Effects of calcium and vitamin D on MLH1 and MSH2 expression in rectal mucosa of sporadic colorectal adenoma patients. *Cancer Epidemiol. Biomarkers Prev.* 19: 1022–1032.

Sroga G.E., L. Karim, W. Colon, and D. Vashishith. 2011. Biochemical characterization of Major bone-matrix proteins using nano-scale size bone samples and proteomics methodology. *Mol. cell Proteomics* 10:MIID.006718.

St-Arnaud, R. 2010. CYP24A1-deficient mice as a tool to uncover a biological activity for vitamin D metabolites hydroxylated at position 24. *J. Steroid Biochem. Mol. Biol.* 121: 254–256.

Staal, A., A.J. van Wijnen, J.C. Birkenhäger, H.A.P. Pols, J. Prahl, H. DeLuca, M.-P. Gaub, J.B. Lian, G.S. Stein, J.P.T.M. van Leeuwen et al. 1996. Distinct conformations of vitamin D receptor/retinoid X receptor-α heterodimers are specified by dinucleotide differences in the vitamin D-responsive elements for the osteocalcin and osteopontin genes. *Mol. Endocrinol.* 10: 1444–1456.

Staudinger, J.L., B. Goodwin, S.A. Jones, D. Hawkins-Brown, K.I. MacKenzie, A. LaTour, Y. Liu, C.D. Klaassen, K.K. Brown, J. Reinhard et al. 2001. The nuclear receptor PXR is a lithocholic acid sensor that protects against liver toxicity. *Proc. Natl. Acad. Sci. USA* 98: 3369–3374.

Stubbs, J.R., S. Liu, W. Tang, J. Zhou, Y. Wang, X. Yao, and L.D. Quarles. 2007. Role of hyperphosphatemia and 1,25-dihydroxyvitamin D in vascular calcification and mortality in fibroblastic growth factor 23 null mice. *J. Am. Soc. Nephrol.* 18: 2116–2124.

Taketani, Y., H. Segawa, M. Chikamori, K. Morita, K. Tanaka, S. Kido, H. Yamamoto, Y. Iemori, S. Tatsumi, N. Tsugawa et al. 1998. Regulation of type II renal Na⁺-dependent inorganic phosphate transporters by 1,25-dihydroxyvitamin D_3. Identification of a vitamin D-responsive element in the human NAPi-3 gene. *J. Biol. Chem.* 273: 14575–14581.

Terpening, C.M., C.A. Haussler, P.W. Jurutka, M.A. Galligan, B.S. Komm, and M.R. Haussler. 1991. The vitamin D-responsive element in the rat bone Gla protein gene is an imperfect direct repeat that cooperates with other cis-elements in 1,25-dihydroxyvitamin D3-mediated transcriptional activation. *Mol. Endocrinol.* 5: 373–385.

Thompson, E.B., and R. Kumar. 2003. DNA binding of nuclear hormone receptors influences their structure and function. *Biochem. Biophys. Res. Commun.* 306: 1–4.

Thompson, P.D., P.W. Jurutka, G.K. Whitfield, S.M. Myskowski, K.R. Eichhorst, C.E. Dominguez, C.A. Haussler, and M.R. Haussler. 2002a. Liganded VDR induces CYP3A4 in small intestinal and colon cancer cells via DR3 and ER6 vitamin D responsive elements. *Biochem. Biophys. Res. Commun.* 299: 730–738.

Thompson, P.D., P.W. Jurutka, G.K. Whitfield, S.M. Myskowski, K.R. Eichhorst, C. Encinas Dominguez, C.A. Haussler, and M.R. Haussler. 2002b. Liganded VDR induces CYP3A4 in small intestinal and colon cancer cells via DR3 and ER6 vitamin D responsive elements. *Biochem. Biophys. Res. Commun.* 299: 730–738.

Thompson, P.D., L.S. Remus, J.-C. Hsieh, P.W. Jurutka, G.K. Whitfield, M.A. Galligan, C. Encinas Dominguez, C.A. Haussler, and M.R. Haussler. 2001. Distinct retinoid X receptor activation function-2 residues mediate transactivation in homodimeric and vitamin D receptor heterodimeric contexts. *J. Mol. Endocrinol.* 27: 211–227.

Thummel, K.E., C. Brimer, K. Yasuda, J. Thottassery, T. Senn, Y. Lin, H. Ishizuka, E. Kharasch, J. Schuetz, and E. Schuetz. 2001. Transcriptional control of intestinal cytochrome P-4503A by 1alpha,25-dihydroxy vitamin D_3. *Mol. Pharmacol.* 60: 1399–1406.

Thurston, R.D., C.B. Larmonier, P.M. Majewski, R. Ramalingam, M. Midura-Kiela, D. Laubitz, A. Vandewalle, D.G. Besselsen, M. Muhlbauer, C. Jobin et al. 2010. Tumor necrosis factor and interferon-gamma down-regulate Klotho in mice with colitis. *Gastroenterology* 138: 1384–1394, 1394, 1381–1382.

Tsujikawa, H., Kurotaki, Y., Fujimori, T., Fukuda, K., and Nabeshima, Y. 2003. Klotho, a gene related to a syndrome resembling human premature aging, functions in a negative regulatory circuit of vitamin D endocrine system. *Mol. Endocrinol.* 17: 2393–2403.

van Den Bemd, G.J., M. Jhamai, A. Staal, A.J. van Wijnen, J.B. Lian, G.S. Stein, H.A. Pols, and J.P. van Leeuwen. 2002. A central dinucleotide within vitamin D response elements modulates DNA binding and transactivation by the vitamin D receptor in cellular response to natural and synthetic ligands. *J. Biol. Chem.* 277: 14539–14546.

Wang, T.T., B. Dabbas, D. Laperriere, A.J. Bitton, H. Soualhine, L.E. Tavera-Mendoza, S. Dionne, M.J. Servant, A. Bitton, E.G. Seidman et al. 2010. Direct and indirect induction by 1,25-dihydroxyvitamin D_3 of the NOD2/CARD15-defensin beta2 innate immune pathway defective in Crohn disease. *J. Biol. Chem.* 285: 2227–2231.

Wang, T.T., L.E. Tavera-Mendoza, D. Laperriere, E. Libby, N.B. MacLeod, Y. Nagai, V. Bourdeau, A. Konstorum, B. Lallemant, R. Zhang et al. 2005. Large-scale in silico and microarray-based identification of direct 1,25-dihydroxyvitamin D_3 target genes. *Mol. Endocrinol.* 19: 2685–2695.

Wang, X., T.T. Wang, J.H. White, and G.P. Studzinski. 2006. Induction of kinase suppressor of RAS-1(KSR-1) gene by 1, alpha25-dihydroxyvitamin D_3 in human leukemia HL60 cells through a vitamin D response element in the 5′-flanking region. *Oncogene* 25: 7078–7085.

Wang, X., T.T. Wang, J.H. White, and G.P. Studzinski. 2007. Expression of human kinase suppressor of Ras 2 (hKSR-2) gene in HL60 leukemia cells is directly upregulated by 1,25-dihydroxyvitamin $D(_3)$ and is required for optimal cell differentiation. *Exp. Cell. Res.* 313: 3034–3045.

Wang, Y. and Z. Sun. 2009. Klotho gene delivery prevents the progression of spontaneous hypertension and renal damage. *Hypertension* 54: 810–817.

Warnmark, A., E. Treuter, A. Wright, and J.A. Gustafsson. 2003. Activation Functions 1 and 2 of Nuclear Receptors: Molecular Strategies for Transcriptional Activation. *Mol. Endocrinol.* 17: 1901–1909.

Watkins, R.E., J.M. Maglich, L.B. Moore, G.B. Wisely, S.M. Noble, P.R. Davis-Searles, M.H. Lambert, S.A. Kliewer, and M.R. Redinbo. 2003. 2.1 A crystal structure of human PXR in complex with the St. John's wort compound hyperforin. *Biochemistry (Mosc)* 42: 1430–1438.

Watkins, R.E., G.B. Wisely, L.B. Moore, J.L. Collins, M.H. Lambert, S.P. Williams, T.M. Willson, S.A. Kliewer, and M.R. Redinbo. 2001. The human nuclear xenobiotic receptor PXR: Structural determinants of directed promiscuity. *Science* 292: 2329–2333.

Wecksler, W.R., H.L. Henry, and A.W. Norman. 1977. Studies on the mode of action of calciferol. Subcellular localization of 1,25-dihydroxyvitamin D_3 in chicken parathyroid glands. *Arch. Biochem. Biophys.* 183: 168–175.

Weissen-Plenz, G., Y. Nitschke, and F. Rutsch. 2008. Mechanisms of arterial calcification: Spotlight on the inhibitors. *Adv. Clin. Chem.* 46: 263–293.

Whitfield, G.K., P.W. Jurutka, C.A. Haussler, J.C. Hsieh, T.K. Barthel, E.T. Jacobs, C. Encinas Dominguez, M.L. Thatcher, and M.R. Haussler. 2005. Nuclear vitamin D receptor: Structure–function, molecular control of gene transcription, and novel bioactions. In *Vitamin D*, ed. D. Feldman, J.W. Pike, and F.H. Glorieux, 219–261. Oxford, U.K.: Elsevier Academic Press.

Whitfield, G.K., S.H. Selznick, C.A. Haussler, J.-C. Hsieh, M.A. Galligan, P.W. Jurutka, P.D. Thompson, S.M. Lee, J.E. Zerwekh, and M.R. Haussler. 1996. Vitamin D receptors from patients with resistance to 1,25-dihydroxyvitamin D_3: Point mutations confer reduced transactivation in response to ligand and impaired interaction with the retinoid X receptor heterodimeric partner. *Mol. Endocrinol.* 10: 1617–1631.

Xie, W., A. Radominska-Pandya, Y. Shi, C.M. Simon, M.C. Nelson, E.S. Ong, D.J. Waxman, and R.M. Evans. 2001. An essential role for nuclear receptors SXR/PXR in detoxification of cholestatic bile acids. *Proc. Natl. Acad. Sci. USA* 98: 3375–3380.

Yu, X.-P., T. Bellido, and S.C. Manolagas. 1995. Downregulation of NF-kB protein levels in activated human lymphocytes by 1,25-dihydroxyvitamin D_3. *Proc. Natl. Acad. Sci. USA* 92: 10990–10994.

Zella, L.A., S. Kim, N.K. Shevde, and J.W. Pike. 2007. Enhancers located in the vitamin D receptor gene mediate transcriptional autoregulation by 1,25-dihydroxyvitamin D_3. *J. Steroid. Biochem. Mol. Biol.* 103: 435–439.

Zhang, J., M.J. Chalmers, K.R. Stayrook, L.L. Burris, Y. Wang, S.A. Busby, B.D. Pascal, R.D. Garcia-Ordonez, J.B. Bruning, M.A. Istrate et al. 2011. DNA binding alters coactivator interaction surfaces of the intact VDR-RXR complex. *Natl. Struct. Mol. Biol.* 18: 556–563.

Zierold, C., H.M. Darwish, and H.F. DeLuca. 1994a. Identification of a vitamin D-response element in the rat calcidiol (25-hydroxyvitamin D_3) 24-hydroxylase gene. *Proc. Natl. Acad. Sci. USA* 91: 900–902.

Zierold, C., H.M. Darwish, and H.F. DeLuca. 1994b. Identification of a vitamin D-responsive element in the rat calcidiol (25-hydroxyvitamin D_3) 24-hydroxylase gene. *Proc. Natl. Acad. Sci. USA* 91: 900–902.

Zinser, G., K. Packman, and J. Welsh. 2002a. Vitamin D_3 receptor ablation alters mammary gland morphogenesis. *Development* 129: 3067–3076.

Zinser, G.M., J.P. Sundberg, and J. Welsh. 2002b. Vitamin D_3 receptor ablation sensitizes skin to chemically induced tumorigenesis. *Carcinogenesis* 23: 2103–2109.

Zou, A., M.G. Elgort, and E.A. Allegretto. 1997. Retinoid X receptor (RXR) ligands activate the human 25-hydroxyvitamin D_3-24-hydroxylase promoter via RXR heterodimer binding to two vitamin D-responsive elements and elicit additive effects with 1,25-dihydroxyvitamin D_3. *J. Biol. Chem.* 272: 19027–19034.

2 Vitamin D Receptor: Genomic and Epigenomic Effects

Prashant K. Singh and Moray J. Campbell

CONTENTS

2.1 INTRODUCTION

2.1.1 TRANSCRIPTIONAL CONTROL OF THE HUMAN GENOME

Complete sequencing in 1998 of the first multicellular organism, *Caenorhabditis elegans*, revealed it to have about 19,000 genes. Surprisingly, completion of the human genome sequence several years later revealed just 4000 more genes [1]. Complexity, therefore, as measured by the number of different cell types and the behaviours of intra- and intercellular networks, clearly does not relate directly to the number of genes. It is revealing that, while the human genome encodes approximately 3000 transcription factors regulating 23,000 genes spread over 3300 mb of DNA, that of *C. elegans* encodes just 600 transcription factors, regulating 19,000 genes spread over 100 mb of DNA. Therefore, transcription factor numbers and gene spacing relate more closely to complexity than gene numbers [2]. Furthermore, the recent identification of genome-wide transcription even from DNA in noncoding regions suggests further aspects of integration to control overall regulation of expression [3,4]. The appreciation of the diversity and regulation of human protein expression is likely to be increased further, e.g., as the impact emerges fully of alternative mRNA splicing and the range and function of noncoding RNA.

Evidence also supporting a transcriptional basis to human complexity emerges from analyses of transcription factors that may have been positively selected for in humans, compared to other primates. Multispecies cDNA expression array studies have suggested stabilizing selection for many genes across primate phyla, but the subset of genes showing selective increase in expression in humans had a notable excess of transcription factors, which is a result consistent with positive selection [5]. Thus, transcriptional control of coding and noncoding genes appears central to human complexity, in health, and most likely in disease.

2.1.2 Epigenetic Events Control the Rate and Magnitude of Transcription

Of the various protein–DNA and RNA–DNA interactions that mediate genome function, those between the histones and DNA are of pre-eminent importance due to both the frequency of the histones and the intimacy of their interactions. 146 bp of DNA is packaged in 13/4 superhelical turns around an octamer of histones (two each of H2A, H2B, H3, and H4) [6], which is a structure that is found in virtually all eukaryotes. Movement of nucleosomes either away or toward transcription-factor-binding sites can, in itself, modulate transcription factor binding, but the nucleosome can also influence genomic functions in more subtle ways.

The four core histones are subject to more than 100 different post-translational modifications (PTMs) to individual amino acids, including acetylation of lysine, methylation of lysine and arginine, and phosphorylation of serine and threonine (reviewed in [7]). The most extensively studied modifications are found along the N-terminal histone tails, which are regions that contain little secondary structure and are exposed on the surface of the nucleosome [6]. However, other studies, which are often by mass spectrometry, have shown that numerous amino acids internal to the nucleosome are subject to modifications that are exactly analogous to those on the histone tails, and they may serve to govern nucleosome mobility [8].

The list of enzymes identified to be involved in setting and removing histone modifications are increasingly numerous; for example, there are 18 histone deacetylases in humans and mice and 28 different methyltransferases known to act on histones, at least *in vitro* [9]. Histone modifications can exert a direct effect on chromatin structure, sometimes by altering the net charge on the histone tail and reducing histone-DNA binding. Alternatively, specific modified residues, or the combinations thereof, can form binding sites for nonhistone proteins that, in turn, influence chromatin structure and function.

These observations formed the basis for the histone code hypothesis. This concept, which was first proposed in 1993, held that these modifications were governed in a coordinated manner and formed a code that mirrored the underlying DNA code to convey heritable information on transcription and expression [10]. Given the rapid expansion of the understanding in the number of histone modifications, their genomic distribution, and their combinatorial manner, it is actually only relatively recently that the true diversity of the range of binding options available and their functional outcomes has become apparent [11]. The strongest evidence that histone modifications at the level of metachromatin architecture form a stable and heritable "histone code" is perhaps seen with X chromosome inactivation (reviewed in [12]). The extent to which similar processes operate to govern the activity of microchromatin contexts, such as gene promoter regions, is an area of debate [13,14]. The regulation of transcription and the patterns of mRNA expression have been related to the expression of these histone modifications through a wide range of correlative and functional studies. In this manner, the status of histone modifications appears to control the transcriptional action of gene loci.

Acetylation of lysine residues of H3 are well established to be associated with gene activation. However, methylated marks can have diverse impact. For example, methylation of H3K4 and H3K9 are distinctive marks of euchromatin and heterochromatin, respectively, and are mutually exclusive. This is illustrated by the control of transcription by the androgen receptor (AR). The histone demethylase enzymes KDM1A/LSD1 and JMJD2C are significant regulators of the AR's ability to govern transcription [15,16]. JMJD2C demethylates H3K9me3, whereas KDM1A/LSD1 demethylates H3K9me2/me1 at promoters such as *PSA* and *KLK2* to remove H3K9 methylation associated with transcriptional silencing. By contrast, KDM1A/LSD1, as a member of the CoREST corepressor complex, demethylates H3K4me2 and me1, with H3K4me3 being associated with transcriptional activation. Therefore, the complex that KDM1A/LSD1 interacts with profoundly alters the transcriptional outputs since demethylation of H3K9 has a gene-activating effect, whereas demethylation of H3K4 has a gene-silencing effect.

This differential regulation of histone methylation has profound implications for transcriptional control. DNA methylation and H3K4 methylation are mutually exclusive, whereas H3K9 methylation is strongly associated with DNA methylation, e.g., through the formation of heterochromatin by HP1 binding and histone deacetylation. Absent of DNA methylation, these interrelationships are highly dynamic, with target gene promoters often poised to be subsequently pushed toward a fully active or a more stably repressed state. For example, CpG island promoter regions of nonexpressed genes do, in fact, show low-level RNA Pol II association and modest transcriptional initiation. It seems that the presence of H3K4 methylation holds these promoters in a chromatin structure that is accessible to the transcriptional machinery, poised to recruit specific transcription factors to drive high-level, efficient transcription. In turn, this prevents H3K9 methylation and DNA methylation. Aberrant DNA methylation of these CpG islands in cancer cells reduces this plasticity and coincides with loss of H3K4 methylation; gain of H3K9 methylation, along with other heterochromatin marks; and stable transcriptional silencing [17]. The distributions of these histone modifications in various cell line models are being organized by various research consortia, e.g., ENCODE [18]. Again, these genome-wide data sets appear to support the idea that these histone marks are strongly associated with genomic architecture, such as gene regions, transcription start sites (TSSs), and enhancer regions where regulatory transcription factors can bind.

Two further points are particularly important; first, the steady-state level of each modification represents a highly dynamic balance between the antagonistic effects of key enzymes (with turnover likely to vary from one part of the genome to another and between cell types). Second, many, if not all, of the enzymes are dependent on or influenced by either metabolites or components present in the intra- or extracellular environment. Thus, the nucleosome, through the array of histone modifications it carries and the enzymes that put them in place, is a finely tuned sensor of the metabolic state of the cell and the composition of its environment. In this manner, it provides a platform through which external environmental and internal variables can influence genomic function. In this manner, epigenetic states are a key modulator of transcriptional capacity and are regulated directly by cell context such as cell cycle status, cell–cell interactions, and cell lineage commitment.

2.1.3 HIGHER ORDER CHROMATIN INTERACTIONS ASSOCIATED WITH TRANSCRIPTION

Another theme that has emerged concerning epigenetic regulation of transcription is higher order chromosomal interactions. It seems that large-scale chromatin rearrangement, through looping, is frequent and widespread; can be inter- or intrachromosomal; and is guided by transcription factors, key pioneer factors, and chromatin-modifying enzymes [19,20]. Improved microscopy techniques have recently shown nascent RNA on the surface of protein dense transcription factories ("gene hubs") that seem to correspond to structures previously termed "nuclear speckles" [21].

A clear example of these interactions has been illustrated in the transcriptional responses of B-cells, where translocation of genes occurs from separate chromosomes and nuclear regions to common sites referred to as transcription factories. These sites contain significant levels of RNA Pol II and other proteins, including factors required for elongation, chromatin remodeling, capping, splicing, and nonsense mediated decay. Recruitment of genes to transcription factories is highly selective, with certain genes and chromosome regions colocalizing far more frequently than expected by chance. Intriguingly, sites of chromosome translocation associated with various cancers often co-localize. For example, *Myc* and *Igh* tend to co-localize, and their fusion, in human lymphoid cells, is a common cause of Burkitt's lymphoma. These rapid movements are associated with movements of the nuclear architecture and involve ATP-dependent mechanisms that involve a chromosome locus that is usually located at the nuclear periphery being rapidly translocated to the interior in a direction perpendicular to the nuclear membrane [22].

2.2   VDR AS A MEMBER OF THE NUCLEAR RECEPTOR SUPERFAMILY

2.2.1   IDENTIFICATION OF THE VDR

Nuclear receptors (NRs) are both receptors and transcription factors. The functions mediated by NRs depend on their translocation to the nucleus either bound to ligand or in a ligand-free state that is subsequently able to bind ligand when within the nucleus. These receptors commonly function within multicomponent transcriptional complexes that selectively regulate target gene expression. Indeed, a great deal of the understanding of gene transcription has arisen from the study of NR-mediated transactivation and transrepression.

In 1985, the primary structure of a cDNA encoding a human glucocorticoid receptor was described [23]. At the same time, cDNA encoding the first human estrogen receptor was also cloned [24]. In the latter study, it was noted that the cloned estrogen receptor had extensive homology with the v-*erb-A* oncogene of the avian erythroblastosis virus. Similarly, a follow-up study by the group that had identified the glucocorticoid receptor also reported homologies with v-*erb-A* [25]. These observations quickly led to the identification of the c-*erbA* locus as the gene encoding the human and chicken thyroid hormone receptors [26]. The next year, in 1987, the cloning of the avian vitamin D receptor (VDR) was described using newly available receptor-specific monoclonal antibodies to recover the appropriate cDNA [27]. The identification that glucocorticoid, estrogen, thyroid, and VDRs were structurally related prompted researchers to seek additional family members. Thus, in just two years, the field went from being unaware of how cells perceived diverse lipophilic signaling hormones to having identified four receptors that displayed cross-specie conservation. Most remarkable was the realization that chemically distinct hormones and other small lipophilic molecules were demonstrably signaling via highly related receptors, providing the first evidence of a new receptor family. Subsequently, it emerged that the NR superfamily is the largest superfamily of human transcription factors.

Phylogenetic classification has defined seven NR subfamilies, and within these, the VDR is in the group-1 subfamily, sharing homology with the liver X receptors (LXRs) and farnesoid X receptor (FXR), and, more distantly, the peroxisome proliferator-activated receptors (PPARs) [28,29]. The receptors within this subfamily preferentially form homo- or heterodimeric complexes, with RXR acting as a common central partner for VDR, PPARs, LXRs, and FXR. Thus, the receptors in the group appear to be all responsive to either bile acid or xenobiotic receptors and therefore widely integrated with bile acid homeostasis and detoxification. In keeping with this capacity, the bile acid lithocholic acid (LCA) has recently been shown to be a potent ligand for the VDR, all be it with lower millimolar affinity [30].

Human NR can also be divided into three groups based on their affinity to their ligands: 1) endocrine receptors (e.g., ERs, AR, VDR, RARs, and TR), which bind their ligands with high affinity (K_d 0.1–1 nM); 2) adopted orphan receptors (e.g., PPARs, LXRs, and FXR), which bind endogenous ligands with low affinity (K_d 1–1000 uM); and 3) orphan receptors (e.g., DAX-1, SF-1, and ROR), for which a ligand either does not exist or has not been discovered. The transcriptional actions of these receptors are finely controlled by several hetero-complexes. With few exceptions, corepressor complexes suppress receptor activity, enhancing their silencing effect in opposition to coactivators that, instead, drive transactivation. Upon ligand binding, NRs undergo conformational changes that lower affinity for corepressor complexes and promote the interaction with coactivator complexes, which allows for transcriptional activation [31]. It is worth noting that, despite their name, NRs are actually found dynamically distributed between different cellular compartments, and indeed, this provides a signaling sensitivity advantage [32].

2.2.2 ACTIVATION OF THE VDR BY LIGAND

$1\alpha25(OH)_2D_3$ is the principal biologically active ligand for the VDR and, in common with most NR ligands, is highly hydrophobic and transported in the aqueous blood stream associated with a specific binding protein (DBP) [33,34]. At the cell membrane, free diffusion across the lipid membrane occurs, although the identification of Megalin as an active transport protein for $25(OH)D_3$ suggests that transport into the cell of vitamin D_3 metabolites may be more tightly regulated than merely by passive diffusion alone [35]. Similarly, for various NRs, it has been predicted that transport across the aqueous cytoplasm by specific transport proteins is an important regulatory mechanism and will most likely apply to the VDR as well [32]. Interestingly, there is also evidence for the VDR to be trafficked actively into the nucleus upon ligand activation, in tandem with heterodimeric partner RXRs [36], each in association with specific importins [37].

In the absence of ligand, the VDR may be distributed throughout the cell, although predominantly located in the nucleus. There is evidence of cytoplasmic expression and cell-membrane-associated VDR that may mediate nongenomic signal transduction responses [38,39]. This is a feature of several NRs, such as the ERα, where the NR is cycled through caveolae at the cell membrane to initiate signal transduction pathways [39,40]. The contribution of these actions to the overall functions of $1\alpha,25(OH)_2D_3$ remains to be clarified fully.

The majority of research to date has addressed a nuclear function for the VDR that binds to the regulatory regions of target genes and therefore is associated with direct regulation of transcription. Structurally, the VDR is uncommon, compared to other NRs, as it does not contain an activation domain at its amino terminus (AF1). In most other receptors, this is an important domain for activation, e.g., for autonomous ligand-independent AF function domain. The VDR instead relies on a domain in the carboxy terminus (AF-2) for activation and other domains for heterodimerization with RXR [41]. The VDR ligand-binding pocket contains hydrophobic residues such as His-305 and -397, which are important in the binding of $1\alpha25(OH)_2D_3$. Ligand binding specifically requires interaction of the hydroxyl group of the A ring at carbon 1 of $1\alpha,25(OH)_2D_3$, which is added by the action of the 1α hydroxylase enzyme (encoded by *CYP27B1*).

Once $1\alpha25(OH)_2D_3$ is bound, it causes a conformational change in the ligand-binding domain (LBD), which allows the C-terminal helix 12 of the AF2 domain to reposition into an active conformation exposing a docking surface for transcriptional coregulators [42–44]. This switch of conformation of the LBD in the presence of ligand is a common feature of all ligand-binding NRs and facilitates the switch of receptor–cofactor interactions. Thus, both the unliganded and liganded VDR associates with a large number of different proteins involved with transcriptional suppression and activation, respectively.

2.2.3 VDR ASSOCIATES WITH CHROMATIN IN LARGE MULTIMERIC COMPLEXES

When located within the nucleus and in the absence of ligand, the VDR exist in an *'apo'* state associated with RXR and corepressors (e.g., NCOR1 and NCOR2/SMRT) [45,46] as part of large complexes (~2.0 MDa) [45,47] and bound to VDR response elements (VDREs). These complexes, in turn, actively recruit a wide range of enzymes that post-translationally modify histone tails, including HDACs and KDMs, and thereby maintain a locally condensed chromatin structure around response element sequences [48–51]. Ligand binding induces a so-called *holo* state, facilitating the association of the VDR-RXR dimer with coactivator complexes. A large number of interacting coactivator proteins have been described, which can be divided into multiple families, including the p160 family, the non-p160 members, and members of the large 'bridging' TRAP/DRIP/ARC complex, which links the receptor complex to the co-integrators CBP/p300 and basal transcriptional machinery [52,53]. These complexes contain and include within them enzymes such as HATs and KATs that antagonize the enzyme corepressor function and open up chromatin.

The specific and complex choreography of the exchange of receptor bound complex has been investigated in the context of the VDR [48,54–59] and other NRs [60–63]. Collectively, these approaches have revealed that the exchange of these factors is highly dynamic and involves cyclical rounds of promoter-specific complex assembly, gene transactivation, complex disassembly, and proteosome-mediated receptor degradation coincident with corepressor complex binding and silencing of transcription. More recent data have emerged to support chromatin looping within the same VDR target gene loci [45]. This gives rise to the characteristic periodicity of NR transcriptional activation and pulsatile mRNA and protein accumulation. These actions appear to be highly specific to individual genes and the separate regulatory VDRE. Therefore, the periodicity of VDR-induced mRNA accumulation of target genes is not shared but rather tends toward patterns that are specific to individual target genes and suggests that promoter-specific complexes combine to determine the precise periodicity [54,55]. There is good evidence that specific histone modifications also determine the assembly of transcription factors on the promoter and control individual promoter transcriptional responsiveness [64–66]. For example, the VDR may recognize basal histone modifications on target gene promoters; functional studies of the SANT motif contained in the corepressor NCOR2/SMRT supports this latter idea [67,68] and again support the cyclical nature of gene regulation.

It remains an enticing prospect that the variability in gene regulatory patterns reflects the organization of different subsets of gene targets, which are perhaps organized by phenotype. Thus, VDR-regulated networks may be differentially governed by the chromatin context at response elements and include near and long-term chromatin associations.

2.2.4 VDR Signal Specificity

2.2.4.1 Interactions with Corepressors and Coactivators

Historically, researchers have tended to consider transcription factor actions in a somewhat monochrome view, with the aim of illuminating complex phenotypes in terms of subsets of regulated genes. For example, the roles of either MYC or AP-1 transcription factors were historically dissected in cancer models, and therefore, their target genes were dissected in terms of control of proliferation. These findings have been revised in light of surveys of genome-binding sites and dissection of biological actions in a broader context (for example, as reviewed in [69,70]). These findings suggest that the functions of a given transcription factor are distilled through interaction with multiple proteins and the normal actions are extremely flexible and integrated with other signaling systems.

Such integration is emerging for the VDR whose function appears pleiotropic. The receptor is detected in virtually all cells of a human and functionally associated with disparate phenotypic effects, from regulating calcium transport to sensing redox potential and DNA damage. Therefore, questions emerge as to what governs temporal regulation of VDR-dependent transcritpomes among different cell types. A high level of specificity over the timing and choice of VDR cofactor interactions may provide a mechanistic basis for signaling specificity. In this manner, the expression and choice of interacting cofactors may determine the choice and timing of gene regulation. A further emerging theme concerns the hierarchical organization: the extent to which VDR is upstream or downstream of other regulatory events.

Of the principal corepressors, it remains to be established to what extent specificity and redundancy occur. The expression, localization, and isoforms of NCOR1 and NCOR2/SMRT corepressors strongly influence the spatio-temporal equilibrium between repressing and activating transcription complexes and transcriptional outputs [71] (reviewed in [68]). These dynamic exchanges between corepressors and coactivators and their epigenetic impact have been illustrated clearly with several members of the NR superfamily, including the AR, VDR, and ERα [45,54,61,72,73], and also with AP-1 members and NF-κB factors (reviewed in [74]).

The principal corepressors NCOR1 and NCOR2/SMRT have been investigated *in vivo*, and these studies revealed specificities with regard to tissue and function with respect to interactions with RAR and PPARs. NCOR1 and NCOR2/SMRT expression, localization, and isoforms have emerged as critical in determining the spatio-temporal equilibrium between the antagonistic actions of the ligand bound and unbound NR complexes and thus determine the dynamics of target gene promoter responsiveness. Knockout of these proteins is embryonically lethal, which is most likely due to the significant number of transcription factors other than NRs that NCOR1 and NCOR2/SMRT interact with such as MYC and FOXO family members. More recently, stem cell components from *Ncor1*[−/−] and *Ncor2/Smrt*[−/−] mice and other studies using conditional approaches have revealed a network of interactions with NRs and other transcription factors [51,75–77]. For example, Ncor1 and Ncor2/Smrt have a Venn diagram distribution of transcriptional targets in neuronal stem cells with both unique and shared targets. Notably, NCOR2/SMRT represses RAR-induced expression of the Jumonji domain containing histone demethylases that are able to de-methylate H3K27me3 and induce genes associated with neuronal differentiation [75]. Similarly, a knock-in approach was used to circumvent the embryonic lethality and allow specific disruption of the interaction of Ncor2/Smrt with NRs. These models revealed dramatically enhanced differentiation rates, notably in adipocyte differentiation mediated by PPARγ [78]. These findings suggest that corepressors significantly regulate transcriptional actions, over both choice and periodicity of target sequence regulation.

A few in vitro studies have examined VDR transcriptional actions in nonmalignant human systems. For example, VDR activation of established target genes in nonmalignant prostate epithelial cells revealed cyclical mRNA and protein accumulation. For one target gene *CDKN1A* [encodes p21[(waf1/cip1)]], NCOR1 and NCOR2/SMRT display both parallel loss and gain of enrichment at specific VDRE on the promoter in response to ligand activation. That is, both corepressor loss and recruitment occurred simultaneously on the same promoter at different regions. The altered NCOR1 enrichment patterns associated with both loss and gain and activating histone modifications collectively appeared to contribute to the oscillation of mRNA [73]. These ligand-induced corepressor associations have now also been established on a genome-wide scale for the related glucocorticoid receptor, where ligand activation induces receptor binding to NCOR1 and NCOR2/SMRT at specific response elements and drives transrepression [79].

Beyond NCOR1 and NCOR2/SMRT the number or list continues to grow of novel corepressor proteins that the VDR interacts with. Compared to the relatively massive size of the corepressors NCOR1 and NCOR2/SMRT, a number of smaller molecules have emerged as showing corepressor function. TRIP15/COPS2/Alien has been demonstrated to interact with the VDR and act as a corepressor in an AF-2 independent manner that may not require the same interactions with HDACs that NCOR1 does [80]. Intriguingly, this protein contributes to the lid subcomplex of the 26S proteasome and thereby potentially links VDR function with the regulation of protein stability [81]. Similarly, SLIRP [82] has also emerged as a repressive factor for the VDR, although, to date, very little is known about the specificity, in terms of tissue and target gene.

Other repressors appear to demonstrate more specific phenotypic specificity. Hairless blocks VDR-mediated differentiation of keratinocytes, whereas addition of $1\alpha,25(OH)_2D_3$ displaces Hairless from the promoter of target genes and recruits coactivators to promote differentiation [83–85]. Similarly, a downstream regulatory element antagonist modulator (DREAM) usually binds to direct repeat response elements in the promoters of target genes to enhance transcription in VDR and RAR target genes, in a calcium-dependent manner, and suggests that specificity arises from the interactions of VDR with further tissue-specific cofactors [86].

A similar level of coactivator specificity is also emerging. Members of the TRAP/DRIP complex were identified independently in association with the VDR and other NRs, including the GR [87,88] and TR [89–91]. The exact specificity of many of the coregulatory factors remains to be established fully, although there are some suggestions that certain coactivators are VDR specific, e.g., NCoA-62 [92]. Similarly, knockout of TRAP220, which has multiple NR interacting domains, has begun to reveal distinct interactions and notably disrupts the ability of the VDR to regulate hematopoietic differentiation [93,94]. In keeping with the skin being a critical target for VDR actions, the specificity of VDR interactions with cofactor complexes has been dissected in detail by Bikle and colleagues who have demonstrated the timing and extent of coactivator binding and established a role for SRC3 during specific stages of keratinocyte differentiation [95,96].

Aside from the established coregulators, some chaperone proteins have been reported to be regulators of VDR mediated transcription. HSP70 down-regulates VDR to repress transcription [97], whereas BAG1L, which is an HSP70 binding protein, has been shown to bind to the VDR and enhances VDR-mediated transcription [98]. Similarly, p23 and HSP90 have been shown to release the VDR/coactivator complex from the promoter of target genes in the presence of $1\alpha,25(OH)_2D_3$ [99]. The association of these HSPs suggests a natural crosstalk with other NRs, such as the AR, that associate with these chaperones in the cytoplasm.

2.2.4.2 PTMs of the VDR Complex

A further level of VDR specificity emerges from considering the range of PTMs that target the VDR complex. PTMs resulting from signal transduction processes, for example, bring about phosphorylation, acetylation, and ubiquitinylation events that are established on the AR [100]. The VDR has been less extensively studied, but crucial roles have emerged for the phosphorylation of serine and threonine residues [101,102]. Subsequently, several residues have been identified that appear to regulate DNA binding and cofactor recruitment. The zinc finger DNA-binding domain is located at the N terminal of the VDR, and adjacent to this domain is the Serine 51 residue. This residue appears crucial for ligand-induced and phosphorylation-dependent transcriptional activation by the VDR. When Ser51 is mutated phosphorylation of the VDR, by protein kinase C (PKC), at least, is all but completely abolished and its transcriptional activity is markedly reduced [103]. It is intriguing that the crucial site of PKC activity is located so close to the DNA-binding domain, but whether there are allosteric or biochemical changes that alter the ability of the VDR to bind DNA remains to be elucidated.

The common NR partner RXR can also be phosphorylated and, as a result, alters recruitment of cofactors to its *holo*-complexes. Ser260 is located within the ligand-binding domain of the RXR and appears crucial for mediating cofactor binding and ligand-induced transcriptional responses. When phosphorylated, Ser260 allows binding between the RXR and VDR but, presumably through allosteric changes to the complex, limits the recruitment of cofactors to the complex [104].

The recruitment of cofactors to the VDR *holo*-complex also appears to be regulated further by the presence of PTMs, e.g., kinase CK-II. The phosphomimic mutant VDRS208D does not increase or decrease VDR-DNA, VDR-RXR, or VDR-SRC interactions, but it does increase the levels of VDR-DRIP205 complexes present. CK-II, which specifically phosphorylates Ser208, enhances $1,25(OH)_2D_3$-induced transactivation of VDR targets [105,106]. In addition, phosphatase inhibitors (okadoic acid), in combination with $1,25(OH)_2D_3$, shifts the cofactor preference from NCOA2/GRIP-1 to TRIP2/DRIP205 [107]. Taken together, these data suggest that the TRIP2/DRIP205 coactivator complex enhances the transcriptional response by VDR and is recruited by CK-II-dependent phosphorylation of the VDR at Ser208.

Interestingly, the NAD-dependent Class-III deacetylase SIRT1 has been shown to act on both p53 [108] and VDR [109] and provides an intriguing link with the metabolic state of the cell [110]. A high NAD/NADH ratio enhances SIRT1 activity, with deacetylation of AR and diminution of its growth-promoting activity. Conversely, low levels of NAD, or high levels of the inhibitor (and SIRT1 product) nicotinamide, suppress SIRT1 activity and hence can enhance the acetylation dependent activities of AR. In this way, SIRT1 may act as a sensor of the redox state of the cell [111] and govern the transcriptional abilities of a number of different factors.

2.2.5 Characterization of Vitamin D Response Elements from Candidate and Genome-Wide Approaches

Specificity of gene regulation also appears to arise from the arrangement and orientation of binding sequences contained within VDRE regions. Simple VDREs are formed by two recognition motifs, and their relative distance and orientation contributes to receptor-binding specificity. Thus, the first identified VDRE was the DR3, which is an imperfect hexameric direct repeat-sequence AGTTCA with a spacer of three nucleotides. In the DR3 configuration, RXR (the heterodimer partner) is believed to occupy the upstream half-site, and VDR (the downstream motif), two half-sites spaced by three nucleotides. Other types of VDREs have since been identified. One such VDRE is a palindromic sequence with a nine-base-pair nucleotide spacer (IR9). This sequence was identified in the human calbindin D9K gene, and like most VDREs, the VDR/RXR binds this sequence in a 5′-RXR-VDR-3′ polarity (reviewed in [112]). More recently, a novel everted repeat sequence with a six-base-pair nucleotide spacer (ER6) has been identified in the gene for *CYP3A4* (an enzyme important in xenobiotic metabolism), in addition to the DR3 already known to be present in this gene [113]. An inverted repeat with no spacer (IR0) has also been identified in the *SULT2A1* gene [114].

Similarly, the ability of VDR to display transrepression, i.e., ligand-dependent transcriptional repression, has received significant interest and reflects emerging themes for other NRs, e.g., PPARs [68,115], and highlights further the hitherto unsuspected flexibility of the VDR to associate with a diverse array of protein factors to adapt function [102,116]. For example, analysis of the avian PTH gene has revealed a ligand-dependent repression of this gene by VDR [117]. The element mediating this effect was identified as a DR3, and since it resulted in transcriptional repression, the motif was referred to as a negative nVDRE. A similar nVDRE has been identified in the human kidney in the *CYP27b1* gene [118]. Interestingly, the VDR does not bind directly to this sequence; binding has been shown to be mediated by an intermediary factor known as a bHLH-type transcription factor VDR interacting repressor (VDIR). It has since been shown that liganded VDR binds to the VDIR and indirectly causes repression through HDAC mechanisms [102].

More recently, larger and integrated responsive regions have been identified, suggesting more intricate control involving integration with other transcription factors, e.g., p53 and C/EBPα, as demonstrated on the promoter/enhancer regions of *CDKN1A* and *SULT2A1*, respectively [54,114]. Thus, the combinatorial actions of the VDR with other TFs most likely go somehow toward explaining the apparent diversity of VDR biological actions. Again, for other NRs (e.g., AR and ERα), more dominant transcription factors, which are so-called pioneer factors, appear to be highly influential in determining choice and magnitude of transcriptional actions [119]. Recently, C/EBP family members have been demonstrated to act in a similar cooperative manner with the related PPARγ [120], and it remains to be established to what extent the VDR interacts with other transcription factors. The aforementioned findings are suggestive of similar mechanisms.

The understanding of the arrangement and location of VDRE has been significantly increased by the application of next-generation sequencing technologies, e.g., ChIP-Seq approaches. Perhaps one of the most well-established NR superfamily members in terms of understanding gene regulatory function is ERα, which has been studied intensively under different conditions, owing to its vital role in the initiation, progression, and therapy of breast cancer. Analysis of binding sequences shows that ERα has the potential to regulate >3% of genes [121] and, perhaps, an equally significant role in regulation of nonprotein-coding regions. However, global ChIP-chip analyses show that ERα binding frequently occurs far from the gene whose expression it potentially controls, with 80% of potential ER response elements (EREs) being more than 10 kb away from their target gene [63]. In the absence of ligand, ERα can bind to ERE as part of large transcriptionally repressive complexes containing corepressor proteins, such as NCOR1, maintaining a locally condensed chromatin structure [61,122]. Prior to entry of estradiol (E2) into cells, the epigenome is essentially primed for particular responses; combinations of specific histone modifications dictate where pioneer factors such as FOXA1 and transcription factors will bind and, thus, which genes are activated [123]. This equilibrium is altered upon addition of ligand, with recruitment of transcription factors and their associated chromatin remodeling enzymes, allowing relaxation of the local chromatin and binding of further factors. For example, the repressive function of ERα complexes depends on the methyltransferases RIZ1, ESET, and Eu-HMTase1 contained within them and their ability to place the repressive H3K9me2 mark on adjacent nucleosomes. HDACs are also crucial components of the repressive complex, potentially facilitating chromatin condensation through reduced acetylation [124].

To date, two comprehensive ChIP-Seq approaches have identified VDRE in two different cell types, i.e., THP1 human monocytic leukemia cells [125] and lymphoblastoid cell lines [126]. Thus, these key studies differed in their choice of cell model and also differed dramatically in the amount of time that the cells were stimulated by $1\alpha25(OH)_2D_3$ (40 min versus 36 h). Setting these differences aside, a number of themes emerge from both studies. In the basal state, the VDR is clearly located in the nucleus bound to chromatin. The number of basal binding sites were broadly similar (around 1000) and were significantly enriched upstream of known genes in the promoter regions, being significantly associated with histone marks that characterize the TSS. Upon stimulation, a similar number of binding sites were also identified (about 2500). Interestingly, analyses of the binding sites in the basal state revealed that only around 30% were classical DR-3 type [125], with other sites showing enrichment for SP1 and ETS transcription-factor-binding sequences. Both studies also revealed that, upon ligand activation, there is a redistribution of VDR to sites that are, principally, of DR-3 type but are located more distally from putative target genes, up to several hundred kilobases away from genes that were both upregulated and repressed. An interesting observation too is the pattern of derepression, i.e., where ligand stimulation causes a loss of VDR binding associated with upregulation of an associated gene. The caveat to this is that, in both studies, the investigators compared ChIP-Seq to microarray studies, and the links established are correlative and not necessarily mechanistic. Setting that concern aside, the redistribution of the VDR upon ligand activation was highly varied, with multiple different patterns of movement identified and appearing different for upregulated genes, compared to downregulated genes. The extent of this complexity is likely to increase as investigators generate time-resolved ChIP-Seq data sets. While the two studies have much in common in terms of the number and distribution patterns of the VDR, the actual binding sites cotargeted in both studies is small, being less than 20%, with only around 5% of genes being in common. A final point to note is that certain binding sites also strongly associated with single nucleotide polymorphism variation and help to build a more complete risk of disease susceptibility [126].

It should also be noted that the majority of research of VDR target genes has focused on mRNA that is translated to protein. A recently emerging theme is that VDR regulate micro-RNA [73,127,128] directly, and these form coregulatory circuits that combine to determine mRNA expression of target genes.

2.3   VDR TRANSCRIPTIONAL NETWORKS

2.3.1   LESSONS FROM TRANSGENIC MURINE MODELS

Key insights into these functions have been gained in *Vdr*-deficient mice [129–131]. The *Vdr* is expressed widely during murine embryonic development in tissues involved in calcium homeostasis and bone development. *Vdr* disruption results in a profound phenotype in these models, which is principally observed postweaning and is associated with the alteration of duodenal calcium absorption and bone mineralization, resulting in hypocalcemia, secondary hyperparathyroidism, osteomalacia, rickets, impaired bone formation, and elevated serum levels of $1\alpha,25(OH)_2D_3$. In parallel, a range of more subtle effects are seen more clearly when the animals are rescued with dietary calcium supplementation and may represent autocrine and noncalcaemic actions. The animals became growth retarded and display alopecia, uterine hypoplasia, impaired ovarian folliculogenesis, reproductive dysfunction, cardiac hypertrophy, and enhanced thrombogenicity.

The *Vdr* is readily detected in keratinocytes, and cotreatment of calcium and $1\alpha,25(OH)_2D_3$ decreases proliferation and promotes differentiation of cultured keratinocytes [132]. The *Vdr* is also detected in outer root sheath and hair follicle bulb, as well as in sebaceous glands [133], and the *Vdr*$^{-/-}$ mice develop hair loss and, ultimately, alopecia totalis, which is associated with large dermal cysts, which, in turn, are not prevented by high-calcium rescue diet. Alopecia arises due to complete failure to initiate anagen, which is the first postnatal hair growth phase. Subsequently, the hair follicles convert into epidermal cysts [134]. Hair follicle formation requires highly coordinated signaling between different cell types, including contributions from the stem cell components; therefore, the alopecia phenotype has attracted significant research interest as it may represent a role in the VDR in stem cell maintenance. Subsequent studies have demonstrated that failure to maintain hair follicles in *Vdr* –/– animals does not actually reflect a loss of follicle stem cells but rather an inability of the primitive progenitor cells to migrate along the follicle at the onset of anagen [135].

Wnt signaling is one of the major processes that regulate postmorphogenic hair follicle development. Interestingly, the development of dermal cysts and increase in sebaceous glands observed in the *Vdr* and hairless –/– mice are also similar to mice expressing keratinocyte-specific disruption to β-catenin [136,137]. These findings have raised the possibility that one function of the *Vdr* may be to coregulate aspects of Wnt signaling, which is a concept supported further by the physical association of VDR in a complex with β-catenin and other Wnt components [138].

Another unexpected finding of the *Vdr*–/– animals was uterine hypoplasia and impaired ovarian function in the females, which leads to dramatically reduced fertility. Similar to the hair phenotype, this was not restored by the rescue diet of high calcium [129]. Estradiol supplementation, however, of the female mice restored uterine function and fertility, and suggests that the fault lies with the inability to generate estrogen. The mammary gland has also been studied extensively in a comprehensive series of experiments by Welsh et al. [139,140] and represents an intriguing tissue where endocrine (calcaemic) and autocrine (antimitotic, prodifferentiative, and proapoptotic) effects of the VDR appear to converge.

These phenotypes underscore the integrated nature of VDR signaling. That is, the biology of hair regeneration and mammary gland function reflects the choreographed actions of VDR, with other NRs, alongside other regulatory processes including Wnt signaling. Disfunction of multiple aspects of this is seen in many cancer phenotypes. More recently, these studies have been complemented by unbiased interrogation of all transcriptional networks that are involved in hematopoietic differentiation, and these approaches identified a significant role for the VDR to cooperate functionally in a module of transcription factors that includes CEBPα and PU-1 to govern granulocyte and monocyte differentiation [141].

2.3.2 VDR Transcriptional Networks in Malignancy

Defining the mechanisms by which the VDR exerts desirable anticancer effects has been an area of significant investigation since the early 1980s. In 1981, Colston et al. were first to demonstrate that $1\alpha,25(OH)_2D_3$ inhibited human melanoma cell proliferation significantly in vitro at nanomolar concentrations [142]. Parallel studies in the same year also found that $1\alpha25(OH)_2D_3$ could induce differentiation in cultured mouse and human myeloid leukemia cells [143,144]. Following these studies, antiproliferative effects have been demonstrated in a wide variety of cancer cell lines, including those from prostate, breast, and colon [145–152]. To identify critical target genes that mediate these actions, comprehensive genome-wide *in silico* and transcriptomic screens have analyzed the antiproliferative VDR transcriptome and revealed broad consensus on certain targets but has also highlighted variability [145,153–155]. This heterogeneity may, in part, reflect experimental conditions, cell line differences, and genuine tissue-specific differences of cofactor expression that alter the amplitude and periodicity of VDR transcriptional actions.

2.3.2.1   Control of Cell Cycle Arrest

A common antiproliferative VDR function is associated with arrest at G_0/G_1 of the cell cycle, coupled with upregulation of a number of cell cycle inhibitors including p21$^{(waf1/cip1)}$ and p27$^{(kip1)}$. Promoter characterization studies have demonstrated a series of VDREs in the promoter/enhancer region of *CDKN1A* [54,156]. By contrast, the regulation of the related CDKI p27$^{(kip1)}$ is mechanistically enigmatic, reflecting both transcriptional and translational regulation, such as enhanced mRNA translation, and attenuating degradative mechanisms [157–160].

The upregulation of p21$^{(waf1/cip1)}$ and p27$^{(kip1)}$ principally mediate G_1 cell cycle arrest, but $1\alpha,25(OH)_2D_3$ has been shown to mediate G_2/M cell cycle arrest in a number of cancer cell lines via direct induction of *GADD45α* [154,161,162]. Again, this regulation appears to combine direct gene transcription and a range of posttranscriptional mechanisms. These studies highlight the difficulty of establishing strict transcriptional effects of the VDR, as a range of posttranscriptional effects act in concert to regulate target protein levels. Concomitant with changes in the cell cycle, there is some evidence that $1\alpha,25(OH)_2D_3$ also induces differentiation, which is most clearly evidenced in myeloid cell lines but also supported by other cell types, and most likely reflects the intimate links that exist between the regulation of the G_1 transition, the expression of CDKIs such as p21$^{(waf1/cip1)}$, and the induction of cellular differentiation [163].

Historically, hematological malignancies combined ease of interrogation with robust classification of cellular differentiation capacity, which were envied by investigators of solid tumors. It is therefore no coincidence that these cell systems generally yielded many important insights for cancer cell biologists, such as chromosomal translocations and instability, and the role of committed adult stem cells.

Indeed, the capacity to readily differentiate in response to external and internal signals has fascinated leukemia researchers as they have sought to understand why leukemia cells appear to fail at certain stages of differentiation. It is within this context that, in the 1980s, investigators [164,165] considered a role for the VDR and the related retinoic acid receptor (RAR) to reactivate dormant differentiation programs in so-called differentiation therapies. Over the following two decades, researchers began to reveal how these receptors instill mitotic restraint and facilitate differentiation programs and how discord over the control and integration of these processes is central to leukemogenesis. Despite these efforts, clinical exploitation of these receptors has largely proved to be equivocal. The one exception to this translational failure has been the exploitation of RAR signaling in patients with acute promyelocytic leukemia. Again, understanding the basic signaling behind this application proved significant to the developing understanding of epigenetic regulation of transcription and the promise of HDAC inhibitors [166].

Against this backdrop, various groups, including that of Studzinski, have worked consistently, exploring mechanisms of resistance to VDR signaling and methods of exploitation, and have recently demonstrated, elegantly, a role for VDR to downregulate miR181a, which, when left unchecked, degrades p27$^{(kip1)}$ [128]. Thus, indirectly, VDR activation elevates the expression of p27$^{(kip1)}$, initiates cell cycle arrest, and commits cells toward differentiation. Transcriptional control of miRNAs and their biological effects is clearly a field of rapid expansion, and members of the NR superfamily are implicated in their regulation [167,168]. A role for the VDR to govern the expression of this regulatory miRNA and, importantly, place its role in a well-understood map of differentiation is highly novel.

Similar integration of miRNA and mRNA was revealed to control the regulation of *CDKN1A* [encodes p21$^{(waf1/cip1)}$]. In nonmalignant prostate epithelial cells, rapid and dynamic patterns of *CDKN1A* mRNA accumulation occurred [73]. These patterns of mRNA accumulation reflected receptor and corepressor exchanges occurring in a unique manner at each of three VDREs. These results support the concept that, in nonmalignant systems, VDR transcriptional responses can be rapid and functional, and faithfully lead to changes in protein. The transition through activated epigenetic states appears to allow the integration of transcriptional signals, e.g., strong cooperation with p53 at 1 h, and suggests an epigenetic basis for the observed cooperation between these two pathways [54,169–172]. Furthermore, these studies in RWPE1 cells revealed that the magnitude of epigenetic modulation was refined by the cell cycle status. ChIP approaches in FACS-sorted cells revealed that G_1 phase cells were characterized by enhanced VDR-induced activating histone modifications (e.g., H3K9ac), with the S and G_2/M phases being largely repressive, e.g., with H3K27me3 enrichment on the *CDKN1A* promoter. Thus, events that were not significant when considering bulk populations emerged with significant clarity when considering each specific cell cycle phase and underscore the fact that bulk culture findings represent an average event of potentially very different populations. While other NRs have been demonstrated to display cell cycle specific phases of activation [173,174], this study revealed an underlying role for differential regulation of histone modifications through the cell cycle to govern these actions. Together, these data suggest the magnitude of *CDKN1A* activation, at least at early time points, is influenced significantly by the stage of the cell cycle. Finally, VDR-dependent coregulation of miR-106b was revealed to modulate the precise timing of *CDKN1A* accumulation and also the expression of p21$^{(waf1/cip1)}$. Together, these data demonstrate that VDR-induced regulation of p21$^{(waf1/cip1)}$ is determined by interplay of histone modifications and miRNA expression that combine in a feed-forward loop and determine the final extent of the cell cycle arrest.

The expression of miRNA contained within feed-forward motifs is disrupted in cancer [175,176]. For example, the cellular advantage to retaining the *MCM-7* promoter selectively in an active state [162] has recently been underscored in prostate cancer by establishing that miR-106b is a proto-oncogene [177]. A strong emerging literature now supports the concept that tumor changes in miRNA regulation are translated to altered serum expression and therefore offers an important diagnostic and prognostic therapeutic window [178–181]. In this manner, the regulation of miRNA such as miR-106b in feed-forward loop motifs may be critical biomarkers to monitor VDR responsiveness. The current studies open up the door to the possibility that serum expression of VDR tumor-regulated miRNA defines molecular phenotypes associated with prostate cancer aggressiveness and responsiveness to vitamin D compound treatment.

2.3.2.2   Sensing DNA Damage

An important and emergent area, in terms of both physiology and therapeutic exploitation, is the role that the liganded VDR appears to play in maintaining genomic integrity and facilitating DNA repair. There appears to be close cooperation between VDR actions and the p53 tumor suppressor pathway. The maintenance of genomic fidelity against a backdrop of self-renewal is central to the normal development and adult function of many tissues, including the mammary and prostate glands, and the colon. For example, in the mammary gland, p53 family members play a role in gland development and maintenance. *P63* –/– animals have an absence of mammary and other epithelial structures, which is associated with failure of lineage commitment (reviewed in [182]), whereas *p53* –/– animals have delayed mammary gland involution, reflecting the *Vdr* –/– animals and wider tumor susceptibility (reviewed in [183]).

The overlap between p53 and VDR appears to extend beyond cellular phenotypes. The *VDR* is a common transcriptional target of both p53 and p63 [170,172], and VDR and p53 share a cohort of direct target genes associated with cell cycle arrest, signal transduction, and programmed cell death including *CDKN1A GADD45A, RB1, PCNA, Bax, IGFBP3, TGFB1/2,* and *EGFR* [54,155,162,184–188]. At the transcriptional level, both VDR heterodimers and p53 tetramers associate, for example, with chromatin remodeling factors CBP/p300 and the SWI/SNF to initiate transactivation [189] By contrast, in the gene-repressive state, VDR and p53 appear to associate with distinct repressor proteins, e.g., p53 with SnoN [190] and VDR with NCOR1, suggesting the possibly association with distinct sets of histone deacetylases. Indeed, *CDKN1A* promoter-dissection studies revealed adjacent p53 and VDR binding sites, suggesting composite responsive regions [54]. Together, these findings suggest that $1\alpha,25(OH)_2D_3$-replete environments enhance p53 signaling to regulate mitosis negatively.

Similarly, the role of $1\alpha,25(OH)_2D_3$ in the skin is also suggestive of its chemopreventive effects. Ultraviolet (UV) light from sun exposure has several effects to the skin; UV-A light induces DNA damage through increasing the level of reactive oxygen species (ROSs), but importantly, UV-B light also catalyzes the conversion of 7-dehydroxycholesterol to 25(OH)-D and induces the expression of VDR.

In addition, antimicrobial and antiinflammatory genes are another subset of VDR targets that are induced by UV radiation. Suppression of the adaptive inflammatory response is thought to be protective for several reasons. Inflamed tissues contain more ROS, which, in turn, can damage DNA and prevent proper function of DNA repair machinery. In addition, the induction of cytokines and growth factors associated with inflammation act to increase the proliferative potential of the cells. NF-κB is a key mediator of inflammation, and the VDR attenuates this process by negatively regulating NF-κB signaling [191]. This control by VDR is underscored by studies, showing *Vdr−/−* mice are more sensitive to chemicals that induce inflammation than their wild-type counterparts [192]. The normally protective effect of inflammation that occurs under other conditions is lost through VDR-mediated suppression but is compensated for by the induction of a cohort of antimicrobial and antifungal genes [193–195]. The induction of antimicrobials not only prevents infection in damaged tissue but can also be cytotoxic for cells with increased levels of anion phospholipids within their membranes, which is a common feature of transformed cells [196]. Finally and most recently, network strategies have been used in different strains of mice with altered sensitivity toward skin cancer. Remarkably, in such unbiased screens, the VDR emerges as a key nodal control point in determining sensitivity toward skin tumors as it regulates both turnover of self-renewal and inflammatory infiltrate [197].

The key question, which is central to exploiting any therapeutic potential of this receptor, is why the VDR should exert such pleiotropic actions. One possible explanation for this pleiotropism is that it represents an adaptation of the skin to UV exposure, coupling the paramount importance of initiating $1\alpha,25(OH)_2D_3$ synthesis with protection of cell and tissue integrity. Thus, VDR actions are able to maximize UV-initiated synthesis of $1\alpha,25(OH)_2D_3$ production while controlling the extent of local inflammation that can result from sun exposure. To compensate for the potential loss of protection associated with immunosuppression, the VDR mediates a range of antimicrobial actions. Equally, local genomic protection is ensured through the upregulation of target genes, which induce G_0/G_1 arrest, cooperation with p53, and induction of cell differentiation. It remains a tantalizing possibility that the functional convergence between p53 family and VDR signaling, which arose in the dermis as an evolutionary adaptation to counterbalance the conflicting physiological requirements of vitamin D synthesis and genome protection, is sustained in epithelial systems, such as the lining of the mammary gland, to protect against genotoxic insults derived from either the environment or local inflammation.

2.4 GENETIC AND EPIGENETIC MECHANISMS OF RESISTANCE TOWARD THE VDR

2.4.1 DOMINANT SIGNAL TRANSDUCTION EVENTS

In terms of distribution, evidence that the normally dynamic flux of the VDR becomes altered in more transformed and aggressive cancer cells is emerging, becoming restricted to the nucleus [198,199]. Findings that the normal transport rates, such as importin-mediated processes, become distorted in malignancy and may result in reduced ability for the VDR to sample $1\alpha,25(OH)_2D_3$ from the cytoplasm. Findings on mechanisms that redistribute the VDR and control its movement are of key significance.

Also reflecting the cooperative and integrated nature of the VDR function, a number of workers have identified mechanisms by which more dominant signaling processes are able either to ablate or attenuate VDR signaling. For example, Munoz et al. have dissected the interrelationships between the VDR, E-Cadherin, and the Wnt signaling pathway in colon cancer cell lines and primary tumors. In these studies, the induction of *CDH1* (encodes E-Cadherin) was seen in subpopulations of SW480 colon cancer cells, which express the VDR and respond to $1\alpha,25(OH)_2D_3$. The VDR thereby limits the transcriptional effects of β-catenin by physically and directly binding it in the nucleus and by upregulating E-cadherin to sequestrate β-catenin in the cytoplasm. In malignancy, these actions are corrupted through downregulation of *VDR* mRNA, which appears to be a direct consequence of binding by the transcriptional repressor SNAIL, which is a key regulator of the epithelial-mesenchyme transition, which is overexpressed in colon cancer [200–202]. Equally underscoring the central importance of β-catenin, it has recently been shown to be posttranslationally modified to act as VDR coactivator and supports a model of checks and balances between these two signaling processes [200,203].

2.4.2 GENETIC RESISTANCE

In cancer, and outside of the very limited pool of mutations reported in the VDR in type-II rickets, the receptor, generally, is neither mutated nor does it appear to be the subject of cytogenetic abnormalities [204]. By contrast, polymorphic variations of the *VDR* have been widely reported. Thus, polymorphisms in the 3′ and 5′ regions of the gene have been described and variously associated with risk of breast, prostate, and colon cancer, although the functional consequences remain to be established clearly. For example, a start codon polymorphism in exon II at the 5′ end of the gene, which was determined using the *fok*-I restriction enzyme, result in a truncated protein. At the 3′ end of the gene, three polymorphisms that do not lead to any change in either the transcribed mRNA or the translated protein have been identified. The first two sequences generate *Bsm*I and *Apa*I restriction sites and are intronic, lying between exons 8 and 9. The third polymorphism, which generates a *Taq*I restriction site, lies in exon 9 and leads to a silent codon change (from ATT to ATC), which both insert an isoleucine residue at position 352. These three polymorphisms are linked to further gene variation, which is a variable-length adenosine sequence within the 3′ untranslated region (3′UTR). The poly(A) sequence varies in length and can be segregated into two groups, i.e., long sequences of 18–24 adenosines or short ones [205–208]. The length of the poly(A) tail can determine mRNA stability [209–211], so the polymorphisms resulting in long poly(A) tails may increase the local levels of the VDR protein.

Multiple studies have addressed the association between *VDR* genotype and cancer risk and progression. In breast cancer, the *Apa*I polymorphism shows a significant association with breast cancer risk, as it indeed has *Bsm*I and the 'L' poly(A) variant. Similarly, the *Apa*I polymorphism is associated with metastases to bone [212,213]. The functional consequences of the *Bsm*I, *Apa*I, and *Taq*I polymorphisms are unclear but, because of genetic linkage, may act as a marker for the poly(A) sequence within the 3′UTR, which, in turn, determine transcript stability. Interestingly, combined polymorphisms and serum 25OH-D levels have been shown to compound breast cancer risk and disease severity further [214].

Earlier studies suggested that polymorphisms in the VDR gene might also be associated with a risk factor of prostate cancer. Ntais and coworkers performed a metaanalysis of 14 published studies with four common gene polymorphisms (*Taq*1, poly A repeat, *Bsm*1, and *Fok*1) in individuals of European, Asian, and African descent. They concluded that these polymorphisms are unlikely to be major determinants of susceptibility to prostate cancer on a wide population basis [215]. Equally, studies in colon cancer have yet to reveal conclusive relationships and may be dependent on the ethnicity of the population studied.

Despite these strong candidate approaches focusing on the VDR, no studies using agnostic GWAS approaches have revealed genetic variants associated with the VDR to be strongly associated with cancer risk or other syndromes in which its actions were implicated. That said, variation in VDRE sequences is emerging as an area by which altered VDR function can impact on disease risk [126].

2.4.3 EPIGENETIC RESISTANCE

In cancer cells, the lack of an antiproliferative response is reflected by suppression of the transcriptional responsiveness of antiproliferative target genes such as *CDKN1A CDKN1B, GADD45A*, and *IGFBPs, BRCA1* [147,162,216,217]. Paradoxically, VDR transactivation of other targets is sustained or even enhanced, as measured by induction of the highly $1\alpha25(OH)_2D_3$-inducible *CYP24* gene [218,219]. Together, these data suggest that the lack of functional VDR alone cannot explain resistance, and instead, the VDR transcriptome is skewed in cancer cells to disfavor antiproliferative target genes. It has been proposed that this apparent $1\alpha25(OH)_2D_3$ insensitivity is the result of epigenetic events that selectively suppress the ability of the VDR to transactivate target genes [220].

The epigenetic basis for such transcriptional discrepancies has been investigated intensively in prostate cancer. There is compelling evidence that histone and DNA methylation processes disrupt NR transcriptional actions, both alone and together. Elevated NCOR2/SMRT acts to suppress VDR regulation of a subset of target mRNAs (e.g., *GADD45A*) [46,162,221] and NCOR1 to disrupt a subset of PPARα/γ actions [222]. One consequence of NCOR1 and NCOR2/SMRT association at target genes is the loss of H3K9ac and accumulation of H3K9me2, allowing the potential for hypermethylation at adjacent CpG regions. Further links exist between NCOR1 and DNA methylation through its interaction with KAISO to regulate DNA methylation [223]. Certainly, a number of key VDR target genes are silenced by increased CpG methylation [224,225]. At high-density regions of CpG methylation, spanning hundreds of base pairs, the entire region acquires H3K9 and –K27 methylation, loses H3K4 methylation, and recruits heterochromatin binding protein 1 (HP1) [17]. The recruitment of HP1 through interaction with MBD1 leads to recruitment of both a H3K9 methylase (KMT1A/SUV39H1) [226] and DNA methyltransferases (DNMTs) [227], which are enzymes that add repressive methylation marks to histones and CpG. DNMT3L and UHFR1 also provide potential links between DNA methylation and the absence of H3K4 methylation and the presence of H3K9 methylation, respectively (reviewed in [228]).

Thus, these processes become self-reinforcing. It is not precisely clear, however, in mammalian cells whether the H3K9 methylation or the high density of CpG methylation is required first to set up this heterochromatic structure. In *Neurospora crassa*, loss of HP1 (which requires H3K9 methylation for binding to chromatin) leads to loss of DNA methylation [229]. This situation describes stable heterochromatic silencing of genomic regions and is in contrast to the dynamic changes at a locus with active epigenetic regulation of transcription in response to NR activation. However, even in such actively regulated regions, dynamic changes in DNA methylation appear to occur. For example, these have been measured in response to NR actions, including ERα, VDR, RARα, and PPARγ, at multiple target genes [230–232].

VDR-resistant prostate cancer cells are associated with elevated levels of NCOR2/SMRT [162,216]; these data indicate that the ratio of *VDR* to corepressor may be critical to determine $1\alpha25(OH)_2D_3$ responsiveness in cancer cells. A siRNA approach toward *NCoR2/SMRT* demonstrated a role for this corepressor to regulate *GADD45α* expression in response to $1\alpha25(OH)_2D_3$. By contrast, knockdown of NCOR1 does not restore antiproliferative responsiveness toward $1\alpha,25(OH)_2D_3$ but does reactivate transcriptional networks governed by PPARs [222].

Parallel studies have demonstrated a similar spectrum of reduced $1\alpha25(OH)_2D_3$ responsiveness between nonmalignant breast epithelial cells and breast cancer cell lines. Again, this was not determined solely by a linear relationship between the levels of $1\alpha25(OH)_2D_3$ and *VDR* expression. Rather, elevated corepressors mRNA levels, notably of *NCoR1*, in ERα negative breast cancer cell lines and primary cultures, were associated with $1\alpha25(OH)_2D_3$ insensitivity [233]. Elevated NCOR1 has also been demonstrated to suppress the VDR responsiveness of bladder cancer cell lines [198] notably toward the VDR ligand LCA [30], suggesting a role for epigenetic disruption of the capacity of cells to sense and metabolize potential genotoxic insults.

The epigenetic lesion rising from elevated NCOR1 can be targeted by cotreatment of either $1\alpha25(OH)_2D_3$ or its analogs, plus the HDAC inhibitors such as trichostatin A to restore the $1\alpha25(OH)_2D_3$ responses of androgen-independent PC-3 cells to levels indistinguishable from control normal prostate epithelial cells. This reversal of $1\alpha25(OH)_2D_3$ insensitivity was associated with reexpression of gene targets associated with the control of proliferation and induction of apoptosis, notably *GADD45A* [147,162,217]. Similarly, targeting in breast cancer cells through cotreatments of $1\alpha25(OH)_2D_3$ with HDAC inhibitors coordinately regulated VDR targets and restored antiproliferative responsiveness [233,234]. Similarly, combinatorial chemistry approaches have been used to combine aspects of the structure of $1\alpha,25(OH)_2D_3$ and HDAC inhibitors into a single molecule that demonstrates very significant potency [235].

Together, these data support the concept that altered patterns of corepressors inappropriately sustains histone deacetylation around the VDRE of specific target gene promoter/enhancer regions and shifts the dynamic equilibrium between *apo* and *holo* receptor conformations to favor transcriptional repression of key target genes. Furthermore, targeting this epigenetic lesion with cotreatments of vitamin D_3 compounds plus HDAC inhibitors generates a temporal window where the equilibrium point between *apo* and *holo* complexes is shifted to sustain a more transcriptionally permissive environment.

These findings compliment a number of parallel studies that have established cooperativity between $1\alpha25(OH)_2D_3$ and butyrate compounds, such as sodium butyrate (NaB) [236–241]. These compounds are short-chain fatty acids produced during fermentation by endogenous intestinal bacteria and have the capacity to act as HDAC inhibitors. Stein and coworkers have identified the effects in colon cancer cells of $1\alpha25(OH)_2D_3$ plus NaB cotreatments to include the coordinate regulation of the VDR itself. Together, these studies underscore further the importance of the dietary-derived milieu to regulate epithelial proliferation and differentiation beyond sites of action in the gut.

2.5 SUMMARY

A highly conserved VDR is found widely throughout metazoans and even in certain noncalcified chordates such as the lamprey (reviewed in [242]). Within prokaryotes, there appears to be the capacity to undertake UV-catalyzed metabolism of cholesterol compounds and suggests that the evolution of vitamin D biochemistry is very ancient. These findings suggest that the VDR system has been adapted to regulate calcium function and retains other functions that are calcium independent and include the capacity to sense the local environment.

The VDR participates in at least three fundamental areas of biology that are required for human health and disrupted in human disease. It participates in the regulation of serum calcium and, by implication, the maintenance of bone integrity; in the control of cell proliferation and differentiation and, by implication, the disruption of these actions in malignancy; and as a modifier of immune responses and, by implication, contributes toward autoimmune diseases [243]. The divergence of these actions may make the VDR a particularly challenging receptor to understand in terms of biology and to exploit therapeutically.

The VDR transcriptional actions reflect a convergence of multiple complexes, the detail of which is still emerging and is centered on genomic and epigenomic regulation and also integrates crosstalk with cellular signaling systems and extends beyond the nucleus and integrates levels of cytoplasmic shuttling. Establishing the specificity of function and selectivity of VDR interactions has to an extent been limited by technical approaches. Unbiased approaches are now required to dissect VDR interactions (in the membrane, cytoplasm, and nucleus) in either individual cells or very pure populations, thereby generating a comprehensive understanding of the spatial temporal network of its interactions.

The next developments will include genomic and epigenomic approaches to defining VDR signaling. The challenge is to model the spatio-temporal actions of the VDR network and, in particular, when and where the VDR exerts critical control over transcription and translation and when this system is at the control of more dominant signaling and transcriptional signals. Such an understanding requires clear awareness of the chromatin architecture and context of the promoter regions (e.g., histone modifications and DNA methylation), genomic organization, gene regulation hierarchies, and $1\alpha,25(OH)_2D_3$-based metabolomic cascades, all within the context of specific cell backgrounds. The ultimate research goal will be to translate this understanding to strategies that can predict the capacity of subsets of VDR actions to be regulated in targeted distinct cell types and exploited in discrete disease settings. The implementation of postgenomic techniques, together with bioinformatics and systems biology methodology, is expected to generate such an integral view, thereby revealing and quantifying the mechanisms by which cells, tissues, and organisms interact with environmental factors such as diet [244,245].

REFERENCES

1. Venter, J. C. et al. 2001. The sequence of the human genome. *Science* 291:1304–1351.
2. Levine, M. and R. Tjian. 2003. Transcription regulation and animal diversity. *Nature* 424:147–151.
3. Hah, N. et al. 2011. A rapid, extensive, and transient transcriptional response to estrogen signaling in breast cancer cells. *Cell* 145:622–634.
4. Core, L. J., J. J. Waterfall, and J. T. Lis. 2008. Nascent RNA sequencing reveals widespread pausing and divergent initiation at human promoters. *Science* 322:1845–1848.
5. Gilad, Y., A. Oshlack, G. K. Smyth, T. P. Speed, and K. P. White. 2006. Expression profiling in primates reveals a rapid evolution of human transcription factors. *Nature* 440:242–245.
6. Chodaparambil, J. V. et al. 2007. A charged and contoured surface on the nucleosome regulates chromatin compaction. *Nature Structural and Molecular Biology* 14:1105–1107.
7. Kouzarides, T. 2007. Chromatin modifications and their function. *Cell* 128:693–705.
8. Cosgrove, M. S. 2007. Histone proteomics and the epigenetic regulation of nucleosome mobility. *Expert Review of Proteomics* 4:465–478.
9. Allis, C. D. et al. 2007. New nomenclature for chromatin-modifying enzymes. *Cell* 131:633–636.
10. Turner, B. M. 1993. Decoding the nucleosome. *Cell* 75:5–8.
11. Goldberg, A. D., C. D. Allis, and E. Bernstein. 2007. Epigenetics: A landscape takes shape. *Cell* 128:635–638.
12. Turner, B. M. 1998. Histone acetylation as an epigenetic determinant of long-term transcriptional competence. *Cell and Molecular Life Sciences* 54:21–31.
13. Jenuwein, T., and C. D. Allis. 2001. Translating the histone code. *Science* 293:1074–1080.
14. Turner, B. M. 2002. Cellular memory and the histone code. *Cell* 111:285–291.
15. Wissmann, M. et al. 2007. Cooperative demethylation by JMJD2C and LSD1 promotes androgen receptor–dependent gene expression. *Nature Cell Biology* 9:347–353.
16. Metzger, E. et al. 2005. LSD1 demethylates repressive histone marks to promote androgen receptor–dependent transcription. *Nature* 437:436–439.
17. Mohn, F. and D. Schubeler. 2009. Genetics and epigenetics: Stability and plasticity during cellular differentiation. *Trends in Genetics* 25:129–136.
18. Birney, E. et al. 2007. Identification and analysis of functional elements in 1% of the human genome by the ENCODE pilot project. *Nature* 447:799–816.
19. Bau, D. et al. 2011. The three-dimensional folding of the alpha-globin gene domain reveals formation of chromatin globules. *Nature Structural and Molecular Biology* 18:107–114.
20. Li, Q., G. Barkess, and H. Qian. 2006. Chromatin looping and the probability of transcription. *Trends in Genetics* 22:197–202.
21. Eskiw, C. H., A. Rapp, D. R. Carter, and P. R. Cook. 2008. RNA polymerase II activity is located on the surface of protein-rich transcription factories. *Journal of Cell Science* 121:1999–2007.
22. Mitchell, J. A. and P. Fraser. 2008. Transcription factories are nuclear subcompartments that remain in the absence of transcription. *Genes and Development* 22:20–25.
23. Hollenberg, S. M. et al. 1985. Primary structure and expression of a functional human glucocorticoid receptor cDNA. *Nature* 318:635–641.

24. Green, S. et al. 1986. Human oestrogen receptor cDNA: Sequence, expression and homology to v-erb-A. *Nature* 320:134–139.
25. Weinberger, C. et al. 1986. The c-erb-A gene encodes a thyroid hormone receptor. *Nature* 324:641–646.
26. Sap, J. et al. 1986. The c-erb-A protein is a high-affinity receptor for thyroid hormone. *Nature* 324:635–640.
27. McDonnell, D. P., D. J. Mangelsdorf, J. W. Pike, M. R. Haussler, and B. W. O'Malley. 1987. Molecular cloning of complementary DNA encoding the avian receptor for vitamin D. *Science* 235:1214–1217.
28. Bookout, A. L. et al. 2006. Anatomical profiling of nuclear receptor expression reveals a hierarchical transcriptional network. *Cell* 126:789–799.
29. Carlberg, C., and T. W. Dunlop. 2006. An integrated biological approach to nuclear receptor signaling in physiological control and disease. *Critical Reviews in Eukaryotic Gene Expression* 16:1–22.
30. Makishima, M. et al. 2002. Vitamin D receptor as an intestinal bile acid sensor. *Science* 296:1313–1316.
31. Cohen, R. N., A. Putney, F. E. Wondisford, and A. N. Hollenberg. 2000. The nuclear corepressors recognize distinct nuclear receptor complexes. *Molecular Endocrinology* 14:900–914.
32. Kolodkin, A. N. et al. 2010. Design principles of nuclear receptor signaling: How complex networking improves signal transduction. *Molecular Systems Biology* 6:446.
33. Imawari, M., K. Kida, and D. S. Goodman. 1976. The transport of vitamin D and its 25-hydroxy metabolite in human plasma. Isolation and partial characterization of vitamin D and 25-hydroxyvitamin D binding protein. *Journal of Clinial Investigation* 58:514–523.
34. Bouillon, R., F. A. Van Assche, H. Van Baelen, W. Heyns, and P. De Moor. 1981. Influence of the vitamin D–binding protein on the serum concentration of 1,25-dihydroxyvitamin D3: Significance of the free 1,25-dihydroxyvitamin D3 concentration. *Journal of Clinial Investigation* 67:589–596.
35. Nykjaer, A. et al. 1999. An endocytic pathway essential for renal uptake and activation of the steroid 25-(OH) vitamin D3. *Cell* 96:507–515.
36. Prufer, K., A. Racz, G. C. Lin, and J. Barsony. 2000. Dimerization with retinoid X receptors promotes nuclear localization and subnuclear targeting of vitamin D receptors. *Journal of Biological Chemistry* 275:41114–41123.
37. Yasmin, R., R. M., Williams, M., Xu, and N. Noy. 2005. Nuclear import of the retinoid X receptor, the vitamin D receptor, and their mutual heterodimer. *Journal of Biological Chemistry* 280:40152–40160.
38. Barsony, J., I. Renyi, and W. McKoy. 1997. Subcellular distribution of normal and mutant vitamin D receptors in living cells: Studies with a novel fluorescent ligand. *Journal of Biological Chemistry* 272:5774–5782.
39. Huhtakangas, J. A., C. J. Olivera, J. E. Bishop, L. P. Zanello, and A. W. Norman. 2004. The vitamin D receptor is present in caveolae-enriched plasma membranes and binds 1 alpha,25(OH)$_2$-vitamin D$_3$ in vivo and in vitro. *Molecular Endocrinology* 18:2660–2671.
40. Boyan, B. D. et al. 2006. Regulation of growth plate chondrocytes by 1,25-dihydroxyvitamin D$_3$ requires caveolae and caveolin-1. *Journal of Bone and Mineral Research* 21:1637–1647.
41. Quack, M. and C. Carlberg. 2000. The impact of functional vitamin D(3) receptor conformations on DNA-dependent vitamin D(3) signaling. *Molecular Pharmacology* 57:375–384.
42. Renaud, J. P. et al. 1995. Crystal structure of the RAR-gamma ligand-binding domain bound to all-trans retinoic acid. *Nature* 378:681–689.
43. Nakabayashi, M. et al. 2008. Crystal structures of rat vitamin D receptor bound to adamantyl vitamin D analogs: Structural basis for vitamin D receptor antagonism and partial agonism. *Jorunal of Medicinal Chemistry* 51:5320–5329.
44. Carlberg, C. and F. Molnar. 2006. Detailed molecular understanding of agonistic and antagonistic vitamin D receptor ligands. *Current Topics in Medicinal Chemistry* 6:1243–1253.
45. Saramaki, A. et al. 2009. Cyclical chromatin looping and transcription factor association on the regulatory regions of the p21 (CDKN1A) gene in response to 1alpha,25-dihydroxyvitamin D3. *Journal of Biological Chemistry* 284:8073–8082.
46. Kim, J. Y., Y. L. Son, and Y. C. Lee. 2009. Involvement of SMRT Corepressor in Transcriptional Repression by the Vitamin D Receptor. *Molecular Endocrinology* 23:251–264.
47. Li, J. et al. 2000. Both corepressor proteins SMRT and N-CoR exist in large protein complexes containing HDAC3. *EMBO Journal* 19:4342–4350.
48. Malinen, M. et al. 2008. Distinct HDACs regulate the transcriptional response of human cyclin-dependent kinase inhibitor genes to trichostatin A and 1{alpha},25-dihydroxyvitamin D$_3$. *Nucleic Acids Research* 36:121–132.
49. Yoon, H. G. et al. 2003. Purification and functional characterization of the human N-CoR complex: The roles of HDAC3, TBL1 and TBLR1. *EMBO Journal* 22:1336–1346.

50. Alenghat, T., J. Yu, and M. A. Lazar. 2006. The N-CoR complex enables chromatin remodeler SNF2H to enhance repression by thyroid hormone receptor. *EMBO Journal* 25:3966–3974.

51. Yu, C. et al. 2005. The nuclear receptor corepressors NCoR and SMRT decrease peroxisome proliferator-activated receptor gamma transcriptional activity and repress 3T3-L1 adipogenesis. *Journal of Biological Chemistry* 280:13600–13605.

52. Oda, Y. et al. 2003. Two distinct coactivators, DRIP/mediator and SRC/p160, are differentially involved in vitamin D receptor transactivation during keratinocyte differentiation. *Molecular Endocrinology* 17:2329–2339.

53. Rachez, C. et al. 2000. The DRIP complex and SRC-1/p160 coactivators share similar nuclear receptor binding determinants but constitute functionally distinct complexes. *Molecular and Cellular Biology* 20:2718–2726.

54. Saramaki, A., C. M. Banwell, M. J. Campbell, and C. Carlberg. 2006. Regulation of the human p21(waf1/cip1) gene promoter via multiple binding sites for p53 and the vitamin D3 receptor. *Nucleic Acids Research* 34:543–554.

55. Vaisanen, S., T. W. Dunlop, L. Sinkkonen, C. Frank, and C. Carlberg. 2005. Spatiotemporal activation of chromatin on the human CYP24 gene promoter in the presence of 1alpha,25-dihydroxyvitamin D(3). *Journal of Molecular Biology* 350:65–77.

56. Zella, L. A., S. Kim, N. K. Shevde, and J.W. Pike. 2006. Enhancers located within two introns of the vitamin D receptor gene mediate transcriptional autoregulation by 1,25-dihydroxyvitamin D_3. *Molecular Endocrinology* 20:1231–1247.

57. Kim, S., N. K. Shevde, and J. W. Pike. 2005. 1,25-Dihydroxyvitamin D_3 stimulates cyclic vitamin D receptor/retinoid X receptor DNA-binding, coactivator recruitment, and histone acetylation in intact osteoblasts. *Journal of Bone and Mineral Research* 20:305–317.

58. Seo, Y. K. et al. 2007. Xenobiotic- and vitamin D-responsive induction of the steroid/bile acid-sulfotransferase Sult2A1 in young and old mice: The role of a gene enhancer in the liver chromatin. *Gene* 386:218–223.

59. Meyer, M. B., M. Watanuki, S. Kim, N. K. Shevde, and J. W. Pike. 2006. The human TRPV6 distal promoter contains multiple vitamin D receptor binding sites that mediate activation by 1,25-dihydroxyvitamin D_3 in intestinal cells. *Molecular Endocrinology* 20:1447–1461.

60. Reid, G. et al. 2003. Cyclic, proteasome-mediated turnover of unliganded and liganded ERalpha on responsive promoters is an integral feature of estrogen signaling. *Molecular Cell* 11:695–707.

61. Metivier, R. et al. 2003. Estrogen receptor-alpha directs ordered, cyclical, and combinatorial recruitment of cofactors on a natural target promoter. *Cell* 115:751–763.

62. Yang, X. et al. 2006. Nuclear receptor expression links the circadian clock to metabolism. *Cell* 126:801–810.

63. Carroll, J. S. et al. 2006. Genome-wide analysis of estrogen receptor binding sites. *Nature Genetics* 38:1289–1297.

64. Shogren-Knaak, M. et al. 2006. Histone H4-K16 acetylation controls chromatin structure and protein interactions. *Science* 311:844–847.

65. Shi, X. et al. 2009. ING2 PHD domain links histone H3 lysine 4 methylation to active gene repression. *Nature* 442:96–99.

66. Varambally, S. et al. 2002. The polycomb group protein EZH2 is involved in progression of prostate cancer. *Nature* 419:624–629.

67. Yu, J., Y. Li, T. Ishizuka, M. G. Guenther, and M. A. Lazar. 2003. A SANT motif in the SMRT corepressor interprets the histone code and promotes histone deacetylation. *EMBO Journal* 22:3403–3410.

68. Rosenfeld, M. G., V. V. Lunyak, and C. K. Glass. 2006. Sensors and signals: A coactivator/corepressor/epigenetic code for integrating signal-dependent programs of transcriptional response. *Genes & Development* 20:1405–1428.

69. Eferl, R. and E. F. Wagner. 2003. AP-1: A double-edged sword in tumorigenesis. *Nature Reviews Cancer* 3:859–868.

70. Watt, F. M., M. Frye, and S. A. Benitah. 2008. MYC in mammalian epidermis: How can an oncogene stimulate differentiation? *Nature Reviews Cancer* 8:234–242.

71. Goodson, M. L., B. A. Jonas, and M. L. Privalsky. 2005. Alternative mRNA splicing of SMRT creates functional diversity by generating corepressor isoforms with different affinities for different nuclear receptors. *Journal of Biological Chemistry* 280:7493–7503.

72. Kang, Z., O. A. Janne, and J. J. Palvimo. 2004. Coregulator recruitment and histone modifications in transcriptional regulation by the androgen receptor. *Molecular Endocrinology* 18:2633–2648.

73. Thorne, J. L. et al. 2011. Epigenetic control of a VDR-governed feed-forward loop that regulates p21(waf1/cip1) expression and function in nonmalignant prostate cells. *Nucleic Acids Research* 39: 2045–2056.

74. Thorne, J. L., M. J. Campbell, and B. M. Turner. 2009. Transcription factors, chromatin and cancer. *International Journal of Biochemistry and Cell Biology* 41:164–175.

75. Jepsen, K. et al. 2007. SMRT-mediated repression of an H3K27 demethylase in progression from neural stem cell to neuron. *Nature* 450:415–419.

76. Alenghat, T. et al. 2008. Nuclear receptor corepressor and histone deacetylase 3 govern circadian metabolic physiology. *Nature* 456:997–1000.

77. Astapova, I. et al. 2008. The nuclear corepressor, NCoR, regulates thyroid hormone action in vivo. *Proceedings of the National Academy of Sciences USA* 105:19544–19549.

78. Nofsinger, R. R. et al. 2008. SMRT repression of nuclear receptors controls the adipogenic set point and metabolic homeostasis. *Proceedings of the National Academy of Sciences USA* 105:20021–20026.

79. Surjit, M. et al. 2011. Widespread negative response elements mediate direct repression by agonist-liganded glucocorticoid receptor. *Cell* 145:224–241.

80. Polly, P. et al. 2000. VDR-Alien: A novel, DNA-selective vitamin D(3) receptor-corepressor partnership. *FASEB Journal* 14:1455–1463.

81. Lykke-Andersen, K. et al. 2003. Disruption of the COP9 signalosome Csn2 subunit in mice causes deficient cell proliferation, accumulation of p53 and cyclin E, and early embryonic death. *Molecular and Cellular Biology* 23:6790–6797.

82. Hatchell, E. C. et al. 2006. SLIRP, a small SRA binding protein, is a nuclear receptor corepressor. *Molecular Cell* 22:657–668.

83. Hsieh, J. C. et al. 2003. Physical and functional interaction between the vitamin D receptor and hairless corepressor, two proteins required for hair cycling. *Journal of Biological Chemistry* 278:38665–38674.

84. Miller, J. et al. 2001. Atrichia caused by mutations in the vitamin D receptor gene is a phenocopy of generalized atrichia caused by mutations in the hairless gene. *Journal of Investigative Dermatology* 117:612–617.

85. Xie, Z., S. Chang, Y. Oda, and D. D. Bikle. 2006. Hairless suppresses vitamin D receptor transactivation in human keratinocytes. *Endocrinology* 147:314–323.

86. Scsucova, S. et al. 2005. The repressor DREAM acts as a transcriptional activator on Vitamin D and retinoic acid response elements. *Nucleic Acids Research* 33:2269–2279.

87. Ding, X. F. et al. 1998. Nuclear receptor-binding sites of coactivators glucocorticoid receptor interacting protein 1 (GRIP1) and steroid receptor coactivator 1 (SRC-1): Multiple motifs with different binding specificities. *Molecular Endocrinology* 12:302–313.

88. Eggert, M. et al. 1995. A fraction enriched in a novel glucocorticoid receptor-interacting protein stimulates receptor-dependent transcription in vitro. *Journal of Biological Chemistry* 270:30755–30759.

89. Zhang, J. and J. D. Fondell. 1999. Identification of mouse TRAP100: A transcriptional coregulatory factor for thyroid hormone and vitamin D receptors. *Molecular Endocrinology* 13:1130–1140.

90. Yuan, C. X., M. Ito, J. D. Fondell, Z. Y. Fu, and R. G. Roeder. 1998. The TRAP220 component of a thyroid hormone receptor-associated protein (TRAP) coactivator complex interacts directly with nuclear receptors in a ligand-dependent fashion. *Proceedings of the National Academy of Sciences USA* 95:7939–7944.

91. Lee, J. W., H. S. Choi, J. Gyuris, R. Brent, and D. D. Moore. 1995. Two classes of proteins dependent on either the presence or absence of thyroid hormone for interaction with the thyroid hormone receptor. *Molecular Endocrinology* 9:243–254.

92. Zhang, C. et al. 2003. Nuclear coactivator-62 kDa/Ski-interacting protein is a nuclear matrix-associated coactivator that may couple vitamin D receptor-mediated transcription and RNA splicing. *Journal of Biological Chemistry* 278:35325–35336.

93. Urahama, N. et al. 2005. The role of transcriptional coactivator TRAP220 in myelomonocytic differentiation. *Genes to Cells* 10:1127–1137.

94. Ren, Y. et al. 2000. Specific structural motifs determine TRAP220 interactions with nuclear hormone receptors. *Molecular and Cellular Biology* 20:5433–5446.

95. Teichert, A. et al. Quantification of the vitamin D receptor-coregulator interaction (dagger). *Biochemistry* (2009).

96. Hawker, N. P., S. D. Pennypacker, S. M. Chang, and D. D. Bikle. 2007. Regulation of human epidermal keratinocyte differentiation by the vitamin D receptor and its coactivators DRIP205, SRC2, and SRC3. *Journal of Investigative Dermatology* 127:874–880.

97. Lutz, W., K. Kohno, and R. Kumar. 2001. The role of heat shock protein 70 in vitamin D receptor function. *Biochemical and Biophysical Research Communications* 282:1211–1219.

98. Guzey, M., S. Takayama, and J. C. Reed. 2000. BAG1L enhances transactivation function of the vitamin D receptor. *Journal of Biological Chemistry* 275:40749–40756.

99. Bikle, D., A. Teichert, N. Hawker, Z. Xie, and Y. Oda. 2007. Sequential regulation of keratinocyte differentiation by 1,25(OH)2D3, VDR, and its coregulators. *Journal of Steroid Biochemistry and Molecular Biology* 103:396–404.

100. Blok, L. J., P. E. de Ruiter, and A. O. Brinkmann. 1996. Androgen receptor phosphorylation. *Endocrine Research* 22:197–219.

101. Hilliard, G. M. T., R. G. Cook, N. L. Weigel, and J. W. Pike. 1994. 1,25-dihydroxyvitamin D3 modulates phosphorylation of serine 205 in the human vitamin D receptor: Site-directed mutagenesis of this residue promotes alternative phosphorylation. *Biochemistry* 33:4300–4311.

102. Murayama, A., M. S. Kim, J. Yanagisawa, K. Takeyama, and S. Kato. 2004. Transrepression by a liganded nuclear receptor via a bHLH activator through coregulator switching. *EMBO Journal* 23:1598–1608.

103. Hsieh, J. C. et al. 1991. Human vitamin D receptor is selectively phosphorylated by protein kinase C on serine 51, a residue crucial to its trans-activation function. *Proceedings of the National Academy of Sciences USA* 88:9315–9319.

104. Macoritto, M. et al. 2008. Phosphorylation of the human retinoid X receptor {alpha} at Serine 260 impairs coactivator(s) recruitment and induces hormone resistance to multiple ligands. *Journal of Biological Chemistry* 283:4943–4956.

105. Arriagada, G. et al. 2007. Phosphorylation at serine 208 of the 1[alpha],25-dihydroxy Vitamin D3 receptor modulates the interaction with transcriptional coactivators. *Journal of Steroid Biochemistry and Molecular Biology* 103:425–429.

106. Jurutka, P. W. et al. 1996. Human vitamin D receptor phosphorylation by casein kinase II at Ser-208 potentiates transcriptional activation. *Proceedings of the National Academy of Sciences USA* 93:3519–3524.

107. Barletta, F., L. P. Freedman, and S. Christakos. 2002. Enhancement of VDR-mediated transcription by phosphorylation: Correlation with increased interaction between the VDR and DRIP205, a subunit of the VDR-interacting protein coactivator complex. *Molecular Endocrinology* 16:301–314.

108. Vaziri, H. et al. 2001. hSIR2(SIRT1) functions as an NAD-dependent p53 deacetylase. *Cell* 107:149–159.

109. An, B. S. et al. 2010. Stimulation of Sirt1-regulated FoxO protein function by the ligand-bound vitamin D receptor. *Molecular and Cellular Biology* 30:4890–4900.

110. Rodgers, J.T. et al. 2005. Nutrient control of glucose homeostasis through a complex of PGC-1alpha and SIRT1. *Nature* 434:113–118.

111. Fulco, M. et al. 2003. Sir2 regulates skeletal muscle differentiation as a potential sensor of the redox state. *Molecular Cell* 12:51–62.

112. Carlberg, C. and S. Seuter. 2007. The vitamin D receptor. *Dermatologic Clinics* 25:515–523, viii.

113. Thompson, P. D. et al. 2002. Liganded VDR induces CYP3A4 in small intestinal and colon cancer cells via DR3 and ER6 vitamin D responsive elements. *Biochemical and Biophysical Research Communications* 299:730–738.

114. Song, C. S. et al. 2006. An essential role of the CAAT/enhancer binding protein-alpha in the vitamin D-induced expression of the human steroid/bile acid-sulfotransferase (SULT2A1). *Molecular Endocrinology* 20:795–808.

115. Chen, C. D. et al. 2004. Molecular determinants of resistance to antiandrogen therapy. *Nature Medicine* 10:33–39.

116. Fujiki, R. et al. 2005. Ligand-induced transrepression by VDR through association of WSTF with acetylated histones. *EMBO Journal* 24:3881–3894.

117. Kim, M.S. et al. 2007. 1Alpha,25(OH)2D3-induced transrepression by vitamin D receptor through E-box-type elements in the human parathyroid hormone gene promoter. *Molecular Endocrinology* 21: 334–342.

118. Turunen, M. M., T. W. Dunlop, C. Carlberg, and S. Vaisanen. 2007. Selective use of multiple vitamin D response elements underlies the 1 alpha,25-dihydroxyvitamin D3-mediated negative regulation of the human CYP27B1 gene. *Nucleic Acids Research* 35:2734–2747.

119. Eeckhoute, J., J. S. Carroll, T. R. Geistlinger, M. I. Torres-Arzayus, and M. Brown. 2006. A cell type–specific transcriptional network required for estrogen regulation of cyclin D1 and cell cycle progression in breast cancer. *Genes & Development* 20:2513–2526.

120. Lefterova, M. I. et al. 2008. PPAR{gamma} and C/EBP factors orchestrate adipocyte biology via adjacent binding on a genome-wide scale. *Genes & Development* 22:2941–2952.

121. Kwon, Y. S. et al. 2007. Sensitive ChIP-DSL technology reveals an extensive estrogen receptor alpha-binding program on human gene promoters. *Proceedings of the National Academy of Sciences USA* 104:4852–4857.

122. Metivier, R. et al. 2004. Transcriptional complexes engaged by apo-estrogen receptor-alpha isoforms have divergent outcomes. *EMBO Journal* 23:3653–3666.

123. Lupien, M. et al. 2008. FoxA1 translates epigenetic signatures into enhancer-driven lineage-specific transcription. *Cell* 132:958–970.

124. Hartman, H. B., J. Yu, T. Alenghat, T. Ishizuka, and M. A. Lazar. 2005. The histone-binding code of nuclear receptor corepressors matches the substrate specificity of histone deacetylase 3. *EMBO Reports* 6:445–451.

125. Heikkinen, S. et al. 2011. Nuclear hormone 1{alpha},25-dihydroxyvitamin D3 elicits a genome-wide shift in the locations of VDR chromatin occupancy. *Nucleic Acids Research* 9:9181–9193.

126. Ramagopalan, S. V. et al. 2010. A ChIP-seq defined genome-wide map of vitamin D receptor binding: Associations with disease and evolution. *Genome Research* 20:1352–1360.

127. Wang, W. L., N. Chatterjee, S. V. Chittur, J. Welsh, and M. P. Tenniswood. 2011. Effects of 1alpha,25 dihydroxyvitamin D_3 and testosterone on miRNA and mRNA expression in LNCaP cells. *Molecular Cancer* 10:58.

128. Wang, X., E. Gocek, C. G. Liu, and G. P. Studzinski. 2009. MicroRNAs181 regulate the expression of p27(Kip1) in human myeloid leukemia cells induced to differentiate by 1,25-dihydroxyvitamin D(3). *Cell Cycle* 8:736–741.

129. Yoshizawa, T. et al. 1997. Mice lacking the vitamin D receptor exhibit impaired bone formation, uterine hypoplasia and growth retardation after weaning. *Nature Genetics* 16:391–396.

130. Li, Y. C. et al. 1997. Targeted ablation of the vitamin D receptor: An animal model of vitamin D-dependent rickets type II with alopecia. *Proceedings of the National Academy of Sciences USA* 94:9831–9835.

131. Van Cromphaut, S. J. et al. 2001. Duodenal calcium absorption in vitamin D receptor-knockout mice: Functional and molecular aspects. *Proceedings of the National Academy of Sciences USA* 98:13324–13329.

132. Bikle, D. D., E. Gee, and S. Pillai. 1993. Regulation of keratinocyte growth, differentiation, and vitamin D metabolism by analogs of 1,25-dihydroxyvitamin D. *Journal of Investigative Dermatology* 101:713–718.

133. Reichrath, J. et al. 1994. Hair follicle expression of 1,25-dihydroxyvitamin D3 receptors during the murine hair cycle. *British Journal of Dermatology* 131:477–482.

134. Sakai, Y., J. Kishimoto, and M. B. Demay. 2001. Metabolic and cellular analysis of alopecia in vitamin D receptor knockout mice. *Journal of Clinial Investigation* 107:961–966.

135. Palmer, H. G., D. Martinez, G. Carmeliet, and F. M. Watt. 2008. The vitamin D receptor is required for mouse hair cycle progression but not for maintenance of the epidermal stem cell compartment. *Journal of Investigative Dermatology* 128:2113–2117.

136. Beaudoin, G. M. 3rd, J. M. Sisk, P. A. Coulombe, and C. C. Thompson. 2005. Hairless triggers reactivation of hair growth by promoting Wnt signaling. *Proceedings of the National Academy of Sciences USA* 102:14653–14658.

137. Thompson, C. C., J. M. Sisk, and G. M. Beaudoin 3rd. 2006. Hairless and Wnt signaling: Allies in epithelial stem cell differentiation. *Cell Cycle* 5:1913–1917.

138. Palmer, H. G., F. Anjos-Afonso, G. Carmeliet, H. Takeda, and F. M. Watt. 2008. The vitamin D receptor is a Wnt effector that controls hair follicle differentiation and specifies tumor type in adult epidermis. *PLoS ONE* 3:e1483.

139. Zinser, G., K. Packman, and J. Welsh, 2002. Vitamin D(3) receptor ablation alters mammary gland morphogenesis. *Development* 129:3067–3076.

140. Zinser, G. M. and J. Welsh. 2004. Accelerated mammary gland development during pregnancy and delayed postlactational involution in vitamin D3 receptor null mice. *Molecular Endocrinology* 18:2208–2223.

141. Novershtern, N. et al. 2011. Densely interconnected transcriptional circuits control cell states in human hematopoiesis. *Cell* 144:296–309.

142. Colston, K., M. J. Colston, and D. Feldman, 1981. 1,25-dihydroxyvitamin D3 and malignant melanoma: The presence of receptors and inhibition of cell growth in culture. *Endocrinology* 108:1083–1086.

143. Miyaura, C. et al. 1981. 1 alpha,25-Dihydroxyvitamin D3 induces differentiation of human myeloid leukemia cells. *Biochemical and Biophysical Research Communications* 102:937–943.

144. Abe, E. et al. 1981. Differentiation of mouse myeloid leukemia cells induced by 1 alpha,25-dihydroxyvitamin D3. *Proceedings of the National Academy of Sciences USA* 78:4990–4994.

145. Palmer, H. G. et al. 2003. Genetic signatures of differentiation induced by 1alpha,25-dihydroxyvitamin D₃ in human colon cancer cells. *Cancer Research* 63:7799–7806.
146. Koike, M. et al. 1997. 19-nor-hexafluoride analogue of vitamin D3: A novel class of potent inhibitors of proliferation of human breast cell lines. *Cancer Research* 57:4545–4550.
147. Campbell, M. J., E. Elstner, S. Holden, M. Uskokovic, and H. P. Koeffler. 1997. Inhibition of proliferation of prostate cancer cells by a 19-nor-hexafluoride vitamin D3 analogue involves the induction of p21waf1, p27kip1 and E-cadherin. *Molecular Endocrinology* 19:15–27.
148. Elstner, E. et al. 1999. Novel 20-epi-vitamin D3 analog combined with 9-cis-retinoic acid markedly inhibits colony growth of prostate cancer cells. *Prostate* 40:141–149.
149. Peehl, D. M. et al. 1994. Antiproliferative effects of 1,25-dihydroxyvitamin D3 on primary cultures of human prostatic cells. *Cancer Research* 54:805–810.
150. Welsh, J. et al. 2002. Impact of the Vitamin D3 receptor on growth-regulatory pathways in mammary gland and breast cancer. *Journal of Steroid Biochemistry and Molecular Biology* 83:85–92.
151. Colston, K. W., U. Berger, and R. C. Coombes. 1989. Possible role for vitamin D in controlling breast cancer cell proliferation. *Lancet* 1:188–191.
152. Colston, K., M. J. Colston, A. H. Fieldsteel, and D. Feldman. 1982. 1,25-dihydroxyvitamin D3 receptors in human epithelial cancer cell lines. *Cancer Research* 42:856–859.
153. Eelen, G. et al. 2004. Microarray analysis of 1alpha,25-dihydroxyvitamin D3-treated MC3T3-E1 cells. *Journal of Steroid Biochemistry and Molecular Biology* 89–90:405–407.
154. Akutsu, N. et al. 2001. Regulation of gene expression by 1alpha,25-dihydroxyvitamin D3 and its analog EB1089 under growth-inhibitory conditions in squamous carcinoma cells. *Molecular Endocrinology* 15:1127–1139.
155. Wang, T. T. et al. 2005. Large-scale *in silico* and microarray-based identification of direct 1,25-dihydroxyvitamin D₃ target genes. *Molecular Endocrinology* 19:2685–2695.
156. Liu, M., M. H. Lee, M. Cohen, M. Bommakanti, and L. P. Freedman. 1996. Transcriptional activation of the Cdk inhibitor p21 by vitamin D3 leads to the induced differentiation of the myelomonocytic cell line U937. *Genes & Development*. 10:142–153.
157. Wang, Q. M., J. B. Jones, and G. P. Studzinski. 1996. Cyclin-dependent kinase inhibitor p27 as a mediator of the G1-S phase block induced by 1,25-dihydroxyvitamin D₃ in HL60 cells. *Cancer Research* 56:264–267.
158. Li, P. et al. 2004. p27(Kip1) stabilization and G(1) arrest by 1,25-dihydroxyvitamin D(3) in ovarian cancer cells mediated through downregulation of cyclin E/cyclin-dependent kinase 2 and Skp1-Cullin-F-box protein/Skp2 ubiquitin ligase. *Journal of Biological Chemistry* 279:25260–25267.
159. Huang, Y. C., J. Y. Chen, and W. C. Hung. 2004. Vitamin D(3) receptor/Sp1 complex is required for the induction of p27(Kip1) expression by vitamin D(3). *Oncogene* 23:4856–4861.
160. Hengst, L. and S. I. Reed, 1996. Translational control of p27Kip1 accumulation during the cell cycle. *Science* 271:1861–1864.
161. Jiang, F., P. Li, A. J. Fornace Jr., S. V. Nicosia, and W. Bai. 2003. G2/M arrest by 1,25-dihydroxyvitamin D3 in ovarian cancer cells mediated through the induction of GADD45 via an exonic enhancer. *Journal of Biological Chemistry* 278:48030–48040.
162. Khanim, F. L. et al. 2004. Altered SMRT levels disrupt vitamin D₃ receptor signalling in prostate cancer cells. *Oncogene* 23:6712–6725.
163. Lubbert, M. et al. 1991. Stable methylation patterns of MYC and other genes regulated during terminal myeloid differentiation. *Leukemia* 5:533–539.
164. Koeffler, H. P. 1983. Induction of differentiation of human acute myelogenous leukemia cells: Therapeutic implications. *Blood* 62:709–721.
165. Studzinski, G. P., A. K. Bhandal, and Z. S. Brelvi. 1986. Potentiation by 1-alpha,25-dihydroxyvitamin D3 of cytotoxicity to HL-60 cells produced by cytarabine and hydroxyurea. *Journal of the National Cancer Institute* 76:641–648.
166. Lin, R. J. et al. 1998. Role of the histone deacetylase complex in acute promyelocytic leukaemia. *Nature* 391:811–814.
167. Song, G. and L. Wang. 2008. Transcriptional mechanism for the paired miR-433 and miR-127 genes by nuclear receptors SHP and ERR gamma. *Nucleic Acids Research* 36:5727–5735.
168. Shah, Y. M. et al. 2007. Peroxisome proliferator-activated receptor alpha regulates a microRNA-mediated signaling cascade responsible for hepatocellular proliferation. *Molecular and Cellular Biology* 27:4238–4247.
169. Ellison, T. I., M. K. Smith, A. C. Gilliam, and P. N. MacDonald. 2008. Inactivation of the vitamin D receptor enhances susceptibility of murine skin to UV-induced tumorigenesis. *Journal of Investigative Dermatology* 128:2508–2517.

170. Maruyama, R. et al. 2006. Comparative genome analysis identifies the vitamin D receptor gene as a direct target of p53-mediated transcriptional activation. *Cancer Research* 66:4574–4583.
171. Lambert, J. R. et al. 2006. Prostate derived factor in human prostate cancer cells: Gene induction by vitamin D via a p53-dependent mechanism and inhibition of prostate cancer cell growth. *Journal of Cellular Physiology* 208:566–574.
172. Kommagani, R., T. M. Caserta, and M. P. Kadakia, 2006. Identification of vitamin D receptor as a target of p63. *Oncogene* 25:3745–3751.
173. Jin, F. and J. D. Fondell. 2009. A novel androgen receptor-binding element modulates Cdc6 transcription in prostate cancer cells during cell-cycle progression. *Nucleic Acids Research* 37:4826–4838.
174. Okada, M. et al. 2008. Switching of chromatin-remodeling complexes for oestrogen receptor-alpha. *EMBO Reports* 9:563–568.
175. Cohen, E. E. et al. 2009. A feedforward loop involving protein kinase Calpha and microRNAs regulates tumor cell cycle. *Cancer Research* 69:65–74.
176. Brosh, R. et al. 2008. p53-Repressed miRNAs are involved with E2F in a feed-forward loop promoting proliferation. *Mol Syst Biol* 4:229.
177. Poliseno, L. et al. 2010. Identification of the miR-106b~25 microRNA cluster as a proto-oncogenic PTEN-targeting intron that cooperates with its host gene MCM7 in transformation. *Sci Signal* 3:29.
178. Resnick, K. E. et al. 2009. The detection of differentially expressed microRNAs from the serum of ovarian cancer patients using a novel real-time PCR platform. *Gynecologic Oncology* 112:55–59.
179. Mitchell, P. S. et al. 2008. Circulating microRNAs as stable blood-based markers for cancer detection. *Proceedings of the National Academy of Sciences USA* 105:10513–10518.
180. Chen, X. et al. 2008. Characterization of microRNAs in serum: A novel class of biomarkers for diagnosis of cancer and other diseases. *Cell Research* 18:997–1006.
181. Brase, J. C. et al. 2011. Circulating miRNAs are correlated with tumor progression in prostate cancer. *International Journal of Cancer* 128:608–616.
182. Barbieri, C. E. and J. A. Pietenpol. 2006. p63 and epithelial biology. *Experimental Cell Research* 312:695–706.
183. Blackburn, A. C. and D. J. Jerry. 2002. Knockout and transgenic mice of Trp53: What have we learned about p53 in breast cancer? *Breast Cancer Research* 4:101–111.
184. Lu, J. et al. 2005. Transcriptional profiling of keratinocytes reveals a vitamin D-regulated epidermal differentiation network. *Journal of Investigative Dermatology* 124:778–785.
185. Yang, L., J. Yang, S. Venkateswarlu, T. Ko, and M. G. Brattain. 2001. Autocrine TGFbeta signaling mediates vitamin D_3 analog-induced growth inhibition in breast cells. *Journal of Cellular Physiology* 188:383–393.
186. Wu, Y., T. A. Craig, W. H. Lutz, and R. Kumar. 1999. Identification of 1 alpha,25-dihydroxyvitamin D3 response elements in the human transforming growth factor beta 2 gene. *Biochemistry* 38:2654–2660.
187. Matilainen, M., M. Malinen, K. Saavalainen, and C. Carlberg. 2005. Regulation of multiple insulin-like growth factor binding protein genes by 1alpha,25-dihydroxyvitamin D_3. *Nucleic Acids Research* 33:5521–5532.
188. Vousden, K. H. and X. Lu. 2002. Live or let die: The cell's response to p53. *Nature Reviews Cancer* 2:594–604.
189. Lee, D. et al. 2002. SWI/SNF complex interacts with tumor suppressor p53 and is necessary for the activation of p53-mediated transcription. *Journal of Biological Chemistry* 277:22330–22337.
190. Wilkinson, D. S. et al. 2005. A direct intersection between p53 and transforming growth factor beta pathways targets chromatin modification and transcription repression of the alpha-fetoprotein gene. *Molecular and Cellular Biology* 25:1200–1212.
191. Szeto, F. L. et al. 2007. Involvement of the vitamin D receptor in the regulation of NF-κB activity in fibroblasts. *Journal of Steroid Biochemistry and Molecular Biology* 103:563–566.
192. Froicu, M. and M. Cantorna. 2007. Vitamin D and the vitamin D receptor are critical for control of the innate immune response to colonic injury. *BMC Immunology* 8:5.
193. Gombart, A. F., N. Borregaard, and H. P. Koeffler. 2005. Human cathelicidin antimicrobial peptide (CAMP) gene is a direct target of the vitamin D receptor and is strongly upregulated in myeloid cells by 1,25-dihydroxyvitamin D_3. *FASEB Journal* 19:1067–1077.
194. Wang, T.-T. et al. 2004. Cutting Edge: 1,25-dihydroxyvitamin D3 is a direct inducer of antimicrobial peptide gene expression. *Journal of Immunology* 173:2909–2912.
195. Mallbris L., D. W. Edstrom, L. Sundblad, F. Granath, and M. Stahle. 2005. UVB upregulates the antimicrobial protein hCAP18 mRNA in human skin. *Journal of Investigative Dermatology* 125:1072–1074.
196. Zasloff, M. 2005. Sunlight, vitamin D, and the innate immune defenses of the human skin. *Journal of Investigative Dermatology* 125:xvi–xvii.

197. Quigley, D. A. et al. 2009. Genetic architecture of mouse skin inflammation and tumour susceptibility. *Nature* 458:505–508.
198. Abedin, S. A. et al. 2009. Elevated NCOR1 disrupts a network of dietary-sensing nuclear receptors in bladder cancer cells. *Carcinogenesis* 30:449–456.
199. Menezes, R. J. et al. 2008. Vitamin D receptor expression in normal, premalignant, and malignant human lung tissue. *Cancer Epidemiology, Biomarkers & Prevention* 17:1104–1110.
200. Pendas-Franco, N. et al. 2008. DICKKOPF-4 is induced by TCF/beta-catenin and upregulated in human colon cancer, promotes tumor cell invasion and angiogenesis and is repressed by 1alpha,25-dihydroxyvitamin D$_3$. *Oncogene* 27:4467–4477.
201. Palmer, H. G. et al. 2004. The transcription factor SNAIL represses vitamin D receptor expression and responsiveness in human colon cancer. *Nature Medicine* 10:917–919.
202. Palmer, H. G. et al. 2001. Vitamin D(3) promotes the differentiation of colon carcinoma cells by the induction of E-cadherin and the inhibition of beta-catenin signaling. *Journal of Cell Biology* 154:369–387.
203. Shah, S. et al. 2006. The molecular basis of vitamin D receptor and beta-catenin cross regulation. *Molecular Cell* 21:799–809.
204. Miller, C. W., R. Morosetti, M. J. Campbell, S. Mendoza, and H. P. Koeffler. 1997. Integrity of the 1,25-dihydroxyvitamin D$_3$ receptor in bone, lung, and other cancers. *Molecular Carcinogenesis* 19:254–257.
205. Guy, M., L. C. Lowe, D. Bretherton-Watt, J. L. Mansi, and K. W. Colston. 2003. Approaches to evaluating the association of vitamin D receptor gene polymorphisms with breast cancer risk. *Recent Results in Cancer Research* 164:43–54.
206. John, E. M., G. G. Schwartz, J. Koo, B. D. Van Den, and S. A. Ingles. 2005. Sun exposure, vitamin D receptor gene polymorphisms, and risk of advanced prostate cancer. *Cancer Research* 65:5470–5479.
207. Ingles, S. A. et al. 1998. Association of prostate cancer with vitamin D receptor haplotypes in African-Americans. *Cancer Research* 58:1620–1623.
208. Ma, J. et al. 1998. Vitamin D receptor polymorphisms, circulating vitamin D metabolites, and risk of prostate cancer in United States physicians. *Cancer Epidemiology, Biomarkers & Prevention* 7:385–390.
209. Gorlach, M., C. G. Burd, and G. Dreyfuss. 1994. The mRNA poly(A)-binding protein: localization, abundance, and RNA-binding specificity. *Experimental Cell Research* 211:400–407.
210. Kim, J. G. et al. 2003. Association between vitamin D receptor gene haplotypes and bone mass in postmenopausal Korean women. *American Journal of Obstetrics and Gynecology* 189:1234–1240.
211. Kuraishi, T., Y. Sun, F. Aoki, K. Imakawa, and S. Sakai. 2000. The poly(A) tail length of casein mRNA in the lactating mammary gland changes depending upon the accumulation and removal of milk. *Biochemical Journal* 347:579–583.
212. Schondorf, T. et al. 2003. Association of the vitamin D receptor genotype with bone metastases in breast cancer patients. *Oncology* 64:154–159.
213. Lundin, A. C., P. Soderkvist, B. Eriksson, M. Bergman-Jungestrom, and S. Wingren. 1999. Association of breast cancer progression with a vitamin D receptor gene polymorphism: Southeast Sweden Breast Cancer Group. *Cancer Research* 59:2332–2334.
214. Guy, M. et al. 2004. Vitamin d receptor gene polymorphisms and breast cancer risk. *Clinical Cancer Research* 10:5472–5481.
215. Ntais, C., A. Polycarpou, and J. P. Ioannidis. 2003. Vitamin D receptor gene polymorphisms and risk of prostate cancer: A meta-analysis. *Cancer Epidemiology, Biomarkers & Prevention* 12:1395–1402.
216. Rashid, S. F. et al. 2001. Synergistic growth inhibition of prostate cancer cells by 1 alpha,25 Dihydroxyvitamin D$_3$ and its 19-nor-hexafluoride analogs in combination with either sodium butyrate or trichostatin A. *Oncogene* 20:1860–1872.
217. Campbell, M. J., A. F. Gombart, S. H. Kwok, S. Park, and H. P. Koeffler. 2000. The antiproliferative effects of 1alpha,25(OH)2D3 on breast and prostate cancer cells are associated with induction of BRCA1 gene expression. *Oncogene* 19:5091–5097.
218. Miller, G. J., G. E. Stapleton, T. E. Hedlund, and K. A. Moffat. 1995. Vitamin D receptor expression, 24-hydroxylase activity, and inhibition of growth by 1alpha,25-dihydroxyvitamin D$_3$ in seven human prostatic carcinoma cell lines. *Clinical Cancer Research* 1:997–1003.
219. Rashid, S. F., J. C. Mountford, A. F. Gombart, and M. J. Campbell. 2001. 1alpha,25-dihydroxyvitamin D$_3$ displays divergent growth effects in both normal and malignant cells. *Steroids* 66:433–440.
220. Campbell, M. J. and L. Adorini. 2006. The vitamin D receptor as a therapeutic target. *Expert Opinion on Therapeutic Targets* 10:735–748.
221. Ting, H. J., B. Y. Bao, J. E. Reeder, E. M. Messing, and Y. F. Lee. 2007. Increased expression of corepressors in aggressive androgen-independent prostate cancer cells results in loss of 1alpha,25-dihydroxyvitamin D$_3$ responsiveness. *Molecular Cancer Research* 5:967–980.

222. Battaglia, S. et al. 2010. Elevated NCOR1 disrupts PPARα/γ signaling in prostate cancer and forms a targetable epigenetic lesion. *Carcinogenesis* 31:1650–1660.
223. Yoon, H. G., D. W. Chan, A. B. Reynolds, J. Qin, and J. Wong. 2003. N-CoR mediates DNA methylation-dependent repression through a methyl CpG binding protein Kaiso. *Molecular Cell* 12:723–734.
224. Yegnasubramanian, S. et al. 2004. Hypermethylation of CpG islands in primary and metastatic human prostate cancer. *Cancer Research* 64:1975–1986.
225. Asatiani, E. et al. 2005. Deletion, methylation, and expression of the NKX3.1 suppressor gene in primary human prostate cancer. *Cancer Research* 65:1164–1173.
226. Fujita, N. et al. 2003. Methyl-CpG binding domain 1 (MBD1) interacts with the Suv39h1-HP1 heterochromatic complex for DNA methylation-based transcriptional repression. *Journal of Biological Chemistry* 278:24132–24138.
227. Esteve, P. O. et al. 2006. Direct interaction between DNMT1 and G9a coordinates DNA and histone methylation during replication. *Genes & Development* 20:3089–3103.
228. Cheng, X. and R. M. Blumenthal. 2010. Coordinated chromatin control: Structural and functional linkage of DNA and histone methylation. *Biochemistry* 49:2999–3008.
229. Freitag, M., P. C. Hickey, T. K. Khlafallah, N. D. Read, and E. U. Selker. 2004. HP1 is essential for DNA methylation in neurospora. *Molecular Cell* 13:427–434.
230. Le May, N. et al. 2010. NER factors are recruited to active promoters and facilitate chromatin modification for transcription in the absence of exogenous genotoxic attack. *Molecular Cell* 38:54–66.
231. Metivier, R. et al. 2008. Cyclical DNA methylation of a transcriptionally active promoter. *Nature* 452:45–50.
232. Kangaspeska, S. et al. 2008. Transient cyclical methylation of promoter DNA. *Nature* 452:112–115.
233. Banwell, C. M. et al. 2006. Altered nuclear receptor corepressor expression attenuates vitamin D receptor signaling in breast cancer cells. *Clinical Cancer Research* 12:2004–2013.
234. Banwell, C. M., L. P. O'Neill, M. R. Uskokovic, and M. J. Campbell. 2004. Targeting 1alpha,25-dihydroxyvitamin D₃ antiproliferative insensitivity in breast cancer cells by cotreatment with histone deacetylation inhibitors. *Journal of Steroid Biochemistry and Molecular Biology* 89–90:245–249.
235. Tavera-Mendoza, L. E. et al. 2008. Incorporation of histone deacetylase inhibition into the structure of a nuclear receptor agonist. *Proceedings of the National Academy of Sciences USA* 105:8250–8255.
236. Costa, E. M. and D. Feldman. 1987. Modulation of 1,25-dihydroxyvitamin D₃ receptor binding and action by sodium butyrate in cultured pig kidney cells (LLC-PK1). *Journal of Bone and Mineral Research* 2:151–159.
237. Gaschott, T. and J. Stein. 2003. Short-chain fatty acids and colon cancer cells: The vitamin D receptor—Butyrate connection. *Recent Results in Cancer Research* 164:247–257.
238. Daniel, C., O. Schroder, N. Zahn, T. Gaschott, and J. Stein. 2004. p38 MAPK signaling pathway is involved in butyrate-induced vitamin D receptor expression. *Biochemical and Biophysical Research Communications* 324:1220–1226.
239. Chen, J. S., D. V. Faller, and R. A. Spanjaard. 2003. Short-chain fatty acid inhibitors of histone deacetylases: Promising anticancer therapeutics? *Current Cancer Drug Targets* 3:219–236.
240. Gaschott, T., O. Werz, A. Steinmeyer, D. Steinhilber, and J. Stein. 2001. Butyrate-induced differentiation of Caco-2 cells is mediated by vitamin D receptor. *Biochemical and Biophysical Research Communications* 288:690–696.
241. Tanaka, Y., K. K. Bush, T. M. Klauck, and P. J. Higgins. 1989. Enhancement of butyrate-induced differentiation of HT-29 human colon carcinoma cells by 1,25-dihydroxyvitamin D₃. *Biochemical Pharmacology* 38:3859–3865.
242. Krasowski, M. D., K. Yasuda, L. R. Hagey, and E. G. Schuetz. 2005. Evolutionary selection across the nuclear hormone receptor superfamily with a focus on the NR1I subfamily (vitamin D, pregnane X, and constitutive androstane receptors). *Nuclear Receptor* 3:2.
243. Adorini, L., K. C. Daniel, and G. Penna. 2006. Vitamin D receptor agonists, cancer and the immune system: An intricate relationship. *Current Topics in Medicinal Chemistry* 6:1297–1301.
244. Westerhoff, H. V. and B. O. Palsson. 2004. The evolution of molecular biology into systems biology. *Nature Biotechnology* 22:1249–1252.
245. Muller, M. and S. Kersten. 2003. Nutrigenomics: Goals and strategies. *Nature Reviews Genetics* 4:315–322.

3 Vitamin D Analogs and Their Clinical Uses

Glenville Jones

CONTENTS

3.1   INTRODUCTION

Vitamin D, its metabolites, and analogs constitute a valuable group of compounds that can be used to regulate gene expression in many cells of the body in functions such as calcium and phosphate homeostasis, as well as cell growth regulation and cell differentiation of a variety of cell types (such as enterocytes, keratinocytes, and epithelial lining cells of vascular, GI, and other ductwork of the body) [1,2]. The parent vitamin (or ultraviolet (UV) light that substitutes for any vitamin D pharmaceutical preparation as a source of the parent vitamin) has been used as a treatment for rickets and osteomalacia since its discovery in the 1920s [3]. The discovery of the principal metabolites, i.e., 25-hydroxyvitamin D_3 (25-OH-D_3; calcidiol) and 1α,25-dihydroxyvitamin D_3 (1α,25-$(OH)_2D_3$; calcitriol), in the early 1970s provided the first generation of vitamin D analogs [4–6], and these have been followed over the past four decades by further generations of analogs featuring modifications of the basic vitamin D structure and culminating in nonsecosteroid vitamin D analogs. With the understanding that the hormonal form, 1α,25-$(OH)_2D_3$, is both a *calcemic agent*, regulating calcium and phosphate transport, and a *cell-differentiating agent*, promoting terminal development of the osteoclast, enterocyte, and keratinocyte [7], it is not surprising that the pharmaceutical industry has striven hard to separate these two properties and thereby develop synthetic vitamin D analogs with specialized "calcemic" and "noncalcemic" (cell-differentiating) uses [8–10]. From this type of research has come several "low-calcemic" agents in the form of calcipotriol, OCT, 19-nor-1α,25-$(OH)_2D_2$, and 1α-OH-D_2, which have found widespread use in dermatology and the treatment of secondary hyperparathyroidism. Newer analogs include not only specialized "selective" vitamin D receptor (VDR) agonists but also VDR antagonists and compounds, which target CYP24, which is a component of the calcitriol metabolism machinery that extends the life of calcitriol within the target cell. These other analogs are thus under development for use in metabolic bone diseases, osteoporosis, and cancer [6,11]. More recently, our concepts of vitamin D have been taken one step further with the realization that, while the bulk of circulating 1α,25-$(OH)_2D_3$ is made in the kidney, some 1α,25-$(OH)_2D_3$ is produced locally by target cells, making this molecule an endocrine and paracrine/autocrine factor [1,2]. This has caused us to revisit the use of 25-OH-D repletion as a method to modulate vitamin D–dependent processes. This review will discuss the spectrum of compounds available, possible clinical uses of these compounds, and their potential mechanisms of action.

3.2 PHARMACOLOGICALLY RELEVANT VITAMIN D COMPOUNDS

Vitamin D compounds can be subdivided into four major groups, which are listed in Tables 3.1 through 3.4 and described here.

3.2.1 VITAMIN D AND ITS NATURAL METABOLITES

Table 3.1 shows the structures of vitamin D_3 and some of its important metabolites. Ironically, vitamin D_3, which is the natural form of vitamin D, is not approved for use as a prescription drug in the United States, but it is found increasingly as an over-the-counter natural food supplement in the United States and is used in both roles in virtually every other country in the world.

During the late 1960s and early 1970s, most of the principal vitamin D metabolites were first isolated and identified by gas-chromatography-mass spectrometry, and then, their exact stereochemical structure was determined [4]. This led to the chemical synthesis of naturally occurring isomer and its testing in various biological assays *in vitro* and *in vivo*. Indeed, all the major metabolites, namely 25-hydroxyvitamin D_3 (25-OH-D_3; Calderol), 1α,25-$(OH)_2D_3$ (Rocaltrol), and 24,25-dihydroxyvitamin D_3 (24R,25-$(OH)_2D_3$; Secalciferol), are/or have been available for use as drugs.

3.2.2 VITAMIN D PRODRUGS

Table 3.2 lists some of the important prodrugs of vitamin D. All of these compounds require a step (or more) of activation *in vivo* before they are biologically active. Included here is vitamin D_2 (ergocalciferol), which is derived by the irradiation of the fungal sterol ergosterol. Since fungi grow in the dark and are rarely exposed to UV light, vitamin D_2 is unlikely to be a natural product. Indeed, vitamin D_2 is hard to detect in nonsupplemented humans eating nonfortified food, although the U.S. mushroom industry is attempting to introduce UV-irradiated mushrooms as a source of vitamin D.

TABLE 3.1
Vitamin D and Its Natural Metabolites

Vitamin D Metabolites [Ring Structure][a]	Side-Chain Structure	Site of Synthesis	Relative VDR-Binding Affinity[b]	Relative DBP-Binding Affinity[c]	References
Vitamin D$_3$ [1]	21 22 24 27 / 20 23 25 26	Skin	~0.001	3,180	213: Mellanby 1919 214: McCollum et al. 1922
25-OH-D$_3$ [1]	—OH	Liver	0.1	66,800	215: Blunt et al. 1968
1α,25-(OH)$_2$D$_3$ [3]	—OH	Kidney	100	100	216: Fraser and Kodiçek 1970 217: Holick et al. 1971
24(R),25-(OH)$_2$D$_3$ [1]	OH / —OH	Kidney	0.02	33,900	218: Holick et al. 1972
1α,24(R),25-(OH)$_3$D$_3$ [3]	OH / —OH	Target tissues[d]	10	21	219: Holick et al. 1973
25(S),26-(OH)$_2$D$_3$ [1]	ıllOH / —OH	Liver	0.02	26,800	220: Suda et al. 1970
25-OH-D$_3$-26,23- lactone [1]	ıllOH / O / O	Kidney	0.01	250,000	221: Horst 1979

Vitamin D Nucleus

CH$_2$ [1]   CH$_3$ [2]   CH$_2$ [3]

HO''''   3   HO''''   3   OH   HO''''   3   1   OH

[a] Structure of the vitamin D nucleus (secosterol ring structure).
[b] Values reproduced from previously published data. (From Stern, P., *Calcified Tissue International*, 33, 1–4, 1981.)
[c] Values reproduced from previously published data. (From Bishop et al. 1994.)
[d] Known target tissues included intestine, bone, kidney, skin, and the parathyroid gland.

TABLE 3.2

Vitamin D Prodrugs

Vitamin D Prodrug [Ring Structure][a]	Side-Chain Structure	Company	Status	Possible Target Diseases	Mode of Delivery	Reference
1α-OH-D$_3$ [3]	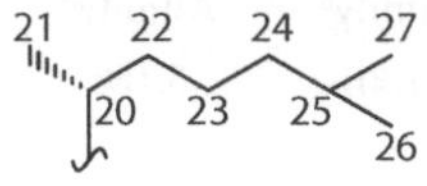 21 22 24 27 / 20 23 25 26	Leo	In use Europe	Osteoporosis	Systemic	23: Barton et al. 1973
1α-OH-D$_2$ [3]	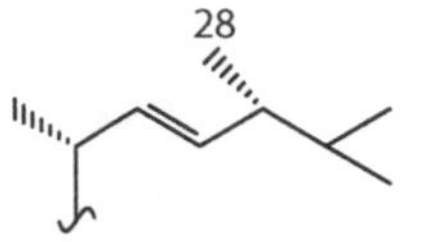 28	Genzyme	In use USA	Secondary Hyperparathyroidism	Systemic	24: Paaren et al. 1978
Dihydrotachysterol [2]	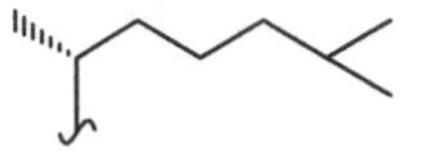	Duphar	Withdrawn	Renal failure	Systemic	26: Jones et al. 1988
Vitamin D$_2$ [1]	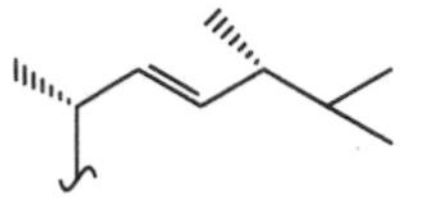	Various	In use USA	Rickets Osteomalacia	Systemic Systemic	3: Park 1940
1α-OH-D$_5$ [3]	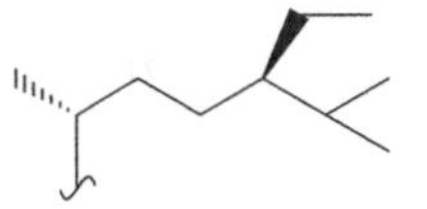	Various	In use USA	Rickets Osteomalacia	Systemic Systemic	25: Mehta et al. 2000

Vitamin D Nucleus

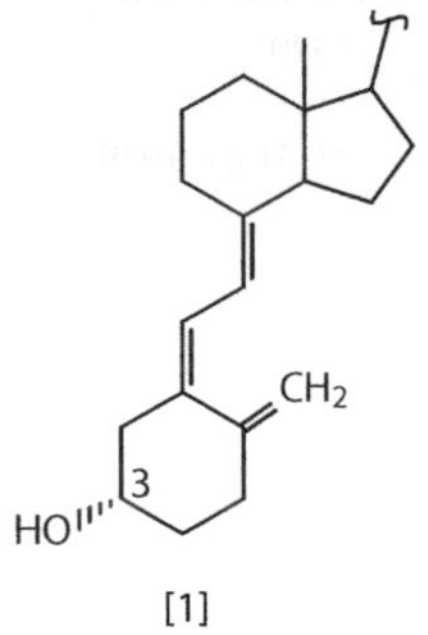

[1] 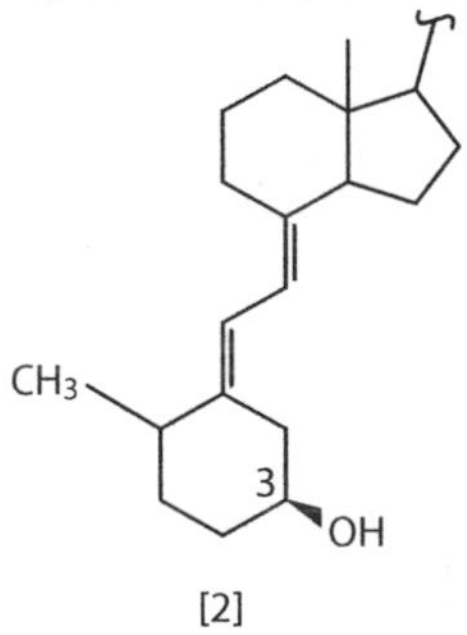   [2] 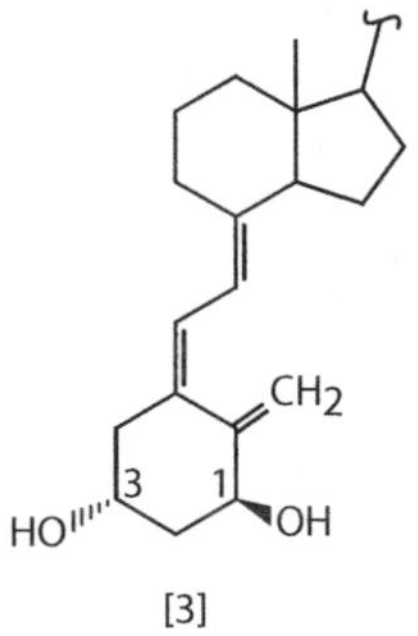   [3]

[a]   Structure of the vitamin D nucleus (secosterol ring structure).

Nevertheless, synthetically produced vitamin D$_2$ can be considered a prodrug since it is used as a substitute for the natural form, vitamin D$_3$, in pharmaceutical preparations or over-the-counter supplements in the United States. Vitamin D$_2$ possesses two specific modifications of the side chain (see Table 3.2), but these differences do not preclude the same series of activation steps as vitamin D$_3$, thus giving rise to 25-OH-D$_2$, 1α,25-(OH)$_2$D$_2$, and 24,25-(OH)$_2$D$_2$ respectively. Recently, there has been much debate in the vitamin D field, particularly in the United States, where vitamin D$_2$ is the sole prescription form available, about the relative utility of vitamin D$_2$ and vitamin D$_3$ to raise the circulating 25-OH-D level [12]. Vitamin D$_3$ preparations have appeared in the United States but under the banner "dietary supplements." Evidence from research studies suggests that oral pharmacological doses of vitamin D$_3$ are significantly more effective than equivalent doses of vitamin

D_2 for increasing the 25-OH-D level into the adequate range (>20 ng/mL) [13,14]; at the same time, there is ample evidence that vitamin D_2 compounds are less toxic than their vitamin D_3 counterparts [15–19]. On the other hand, recent studies using more physiological daily oral dosing of vitamins D_2 and D_3 suggests bioequivalence [20–22], and pediatricians view the antirachitic activity of both as roughly equal [3].

25-OH-D_3 was developed and approved as the pharmaceutical preparation Calderol in the 1970s by Upjohn, which was later acquired by Organon, but was withdrawn in the United States in the early 2000s. Two other prodrugs, i.e., 1α-OH-D_3 and 1α-OH-D_2, were synthesized [23,24] as alternative sources of $1\alpha,25$-$(OH)_2D_3$ and $1\alpha,25$-$(OH)_2D_2$, respectively, which circumvent the renal 1α-hydroxylase enzyme, which, in turn, was shown to be tightly regulated and prone to damage in renal disease. The prodrug 1α-OH-D_5 has been in clinical trials for the treatment of breast cancer [25].

The final prodrug in the list, dihydrotachysterol (DHT), can be viewed as the original vitamin D analog developed in the 1930s. At one time, DHT was believed to be "active" when converted to 25-OH-DHT by virtue of its A-ring being rotated 180° such that the 3β-hydroxyl function assumes a pseudo-1α-hydroxyl position [26]. The mechanism of action of DHT has become less clear with the description of the extrarenal metabolism of 25-OH-DHT to $1\alpha,25$-$(OH)_2$-DHT and $1\beta,25$-$(OH)_2$-DHT, which are two further metabolites that have greater biological activity than either 25-OH-DHT or DHT itself [27].

3.2.3 CALCITRIOL ANALOGS

Table 3.3 lists some of the most promising vitamin D analogs of $1\alpha,25$-$(OH)_2D_3$ approved by various governmental agencies, those currently under development by various industrial/or university research groups, and a few analogs that have been abandoned at various stages of the development process. Since the number of vitamin D analogs synthesized now lists in the thousands, the table is provided mainly to give a sample of the structures experimented with so far, the worldwide scope of the companies involved, and the broad spectrum of target diseases and uses. While Table 3.3 might give the impression that specific vitamin D analogs have been developed to fill particular disease niches, this is rarely the case, and some anlogs have found utility in the treatment of several diseases.

Early generations of calcitriol analogs included molecules with fluorine atoms placed at metabolically vulnerable positions in the side chain and resulted in highly stable and potent "calcemic" agents, such as $26,27$-F_6-$1\alpha,25$-$(OH)_2D_3$ (falecalcitriol). Later generations of analogs focused on features that make the molecule more susceptible to clearance, such as in calcipotriol (MC903), where a C22–C23 double bond, a 24-hydroxyl function, and a cyclopropane ring have been introduced into the side chain or in 22-oxacalcitriol (OCT), where the 22-carbon has been replaced with an oxygen atom. Both modifications have given rise to highly successful analogs marketed in Europe and Japan, respectively [28,29].

The C-24 position is a favorite site for modification, and numerous analogs contain 24-hydroxyl groups, e.g., $1\alpha,24(S)$-$(OH)_2D_2$ and $1\alpha,24(R)$-$(OH)_2D_3$ [30]. Other analogs contain a combination of multiple changes in the side chain, including unsaturation, 20-epimerization, 22-oxa replacement, homologation in the side chain, or terminal methyl groups. The resultant molecules such as EB1089 and KH1060 have attracted strong attention of researchers because of their increased potency *in vitro* and were pursued as possible anticancer and immunomodulatory compounds, respectively, but their development seems to have stalled.

A few attempts have been made to modify the nucleus of calcitriol. The Roche compound $1\alpha,25$-$(OH)_2$-16-ene-23-yne-D_3, which has been touted as an antitumor compound *in vivo*, possesses a D-ring double bond [31]. Declercq and Bouillon have made a 14-epi,19-nor-23-yne derivative with the same 23-yne side chain, which also holds promise in cancer therapy (Table 3.3) [32], and the same

TABLE 3.3
Analogs of 1α,25-(OH)₂D₃

Vitamin D Analog [Ring Structure][a]	Side Chain Structure	Company	Status	Possible Target Diseases	Mode of Delivery	Reference
Calcitriol, 1α,25-(OH)₂D₃ [3]	(side chain; positions 20, 21, 22, 23, 24, 25, 26, 27, OH)	Roche, Duphar	In use worldwide	Hypocalcemia / Psoriasis	Systemic / Topical	222: Baggiolini et al. 1982
26,27-F₆-1α,25-(OH)₂D₃ [3]	(side chain; CF₃, OH, CF₃)	Sumitomo-Taisho	In use Japan	Osteoporosis / Hypoparathyroidism	Systemic / Systemic	223: Kobayashi et al. 1982
19-Nor-1α,25-(OH)₂D₂ [5]	(side chain; 28, OH)	Abbott	In use USA	Secondary hyperparathyroidism	Systemic	224: Perlman et al. 1990
22-Oxacalcitriol (OCT) [3]	(side chain; O, OH)	Chugai	In use Japan	Secondary hyperparathyroidism / Psoriasis	Systemic / Topical	225: Murayama et al. 1986
Calcipotriol (MC903) [3]	(side chain; OH, cyclopropyl)	Leo	In use worldwide	Psoriasis / Cancer	Topical / Topical	226: Calverley 1987
EB1089 [3]	(side chain; 27a, OH, 26a, 24a)	Leo	Clinical trials	Cancer	Systemic	227: Binderup et al. 1991
20-epi-1α,25-(OH)₂D₃ [3]	(side chain; OH)	Leo	Preclinical	Immune diseases	Systemic	228: Calverley et al. 1991
2-methylene-19-nor-20-epi-1α,25-(OH)₂D₃ (2MD) [7]	(side chain; OH)	Deltanoids	Preclinical	Osteoporosis	Systemic	35: Shevde et al. 2002

Name	Company	Status	Indication	Route	Reference
BXL-628 (formerly Ro-269228) [8]	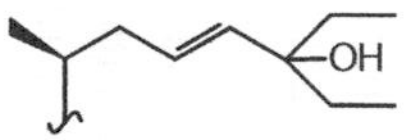 Bioxell	Clinical trials	Prostate cancer	Systemic	229: Marchiani et al. 2006
14-epi-19-nor-23-yne, 1α,25-(OH)$_2$ D$_3$ (TX522) [6]	 Hybrigenics	Preclinical	Cancer	Systemic	32: Eelen et al. 2008
ED71 (Eldecalcitol) [4]	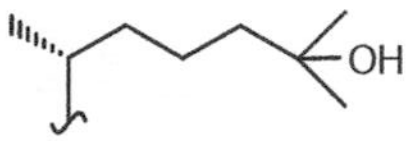 Chugai	Clinical trials	Osteoporosis	Systemic	73: Nishii et al. 1993
1α,24(S)-(OH)$_2$D$_2$ [3]	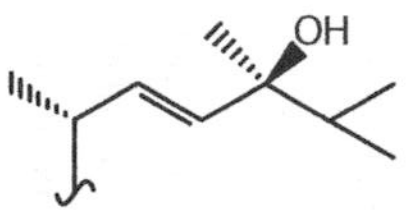 Genzyme	Preclinical	Psoriasis	Topical	30: Strugnell et al. 1995
1α,24(R)-(OH)$_2$D$_3$ (TV-02) [3]	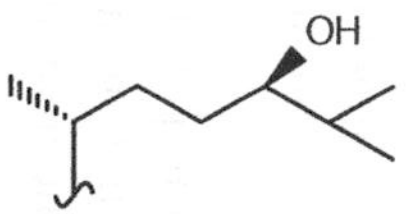 Teijin	In use Japan	Psoriasis	Topical	230: Morisaki et al. 1975

Vitamin D Nucleus

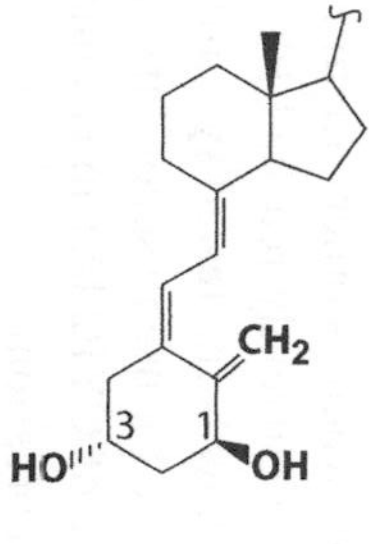

[3] 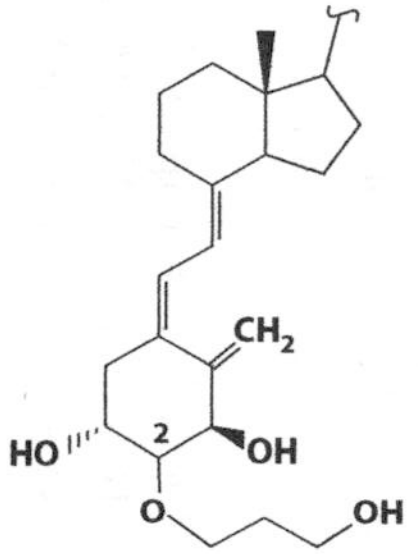 [4] 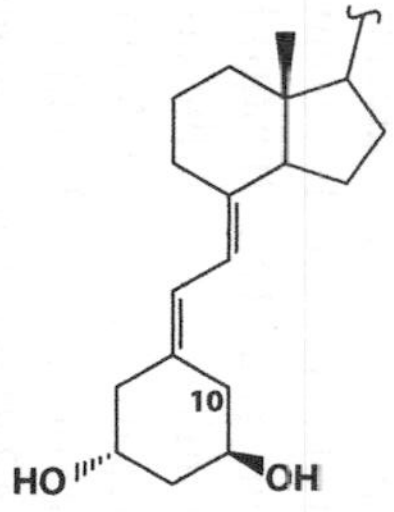 [5] 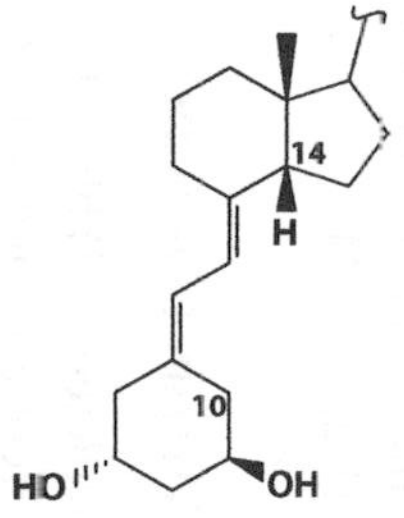 [6] 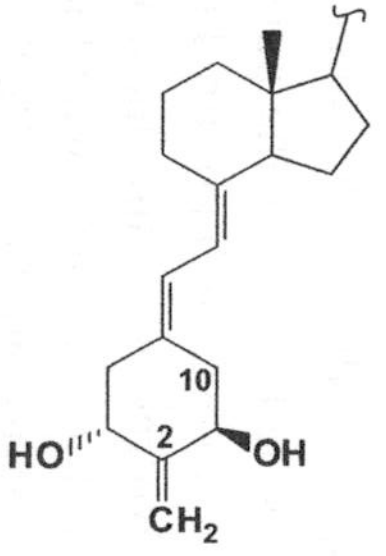 [7] 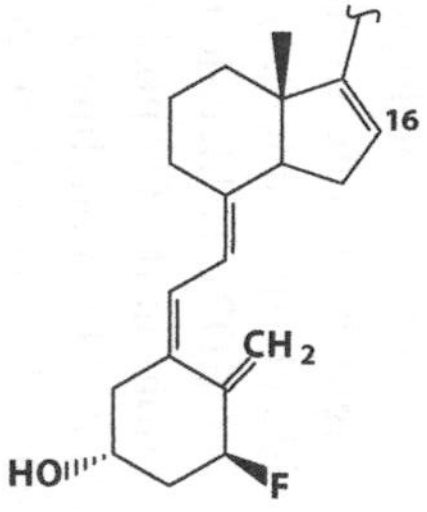 [8]

a Structure of the vitamin D nucleus (secosterol ring structure).

researchers have introduced a series of biologically active analogs without one or the other of the C/D rings but with a rigid backbone to maintain the spatial arrangement of the A-ring hydroxyl groups and the side chain (Table 3.4) [33]. Relatively recently, the A-ring-substituted 2-hydroxypropoxy-derivative ED71 (Eldecalcitol) has been launched as an antiosteoporosis drug. Other bulky modifications at the C2 position of the A-ring are accommodated well by the vitamin D receptor, as indicated by modeling and biological activity studies [34,35]. The Abbott compound 19-nor-1α,25-$(OH)_2D_2$ (Zemplar) lacks a 19-methylene group and is roughly based on the *in vivo* active metabolite 1α,25-$(OH)_2DHT_2$ formed from DHT, which retains biological activity, although the C-19 methylene is replaced by a C-19 methyl. Many other compounds have been developed with rigid or altered *cis*-triene structures [36] or modifications of the 1α, 3β-, or 25-hydroxyl functions, not only for the purpose of developing active molecules for use as drugs but also to allow us to establish minimal requirements for biological activity in structure/activity studies [8,9]. Two recent compounds, i.e., Bioxell's BXL-628 and Deltanoids' 2-MD, combine modifications in the side chain with those in the nucleus. BXL-628 combines 1-fluorination; 16-ene and 23-ene unsaturations; 26,27-homologation; and 20-epimerization, which are all found in earlier generations of analogs, to make a antiproliferative agent currently in clinical trials for the treatment of prostate cancer and prostatitis [37,38]. Likewise, 2-MD, which is touted as being bone specific, combines a novel 2-methylene substitution and the 19-nor feature with side chain 20-epimerization [35].

3.2.4 MISCELLANEOUS VITAMIN D ANALOGS AND ASSOCIATED DRUGS

A series of compounds depicted in Table 3.4 is the substituted biphenyls, which was originally developed by Ligand, representing nonsteroidal scaffolds selected by high-throughput screening, which show weak VDR binding but good transactivation through VDRE-driven vitamin D-dependent genes and produce hypercalcemia *in vivo* [39]. This family has recently been extended by the synthesis of some highly potent tissue-selective nonsecosteroidal VDR modulators with nanomolar affinity (e.g., LY2109866) by a research group at Eli Lilly [40]. This is the first class of vitamin D mimics that lack the conventional *cis*-triene secosteroid structure while maintaining the spatial separation of the A-ring and side-chain hydroxyl functions needed to bind to certain key residues of the ligand-binding pocket of the VDR. Although these nonsecosteroidal compounds exhibit a 270-fold improvement of the therapeutic index over calcitriol in animal models, they are still to be tested clinically. On the contrary, Table 3.4 also shows the structures of two different classes of VDR/cacitriol antagonists made by Teijin and Schering, respectively. The former compounds, most notably TEI-9647, are dehydration products of the natural metabolite 1α,25-$(OH)_2D_3$-26,23-lactone (Table 3.1) and have found clinical utility in the treatment of Paget's disease [41,42].

Another group of compounds that impact the vitamin D field that are under development are the CYP24A1 inhibitors. By blocking CYP24A1, which is the main catabolic pathway within the vitamin D target cell, these agents extend the life of the natural agonist, calcitriol, giving rise to longer lasting biological effect [43]. Sandoz/Novartis developed a group of azole molecules that have greater specificity toward CYP24A1 and CYP27B1 from the general cytochrome P450 (CYP) inhibitor, ketoconazole, which showed utility in blocking cell proliferation *in vitro*, but these compounds were discontinued after early clinical trials [44]. Cytochroma has developed a group of CYP24A1 inhibitors synthesized by Gary Posner based on vitamin D templates; some of these are pure CYP24A1 inhibitors, whereas others are mixed VDR agonist/CYP24A1 inhibitors (Table 3.4). Some of these drugs have currently reached Phase IIB human clinical trials for the treatment of psoriasis [45,46] and are now being tested systemically in the treatment of secondary hyperparathyroidism [47]. Their promise for use in secondary hyperparathyroidism presumably stems from their ability to counter the role of CYP24A1 in attenuating the effect of calcitriol on preproparathyroid hormone (PTH) gene suppression.

TABLE 3.4
Miscellaneous Vitamin D Compounds

Name	Structure	Name	Structure
LG190090 **Ligand Pharmaceuticals** Nonsteroidal VDR agonist 39: Boehm et al. 1999		**LY2108491** **Eli Lilly** Nonsteroidal VDR agonist 40: Ma et al. 2006	
TEI-9647 **Teijin** VDR antagonist Dehydration product of $1\alpha,25(R)$-$(OH)_2D_3$-$26,23(S)$-lactone 41: Saito and Kittaka 2006 231: Ochiai et al. 2005 232: Toell et al. 2001		**ZK159222** **Schering** VDR antagonist 232: Toell et al. 2001	
SDZ 89-443 **Sandoz/Novartis** P450 inhibitor 210: Schuster et al. 2003		**VID400** **Sandoz/Novartis** P450 inhibitor 210: Schuster et al. 2003	

(continued)

TABLE 3.4 (Continued)
Miscellaneous Vitamin D Compounds

Name	Structure	Name	Structure
CTA018 (MT2832) Cytochroma VDR agonist/ CYP24A1 inhibitor 45: Posner et al. 2009		**CTA091** Cytochroma CYP24A1 inhibitor 46: Posner et al. 2009	
KS 176 C-ring modified VDR agonist 33: Verstuyf et al. 2000		**SL 117** E-ring modified VDR agonist 33: Verstuyf et al. 2000	

3.3 CLINICAL APPLICATIONS OF VITAMIN D COMPOUNDS

The clinical usefulness of vitamin D analogs has been reviewed comprehensively elsewhere [1,5,11,48]. This review summarizes, updates, and highlights some of this information.

3.3.1 RICKETS, OSTEOMALACIA, AND VITAMIN D INSUFFICIENCY

When the nutritional basis of rickets and osteomalacia became apparent in the first half of the 20th century, vitamin D (particularly the less expensive substitute, vitamin D_2) became the treatment of choice for these diseases. Food fortification in the form of vitamin D supplements to milk, margarine, and bread replaced much of the need for therapeutic vitamin D to abolish overt rickets and osteomalacia. In fact, since then, vitamin D deficiency rickets (defined in [48] as plasma 25-OH-D levels below 12 ng/mL or 30 nmol/L) has become very uncommon in North America because vitamin D fortification is required by law, whereas it was quite common before the practice became mandatory and is still more prevalent in the world, where food fortification is not required. On the other hand, mild vitamin D deficiency (defined in [48] as plasma 25-OH-D levels in the ranges of 12–20 ng/mL or 30–50 nmol/L) remains common in the general population, especially in the winter months. Some are claiming that levels of 25-OH-D below 30 ng/mL or 75 nmol/L are correlated with poor outcomes in several health-related areas, including optimal bone mineral density, various types of cancer, autoimmune diseases, infections, and cardiovascular disease [49,50].

Vitamin D deficiency is still quite prevalent in the elderly and is usually treated with modest doses of ~800 IU of vitamin D [51], although these are not always successful, presumably due to confounding disease complications [52]. In recent years, several world, continent-wide, and national food agencies and societies have reviewed population needs and proposed new guidelines, in some cases raising the recommendations for vitamin D intake for specific age groups (particularly for those in the neonatal, elderly, or postmenopausal categories) to try to ensure adequate intakes, irrespective of geographical, dietary, and sun exposure differences [48]. However, the need for the use of expensive pharmaceutical prescription drugs containing calcitriol or its analogs to cure simple rickets and osteomalacia is not warranted for the otherwise-healthy general population.

Although many of the hallmarks of rickets and osteomalacia are successfully relieved by doses of vitamin D in the range of 600–800 IU/day (15–20 µg/day) recommended in the recent 2011 Insitute of Medicine (IOM) report [48], critics point to epidemiological data that suggest that the current recommended dietary reference ranges (also known as DRIs) do not result in plasma 25-OH-D levels >40 ng/mL, which correlate with maximal bone mineral density [1,53] or the other health benefits of vitamin D [48,49]. Consequently, there has been much recent debate over the optimal level of vitamin D intakes, and this has led to a view among some experts that vitamin D intakes might need to be increased above 1500 IU/day [54] and possibly higher [55] in order to achieve target plasma 25-OH-D levels >40 ng/mL. However, the IOM [48] questions the scientific basis for these claims and points to the fact that recommendations for general populations should be based on current evidence and not opinions.

3.3.2 OSTEOPOROSIS

The etiology of this disease is complex and likely to be multifactorial [56,57]. With the demonstration that ovariectomy and estrogen deficiency results in enhanced production of osteoclastogenic cytokines such as interleukin-6, TNF-α, and interleukin-1, as well as cytokine-mediated osteoclast recruitment and increased bone resorption, has come a clearer understanding of the molecular processes underlying postmenopausal osteoporosis [58,59]. Theories focusing on osteoblast/osteoclast communication led to the discovery of RANK; its ligand RANKL and the decoy receptor, i.e., osteoprotegerin; and an understanding of how agents such as vitamin D or $1\alpha,25\text{-}(OH)_2D_3$ can influence osteoclastogenesis and bone resorption [60–62]. There have been frequent claims that levels of $1\alpha,25$-$(OH)_2D_3$ are low in osteoporosis and lead to a fall in intestinal calcium absorption [63]. In addition, there has been a long-standing debate over different VDR genotypes correlating with bone mineral density [64,65], and this can be widened to implicate variants of any component of the vitamin D machinery in a genetic basis for susceptibility to osteoporosis. Recently, a genome-wide study of 30,000 Caucasians showed that polymorphisms of four vitamin D-related genes, i.e., DBP, CYP2R1, CYP24A1, and 7-DHC-reductase, contribute to variations in serum 25-OH-D levels [66] and thus may affect vitamin D sufficiency and bone health.

As a consequence, it not surprising that vitamin D and vitamin D analogs have been tried in attempts to slow down bone loss and reduce fracture rates in elderly patients with osteopenia and osteoporosis. Small doses of vitamin D (800–1000 IU) have proven effective in treating vitamin D deficiency and osteopenia in elderly populations, especially in combination with supplemental calcium, by increasing bone mineral density and reducing fracture rates [51,52,67]. However, the use and effectiveness of active vitamin D metabolites in the treatment of osteoporosis remains controversial. Clinical trials of 1α-OH-D_3 [68], 1α-OH-D_2 [69], and $1\alpha,25$-$(OH)_2D_3$ [70–72] have been undertaken in this condition. The experience seems to have been that while benefits have been observed in terms of reductions in vertebral deformities in the second and third years of longer studies [72], there is no evidence that vitamin D analogs offer advantages over vitamin D or calcium in the treatment of osteoporosis [67].

In North America, where dietary Ca intakes and absorption rates are higher, the use of active vitamin analogs has led to intolerable side effects and the discontinuation of the use of $1\alpha,25$-$(OH)_2D_3$ and 1α-OH-D_3 for the treatment of osteoporosis. In the U.K., Australia, Italy, Japan, New Zealand, and 16 other countries in rest of the world, these drugs are approved, and their side effects are tolerated. Nevertheless, some pharmaceutical companies have sought to develop "milder" but "longer lived" calcitriol analogs for use in osteoporosis. ED-71 represents such an analog, which, by virtue of an A-ring substituent at C-2 and tighter binding affinity to DBP, has a longer $t_{1/2}$ in the plasma [73]. ED-71, which was marketed under the brand name Eldecalcitol, has performed well at restoring bone mass without causing hypercalcemia in long-term studies involving ovariectomized rats [74] and in Phase-I and Phase-II clinical trials [75] and received approval for use in osteoporosis in Japan in January 2011. Another bone-specific analog with potential for treatment of osteoporosis, 2-MD [34], which stimulates bone formation *in vivo* in an ovariectomized rat model. 2-MD recently underwent a randomized double-blind placebo-controlled trial in osteopenic women, and although daily oral treatment with 2-MD caused a marked increase in bone formation markers, it also increased bone resorption, thereby increasing bone remodeling but not net increases in bone mineral density [76].

It should not go unmentioned that there have also been periodic claims that there is a bone-specific role for $24R,25$-$(OH)_2D_3$ in normal bone mineralization that has resurfaced with the finding that the *Cyp24A1*-knockout mouse, which lacks any 24-hydroxylated vitamin D metabolites, exhibits slow bone fracture healing [77]. Thus, it remains a possibility that any vitamin D analog may act on bone by generating a 24-hydroxylated metabolite.

3.3.3 SECONDARY HYPERPARATHYROIDISM AND RENAL OSTEODYSTROPHY

Chronic renal disease-mineral and bone disorder (CKD-MBD) [78] is accompanied by the gradual loss of renal 25-OH-D_3-1α-hydroxylase (CYP27B1) activity over the five-stage natural history of the disease, which culminates in dialysis (stage 5D). As early as stage 2 of CKD-MBD, the 1α-hydroxylase activity declines, leading to reduced plasma levels of $1\alpha,25$-$(OH)_2D_3$, which result in hypocalcemia and secondary hyperparathyroidism. Unchecked, these biochemical events, together with the other sequelae of renal failure such as phosphate retention, can result in renal osteodystrophy. Active vitamin D analogs, such as 1α-OH-D_3 and $1\alpha,25$-$(OH)_2D_3$, raise plasma Ca^{2+} concentrations and, in addition, lower PTH levels by direct suppression of PTH gene transcription at the level of the PTH gene promoter. Slatopolsky et al. [79] showed that intravenous infusion of "active" vitamin D preparations results in a more effective suppression of plasma PTH levels without such a profound increase in plasma $[Ca^{2+}]$ in end-stage renal disease (ESRD). Subsequent work has employed "low-calcemic" vitamin D analogs such as OCT, 19-nor-1α,25-$(OH)_2D_2$ or 1α-OH-D_2 as oral or intravenous substitutes for the more calcemic natural hormone [80,81] at various stages of CKD-MBD from stages 3 to 5D. The FDA has approved both oral and intravenous forms of the drugs for the treatment of secondary hyperparathyroidism at stages 3 and 4 of CKD-MBD and in hemodialysis/peritoneal dialysis patients.

In 2003, a body of nephrologists released guidelines [82] recommended earlier, more aggressive use of vitamin D preparations, and "active" vitamin D analogs in the treatment of secondary hyperparathyroidism in CKD-MBD. KDOQI guidelines suggested that treatment as early as stage 3 (GFR <60) might benefit the patient by limiting the extreme rises in plasma PTH levels and preventing the parathyroid gland resistance to vitamin D treatment often observed in ESRD. KDOQI guidelines also recognized the high frequency of vitamin D deficiency (25-OH-D < 10 ng/mL) and vitamin D insufficiency (KDOQI defined this as 25-OH-D 10–30 ng/mL) in the CKD-MBD and ESRD populations [83,84] and made the opinion-based recommendation to make an initial attempt at vitamin D repletion with escalating doses of vitamin D_2 prior to administration of "active" vitamin D analog replacement therapy. This initial intervention to boost 25-OH-D levels has proven to be successful in Stage-3 CKD-MBD patients, as evidenced by increased 1α,25-$(OH)_2D$ and mild PTH suppression, but the strategy fails to produce the desired effects in Stage-4 CKD-MBD patients due to reduced renal 1α-hydroxylase activity [85,86]. Currently, both oral and intravenous formulations of various active vitamin D analogs are available for use in stage-3, -4, and -5 patients to take over when vitamin D repletion fails to regulate PTH levels.

The emergence of the potential importance of the extra-renal 1α-hydroxylase in normal human physiology has led to a reevaluation of the vitamin D repletion and "active" hormone replacement arms of the CKD-MBD therapy [2,87]. The value of the vitamin D repletion is now seen as providing the substrate 25-OH-D for both the renal 1α-hydroxylase, which is the main determinant of circulating 1α,25-$(OH)_2D_3$, and the extra-renal 1α-hydroxylase, which is postulated to augment 1α,25-$(OH)_2D_3$ synthesis for local or paracrine actions around the body. The latest KDIGO guidelines endorse the combined use of both [78]. While the decline of the renal 1α-hydroxylase enzyme during CKD-MBD is well established, the fate of the extrarenal 1α-hydroxylase in the face of uremia is largely a matter of conjecture. Evidence from anephric patients treated with large doses of 25-OH-D_3 [86] suggests that the extrarenal enzyme survives in CKD patients, arguing that provision of a source of 25-OH-D to vitamin D–deficient patients throughout all stages of CKD-MBD is warranted [88]. It also argues for the more judicious use of "active" vitamin D analogs as hormone replacement therapy layered on top of conventional vitamin D repletion therapy. Early attempts at this type of combined vitamin D/"active" vitamin D analog approach in a pediatric population have resulted in a more efficient PTH control without many of the usual problems of soft-tissue calcification observed in patients treated only with "active" vitamin D analogs [89,90].

Discussion of the optimal vitamin D therapy of CKD patients has refocused attention on the underlying mechanisms for the decline of the renal 1α-hydroxylase activity in this condition. For three decades, it has been widely assumed that the serum 1α,25-$(OH)_2D_3$ declines because of enzyme loss in the proximal tubular cells. Recent progress, including the elucidation of the role of fibroblastlike growth factor 23 (FGF-23) in the phosphate homeostatic loop, has shed new light on the role of increasing PO_4 and FGF-23 levels in causing dysfunctional vitamin D metabolism in the course of renal disease [91,92]. Since the known biological effects of FGF-23 include downregulation of the renal 1α-hydroxylase (CYP27B1) and up-regulation of the catabolic 24-hydroxylase (CYP24A1), the possibility exists that the rising FGF-23 level contributes to the fall in serum 1α,25-$(OH)_2D_3$ by reducing its synthesis and increasing its degradation. Such a scenario opens the door to the use of CYP24A1 inhibitors in renal disease.

3.4   HYPERPROLIFERATIVE CONDITIONS: PSORIASIS AND CANCER

The demonstration that $1\alpha,25\text{-}(OH)_2D_3$ is an antiproliferative prodifferentiating agent for certain cell types *in vivo*, and many cell lines *in vitro* suggested that vitamin D analogs might offer some relief in the treatment of hyperproliferative disorders such as psoriasis and cancer. Early psoriasis trials with systemic $1\alpha,25\text{-}(OH)_2D_3$ were moderately successful but plagued with hypercalcemic side effects. Modifications to the protocol included the following:

1. Administration of calcitriol overnight when intestinal concentrations of $[Ca^{2+}]$ were low
2. Substitution of "low-calcemic" analogs for the calcitriol

According to Holick [93], oral calcitriol is an effective treatment for psoriasis when administered using an overnight protocol. However, by far, the most popular treatment for psoriasis is the topical administration of the "low-calcemic" analog calcipotriol formulated as an ointment [28]. When given orally, calcipotriol is ineffective due to the fact that it is rapidly broken down [94]. When given topically as an ointment, calcipotriol survives long enough to cause improvement in more than 75% of patients [95]. Both $1\alpha,25\text{-}(OH)_2D_3$ and calcipotriol are effective in psoriasis because they block hyperproliferation of keratinocytes, increase differentiation of keratinocytes, and help suppress local inflammatory factors through their immunomodulatory properties. Calcipotriol has been marketed worldwide for use in psoriasis for close to 20 years, most recently in a combination therapy formulation with a corticosteroid.

The literature contains much speculation regarding the exploitation of the antiproliferative properties of vitamin D and its analogs in the prevention and treatment of cancer. While dietary vitamin D may have some value in cancer prevention, most experts feel that vitamin D analogs, with their altered balance of calcemic and antiproliferative properties, offer the best chance of success in cancer treatment. Several thousands of vitamin D analogs have been tested *in vitro* and *in vivo* with some degree of success in controlling the growth of tumor cells offering potential for use as anticancer drug therapies (reviewed extensively in [11]). Many vitamin D compounds are extremely effective antiproliferative or prodifferentiation agents *in vitro* using a variety of mechanisms involving gene expression of cell division and proapototic genes to produce their effects. Preclinical studies in laboratory animals have also resulted in promising data [11]. For example, in mice inoculated with fulminant leukemia, moderate leukemia, or slowly progressive leukemia, the Roche compound $1\alpha,25\text{-}(OH)_2\text{-}16\text{-}ene\text{-}23\text{-}yne\text{-}D_3$ administered at 1.6 µUg/qod was significantly more effective than 0.1 µUg/qod $1\alpha,25\text{-}(OH)_2D_3$ at increasing survival time, even though the $1\alpha,25\text{-}(OH)_2D_3$-treated group developed mild hypercalcemia and the analog-treated animals remained normocalcemic [96]. With the analog EB1089, the promising antiproliferative effects observed *in vitro* and in the NMU-induced mammary tumor and in LNCaP prostate cancer xenograft models [97,98] were also extended into the clinic. Early trials in limited numbers of breast cancer patients have been followed up with more extensive ongoing Phase-II and Phase-III clinical trials in a number of different cancers [99–101]. Several other analogs have entered clinical trials for the treatment of a variety of hyperproliferative diseases, usually involving VDR-positive tumors (see [11]). Many trials are still ongoing, including the testing of BXL-628 (Table 3.3) in prostate-related diseases [37].

Despite the enormous promise of vitamin D analogs as anti-cancer agents, this has yet to result in an approved vitamin D analog for use in any type of cancer [11]. The principal problem in anticancer studies involving orally administered vitamin D compounds is hypercalcemia. Although the newer analogs appear to be less calcemic than calcitriol itself, they still retain some ability to raise serum calcium; they are not "noncalcemic" as is sometimes claimed. One attempt to overcome this problem has been high-dose intermittent therapy (large doses of vitamin D analog administered weekly), which appear to result in less hypercalcemic episodes. Another problem emerging from experience with clinical trials of vitamin D analogs is that effective doses needed to retard cell growth (~1 nM or higher) cannot be attained *in vivo* due to low bioavailability [102–104]. One of the principal determinants of tumor cell vitamin D analog levels is the catabolic enzyme CYP24A1, which is upregulated in vitamin D target cells and limits the effective drug concentration reached. Thus, another approach to effective vitamin D therapy in cancer patients is the potential use of CYP24A1-inhibitors (Table 3.4) with or without calcitriol or one of its analogs. Nevertheless, it remains uncertain whether we will ever develop a vitamin D–based therapy that minimizes the hypercalcemic episodes while retaining sufficient antiproliferative activity to be valuable in slowing tumor growth.

3.5 IMMUNOSUPPRESSION

The immunosuppressive properties of $1\alpha,25$-$(OH)_2D_3$ and its analogs have been the subject of several excellent reviews [1,105–107]. $1\alpha,25$-$(OH)_2D_3$ is believed to work by regulation of the expression of various cytokines, particularly suppressing proinflammatory cytokines and promoting other cytokines, thereby raising the Th2/Th1 ratio. The hormone also stimulates the innate immune system by promoting the transcription of a natural bacterial peptide, cathelicidin (LL-37), which kills Mycobacterium tuberculosis, resulting in increased resistance to tuberculosis [108–110]. The spectrum of effects exhibited by $1\alpha,25$-$(OH)_2D_3$ and its analogs on the immune system results in beneficial effects on a wide variety of autoimmune diseases. Researchers have demonstrated the ability of calcitriol to suppress the onset of experimental encephalitis [111] and Type-I diabetes in NOD mice [112] and to work synergistically with cyclosporine to provide immunosuppression in transplantation medicine [113]. The latter development has led to some optimism that coadministration of a vitamin D analog with cyclosporine can reduce the dosage of the latter drug and minimize the serious side effects associated with its use. Several studies [112,114,115] have focused on the immunosuppressive effects of Leo drugs, i.e., KH1060 and 20-epi-$1\alpha,25$-$(OH)_2D_3$, both of which contain the 20-S side-chain configuration. Recent generations of compounds such as BXL-628 that contain multiple modifications found in the Leo Pharma drugs are being tested in prostatitis, which is an inflammation of the prostate [38]. Again, it remains unclear if analogs that show promise in immunological studies will prove to be effective immunomodulators in the clinic.

3.6 CARDIOVASCULAR DISEASE

Rather unexpectedly, the generation of the VDR-knockout mouse led to the discovery that renin gene expression was under the negative regulation of $1\alpha,25$-$(OH)_2D_3$ [116], the VDR-knockout mouse showing threefold higher expression, compared with wild-type mice, and, as a result, exhibited hypertension. Another "cardiovascular" gene regulated by $1\alpha,25$-$(OH)_2D_3$ is atrial naturetic peptide [117]. One consequence of these findings is a retrospective examination of clinical databases for the association of vitamin D status (serum 25-OH-D) and hypertension. It has been reported in a large 110,000 subject study that the highest relative risk of incident hypertension in both men and women studied exists in those in the lowest decile for plasma 25-OH-D [118].

The connection of vitamin D with cardiovascular disease does not end with its role in hypertension. Work with various uremic rodent models in which there is secondary hyperparathyroidism and accelerated vascular calcification suggests that there are protective effects of calcitriol (and/or 25-OH-D) on the health of the vascular system by suppression of the inflammatory process involved in the development of atherosclerosis, by antiproliferative effects on myocardial cell hypertrophy, and by direct suppressive effects on vascular epithelial cell gene expression of calcification genes, e.g., osteocalcin, Runx-2, and osterix [119–122]. Thus, the emerging concept is that, rather than accelerating renal failure, vitamin D analogs when used at clinically relevant doses are protective to the vasculature and not deleterious, as once believed.

Indeed, this interpretation of the animal data is supported by clinical data from experience of the use of vitamin D analogs in CKD patients, where mortality from CVD causes are threefold higher than in the normal population and constitute one of the major causes of death. Mortality rates are also twofold higher in untreated CKD patients than in vitamin D analog-treated patients [123], a relationship that further widens when one compares vitamin D_2 analogs to calcitriol [124]. Adding further to this story is the finding that those with vitamin D deficiency, as evidenced by serum 25-OH-D levels < 15 ng/mL and hypertension, are also at increased risk of CVD, suggesting that this role of calcitriol is mediated by local conversion of 25-OH-D into calcitriol [125].

3.7 MECHANISM OF ACTION OF VITAMIN D ANALOGS

3.7.1 CRITERIA THAT INFLUENCE PHARMACOLOGICAL EFFECTS OF VITAMIN D COMPOUNDS

It is widely accepted that vitamin D analogs work through a VDR-mediated mechanism. Several decades of work on the vitamin D signal cascade have identified the proteins including VDR, DBP, and various CYPs (Figure 3.1) that play a role in transduction of the biological effects of vitamin D inside the body and inside the target cell. Not surprisingly, this perspective allows us to explore how specific vitamin D analogs manage to interact with these proteins and survive in the body in order to mimic some or all of the actions of $1\alpha,25\text{-}(OH)_2D_3$. These factors are discussed briefly here.

3.7.1.1 Activating Enzymes

It has been shown using *in vitro* models that some vitamin D compounds lacking 1α-hydroxylation (e.g., $24(R),25\text{-}(OH)_2D_3$) are capable of interacting with the VDRs and transactivating reporter genes but this occurs only at high concentrations of ligand [126]. It seems unlikely that these concentrations will be reached *in vivo*, except in hypervitaminosis D [127]. In fact, the recent demonstration in wild-type and VDR-knockout mice that hypercalcemia appears at similar toxic doses of vitamin D_3 suggests that it is 25-OH-D and not $1\alpha,25\text{-}(OH)_2D_3$, which is the metabolite that triggers excessive gene expression in hypervitaminosis D [128]. However, most vitamin D analogs are not used at concentrations required in hypervitaminosis D. Consequently, most of the compounds described in Tables 3.1 and 3.2 lack vitamin D biological activity *unless* they are activated *in vivo*. This is particularly the case for the parent vitamin D_3 itself, for its main circulating form 25-OH-D_3 or for

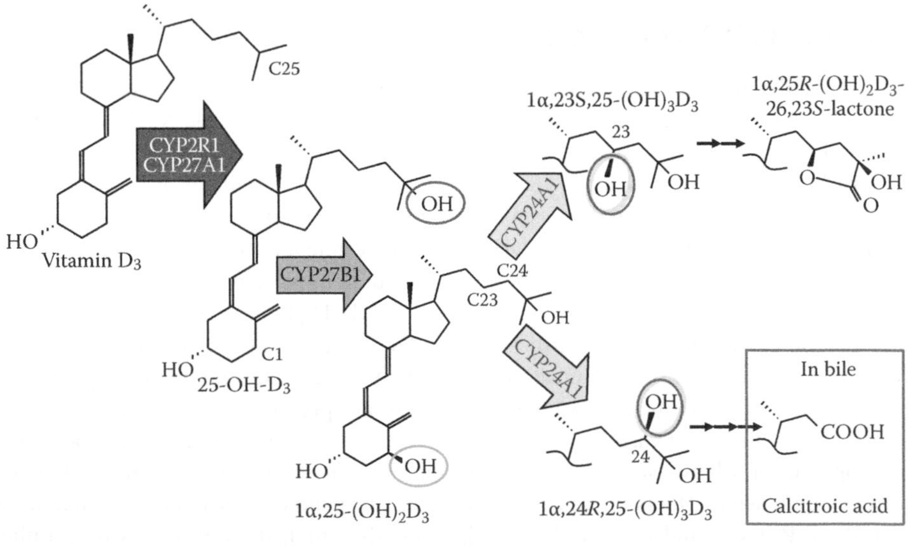

FIGURE 3.1 Enzymes involved in vitamin D metabolism. Cytochrome P450-containing enzymes involved in activation (CYP2R1, CYP27A1, and CYP27B1) and inactivation (CYP24A1) of vitamin D. Some of the same enzymes are involved in the activation of prodrugs (see Table 3.2). The catabolic enzyme CYP24A1 is also implicated in the inactivation of many of the clinically approved vitamin D analogs. Nonspecific enzymes (e.g., CYP3A4) are also believed to play a role in vitamin D analog clearance.

any of the prodrugs listed in Table 3.2. Vitamins D_2 and D_3 depend on both the liver 25-hydroxylase and kidney 1α-hydroxylase enzyme systems in order to be activated, whereas most prodrugs require only a single step of activation. Indeed, the 1α-OH-D drugs were designed to overcome the tightly regulated 1α-hydroxylase step, which is easily damaged in chronic renal failure.

In essence, prodrugs depend on the weakly regulated 25-hydroxylase step in the liver for activation. The cytochrome P450 originally thought to be responsible for 25-hydroxylation of vitamin D_3, i.e., CYP27A, has been cloned and shown to be a bifunctional polypeptide that can execute both activation of vitamin D_3 and the 27-hydroxylation of cholesterol during bile acid biosynthesis [129]. However, the CYP27A enzyme has a relatively low affinity for vitamin D, does not 25-hydroxylate vitamin D_2, and, when mutated, results in cerebrotendinous xanthomatosis and not rickets. Accordingly, another one of the several candidate P450s [43] known to carry out this step may be the "physiologically relevant" 25-hydroxylase. Indeed, CYP2R1 [130], which is a high-affinity microsomal enzyme with known human mutations that cause rickets, has been recently shown to 25-hydroxylate the prodrug 1α-OH-D_2 [131]. Recently, this enzyme was crystallized with several vitamin D substrates in the active site, making it likely that it is the relevant isoform [132]. Certainly, a recent genome-wide study of the determinants of serum 25-OH-D is consistent with this idea since it identified four relevant genes, including CYP2R1 and the others being DBP, CYP24A1, and 7-dehydrocholesterol reductase [66].

However, it is clear that the mitochondrial CYP27A can efficiently 25-hydroxylate 1α-OH-D_3 to give $1\alpha,25$-$(OH)_2D_3$ [133] and is present in a variety of tissues, as well as the liver (e.g., kidney and bone). Studies using cultured bone cells and even keratinocytes *in vitro* are able to demonstrate synthesis of $1\alpha,25$-$(OH)_2D_3$ from 1α-OH-D_3 [134] or $1\alpha,24$-$(OH)_2D_2$ from 1α-OH-D_2 [135]. If these findings can be extrapolated to the *in vivo* situation, the implications of this work are that in CYP27A, vitamin D target cells may have some ability to synthesize the active form from a prodrug *without* the need for the hormone to enter the bloodstream.

The ability of extrarenal tissues to 1α-hydroxylate various 25-hydroxylated metabolites and analogs has always been a controversial story. However, it was widely accepted that extrarenal 1α-hydroxylase activity is associated with certain granulomatous conditions (e.g., sarcoidosis) [110,136]. Currently, there is little information for why the enyme is overexpressed in sarcoidosis. In sarcoid patients, 25-OH-D can be converted to $1\alpha,25$-$(OH)_2D$, which is a step that, unlike the renal case, is not subject to tight regulation and is thus potentially more likely to result in hypercalcemia. Exposure of such patients to sunlight or administration of 25-OH-D can result in excessive plasma levels of $1\alpha,25$-$(OH)_2D$. Following the cloning of the cytochrome P450 representing the 1α-hydroxylase (CYP27B1) [137,138], it was quickly shown that CYP27B1 can be expressed extrarenally in skin and lung cancer cells [139,140]. This has been extended over the past decade with further studies of CYP27B1 mRNA levels using real-time PCR and specific anti-CYP27B1 antibodies [141] to show the widespread distribution of this enzyme in many normal tissues, as well as pathological situations. As alluded to earlier, the concept of the extrarenal 1α-hydroxylase suggests that this enzyme plays an important physiological and pathological role [1,2], and this has, in turn, raised the level of importance given to ensuring maintenance of adequate 25-OH-D levels by vitamin D or direct 25-OH-D_3 supplementation rather than just calcitriol hormone replacement.

Most of the calcitriol analogs listed in Table 3.3 are thought to be active as such, not requiring any step of activation prior to their action on the transcriptional machinery or in nongenomic pathways. It remains a theoretical possibility though that the biological activity of one of these parent analogs could be altered by enzyme systems *in vivo*, either by the generation of a more potent metabolite or by giving rise to a less active but more long-lived catabolite.

3.7.1.2 Vitamin D–Binding Protein

The vitamin D–binding protein (DBP) provides transport for all lipid-soluble vitamin D compounds, from vitamin D to $1\alpha,25$-$(OH)_2D_3$, so it is not surprising that it also carries vitamin D analogs. Most of the analogs of calcitriol designed to date contain modifications to the side chain, and this is usually detrimental to binding to DBP. Several analogs, e.g., calcipotriol, OCT, and 19-nor-1,25-$(OH)_2D_2$, have very weak affinities for DBP and have been reduced by as much as two to three orders of magnitude relative to $1\alpha,25$-$(OH)_2D_3$. This property has important implications on metabolic clearance rates, delivery to target cells, and tissue distribution [142,143]. Detailed studies with one analog, OCT, have shown it to bind primarily to β-lipoprotein and exhibit an abnormal tissue distribution *in vivo*, with abnormally high concentrations (ng/g tissue) in the parathyroid gland [144]. It was thus proposed that this unusual distribution may make OCT a useful systemically administered drug with a selective advantage in the treatment of hyperparathyroidism. Another vitamin D analog with a modified side chain is 20-epi-$1\alpha,25$-$(OH)_2D_3$, where the 20-*S* configuration of the side chain is opposite to the normal 20-*R* configuration. The DBP binding affinity of this analog is virtually unmeasurable because it does not displace [^{3}H]25-OH-D_3 from the plasma binding protein [145]. Confirmation that this is indeed the case comes from GH-reporter gene transactivation assays, where 20-epi-$1\alpha,25$-$(OH)_2D_3$ transactivates equally well in COS cells incubated in the presence and absence of fetal calf serum (as a source of DBP). On the other hand, $1\alpha,25$-$(OH)_2D_3$-induced GH reporter gene expression is sensitive to DBP in the external growth medium, requiring twofold less hormone in the absence of DBP as in its presence [145]. It therefore appears that analogs that bind DBP less well than $1\alpha,25$-$(OH)_2D_3$ derive a target cell advantage over the natural hormone *if they are able to find alternative plasma carrier proteins to transport them to their target cells.* However, these same alternative plasma carriers presumably result in changes in the tissue distribution and hepatic clearance of analogs over the natural metabolites of vitamin D. Studies in the DBP-knockout mouse [146] suggest that 25-OH-D_3 clearance is more rapid in the absence of DBP. The author is unaware of any studies of the effects of vitamin D analogs in the DBP-knockout mouse, although the availability of this model offers a unique approach to the study of alternate vitamin D analog transport mechanisms in an *in vivo* setting.

3.7.1.3   Vitamin D Receptor/RXR/VDRE Interactions

Three decades of work have established that $1\alpha,25$-$(OH)_2D_3$ is able to work through a VDR-mediated genomic mechanism to stimulate transcriptional activity at vitamin D–dependent genes (see other chapters in this book). Cloning of the VDR and elucidation of the 3-D structure of its ligand-binding domain have provided a huge boost to delineating the precise conformational changes that take place when the natural ligand binds to the VDR [147,148] and the nature of the postligand-binding transcriptional events that occur thereafter, particularly the nature of the coactivator proteins involved [149,150]. Basic knowledge of the mechanism of action of vitamin D has been aided by the opportunity to observe the lack of effects of calcitriol and its analogs in the VDR-knockout mouse [151,152]. These studies have largely refuted claims of alternative non-VDR-mediated mechanisms to produce physiologically relevant effects that might complicate our understanding of the pharmacological effects of vitamin D analogs.

Much evidence exists to support the viewpoint that vitamin D analogs mimic $1\alpha,25$-$(OH)_2D_3$ and use a genomic mechanism. In early work, Stern [153] showed that there exists a strong correlation between chick intestinal VDR binding of an analog and its potency in a [^{45}Ca] rat bone resorption assay, suggesting that a vitamin D analog is only as good as its affinity for the VDR. More recent work has suggested that this is a highly simplified viewpoint and that VDR binding affinity may not even be the major factor, with transactivation activity stemming from a series of parameters such as conformation of the ligand/VDR complex, binding of the RXR partner, stability of the VDR/RXR/ ligand complex, or even the nature of the coactivator proteins recruited to the complex. Recent data from the "superagonist" analogs [32,33,154] suggest that the 20-epi-analogs including KH1060 are only approximately one order of magnitude more potent than $1\alpha,25$-$(OH)_2D_3$ in gene transactivation assays and in differentiation assays. Thus, it appears that the quantitative advantages originally claimed for calcitriol "superagonists" are modest at best and may be partially explained by other factors such as differences in DBP binding or metabolic clearance rates [155,156].

Perhaps, more important is whether analogs can be qualitatively different from $1\alpha,25\text{-}(OH)_2D_3$ in their actions and work selectively in either calcium and phosphate homeostasis or cell differentiation roles. Freedman's group reported that the ability of various analogs to transactivate vitamin D-dependent genes or to stimulate differentiation of cells is best correlated with their ability to recruit the coactivator, i.e., DRIP-205, one of the many components of the DRIP complex isolated by Freedman's group [157,158]. Among the other coactivators/transcription factors implicated in vitamin D analog action is GRIP-1(TIF-2), which has been purported to have a particular propensity to interact with the analog OCT [159]. In another study by Issa et al. [160], a broad panel of vitamin D analogs showed that GRIP-1 was more consistently recruited at levels closer to that of $1\alpha,25\text{-}(OH)_2D_3$ than was another coactivator AIB-1. Work by Peleg et al. [161] offers an insight into the purported bone tissue selectivity of the Roche analog Ro 26-9228 (Table 3.3 renamed BXL-628) by showing recruitment of GRIP-1 in osteoblasts but not in CaCo-2 colon cancer cells, although paradoxically, BXL-628 is now being pursued clinically in prostatic disease rather than osteoporosis. Nevertheless, it appears that there is a fairly strong basis for the hypothesis that differences in the biopotency advantage of certain vitamin D analogs over $1\alpha,25\text{-}(OH)_2D_3$ are due in part to changes in the recruitment of the RXR dimerization partner and/or coactivators (e.g., [33]), but there is no consensus on which of these coactivator proteins is the important one or if these different coactivators can explain tissue/cell selectivity. Work using CHIP assays [162], which show temporal changes in coactivator recruitment at vitamin D-dependent gene promoters, may aid in our understanding of this complex transcriptional story.

3.7.1.4 Target Cell Catabolic Enzymes

In recent years, much evidence has accumulated to support the hypothesis that $1\alpha,25\text{-}(OH)_2D_3$ is subject to target cell catabolism and side-chain cleavage to calcitroic acid via a 24-oxidation pathway [163]. The cloning of CYP24A1, with the cytochrome P450 involved, has confirmed that it is vitamin D–inducible since its gene promoter contains a VDRE, carries out multiple steps in the side-chain modification process, and is present in most (if not all) vitamin D target cells [164,6,43]. We have postulated that the purpose of this catabolic pathway is to desensitize the target cell to continuing hormonal stimulation by $1\alpha,25\text{-}(OH)_2D_3$ [165]. Support for this hypothesis came from St. Arnaud's group when they engineered a CYP24A1-knockout mouse, which shows 50% lethality at weaning, with death resulting from hypercalcemia and nephrocalcinosis [166]. Surviving mice show an inability to rapidly clear a bolus dose of $1\alpha,25\text{-}(OH)_2D_3$ from the bloodstream and tissues [167] and a metabolic bone disease that is reminiscent of the excessive osteoid bone pathology observed in rodents given excessive amounts of $1\alpha,25\text{-}(OH)_2D_3$ [168]. The bone defect is probably due to excessive $1\alpha,25\text{-}(OH)_2D_3$ levels since crossing the CYP24A1-knockout mouse with the VDR-knockout mouse results in a phenotype without the bone defect [169], although, as stated earlier, the bone lesion can also be relieved by $24(R),25\text{-}(OH)_2D_3$ administration [77]. Recently, Schlingmann et al. [170] have reported the human equivalent of the CYP24A1-knockout mouse represented by patients with idiopathic infantile hypercalcemia who show inactivating coding-sequence mutations of the CYP24A1 gene. The disease reinforces our view that CYP24A1 exists in the kidney and target cells to degrade $25\text{-}OH\text{-}D_3$ and $1\alpha,25\text{-}(OH)_2D_3$ to inactive products since these patients exhibit elevated vitamin D metabolites and hypercalcemia. Given the demonstrated importance of CYP24A1 to $1\alpha,25\text{-}(OH)_2D_3$ clearance, one must ask the question of whether vitamin D analogs might be subject to the same catabolic processes? Is CYP24A1 largely responsible for their observed pharmacokinetics? If not, what other drug-catabolizing systems are present within vitamin D target cells to inactivate the vitamin D analog?

Certainly, there are vitamin D analogs such as calcipotriol, OCT, EB1089, and KH1060 that are metabolized by vitamin D target cells to clearly defined and unique metabolites [171–173,155], which resemble products of the 24-oxidation pathway for $1\alpha,25\text{-}(OH)_2D_3$ or which are unique to the particular analog. Furthermore, some of these metabolites are products only of vitamin D target cells and are vitamin D inducible, implying that CYP24A1 is involved in their formation, but this has been confirmed with some analogs such as calcipotriol [131]. However, in the case of several analogs blocked at C-24 and subject to metabolism elsewhere on the side chain, the direct involvement of CYP24A1 is strongly implicated or proven. These include 23-hydroxylation of 26,27-hexafluoro-$1\alpha,25$-$(OH)_2D_3$ [174]; 26-hydroxylation of 24-difluro-$1\alpha,25$-$(OH)_2D_3$ [175]; 26-hydroxylation of $1\alpha,25$-$(OH)_2$-16ene-23yne-D_3 [176]; and 26- and 28-hydroxylation of $1\alpha,25$-$(OH)_2D_2$ [177,178]. Since many of these same products are observed *in vitro* and *in vivo*, and since pharmacokinetic parameters often parallel target cell metabolic parameters [143,179], one concludes that target cell metabolism of vitamin D analogs must contribute to the pharmacokinetics and biological activity observed *in vitro* and *in vivo*. Even studies such as that of Eelen et al. [32], which purport to show changes in VDR-mediated gene expression at the coactivator level, show that use of a CYP24A1 inhibitor VID-400 (Table 3.4) in narrowing potency differences between 23-yne analogs and $1\alpha,25$-$(OH)_2D_3$ by blocking catabolism of the latter molecule and allowing the evaluation of the potency issue without metabolic considerations. Unfortunately, this has not always been the approach, and there is little doubt that the poor performance of some promising vitamin D analogs during *in vivo* testing is due to their poor metabolic stability. Accordingly, greater attention to the metabolic potential of *in vitro* testing systems and/or greater use of defined target cell (and hepatic) metabolic systems is warranted.

One factor regarding target cell metabolism considered in recent years is the possibility that vitamin D analogs might be *activated* rather than *catabolized* by the same enzymes [180,181]. While this is potentially more important for prodrugs (Table 3.2), the generation of large numbers of metabolites from such analogs as KH1060 [155] or the formation of long-lived metabolites such as 26,27-hexafluoro-$1\alpha,23,25$-$(OH)_3D_3$ from 26,27-hexafluoro-$1\alpha,25$-$(OH)_2D_3$ [174] complicates the picture. In most cases, however, this issue can be resolved on pharmacokinetic grounds.

Lastly, some of the calcitriol analogs with modifications in the vicinity of C23, namely 20-epi-$1\alpha,25$-$(OH)_2D_3$ [145], $1\alpha,25$-$(OH)_2$-16ene-D_3 [182], and 20-methyl-$1\alpha,25$-$(OH)_2D_3$ [183], undergo 24-oxidation pathway metabolism that stalls at the level of the 24-oxo-metabolite seemingly because the enzyme CYP24A1 cannot efficiently carry out the usual 23-hydroxylation step and complete the catabolic sequence to the inactive calcitroic acid. The consequence, at least *in vitro*, is that the 24-oxo-metabolite accumulates, and there have been claims that this metabolite retains significant biological activity [181,182]. Recently, this hypothesis has received a boost with the work of Zella et al. [156], who have found that the "superagonist" 20-epi-$1\alpha,25$-$(OH)_2D_3$ exhibits a prolonged duration of action on intestinal calcium regulating genes selectively, and these researchers have proposed that this advantage over $1\alpha,25$-$(OH)_2D_3$ stems from a reduction in its catabolic rate specifically in that tissue.

3.7.1.5 Hepatic Clearance or Nonspecific Metabolism

The poor DBP binding properties of many side-chain modified calcitriol analogs opens up the possibility of alternative plasma carriers and accelerated degradation. The liver plays a major role in such metabolic clearance and a small number of detailed studies performed to date have included *in vitro* incubation with liver preparations. Calcipotriol [184], OCT [172], EB1089 [185], and KH1060 [186] are all subject to metabolism by liver enzymes. One such liver enzyme that is capable of 23- and 24-hydroxylation of $1\alpha,25$-$(OH)_2D_3$, and possibly some of its analogs is the abundant general cytochrome P450, CYP3A4 [187]. Indeed, this enzyme is upregulated by $1\alpha,25$-$(OH)_2D_3$ in duodenum, suggesting that a physiologically-relevant loop exists [188]. Since over the years there have been frequent reports of drug-induced osteomalacia associated with coincidental use of anticonvulsants (e.g., diphenylhydantoin) or barbiturates and vitamin D preparations [e.g., 189], the direct association between CYP3A4 and $1\alpha,25$-$(OH)_2D_3$ is potentially important to explain the putative accelerated clearance of vitamin D metabolites [190]. One such phenomenon that might be explained by intestinal CYP3A4 action is the purported lower toxicity of vitamin D_2 compounds, compared to their D_3 counterparts, which were referred to earlier. Work using microsomes from an intestinal cell line and supersomes enriched in recombinant human CYP3A4 catabolize $1\alpha,25$-$(OH)_2D_2$ at a significantly faster rate than $1\alpha,25\text{-}(OH)_2D_3$ [191]. The implication of this finding is that $1\alpha,25\text{-}(OH)_2D_2$ and possibly other synthetic analogs, such as the mixed VDR agonist/CYP24A1 inhibitor CTA-018 [47], are likely to be selectively broken down in the intestine, potentially reducing their gene expression effects on intestinal calcium and phosphate absorption but not on other tissues.

These cytochrome P450 enzymes give rise to intermediate polarity or truncated metabolites, which can be further glucuronidated and excreted in bile (e.g., OCT [192]). A recent study has defined UGT1A3 as the isoform of UDP-glucuronosyltransferase involved in glucuronidation of the 23-hydroxylated metabolite of the analog $26,27\text{-}F_6\text{-}1\alpha,25\text{-}(OH)_2D_3$ [193].

Few, if any, studies have separately considered the *rate* of catabolism or glucuronidation relative to $1\alpha,25\text{-}(OH)_2D_3$. However, data are available comparing the *in vivo* rate of metabolic clearance of vitamin D analogs to $1\alpha,25\text{-}(OH)_2D_3$, although inevitably, this probably measures a few *in vitro* parameters, such as the rate of both hepatic and target cell metabolism, in addition to the affinity of DBP binding within a single *in vivo* parameter. Thus, in lieu of detailed *in vitro* metabolic analyses, the $t1/2$ of the vitamin D analog is a useful term for indicating the general survival of the vitamin D drug *in vivo*. Data for this parameter have been published for some of the most interesting analogs [143].

One controversial nonspecific route of metabolism described in the vitamin D literature is 3-epimerization [194]. That 3-epimeric metabolites of vitamin D compounds are formed is not in question, with the most notable example being $3\text{-}epi\text{-}25\text{-}OH\text{-}D_3$ detected in ample concentrations in pediatric serum samples by LC-MS/MS [195]. The controversy surrounds whether 3-epimers are an artifact of sample preparation; if indeed they are found *in vivo*, then what is the nature of the enzyme systems involved? The author of this chapter is credited with the generation of the first 3-epimer in the form of $3epi\text{-}1,25\text{-}(OH)_2D_3$ by incubating very high concentrations of $1,25\text{-}(OH)_2D_3$ with cultured bone cells *in vitro* [196]. However, the 3-epimeric forms were first identified by Satya Reddy's group. Reddy has shown that various *in vitro* cultured cell systems generate differential amounts of 3-epimeric forms [197], 3-epimeric forms can be further metabolized to 1α- and 24-hydroxylated products, 3-epimeric forms of calcitriol analogs can also be generated *in vitro*, and 3-epimers retain biological activity [194]. However, so far, the nature of the putative 3-epimerase has not been determined and thus whether it is specific for vitamin D compounds. In conclusion, both the physiological and pharmacological importance of 3-epimersation to vitamin D and its analogs remain obscure.

3.7.1.6 Putative Nongenomic Non-VDR-Mediated Actions of Vitamin D and Its Analogs

The nongenomic and non-VDR mediated actions of $1\alpha,25\text{-}(OH)_2D_3$ have been reviewed extensively in Chapter 11 of this book and elsewhere [9,198,199]. The membrane VDR initially described by Nemere et al. [200] and first identified as annexin II [201] was proposed to be involved in mediating rapid nongenomic effects. In the past decade, further attempts to purify and identify the putative membrane receptor have resulted in identification of a membrane-associated rapid response system $(1,25\text{-}(OH)_2D_3\text{-}MARRS)$ in chick intestinal cells [202], which may explain rapid nongenomic actions [198,199]. Recently, the $1,25\text{-}(OH)_2D_3\text{-}MARRS$ membrane receptor from chondrocytes was identified as a protein disulphide isomerase [203], but exactly what this protein does is not clear. Furthermore, at this point in time, little work has been performed on the specificity of the vitamin D–binding site of membrane VDR/annexin II or the $1,25\text{-}(OH)_2D_3\text{-}MARRS$ complex, and thus, the possibility that the nongenomic actions/membrane VDR might explain vitamin D analog actions seems speculative.

3.7.2 SUMMARY OF FACTORS WHICH MODULATE POTENCY OR SPECIFICITY OF VITAMIN D ANALOGS

Figure 3.2 contains a list of the factors that are thought to modulate the potency or specificity of vitamin D analogs and therefore determine if a specific analog is likely to have a greater calcemic activity or procell differentiating/antiproliferatating activity. Although this list offers clues as to

THE VITAMIN D DECATHLON/ VITAMIN D ANALOG PENTATHLON

STEP 1: Intestinal Absorption from Diet OR Synthesis in the Skin

STEP 2: Transport on DBP or Chylomicrons to Liver OR Stores

STEP 3: Liver 25-Hydroxylation

STEP 4: Transport on DBP to Kidney

STEP 5: 1α- (or 24-)Hydroxylation in Kidney to form active 1α,25-(OH)$_2$D$_3$ (or inactive 24,25-(OH)$_2$D$_3$)

STEP 6: Transport on DBP to Target Cell

STEP 7: Binding of vitamin D ligand to VDR/RXR Heterodimer in Target Cell

STEP 8: Conformational change of VDR & Binding of Coactivators to modulate Gene Transcription

STEP 9: Inactivation of vitamin D ligand by target cell CYP24A1

STEP 10: Excretion of 1α,25-(OH)$_2$D$_3$ catabolite, calcitroic acid or the vitamin D analog product in bile

Underlined steps apply to calcitriol analogs

FIGURE 3.2 Vitamin D decathlon/vitamin D analog pentathlon. The 10 steps involved in the vitamin D signal cascade from absorption of vitamin D in the intestine of synthesis in the skin to excretion of calcitroic acid in bile. Many of the steps involve vitamin D–specific proteins, i.e., DBP, VDR, CYPs, although more general transcription factors such as RXR, coactivator proteins, and general cytochromes P450 are sometimes involved. Except for vitamin D$_2$, vitamin D prodrugs (Table 3.2) use only a subset of the activation steps. The calcitriol analogs (Table 3.3), which are active as administered, require only five or six of these steps.

how to design analogs to meet certain needs, the truth is that most current analogs were identified by screening libraries of hundreds or thousands of compounds and not by rational drug design. Furthermore, despite claims in the literature, no pharmaceutical company has succeeded in the development of an antiproliferative or immunomodulatory vitamin D analog without some calcemic activity. Thus, the field of making specific analogs for specific applications remains an art rather than a science.

3.8 FUTURE PROSPECTS

A number of researchers remain optimistic that the unraveling of the genomic (or nongenomic) mechanism of action of 1α,25-(OH)$_2$D$_3$ will reveal new approaches by which the vitamin D signaling cascade can be exploited. Certainly, the significant progress made in characterizing the coactivator proteins and the rest of the transcriptional apparatus will continue. One is able to predict fairly confidently from success in related steroid hormone fields that a fully functional vitamin D–dependent *in vitro* reconstituted VDR-RXR transcriptional system, devoid of the complications of metabolic enzymes, will be the perfect model to test the transactivation activity of future vitamin D analogs. It seems likely that this approach will allow us to dissect out the exact features that give certain analogs a transcriptional advantage to provide increased potency and/or selectivity over 1α,25-(OH)$_2$D$_3$.

Studies of the vitamin D–binding pockets of VDR, DBP, and the three (or more) vitamin D–related cytochrome P450s (especially CYP2R1 and CYP24A1) will continue to be a major goal now that all these specific proteins have been cloned, overexpressed, and crystallized. While the ligand-binding domains of the nuclear receptors have been studied, the full-length proteins are beyond the current limits of NMR or X-ray crystallography. It is also likely that technical problems with these procedures will be overcome shortly and that the full-length proteins can be tackled. The work of the Moras group [147] on the ligand-binding domain of the VDR will be extended to the interaction with coactivators, and there will also be a growing focus on the other major proteins in the vitamin D signal transduction pathway.

The wide availability of recombinant proteins for hundreds of cytochromes P450 from species across the phylogenetic tree, including 58 CYPs in the human genome, has allowed for the elucidation of some crystal structures (human CYP2R1 and rat CYP24A1) [132,204] and also homology modeling studies of the enzymes involved in vitamin D metabolism [205–207]. CYP2R1 was the first vitamin D–related cytochrome P450 to have its crystal structure solved, but it is a microsomal enzyme that is quite unrelated to the mitochondrial isoforms, which perform the later steps of vitamin D metabolism [132]. The newly released crystal structure of rat CYP24A1 [204] lacks a vitamin D substrate, and so it is limited in what it reveals about substrate binding. However, the crystal stucture does confirm the importance of several amino acid residues postulated from homology modeling and mutagenesis studies (Figure 3.3) [205–207]. These crystal structure and homology models of CYP24A1 are consistent on the residues that line the substrate-binding pocket and contact the side chain (e.g., Ala326) [208] and ultimately help determine the specificity/hydroxylation site. The availability of structural information from the crystal structure of CYP24A1 will likely allow for new approaches for the design and testing of vitamin D analogs. Reddy's group recently published work [209] comparing *in silico* docking of known vitamin D analogs in the substrate-binding domain of rat CYP24A1 with the side-chain hydroxylation activity of the same analogs in a CYP24A1 reconstitution assay. No doubt that these are the first attempts at rational vitamin D analog drug design using the CYP24A1 enzyme structure to predict metabolic stability.

Even without this rational approach, access to full-length CYP24A1 and CYP27B1 has also permitted a more efficient search for potential inhibitors of both. Such specific inhibitors of CYP24A1

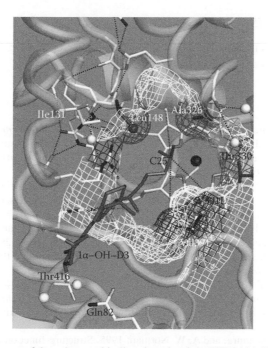

FIGURE 3.3 Crystal structure of the substrate-binding domain of rat CYP24A1. Illustrated are the key active site residues, many of which have been substituted in mutagenesis studies. As shown in Figure 3.1, CYP24A1 is a versatile enzyme that is capable of 24- or 23-hydroxylation of the vitamin D side chain. Vitamin D analogs with their side chain and nuclear modifications are often blocked at C-23 or C-24. Again, the versatility of CYP24A1 is revealed because the literature documents alternate hydroxylations at C-26, C-27, and C-28 (in vitamin D_2 compounds). The availability of the active site structure should allow for rational drug design.

and CYP27B1 [44–47,210,211] may be of value in blocking $1\alpha,25\text{-}(OH)_2D_3$ catabolism or synthesis in certain clinical conditions where excessive breakdown or production is supected. In general, modeling of VDR and cytochromes P450 is expected to lead to more rational vitamin D analog design to take advantage of structural idiosyncrasies of all of these key proteins. Meanwhile, the not-so-rational synthesis of new analogs is likely to continue.

The list of clinical applications for vitamin D analogs continues to increase [1,6]. These applications have been further rationalized with the availability of VDR knockout mice to demonstrate vitamin D ligand-dependent processes [151,152]. Elucidation of the mechanism by which $1\alpha,25\text{-}(OH)_2D_3$ and its analogs regulate the cell cycle and proliferation remains an important priority of the field [11,212]. Current applications of vitamin D analogs still fall mainly into calcium-related and cell-proliferative/differentiating arenas, but the "rediscovery" of the broad consequences of vitamin D deficiency has reinvigorated the whole field. The goal of developing analogs that can completely separate the "calcemic" and "nonclassical" properties of $1\alpha,25\text{-}(OH)_2D_3$ has not yet been fully realized. However, some promising compounds have been synthesized, and interesting idiosyncrasies of their biological actions have surfaced (e.g., tissue, cell, gene, and VDRE differences). It remains to be seen if these differences can be exploited. On the other hand, it must be stated that if vitamin D analogs work *only* through a VDR-mediated genomic mechanism, it is difficult to appreciate how the "calcemic" properties of $1\alpha,25\text{-}(OH)_2D_3$ can ever be fully resolved from the "cell-differentiating" properties, given that pharmacokinetic differences have provided only a partial separation. On a more optimistic front, it can be stated that there are now many vitamin D analogs such as calcipotriol, OCT, $1\alpha,24\text{-}(OH)_2D_3$, $1\alpha\text{-}OH\text{-}D_2$, $19\text{-nor-}1\alpha,25\text{-}(OH)_2D_2$, and ED-71 approved for use in the treatment of various clinical conditions worldwide, and this list continues to grow.

ACKNOWLEDGMENTS

The author acknowledges the contributions of Dr. David Prosser and Dr. Martin Kaufmann to the compilation of the figures and tables generated for this chapter.

REFERENCES

1. Holick, M. F. 2007. Vitamin D deficiency. *New England Journal of Medicine* 357:266–281.
2. Jones, G. 2007. Expanding role for vitamin D in chronic kidney disease: Importance of blood 25-OH-D levels and extra-renal 1α-hydroxylase in the classical and nonclassical actions of $1\alpha,25$-dihydroxyvitamin D_3. *Seminars in Dialysis* 20:316–324.
3. Park, E. A. 1940. The therapy of rickets. *The Journal of the American Medical Association* 94:370–379.
4. De Luca, H. F. 1988. The vitamin D story: A collaborative effort of basic science and clinical medicine. *The FASEB Journal* 2:224–236.
5. Jones, G. and M. J. Calverley. 1993. A dialogue on analogues: Newer vitamin D drugs for use in bone disease, psoriasis, and cancer. *Trends in Endocrinology and Metabolism* 4:297–303.
6. Jones, G., S. Strugnell, and H. F. De Luca. 1998. Current understanding of the molecular actions of vitamin D. *Physiological Reviews* 78:1193–1231.
7. Miyaura, C., E. Abe, T. Kuribayashi et al. 1981. $1\alpha,25$-dihydroxyvitamin D3 induces differentiation of human myeloid leukemia cells. *Biochemical and Biophysical Research Communications* 102:937–943.
8. Calverley, M. J. and G. Jones. Vitamin D. 1992. In: *Antitumour Steroids*. Ed: Blickenstaff, R. T. Toronto: Academic Press. 193–270.
9. Bouillon, R., W. H. Okamura, and A. W. Norman. 1995. Structure-function relationships in the vitamin D endocrine system. *Endocrine Reviews* 16:200–257.
10. Jones, G. 2008. Pharmacological mechanisms of therapeutics—Chapter 83: Vitamin D and analogues. In: *Principles of Bone Biology*, 3rd Edition. Eds: Bilezikian J., L. Raisz, and G. Rodan. San Diego: Academic Press Inc.; pp. 1777–1799.
11. Masuda, S. and G. Jones. 2006. The promise of vitamin D analogs in the treatment of hyperproliferative conditions. *Molecular Cancer Therapeutics* 5:797–808.

12. Vieth, R. 2005. Chapter 61: The pharmacology of vitamin D, including fortification strategies. In: Vitamin D, 2nd Edition. Eds: Feldman D., J. W. Pike, and F. H. Glorieux. New York: Elsevier Academic Press. 995–1015.

13. Trang, H. M., D. E. Cole, L. A. Rubin et al. 1998. Evidence that vitamin D_3 increases serum 25-hydroxy-vitamin D more efficiently than does vitamin D_2. *American Journal of Clinical Nutrition* 68:854–858.

14. Armas, L. A., B. W. Hollis, and R. P. Heaney. 2004. Vitamin D_2 is much less effective than vitamin D_3 in humans. *The Journal of Clinical Endocrinology & Metabolism* 89:5387–5391.

15. Roborgh, J. R. and T. de Man. 1960. The hypercalcemic activity of dihydrotachysterol-2 and dihydro-tachysterol-3 and of the vitamins D2 and D3: Comparative experiments in rats. *Biochemical Pharmacology* 2:1–6.

16. Roborgh, J. R. and T. de Man. 1960. The hypercalcemic activity of dihydrotachysterol-2 and dihydro-tachysterol-3 and of the vitamins D2 and D3 after intravenous injection of the aqueous preparations—Part II: Comparative experiments on rats. *Biochemical Pharmacology* 3:277–282.

17. Hunt, R. D., F. G. Garcia, and R. J. Walsh. 1972. A comparison of the toxicity of ergocalciferol and cho-lecalciferol in rhesus monkeys (Macacamulatta). *Journal of Nutrition* 102:975–986.

18. Sjöden, G., C. Smith, U. Lindgren et al. 1985. 1α-Hydroxyvitamin D2 is less toxic than 1α-hydroxyvitamin D3 in the rat. *Proceedings of the Society for Experimental Biology and Medicine* 178:432–436.

19. Weber, K., M. Goldberg, M. Stangassinger et al. 2001. 1α-hydroxyvitamin D2 is less toxic but not bone selective relative to 1α-hydroxyvitamin D3 in ovariectomized rats. *Journal of Bone and Mineral Research* 16:639–651.

20. Rapuri, P. B., J. C. Gallagher, and G. Haynatzki. 2004. Effect of vitamins D2 and D3 supplement use on serum 25,OHD concentration in elderly women in summer and winter. *Calcified Tissue International* 74:150–156.

21. Holick, M. F., R. M., Biancuzzo, T. C. Chen et al. 2008. Vitamin D2 is as effective as vitamin D3 in maintaining circulating concentrations of 25-hydroxyvitamin D. *The Journal of Clinical Endocrinology & Metabolism* 93:677–681.

22. Thacher, T. D., M. O. Obadofin, K. O. O'Brien et al. 2009. The effect of vitamin D2 and vitamin D3 on intestinal calcium absorption in Nigerian children with rickets. *The Journal of Clinical Endocrinology & Metabolism* 94:3314–3321.

23. Barton D. H., R. H. Hesse, M. M. Pechet et al. 1973. A convenient synthesis of 1α-hydroxy-vitamin D_3. *Journal of the American Chemical Society* 95:2748–2749.

24. Paaren, H. E., D. E. Hamer, H. K. Schnoes et al. 1978. Direct C-1 hydroxylation of vitamin D compounds: Convenient preparation of 1α-hydroxyvitamin D_3, 1α, 25-dihydroxyvitamin D_3, and 1α-hydroxyvitamin D_2. *Proceedings of the National Academy of Sciences USA* 75:2080–2081.

25. Mehta, R., M. Hawthorne, L. Uselding et al. 2000. Prevention of N-methyl-N-nitrosourea-induced mam-mary carcinogenesis in rats by 1alpha-hydroxyvitamin D5. *Journal of the National Cancer Institute* 92:1836–1840.

26. Jones, G., N. Edwards, D. Vriezen et al. 1988. Isolation and identification of seven metabolites of 25-hydroxydihydrotachysterol$_3$ formed in the isolated perfused rat kidney: A model for the study of side-chain metabolism of vitamin D. *Biochemistry* 27:7070–7079.

27. Qaw, F., M. J. Calverley, N. J. Schroeder et al. 1993. *In vivo* metabolism of the vitamin D analog, dihy-drotachysterol: Evidence for formation of 1α,25-and 1β,25-dihydroxy-dihydrotachysterol metabolites and studies of their biological activity. *Journal of Biological Chemistry* 268:282–292.

28. Kragballe, K. 1992. Vitamin D analogues in the treatment of psoriasis. *Journal of Cellular Biochemistry* 49:46–52.

29. Abe-Hashimoto, J., T. Kikuchi, T. Matsumoto et al. 1993. Antitumor effect of 22-oxa-calcitriol, a noncal-cemic analogue of calcitriol, in athymic mice implanted with human breast carcinoma and its synergism with tamoxifen. *Cancer Research* 53:2534–2537.

30. Strugnell, S., V. Byford, H. L. J. Makin et al. 1995. 1α,24(S)-dihydroxyvitamin D_2: A biologically active product of 1α-hydroxyvitamin D_2 made in the human hepatoma, Hep3B. *Biochemical Journal* 310:233–241.

31. Baggiolini, E. G., J. J. Partridge, S.-J. Shiuey et al. 1989. Cholecalciferol 23-yne derivatives, their phar-maceutical compositions, their use in the treatment of calcium related diseases, and their antitumor activ-ity. US 4,804,502. *Chemical Abstracts* 111:58160d [Abstract].

32. Eelen, G., N. Valle, Y. Sato et al. 2008. Superagonistic fluorinated vitamin D_3 analogs stabilize helix 12 of the vitamin D receptor. *Chemistry & Biology* 15:1029–1034.

33. Verstuyf, A., L. Verlinden, E. van Etten et al. 2000. Biological activity of CD-ring modified 1α,25-dihydroxyvitamin D analogues: C-ring and five-membered D-ring analogues. *Journal of Bone and Mineral Research* 15:237–252.

34. Suhara, Y., K. I. Nihei, M. Kurihara et al. 2001. Efficient and versatile synthesis of novel 2α-substituted 1α,25-dihydroxyvitamin D_3 analogues and their docking to vitamin D receptors. *The Journal of Organic Chemistry* 266:8760–8671.

35. Shevde, N. K., L. A. Plum, M. Clagett-Dame et al. 2002. A potent analog of 1α,25-dihydroxyvitamin D_3 selectively induces bone formation. *Proceedings of the National Academy of Sciences USA* 99: 13487–13491.

36. Okamura, W. H., M. M. Midland, A. W. Norman et al. 1995. Biochemical significance of the 6-s-*cis* conformation of the steroid hormone 1α,25-dihydroxyvitamin D_3 based on the provitamin D skeleton. *Annals of the New York Academy of Sciences* 761:344–348.

37. Crescioli, C., P., Ferruzzi, A. Caporali et al. 2004. Inhibition of prostate cell growth by BXL-628, a calcitriol analogue selected for a phase II clinical trial in patients with benign prostate hyperplasia. *European Journal of Endocrinology* 150:591–603.

38. Adorini, L., G. Penna, S. Amuchastegui et al. 2007. Inhibition of prostate growth and inflammation by the vitamin D receptor agonist BXL-628 (elocalcitol). *The Journal of Steroid Biochemistry and Molecular Biology* 103:689–693.

39. Boehm, M. F., P. Fitzgerald, A. Zou et al. 1999. Novel nonsecosteroidal vitamin D mimics exert VDR-modulating activities with less calcium mobilization than 1,25-dihydroxyvitamin D_3. *Chemistry & Biology* 6:265–275.

40. Ma, Y., B. Khalifa, Y. K. Yee et al. 2006. Identification and characterization of noncalcemic, tissue-selective, nonsecosteroidal vitamin D receptor modulators. *Journal of Clinical Investigation* 116: 892–904.

41. Ishizuka, S., N. Kurihara, S. V. Reddy et al. 2005. (23S)-25-Dehydro-1α-hydroxyvitamin D_3-26,23-lactone, a vitamin D receptor antagonist that inhibits osteoclast formation and bone resorption in bone marrow cultures from patients with Paget's disease. *Endocrinology* 146:2023–2030.

42. Saito, N. and A. Kittaka. 2006. Highly potent vitamin D receptor antagonists: Design, synthesis, and biological evaluation. *ChemBioChem* 7:1479–1490.

43. Prosser, D. E. and G. Jones. 2004. Enzymes involved in the activation and inactivation of vitamin D. *Trends in Biochemical Sciences* 29:664–673.

44. Schuster, I., H. Egger, N. Astecker et al. 2001. Selective inhibitors of CYP24: Mechanistic tools to explore vitamin D metabolism in human keratinocytes. *Steroids* 66:451–462.

45. Posner, G. H., K. R. Crawford, H. W. Yang et al. 2004. Potent low-calcemic selective inhibitors of CYP24 hydroxylase: 24-sulphone analogs of the hormone 1α,25-dihydroxyvitamin D_3. *The Journal of Steroid Biochemistry and Molecular Biology* 89–90:5–12.

46. Kahraman, M., S. Sinishtaj, P. M. Dolan et al. 2004. Potent, selective and low-calcemic inhibitors of CYP24 hydroxylase: 24-sulfoximine analogues of the hormone 1α,25-dihydroxyvitamin D_3. *Journal of Medicinal Chemistry* 47:6854–6863.

47. Posner, G., C. Helvig, D. Cuerrier et al. 2010. Vitamin D analogues targeting CYP24 in chronic kidney disease. *Journal of Steroid Biochemistry and Molecular Biology* 121(1–2):13–19.

48. Institute of Medicine of the National Academy of Sciences. 2011. *Dietary Reference Intakes for Vitamin D and Calcium*. The National Academies Press, Washington, DC. 1–1115.

49. Bischoff-Ferrari, H. A., E. Giovannucci, W. C. Willett et al. 2006. Estimation of optimal serum concentrations of 25-hydroxyvitamin D for multiple health outcomes. *American Journal of Clinical Nutrition* 84:18–28.

50. Bischoff-Ferrari, H. A. 2008. Optimal serum 25-hydroxy vitamin D levels for multiple health outcomes. *Advances in Experimental Medicine and Biology* 624:55–71.

51. Chapuy, M. C., M. E. Arlot, F. Duboeuf et al. 1992. Vitamin D_3 and calcium to prevent hip fractures in the elderly women. *New England Journal of Medicine* 327:1637–1642.

52. Yendt, E., K. A. Kovacs, and G. Jones. 2008. Secondary hyperparathyroidism in primary osteoporosis and osteopenia: Optimizing calcium and vitamin D intakes to levels recommended by expert panels may not be sufficient for correction. *Clinical Endocrinology* 69:855–863.

53. Bischoff-Ferrari, H. A., T. Dietrich, E. J. Orav et al. 2003. Positive association between 25-hydroxy vitamin D levels and bone mineral density: A population-based study of younger and older adults. *American Journal of Medicine* 116:634–639.

54. Dawson-Hughes, B., R. P. Heaney, M. F. Holick et al. 2005. Estimates of optimal vitamin D status. *Osteoporosis International* 16:713–716.

55. Heaney, R. P. and M. F. Holick. 2011. Why the IOM recommendations for vitamin D are deficient. *Journal of Bone and Mineral Research* 26:455–457.

56. Riggs, B. L. and L. J. Melton III. 1983. Evidence for two distinct syndromes of involutional osteoporosis. *American Journal of Medicine* 75:899–901.
57. Nordin, B. E. C. 1997. Calcium and osteoporosis. *Nutrition* 13:664–686.
58. Manolagas, S. C. and R. L. Jilka. 1995. Bone marrow, cytokines, and bone remodeling: Emerging insights into the pathophysiology of osteoporosis. *New England Journal of Medicine* 332:305–311.
59. Riggs, B. L., S. Khosla, L. J. Melton III. 1998. A unitary model for involutional osteoporosis: Estrogen deficiency causes both type I and type II osteoporosis in postmenopausal women and contributes to bone loss in aging men. *Journal of Bone and Mineral Research* 13:763–773.
60. Aubin, J. E. and E. Bonnelye. 2000. Osteoprotegerin and its ligand: A new paradigm for regulation of osteoclastogenesis and bone resorption. *Osteoporosis International* 11:905–913.
61. Teitelbaum, S. L. and F. P. Ross. 2003. Genetic regulation of osteoclast development and function. *Nature Reviews Genetics* 4:638–649.
62. Manolagas, S. C. 2010. From estrogen-centric to aging and oxidative stress: A revised perspective on the pathogenesis of osteoporosis. *Endocrine Reviews* 31:266–300.
63. Eastell, R. and B. L. Riggs. 2005. Chapter 67: Vitamin D and osteoporosis. In: *Vitamin D*, 2nd Edition. Eds: Feldman, D., W. Pike, and F. Glorieux. San Diego: Academic Press. 1101–1120.
64. Morrison, N. A., J. C. Qi, A. Tokita et al. 1994. Prediction of bone density from vitamin D receptor alleles. *Nature* 367:284–287.
65. Uitterlinden, A. G. 2011. Chapter 56: Genetics of the vitamin D endocrine system. In: *Vitamin D*, 3rd Edition. Eds: Feldman, D., W. Pike, and J. S. Adams. San Diego: Academic Press. 1025–1040.
66. Wang, T. J., F. Zhang, J. B. Richards et al. 2010. Common genetic determinants of vitamin D insufficiency: A genome-wide association study. *Lancet* 376:180–188.
67. Eberling, P. R. and J. A. Eisman. 2011. Chapter 61: Vitamin D and osteoporosis. In: *Vitamin D*, 3rd Edition. Eds: Feldman, D., W. Pike, and J. S. Adams. San Diego: Academic Press. 1129–1144.
68. Orimo, H., M. Shiraki, T. Hayashi et al. 1987. Reduced occurrence of vertebral crush fractures in senile osteoporosis treated with 1α(OH)-vitamin D_3. *Journal of Bone and Mineral Research* 3:47–52.
69. Gallagher, J. C., C. W. Bishop, J. C. Knutson et al. 1994. Effects of increasing doses of 1α-hydroxyvitamin D_2 on calcium homeostasis in postmenopausal osteopenic women. *Journal of Bone and Mineral Research* 9:607–614.
70. Gallagher, J. C., B. L. Riggs, R. R. Recker, and D. Goldgar. 1989. The effect of calcitriol on patients with postmenopausal osteoporosis with special reference to fracture frequency. *Proceedings of the Society for Experimental Biology and Medicine* 191:287–292.
71. Ott, S. and C. H. Chesnut. 1989. Calcitriol treatment is not effective in postmenopausal osteoporosis. *Annals of Internal Medicine* 110:267–274.
72. Tilyard, M. W., G. F. S. Spears, J. Thomson, and S. Dovey. 1992. Treatment of postmenopausal osteoporosis with calcium. *New England Journal of Medicine* 326:357–362.
73. Nishii, Y., K. Sato, and T. Kobayashi. 1993. The development of vitamin D analogues for the treatment of osteoporosis. *Osteoporosis International* 1(Suppl.):S190–S193.
74. Okano, T., N. Tsugawa, S. Masuda et al. 1991. A novel synthetic vitamin D_3 analogue, 2-β-(3-hydroxypropoxy)-calcitriol (ED-71): Its biological activities and pharmacological effects on calcium metabolism. *Contributions to Nephrology* 91:116–122.
75. Matsumoto, T. and N. Kubodera. 2007. ED-71, a new active vitamin D_3, increases bone mineral density regardless of serum 25(OH)D levels in osteoporotic subjects. *The Journal of Steroid Biochemistry and Molecular Biology* 103:584–586.
76. De Luca, H. F., W. Bedale, N. Binkley, J. C. Gallagher, M. Bolognese, and M. Peacock. 2011. The vitamin D analog 2-MD increase bone turnover but not BMD in postmenopausal women with osteopenia: Results of a 1-year, phase 2, double-blind, placebo-controlled, randomized clinical trial. *Journal of Bone and Mineral Research* 26:538–545.
77. St-Arnaud, R., A. Arabian, V. W. Yu et al. 2008. 1α,24(S)(OH)2D2 normalizes bone morphology and serum parathyroid hormone without hypercalcemia in 25-hydroxyvitaminD-1-hydroxylase(CYP27B1)-deficient mice: An animal model of vitamin D deficiency with secondary hyperparathyroidism. *Journal of Endocrinological Investigation* 31:711–717.
78. Moe, S. M. and T. Drueke. 2008. Kidney Disease: Improving Global Outcomes (KDIGO). *Clinical Journal of the American Society of Nephrology* 3(Suppl 3):S127–S130.
79. Delmez, J. A., C. Tindira, P. Grooms et al. 1989. Parathyroid hormone suppression by intravenous 1,25-dihydroxyvitamin D: A role for increased sensitivity to calcium. *Journal of Clinical Investigation* 83:1349–1355.

80. Brown, A. J., C. R. Ritter, J. L. Finch et al. 1989. The noncalcemic analogue of vitamin D, 22-oxacalcitriol, suppresses parathyroid hormone synthesis and secretion. *Journal of Clinical Investigation* 84:728–732.
81. Slatopolsky, E., J. Finch, C. Ritter et al. 1995. A new analog of calcitriol, 19-nor-1,25-$(OH)_2D_2$, suppresses PTH secretion in uremic rats in the absence of hypercalcemia. *Journal of Bone and Mineral Research* 10:S167 [Abstract].
82. National Kidney Foundation Disease Outcomes Quality Initiative (KDOQI). 2003. Clinical practise guidelines for bone metabolism and disease in chronic kidney disease. *American Journal of Kidney Diseases* 42(Suppl 3):S1–S202.
83. Gonzalez, E. A., A. Sachdeva, D. A. Oliver, and K. J. Martin. 2004. Vitamin D insufficiency and deficiency in chronic kidney disease: A single center observational study. *American Journal of Nephrology* 24:503–510.
84. Levin, A., G. L. Bakris, M. Molitch, M. Smulders, J. Tian, L. A. Williams, and D. L. Andress. 2007. Prevalence of abnormal serum vitamin D, PTH, calcium, and phosphorus in patients with chronic kidney disease: Results of the study to evaluate early kidney disease. *Kidney International* 71:31–38.
85. Al-Aly, Z., R. A. Qazi, E. A. González, A. Zeringue, and K. J. Martin. 2007. Changes in serum 25-hydroxy vitamin D and plasma intact PTH levels following treatment with ergocalciferol in patients with CKD. *American Journal of Kidney Diseases* 50:59–68.
86. Zisman, A. L., M. Hristova, L. T. Ho, and S. M. Sprague. 2007. Impact of ergocalciferol treatment of vitamin D deficiency on serum parathyroid hormone concentrations in chronic kidney disease. *American Journal of Nephrology* 27:36–43.
87. Dusso, A., S. Lopez-Hilker, N. Rapp, and E. Slatopolsky. 1988. Extrarenal production of calcitriol in chronic renal failure. *Kidney International* 34:368–375.
88. Jones, G. 2010. Editorial: Why dialysis patients need combination therapy with cholecalciferol and a calcitriol analog. *Seminars in Dialysis* 23:239–243.
89. Briese, S., S. Wiesner, J. C. Will et al. 2006. Arterial and cardiac disease in young adults with childhood-onset end-stage renal disease-impact of calcium and vitamin D therapy. *Nephrology Dialysis Transplantation* 21:1906–1914.
90. Fournier, A., L. Harbouche, J. Mansour, and I. Shahapuni. 2007. Impact of calcium and vitamin D therapy on arterial and cardiac disease in young adults with childhood-onset end stage renal disease. *Nephrology Dialysis Transplantation* 22:956–957.
91. Petkovich, M. P. and G. Jones. 2011. CYP24A1 and chronic kidney disease. *Current Opinion in Nephrology and Hypertension* 20:337–344.
92. Dusso, A. S. and E. Slatopolsky. 2011. Chapter 70: Vitamin D and renal disease. In: *Vitamin D*, 3rd Edition. Eds: Feldman, D., W. Pike, and J. S. Adams. San Diego: Academic Press. 1325–1357.
93. Holick, M. F. 1995. Noncalcemic actions of 1,25-dihydroxyvitamin D_3 and clinical applications. *Bone* 17:107S–111S.
94. Binderup, L. and E. Bramm. 1988. The vitamin D story: A collaborative effort of basic science and clinical medicine. *Biochemical Pharmacology* 37:889–895.
95. Kragballe, K., B. T. Gjertsen, D. De Hoop et al. 1991. Double-blind, right/left comparison of calcipotriol and betamethasone valerate in treatment of psoriasis vulgaris. *Lancet* 337:193–196.
96. Zhou, J.-Y., A. W. Norman, D. L. Chen et al. 1990. 1,25-Dihydroxy-16-ene-23-yne-vitamin D_3 prolongs survival time of leukemic mice. *Proceedings of the National Academy of Sciences USA* 1990;87:3929–3932.
97. Colston, K. W., G. Pirianov, E. Bramm et al. 2003. Effects of Seocalcitol (EB1089) on nitrosomethyl urea-induced rat mammary tumors. *Breast Cancer Research and Treatment* 80:303–311.
98. Blutt, S. E., T. C. Polek, L. V. Stewart et al. 2000. A calcitriol analogue, EB1089, inhibits the growth of LNCaP tumors in nude mice. *Cancer Research* 60:779–782.
99. Gulliford, T., J. English, K. W. Colston et al. 1998. A phase I study of the vitamin D analogue EB 1089 in patients with advanced breast and colorectal cancer. *British Journal of Cancer* 78:6–13.
100. Evans, T. R., K. W. Colston, F. J. Lofts et al. 2002. A phase II trial of the vitamin D analogue Seocalcitol (EB1089) in patients with inoperable pancreatic cancer. *British Journal of Cancer* 86:680–685.
101. Dalhoff, K., J. Dancey, L. Astrup et al. 2003. A phase II study of the vitamin D analogue, Seocalcitol in patients with inoperable hepatocellular carcinoma. *British Journal of Cancer* 89:252–257.
102. Beer, T. M., A. Myrthue, M. Garzotto. 2004. Randomized study of high-dose pulse calcitriol or placebo prior to radical prostatectomy. *Cancer Epidemiology, Biomarkers and Prevention* 13:2225–2232.
103. Trump, D. L., P. A. Hershberger, R. J. Bernardi et al. 2004. Antitumor activity of calcitriol: Preclinical and clinical studies. *The Journal of Steroid Biochemistry and Molecular Biology* 89–90:519–526.
104. Deeb, K. K., D. L. Trump, and C. S. Johnson. 2007. Vitamin D signaling pathways in cancer: Potential for anticancer therapeutics. *Nature Reviews Cancer* 7:684–700.

105. Van Etten, E., D. D. Branisteanu, A. Verstuyf et al. 2000. Analogs of 1,25-dihydroxyvitamin D_3 as dose-reducing agents for classical immunosuppressants. *Transplantation* 69:1932–1942.
106. Mathieu, C. and L. Adorini. 2002. The coming of age of 1,25-dihydroxyvitamin D3 analogs as immuno-modulatory agents. *Trends in Molecular Medicine* 8:174–179.
107. Baeke, F., E. V. Etten, L. Overbergh, and C. Mathieu. 2007. Vitamin D3 and the immune system: Maintaining the balance in health and disease. *Nutrition Research Reviews* 20:106–118.
108. Wang, T. T., F. P. Nestel, V. Bourdeau et al. 2004. Cutting edge: 1,25-dihydroxyvitamin D3 is a direct inducer of antimicrobial peptide gene expression. *The Journal of Immunology* 173:2909–2912.
109. Liu, P. T., S. Stenger, H. Li et al. 2006. Toll-like receptor triggering of a vitamin D-mediated human antimicrobial response. *Science* 311:1770–1773.
110. Adams, J. S. and M. Hewison. 2010. Update in vitamin D. *The Journal of Clinical Endocrinology & Metabolism* 95:471–478.
111. Lemire, J. M. and C. D. Archer. 1991. 1,25-Dihydroxyvitamin D_3 prevents the *in vivo* induction of murine experimental autoimmune encephalomyelitis. *Journal of Clinical Investigation* 87:1103–1107.
112. Mathieu, C., M. Waer, K. Casteels et al. 1995. Prevention of type I diabetes in NOD mice by nonhypercalce-mic doses of a new structural analog of 1,25-dihydroxyvitamin D_3, KH1060. *Endocrinology* 136:866–872.
113. Mathieu, C., R. Bouillon, O. Rutgeerts et al. 1994. Potential role of 1,25(OH)$_2$ vitamin D_3 as a dose-reducing agent for cyclosporine and FK 506. *Transplantation Proceedings* 26:3130.
114. Mathieu, C., J. Laureys, M. Waer, and R. Bouillon. 1994. Prevention of autoimmune destruction of trans-planted islets in spontaneously diabetic NOD mice by KH1060, a 20-epi analog of vitamin D: Synergy with cyclosporine. *Transplantation Proceedings* 26:3128–3129.
115. Veyron, P., R. Pamphile, L. Binderup et al. 1993. Two novel vitamin D analogues, KH 1060 and CB 966, prolong skin allograft survival in mice. *Transplant Immunology* 1:72–76.
116. Li, Y. C., J. Kong, M. Wei et al. 2002. 1,25-Dihydroxy vitamin D3 is a negative endocrine regulator of the renin-angiotens in system. *Journal of Clinical Investigation* 110:229–238.
117. Xiang, W., J. Kong, S. Chen et al. 2005. Cardiac hypertrophy in vitamin D receptor knockout mice: Role of the systemic and cardiacrenin-angiotens in systems. *American Journal of Physiology—Endocrinology and Metabolism* 288:E125–32.
118. Forman, J. P., E. Giovannucci, M. D. Holmes et al. 2007. Plasma 25-hydroxy vitamin D levels and risk of incident hypertension. *Hypertension* 49:1063–1069.
119. Levin, A. and Y. C. Li. 2005. Vitamin D and its analogues: Do they protect against cardiovascular disease in patients with kidney disease? *Kidney International* 68:1973–1981.
120. Mizobuchi, M., J. L. Finch, D. R. Martin et al. 2007. Differential effects of vitamin D receptor activators on vascular calcification in uremic rats. *Kidney International* 72:709–715.
121. Mathew, S., R. J. Lund, L. R. Chaudhary et al. 2008. Vitamin D receptor activators can protect against vascular calcification. *Journal of the American Society of Nephrology* 19:1509–1519.
122. Judd, S. E. and V. Tangpricha. 2009. Vitamin D deficiency and risk for cardiovascular disease. *The American Journal of the Medical Sciences* 338:40–44.
123. Teng, M., M. Wolf, M. N. Ofsthun et al. 2005. Activated injectable vitamin D and hemodialysis survival: A historical cohort study. *Journal of the American Society of Nephrology* 16:1115–1125.
124. Tentori, F., W. C. Hunt, C. A. Stidley et al. 2006. Mortality risk among hemodialysis patients receiving different vitamin D analogs. *Kidney International* 70:1858–1865.
125. Wang, T. J., M. J. Pencina, S. L. Booth et al. 2008. Vitamin D deficiency and risk of cardiovascular dis-ease. *Circulation* 117:503–511.
126. Uchida, M., K. Ozono, and J. W. Pike. 1994. Activation of the human osteocalcin gene by 24R,25-dihydroxyvitamin D_3 occurs through the vitamin D receptor and the vitamin D-responsive element. *Journal of Bone and Mineral Research* 9:1981–1987.
127. Jones, G. 2008. Pharmacokinetics of vitamin D toxicity. *The American Journal of Clinical Nutrition* 88(Suppl):582S–586S.
128. De Luca, H. F., J. M. Prahl, and L. A. Plum. 2011. 1,25-Dihydroxy vitamin D is not responsible for toxicity caused by vitamin D or 25-hydroxy vitamin D. *Archives of Biochemistry and Biophysics* 505:226–230.
129. Okuda, K. I., E. Usui, and Y. Ohyama. 1995. Recent progress in enzymology and molecular biology of enzymes involved in vitamin D metabolism. *The Journal of Lipid Research* 36:1641–1652.
130. Cheng, J. B., M. A. Levine, N. H. Bell et al. 2004. Genetic evidence that the human CYP2R1 enzyme is a key vitamin D 25-hydroxylase. *Proceedings of the National Academy of Sciences USA* 101:7711–7715.
131. Jones, G., V. Byford, S. West et al. 2006. Hepatic activation and inactivation of clinically relevant vitamin D analogs and prodrugs. *Anticancer Research* 26:2589–2596.

132. Strushkevich, N., S. A. Usanov, A. N. Plotnikov, G. Jones, and H.-W. Park. 2008. Structural Analysis of CYP2R1 in complex with vitamin D_3. *Journal of Molecular Biology* 380:95–106.

133. Guo, Y.-D., S. Strugnell, D. W. Back et al. 1993. Transfected human liver cytochrome P-450 hydroxylates vitamin D analogs at different side-chain positions. *Proceedings of the National Academy of Sciences USA* 90:8668–8672.

134. Ichikawa, F., K. Sato, M. Nanjo et al. 1995. Mouse primary osteoblasts express vitamin D_3 25-hydroxylase mRNA and convert 1α-hydroxyvitamin D_3 into 1α,25-dihydroxyvitamin D_3. *Bone* 16:129–135.

135. Masuda, S., S. Strugnell, J. C. Knutson et al. 2006. Evidence for the activation of 1α-hydroxyvitamin D_2 by 25-hydroxyvitaminD-24-hydroxylase: Delineation of pathways involving 1α,24-dihydroxyvitamin D_2 & 1α,25-dihydroxyvitamin D_2. *Biochimica et Biophysica Acta* 1761:221–234.

136. Adams, J. S. and M. A. Gacad. 1985. Characterization of 1α-hydroxylation of vitamin D_3 sterols by cultured alveolar macrophages from patients with sarcoidosis. *The Journal of Experimental Medicine* 161:755–765.

137. St. Arnaud, R., S. Messerlian, J. M. Moir et al. 1997. The 25-hydroxyvitamin D 1-α-hydroxylase gene maps to the pseudovitamin D-deficiency rickets (PDDR) disease locus. *Journal of Bone and Mineral Research* 12:1552–1559.

138. Takeyama, K., S. Kitanaka, T. Sato et al. 1997. 25-Hydroxyvitamin D_3 1α-hydroxylase and vitamin D synthesis. *Science* 277:1827–1830.

139. Fu, G. K., D. Lin, M. Y. Zhang et al. 1997. Cloning of human 25-hydroxyvitamin D-1α-hydroxylase and mutations causing vitamin D-dependent rickets type 1. *Molecular Endocrinology* 11:1961–1970.

140. Jones, G., H. Ramshaw, A. Zhang et al. 1999. Expression and activity of vitamin D-metabolizing cytochrome P450s (CYP1α and CYP24) in human non-small-cell lung carcinomas. *Endocrinology* 140:3303–3310.

141. Hewison, M. and J. Adams. 2011. Chapter 45: Extrarenal 1α-hydroxylase. In: *Vitamin D*, 3rd Edition. Eds: Feldman, D., W. Pike, J. S. Adams. San Diego: Academic Press. 777–804.

142. Bouillon, R., K. Allewaert, D. Z. Xiang et al. 1991. Vitamin D analogs with low affinity for the vitamin D binding protein: Enhanced *in vitro* and decreased *in vivo* activity. *Journal of Bone and Mineral Research* 6:1051–1057.

143. Kissmeyer, A.-M., I. S. Mathiasen, S. Latini et al. 1995. Pharmacokinetic studies of vitamin D, analogues: Relationship to vitamin D binding protein (DBP). *Endocrine* 3:263–266.

144. Tsugawa, N., T. Okano, S. Masuda et al. 1991. A novel vitamin D_3 analogue, 22-oxacalcitriol (OCT): Its different behavior from calcitriol in plasma transport system. In: *Vitamin D: Gene Regulation Structure-Function Analysis and Clinical Application*. Eds: Norman, A. W., R. Bouillon, M. Thomasset. Berlin: De Gruyter. 312–313.

145. Dilworth, F. J., M. J. Calverley, H. L. J. Makin, and G. Jones. 1994. Increased biological activity of 20-epi-1,25-dihydroxyvitamin D_3 is due to reduced catabolism and altered protein binding. *Biochemical Pharmacology* 47:987–993.

146. Safadi, F. F., P. Thornton, H. Magiera et al. 1999. Osteopathy and resistance to vitamin D toxicity in mice null for vitamin D binding protein. *Journal of Clinical Investigation* 103:239–251.

147. Rochel, N., G. Tocchini-Valentini, P. F. Egea et al. 2001. Functional and structural characterization of the insertion region in the ligand binding domain of the vitamin D nuclear receptor. *European Journal of Biochemistry* 268:971–979.

148. Rochel, N. and D. Moras. 2011. Chapter 9: Structural basis or ligand activity in VDR. In: *Vitamin D*, 3rd Edition. Eds: Feldman, D., W. Pike, and J. S. Adams. San Diego: Academic Press. 171–191.

149. Rachez, C. and L. P. Freedman. 2000. Mechanisms of gene regulation by vitamin D_3 receptor: A network of coactivator interactions. *Gene* 246:9–21.

150. Kato, S. 2000. Molecular mechanism of transcriptional control by nuclear vitamin receptors. *British Journal of Nutrition* 84:229–233.

151. Bouillon, R., G. Carmeliet, L. Verlinden, E. van Etten, A. Verstuyf, H. F. Luderer, L. Lieben, C. Mathieu, and M. Demay. 2008. Vitamin D and human health: Lessons from vitamin D receptor null mice. *Endocrine Reviews* 29:726–776.

152. Hendy, G. N., R. Kremer, and D. Goltzman. 2011. Chapter 33: Contributions of genetically modified mouse models to understanding the physiology and pathophysiology of the 25-hydroxyvitamin D-1α-hydroxylase (1α-OH-ase) and the vitamin D receptor (VDR). In: *Vitamin D*, 3rd Edition. Eds: Feldman, D., W. Pike, and J. S. Adams. San Diego: Academic Press. 583–607.

153. Stern, P. 1981. A monolog on analogs: *In vitro* effects of vitamin D metabolites and consideration of the mineralisation question. *Calcified Tissue International* 33:1–4.

154. Yang, W. and L. P. Freedman. 1999. 20-Epi analogues of 1,25-dihydroxyvitamin D_3 are highly potent inducers of DRIP coactivator complex binding to the vitamin D_3 receptor. *Journal of Biological Chemistry* 274:16838–16845.

155. Dilworth, F. J., G. R. Williams, A.-M. Kissmeyer et al. 1997. The vitamin D analog, KH1060 is rapidly degraded both *in vivo* and *in vitro* via several pathways: Principal metabolites generated retain significant biological activity. *Endocrinology* 138:5485–5496.
156. Zella, L. A., M. B. Meyer, R. D. Nerenz, and J. W. Pike. 2009. The enhanced hypercalcemic response to 20-epi-1,25-dihydroxy vitamin D3 results from a selective and prolonged induction of intestinal calcium-regulating genes. *Endocrinology* 150:3448–3456.
157. Cheskis, B., B. D. Lemon, M. Uskokovic, P. T. Lomedico, and L. P. Freedman. 1995. Vitamin D_3-retinoid X receptor dimerization, DNA binding, and transactivation are differentially affected by analogs of 1,25-dihydroxyvitamin D_3. *Molecular Endocrinology* 9:1814–1824.
158. Rachez, C., B. D. Lemon, Z. Suldan et al. 1999. Ligand-dependent transcription activation by nuclear receptors requires the DRIP complex. *Nature* 398:824–828.
159. Takeyama, K., Y. Masuhiro, H. Fuse et al. 1999. Selective interaction of vitamin D receptor with transcriptional coactivators by a vitamin D analog. *Molecular and Cellular Biology* 19:1049–1055.
160. Issa, L. L., G. M. Leong, R. L. Sutherland, and J. A. Eisman. 2000. Vitamin D analogue-specific recruitment of vitamin D receptor coactivators. *Journal of Bone and Mineral Research* 17:879–890.
161. Peleg, S., A. Ismail, M. R. Uskokovic, and Z. Avnur. 2003. Evidence for tissue- and cell-type selective activation of the vitamin D receptor by Ro-26-9228, a noncalcemic analog of vitamin D_3. *Journal of Cellular Biochemistry* 88:267–273.
162. Pike, J. W., L. A. Zella, M. B. Meyer et al. 2007. Molecular actions of 1,25-dihydroxy vitamin D3 on genes involved in calcium homeostasis. *Journal of Bone and Mineral Research* 22 Suppl 2:V16-9.
163. Makin, G., D. Lohnes, V. Byford, R. Ray, and G. Jones. 1989. Target cell metabolism of 1,25-dihydroxyvitamin D_3 to calcitroic acid: Evidence for a pathway in kidney and bone involving 24-oxidation. *Biochemical Journal* 262:173–180.
164. Akiyoshi-Shibata, M., T. Sakaki, Y. Ohyama et al. 1994. Further oxidation of hydroxycalcidiol by calcidiol 24-hydroxylase: A study with the mature enzyme expressed in *Escherichia coli*. *European Journal of Biochemistry* 224:335–343.
165. Lohnes, D. and G. Jones. 1992. Further metabolism of 1α,25-dihydroxyvitamin D_3 in target cells. *Journal of Nutritional Science and Vitaminology*, Special Issue:75–78.
166. St. Arnaud, R. 1991. Targeted inactivation of vitamin D hydroxylases in mice. *Bone* 25:127–129.
167. Masuda. S., V. Byford, A. Arabian et al. 2005. Altered pharmacokinetics of 1α,25-dihydroxyvitamin D_3 and 25-hydroxyvitamin D_3 in the blood and tissues of the 25-hydroxyvitamin D-24-hydroxylase (CYP24A1) null mouse. *Endocrinology* 146:825–834.
168. Hock, J. M., M. Gunness-Hey, J. Poser et al. 1986. Stimulation of undermineralized matrix formation by 1,25-dihydroxyvitamin D_3 in long bones of rats. *Calcified Tissue International* 38:79–86.
169. St. Arnaud, R., A. Arabian, R. Travers et al. 2000. Deficient mineralization of intramembranous bone in vitamin D-24-hydroxylase-ablated mice is due to elevated 1,25-dihydroxyvitamin D and not to the absence of 24,25-dihydroxyvitamin D. *Endocrinology* 141:2658–2666.
170. Schlingmann, K. P., M. Kaufmann, S. Weber, A. Irwin, C. Goos, A. Wassmuth, U. John, J. Misselwitz, G. Klaus, E. Kuwertz-Broking, H. Fehrenbach, A. M. Wingen, T. Guran, T. Akcay, J. G. Hoenderop, R. J. Bindels, D. E. Prosser, G. Jones, and M. Konrad. 2011. Mutations of CYP24A1 and idiopathic infantile hypercalcemia. *The New England Journal of Medicine* 365:410–421.
171. Masuda, S., S. Strugnell, M. J. Calverley et al. 1994. *In vitro* metabolism of the antisporiatic vitamin D analog, calcipotriol, in two cultured human keratinocyte models. *Journal of Biological Chemistry* 269:4794–4803.
172. Masuda, S., V. Byford, R. Kremer et al. 1996. *In vitro* metabolism of the vitamin D analog, 22-oxacalcitriol, using cultured osteosarcoma, hepatoma and keratinocyte cell lines. *Journal of Biological Chemistry* 271:8700–8708.
173. Shankar, V. N., F. J. Dilworth, H. L. J. Makin et al. 1997. Metabolism of the vitamin D analog EB1089 by cultured human cells: Redirection of hydroxylation site to distal carbons of the side chain. *Biochemical Pharmacology* 53:783–793.
174. Sasaki, H., H. Harada, Y. Handa et al. 1995. Transcriptional activity of a fluorinated vitamin D analog on VDR-RXR-mediated gene expression. *Biochemistry* 34:370–377.
175. Miyamoto, Y., T. Shinki, K. Yamamoto et al. 1997. 1α,25-Dihydroxyvitamin D_3-24-hydroxylase (CYP24) hydroxylates the carbon at the end of the side chain (C-26) of the C-24-fluorinated analog of 1α,25-dihydroxyvitamin D_3. *Journal of Biological Chemistry* 272:14115–14119.
176. Satchell, D. P. and A. W. Norman. 1995. Metabolism of the cell differentiating agent 1α,25(OH)$_2$-16-one-23-yne vitamin D_3 by leukemic cells. *The Journal of Steroid Biochemistry and Molecular Biology* 57:117–124.

177. Rao, D. S., M. L. Siu-Caldera, M. R. Uskokovic et al. 1999. Physiological significance of C-28 hydroxylation in the metabolism of 1α,25-dihydroxyvitamin D₂. *Archives of Biochemistry and Biophysics* 368:319–328.
178. Shankar, V. N., A. E. Propp, N. S. Schroeder et al. 2001. *In vitro* metabolism of 19-nor-1α,25-(OH)₂D₃ in cultured cell lines: Inducible synthesis of lipid- and water-soluble metabolites. *Archives of Biochemistry and Biophysics* 387:297–306.
179. Jones, G. 2010. Vitamin D analogs. *Endocrinology and Metabolism Clinics of North America* 39:447–472.
180. Siu-Caldera, M. L., J. W. Clark, A. Santos-Moore et al. 1996. 1α,25-dihydroxy-24-oxo-16-ene vitamin D3, a metabolite of a synthetic vitamin D3 analog, 1α,25-dihydroxy-16-ene vitamin D₃, is equipotent to its parent in modulating growth and differentiation of human leukemic cells. *The Journal of Steroid Biochemistry and Molecular Biology* 59(5-6):405–412.
181. Swami, S., X. Y. Zhao, S. Sarabia et al. 2003. A low-calcemic vitamin D analog (Ro25-4020) inhibits the growth of LNCaP human prostate cancer cells with increased potency by producing an active 24-oxo metabolite (Ro29-9970). *Recent Results in Cancer Research* 164:349–352.
182. Siu-Caldera, M. L., H. Sekimoto, S. Peleg et al. 1999. Enhanced biological activity of 1α,25-dihydroxy-20-epi-vitamin D₃, the C-20 epimer of 1α,25-dihydroxyvitamin D₃, is in part due to its metabolism into stable intermediary metabolites with significant biological activity. *The Journal of Steroid Biochemistry and Molecular Biology* 71:111–121.
183. Shankar, V. N., V. Byford, D. E. Prosser et al. 2001. Metabolism of a 20-methyl substituted series of vitamin D analogs by cultured human cells: Apparent reduction of 23-hydroxylation of the side chain by 20-methyl group. *Biochemical Pharmacology* 61:893–902.
184. Sorensen, H., L. Binderup, M. J. Calverley et al. 1990. *In vitro* metabolism of calcipotriol (MC 903), a vitamin D analogue. *Biochemical Pharmacology* 39:391–393.
185. Kissmeyer, A.-M., E. Binderup, L. Binderup et al. 1997. The metabolism of the vitamin D analog EB 1089: Identification of *in vivo* and *in vitro* metabolites and their biological activities. *Biochemical Pharmacology* 53:1087–1097.
186. Rastrup-Anderson, N., F. A. Buchwald, and G. Grue-Sorensen. 1992. Identification and synthesis of a metabolite of KH1060, a new potent 1α,25-dihydroxyvitamin D₃ analogue. *Bioorganic & Medicinal Chemistry Letters* 2:1713–1716.
187. Xu, Y., T. Hashizume, M. C. Shuhart et al. 2005. Intestinal and hepatic CYP3A4 catalyze hydroxylation of 1α,25-dihydroxyvitamin D₃: Implications for drug-induced osteomalacia. *Molecular Pharmacology* 69:56–65.
188. Thummel, K. E., C. Brimer, K. Yasuda et al. 2001. Transcriptional control of intestinal cytochrome P-450 3A by 1α,25-dihydroxyvitamin D₃. *Molecular Pharmacology* 60:1399–1406.
189. Onodera, K., A. Takahashi, H. Mayanagi et al. 2001. Phenytoin-induced bone loss and its prevention with alfacalcidol or calcitriol in growing rats. *Calcified Tissue International* 69:109–116.
190. Gascon-Barre, M., J. P. Villeneuve, and L. H. Lebrun. 1984. Effect of increasing doses of phenytoin on the plasma 25-hydroxyvitamin D and 1,25-dihydroxyvitamin D concentrations. *Journal of the American College of Nutrition* 3:45–50.
191. Helvig, C., D. Cuerrier, A. Kharebov et al. 2008. Comparison of 1,25-dihydroxyvitamin D₃ and calcitriol effects in an adenine-induced uremic model of CKD reveals differential control over calcium and phosphate. *American Society for Bone and Mineral Research* 23:S357 (Abstract).
192. Kobayashi, T., T. Okano, N. Tsugawa et al. 1991. Metabolism and transporting system of 22-oxacalcitriol. *Contributions to Nephrology* 91:129–133.
193. Kasai, N., T. Sakaki, R. Shinkyo et al. 2005. Metabolism of 26,26,26,27,27,27-F₆-1α,23S,25-trihydroxyvitamin D₃ by human UDP-glucuronosyltransferase 1A3. *Drug Metabolism and Disposition* 33:102–107.
194. Bischof, M. G., M. L. Siu-Caldera, A. Weiskopf, P. Vouros, H. S. Cross, M. Peterlik, and G. S. Reddy. 1998. Differentiation-related pathways of 1alpha,25-dihydroxychole calciferol metabolism in human colonadeno carcinoma-derived Caco-2cells: Production of 1alpha,25-dihydroxy-3epi-cholecalciferol. *Experimental Cell Research* 241:194–201.
195. Singh, R. J., R. L. Taylor, G. S. Reddy, and S. K. Grebe. 2006. C-3 epimer scan account for a significant proportion of total circulating 25-hydroxyvitamin D in infants, complicating accurate measurement and interpretation of vitamin D status. *The Journal of Clinical Endocrinology & Metabolism* 91:3055–3061.
196. Miller, B. E., D. P. Chin, and G. Jones. 1990. 1,25-dihydroxyvitamin D₃ metabolism in a human osteosarcoma cell line and human bone cells. *Journal of Bone and Mineral Research* 5:597–607.

197. Siu-Caldera, M. L., H. Sekimoto, A. Weiskopf, P. Vouros, K. R. Muralidharan, W. H. Okamura, J. Bishop, A. W. Norman, M. R. Uskoković, I. Schuster, and G. S. Reddy. 1999. Production of $1\alpha,25$-dihydroxy-3-epi-vitamin D_3 in two rat osteosarcoma cell lines (UMR106andROS17/2.8):existence of the C-3 epimerization pathway in ROS17/2.8 cells in which the C-24 oxidation pathway is not expressed. *Bone* 24:457–463.

198. Norman, A. W., I. Nemere, L. Zhou et al. 1992. $1,25(OH)_2$-vitamin D_3, a steroid hormone that produces biological effects via both genomic and nongenomic pathways. *The Journal of Steroid Biochemistry and Molecular Biology* 41:231–240.

199. Norman, A. W. 2011. Chapter 15: Vitamin D sterol/VDR conformational dynamics and nongenomic responses. In: *Vitamin D*, 3rd Edition. Eds: Feldman, D., W. Pike, and J. S. Adams. San Diego: Academic Press. 271–300.

200. Nemere, I., M. C. Dormanen, M. W. Hammond et al. 1994. Identification of a specific binding protein for $1\alpha,25$-dihydroxyvitamin D_3 in basal-lateral membranes of chick intestinal epithelium and relationship to transcaltachia. *Journal of Biological Chemistry* 269:23750–23756.

201. Baran, D. T., J. M. Quail, R. Ray et al. 2000. Annexin II is the membrane receptor that mediates the rapid actions of $1\alpha,25$-dihydroxyvitamin D_3. *Journal of Cellular Biochemistry* 78:34–46.

202. Rohe, B., S. E. Safford, I. Nemere, and M. C. Farach-Carson. 2005. Identification and characterization of 1,25D3-membrane-associated rapid response, steroid (1,25D3-MARRS)-binding protein in rat IEC-6 cells. *Steroids* 70:458–463.

203. Chen, J., R. Olivares-Navarrete, Y. Wang, T. R. Herman, B. D. Boyan, and Z. Schwartz. 2010. Protein-disulfide isomerase-associated3 (Pdia3) mediates the membrane response to 1,25-dihydroxy vitamin D_3 in osteoblasts. *Journal of Biological Chemistry* 285:37041–37050.

204. Annalora, A. J., D. B. Goodin, W. X. Hong, Q. Zhang, E. F. Johnson, and C. D. Stout. 2010. Crystal structure of CYP24A1, amitochondrial cytochrome P450 involved in vitamin D metabolism. *Journal of Molecular Biology* 396:441–451.

205. Prosser, D. E., Y.-D. Guo, K. R. Geh et al. 2006. Molecular modeling of CYP27A1 and site-directed mutational analyses affecting vitamin D hydroxylation. *Biophysical Journal* 90:1–21.

206. Hamamoto, H., T. Kusudo, N. Urushino et al. 2006. Structure-function analysis of vitamin D 24-hydroxylase (CYP24A1) by site-directed mutagenesis: Amino acid residues responsible for species-based difference of CYP24A1 between humans and rats. *Molecular Pharmacology* 70:120–128.

207. Masuda, S., D. Prosser, Y.-D. Guo et al. 2007. Generation of a homology model for the human cytochrome P450, CYP24A1, and the testing of putative substrate binding residues by site-directed mutagenesis and enzyme activity studies. *Archives of Biochemistry and Biophysics* 460:177–191.

208. Prosser, D., M. Kaufmann, B. O'Leary et al. 2007. Single A326G mutation converts hCYP24A1 from a 25-OH-D_3-24-hydroxylase into -23-hydroxylase generating $1\alpha,25$-$(OH)_2D_3$-26,23-lactone. *Proceedings of the National Academy of Sciences USA* 104:12673–12678.

209. Rhieu, S. Y., A. J. Annalora, R. M. Gathungu, P. Vouros, M. R. Uskokovic, I. Schuster, G. T. Palmore, and G. S. Reddy. 2011. A new insight into the role of rat cytochrome P45024A1 in metabolism of selective analogs of $1\alpha,25$-dihydroxy vitamin D_3. *Archives of Biochemistry and Biophysics* 509:33–43.

210. Schuster, I., H. Egger, P. Nussbaumer, and R. T. Kroemer. 2003. Inhibitors of vitamin D hydroxylases: Structure-activity relationships. *Journal of Cellular Biochemistry* 88:372–380.

211. Muralidharan, K. R., M. Rowland-Goldsmith, A. S. Lee et al. 1997. Inhibitors of 25-hydroxyvitamin D_3-1α-hydroxylase: Thiavitamin D analogs and biological evaluation. *The Journal of Steroid Biochemistry and Molecular Biology* 62:73–78.

212. Studzinski, G. P., E. Gocek, and M. Danilenko. 2011. Chapter 84: Vitamin D Effects on Differentiation and the Cell Cycle. In: *Vitamin D*, 3rd Edition. Eds.: Feldman, D., W. Pike, and J. S. Adams. San Diego: Academic Press. 1625–1656.

213. Mellanby, E. and M. D. Cantag. 1919. Experimental investigation on rickets. *Lancet* 196:407–412.

214. McCollum, E. V., N. Simmonds, J. E. Becker, and P. G. Shipley. 1922. Studies on experimental rickets. XXI. An experimental demonstration of the existence of a vitamin which promotes calcium deposition. *Journal of Biological Chemistry* 53:293–312.

215. Blunt, J. W., H. F. De Luca, and H. K. Schnoes. 1968. 25-Hydroxycholecalciferol. A biologically active metabolite of vitamin D_3. *Biochemistry* 7:3317–3322.

216. Fraser, D. R. and E. Kodicek. 1970. Unique biosynthesis by kidney of a biologically active vitamin D metabolite. *Nature* 228:764–766.

217. Holick, M. F., H. K. Schnoes, H. F. De Luca et al. 1971. Isolation and identification of 1,25-dihydroxycholecalciferol: A metabolite of vitamin D active in intestine. *Biochemistry* 10:2799–2804.

218. Holick, M. F., H. K. Schnoes, H. F. De Luca et al. 1972. Isolation and identification of 24,25-dihydroxy-cholecalciferol: A metabolite of vitamin D_3 made in the kidney. *Biochemistry* 11:4251–4255.
219. Holick, M. F., A. Kleiner-Bossaller, H. K. Schnoes et al. 1973. 1,24,25-Trihydroxyvitamin D_3. A metabolite of vitamin D_3 effective on intestine. *Journal of Biological Chemistry* 248:6691–6696.
220. Suda, T., H. F. De Luca, H. K. Schnoes et al. 1970. 25,26-dihydroxyvitamin D_3, a metabolite of vitamin D_3 with intestinal transport activity. *Biochemistry* 9:4776–4780.
221. Horst, R. L. 1979. *Biochemical and Biophysical Research Communications* 89:286–293.
222. Baggiolini, E. G., P. M. Wovkulich, J. A. Iacobelli et al. 1982. Preparation of 1-alpha hydroxylated vitamin D metabolites by total synthesis. In: *Vitamin D: Chemical, Biochemical and Clinical Endocrinology of Calcium Metabolism.* Eds.: Norman, A. W., K. Schaefer, D. von Herrath, H.-G. Grigoleit. Berlin: De Gruyter. 1089–1100.
223. Kobayashi, Y., T. Taguchi, S. Mitsuhashi et al. 1982. Studies on organic fluorine compounds—XXXIX: Studies on steroids. LXXIX: Synthesis of 1α,25-dihydroxy-26,26,26,27,27,27-hexaflurovitamin D_3. *Chemical and Pharmaceutical Bulletin (Tokyo)* 30:4297–4303.
224. Perlman, K. L., R. R. Sicinski, H. K. Schnoes, H. F. De Luca. 1990. 1α,25-Dihydroxy-19-nor-vitamin D_3, a novel vitamin D-related compound with potential therapeutic activity. *Tetrahedron Letters* 31:1823–1824.
225. Murayama, E., K. Miyamoto, N. Kubodera et al. 1986. Synthetic studies of vitamin D analogues—Part VIII: Synthesis of 22-oxavitamin D_3 analogues. *Chemical and Pharmaceutical Bulletin (Tokyo)* 34:4410–4413.
226. Calverley, M. J. 1987. Synthesis of MC-903, a biologically active vitamin D metabolite analog. *Tetrahedron* 43:4609–4619.
227. Binderup, E., M. J. Calverley, and L. Binderup. 1991. Synthesis and biological activity of 1α-hydroxylated vitamin D analogues with polyunsaturated side chains. In: *Vitamin D: Proceedings of the Eighth Workshop on Vitamin D*, Paris, France. Eds.: Norman, A. W., R. Bouillon, and M. Thomasset. De Gruyter: Berlin. 192–193.
228. Calverley, M. J., E. Binderup, and L. Binderup. 1991. The 20-epi modification in the vitamin D series: Selective enhancement of "nonclassical" receptor-mediated effects. In: *Vitamin D: Proceedings of the Eighth Workshop on Vitamin D*, Paris, France. Eds.: Norman, A. W., R. Bouillon, and M. Thomasset. Berlin: De Gruyter. 163–164.
229. Marchiani, S., L. Bonaccorsi, P. Ferruzzi et al. 2006. The vitamin D analogue BXL-628 inhibits growth factor-stimulated proliferation and invasion of DU145 prostate cancer cells. *Journal of Cancer Research and Clinical Oncology* 132:408–416.
230. Morisaki, M., N. Koizumi, N. Ikekawa et al. 1975. Synthesis of active forms of vitamin D. Part IX. Synthesis of 1α,24-dihydroxycholecalciferol. *Journal of the Chemical Society—Perkins Transactions* 1(1):1421–1424.
231. Ochiai, E., D. Miura, H. Eguchi et al. 2005. Molecular mechanism of the vitamin D antagonistic actions of (23S)-25-dehydro-1alpha-hydroxyvitamin D_3-26,23-lactone depends on the primary structure of the carboxyl-terminal region of the vitamin D receptor. *Molecular Endocrinology* 19:1147–1157.
232. Toell, A., M. M. Gonzalez, D. Ruf et al. 2001. Different molecular mechanisms of vitamin D_3 receptor antagonists. *Molecular Pharmacology* 59:1478–1485.
233. Bishop, J. E., E. D. Collins, W. H. Okamura, and A. W. Norman. 1994. Profile of ligand specificity of the vitamin D binding protein for 1α,25-dihydroxyvitamin D_3 and its analogs. *Journal of Bone and Mineral Research* 9:1277–1288.

4 Extrarenal CYP27B1 and Vitamin D Physiology

Martin Hewison

CONTENTS

4.1 INTRODUCTION

In the last 10 years, there has been a remarkable renaissance in vitamin D research. Two key concepts have underpinned this renewed interest in the health benefits of vitamin D. First is the continuing debate on the worldwide prevalence of vitamin D insufficiency [1] and how optimal vitamin D status can be safely achieved through conventional exposure to sunlight and dietary intake [2]. Second is the potential for vitamin D to promote health benefits beyond its classical effects on the skeleton [3–6]. Following a recent data review, the Institute of Medicine (IOM) has issued statements aimed at addressing some of the key questions concerning our new perspective on vitamin D and human health [7]. The Recommended Dietary Allowance of vitamin D for all age groups has been elevated based on bone responses to vitamin D. However, the IOM report also recognized the need for further research to better define other "nonclassical" health benefits of vitamin D. The latter reflects the accumulation of recent data describing anticancer, immunomodulatory, angiogenic, and antihypertensive actions of vitamin D [3,5,8,9]. Central to this new perspective on vitamin D has been its proposed localized actions, with target cells expressing both the intracellular vitamin D receptor (VDR) and the enzyme that synthesizes the active form of vitamin D, i.e., 1,25-hydroxyvitamin D (1,25(OH)$_2$D), from precursor 25-hydroxyvitamin D (25OHD). In classical vitamin D endocrinology, the vitamin activating enzyme 25-hydroxyvitamin D-1α-hydroxylase (1α-hydroxylase) is expressed predominantly in the kidneys, where it acts to support systemic levels of 1,25(OH)$_2$D. However, in extrarenal tissues, the enzyme appears to act at a more localized level promoting nonclassical actions in a site-specific manner. In this setting, it is proposed that biological responses to vitamin D will be more dependent on the availability of substrate 25OHD for 1α-hydroxylase or, in other words, serum vitamin D status. Thus, extrarenal 1α-hydroxylase may be a pivotal factor in defining the impact of vitamin D on the broader features of human health. The following chapter compares the basic biochemistry and physiology of vitamin D metabolism in extrarenal tissues, with emphasis on the differential regulation of this enzyme relative to renal 1α-hydroxylase and the possible biological impact of the enzyme on diverse peripheral tissues.

4.2 RENAL EXPRESSION OF 1α-HYDROXYLASE AND CLASSICAL VITAMIN D PHYSIOLOGY

In classical vitamin D endocrinology, the actions of vitamin D are mediated, following renal generation of 1,25(OH)$_2$D and its subsequent action on distal tissues expressing the VDR. Synthesis of 1,25(OH)$_2$D appears to occur specifically in the epithelial cells of the proximal tubule [10], although 1α-hydroxylase expression [11] and activity [12] have been described in cells from the distal part of the nephron. Expression of renal 1α-hydroxylase and associated synthesis of 1,25(OH)$_2$D is conventionally stimulated by parathyroid hormone (PTH), which stimulates transcription of the gene for 1α-hydroxylase (CYP27B1) under conditions of low extracellular calcium (see Figure 4.1). The

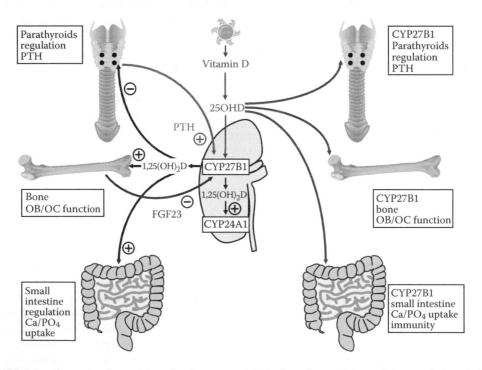

FIGURE 4.1   (See color insert.) Renal and extrarenal 1α-hydroxylase activity and the regulation of classical vitamin D function. The schematic shows the renal conversion of 25OHD to 1,25(OH)$_2$D via the enzyme 1α-hydroxylase, its subsequent systemic effects, and its regulation by endocrine factors such as PTH, fibroblast growth factor 23 (FGF23), and 24-hydroxylase (CYP24A1). These actions depicted on the left-hand side of the schematic are contrasted with proposed extrarenal actions of vitamin D, mediated via target tissue expression of CYP27B1 and localized intracrine synthesis of 1,25(OH)$_2$D from 25OHD.

1,25(OH)$_2$D produced by renal 1α-hydroxylase activity then enters the circulation and acts as classical steroid hormone, enhancing calcium and phosphate uptake by the gastrointestinal (GI) tract and promoting skeletal homeostasis through effects on cells such as bone-forming osteoblasts [13]. The efficacy of this system is dependent first on the expression of functional receptors for 1,25(OH)$_2$D (VDR) in target intestinal, bone, and parathyroid cells but also requires sufficient 1,25(OH)$_2$D to promote VDR signaling [14]. Once calcium homeostasis has been reestablished, 1,25(OH)$_2$D is then able to complete feedback regulation of this system by suppressing transcription of PTH in parathyroid cells. Additional feedback regulation of renal 1α-hydroxylase is provided by another endocrine factor, i.e., fibroblast growth factor 23 (FGF23), which acts to suppress expression of CYP27B1 and therefore inhibits renal synthesis of 1,25(OH)$_2$D [15]. FGF23, which is essential for normal phosphate homeostasis [16], is produced in bone and is induced by 1,25(OH)$_2$D [17]. FGF23 acts as an important feedback control pathway for control of renal endocrine synthesis of 1,25(OH)$_2$D, but an additional mechanism is also provided by the vitamin D catabolic enzyme vitamin D-24-hydroxylase (24-hydroxylase). The gene for which (CYP24A1) is expressed in the kidneys and is induced by 1,25(OH)$_2$D, with resulting 24-hydroxylase activity catalyzing conversion of 25OHD and 1,25(OH)$_2$D to less active metabolites [18].

The combined endocrine actions of PTH, FGF23, and CYP24A1 mean that serum levels of 1,25(OH)$_2$D are not usually correlated with circulating concentrations of its precursor 25OHD. Nevertheless, certain facets of vitamin D endocrinology appear to be strongly influenced by serum 25OHD, suggesting extrarenal modes of action. Recent studies have shown that, in addition to expressing VDR, cells from the parathyroid glands [19] and bone-forming osteoblasts [20] exhibit significant levels of 1α-hydroxylase activity. It has therefore been proposed that some calciotropic effects of vitamin D are due to localized intracrine synthesis of 25OHD to 1,25(OH)$_2$D and subsequent signaling via VDR in the same cells (Figure 4.1). The key distinction between the systemic and localized actions of vitamin D in parathyroid glands and bone is that the latter does not appear to be subject to endocrine regulation by PTH and FGF23 but is modulated instead by factors that are associated with the function of the target tissue itself. In this setting, the intracrine synthesis of 1,25(OH)$_2$D will be more dependent on the availability of substrate for 1α-hydroxylase, i.e., the circulating level of 25OHD. The conclusion that extrarenal synthesis of 1,25(OH)$_2$D will be more reflective of the vitamin D status of any given individual is a pivotal concept in the recently revised perspective on vitamin D and human health, in which a variety of diseases have been associated with impaired vitamin D status. The remainder of this chapter will detail the diversity of tissues, where extrarenal expression of CYP27B1/1α-hydroxylase has been described, and the implications of this for tissue-specific functions of vitamin D.

4.3 EXTRARENAL SYNTHESIS OF 1,25(OH)$_2$D IN NORMAL PHYSIOLOGY

Although extrarenal synthesis of 1,25(OH)$_2$D was first described more than 25 years ago, the key advance in linking this to normal human physiology arose from the cloning of the gene for 1α-hydroxylase (CYP27B1) [21]. This enabled a much more comprehensive appraisal of the tissue distribution of this enzyme than was previously available but also showed that both renal and extrarenal activities of 1α-hydroxylase were due to the same gene product, with the human CYP27B1 cDNA being cloned from keratinocytes, which is a recognized extrarenal source of 1,25(OH)$_2$D [22]. These observations have simplified the development of probes and antisera to characterize both renal and extrarenal expression of the enzyme [11,23]. The next section details some of the key extrarenal tissues that have been reported to express 1α-hydroxylase. Dysregulated extrarenal CYP27B1 leading to elevated serum levels of 1,25(OH)$_2$D will be discussed in later parts of the chapter.

4.3.1 SKIN

As outlined previously, the human cDNA for CYP27B1 was cloned from skin keratinocytes, and these cells were also used to demonstrate impaired synthesis of 1,25(OH)$_2$D in patients with the hereditary disorder vitamin D–dependent rickets type 1 [22]. Prior studies *in vitro* demonstrated expression and activity of 1α-hydroxylase in these cells, which appears to be dependent on their stage of development or differentiation [24,25]. Proliferating keratinocytes, which are similar to those found in the stratum basalis of the epidermis, show relatively high levels of 1,25(OH)$_2$D production, but this declines as keratinocytes differentiate toward cornified envelope precursor cells [26]. Decreased synthesis of 1,25(OH)$_2$D by keratinocytes is accompanied by increased activity of the catabolic enzyme 24-hydroxylase, along with decreased expression of VDR [25]. These changes in vitamin D metabolism appear to occur before any changes in epidermal differentiation markers, suggesting a role for locally produced 1,25(OH)$_2$D in the differentiation of keratinocytes within the normal epidermis. Analysis of mice, in which the Cyp27b1 gene has been replaced by a LacZ reporter cDNA, suggests that the enzyme is poorly expressed in the skin [27]. Nevertheless, other studies using mice with gene ablation of Cyp27b1 have confirmed that synthesis of 1,25(OH)$_2$D is required for normal differentiation and function of epidermal keratinocytes [28,29]. In the absence of a tissue-specific Cyp27b1 knockout, the importance of epidermal 1,25(OH)$_2$D production versus renal synthesis of the hormone in mediating its effects on the skin has yet to be confirmed. However, studies *in vitro* using keratinocytes with either knockout [30] or overexpression [31] of Cyp27b1 have underlined the importance of autocrine synthesis of 1,25(OH)$_2$D in promoting antiproliferation and differentiation effects in these cells.

4.3.2 IMMUNE SYSTEM

One of the initial observations linking vitamin D and the immune system was the ability of 1,25(OH)$_2$D to stimulate differentiation of precursor monocytes to more mature phagocytic macrophages *in vitro* [32–35]. This suggested that monocytes/macrophages are important target cells for vitamin D. Further studies showed that these cells were also able to convert 25OHD to 1,25(OH)$_2$D *in vitro* [36], most notably following activation of normal monocytes by pathogen-associated molecular patterns (PAMPs) [37] or cytokines [38]. *In vitro*, the ability of human monocytes to synthesize 1,25(OH)$_2$D increases as the cells differentiated toward a more mature phenotype, although, paradoxically, this is associated with decreased expression of VDR [39]. The overarching conclusion from these observations was that localized synthesis of 1,25(OH)$_2$D, coupled with expression of endogenous VDR, supported an intracrine mode of action for vitamin D in normal human monocytes [40]. However, the functional significance of this was only defined much later in studies showing that monocyte VDR and CYP27B1 are specifically induced by PAMPs derived from *Mycobacterium tuberculosis* (*M. tb*) [41]. Specifically, the toll-like receptor 2 (TLR2) ligand and *M. tb* PAMP 19 kDa lipoprotein was able to induce intracrine responses to vitamin D, combined with treatment with 19-kDa lipoprotein and 25OHD stimulated monocyte expression of the antibacterial protein cathelicidin (LL37), thereby promoting killing of *M. tb* [41]. Crucially, this response appears to be directly dependent on the 25OHD status of the donor serum used for monocyte culture [41,42]. These observations provide the first clear evidence of a physiological role for serum 25OHD as a regulator of innate immunity, with CYP27B1 acting as the pivotal activation step.

Subsequent studies have expanded the spectrum of monocyte innate immune responses regulated via intracrine activation of vitamin D. These include the induction of other antibacterial proteins such as β-defensin 2 (DEFB4), which appears to require local generation of 1,25(OH)$_2$D, in conjunction with the transcription factor nuclear factor-kappa B (NF-κB) [43,44]. The latter may be generated by a variety of mechanisms, including responses to the inflammatory cytokine interleukin-1 (IL-1) [43] and the intracellular pathogen recognition receptor (PRR) NOD2 [44]. Vitamin D induces expression of both LL37 due to direct transcriptional regulation of these genes by 1,25(OH)$_2$D$_3$-liganded VDR acting via functional vitamin D response elements (VDREs) within the proximal promoters of the LL37 and DEFB4 genes [45,46]. For LL37, this VDRE occurs within a small interchangeable nuclear element sequence, which only appears to be present in the gene promoter of higher primates. Thus, vitamin D regulation of this innate immune response is likely to have been a relatively recent evolutionary event [46]. Despite this, monocytes from lower primates such as mice express a functional 1α-hydroxylase, suggesting alternative targets for intracrine synthesized 1,25(OH)$_2$D in these animals. This may include effects on reactive oxygen species (ROSs), which can act as bacteriocides. Previous studies have reported that macrophages infected with *M. tb* in the presence of 1,25(OH)$_2$D produce high levels of superoxide anions via the NADPH oxidase system [47]. More recently, attention has focused on nitric oxide (NO), which is produced by macrophages as part of the innate immune response to infection. In addition to its established vascular activity, NO can function as a ROS and thus exert bacteriocidal effects [48]. The NO pathway appears to play a pivotal role in mouse responses to *M. tb* infection [49], but its importance to humans is less clear.

In recent years, attention has also focused on the environment in which bacterial killing takes place. The strongly acidic nature of the fused lysosome-phagosome that encapsulate intracellular bacteria is important for their subsequent killing but may also involve incorporation of specific antibacterial factors such as LL37 and DEFB4 [50]. The efficacy of such a mechanism is far from clear, and attention has now focused on autophagy as a process that may enhance the management of phagocytosed pathogens. Autophagy is a cellular mechanism common to all eukaryotic organisms that involves membrane encapsulation of organelles or cell proteins in an autophagosome prior to fusion with lysosomes and degradation of the autolysosomal contents. In addition to its well-recognized function as a pivotal factor in the maintenance of cytosolic homeostasis [51], autophagy also plays a key role in cellular response to infection, with pathogens contained in autophagosomes being either eliminated or degraded prior to presentation to PRRs [52,53]. Recent studies have shown that TLR2-induction of CYP27B1 is associated with enhanced autophagy, indicating that this mechanism is also likely to be strongly influenced by bioavailability of 25OHD and localized generation of 1,25(OH)$_2$D [54].

In addition to normal monocytes, several other related immune cell types are known to exhibit 1α-hydroxylase expression and activity. Antigen-presenting dendritic cells (DCs) are part of the innate immune system, but they play an important role in mediating the effects of vitamin D on cells from the adaptive immune system, such as T lymphocytes (T-cells) and B lymphocytes (B-cells). As such, the effects of vitamin D on DCs are central to its subsequent effects on autoimmune activity and host-graft rejection. DCs express VDR [55], but like monocytes, they also express 1α-hydroxylase, suggesting a similar intracrine mode of action [56,57]. Monocyte-derived DCs show baseline 1α-hydroxylase expression and activity that increases as they differentiate toward a mature phenotype [56]. Localized conversion of 25OHD to 1,25(OH)$_2$D by these cells has been shown to suppress DC maturation, leading, in turn, to inhibition of T-cell proliferation [56]. In contrast to CYP27B1 expression, mature DCs express fewer VDR than immature DCs [56], so that mature antigen-presenting DCs will be less sensitive to the suppressive effects of 1,25(OH)$_2$D. Conversely, the high levels of 1,25(OH)$_2$D synthesized by mature DCs will be able to act on VDR-enriched immature DCs and thus prevent their further maturation [58]. This way, the 1α-hydroxylase in DCs may act in a paracrine rather than intracrine fashion by enabling initial antigen presentation to T-cells by mature DCs while preventing sustained maturation of DCs and overstimulation of adaptive immune responses. CYP27B1 expression in DCs has also been linked to T-cell homing by promoting interaction between chemokine ligand 27 and its receptor, chemokine receptor 10 [59]. In this case, it was suggested that T-cells express CYP27B1 and therefore provide a local source of 1,25(OH)$_2$D. Similar expression of 1α-hydroxylase has also been reported for B-cells [60], suggesting that a range of lymphocyte responses to vitamin D may be mediated via intracrine rather than paracrine metabolism of vitamin D.

4.3.3 PLACENTA

The placenta was one of the first extrarenal sites reported to exhibit extrarenal activity of 1α-hydroxylase, with the enzyme being detectable in both maternal decidua and fetal trophoblast [61,62]. Subsequent studies have shown that expression of CYP27B1 is higher in first and second trimester tissues and parallels similar patterns of expression for VDR [63,64]. By contrast, expression of the vitamin D catabolic enzyme 24-hydroxylase decreases in the first- and second-trimester placentas. This appears to be due to extensive methylation of the CYP24A1 gene in placental tissue, leading to transcriptional silencing of the enzyme [65]. This way, CYP24A1 gene methylation provides an additional mechanism for maximal local generation of $1,25(OH)_2D_3$ in the placenta by suppressing catabolism of $1,25(OH)_2D$. Initial studies linked the high level of 1α-hydroxylase activity in the placenta to the elevated serum levels of $1,25(OH)_2D$ that are characteristic of the first trimester of pregnancy. However, analysis of 1α-hydroxylase-deficient animals and an anephric pregnant woman indicate that this is not likely to be the case [66]. Instead, the presence of VDR in the placenta is consistent with an intracrine mode of vitamin D action at the fetal–maternal interface, similar to that observed in the immune system [67]. Indeed, an immunomodulatory role for placental activation of vitamin D has been proposed [68]. This stems, in part, from the recognized immune activity within the wide variety of cells that make up the placenta as a whole. Both maternal and fetal cells are involved in mediating innate [69–71] and adaptive [72,73] immune responses. In particular, the placenta and reproductive tissues express a wide range of antibacterial [74–76] and antiviral factors [77], which may act as targets for the high levels of local $1,25(OH)_2D$ production within the placenta. Experiments using human placental tissue and cells have shown that 25OHD and $1,25(OH)_2D$ induce expression of LL37 in both decidua [78] and trophoblast [79], leading to enhanced bacterial killing [79]. Similar studies have also highlighted the antiinflammatory action of 25OHD in placental cells, following local activation via CYP27B1 [78]. The significance of this has been further endorsed by *in vivo* data using mice with ablation of the CYP27B1 gene. Knockout of mouse Cyp27b1 within the fetal component of the placenta alone was sufficient to greatly exacerbate inflammatory responses to the PAMP lipopolysaccharide (LPS) [80].

The above *in vitro* and animal data strongly suggest that a key function of 1α-hydroxylase in the placenta is to support antibacterial and antiinflammatory actions. It is therefore interesting to note studies showing that maternal vitamin D insufficiency is associated with increased rates of bacterial vaginosis in the first trimester of pregnancy [81]. Other reports have described association between impaired vitamin D status in pregnant women and risk of maternal to child transmission of human immunodeficiency virus (HIV) [82]. It is hypothesized that the lower levels of HIV transmission under conditions of vitamin D sufficiency may be due to improved innate immune response to infection in these women. Given the fundamental role of the placenta in vertical transmission of HIV from mother to the fetus [83], it is possible that vitamin D–induced innate immune responses within the placenta plays a role in combating viral infection during pregnancy. Other groups have reported altered levels of 1α-hydroxylase in placentas from preeclampsia pregnancies [84,85], and this may be linked to epidemiology, indicating that vitamin D deficiency significantly increases the risk of preeclampsia [86–88].

4.3.4 PROSTATE AND BREAST

Expression of 1α-hydroxylase has been reported in human prostate cells [89], suggesting that locally synthesized $1,25(OH)_2D$ may act in an intracrine fashion to regulate prostate cell proliferation. Subsequent studies using primary cultures and cell lines showed that 1α-hydroxylase activity is lower in prostate cancer cells relative to normal prostate cells [90,91]. This provides an explanation for the ability of nonneoplastic prostate cells to show decreased cell proliferation when treated with either $1,25(OH)_2D$ or 25OHD, whereas prostate cancer cells appear to be only responsive to $1,25(OH)_2D$ [91]. Overall, these observations support a possible role for vitamin D as an endogenous generator of chemopreventative $1,25(OH)_2D$ in cancer [92,93].

Analysis of paired normal and neoplastic biopsy tissue from a cohort of women with breast cancer revealed that, although CYP27B1 mRNA was detectable in normal breast tissue, expression was much higher in breast tumors [94]. Similar observations have been made by other groups [95,96], but other groups have shown no change in CYP27B1 mRNA expression [96]. Immunohistochemical analysis of 1α-hydroxylase protein endorses mRNA data showing increased levels of the enzyme in tumors but also indicates that expression of the enzyme is not restricted to tumor cells but is also a feature of the inflammatory infiltrate associated with breast tumors [94]. Thus, *in vivo*, the expression and function of 1α-hydroxylase in tumors may be complicated by the heterogeneous composition of breast tumors, notably the extensive presence of cells from the immune system. Another observation arising from studies of normal breast tissue and tumors is that expression of CYP24A1 is elevated in a vitamin D–independent fashion in tumors [94]. Dysregulation of vitamin D metabolism such as this is likely to disrupt any beneficial effects of locally synthesized $1,25(OH)_2D$ within breast tissue.

In contrast to tumor biopsies, breast cancer cell lines show lower levels of CYP27B1 expression relative to nonmalignant lines [94,97]. This may reflect the selection process by which malignant cells are isolated, with low expression of CYP27B1 and associated production of antiproliferative $1,25(OH)_2D$ being disadvantageous to cell propagation. Alternatively, this may simply be a reflection of the high levels of CYP24A1 commonly observed in these cell lines. Oncogenic transformation of normal human breast epithelial cells leads to decreased expression of VDR and CYP27B1, suppression of the local generation of $1,25(OH)_2D$, and impaired cell responses to 25OHD [98]. Despite this, breast cancer cells appear to be responsive to 25OHD [99], possibly as a consequence of megalin-mediated uptake of 25OHD and vitamin D–binding protein (DBP), as observed in the kidneys [100]. *Ex vivo* and *in vitro* studies of breast 1α-hydroxylase are supported by analysis of mouse models, which has demonstrated expression of mouse Cyp27b1 in normal breast tissue [101], suggesting that locally synthesized $1,25(OH)_2D$ contributes to normal breast tissue development in these animals.

4.3.5 BONE

The skeleton has long been considered a key target for the endocrine actions of $1,25(OH)_2D$, despite early studies describing the synthesis of $1,25(OH)_2D$ in primary cultures of human and mouse osteoblasts [102–104]. As with the parathyroid glands, recent reports have provided more comprehensive evidence for the presence of 1α-hydroxylase activity in bone [20,105,106]. These studies showed that local conversion of 25OHD to $1,25(OH)_2D$ by primary cultures of human osteoblasts and osteoblastic cell lines promotes intracrine regulation of osteoblast differentiation and function, supporting the overall hypothesis that vitamin D–mediated effects on the skeleton are not exclusively endocrine in nature [107,108]. Similar intracrine activation of vitamin D has also been reported for chondrocytes, with transforming growth factor β1 (TGFβ1) stimulating expression of 1α-hydroxylase [109,110] in a similar fashion to that described for keratinocytes [111]. These studies *ex vivo* and *in vitro* have been extended to include analysis of chondrocyte-specific Cyp27b1 knockout mice that present a wide range of bone and growth plate abnormalities [112]. By contrast, mice with transgenic overexpression of chondrocytic Cyp27b1 show the opposite phenotype to those with gene ablation [113]. A possible role for localized activity of 1α-hydroxylase in bone is provided by recent studies of bone mineral density (BMD) in healthy adults, which showed close correlation between BMD and serum levels of 'free' 25OHD [114]. This observation suggests that the effects of 25OHD on bone are independent of serum $1,25(OH)_2D$ and may therefore involve local intracrine effects (see Figure 4.1).

4.3.6 ENDOCRINE GLANDS AND REPRODUCTIVE TISSUES

In classical calcium endocrinology, the secretion of PTH by the parathyroid glands is suppressed by $1,25(OH)_2D$ produced via 1α-hydroxylase activity in the kidneys. However, it now appears that 1α-hydroxylase is also present in the parathyroid glands themselves and may thus facilitate intracrine regulation of PTH (see Figure 4.1). CYP27B1 mRNA has been detected in normal parathyroid gland tissue, as well as parathyroid adenomas and carcinomas [115,116], and studies using bovine parathyroid cells showing that 25OHD can suppress PTH secretion in a similar fashion to that observed with active $1,25(OH)_2D$ [19]. The parathyroid cells used in the latter study showed only a low rate of conversion of 25OHD to $1,25(OH)_2D$, and addition of a 1α-hydroxylase inhibitor did not suppress the effects of 25OHD on PTH secretion [19]. This suggests that the effects of 25OHD did not involve localized conversion to 1,25(OH)2D, but other reports have reached the opposite conclusion, showing that suppression of PTH by 25OHD could indeed be blocked using the 1α-hydroxylase inhibitor ketoconazole [117]. Although these observations support a role for extrarenal 1α-hydroxylase in mediating the calcium/bone effects of low vitamin D (25OHD) status, a possible role for parathyroid 1α-hydroxylase as a factor in parathyroid tumor formation has also been proposed. Specifically, it has been hypothesized that parathyroid CYP27B1 may act to suppress tumor suppressor genes in this organ by enhancing localized concentrations of antiproliferative 1,25(OH)$_2$D [118]. In this regard, it is interesting to note that FGF23, which suppresses renal 1α-hydroxylase activity, appears to enhance CYP27B1 in cultured bovine parathyroid cell [119]. The potential effect of FGF23 on parathyroid adenomas remains unclear.

Protein for 1α-hydroxylase has been detected in other endocrine organs such as the pancreas [23,120]. As the VDR is also expressed in these different models, an intracrine mode of action has been proposed for pancreatic response to vitamin D, with possible effects on normal pancreatic physiology [121], and risk of diabetes [122]. Expression of 1α-hydroxylase has also been described for normal thyroid tissue and papillary thyroid carcinomas [123], with immunohistochemistry suggesting that the enzyme and VDR are more strongly expressed in carcinoma tissue [123]. Other immunohistochemical analyses have shown that 1α-hydroxylase is present in the adrenal medulla [23], suggesting a possible secretory function for intracrine 1,25(OH)$_2$D. By contrast, no expression of CYP27B1 was observed in the adrenal cortex, despite this being the location of many other CYP450 enzymes involved in steroidogenesis [23].

Within male reproductive tissues, CYP27B1 expression has been reported for testes in general [22,124] and, more specifically in epididymis, seminal vesicle, prostate, and spermatozoa [124]. Thus, local synthesis of 1,25(OH)$_2$D may influence spermatogenesis, although effects of vitamin D on male and female gonadal estrogen synthesis have also been proposed [125]. Vitamin D has been shown to regulate expression of enzymes involved in estrogen metabolism such as aromatase and 17β-hydroxysteroid dehydrogenase [126], suggesting a possible role in the regulation of ovarian function. This is supported by expression studies documenting expression of 1α-hydroxylase in normal ovaries [127,128], ovarian carcinomas [129], and dysgerminomas [128]. In carcinomas and dysrgerminomas, expression of 1α-hydroxylase is elevated relative to normal ovary tissue, with some dysgerminoma patients showing elevated serum levels of 1,25(OH)$_2$D and hypercalcemia [128]. Common with observations for breast tumors and granulomatous diseases, overexpression of 1α-hydroxylase in dysgerminomas is associated with infiltrating tumor macrophages, suggesting an immune component to this disease [128]. Expression of 1α-hydroxylase has also been reported in endometrial tissue, with higher levels of the enzyme being associated with endometriosis [130] and cervical carcinomas [129].

4.3.7 GI TRACT

Common with bone and parathyroid cells, the GI tract is an important target for endocrine 1,25(OH)$_2$D, which can act on the abundant VDR expressed at this site. Another similarity between bone, parathyroid cells and the GI tract is the increasing evidence for more localized metabolism of vitamin D in these tissues. *In vitro*, the expression and activity of 1α-hydroxylase in colonic cell models appears to be highly dependent on the proliferation and differentiation status of specific cell clones [131], with adenocarcinoma-derived Caco-2 and HT-29 human colonic cells showing responses to 25OHD [132,133]. Protein and mRNA for 1α-hydroxylase is also detectable in normal colon tissue and tumors, with expression being increased in moderate to high differentiated colon tumors but lost in highly differentiated tumors [134]. Thus, the capacity for local colonic synthesis of 1,25(OH)$_2$D appears to be enhanced in early tumor initiation, thereby providing an intracrine target for supplementary 25OHD [135,136]. Conversely, following establishment of a tumor, this mechanism becomes corrupted and may thus negate the potential benefits of vitamin D supplementation [137]. Unlike many other extrarenal tissues, characterization of 1α-hydroxylase expression in colon tumors has been complemented by analysis of vitamin D metabolites from freshly isolated tissues [134]. In these studies, the level of 1,25(OH)$_2$D in colonic tumors was inversely related to activity of catabolic 24-hydroxylase [134], indicating a similar pattern of vitamin D metabolism to that reported for breast tumors [94].

Protein and mRNA for 1α-hydroxylase is also detectable in mouse [138] and human [23] colonic epithelial cells. Expression of the enzyme is elevated in GI tissue from patients with Crohn's disease, which is a form of inflammatory bowel disease (IBD), but this appears to be due to increased levels of CYP27B1 in disease-affected granulomatous tissue [139]. Similar observations have also been made for mice with experimental forms of IBD, where increased expression of CYP27B1 was noted in colonic lymphomatous tissue [138]. Common with VDR knockout mice, ablation of the Cyp27b1 gene has been shown to increase susceptibility to experimental IBD [138], although this may be due to suppression of both renal and extrarenal synthesis of 1,25(OH)$_2$D. Recent data have shown that vitamin D deficiency also increases the severity of experimental IBD in mice [140]. In this study, mice exhibited almost undetectable levels of serum 25OHD, but their circulating levels of 1,25(OH)$_2$D remained within the normal range, suggesting a greater role for intracrine 1α-hydroxylase activity in protecting against IBD in these animals. In vitamin D–deficient mice, low serum 25OHD appears to predispose mice to the onset of IBD as a result of decreased expression of the colonic antimicrobial angionenin-4, which is a key regulator of tissue invasion by enteric bacteria [141]. Vitamin D may therefore play a pivotal role in maintaining innate immune surveillance of enteric bacteria within the GI tract, with this being compromised under conditions of vitamin D insufficiency. Recent studies have implicated aberrant innate immune handling of enteric microbiota as an initiator of the adaptive immune damage associated with Crohn's disease [142]. It is therefore possible to speculate that the association between vitamin D and Crohn's disease may involve both the activation of innate immune responses and the suppression of adaptive immunity and associated inflammation.

4.4 MECHANISMS FOR THE REGULATION OF EXTRARENAL 1α-HYDROXYLASE

Early studies using extrarenal sources of 1α-hydroxylase indicated that macrophage synthesis of 1,25(OH)$_2$D is not subject to the sensitive feedback regulation characteristic of its renal counterpart. One possible explanation is that renal and extrarenal synthesis of 1,25(OH)$_2$D is catalyzed by distinct enzymes. However, the cloning of the gene for CYP27B1 showed that a single gene product is responsible for 1α-hydroxylase activity. Following the initial cloning of the mouse Cyp27b1 gene from renal tissue [21] it is notable that the human homolog (CYP27B1) was isolated from keratinocytes, which is an extrarenal source of 1,25(OH)$_2$D [22]. The identical gene product in renal and extrarenal tissues strongly supported the presence of a single but differentially regulated 1α-hydroxylase protein within specific tissue. This was endorsed by studies showing that monocytes from patients with CYP27B1 mutations associated with vitamin D–dependent rickets type 1 exhibited lower levels of 1,25(OH)$_2$D production than observed in normal subjects [143]. DNA and amino acid sequence information for CYP27B1 has also facilitated the development of specific antisera and probes for 1α-hydroxylase. This has further emphasized the widespread tissue-distribution of the enzyme but has also underlined the identity between the renal and extrarenal CYP27B1 [11,23].

4.4.1 BIOCHEMISTRY AND REGULATION OF EXTRARENAL 1α-HYDROXYLASE

In both renal and extrarenal tissues, 1α-hydroxylase functions as a mitochondrial mixed function oxidase with cytochrome P450 activity [144]. Using reconstituted mitochondrial extracts, it has been shown that electron transfer to the cytochrome P450 and subsequent insertion of an oxygen atom in the substrate 25OHD require the following: flavoprotein, ferredoxin reductase, an electron source, and molecular oxygen [144]. Both renal and extrarenal 1α-hydroxylase requires a secosterol (vitamin D sterol molecule with an open B-ring) as substrate [145], and in both cases, the enzyme has a particular affinity for secosterols bearing a carbon-25 hydroxy group (i.e., 25OHD and 24,25-dihydroxyvitamin D (24,25(OH)$_2$D)) [145,146]. The calculated Km (affinity) of the 1α-hydroxylase in pulmonary alveolar macrophages is in the range of 50–100 nM for these two substrates, which is similar to that observed in renal proximal tubule cells [145,146]. Both renal and extrarenal 1α-hydroxylase activities are inhibited by naphthoquinones, which are molecules that compete with reductase for donated electrons, and by imidazoles, which compete with the enzyme for receipt of O_2 [147]. The availability of cDNA sequences for CYP27B1 has shed more light on the catalytic properties of the enzyme [21,22,148,149] but has so far failed to provide a clear mechanism for the differential regulation of $1,25(OH)_2D$ production in renal and extrarenal tissues.

As outlined earlier in the chapter, there appear to be four major endocrine regulators of the renal 1α-hydroxylase: the serum concentration of calcium and phosphate; PTH; FGF23; and $1,25(OH)_2D$ itself (Figure 4.1). However, 1α-hyroxylase at extrarenal sites appears to be unaffected by the stimulatory effects of PTH and phosphate [36,147]. Cells such as macrophages do not express abundant levels of PTH receptors [150], and there is no evidence of responses to PTH or PTHrP in these cells. Similarly, the macrophage 1α-hydroxylase does not appear to be influenced by changes in extracellular phosphate [147]. Calcium effects on extrarenal 1α-hydroxylase activity have been described but do not appear to be consistent with the renal enzyme: extracellular calcium has been shown to inhibit renal 1α-hydroxylase [151], whereas, in macrophages, calcium ionophores stimulate $1,25(OH)_2D$ production [152]. Effects of FGF23 on extrarenal 1α-hydroxylase activity have yet to be documented but are likely to be extremely important, given the extremely high circulating levels of FGF23 reported for many patients with chronic kidney disease (CKD). In the meantime, the general conclusion is that the key endocrine systems involved in regulating renal 1α-hydroxylase activity are not utilized at extrarenal sites. Furthermore, there is no clear evidence that extrarenal synthesis of $1,25(OH)_2D$ is influenced by other endocrine factors such as estrogen, prolactin, and growth hormone, which have been reported to stimulate renal 1α-hydroxylase [153–155]. It therefore seems likely that synthesis of $1,25(OH)_2D$ at many extrarenal sites will be defined by factors that are distinct from those involved within the kidney (see Figure 4.2). These are discussed in the succeeding sections of the chapter.

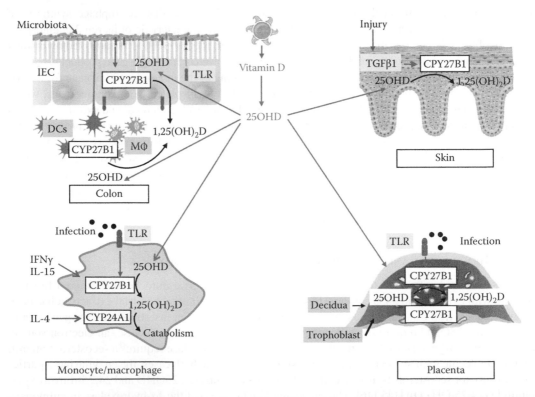

FIGURE 4.2   (See color insert.) Expression and regulation of 1α-hydroxylase in extrarenal tissues. The schematic shows key tissues known to express 1α-hydroxylase outside the kidney. The expression and putative function of 1α-hydroxylase (CYP27B1) is shown for colon, skin, placenta, and monocytes/macrophages. Prominent regulators of CYP27B1 expression at each site are shown as red arrows.

4.4.2 CYP24A1 and Extrarenal Synthesis of 1,25(OH)₂D

In addition to PTH and FGF23, the other major determinant of renal 1,25(OH)$_2$D production is the catabolic enzyme 24-hydroxylase. Like 1α-hydroxylase, 24-hydroxylation of vitamin D metabolites is due to a mitochondrial CYP450 [156–158] encoded by the CYP24A1/Cyp24a1 gene [159,160]. Expression of CYP24A1 is stimulated in kidney cells by 1,25(OH)$_2$D, whereas PTH inhibits transcription of CYP24A1 and the synthesis of 24,25(OH)$_2$D from 25OHD [161]. Because 24-hydroxylase is coexpressed in the kidney along with the 1α-hydroxylase, the first point of regulation involves the bioavailability of 25OHD as a substrate [162]. Although renal 24-hydroxylase has a lower affinity for 25OHD than renal 1α-hydroxylase, it has a higher capacity for the substrate [146]. Therefore, when CYP24A1 expression is induced by systemic or locally produced 1,25(OH)$_2$D, or by low serum PTH levels, the 24-hydroxylase can actively compete with 1α-hydroxylase for the substrate 25OHD. CYP24A1 can also influence circulating concentrations of 1,25(OH)$_2$D by catalyzing 24-hydroxylation of this metabolite leading, in turn, to the generation of nonbiologically active water-soluble excretable metabolites. Although both 25OHD and 1,25(OH)$_2$D are metabolized by 24-hydroxylase [149], the latter appears to be the preferred substrate [158].

Expression and activity of 24-hydroxylase appear to be common to all VDR-expressing cell types [163] and may play a key role in defining the functional efficacy of extrarenal synthesis of 1,25(OH)$_2$D. For example, recent studies have shown that the cytokine interleukin-4 (IL-4) suppresses 25OHD-induced antibacterial activity [164]. This does not involve effects on CYP27B1 expression but, instead, appears to be due to enhanced vitamin D catabolism. IL-4-treated monocytes showed increased activity of 24-hydroxylase, but paradoxically, this was not coincident with increased expression of CYP24A1 [164]. The explanation for this is still unclear but may involve regulation of another CYP enzyme acting as a decoy for the 24-hydroxylase. Beyond this observation with IL-4, studies using monocytes [40] and macrophages [145] have shown relatively low basal 24-hydroxylase activity, following treatment with added 1,25(OH)$_2$D. This contrasts the sensitive induction of 24-hydroxylase activity in renal tubular epithelial cells and other epithelial cells [165] and occurs despite the fact that expression of mRNA for CYP24A1 in monocytes and macrophages is readily induced by 1,25(OH)$_2$D. An explanation for this provided by studies showing that, in addition to the conventional 24-hydroxylase enzyme, monocytes/macrophages also express a splice variant form of the 24-hydroxylase protein [166]. This splice variant protein termed CYP24-SV was shown to be a more potent suppressor of macrophage 1,25(OH)$_2$D$_3$ production than CYP24A1 itself [166], although this does not appear to be due to any catabolic activity of the enzyme. Truncation of the CYP24-SV protein involves alternative exon splicing that removes exons 1 and 2 from CYP24A1. This part of CYP24A1 encodes a mitochondrial targeting sequence, and as a result, expression of CYP24A1 is restricted to the cytoplasm, thereby eliminating any catalytic activity. However, because CYP24-SV retains the substrate binding domain of CYP24A1, it still has the potential to bind 25OHD or 1,25(OH)$_2$D, and may therefore act as a decoy for mitochondrial 1α-hydroxylase or 24-hydroxylase.

Molecular modeling of human CYP24A1 and CYP24-SV based on the crystal structures of CYP3A4 and CYP2C8, respectively, has highlighted significant alterations in the CYP24-SV structure, particularly within helical regions adjacent to the heme-binding domain [167]. The resulting changes in substrate binding pocket hydrophobicity immediately above the heme-binding domain may lead to changes in substrate binding and a shift in preference for 25OHD rather than 1,25(OH)$_2$D. This provides a potential explanation for the potent effects of CYP24-SV on macrophage metabolism of 25OHD, with the splice variant functioning to attenuate synthesis of 1,25(OH)$_2$D. However, in cells such as macrophages that have a high capacity for 1,25(OH)$_2$D production, CYP24-SV may also bind 1,25(OH)$_2$D, thereby limiting its access to the VDR. This way, CYP24-SV may provide a metabolically economical alternative to 24-hydroxylation of 1,25(OH)$_2$D. Moreover, such a mechanism provides a potential explanation for accumulation and possible systemic spillover of 1,25(OH)$_2$D in some extrarenal settings, as detailed in the final sections of this chapter.

4.4.3 REGULATION OF MONOCYTE 1α-HYDROXYLASE

Studies using cultured monocytes and DCs have shown that 150-nM 25OHD is as effective as 100-nM 1,25(OH)$_2$D in stimulating changes in monocyte or DC phenotype markers. This occurs despite the fact that intracellular accumulation of 1,25(OH)$_2$D by monocyte intracrine metabolism is 30 times lower than that obtained with exogenously added 1,25(OH)$_2$D [168]. The conclusion from this observation is that extrarenal 1α-hydroxylase provides a sensitive mechanism for mediating cellular responses to vitamin D. However, the key question remaining is whether or not extrarenal 1α-hydroxylase is sufficiently abundant *in vivo* to fulfill its *in vitro* potential. Studies using Cyp27b1 knockout mice observed that active transcription of enzyme was only evident in the kidney and the placenta [27], suggesting that, in healthy mice, expression of 1α-hydroxylase is very limited without additional stimulus. The next section will consider the various factors that may contribute to the regulation of CYP27B1 expression and activity in extrarenal tissues.

Initial studies of macrophages from alveolar fluid indicated that the cytokine IFNγ is a potent stimulator of 1,25(OH)$_2$D in these cells [77]. However, it is now clear that other immunomodulators are also able to stimulate macrophage 1α-hydroxylase, including other cytokines such as tumor necrosis factor α (TNFα) (79,43), interleukin-2 (IL-2), and IL-15 [169]. The precise mechanism by which IL-15 promotes CYP27B1 transcription in monocytes has yet to be defined, but data from other cell models indicated that the JAK-STAT pathway is the most likely candidate [170]. The ability of IL-15 and IFNγ to induce CYP27B1 expression independent of PRRs such as the TLR system suggests that monocyte synthesis of 1,25(OH)$_2$D is not exclusively regulated by innate immune activity. IFNγ is produced by natural killer (NK) and NK T cells, as well as CD4 and CD8-positive T-cells [171]. Conversely, T-cells expressing factors such as IFNγ are known to stimulate IL-15 synthesis by DCs [172]. The induction of monocyte CYP27B1 activity by either of these cytokines may therefore provide a mechanism by which the adaptive immune system is able to enhance innate immune response to infection. This is likely to be important in view of the fact that innate immunity alone is not sufficient to combat infectious diseases, as illustrated by the high rates of TB infection in people who are positive for the HIV.

At a molecular level, the CYP27B1 promoter region includes putative AP-1 and nuclear factor-kappa B (NF-κB) binding sites, which are potential targets for cytokine regulation of 1α-hydroxylase [173,174]. IFNγ signals via the janus kinase 1 (JAK1) and JAK2 pathways with subsequent phosphorylation of signal transducers and activators of transcription 1 alpha (STAT1α) and transregulation of target genes via cis-acting promoter elements [175]. However, the JAK/STAT pathway is essential for the effects of many cytokines and growth factors including IL-2 and IL-15 [170,176]. In macrophages, the JAK/STAT system may also interact with other signaling pathways, including p38 mitogen-activated protein kinase (MAPK) and NF-κB [177,178]. Molecular analysis of the induction of monocyte CYP27B1 expression by IFNγ and LPS confirmed that multiple signal transduction pathways are involved including phosphorylation of C/EBPβ [179]. Binding of the latter to recognition sites in the CYP27B1 gene promoter appears to be necessary for induction of the enzyme by IFNγ and LPS [180]. Subsequent studies using mice with gene ablation of components in the IFNγ signaling pathway have confirmed the importance of C/EBPβ and STA1α as key regulators of 1α-hydroxylase activity in monocytes [181].

As detailed earlier in the chapter, a pivotal observation linking intracrine vitamin D metabolism and immune function was the induction of monocyte CYP27B1 by the *M. tb* PAMP 19 kDa lipoprotein, a ligand for TLR2 [41]. Similar observations have also been made for the Gram+ PAMP LPS [42]. Responses to LPS are mediated via a membrane receptor complex involving the monocyte marker cluster of differentiation 14 (CD14), lymphocyte antigen 96 (MD2), and the PRR TLR4 [182]. Binding of ligand LPS to the TLR4 complex activates MAPK, NF-κB, and interferon regulatory factor, and all three of these appear to be involved in LPS regulation of 1α-hydroxylase [180].

However, there are now known to be 13 TLRs (TLR1–TLR13) in humans. These PRRs are expressed by many cell types and can respond to a variety of pathogenic stimuli [183], suggesting that TLRs and pathogen recognition may be a universal mechanism for enhancing extrarenal synthesis of $1,25(OH)_2D$.

PAMPs such as LPS and cytokines such as IFNγ commonly activate different signal transduction pathways, but there is potential for crosstalk between these pathways. Notably, IFNγ and LPS are effective stimulators of NO synthesis in macrophages, suggesting a possible role for NO production in the extrarenal synthesis of $1,25(OH)_2D$ [184–186]. Generation of NO in the macrophage is under the control of the inducible NO synthase (iNOS) [187], which is transcriptionally regulated by IFNγ and LPS [188,189]. Furthermore, NO has established inhibitory effects on other cytochrome P450s [190,191] that suggest a possible link with the enzymes involved in vitamin D metabolism. Indeed, NO may act as an alternative to NADPH as a source of unpaired electrons for the 1α-hydroxylase reaction in macrophages [184,185].

4.4.4 REGULATION OF EXTRARENAL 1α-HYDROXYLASE IN CELLS OTHER THAN MACROPHAGES

Although most of our current knowledge of the factors involved in regulating extrarenal 1α-hydroxylase has stemmed from studies of the enzyme in monocytes and macrophages, it is important to recognize that other cells from the immune system may also actively synthesize $1,25(OH)_2D$. As outlined previously, transcripts for CYP27B1 have been reported in T cells [59], B cells [60], and DCs [56,57,59]. Expression and activity of 1α-hydroxylase is increased in DCs as they differentiate toward a mature antigen-presenting phenotype [56,57]. However, this can be further enhanced by several factors including LPS, IFNγ, TNFα, and polyinosinic:polycytidylic acid (poly I:C), which is a ligand for TLR3 that is structurally similar to double stranded DNA [56]. Other studies have confirmed the induction of DC 1α-hydroxylase by TLR3, with this being linked to enhanced lymphoid trafficking of these cells [192]. It would therefore appear that macrophage/DC 1α-hydroxylase can be regulated by TLRs associated with Gram +VE (TLR2) and Gram –VE (TLR4) bacteria, as well as viruses (TLR3). DC synthesis of $1,25(OH)_2D$ is also regulated by the TLR4-ligand monophosphoryl lipid A (MPLA) [193]. However, unlike LPS, which utilizes the Myd88 adapter for TLR4 signaling [194], MPLA appears to signal via activation of the Toll-IL-1R domain-containing adapter-inducing IFN-β signaling pathway [195,196]. In view of the properties of MPLA as a vaccine adjuvant, these data underline the potential importance of DC 1α-hydroxylase as a regulatory mechanism for antigen presentation while also underlining the complexity of pathways involved in regulating expression of 1α-hydroxylase in immune cells.

With the exception of macrophages and the placenta, extrarenal 1α-hydroxylase has been most well studied in epidermal keratinocytes. Like monocytes, keratinocytes can respond to TLRs such as TLR2, but, because keratinocytes express low levels of TLRs, initial 'priming' of these receptors is required to facilitate pathogen-sensing and induction of 1α-hydroxylase. Basal expression of keratinocyte 1α-hydroxylase is stimulated by TGF-β1, and signaling via this growth factor appears to be sufficient for induction of localized synthesis of $1,25(OH)_2D$. This, in turn, leads to enhanced expression of TLR2, which can then support PAMP-PRR activity similar to that described for macrophages [111]. In human skin, expression of TGF-β1 is a feature of tissue injury, and thus, induction of epidermal 1α-hydroxylase activity may be part of a mechanism linking wound repair with vitamin D–induced innate immunity [111]. In contrast to monocytes and keratinocytes, $1,25(OH)_2D$ appears to be constitutively synthesized by cells from the placenta [61,62], with expression of CYP27B1 and VDR being profoundly elevated in maternal decidua and fetal trophoblast early in the first trimester of pregnancy. However, studies using mouse placentas have shown that expression of Cyp27b1 is further elevated, following *in vivo* challenge with LPS during pregnancy [80]. It therefore seems likely that the high baseline production of $1,25(OH)_2D$ by placental cells can be further enhanced following infection or other immune challenge.

4.5 HUMAN DISEASES ASSOCIATED WITH EXTRARENAL OVERPRODUCTION OF 1,25(OH)$_2$D

Extrarenal expression of 1α-hydroxylase provides an efficient mechanism for mediating the non-classical actions of vitamin D in a variety of tissues. However, it is important to recognize that the initial observations describing extrarenal production of 1,25(OH)$_2$D arose from pathological conditions in which extrarenal 1α-hydroxylase is dysregulated. Since then, more than 30 conditions have been described, in which elevated circulating levels of 1,25(OH)$_2$D associated with extrarenal expression of 1α-hydroxylase are thought to be the cause of patient hypercalciuria or hypercalcemia [197]. In some cases, aberrant calcium homeostasis has been reported to be associated with only 'high normal' concentrations of serum 1,25(OH)$_2$D [197]. Nevertheless, this would still represent an inappropriate elevation of 1,25(OH)$_2$D, given that suppression of PTH levels is common to most of these diseases. The elevation of serum 1,25(OH)$_2$D and associated hypercalciuria/hypercalcemia is likely to be highly dependent on the availability of substrate 25OHD or, in other words, the vitamin D status of the individual. This was illustrated by recent studies of a patient with giant cell polymyositis (GCP) with low serum calcium levels who was prescribed vitamin D supplementation therapy and then went on to develop increased serum 1,25(OH)$_2$D and hypercalemia [197]. In this setting, extrarenal 1α-hydroxylase is functional with effective conversion of 25OHD to 1,25(OH)$_2$D. However, the downstream effects of this are absent, diminished, or simply unable to counter the established inflammatory disease. The next section of the chapter details the specific disease scenarios associated with aberrant extrarenal synthesis of 1,25(OH)$_2$D.

4.5.1 Extrarenal 1α-Hydroxylase and Sarcoidosis

A link between vitamin D and sarcoidosis was first recognized 80 years ago [198], with subsequent studies showing that the hypercalcemia of sarcoidosis, which was characterized by increased intestinal calcium absorption and bone resorption, was similar to that described for exogenous vitamin D intoxication [199]. Seasonal evaluation of serum calcium levels in patients with sarcoidosis showed that there was significant increase in the mean serum calcium concentration in 345 patients with sarcoidosis from winter to summer but no such change in more than 12,000 control subjects [200]. This observation was confirmed by later studies, in which the serum calcium concentration in patients with active sarcoidosis was increased upon whole body exposure to ultraviolet (UV) light irradiation [201]. It was initially hypothesized that the clinical abnormality in calcium balance in patients with active sarcoidosis resulted from increased target organ responsiveness to vitamin D [202]. However, subsequent studies revealed that the hypercalcemia of sarcoidosis was, in fact, due to increased circulating concentrations of active 1,25(OH)$_2$D [203–206], with this being produced outside the kidney [207]. Shortly after these studies, using alveolar cells from sarcoidosis patients showed that macrophages were the extrarenal source of 1,25(OH)$_2$D in these patients [208]. Unequivocal structural characterization of the metabolite as 1,25(OH)$_2$D was later obtained by the same investigators [209,210].

As described in detail earlier in this chapter, the synthesis of 1,25(OH)$_2$D by the renal 1α-hydroxylase is normally strictly regulated, with levels of the hormone product being some 1000-fold lower than substrate 25OHD. By contrast, it is now clear that endogenous 1,25(OH)$_2$D production in patients with sarcoidosis is dysregulated and not bound by the same set of endocrine factors known to regulate 1,25(OH)$_2$D synthesis in the kidney [211]. Hypercalcemic patients with sarcoidosis possess high or inappropriately elevated serum 1,25(OH)$_2$D levels, despite having low serum PTH levels and relatively elevated serum phosphate [212,213]. Serum 1,25(OH)$_2$D concentrations in patients with active sarcoidosis are also sensitive to increases in substrate 25OHD, which is not commonly observed in normal healthy subjects [205]. This provides an explanation for the long-recognized association of hypercalciuria and/or hypercalciuria in sarcoidosis patients in the summer months or following holidays to geographic locations at lower latitudes than those at which the patient normally resides [205,214]. This link between increased cutaneous vitamin D synthesis and the development of clinical abnormalities of calcium balance can be replicated by oral administration of vitamin D [205,213,215]. Production of 1,25(OH)$_2$D in patients with sarcoidosis is unusually sensitive to inhibition by drugs that do not affect renal 1α-hydroxylation at the same doses [216]. Antiinflammatory concentrations of glucocorticoids have long been recognized as effective treatment for sarcoidosis-associated hypercalcemia, with dramatic effects on serum 1,25(OH)$_2$D levels [206,215,217]. By contrast, administration of glucocorticoids to patients without sarcoidosis is not normally associated with a reduction in serum 1,25(OH)$_2$D or calcium. Chloroquine and its hydroxylated analog, hydroxychloroquine, are examples of pharmaceutical agents that appear to act preferentially on the extrarenal vitamin D-1α-hydroxylase [218–221].

4.5.2   Extrarenal 1α-Hydroxylase in Other Granuloma-Forming Diseases

Of the other human granuloma-forming diseases reported to be associated with extrarenal 1,25(OH)$_2$D-mediated hypercalcemia, TB is the most commonly reported. Hypercalcemia has been recognized as a complication of *M. tb* infection for many years and was eventually linked to overproduction of extrarenal 1,25(OH)$_2$D [222–225]. In common with sarcoidosis, circulating 1,25(OH)$_2$D in TB patients is synthesized by disease-activated macrophages [226,227], with this being abnormally responsive to small changes substrate 25OHD [228], and subject to suppression by glucocorticoids [229,230]. The prevalence of hypercalcemia in TB patients has been reported to be as high as 26% [231].

Hypercalciuria or overt hypercalcemia has also been observed for infectious diseases characterized by widespread granuloma formation and macrophage proliferation in affected tissues. These include leprosy [232,233], disseminated candidiasis [234], crytococcosis [235], histoplasmosis [236,237], and coccidioidomycosis [238]. Hypercalcemia in most of these conditions has been associated with elevated serum concentrations of 1,25(OH)$_2$D, but the prevalence and incidence of hypercalcemia and hypercalciuria associated with these diseases is less clear. However, it seems likely that the dysregulated vitamin D metabolism and action associated with these diseases will increase in frequency as the number of immunocompromised patients, especially those with AIDS, increases worldwide.

Extrarenal overproduction of 1,25(OH)$_2$D has also been documented in patients with silicone-induced granulomata [239], eosinophilic granuloma [240], Wegener's granulomatosis [241], Langerhans cell histiocytosis [242], and the systemic granulomatous disease giant cell polymyosis [197]. In addition, 1,25(OH)$_2$D-mediated hypercalcemia has been observed in newborn infants suffering from subcutaneous fat necrosis [243,244], which is a transient disorder associated with birth trauma and characterized histopathologically by the proliferation of "foreign body-type" giant cells around cholesterol-shaped crystals in necrotizing, subcutaneous adipose tissue. Finally, elevated serum levels of 1,25(OH)$_2$D and associated hypercalcemia has been reported in patients with Crohn's disease [139,245,246]. The possible impact of extrarenal 1α-hydroxylase in this clinical situation is an exciting new development not only because of the prevalence of Crohn's disease particularly in developed countries [247] but also because of several recent reports that have documented expression of 1α-hydroxylase along the GI tract [23,95,134,248]. In the case of patients with Crohn's disease presenting with elevated serum 1,25(OH)$_2$D, it has been possible to use tissue biopsy material to show that the increased activation of 25OHD is due to enhanced expression of 1α-hydroxylase in areas of the colon with extensive granulomatous disease [139].

4.5.3   Extrarenal 1α-Hydroxylase, Malignant Lymphoproliferative Disorders, and Autoimmune Disease

Overproduction of 1,25(OH)2D by extrarenal 1α-hydroxylase is not confined to patients with granuloma-forming diseases but can also be observed in patients with lymphoproliferative neoplasms [249–252]. Indeed, extrarenal overproduction of 1,25(OH)$_2$D is the most common cause of hypercalciuria and hypercalcemia in patients with non-Hodgkin and Hodgkin lymphoma [253,254], especially in patients with B-cell neoplasms [252]. As is the situation with hypercalciuric/calcemic patients with sarcoidosis or other granuloma-forming disease and elevated circulating 1,25(OH)$_2$D levels, serum concentrations of PTH are suppressed and PTHrP normal (i.e., not elevated) in lymphoma patients, indicative of dysregulated overproduction of 1,25(OH)$_2$D. Clinical studies of hypercalcemic patients with lymphoma pre- and postsuccessful antitumor therapy [252–255] are compatible, with either the tumor being an immediate source of 1,25(OH)$_2$D or the source of a factor that stimulates the production of 1,25(OH)$_2$D in the kidney or other inflammatory cells, with the latter being the most likely mechanism. Analysis of a patient with hypercalcemia and raised circulating levels of 1,25(OH)$_2$D associated with a splenic B-cell lymphoma showed that the elevated serum 1,25(OH)$_2$D and calcium were corrected, following resection of the spleen [256]. Subsequent immunonhistochemical analysis of this tissue revealed increased expression of 1α-hydroxylase in macrophages adjacent to the tumor but not in the tumor itself [256]. The nature of the tumor-derived factor that stimulates macrophage 1α-hydroxylase in lymphomas remains to be determined and may include a wide range of cytokines and other immunogenic factors.

Reports describing circulating levels of 1,25(OH)$_2$D in patients with inflammatory autoimmune diseases such as rheumatoid arthritis have been variable and include enhanced [257], decreased [258,259], or unchanged circulating levels [260] of the hormone. In other related diseases such as systemic lupus erythematosus (SLE), circulating levels of 1,25(OH)$_2$D do not appear to be affected by disease status [260], even though SLE patients have been reported to exhibit low serum levels of 25OHD [261–263]. Although patients with inflammatory disease may not present with the same elevated circulating levels of 1,25(OH)$_2$D that are often characteristic of those with granulomatous disease, this does not detract from possible localized expression of extrarenal 1α-hydroxylase. Substrate-dependent accumulation of 1,25(OH)$_2$D has been described in the synovial fluid of patients with 'inflammatory' arthritis [264,265]. These observations have prompted suggestion that local accumulation of 1,25(OH)$_2$D within the synovium may contribute to periarticular bone loss in patients with rheumatoid arthritis. Other diseases associated with membrane inflammation include peritonitis. This involves infectious or noninfectious inflammation of the serous membrane of the abdominal cavity. Infectious peritonitis is frequently observed in patients with CKD receiving peritoneal dialysis. Peritoneal macrophages from these patients have been shown to metabolize 25OHD to 1,25(OH)$_2$D in vitro [266–268]. The impact of this localized (peritoneal) synthesis of 1,25(OH)$_2$D on circulating levels of the hormone is less easy to delineate than for other diseases such as sarcoidosis because CKD patients have underlying impairment of renal 1α-hydroxylase activity and frequently present with low vitamin D status. Nevertheless, some reports have suggested that serum levels of 1,25(OH)$_2$D in CKD patients receiving peritoneal dialysis are higher than those observed for patients receiving conventional hemodialysis [266].

4.6   SUMMARY AND CONCLUSIONS

Although the existence of extrarenal production of 1,25(OH)$_2$D has been recognized for many years, it is only recently that this facet of vitamin D metabolism has been incorporated into normal vitamin D physiology. Previously, the extrarenal 1α-hydroxylase was considered to be a manifestation of granulomatous or inflammatory disease, and this remains a significant health issue in the small number of patients within the spectrum of diseases known to be associated with overproduction of 1,25(OH)$_2$D. However, RNA and protein analyses have shown that CYP27B1 is expressed in a wide range of tissues not affected by disease, suggesting a normal function for extrarenal activation of vitamin D. The precise relationship between this and actual localized synthesis of 1,25(OH)$_2$D has yet to be determined, in part because of the difficulty in analyzing the small tissue concentrations of this hormone. This is further complicated by the fact that 25OHD circulates bound to the DBP, which has higher affinity for 25OHD than 1,25(OH)$_2$D. Recent studies by our group have shown that DBP attenuates intracrine responses to 25OHD in monocytes [269], indicating that concentrations or genotypic variations in DBP may play a pivotal role in the realization of extrarenal 1α-hydroxylase activity within specific tissues.

Irrespective of the mechanism by which 25OHD enters cells, it seems likely that the localized generation of 1,25(OH)$_2$D at extrarenal sites will be low, compared to that observed in the kidneys. However, studies *in vitro* suggest that this may be sufficient to sensitively induce responses via the VDR, particularly under conditions known to stimulate expression of CYP27B1. The most well studied example of this is the TLR-mediated induction of 1α-hydroxylase in cells that encounter infectious challenge. Similar responses may also occur in a noninfection setting, thereby enhancing extrarenal expression of CYP27B1 and facilitating the production of physiologically active levels of 1,25(OH)$_2$D within specific tissues. As our understanding of this improves, it seems likely that the biological actions of vitamin D will continue to expand, providing further impetus for clinical studies of the health benefits of this important natural molecule.

REFERENCES

1. Holick, M. F. 2007. Vitamin D deficiency. *N Engl J Med* 357:266–281.
2. Holick, M. F. 2009. Vitamin D status: measurement, interpretation, and clinical application. *Ann Epidemiol* 19:73–78.
3. Adams, J. S. and M. Hewison. 2008. Unexpected actions of vitamin D: new perspectives on the regulation of innate and adaptive immunity. *Nat Clin Pract Endocrinol Metab* 4:80–90.
4. Carlberg, C. and S. Seuter. 2009. A genomic perspective on vitamin D signaling. *Anticancer Res* 29:3485–3493.
5. Spina, C. S., V. Tangpricha, M. Uskokovic, L. Adorinic, H. Maehr, and M. F. Holick. 2006. Vitamin D and cancer. *Anticancer Res* 26:2515–2524.
6. Holick, M. F. 2004. Vitamin D: importance in the prevention of cancers, type 1 diabetes, heart disease, and osteoporosis. *Am J Clin Nutr* 79:362–371.
7. Ross, A. C., J. E. Manson, S. A. Abrams, J. F. Aloia, P. M. Brannon, S. K. Clinton, R. A. Durazo-Arvizu, J. C. Gallagher, R. L. Gallo, G. Jones, C. S. Kovacs, S. T. Mayne, C. J. Rosen, and S. A. Shapses. 2011. The 2011 report on dietary reference intakes for calcium and vitamin D from the Institute of Medicine: what clinicians need to know. *J Clin Endocrinol Metab* 96:53–58.
8. Krishnan, A. V. and D. Feldman. 2011. Mechanisms of the anti-cancer and anti-inflammatory actions of vitamin D. *Annu Rev Pharmacol Toxicol* 51:311–336.
9. Li, Y. C., G. Qiao, M. Uskokovic, W. Xiang, W. Zheng, and J. Kong. 2004. Vitamin D: A negative endocrine regulator of the renin-angiotensin system and blood pressure. *J Steroid Biochem Mol Biol* 89–90:387–392.
10. Brunette, M. G., M. Chan, C. Ferriere, and K. D. Roberts. 1978. Site of 1,25(OH)$_2$ vitamin D$_3$ synthesis in the kidney. *Nature* 276:287–289.
11. Zehnder, D., R. Bland, E. A. Walker, A. R. Bradwell, A. J. Howie, M. Hewison, and P. M. Stewart. 1999. Expression of 25-hydroxyvitamin D3-1alpha-hydroxylase in the human kidney. *J Am Soc Nephrol* 10:2465–2473.
12. Bland, R., D. Zehnder, S. V. Hughes, P. M. Ronco, P. M. Stewart, and M. Hewison. 2001. Regulation of vitamin D-1alpha-hydroxylase in a human cortical collecting duct cell line. *Kidney Int* 60:1277–1286.
13. Hewison, M., D. Zehnder, R. Bland, and P. M. Stewart. 2000. 1alpha-Hydroxylase and the action of vitamin D. *J Mol Endocrinol* 25:141–148.
14. Haussler, M. R., C. A. Haussler, L. Bartik, G. K. Whitfield, J. C. Hsieh, S. Slater, and P. W. Jurutka. 2008. Vitamin D receptor: molecular signaling and actions of nutritional ligands in disease prevention. *Nutr Rev* 66:S98–S112.
15. Shimada, T., M. Kakitani, Y. Yamazaki, H. Hasegawa, Y. Takeuchi, T. Fujita, S. Fukumoto, K. Tomizuka, and T. Yamashita. 2004. Targeted ablation of Fgf23 demonstrates an essential physiological role of FGF23 in phosphate and vitamin D metabolism. *J Clin Invest* 113:561–568.
16. Razzaque M. S. 2009. The FGF23-Klotho axis: endocrine regulation of phosphate homeostasis. *Nat Rev Endocrinol* 5:611–619.
17. Prie, D. and G. Friedlander. 2010. Reciprocal control of 1,25-dihydroxyvitamin D and FGF23 formation involving the FGF23/Klotho system. *Clin J Am Soc Nephrol* 5:1717–1722.

18. Ohyama. Y. and T. Yamasaki. 2004. Eight cytochrome P450s catalyze vitamin D metabolism. *Front Biosci* 9:3007–3018.
19. Ritter, C. S., H. J. Armbrecht, E. Slatopolsky, and A. J. Brown. 2006. 25-hydroxyvitamin D(3) suppresses PTH synthesis and secretion by bovine parathyroid cells. *Kidney Int* 70:654–659.
20. van Driel, M., M. Koedam, C. J. Buurman, M. Hewison, H. Chiba, A. G. Uitterlinden, H. A. Pols, and J. P. van Leeuwen. 2006. Evidence for auto/paracrine actions of vitamin D in bone: 1alpha-hydroxylase expression and activity in human bone cells. *FASEB J* 20:2417–2419.
21. Takeyama, K., S. Kitanaka, T. Sato, M. Kobori, J. Yanagisawa, and S. Kato. 1997. 25-Hydroxyvitamin D3 1alpha-hydroxylase and vitamin D synthesis. *Science* 277:1827–1830.
22. Fu, G. K., D. Lin, M. Y. Zhang, D. D. Bikle, C. H. Shackleton, W. L. Miller, and A. A. Portale. 1997. Cloning of human 25-hydroxyvitamin D-1 alpha-hydroxylase and mutations causing vitamin D-dependent rickets type 1. *Mol Endocrinol* 11:1961–1970.
23. Zehnder, D., R. Bland, M. C. Williams, R. W. McNinch, A. J. Howie, P. M. Stewart, and M. Hewison. 2001. Extrarenal expression of 25-hydroxyvitamin d(3)-1 alpha-hydroxylase. *J Clin Endocrinol Metab* 86:888–894.
24. Bikle, D. D., S. Pillai, E. Gee, and M. Hincenbergs. 1989. Regulation of 1,25-dihydroxyvitamin D production in human keratinocytes by interferon-gamma. *Endocrinology* 124:655–660.
25. Pillai, S., D. D. Bikle, and P. M. Elias. 1988. 1,25-Dihydroxyvitamin D production and receptor binding in human keratinocytes varies with differentiation. *J Biol Chem* 263:5390–5395.
26. Pillai, S., D. D. Bikle, and P. M. Elias. 1988. Vitamin D and epidermal differentiation: evidence for a role of endogenously produced vitamin D metabolites in keratinocyte differentiation. *Skin Pharmacol* 1:149–160.
27. Vanhooke, J. L., J. M. Prahl, C. Kimmel-Jehan, M. Mendelsohn, E. W. Danielson, K. D. Healy, and H. F. Deluca. 2006. CYP27B1 null mice with LacZreporter gene display no 25-hydroxyvitamin D3-1{alpha}-hydroxylase promoter activity in the skin. *Proc Natl Acad Sci USA* 103:75–80.
28. Bikle, D. D., S. Chang, D. Crumrine, H. Elalieh, M. Q. Man, O. Dardenne, Z. Xie, R. S. Arnaud, K. Feingold, and P. M. Elias. 2004. Mice lacking 25OHD 1alpha-hydroxylase demonstrate decreased epidermal differentiation and barrier function. *J Steroid Biochem Mol Biol* 89–90:347–353.
29. Bikle, D. D., S. Chang, D. Crumrine, H. Elalieh, M. Q. Man, E. H. Choi, O. Dardenne, Z. Xie, R. S. Arnaud, K. Feingold, and P. M. Elias. 2004. 25 Hydroxyvitamin D 1 alpha-hydroxylase is required for optimal epidermal differentiation and permeability barrier homeostasis. *J Invest Dermatol* 122:984–992.
30. Huang, D. C., V. Papavasiliou, J. S. Rhim, R. L. Horst, R. Kremer. 2002. Targeted disruption of the 25-hydroxyvitamin D₃ 1alpha-hydroxylase gene in ras-transformed keratinocytes demonstrates that locally produced 1alpha,25-dihydroxyvitamin D₃ suppresses growth and induces differentiation in an autocrine fashion. *Mol Cancer Res* 1:56–67.
31. Flanagan, J. N., L. W. Whitlatch, T. C. Chen, X. H. Zhu, M. T. Holick, X. F. Kong, and M. F. Holick. 2001. Enhancing 1 alpha-hydroxylase activity with the 25-hydroxyvitamin D-1 alpha-hydroxylase gene in cultured human keratinocytes and mouse skin. *J Invest Dermatol* 116:910–914.
32. Abe, E., C. Miyaura, H. Sakagami, M. Takeda, K. Konno, T. Yamazaki, S. Yoshiki, and T. Suda. 1981. Differentiation of mouse myeloid leukemia cells induced by 1 alpha,25-dihydroxyvitamin D₃. *Proc Natl Acad Sci USA* 78:4990–4994.
33. Abe, E., C. Miyaura, H. Tanaka, Y. Shiina, T. Kuribayashi, S. Suda, Y. Nishii, H. F. De Luca, T. Suda. 1983. 1 alpha,25-dihydroxyvitamin D3 promotes fusion of mouse alveolar macrophages both by a direct mechanism and by a spleen cell-mediated indirect mechanism. *Proc Natl Acad Sci USA* 80:5583–5587.
34. Tanaka, H., E. Abe, C. Miyaura, Y. Shiina, T. Suda. 1983. 1 alpha,25-dihydroxyvitamin D3 induces differentiation of human promyelocytic leukemia cells (HL-60) into monocyte-macrophages, but not into granulocytes. *Biochem Biophys Res Commun* 117:86–92.
35. Koeffler, H. P., T. Amatruda, N. Ikekawa, Y. Kobayashi, and H. F. De Luca. 1984 Induction of macrophage differentiation of human normal and leukemic myeloid stem cells by 1,25-dihydroxyvitamin D3 and its fluorinated analogues. *Cancer Res* 44:5624–5628.
36. Reichel, H., H. P. Koeffler, R. Barbers, and A. W. Norman. 1987. Regulation of 1,25-dihydroxyvitamin D₃ production by cultured alveolar macrophages from normal human donors and from patients with pulmonary sarcoidosis. *J Clin Endocrinol Metab* 65:1201–1209.
37. Reichel, H., H. P. Koeffler, J. E. Bishop, and A. W. Norman. 1987. 25-Hydroxyvitamin D3 metabolism by lipopolysaccharide-stimulated normal human macrophages. *J Clin Endocrinol Metab* 64:1–9.
38. Koeffler, H. P., H. Reichel, J. E. Bishop, and A. W. Norman. 1985. gamma-Interferon stimulates production of 1,25-dihydroxyvitamin D₃ by normal human macrophages. *Biochem Biophys Res Commun* 127:596–603.

39. Kreutz, M., R. Andreesen, S. W. Krause, A. Szabo, E. Ritz, and H. Reichel. 1993. 1,25-dihydroxyvitamin D3 production and vitamin D3 receptor expression are developmentally regulated during differentiation of human monocytes into macrophages. *Blood* 82:1300–1307.

40. Hewison, M., S. Barker, A. Brennan, J. Nathan, D. R. Katz, and J. L. O'Riordan. 1989. Autocrine regulation of 1,25-dihydroxycholecalciferol metabolism in myelomonocytic cells. *Immunology* 68:247–252.

41. Liu, P. T., S. Stenger, H. Li, L. Wenzel, B. H. Tan, S. R. Krutzik, M. T. Ochoa, J. Schauber, K. Wu, C. Meinken, D. L. Kamen, M. Wagner, R. Bals, A. Steinmeyer, U. Zugel, R. L. Gallo, D. Eisenberg, M. Hewison, B. W. Hollis, J. S. Adams, B. R. Bloom, and R. L. Modlin. 2006. Toll-like receptor triggering of a vitamin D-mediated human antimicrobial response. *Science* 311:1770–1773.

42. Adams, J. S., S. Ren, P. T. Liu, R. F. Chun, V. Lagishetty, A. F. Gombart, N. Borregaard, R. L. Modlin, and M. Hewison. 2009. Vitamin D-directed rheostatic regulation of monocyte antibacterial responses. *J Immunol* 182:4289–4295.

43. Liu, P. T., M. Schenk, V. P. Walker, P. W. Dempsey, M. Kanchanapoomi, M. Wheelwright, A. Vazirnia, X. Zhang, A. Steinmeyer, U. Zugel, B. W. Hollis, G. Cheng, and R. L. Modlin. 2009. Convergence of IL-1beta and VDR activation pathways in human TLR2/1-induced antimicrobial responses. *PLoS One* 4:e5810

44. Wang, T. T., B. Dabbas, D. Laperriere, A. J. Bitton, H. Soualhine, L. E. Tavera-Mendoza, S. Dionne, M. J. Servant, A. Bitton, E. G. Seidman, S. Mader, M. A. Behr, and J.H. White. 2010. Direct and indirect induction by 1,25-dihydroxyvitamin D_3 of the NOD2/CARD15-beta defensin 2 innate immune pathway defective in Crohn's disease. *J Biol Chem* 285:2227–2231.

45. Wang, T. T., F. P. Nestel, V. Bourdeau, Y. Nagai, Q. Wang, J. Liao, L Tavera-Mendoza, R. Lin, J. W. Hanrahan, S. Mader, and J. H. White. 2004. Cutting edge: 1,25-dihydroxyvitamin D_3 is a direct inducer of antimicrobial peptide gene expression. *J Immunol* 173:2909–2912.

46. Gombart, A. F., N. Borregaard, H. P. Koeffler. 2005. Human cathelicidin antimicrobial peptide (CAMP) gene is a direct target of the vitamin D receptor and is strongly up-regulated in myeloid cells by 1,25-dihydroxyvitamin D3. *FASEB J* 19:1067–1077.

47. Sly, L. M., M. Lopez, W. M. Nauseef, and N. E. Reiner. 2001. 1alpha,25-Dihydroxyvitamin D3-induced monocyte antimycobacterial activity is regulated by phosphatidylinositol 3-kinase and mediated by the NADPH-dependent phagocyte oxidase. *J Biol Chem* 276:35482–35493.

48. Kohchi, C., H. Inagawa, T. Nishizawa, and G. Soma. 2009. ROS and innate immunity. *Anticancer Res* 29:817–821.

49. Chan, J., Y. Xing, R. S. Magliozzo, and B. R. Bloom. 1992. Killing of virulent Mycobacterium tuberculosis by reactive nitrogen intermediates produced by activated murine macrophages. *J Exp Med* 175:1111–1122.

50. Sorensen, O. E., P. Follin, A. H. Johnsen, J. Calafat, G. S. Tjabringa, P. S. Hiemstra, and N. Borregaard. 2001. Human cathelicidin, hCAP-18, is processed to the antimicrobial peptide LL-37 by extracellular cleavage with proteinase 3. *Blood* 97:3951–3959.

51. Klionsky, D. J. and S. D. Emr. 2000. Autophagy as a regulated pathway of cellular degradation. *Science* 290:1717–1721.

52. Gutierrez, M. G., S. S. Master, S. B. Singh, G. A. Taylor, M. I. Colombo, and V. Deretic. 2004. Autophagy is a defense mechanism inhibiting BCG and Mycobacterium tuberculosis survival in infected macrophages. *Cell* 119:753–766.

53. Deretic, V. and B. Levine. 2009. Autophagy, immunity, and microbial adaptations. *Cell Host Microbe* 5:527–549.

54. Shin, D. M., J. M. Yuk, H. M. Lee, S. H. Lee, J. W. Son, C. V. Harding, J. M. Kim, R L. Modlin, and E. K. Jo. 2011. Mycobacterial Lipoprotein Activates Autophagy via TLR2/1/CD14 and a Functional Vitamin D Receptor Signaling. *Cell Microbiol*

55. Brennan, A., D. R. Katz, J. D. Nunn, S. Barker, M. Hewison, L. J. Fraher, and J. L. O'Riordan. 1987. Dendritic cells from human tissues express receptors for the immunoregulatory vitamin D3 metabolite, dihydroxycholecalciferol. *Immunology* 61:457–461.

56. Hewison, M., L. Freeman, S. V. Hughes, K. N. Evans, R. Bland, A. G. Eliopoulos, M. D. Kilby, P. A. Moss, and R. Chakraverty. 2003. Differential regulation of vitamin D receptor and its ligand in human monocyte-derived dendritic cells. *J Immunol* 170:5382–5390.

57. Fritsche, J., K. Mondal, A. Ehrnsperger, R. Andreesen, and M. Kreutz. 2003. Regulation of 25-hydroxyvitamin D3-1 alpha-hydroxylase and production of 1 alpha,25-dihydroxyvitamin D3 by human dendritic cells. *Blood* 102:3314–3316.

58. Hewison, M., D. Zehnder, R. Chakraverty, and J. S. Adams. 2004. Vitamin D and barrier function: a novel role for extra-renal 1 alpha-hydroxylase. *Mol Cell Endocrinol* 215:31–38.

59. Sigmundsdottir, H., J. Pan, G. F. Debes, C. Alt, A. Habtezion, D. Soler, and E. C. Butcher. 2007. DCs metabolize sunlight-induced vitamin D_3 to 'program' T cell attraction to the epidermal chemokine CCL27. *Nat Immunol* 8:285–293.
60. Chen, S., G. P. Sims, X. X. Chen, Y. Y. Gu, S. Chen, and P. E. Lipsky. 2007. Modulatory effects of 1,25-dihydroxyvitamin d3 on human B cell differentiation. *J Immunol* 179:1634–1647.
61. Gray, T. K., G. E. Lester, and R. S. Lorenc. 1979. Evidence for extra-renal 1 alpha-hydroxylation of 25-hydroxyvitamin D3 in pregnancy. *Science* 204:1311–1313.
62. Weisman, Y., A. Harell, S. Edelstein, M. David, Z. Spirer, and A. Golander. 1979. 1 alpha, 25-Dihydroxyvitamin D3 and 24,25-dihydroxyvitamin D_3 in vitro synthesis by human decidua and placenta. *Nature* 281:317–319.
63. Zehnder, D., K. N. Evans, M. D. Kilby, J. N. Bulmer, B. A. Innes, P. M. Stewart, and M. Hewison. 2002. The ontogeny of 25-hydroxyvitamin D(3) 1alpha-hydroxylase expression in human placenta and decidua. *Am J Pathol* 161:105–114.
64. Evans, K. N., J. N. Bulmer, M. D. Kilby, and M. Hewison. 2004. Vitamin D and placental-decidual function. *J Soc Gynecol Investig* 11:263–271.
65. Novakovic, B., M. Sibson, H. K. Ng, U. Manuelpillai, V. Rakyan, T. Down, S. Beck, T. Fournier, D. Evain-Brion, E. Dimitriadis, J. M. Craig, R. Morley, and R. Saffery. 2009. Placenta-specific methylation of the vitamin D 24-hydroxylase gene: implications for feedback autoregulation of active vitamin D levels at the fetomaternal interface. *J Biol Chem* 284:14838–14848.
66. Kovacs, C. S. and H. M. Kronenberg. 1997. Maternal-fetal calcium and bone metabolism during pregnancy, puerperium, and lactation. *Endocr Rev* 18:832–872.
67. Bruns, M. E. and D. E. Bruns. 1983. Vitamin D metabolism and function during pregnancy and the neonatal period. *Ann Clin Lab Sci* 13:521–530.
68. Rebut-Bonneton, C. and J. Demignon. 1991. Effects of 1,25-dihydroxyvitamin D_3 on in vitro lymphocyte reactions: arguments for a role at the maternofetal interface. *Gynecol Obstet Invest* 32:134–138.
69. Sacks, G., I. Sargent, and C. Redman. 2000. Innate immunity in pregnancy. *Immunol Today* 21:200–201.
70. Sacks, G., I. Sargent, and C. Redman. 1999. An innate view of human pregnancy. *Immunol Today* 20:114–118.
71. Guleria, I. and J. W. Pollard. 2000. The trophoblast is a component of the innate immune system during pregnancy. *Nat Med* 6:589–593.
72. Zenclussen, A. C., A. Schumacher, M. L. Zenclussen, P. Wafula, and H. D. Volk 2007. Immunology of pregnancy: Cellular mechanisms allowing fetal survival within the maternal uterus. *Expert Rev Mol Med* 9:1–14.
73. Laskarin, G., U. Kammerer, D. Rukavina, A. W. Thomson, N. Fernandez, S. M. Blois. 2007. Antigen-presenting cells and materno-fetal tolerance: An emerging role for dendritic cells. *Am J Reprod Immunol* 58:255–267.
74. King, A. E., R. W. Kelly, J. M. Sallenave, A. D. Bocking, and J. R. Challis. 2007. Innate immune defenses in the human uterus during pregnancy. *Placenta* 28:1099–1106.
75. King, A. E., A. Paltoo, R. W. Kelly, J. M. Sallenave, A. D. Bocking, and J. R. Challis. 2007. Expression of natural antimicrobials by human placenta and fetal membranes. *Placenta* 28:161–169.
76. Svinarich, D. M., N A. Wolf, R. Gomez, B. Gonik, and R. Romero. 1997. Detection of human defensin 5 in reproductive tissues. *Am J Obstet Gynecol* 176:470–475.
77. Abrahams, V. M., T. M. Schaefer, J. V. Fahey, I. Visintin, J. A. Wright, P. B. Aldo, R. Romero, C. R. Wira, and G. Mor. 2006. Expression and secretion of antiviral factors by trophoblast cells following stimulation by the TLR-3 agonist, Poly(I : C). *Hum Reprod* 21:2432–2439.
78. Evans, K. N., L. Nguyen, J. Chan, B. A. Innes, J. N. Bulmer, M. D. Kilby, and M. Hewison. 2006. Effects of 25-hydroxyvitamin D3 and 1,25-dihydroxyvitamin D3 on cytokine production by human decidual cells. *Biol Reprod* 75:816–822.
79. Liu, N., A. T. Kaplan, J. Low, L. Nguyen, G. Y. Liu, O. Equils, and M. Hewison. 2009. Vitamin D induces innate antibacterial responses in human trophoblasts via an intracrine pathway. *Biol Reprod* 80:398–406.
80. Liu, N. Q., A. T. Kaplan, V. Lagishetty, Y. B. Ouyang, Y. Ouyang, C. F. Simmons, O. Equils, and M. Hewison. 2011. Vitamin D and the regulation of placental inflammation. *J Immunol* 186:5968–5974.
81. Bodnar, L. M., M. A. Krohn, and H. N. Simhan. 2009. Maternal vitamin D deficiency is associated with bacterial vaginosis in the first trimester of pregnancy. *J Nutr* 139:1157–1161.
82. Mehta, S., D. J. Hunter, F. M. Mugusi, D. Spiegelman, K. P. Manji, E. L. Giovannucci, E. Hertzmark, G. I. Msamanga, and W. W. Fawzi. 2009. Perinatal outcomes, including mother-to-child transmission of HIV, and child mortality and their association with maternal vitamin D status in Tanzania. *J Infect Dis* 200:1022–1030.

83. Al-Husaini, A. M. 2009. Role of placenta in the vertical transmission of human immunodeficiency virus. *J Perinatol* 29:331–336.

84. Fischer, D., A. Schroer, D. Ludders, T. Cordes, B. Bucker, J. Reichrath, and M. Friedrich. 2007. Metabolism of vitamin D3 in the placental tissue of normal and preeclampsia complicated pregnancies and premature births. *Clin Exp Obstet Gynecol* 34:80–84.

85. Diaz, L., C. Arranz, E. Avila, A. Halhali, F. Vilchis, and F. Larrea. 2002. Expression and activity of 25-hydroxyvitamin D-1 alpha-hydroxylase are restricted in cultures of human syncytiotrophoblast cells from preeclamptic pregnancies. *J Clin Endocrinol Metab* 87:3876–3882.

86. Bodnar, L. M., J. M. Catov, H. N. Simhan, M. F. Holick, R. W. Powers, and Roberts. J. M. 2007. Maternal vitamin D deficiency increases the risk of preeclampsia. *J Clin Endocrinol Metab*

87. Bodnar, L. M., H. N. Simhan, R. W. Powers, M. P. Frank, E. Cooperstein, and J. M. Roberts. 2007. High prevalence of vitamin D insufficiency in black and white pregnant women residing in the northern United States and their neonates. *J Nutr* 137:447–452.

88. Hypponen, E. 2005. Vitamin D for the prevention of preeclampsia? A hypothesis. *Nutr Rev* 63: 225–232.

89. Schwartz, G. G., L. W. Whitlatch, T. C. Chen, B. L. Lokeshwar, and M. F. Holick. 1998. Human prostate cells synthesize 1,25-dihydroxyvitamin D3 from 25-hydroxyvitamin D3. *Cancer Epidemiol Biomarkers Prev* 7:391–395.

90. Whitlatch, L. W., M. V. Young, G. G. Schwartz, J. N. Flanagan, K. L. Burnstein, B. L. Lokeshwar, F. S. Rich, M. F. Holick, amd T. C. Chen. 2002. 25 Hydroxyvitamin D-1alpha-hydroxylase activity is diminished in human prostate cancer cells and is enhanced by gene transfer. *J Steroid Biochem Mol Biol* 81:135–140.

91. Hsu, J. Y., D. Feldman, J. E. McNeal, and D. M. Peehl. 2001. Reduced 1alpha-hydroxylase activity in human prostate cancer cells correlates with decreased susceptibility to 25-hydroxyvitamin D3-induced growth inhibition. *Cancer Res* 61:2852–2856.

92. Barreto, A. M., G. G. Schwartz, R. Woodruff, and S. D. Cramer. 2000. 25-Hydroxyvitamin D3, the prohormone of 1,25-dihydroxyvitamin D3, inhibits the proliferation of primary prostatic epithelial cells. *Cancer Epidemiol Biomarkers Prev* 9:265–270.

93. Polek, T. C. and N. L. Weigel. 2002. Vitamin D and prostate cancer. *J Androl* 23:9–17.

94. Townsend, K., C. M. Banwell, M. Guy, K. W. Colston, J. L. Mansi, P. M. Stewart, M. J. Campbell, and M. Hewison. 2005. Autocrine metabolism of vitamin D in normal and malignant breast tissue. *Clin Cancer Res* 11:3579–3586.

95. McCarthy, K., C. Laban, S. A. Bustin, W. Ogunkolade, S. Khalaf, R. Carpenter, and P. J. Jenkins. 2009. Expression of 25-hydroxyvitamin D-1-alpha-hydroxylase, and vitamin D receptor mRNA in normal and malignant breast tissue. *Anticancer Res* 29:155–157.

96. Segersten, U., P. K. Holm, P. Bjorklund, O. Hessman, H. Nordgren, L. Binderup, G. Akerstrom, P. Hellman, and G. Westin. 2005. 25-Hydroxyvitamin D3 1alpha-hydroxylase expression in breast cancer and use of non-1alpha-hydroxylated vitamin D analogue. *Breast Cancer Res* 7:R980–986.

97. Colston, K. W., L. C. Lowe, J. L. Mansi, and M. J. Campbell. 2006. Vitamin D status and breast cancer risk. *Anticancer Res* 26:2573–2580.

98. Kemmis, C. M. and J. Welsh. 2008. Mammary epithelial cell transformation is associated with deregulation of the vitamin D pathway. *J Cell Biochem* 105:980–988.

99. Rowling, M. J., C. M. Kemmis, D. A. Taffany, and J. Welsh. 2006. Megalin-mediated endocytosis of vitamin D binding protein correlates with 25-hydroxycholecalciferol actions in human mammary cells. *J Nutr* 136:2754–2759.

100. Nykjaer, A., D. Dragun, D. Walther, H. Vorum, C. Jacobsen, J. Herz, F. Melsen, E. I. Christensen, and T. E. Willnow. 1999. An endocytic pathway essential for renal uptake and activation of the steroid 25-(OH) vitamin D3. *Cell* 96:507–515.

101. Welsh, J. 2004. Vitamin D and breast cancer: Insights from animal models. *Am J Clin Nutr* 80:1721S–1724S.

102. Turner, R. T., J. E. Puzas, M. D. Forte, G. E. Lester, T. K. Gray, G. A. Howard, and D. J. Baylink. 1980. *In vitro* synthesis of 1 alpha,25-dihydroxycholecalciferol and 24,25-dihydroxycholecalciferol by isolated calvarial cells. *Proc Natl Acad Sci USA* 77:5720–5724.

103. Howard, G. A., R. T. Turner, D. J. Sherrard, and D. J. Baylink. 1981. Human bone cells in culture metabolize 25-hydroxyvitamin D3 to 1,25-dihydroxyvitamin D3 and 24,25-dihydroxyvitamin D3. *J Biol Chem* 256:7738–7740.

104. Pols, H. A., H. P. Schilte, P. J. Nijweide, T. J. Visser, and J. C. Birkenhager. 1984. The influence of albumin on vitamin D metabolism in fetal chick osteoblast-like cells. *Biochem Biophys Res Commun* 125:265–272.

105. Atkins, G. J., P. H. Anderson, D. M. Findlay, K. J. Welldon, C. Vincent, A. C. Zannettino, P. D. O'Loughlin, and H. A. Morris. 2007. Metabolism of vitamin D3 in human osteoblasts: evidence for autocrine and paracrine activities of 1 alpha,25-dihydroxyvitamin D3. *Bone* 40:1517–1528.

106. Somjen, D., S. Katzburg, N. Stern, F. Kohen, O. Sharon, R. Limor, N. Jaccard, D. Hendel, and Y. Weisman. 2007. 25 hydroxy-vitamin D(3)-1alpha hydroxylase expression and activity in cultured human osteoblasts and their modulation by parathyroid hormone, estrogenic compounds and dihydrotestosterone. *J Steroid Biochem Mol Biol* 107:238–244.

107. Anderson, P. H. and G. J. Atkins. 2008. The skeleton as an intracrine organ for vitamin D metabolism. *Mol Aspects Med* 29:397–406.

108. St-Arnaud. R. 2008. The direct role of vitamin D on bone homeostasis. *Arch Biochem Biophys* 473:225–230.

109. Pedrozo, H. A., B. D. Boyan, J. Mazock, D. D. Dean, R. Gomez, and Z. Schwartz. 1999. TGFbeta1 regulates 25-hydroxyvitamin D3 1alpha- and 24-hydroxylase activity in cultured growth plate chondrocytes in a maturation-dependent manner. *Calcif Tissue Int* 64:50–56.

110. Weber, L., U. Hugel, J. Reichrath, H. Sieverts, O. Mehls, and G. Klaus. 2003. Cultured rat growth plate chondrocytes express low levels of 1alpha-hydroxylase. *Recent Results Cancer Res* 164:147–149.

111. Schauber, J., R. A. Dorschner, A. B. Coda, A. S. Buchau, P. T. Liu, D. Kiken, Y. R. Helfrich, S. Kang, H. Z. Elalieh, A. Steinmeyer, U. Zugel, D. D. Bikle, R. L. Modlin, and R. L. Gallo. 2007. Injury enhances TLR2 function and antimicrobial peptide expression through a vitamin D-dependent mechanism. *J Clin Invest* 117:803–811.

112. St. Arnaud, R. O. Dardenne, J. Prud'homme, S. A. Hacking, and F. H. Glorieux. 2003. Conventional and tissue-specific inactivation of the 25-hydroxyvitamin D-1alpha-hydroxylase (CYP27B1). *J Cell Biochem* 88:245–251.

113. Naja, R. P., O. Dardenne, A. Arabian, and R. St. Arnaud. 2009. Chondrocyte-specific modulation of Cyp27b1 expression supports a role for local synthesis of 1,25-dihydroxyvitamin D3 in growth plate development. *Endocrinology* 150:4024–4032.

114. Powe, C. E., C. Ricciardi, A. H. Berg, D. Erdenesanaa, G. Collerone, E. Ankers, J. Wenger, S. A. Karumanchi, R. Thadhani, and I. Bhan. 2011. Vitamin D–binding protein modifies the vitamin D-bone mineral density relationship. *J Bone Miner Res* 26:1609–1616.

115. Segersten, U., P. Correa, M. Hewison, P. Hellman, H. Dralle, T. Carling, G. Akerstrom, and G. Westin. 2002. 25-hydroxyvitamin D(3)-1alpha-hydroxylase expression in normal and pathological parathyroid glands. *J Clin Endocrinol Metab* 87:2967–2972.

116. Correa, P., U. Segersten, P. Hellman, G. Akerstrom, G. Westin. 2002. Increased 25-hydroxyvitamin D3 1alpha-hydroxylase and reduced 25-hydroxyvitamin D3 24-hydroxylase expression in parathyroid tumors—new prospects for treatment of hyperparathyroidism with vitamin D. *J Clin Endocrinol Metab* 87:5826–5829.

117. Kawahara, M., Y. Iwasaki, K. Sakaguchi, T. Taguchi, M. Nishiyama, T. Nigawara, M. Tsugita, M. Kambayashi, T. Suda, and K. Hashimoto. 2008. Predominant role of 25OHD in the negative regulation of PTH expression: clinical relevance for hypovitaminosis D. *Life Sci* 82:677–683.

118. Lauter, K. and A. Arnold. 2009. Analysis of CYP27B1, encoding 25-hydroxyvitamin D-1alpha-hydroxylase, as a candidate tumor suppressor gene in primary and severe secondary/tertiary hyperparathyroidism. *J Bone Miner Res* 24:102–104.

119. Krajisnik, T., P. Bjorklund, R. Marsell, O. Ljunggren, G. Akerstrom, K. B. Jonsson, G. Westin, and T. E. Larsson. 2007. Fibroblast growth factor-23 regulates parathyroid hormone and 1alpha-hydroxylase expression in cultured bovine parathyroid cells. *J Endocrinol* 195:125–131.

120. Bland, R., D. Markovic, C. E. Hills, S. V. Hughes, S. L. Chan, P. E. Squires, and M. Hewison. 2004. Expression of 25-hydroxyvitamin D3-1alpha-hydroxylase in pancreatic islets. *J Steroid Biochem Mol Biol* 89–90:121–125.

121. Leung, P. S. and Q. Cheng. 2010. The novel roles of glucagon-like peptide-1, angiotensin II, and vitamin D in islet function. *Adv Exp Med Biol* 654:339–361.

122. Mathieu, C., C. Gysemans, A. Giulietti, and R. Bouillon. 2005. Vitamin D and diabetes. *Diabetologia* 48:1247–1257.

123. Khadzkou, K., P. Buchwald, G. Westin, H. Dralle, G. Akerstrom, and P. Hellman. 2006. 25-hydroxyvitamin D$_3$ 1alpha-hydroxylase and vitamin D receptor expression in papillary thyroid carcinoma. *J Histochem Cytochem* 54:355–361.

124. Blomberg Jensen, M., J. E. Nielsen, A. Jorgensen, E. Rajpert De Meyts, D. M. Kristensen, N. Jorgensen, N. E. Skakkebaek, A. Juul, and H. Leffers. 2010. Vitamin D receptor and vitamin D metabolizing enzymes are expressed in the human male reproductive tract. *Hum Reprod* 25:1303–1311.

125. Kinuta, K., H. Tanaka, T. Moriwake, K. Aya, S. Kato, and Y. Seino. 2000. Vitamin D is an important factor in estrogen biosynthesis of both female and male gonads. *Endocrinology* 141:1317–1324.
126. Hughes, S. V., E. Robinson, R. Bland, H. M. Lewis, P. M. Stewart, and M. Hewison. 1997. 1,25-dihydroxyvitamin D_3 regulates estrogen metabolism in cultured keratinocytes. *Endocrinology* 138:3711–3718.
127. Fischer, D., M. Thome, S. Becker, T. Cordes, K. Diedrich, M. Friedrich, and M. Thill. 2009. Expression of 25-hydroxyvitamin D_3-24-hydroxylase in benign and malignant ovarian cell lines and tissue. *Anticancer Res* 29:3635–3639.
128. Evans, K. N., H. Taylor, D. Zehnder, M. D. Kilby, J. N. Bulmer, F. Shah, J. S. Adams, and M. Hewison. 2004. Increased expression of 25-hydroxyvitamin D-1alpha-hydroxylase in dysgerminomas: A novel form of humoral hypercalcemia of malignancy. *Am J Pathol* 165:807–813.
129. Friedrich, M., L. Rafi, T. Mitschele, W. Tilgen, W. Schmidt, and J. Reichrath. 2003. Analysis of the vitamin D system in cervical carcinomas, breast cancer and ovarian cancer. *Recent Results Cancer Res* 164:239–246.
130. Agic, A., H. Xu, C. Altgassen, F. Noack, M. M. Wolfler, K. Diedrich, M. Friedrich, R. N. Taylor, and D. Hornung. 2007. Relative expression of 1,25-dihydroxyvitamin D3 receptor, vitamin D 1 alpha-hydroxylase, vitamin D 24-hydroxylase, and vitamin D 25-hydroxylase in endometriosis and gynecologic cancers. *Reprod Sci* 14:486–497.
131. Bareis, P., E. Kallay, M. G. Bischof, G. Bises, H. Hofer, C. Potzi, T. Manhardt, R. Bland, H. S. Cross. 2002. Clonal differences in expression of 25-hydroxyvitamin $D(_3)$-1alpha-hydroxylase, of 25-hydroxyvitamin $D(_3)$-24-hydroxylase, and of the vitamin D receptor in human colon carcinoma cells: Effects of epidermal growth factor and 1alpha,25-dihydroxyvitamin $D(_3)$. *Exp Cell Res* 276:320–327.
132. Bischof, M. G., M. L. Siu-Caldera, A. Weiskopf, P. Vouros, H. S. Cross, M. Peterlik, and G. S. Reddy. 1998. Differentiation-related pathways of 1 alpha,25-dihydroxycholecalciferol metabolism in human colon adenocarcinoma-derived Caco-2 cells: Production of 1 alpha,25-dihydroxy-3epi-cholecalciferol. *Exp Cell Res* 241:194–201.
133. Lagishetty, V., R. F. Chun, N. Q. Liu, T. S. Lisse, J. S. Adams, and M. Hewison. 1alpha-Hydroxylase and innate immune responses to 25-hydroxyvitamin D in colonic cell lines. *J Steroid Biochem Mol Biol*
134. Bareis, P., G. Bises, M. G. Bischof, H. S. Cross, and M. Peterlik. 2001. 25-hydroxy-vitamin D metabolism in human colon cancer cells during tumor progression. *Biochem Biophys Res Commun* 285:1012–1017.
135. Kallay, E., G. Bises, E. Bajna, C. Bieglmaycr, W. Gerdenitsch, I. Steffan, S. Kato, H. J. Armbrecht, and H. S. Cross. 2005. Colon-specific regulation of vitamin D hydroxylases—A possible approach for tumor prevention. *Carcinogenesis* 26:1581–1589.
136. Cross, H. S., G. Bises, D. Lechner, T. Manhardt, and E. Kallay. 2005. The Vitamin D endocrine system of the gut: Its possible role in colorectal cancer prevention. *J Steroid Biochem Mol Biol* 97:121–128.
137. Peterlik, M. and H. S. Cross. 2006. Dysfunction of the vitamin D endocrine system as common cause for multiple malignant and other chronic diseases. *Anticancer Res* 26:2581–2588.
138. Liu, N., L. Nguyen, R. F. Chun, V. Lagishetty, S. Ren, S. Wu, B. Hollis, H. F. De Luca, J. S. Adams, and M. Hewison. 2008. Altered endocrine and autocrine metabolism of vitamin d in a mouse model of gastrointestinal inflammation. *Endocrinology* 149:4799–4808.
139. Abreu, M. T., V. Kantorovich, E. A. Vasiliauskas, U. Gruntmanis, R. Matuk, K. Daigle, S. Chen, D. Zehnder, Y. C. Lin, H. Yang, M. Hewison, and J. S. Adams. 2004. Measurement of vitamin D levels in inflammatory bowel disease patients reveals a subset of Crohn's disease patients with elevated 1,25-dihydroxyvitamin D and low bone mineral density. *Gut* 53:1129–1136.
140. Lagishetty, V., A. V. Misharin, N. Q. Liu, T. S. Lisse, R. F. Chun, Y. Ouyang, S. M. McLachlan, J. S. Adams, and M. Hewison. 2010. Vitamin D deficiency in mice impairs colonic antibacterial activity and predisposes to colitis. *Endocrinology* 151:2423–2432.
141. Hooper, L. V., T. S. Stappenbeck, C. V. Hong, and J. I. Gordon. 2003. Angiogenins: A new class of microbicidal proteins involved in innate immunity. *Nat Immunol* 4:269–273.
142. Packey, C. D. and R. B. Sartor. 2009. Commensal bacteria, traditional and opportunistic pathogens, dysbiosis and bacterial killing in inflammatory bowel diseases. *Curr Opin Infect Dis* 22:292–301.
143. Smith, S. J., A. K. Rucka, J. L. Berry, M. Davies, S. Mylchreest, C. R. Paterson, D. A. Heath, M. Tassabehji, A. P. Read, A. P. Mee, and E. B. Mawer. 1999. Novel mutations in the 1alpha-hydroxylase (P450c1) gene in three families with pseudovitamin D-deficiency rickets resulting in loss of functional enzyme activity in blood-derived macrophages. *J Bone Miner Res* 14:730–739.
144. Shany, S., S. Y. Ren, J. E. Arbelle, T. L. Clemens, and J. S. Adams. 1993. Subcellular localization and partial purification of the 25-hydroxyvitamin D_3 1-hydroxylation reaction in the chick myelomonocytic cell line HD-11. *J Bone Miner Res* 8:269–276.

145. Reichel, H., H. P. Koeffler, A. W. Norman. 1987. Synthesis in vitro of 1,25-dihydroxyvitamin D$_3$ and 24,25-dihydroxyvitamin D$_3$ by interferon-gamma-stimulated normal human bone marrow and alveolar macrophages. *J Biol Chem* 262:10931–10937.

146. Fraser, D. R. 1980. Regulation of the metabolism of vitamin D. *Physiol Rev* 60:551–613.

147. Adams, J. S., S. Y. Ren, J. E. Arbelle, N. Horiuchi, R. W. Gray, T. L. Clemens, and S. Shany. 1994. Regulated production and intracrine action of 1,25-dihydroxyvitamin D$_3$ in the chick myelomonocytic cell line HD-11. *Endocrinology* 134:2567–2573.

148. Nakamura, Y., T. A. Eto, T. Taniguchi, K. Miyamoto, J. Nagatomo, H. Shiotsuki, H. Sueta, S. Higashi, K. I. Okuda, and T. Setoguchi. 1997. Purification and characterization of 25-hydroxyvitamin D$_3$ 1alpha-hydroxylase from rat kidney mitochondria. *FEBS Lett* 419:45–48.

149. Inouye, K. and Sakaki, T. 2001. Enzymatic studies on the key enzymes of vitamin D metabolism; 1 alpha-hydroxylase (CYP27B1) and 24-hydroxylase (CYP24). *Biotechnol Annu Rev* 7:179–194.

150. Hakeda, Y., K. Hiura, T. Sato, R. Okazaki, T. Matsumoto, E. Ogata, R. Ishitani, and M. Kumegawa. 1989. Existence of parathyroid hormone binding sites on murine hemopoietic blast cells. *Biochem Biophys Res Commun* 163:1481–1486.

151. Bland, R., E. A. Walker, S. V. Hughes, P. M. Stewart, and M. Hewison. 1999. Constitutive expression of 25-hydroxyvitamin D$_3$-1alpha-hydroxylase in a transformed human proximal tubule cell line: evidence for direct regulation of vitamin D metabolism by calcium. *Endocrinology* 140:2027–2034.

152. Yuan, J. Y., A. J. Freemont, E. B. Mawer, and M. E. Hayes. 1992. Regulation of 1 alpha, 25-dihydroxyvitamin D3 synthesis in macrophages from arthritic joints by phorbol ester, dibutyryl-cAMP and calcium ionophore (A23187). *FEBS Lett* 311:71–74.

153. Henry, H. L. 1981. 25(OH)D$_3$ metabolism in kidney cell cultures: lack of a direct effect of estradiol. *Am J Physiol* 240:E119–E124.

154. Adams, N. D., T. L. Garthwaite, R. W. Gray, T. C. Hagen, and J. Lemann Jr. 1979. The interrelationships among prolactin, 1,25-dihydroxyvitamin D, and parathyroid hormone in humans. *J Clin Endocrinol Metab* 49:628–630.

155. Brixen, K., H. K. Nielsen, R. Bouillon, A. Flyvbjerg, and L. Mosekilde. 1992. Effects of short-term growth hormone treatment on PTH, calcitriol, thyroid hormones, insulin and glucagon. *Acta Endocrinol (Copenh)* 127:331–336.

156. Henry, H. L. 1992. Vitamin D hydroxylases. *J Cell Biochem* 49:4–9.

157. Omdahl, J. L., H. A. Morris, and B. K. May. 2002. Hydroxylase enzymes of the vitamin D pathway: expression, function, and regulation. *Annu Rev Nutr* 22:139–166.

158. Omdahl, J. L., E. A. Bobrovnikova, S. Choe, P. P. Dwivedi, and B. K. May. 2001. Overview of regulatory cytochrome P450 enzymes of the vitamin D pathway. *Steroids* 66:381–389.

159. Ohyama, Y., M. Noshiro, and K. Okuda. 1991. Cloning and expression of cDNA encoding 25-hydroxyvitamin D$_3$ 24-hydroxylase. *FEBS Lett* 278:195–198.

160. Chen, K. S., J. M. Prahl, and H. F. De Luca. 1993. Isolation and expression of human 1,25-dihydroxyvitamin D$_3$ 24-hydroxylase cDNA. *Proc Natl Acad Sci USA* 90:4543–4547.

161. Zierold, C., J. A. Mings, and H. F. De Luca. 2003. Regulation of 25-hydroxyvitamin D$_3$-24-hydroxylase mRNA by 1,25-dihydroxyvitamin D$_3$ and parathyroid hormone. *J Cell Biochem* 88:234–237.

162. Henry, H. L. 1979. Regulation of the hydroxylation of 25-hydroxyvitamin D$_3$ in vivo and in primary cultures of chick kidney cells. *J Biol Chem* 254:2722–2729.

163. Ebert, R., N. Schutze, J. Adamski, and F. Jakob. 2006. Vitamin D signaling is modulated on multiple levels in health and disease. *Mol Cell Endocrinol* 248:149–159.

164. Edfeldt, K., P. T. Liu, R. Chun, M. Fabri, M. Schenk, M. Wheelwright, C. Keegan, S. R. Krutzik, J. S. Adams, M. Hewison, and R. L. Modlin. 2010. T-cell cytokines differentially control human monocyte antimicrobial responses by regulating vitamin D metabolism. *Proc Natl Acad Sci USA* 107:22593–22598.

165. Xie, Z., S. J. Munson, N. Huang, A. A. Portale, W. L. Miller, and D. D. Bikle. 2002. The mechanism of 1,25-dihydroxyvitamin D$_{(3)}$ autoregulation in keratinocytes. *J Biol Chem* 277:36987–36990.

166. Ren, S., L. Nguyen, S. Wu, C. Encinas, J. S. Adams, and M. Hewison. 2005. Alternative splicing of vitamin D-24-hydroxylase: a novel mechanism for the regulation of extrarenal 1,25-dihydroxyvitamin D synthesis. *J Biol Chem* 280:20604–20611.

167. Adams, J. S., H. Chen, R. Chun, S. Ren, S. Wu, M. Gacad, L. Nguyen, J. Ride, P. Liu, R. Modlin, and M. Hewison. 2007. Substrate and enzyme trafficking as a means of regulating 1,25-dihydroxyvitamin D synthesis and action: the human innate immune response. *J Bone Miner Res* 22(Suppl 2):V20–V24.

168. Hewison, M., F. Burke, K. N. Evans, D. A. Lammas, D. M. Sansom, P. Liu, R. L. Modlin, and J. S. Adams. 2007. Extra-renal 25-hydroxyvitamin D3-1alpha-hydroxylase in human health and disease. *J Steroid Biochem Mol Biol* 103:316–321.

169. Krutzik, S. R., M. Hewison, P. T. Liu, J. A. Robles, S. Stenger, J. S. Adams, and R. L. Modlin. 2008. IL-15 links TLR2/1-induced macrophage differentiation to the vitamin D-dependent antimicrobial pathway. *J Immunol* 181:7115–7120.

170. Waldmann, T. A. 2006. The biology of interleukin-2 and interleukin-15: implications for cancer therapy and vaccine design. *Nat Rev Immunol* 6:595–601.

171. Schoenborn, J. R. and C. B. Wilson. 2007. Regulation of interferon-gamma during innate and adaptive immune responses. *Adv Immunol* 96:41–101.

172. Josien, R., B. R. Wong, H. L. Li, R. M. Steinman, and Y. Choi. 1999. TRANCE, a TNF family member, is differentially expressed on T cell subsets and induces cytokine production in dendritic cells. *J Immunol* 162:2562–2568.

173. Murayama, A., K. Takeyama, S. Kitanaka, Y. Kodera, T. Hosoya, and S. Kato. 1998. The promoter of the human 25-hydroxyvitamin D$_3$ 1 alpha-hydroxylase gene confers positive and negative responsiveness to PTH, calcitonin, and 1 alpha,25(OH)2D3. *Biochem Biophys Res Commun* 249:11–16.

174. Kong, X. F., X. H. Zhu, Y. L. Pei, D. M. Jackson, and M. F. Holick. 1999. Molecular cloning, characterization, and promoter analysis of the human 25-hydroxyvitamin D$_3$-1alpha-hydroxylase gene. *Proc Natl Acad Sci USA* 96:6988–6993.

175. Platanias, L. C. and E. N. Fish. 1999. Signaling pathways activated by interferons. *Exp Hematol* 27:1583–1592.

176. Imada, K. and W. J. Leonard. 2000. The Jak-STAT pathway. *Mol Immunol* 37:1–11.

177. Ishihara, K. and T. Hirano. 2002. Molecular basis of the cell specificity of cytokine action. *Biochim Biophys Acta* 1592:281–296.

178. Xaus, J., M. Comalada, A. F. Valledor, M. Cardo, C. Herrero, C. Soler, J. Lloberas, and A. Celada. 2001. Molecular mechanisms involved in macrophage survival, proliferation, activation or apoptosis. *Immunobiology* 204:543–550.

179. Stoffels, K., L. Overbergh, A. Giulietti, L. Verlinden, R. Bouillon, and C. Mathieu. 2006. Immune regulation of 25-hydroxyvitamin-d(3)-1alpha-hydroxylase in human monocytes. *J Bone Miner Res* 21: 37–47.

180. Stoffels, K., L. Overbergh, A. Giulietti, L. Verlinden, R. Bouillon, and C. Mathieu. 2006. Immune regulation of 25-hydroxyvitamin-D$_3$-1alpha-hydroxylase in human monocytes. *J Bone Miner Res* 21: 37–47.

181. Stoffels, K., L. Overbergh, R. Bouillon, and C. Mathieu. 2007. Immune regulation of 1alpha-hydroxylase in murine peritoneal macrophages: Unraveling the IFNgamma pathway. *J Steroid Biochem Mol Biol* 103:567–571.

182. Medzhitov, R. and C. Janeway Jr. 2000. The Toll receptor family and microbial recognition. *Trends Microbiol* 8:452–456.

183. Trinchieri, G. and A. Sher. 2007. Cooperation of Toll-like receptor signals in innate immune defence. *Nat Rev Immunol* 7:179–190.

184. Adams, J. S., S. Y. Ren, J. E. Arbelle, T. L. Clemens, and S. Shany. 1994. A role for nitric oxide in the regulated expression of the 25-hydroxy-vitamin D-1-hydroxylation reaction in the chick myelomonocytic cell line HD-11. *Endocrinology* 134:499–502.

185. Adams, J. S., S. Y. Ren, J. E. Arbelle, S. Shany, and M. A. Gacad. 1995. Coordinate regulation of nitric oxide and 1,25-dihydroxyvitamin D production in the avian myelomonocytic cell line HD-11. *Endocrinology* 136:2262–2269.

186. Adams, J. S. and S. Y. Ren. 1996. Autoregulation of 1,25-dihydroxyvitamin D synthesis in macrophage mitochondria by nitric oxide. *Endocrinology* 137:4514–4517.

187. Nathan, C. and Q. W. Xie. 1994. Nitric oxide synthases: roles, tolls, and controls. *Cell* 78:915–918.

188. Salkowski, C. A., G. Detore, R. McNally, N. van Rooijen, and S. N. Vogel. 1997. Regulation of inducible nitric oxide synthase messenger RNA expression and nitric oxide production by lipopolysaccharide in vivo: the roles of macrophages, endogenous IFN-gamma, and TNF receptor-1-mediated signaling. *J Immunol* 158:905–912.

189. Alley, E. W., W. J. Murphy, and S. W. Russell. 1995. A classical enhancer element responsive to both lipopolysaccharide and interferon-gamma augments induction of the iNOS gene in mouse macrophages. *Gene* 158:247–251.

190. Khatsenko, O. G., S. S. Gross, A. B. Rifkind, and J. R. Vane. 1993. Nitric oxide is a mediator of the decrease in cytochrome P450-dependent metabolism caused by immunostimulants. *Proc Natl Acad Sci USA* 90:11147–11151.

191. Stadler, J., J. Trockfeld, W. A. Schmalix, T. Brill, J. R. Siewert, H. Greim, and J. Doehmer. 1994. Inhibition of cytochromes P4501A by nitric oxide. *Proc Natl Acad Sci USA* 91:3559–3563.

192. Enioutina, E. Y., D. Bareyan, and R. A. Daynes. 2008. TLR ligands that stimulate the metabolism of vitamin D3 in activated murine dendritic cells can function as effective mucosal adjuvants to subcutaneously administered vaccines. *Vaccine* 26:601–613.
193. Enioutina, E. Y., D. Bareyan, and R. A. Daynes. 2009. TLR-induced local metabolism of vitamin D_3 plays an important role in the diversification of adaptive immune responses. *J Immunol* 182:4296–4305.
194. Takeda, K. and S. Akira. 2005. Toll-like receptors in innate immunity. *Int Immunol* 17:1–14.
195. Casella, C. R. and T. C. Mitchell. 2008. Putting endotoxin to work for us: Monophosphoryl lipid A as a safe and effective vaccine adjuvant. *Cell Mol Life Sci* 65:3231–3240.
196. Mata-Haro, V., C. Cekic, M. Martin, P. M. Chilton, C. R. Casella, and T. C. Mitchell. 2007. The vaccine adjuvant monophosphoryl lipid A as a TRIF-biased agonist of TLR4. *Science* 316:1628–1632.
197. Kallas, M., F. Green, M. Hewison, C. White, and G. Kline. 2010. Rare causes of calcitriol-mediated hypercalcemia: a case report and literature review. *J Clin Endocrinol Metab* 95:3111–3117.
198. Harrell, G. T. and S. Fisher. 1939. Blood chemical changes in Boeck's sarcoid with particular reference to protein, calcium and phosphatase values. *J Clin Invest* 18:687–693.
199. Albright, F., E. L. Carroll, E. F. Dempsey, and P. H. Henneman. 1956. The cause of hypercalcuria in sarcoid and its treatment with cortisone and sodium phytate. *J Clin Invest* 35:1229–1242.
200. Taylor, R. L., H. J. Lynch Jr., W. G. Wysor Jr. 1963. Seasonal influence of sunlight on the hypercalcemia of sarcoidosis. *Am J Med* 34:221–227.
201. Dent, C. E., F. V. Flynn, and J. D. Nabarro. 1953. Hypercalcaemia and impairment of renal function in generalized sarcoidosis. *Br Med J* 2:808–810.
202. Bell, N. H. G. J. J. and F. C. Barter. 1964. Abnormal calcium absorption in sarcoidosis: evidence for increased sensitivity to vitamin D. *Am J Med* 36:500–513.
203. Holick, M. F. 1990. The use and interpretation of assays for vitamin D and its metabolites. *J Nutr* 120(Suppl 11):1464–1469.
204. Bell, N. H., P. H. Stern, E. Pantzer, T. K. Sinha, and H. F. De Luca. 1979. Evidence that increased circulating 1 alpha, 25-dihydroxyvitamin D is the probable cause for abnormal calcium metabolism in sarcoidosis. *J Clin Invest* 64:218–225.
205. Papapoulos, S. E., T. L. Clemens, L. J. Fraher, I. G. Lewin, L. M. Sandler, and J. L. O'Riordan. 1979. 1, 25-dihydroxycholecalciferol in the pathogenesis of the hypercalcaemia of sarcoidosis. *Lancet* 1:627–630.
206. Stern, P. H., J. De Olazabal, and N. H. Bell. 1980. Evidence for abnormal regulation of circulating 1 alpha,25-dihydroxyvitamin D in patients with sarcoidosis and normal calcium metabolism. *J Clin Invest* 66:852–855.
207. Barbour, G. L., J. W. Coburn, E. Slatopolsky, A. W. Norman, and R. L. Horst. 1981. Hypercalcemia in an anephric patient with sarcoidosis: evidence for extrarenal generation of 1,25-dihydroxyvitamin D. *N Engl J Med* 305:440–443.
208. Adams, J. S., O. P. Sharma, M. A. Gacad, and F. R. Singer. 1983. Metabolism of 25-hydroxyvitamin D_3 by cultured pulmonary alveolar macrophages in sarcoidosis. *J Clin Invest* 72:1856–1860.
209. Adams, J. S. and M. A. Gacad. 1985. Characterization of 1 alpha-hydroxylation of vitamin D_3 sterols by cultured alveolar macrophages from patients with sarcoidosis. *J Exp Med* 161:755–765.
210. Adams, J. S., F. R. Singer, M. A. Gacad, O. P. Sharma, M. J. Hayes, P. Vouros, and M. F. Holick. 1985. Isolation and structural identification of 1,25-dihydroxyvitamin D_3 produced by cultured alveolar macrophages in sarcoidosis. *J Clin Endocrinol Metab* 60:960–966.
211. Bell, N. H. 1991. Endocrine complications of sarcoidosis. *Endocrinol Metab Clin North Am* 20:645–654.
212. Basile, J. N., Y. Liel, J. Shary, and N. H. Bell. 1993. Increased calcium intake does not suppress circulating 1,25-dihydroxyvitamin D in normocalcemic patients with sarcoidosis. *J Clin Invest* 91:1396–1398.
213. Sandler, L. M., C. G. Winearls, L. J. Fraher, T. L. Clemens, R. Smith, and J. L. O'Riordan. 1984. Studies of the hypercalcaemia of sarcoidosis: effect of steroids and exogenous vitamin D_3 on the circulating concentrations of 1,25-dihydroxy vitamin D_3. *Q J Med* 53:165–180.
214. Cronin, C. C., S. F. Dinneen, M. S. O'Mahony, C. P. Bredin, and D. J. O'Sullivan. 1990. Precipitation of hypercalcaemia in sarcoidosis by foreign sun holidays: report of four cases. *Postgrad Med J* 66:307–309.
215. Insogna, K. L., B. E. Dreyer, M. Mitnick, A. F. Ellison, and A. E. Broadus. 1988. Enhanced production rate of 1,25-dihydroxyvitamin D in sarcoidosis. *J Clin Endocrinol Metab* 66:72–75.
216. Shulman, L. E., E. H. Schoenrich, and A. M. Harvey. 1952. The effects of adrenocorticotropic hormone (ACTH) and cortisone on sarcoidosis. *Bull Johns Hopkins Hosp* 91:371–415.
217. Anderson, J., C. Harper, C. E. Dent, and G. R. Philpot. 1954. Effect of cortisone on calcium metabolism in sarcoidosis with hypercalcaemia, possibly antagonistic actions of cortisone and vitamin D. *Lancet* 267:720–724.

218. O'Leary, T. J., G. Jones, A. Yip, D. Lohnes, M. Cohanim, and E. R. Yendt. 1986. The effects of chloroquine on serum 1,25-dihydroxyvitamin D and calcium metabolism in sarcoidosis. *N Engl J Med* 315:727–730.
219. Barre, P. E., M. Gascon-Barre, J. L. Meakins, and D. Goltzman. 1987. Hydroxychloroquine treatment of hypercalcemia in a patient with sarcoidosis undergoing hemodialysis. *Am J Med* 82:1259–1262.
220. Adams. J. S., O. P. Sharma, M. M. Diz, and D. B. Endres. 1990. Ketoconazole decreases the serum 1,25-dihydroxyvitamin D and calcium concentration in sarcoidosis-associated hypercalcemia. *J Clin Endocrinol Metab* 70:1090–1095.
221. Adams, J. S., M. M., and O. P. Sharma. 1989. Effective reduction in the serum 1,25-dihydroxyvitamin D and calcium concentration in sarcoidosis-associated hypercalcemia with short-course chloroquine therapy. *Ann Intern Med* 111:437–438.
222. Felsenfeld, A. J., M. K. Drezner, and F. Llach. 1986. Hypercalcemia and elevated calcitriol in a maintenance dialysis patient with tuberculosis. *Arch Intern Med* 146:1941–1945.
223. Gkonos, P. J., R. London, and E. D. Hendler. 1984. Hypercalcemia and elevated 1,25-dihydroxyvitamin D levels in a patient with end-stage renal disease and active tuberculosis. *N Engl J Med* 311:1683–1685.
224. Epstein, S., P. H. Stern, N. H. Bell, I. Dowdeswell, and R. T. Turner. 1984. Evidence for abnormal regulation of circulating 1 alpha, 25-dihydroxyvitamin D in patients with pulmonary tuberculosis and normal calcium metabolism. *Calcif Tissue Int* 36:541–544.
225. Bell, N. H., J. Shary, S. Shaw, and R. T. Turner. 1985. Hypercalcemia associated with increased circulating 1,25 dihydroxyvitamin D in a patient with pulmonary tuberculosis. *Calcif Tissue Int* 37:588–591.
226. Cadranel, J., M. Garabedian, B. Milleron, H. Guillozo, G. Akoun, and A. J. Hance. 1990. 1,25(OH)2D2 production by T lymphocytes and alveolar macrophages recovered by lavage from normocalcemic patients with tuberculosis. *J Clin Invest* 85:1588–1593.
227. Cadranel, J. L., M. Garabedian, B. Milleron, H. Guillozzo, D. Valeyre, F. Paillard, G. Akoun, and A. J. Hance. 1994. Vitamin D metabolism by alveolar immune cells in tuberculosis: correlation with calcium metabolism and clinical manifestations. *Eur Respir J* 7:1103–1110.
228. Isaacs, R. D., G. I. Nicholson, and I. M. Holdaway. 1987. Miliary tuberculosis with hypercalcaemia and raised vitamin D concentrations. *Thorax* 42:555–556.
229. Shai, F., R. K. Baker, J. R. Addrizzo, and S. Wallach. 1972. Hypercalcemia in mycobacterial infection. *J Clin Endocrinol Metab* 34:251–256.
230. Braman, S. S., A. L. Goldman, and M. I. Schwarz. 1973. Steroid-responsive hypercalcemia in disseminated bone tuberculosis. *Arch Intern Med* 132:269–271.
231. Need, A. G., P. J. Phillips, F. Chiu, and H. Prisk. 1980. Hypercalcaemia associated with tuberculosis. *Br Med J* 280:831
232. Hoffman, V. N. and O. M. Korzeniowski. 1986. Leprosy, hypercalcemia, and elevated serum calcitriol levels. *Ann Intern Med* 105:890–891.
233. Ryzen, E., T. H. Rea, and F. R. Singer. 1988. Hypercalcemia and abnormal 1,25-dihydroxyvitamin D concentrations in leprosy. *Am J Med* 84:325–329.
234. Kantarjian, H. M., M. F. Saad, E. H. Estey, R. V. Sellin, and N. A. Samaan. 1983. Hypercalcemia in disseminated candidiasis. *Am J Med* 74:721–724.
235. Spindel, S. J., R. J. Hamill, P. R. Georghiou, C. E. Lacke, L. K. Green, and L. E. Mallette. 1995. Case report: Vitamin D–mediated hypercalcemia in fungal infections. *Am J Med Sci* 310:71–76.
236. Murray, J. J. and C. R. Heim. 1985. Hypercalcemia in disseminated histoplasmosis. Aggravation by vitamin D. *Am J Med* 78:881–884.
237. Walker, J. V., D. Baran, N. Yakub, and R. B. Freeman. 1977. Histoplasmosis with hypercalcemia, renal failure, and papillary necrosis: Confusion with sarcoidosis. *JAMA* 237:1350–1352.
238. Parker, M. S., S. Dokoh, J. M. Woolfenden, and H. W. Buchsbaum. 1984. Hypercalcemia in coccidioidomycosis. *Am J Med* 76:341–344.
239. Kozeny, G. A., A. L. Barbato, V. K. Bansal, L. L. Vertuno, and J. E. Hano. 1984. Hypercalcemia associated with silicone-induced granulomas. *N Engl J Med* 311:1103–1105.
240. Jurney, T. H. 1984. Hypercalcemia in a patient with eosinophilic granuloma. *Am J Med* 76:527–528.
241. Edelson, G. W., G. B. Talpos, and H. G. Bone 3rd. 1993. Hypercalcemia associated with Wegener's granulomatosis and hyperparathyroidism: Etiology and management. *Am J Nephrol* 13:275–277.
242. Al-Ali H., A. A. Yabis, E. Issa, Z. Salem, A. Tawil, N. Khoury, and H. Fuleihan Gel. 2002. Hypercalcemia in Langerhans' cell granulomatosis with elevated 1,25 dihydroxyvitamin D (calcitriol) level. *Bone* 30:331–334.
243. Cook, J. S., M. S. Stone, and J. R. Hansen. 1992. Hypercalcemia in association with subcutaneous fat necrosis of the newborn: Studies of calcium-regulating hormones. *Pediatrics* 90:93–96.

244. Farooque, A., C. Moss, D. Zehnder, M. Hewison, and N. J. Shaw. 2009. Expression of 25-hydroxyvitamin D3-1alpha-hydroxylase in subcutaneous fat necrosis. *Br J Dermatol* 160:423–425.
245. Bosch, X. 1998. Hypercalcemia due to endogenous overproduction of 1,25-dihydroxyvitamin D in Crohn's disease. *Gastroenterology* 114:1061–1065.
246. Tuohy, K. A. and T. I. Steinman. 2005. Hypercalcemia due to excess 1,25-dihydroxyvitamin D in Crohn's disease. *Am J Kidney Dis* 45:e3–e6.
247. Jahnsen, J., J. A. Falch, P. Mowinckel, and E. Aadland. 2002. Vitamin D status, parathyroid hormone and bone mineral density in patients with inflammatory bowel disease. *Scand J Gastroenterol* 37:192–199.
248. Tangpricha, V., J. N. Flanagan, L. W. Whitlatch, C. C. Tseng, T. C. Chen, P. R. Holt, M. S. Lipkin, and M. F. Holick. 2001. 25-hydroxyvitamin D-1alpha-hydroxylase in normal and malignant colon tissue. *Lancet* 357:1673–1674.
249. Zaloga, G. P., C. Eil, and C. A. Medbery. 1985. Humoral hypercalcemia in Hodgkin's disease. Association with elevated 1,25-dihydroxycholecalciferol levels and subperiosteal bone resorption. *Arch Intern Med* 145:155–157.
250. Rosenthal, N., K. L. Insogna, J. W. Godsall, L. Smaldone, J. A. Waldron, and A. F. Stewart. 1985. Elevations in circulating 1,25-dihydroxyvitamin D in three patients with lymphoma-associated hypercalcemia. *J Clin Endocrinol Metab* 60:29–33.
251. Davies, M., E. B. Mawer, M. E. Hayes, and G. A. Lumb. 1985. Abnormal vitamin D metabolism in Hodgkin's lymphoma. *Lancet* 1:1186–1188.
252. Adams, J. S., M. Fernandez, M. A. Gacad, P. S. Gill, D. B. Endres, S. Rasheed, and F. R. Singer. 1989. Vitamin D metabolite-mediated hypercalcemia and hypercalciuria patients with AIDS- and non-AIDS-associated lymphoma. *Blood* 73:235–239.
253. Seymour, J. F., R. F. Gagel, F. B. Hagemeister, M. A. Dimopoulos, and F. Cabanillas. 1994. Calcitriol production in hypercalcemic and normocalcemic patients with non-Hodgkin lymphoma. *Ann Intern Med* 121:633–640.
254. Seymour, J. F. and R. F. Gagel. 1993. Calcitriol: The major humoral mediator of hypercalcemia in Hodgkin's disease and non-Hodgkin's lymphomas. *Blood* 82:1383–1394.
255. Davies, M., M. E. Hayes, J. A. Yin, J. L. Berry, and E. B. Mawer. 1994. Abnormal synthesis of 1,25-dihydroxyvitamin D in patients with malignant lymphoma. *J Clin Endocrinol Metab* 78:1202–1207.
256. Hewison, M., V. Kantorovich, H. R. Liker, A. J. Van Herle, P. Cohan, D. Zehnder, and J. S. Adams. 2003. Vitamin D-mediated hypercalcemia in lymphoma: evidence for hormone production by tumor-adjacent macrophages. *J Bone Miner Res* 18:579–582.
257. Gates, S., J. Shary, R. T. Turner, S. Wallach, and N. H. Bell. 1986. Abnormal calcium metabolism caused by increased circulating 1,25-dihydroxyvitamin D in a patient with rheumatoid arthritis. *J Bone Miner Res* 1:221–226.
258. Kroger, H., I. M. Penttila, and E. M. Alhava. 1993. Low-serum vitamin D metabolites in women with rheumatoid arthritis. *Scand J Rheumatol* 22:172–177.
259. Oelzner, P., A. Muller, F. Deschner, M. Huller, K. Abendroth, G. Hein, and G. Stein. 1998. Relationship between disease activity and serum levels of vitamin D metabolites and PTH in rheumatoid arthritis. *Calcif Tissue Int* 62:193–198.
260. Muller, K., N. J. Kriegbaum, B. Baslund, O. H. Sorensen, M. Thymann, and K. Bentzen. 1995. Vitamin D3 metabolism in patients with rheumatic diseases: low serum levels of 25-hydroxyvitamin D3 in patients with systemic lupus erythematosus. *Clin Rheumatol* 14:397–400.
261. Cutillas-Marco, E., M. Morales-Suarez-Varela, A. Marquina-Vila, and W. Grant. 2010. Serum 25-hydroxyvitamin D levels in patients with cutaneous lupus erythematosus in a Mediterranean region. *Lupus* 19:810–814.
262. Cutolo, M. and K. Otsa. 2008. Review: Vitamin D, immunity and lupus. *Lupus* 17:6–10.
263. Kamen, D. and C. Aranow. 2008. Vitamin D in systemic lupus erythematosus. *Curr Opin Rheumatol* 20:532–537.
264. Hayes, M. E., J. Denton, A. J. Freemont, and E. B. Mawer. 1989. Synthesis of the active metabolite of vitamin D, 1,25(OH)2D3, by synovial fluid macrophages in arthritic diseases. *Ann Rheum Dis* 48: 723–729.
265. Mawer, E. B., M. E. Hayes, P. E. Still, M. Davies, G. A. Lumb, J. Palit, and P. J. Holt. 1991. Evidence for nonrenal synthesis of 1,25-dihydroxyvitamin D in patients with inflammatory arthritis. *J Bone Miner Res* 6:733–739.
266. Hayes M. E., O'Donoghue D. J., Ballardie F. W., and E. B. Mawer. 1987. Peritonitis induces the synthesis of 1 alpha,25-dihydroxyvitamin D3 in macrophages from CAPD patients. FEBS Lett 220:307–310.

267. Levy, R., J. Klein, T. Rubinek, M. Alkan, S. Shany, and C. Chaimovitz. 1990. Diversity in peritoneal macrophage response of CAPD patients to 1,25-dihydroxyvitamin D3. *Kidney Int* 37:1310–1315.
268. Shany, S., J. Rapoport, I. Zuili, A. Gavriel, N. Lavi, and C. Chaimovitz. 1991. Metabolism of 25-OH-vitamin D3 by peritoneal macrophages from CAPD patients. *Kidney Int* 39:1005–1011.
269. Chun, R. F., A. L. Lauridsen, L. Suon, L. A. Zella, J. W. Pike, R. L. Modlin, A. R. Martineau, R. J. Wilkinson, J. Adams, and M. Hewison. 2010. Vitamin D-binding protein directs monocyte responses to 25-hydroxy- and 1,25-dihydroxyvitamin D. *J Clin Endocrinol Metab* 95:3368–3376.

Section II

Oxidative Stress

5 Vitamin D and Oxidative Stress

Huei-Ju Ting and Yi-Fen Lee

CONTENTS

5.1 INTRODUCTION

The beneficial effects of vitamin D in several diseases, such as preventing bone loss, protecting neurons from degeneration, boosting immune response, preventing diabetes/diabetic toxicity, and acting as an anticancer agent, are well documented [1–6]. This encourages vitamin D supplementation in the general healthy population. Vitamin D maintains calcium homeostasis and participates in multiple cellular processes, such as differentiation, proliferation, and metabolism, through both nongenomic action and its cognate receptor (VDR)-mediated transcriptional regulation [6,7]. Emerging evidence supports that maintenance of cellular reductive-oxidative (redox) homeostasis can account for many of these vitamin D actions, yet the role of vitamin D in regulating cellular oxidative status remains controversial. This conflict raises concerns since improper levels of oxidative stress can lead to catastrophic outcomes. Therefore, in this chapter, we intend to cover the studies on the effect and mechanism of vitamin D in regulating cellular oxidative stress responses. The following summarizes recent discoveries from studies investigating the anti- and prooxidation effects of vitamin D and the underlying mechanisms through which vitamin D mediates both aspects. There are many forms of vitamin D applied in these studies, including the precursor forms, vitamin D_2 (VD2), vitamin D_3 (VD3), 25-hydroxyvitamin D_3 (25-VD), the active form of vitamin D_3 (1alpha,25-dihydroxyvitamin D_3, 1,25-VD), and low-calcemic active analogs. Furthermore, the systems, including cell types, application route, and dosages of vitamin D metabolites, and assays measuring oxidative stress applied in each study are varied and therefore should be compared and discussed. Results from these studies might lead to the establishment of guidelines for vitamin D supplementation, so that appropriate dosages and forms of vitamin D can be applied to individual disease types.

5.2 OXIDATIVE STRESS

5.2.1 SOURCE OF OXIDATIVE STRESS

Oxidative stress occurs following the endogenous imbalance of cellular redox status or exogenous oxidative insults. There are several endogenous sources of oxidative stress, including reactive oxygen species (ROS) produced from metabolic activity such as oxidative phosphorylation, p450 metabolism, peroxisomes, and mitochondria transporting electrons from oxidizable donors to oxygen to produce energy [8]. Hyperactive mitochondria during nutrition overload can increase the activity of the electron transport chain to accumulate ROS derived from superoxide anion radical ($O_2 \cdot^-$), including hydroperoxyl radical ($HO_2 \cdot$), hydrogen peroxide (H_2O_2), and hydroxyl radical ($\cdot OH$) [9,10]. Inflammatory responses, such as the innate immune response, recruit macrophages and granulocytes to produce superoxides via NADPH oxidase (NOX) and nitric oxide ($NO \cdot^-$) by nitric oxide synthase (NOS) [11]. Following ischemia, reperfusion of circulating blood brings in oxygen and inflammatory blood cells that generate free radicals in response to tissue damage, therefore causing oxidative injury [12].

Exogenous sources of oxidative stress include xenobiotics that, once metabolized in cells, can generate free radicals, initiating a cascade of electron exchanges to oxygen generating ROS [13]. Those causing oxidative injury of DNA to mutate the genome are particularly well known as carcinogens. Some metals or chemicals can selectively accumulate or be metabolized in neurons to chronically increase neuronal oxidative status, causing neurological disease [14,15]. Certain chemotherapy agents can produce free radicals in proliferating cells to elicit oxidative damage and inhibit cancer cell growth [16,17]. These agents will also damage normal cells with high proliferating rates, such as hematopoietic cells. Aside from xenobiotics, cells exposed to ultraviolet (UV) or ionizing radiation receive radioactive energy that disturbs electrons to produce free radicals that result in ROS generation [18,19].

5.2.2 REDOX BALANCE MAINTENANCE

The cellular oxidative status is maintained by balancing the level of oxidative source exposure/ activity and the level of antioxidant amount/activity. Therefore, decreasing exposure to radiation and xenobiotics, uncoupling mitochondria potential, suppressing enzyme activity, and suppressing inflammatory response can reduce oxidative status. NOX, which is the enzyme that produces superoxide anion, is abundant in macrophages and granulocytes and is also present in a lesser amount in other types of cells. Its expression is regulated by cytokines, growth factor receptor tyrosine kinase, and ligands of G-protein-coupled receptors [20]. Inducible NOS, which is the enzyme that produces nitric oxide, is expressed in macrophages and induced by cytokines IL-1 and TNFα [21]. The activity of these enzymes should be tightly controlled to maintain redox balance.

On the other hand, modulating antioxidative mechanisms also controls cellular oxidative status. There are two types of antioxidants: enzymatic and nonenzymatic [8]. Enzymatic antioxidants of superoxides are superoxide dismutases (SOD), including cytosolic Cu/Zn-SOD (SOD1) and mitochondrial Mn-SOD (SOD2) that dismutate superoxides to produce H_2O_2. The enzymatic antioxidants of H_2O_2 are peroxisomal/mitochondria enzyme catalase (CAT) and glutathione (GSH)-homeostasis enzymes [glutathione peroxidase (GPX) and glutathione reductase (GR)]. H_2O_2 can also be reduced by peroxiredoxins (PRX) to generate oxidized peroxiredoxins, which are then reduced by the thioredoxin (TRx) system. The redox pair of oxidized and reduced TRx is catalyzed by TRx reductase (TR) utilizing NADPH as a reducing source (Figure 5.1). Vitamin D upregulated protein 1 (VDUP1), which is also known as TRx-binding protein-2 (TXNIP), negatively regulates TRx in transforming H_2O_2 to H_2O to increase oxidative status [22]. Nonenzymatic antioxidants are mostly from dietary intake, such as vitamin E, vitamin C, β-carotene, HDL-cholesterol, magnesium (Mg), calcium (Ca), and copper (Cu). Overall, the activity of these proteins and levels of nonenzymatic oxidative sources and antioxidants maintain redox balance in cells.

The expression of SOD is low in some tissues, such as the heart, which are therefore more susceptible to superoxide mediated oxidative stress. The regulatory factors of Cu/Zn-SOD expression include oxidants, metal ions, heat shock proteins, and shear stress in aortic endothelial cells [23]. Zinc deficiency can increase oxidative stress by impairing the activity of Cu/Zn-SOD. The expression and/or activity of Mn-SOD are regulated by p53, oxidants, cytokines (TNFα, IFNγ, IL-1, IL-6), endotoxin, redox regulators, and thiol-reducing agents [24]. The H_2O_2 metabolizing enzyme, CAT, is expressed abundantly in certain tissues, including the liver, kidney, and erythrocytes, but low in connective tissues and a subset of macrophages, thus determining the susceptibility of these tissues to oxidative stress. CAT levels can be regulated by the transcription factors NF-Y, sp1, and PPARγ [25,26]. The expression of CAT is decreased in tumors, liver transplant patients, and aging, and may increase the oxidative status involved in these processes. The GSH pool is a major antioxidant source; therefore, its synthesis is critical in maintaining cellular oxidative status. Two enzymes are responsible for the biosynthesis of GST, i.e., glutamate-cysteine ligase and GSH synthase; γ-glutamyl transpeptidase (GGT) transport GSH between cells is important in replenishing the GSH pool. GSH deficiency caused by chemically depleting GSH and the deficiency of GSH biosynthesis enzymes can manifest increasing cellular oxidative status [27]. The H_2O_2 metabolizing system utilizing GSH consists of two enzymes: GR and GPX. The classical selenocysteine protein GPX is transcriptionally or posttranscriptionally regulated by p53, c-Abl, Arg, and selenium [24,28,29]. Selenium can posttranscriptionally regulate GPX activity by both incorporating into GPX protein and regulating

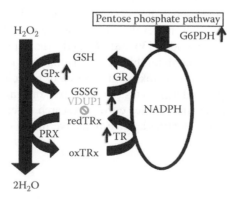

FIGURE 5.1 Vitamin D regulation of proteins in GSH and TRx antioxidation system. Vitamin D regulates several enzymes in the antioxidative pathway that reduces H_2O_2 to H_2O (small arrow). Among them, the induction of G6PDH, GPx, and TR mediates antioxidative effect of vitamin D, whereas induction of VDUP1 mediates the prooxidative effect of vitamin D.

mRNA through the 3′-untranslated region [24]. An oxidative-stress-induced transcription factor, Nrf2, can upregulate antioxidant response element-responsive genes including several antioxidant enzymes (CAT, GPX, and GCLC), and the detoxification enzymes, HMOX-1 and NQO1, can also scavenge superoxide anions [30–32]. Both selenium and Nrf2 can also regulate the activity and expression of TR, which is a selenium containing flavoprotein oxidoreductase that can reduce disulfide in thioredoxins. Steroid hormones, such as estrogen and vitamin D, can regulate TR expression in the heart, osteoblasts, and monocytes [33]. PRX is a protein family consisting of three subclasses with specific expression in organelles. Its expression can be transcriptionally controlled by Nrf2 and posttranscriptionally regulated by Cdc2 phosphorylation, redox-state-controlled oligomerization, proteolysis, and ligand binding [34]. The regulation of these antioxidant proteins is critical in maintaining the cellular redox state.

Cellular redox state can be measured by directly detecting the amount of ROS (superoxide, nitrite, or H_2O_2) or oxidative damage (lipid peroxidation, 8-OH-dG, or cell death). Indirectly, the activity of the antioxidative enzymes (SOD, CAT, and GPX) and stress-activated proteins [heat shock proteins (HSPs) and stress activated kinase (SAK)] provides information suggesting the trend of the redox status but does not represent a definitive level of the oxidative status. Increasing expression of antioxidative enzymes or responsive proteins can be interpreted not only as a response to increased oxidative stress but also as preventing oxidative stress to protect cells from oxidative damage. The most common index of redox status is the ratio of reduced glutathione (GSH) versus oxidized glutathione disulfide (GSSG) (Figure 5.1). Glutathione peroxidase (GPX) catalyzes the reduction of H_2O_2 by utilizing reduced GSH as substrate to generate GSSG. The reduction in GSSG is catalyzed by GR in a NADPH-dependent pathway to regenerate GSH. Therefore, the ratio of reduced GSH versus GSSG represents the status of the thiol pool available for maintaining redox balance. Both GSH and TRx systems depend on NADPH to replenish reduced GSH and TRx. Glucose-6-phosphate dehydrogenase (G6PDH) is the rate-limiting enzyme of the pentose phosphate pathway, which produces NADPH, which, in turn, is the critical coenzyme for replenishing the reduced thiol (GSH and TRx) pool for maintaining the redox homeostasis.

5.3 ANTIOXIDANT ROLE OF VITAMIN D

The antioxidative effect of vitamin D was discovered first in protecting ROS-induced neurotoxicity by regulating proteins reducing oxidative stress [35]. Oxidative injury in neurons is a dominant cause of neurological disease; therefore, several studies investigated the neuroprotective effect by systemically or locally supplementing with VD3. The hypercalcemic side effect of vitamin D raises concerns of using vitamin D supplementation to prevent oxidative injury; therefore, vitamin D analogs with low calcemic effect were developed, and their antioxidative effects were proven to be beneficial in preclinical disease models [36–38]. Recently, the antioxidative effect of vitamin D was found to be mediated partly through inhibition of inflammatory responses such as suppressing immune cell infiltration by attenuating cytokine production or reducing the inducible nitric oxide synthase (iNOS) in macrophages and astrocytes to reduce free radical formation from nitric oxide derivatives [39,40]. The following summarizes current research revealing the antioxidative effect of vitamin D and its analogs in protecting against oxidative injury occurring during toxic insult exposure or disease progression.

5.3.1 CANCER

The chemopreventive effect of vitamin D in tumorigenesis and cancer progression is well documented in several types of cancers [6]. Vitamin D regulates cellular function mainly through gene transcription. Gene-profiling studies have been performed on numerous types of cancer cells (prostate, breast, ovarian, colorectal, squamous cell carcinoma, and leukemia). Among many functional pathways of identified target genes, oxidative stress is one common cellular process found to be regulated by 1,25-VD in different types of cancers. This supports the importance of regulating the oxidative stress status in mediating the chemopreventive effect of vitamin D in cancer.

5.3.1.1 Colorectal Cancer

The chemopreventive effect of 1,25-VD in colon carcinogenesis was first suggested by epidemiological studies showing that low serum levels of 25-VD are associated with colon cancer [41]. The antiproliferative and prodifferentiation effect of 1,25-VD in colon cancer cell lines led researchers to investigate the role of VDR in colon carcinogenesis in VDR deficient (VDRKO) mice [42,43]. The finding that the VDRKO mouse has increased oxidative DNA damage marker 8-hydroxy-2-deoxyguanosine (8-OHdG) levels and proliferation suggests the anti-oxidative stress is one potential mechanism through which vitamin D prevents carcinogenesis. Human clinical trial supplementation with vitamin D (800 IU/d) in patients with colorectal adenoma for six months showed decreased 8-OHdG levels in patients with higher baseline levels of VDR expression [44]. These results further support the antioxidative effect of 1,25-VD/VDR signaling mediating chemoprevention in colorectal carcinogenesis.

5.3.1.2 Breast Cancer

The chemopreventive role of vitamin D/VDR in breast cancer has been suggested in epidemiological studies and proven in preclinical cell and animal studies [45]. The hypercalcemic side effect of 1,25-VD has hindered its use as a chemopreventive reagent in the healthy population. The finding that the prohormone form of 1,25-VD, 25-VD, is present at a higher level than 1,25-VD in the serum and can directly activate VDR and prevent mammary gland carcinogenesis, suggesting potential application of 25-VD as a chemopreventive reagent [46,47]. The connection between oxidative stress and carcinogenesis elicited a study exploring the protective effect of 25-VD against oxidative stress. Treatment with 25-VD indeed protected cells from cell death induced by H_2O_2 in a nonmalignant breast epithelial cell line, i.e., MCF12F [48]. The protective mechanism is suspected to involve counterregulation of stress responsive microRNAs, such as miR-182, which inhibit cell proliferative/apoptotic-related gene expression. The effect of 25-VD on cellular redox status in this study, however, was not examined. Therefore, it demands further investigation to determine whether 25-VD exerts antioxidative effects to protect cells from H_2O_2 challenges.

5.3.1.3 Prostate Cancer

The chemopreventive effect of vitamin D in prostate cancer tumorigenesis is supported by epidemiological and preclinical studies [49]. Accumulated genetic mutation from oxidative stress can lead to tumorigenesis [50]. The antioxidative function of vitamin D in mediating a cancer preventive effect is suggested by gene-profiling studies showing that 1,25-VD induces oxidative stress response genes in normal/nonmalignant prostate epithelial cells [51,52]. In one study, treatment with 1,25-VD (50 nM) induced thioredoxin reductase 1 (TR1), which reduces thioredioxin to maintain reductive status (Figure 5.1). Interestingly, 1,25-VD induces metallothionein isoforms specifically in normal but not cancer cells [52]. In addition to the protective role of metallothionein in oxidative stress, its role in controlling homeostasis of intracellular zinc levels can convert the zinc versus redox signals in cells [53]. Consistent with this, a recent study profiling gene expression in 1,25-VD-treated RWPE-1, which is a nonmalignant human prostate epithelial cell line, also found the induction of TR1 [51]. Moreover, 1,25-VD stimulated the antioxidant pathway that includes oxidative stress responsive transcription factors (Nrf2, JUNB, and FOS), and downstream antioxidants (HMOX1 and GPX3) [51,52]. The induction of G6PDH was identified in both gene profiling and promoter studies in 1,25-VD-treated RWPE-1 [52,54]. The direct regulation of G6PDH was proven to be important for the protective effect of 1,25-VD, increasing cell survival and reducing ROS production, in nonmalignant prostate epithelial cells upon H_2O_2 challenge. Interestingly, this effect of inducing G6PDH is specific to nonmalignant cells but not prostate cancer cell lines, which are not protected from oxidative stress by vitamin D. Overall, the chemopreventive effect of vitamin D in prostate tumorigenesis could be mediated by protecting cells against oxidative stress. This protective effect of vitamin D could be a result of regulating multiple antioxidant mechanisms including glutathione metabolism, GSH-GSSH redox cycle, theoredoxin redox cycle, and metallothionein expression.

5.3.1.4 Liver Cancer

The chemopreventive effect of 1,25-VD is demonstrated in multiple carcinogen-induced liver cancer models in rats, including diethylnitrosamine (DEN), 3,-methyl-4-dimethyl-amino-azobenzene (3,-Met-DAB), and 2-acetylaminofluorene (2-AAF) [55–58]. The carcinogen-induced DNA damage is measured by chromosomal aberrations (CAs), DNA strand breaks, and specific DNA adducts in rat liver. In 1,25-VD-treated rats, effective suppression of hyperplastic nodule formation and early CAs and DNA damage during the process of rat hepatocarcinogenesis was observed [55,57]. Carcinogen-induced changes of reduced GSH concentration, GGT activity, glutathione S-transferase (GST) activity, and glutathione peroxidase (GPX) activity can be controlled by 1,25-VD treatment [58]. Decreased lipid peroxidation marker (MDA), increased expression of SOD, and maintained level of reduced GSH in the 1,25-VD-treated group during hepatocarcinogenesis suggest that the antioxidative effect of 1,25-VD mediates the chemopreventive function [58].

5.3.2 Alcoholic Myopathy

The cause of muscle atrophy in alcoholic patients includes hormone alteration, decreasing protein synthesis, and increasing muscle fiber apoptosis [59]. Vitamin D deficiency is observed in these patients and ethanol-fed animals, linking vitamin D deficiency and muscle atrophy. Aside from promoting proliferation and differentiation of muscle fibers, the antioxidative effect of vitamin D could eliminate oxidative damage, such as cholesterol hydroperoxidation, MDA protein adducts, and mitochondria DNA depletion, which is found in alcoholic patients' muscles [60]. In ethanol-fed rats, vitamin D deficiency is associated with decreased type-II muscle fiber area, decreased GPX, and increased MDA levels. Therefore, loss of the antioxidant effect of vitamin D might participate in the process of alcoholic myopathy.

5.3.3 Cardiovascular Disease

5.3.3.1 CKD-Related Cardiovascular Disease

Myocardial damage in chronic kidney disease (CKD) patients results in cardiovascular disease and, eventually, morbidity. In uremic rats that mimic CKD, the vitamin D analog paricalcitol prevents myocardial hypertrophy [37,38]. Underlying mechanisms include antiinflammatory and antifibrotic effects, and inhibiting overactivity of the rennin–angiotensin–aldosterone system by vitamin D. Angiotensin II can induce NOX activity, which synthesizes superoxide that increases oxidative stress in the cardiovascular system. Oxidative damage of protein, DNA, and membrane lipid peroxidation can lead to cardiovascular dysfunction. The protective effect of vitamin D in CKD-related cardiomyopathy could be mediated through reduction in cardiac oxidative stress. One study indeed demonstrated that altered activities of NOX, GSH levels, and SOD activity in the heart, triggered by uremic surgery, are attenuated by paricalcitol [37]. This suggests the antioxidative role of vitamin D in protecting CKD-related cardiac myopathy.

5.3.3.2 Atherosclerosis

Vitamin D deficiency is associated with high risk of atherosclerosis. Hypovitaminosis D could increase insulin resistance that causes endothelial dysfunction, which is the early marker of atherosclerosis. A study examined endothelial function in human subjects with normal vitamin D levels and vitamin D deficiency before and after vitamin D replacement therapy. Endothelial function was positively correlated, whereas lipid oxidation levels were inversely correlated with serum 25-VD levels between control and vitamin D–deficient subjects. Serum 25-VD levels were elevated from an average of 20–120 nM in vitamin D–deficient subjects after monthly intramuscular injections of 300,000 IU VD3 for three months [61]. This result demonstrated that, in vitamin D–supplemented subjects, lipid peroxidation levels were reduced, and endothelial function was improved. This indicates that antioxidant properties of vitamin D in the endothelium can prevent peripheral vascular diseases, such as atherosclerosis [61]. In atherosclerotic mice with ApoE deficiency, oxidative and inflammatory injuries are increased [40]. After treatment with the vitamin D analog, paracalcitol, lipid peroxidation is decreased and reduced GSH levels are increased that lead to attenuation of oxidative injury in the aorta of the mice [40]. In addition, decreased expression of NOX and increased expression of Cu/Zn-SOD proteins in the aorta of paracalcitol-treated ApoE-deficient mice ameliorated superoxide levels. Furthermore, paracalcitol decreased the levels of inflammatory cytokines (MCP-1 and TNFα) and iNOS in the aorta, indicating decreased inflammatory injury. These studies demonstrate that vitamin D prevents oxidative injury in the aorta endothelium, therefore suggesting that supplementation with vitamin D may delay the process of atherosclerosis.

5.3.4 DIABETES

Vitamin D–deficient patients have decreased insulin synthesis and secretion, and vice versa, diabetic patients are likely to be vitamin D deficient, suggesting that vitamin D participates in the development of diabetes [4]. The preventive effect of vitamin D in type I diabetes has been demonstrated in experimental models [62]. Since the oxidative stress generated by inflammatory responses and hyperglycemia participates in the development of diabetes, the antioxidative effect of vitamin D was examined in an STZ-induced diabetes rat model. Oxidative markers, including SOD and catalase in erythrocytes, were reduced in the vitamin D–treated diabetic group, whose blood glucose regulation and insulin secretion were reversed to normal [63]. Furthermore, infiltrating immune cells secreting inflammatory products, including cytokines, and oxidative stress can damage pancreatic islets, leading to the development of type I diabetes and islet allograft rejection. In a study, vitamin D treatment in pancreatic islets damaged by inflammatory cytokines (IL-1β+TNFα+IFNγ) returns insulin secretion and cell viability to normal levels [39]. Here, the antioxidative stress effect of vitamin D was mediated by the suppression of cytokine-induced nitric oxide production. Oxidative damage remains an obstacle for maintaining pancreatic islet function after allograft transplantation in diabetic patients. The antioxidant effect of VD3 in protecting pancreatic cell viability and function was suggested in studies showing that VD3 and vitamin E treatment yields larger and more intact beta-cell clusters, a longer period of preservation of islet function, and increased insulin secretion of pancreatic islets. However, there was no measurement of oxidative damage and antioxidant level provided [64].

Hyperglycemia in diabetes patients generates more free radicals from glucose autooxidation. This increased level of free radicals results in oxidative stress that causes diabetic toxicity [65,66]. The preventive and therapeutic effect of 1,25-VD in diabetic toxicity of testes, liver, and kidney has been demonstrated in the alloxan-induced diabetic rat model. 1,25-VD treatment before and after alloxan injection can increase insulin levels and correct blood glucose and liver glycogen levels [67,68]. In these studies, the altered antioxidant level and activity in diabetic rat, including SOD, CAT, nonenzymatic antioxidants, and lipid peroxidation levels, are reversed by 1,25-VD. Overall, vitamin D prevents oxidative damage-induced diabetes, preserves pancreatic islets function, and reduces diabetic toxicity, possibly through its antioxidant effect.

5.3.5 SKIN

Vitamin D precursor, 7-dehydrocholesterol, is metabolized to VD3 by photochemical reaction in the epidermal layer by sunlight. It is therefore believed that local 1,25-VD levels can exceed physiological concentrations. Fully equipped with vitamin D signaling machinery, it is believed that local 1,25-VD-activated VDR can protect skin from environmental insults and physiological stress-induced redox imbalance, such as ultraviolet irradiation, chemotherapy, and inflammation. Several studies examine and support the protective effect of 1,25-VD against oxidative stress in skin [69]. Treatment with 1,25-VD starting at 3 nM can protect the keratinocyte cell line, HaCaT, from H_2O_2-induced cytotoxicity, possibly through inhibiting stress activated protein kinase JNK induced apoptosis [69,70]. Studies also find 1,25-VD protects keratinocytes from UV-induced death and thymine dimers, both *in vitro* and *in vivo*, possibly via the induction of metallothionein and suppression of NO^- production [71,72]. Therefore, this protective effect of vitamin D against oxidative damage in skin is through induction of antiapoptosis signaling and scavenging oxidative sources.

5.3.6 NEURON

Since the finding that vitamin D can upregulate the synthesis of nerve growth factor two decades ago, extended studies further found that vitamin D can stimulate the synthesis of neurotrophins and neurotransmitters. Thereafter, the neuroprotective effect of vitamin D was proposed and investigated. Summarized in a review article [2], the neuroprotective effect of vitamin D was observed in animal models of ischemia brain injury, diabetic neuropathy, and autoimmune central nervous system (CNS)-degenerating disease. The antioxidative effect of vitamin D in neuroprotection is suggested by several findings. First, vitamin D can inhibit synthesis of iNOS that produces nitric oxide to cause neuronal damage. Second, vitamin D can induce expression and activity of GGT that participates in the glutathione cycle between neurons and astrocytes. Third, vitamin D increases glutathione levels to protect neurons. The antioxidant effect of vitamin D in neuroprotection was also demonstrated in many other types of insults, including cyanide, zinc, H_2O_2, and iron-induced neuronal injury [35,36,73–75]. These insults stimulate oxidative stress to cause CNS, dopaminergic, and noradrenergic neuron damage. Local infusion, systemic, and *in vitro* culture treatment with 1,25-VD can reduce lipid peroxidation and ROS levels to protect against the neurotoxicity of these insults. Combination treatment of neuroprotective agents with vitamin D was investigated for the ability to protect against ischemic cerebral vascular disease [76–78]. A combination of vitamin D and DHA, a blood-brain barrier transportable form of vitamin C, decreased ischemia/reperfusion induced lipid peroxidation possibly through inducing GSH and SOD activity in the cortex and corpus striatum (CS) in a middle cerebral artery occlusion/reperfusion model [76]. The combination of vitamin D and estrogen-reduced focal cortical ischemia induced expression of HSP27 and HSP32, which are two oxidative stress markers in the brain, supporting the antioxidative effect of vitamin D in neuroprotection [77,78].

5.4 PROOXIDANT ROLE OF VITAMIN D

5.4.1 CANCER

Vitamin D potentiates the cancer killing effect of cytokines (TNFα, IL10, and IL-6) and of chemotherapy drugs (doxorubicin, quinine, and H_2O_2) in breast cancer, leukemia, and renal carcinoma. The fact that ROS generation is one common feature of these cytotoxic agents leads to investigation of prooxidation effects as an underlying mechanism mediating the anticancer effects of vitamin D.

5.4.1.1 Breast Cancer

1,25-VD reduces the growth and viability of cancer cells by cell cycle arrest and apoptosis. The anticancer-promoting effect of 1,25-VD on several cytotoxic agents including doxorubicin, cytokines (TNFα, IL-1, and IL-6), H_2O_2, and menadione in breast cancer has been explored [79–84]. The synergistic or additive effect of 1,25-VD in promoting the cytotoxic effect of these agents can be diminished by ROS scavengers, suggesting the involvement of ROS. In studies combining 1,25-VD with TNFα or doxorubicin, which are two agents involving superoxide-mediated cytotoxicity, the enhancing effect of 1,25-VD is linked to reducing the superoxide-metabolizing capacity. SOD is the first line of antioxidative defense system protecting cells from superoxide induced cytotoxicity. Treatment with 1,25-VD can decrease the mRNA and protein level of Cu/Zn SOD at 96 h [81]. Pretreatment with 1,25-VD for 48 h can enhance the cytotoxicity of doxorubicin. Therefore, it is assumed that 1,25-VD can enhance the cytotoxicity of doxorubicin by attenuating the activity of Cu/Zn SOD metabolizing superoxide. However, the cellular level of superoxide or derivatives was not measured. The mechanism of 1,25-VD downregulation of Cu/Zn SOD and whether this regulation results in oxidative stress in cells, however, remains unclear. The prooxidation effect of vitamin D was found to be associated with its apoptotic effect in the breast cancer cell line MCF-7 [85]. The production of ROS superoxide increased after four days of treatment with 100-nM 1,25-VD. In this study, the level of cytosolic cytochrome c increased during day 2 after treatment. Therefore, the increased level of superoxide was the consequence of apoptotic events where impairment of proton flow results in incomplete reduction of molecular oxygen in mitochondria. Further studies examined the cellular level of glutathione at five days after 1,25-VD treatment and found that depletion of GSH was consistent with increased oxidative stress induced by 1,25-VD. The level of VDUP1, which is induced by 1,25-VD in HL-60 and can deplete thioredoxin, was however not affected by 1,25-VD in the MCF-7 cell line. Interestingly, the level of the reduced form of thioredoxin and its nuclear translocation were increased at five days after 1,25-VD treatment [79]. This is in contrast to the overall increase in cellular oxidative stress level on the same treatment. Therefore, one can assume that the apoptotic pathway regulated by 1,25-VD contributing to the increase in ROS and depletion of GSH is dissociated from the antioxidative pathway regulated by 1,25-VD that increases the reduced form of thioredoxin [79]. In another study, pretreatment with 1,25-VD for two days was required to promote the cytotoxicity of H_2O_2. It was found that the H_2O_2 metabolism and caspase activity were not involved in the cytotoxic-promoting effect of 1,25-VD [82]. Instead, the mitochondrial membrane potential dropped significantly earlier and faster after H_2O_2 challenge when cells were pretreated with 1,25-VD for two days. This supports the premise that the disruption of mitochondrial function after 1,25-VD treatment and induction of apoptosis is a major event promoting sensitivity to cytotoxic agents [82]. Combining 1,25-VD with the GSH-depleting compound menadione increased ROS levels in MCF-7, whereas the activity of antioxidant enzymes SOD, GPX, and CAT were induced by 1,25-VD [80]. Although it was suggested that this induction of antioxidant enzymes is a compensation response of cells facing oxidative stress, it is also possible the antioxidant enzymes are directly upregulated by 1,25-VD. However, this upregulation is not sufficient to combat the ROS produced by disrupted mitochondrial function during apoptosis. Determining the mechanism by which 1,25-VD upregulates antioxidant enzymes and if this upregulation would be diminished in the presence of ROS scavengers will clarify the role of 1,25-VD in regulating cellular redox balance. Overall, the prooxidative effect of 1,25-VD in MCF-7 cells is a late event that can be the result of inhibition of Cu/Zn SOD activity and/or the consequence of apoptosis related mitochondria dysfunction.

5.4.1.2 Leukemia

Vitamin D treatment promotes monocytic differentiation of human myeloid leukemia HL-60 cells [86]. Investigation of the prodifferentiative mechanism of vitamin D identified the vitamin D upregulated protein (VDUP1) from HL-60 [87]. VDUP-1 can bind and negatively regulate TRx function [88]. TRx participates in multiple biological responses, including proliferation, antiapoptosis, and gene expression through various mechanisms including redox regulation of thiol-disulfide bonds, as a hydrogen donor for riobonucleotide reductase during DNA synthesis, ASK1 kinase inhibition, and promotion of transcription factor DNA binding [89]. Therefore, VDUP1 is expected to affect these biological processes involving TRx. Cells overexpressing VDUP1 are more vulnerable to oxidative stress, suggesting that treatment with 1,25-VD can increase the cytotoxicity of oxidative stress in HL-60 cells; however, this was not examined [90]. The increase in ROS production in 1,25-VD- and EB1089-treated HL-60 cells that was measured at 72 h could be the consequence of apoptotic events rather than depletion of TRx by VDUP1 [91,92]. Whether 1,25-VD-induced VDUP1 expression results in the increase in ROS levels in HL-60 cells needs further investigation. The antigrowth and proapoptotic effect of vitamin D_2 (ergocalciferol), which is a natural VD3 analog abundant in plants, was observed in the HL-60 cell line. Treatment with 60-μM VD2 can trigger apoptosis within 12 h, which is associated with GSH depletion occurring within 30 min after vitamin D treatment and ROS (H_2O_2 and superoxide) production occurring in 60 min. ROS production is possibly the primary cause of apoptosis in vitamin D_2–treated HL-60 since CAT and NAC can prevent apoptosis [93]. However, it is not clear whether this high concentration of vitamin D_2 is achievable *in vivo* without side effects.

5.4.2 INFLAMMATION-RELATED DISEASE

5.4.2.1 Obesity

The role of oxidative stress in metabolic syndrome including obesity is established [94]. Excessive uptake of lipid or carbohydrate caused by metabolic overload in adipocytes increases the utilization of the mitochondria electron transport chain. The unusually active mitochondria generate excess amounts of ROS, stimulating oxidative-stress-activated kinases that can directly or indirectly interfere with insulin signaling by increasing cytokine production [94]. Rupture of hypertrophied adipocytes attracts macrophages that further produce more ROS in adipose tissues. This vicious cycle of oxidative stress and inflammation in adipose tissues leads to deregulation of metabolism and, thereafter, obesity [94]. Recent studies support that vitamin D can promote oxidative stress-inflammatory action in adipose tissue. In a mouse obesity model, high-sucrose/high-fat-diet-induced obesity was associated with increasing oxidative stress. A high-calcium diet in this mouse model can decrease ROS and cytokine generation in adipose tissue, which is accompanied with decreased levels of plasma 1,25-VD [95]. Treatment of differentiated 3T3-L1 adipocytes with 1,25-VD increased cytokine production [96]. Therefore, it was implied that the prooxidative effect of 1,25-VD contributes to cytokine production and that a high-calcium diet reversed this effect by reducing plasma 1,25-VD levels. Further investigation found that 1,25-VD can regulate cytokine production in both adipocytes and macrophages, and this regulation is even more profound in adipocyte-macrophage cocultures [97]. The induction of cytokines by 1,25-VD was found to be dependent on calcium and mitochondria uncoupling [97]. Vitamin D increases intracellular calcium through nongenomic regulation and inhibits uncoupling protein 2 (UCP2) expression through VDR-mediated transcriptional regulation in adipocytes [97]. The fact that cellular calcium levels and mitochondria uncoupling can modulate oxidative stress levels further supports the prooxidative effect of vitamin D in adipocytes. Under the challenge of glucose, adipocytes proliferate and produce ROS accompanied by induction of intracellular calcium levels and reduction of UCP2 [98]. Suppression of calcium influx can decrease ROS production and inhibition of mitochondria uncoupling further increases ROS production. Cotreatment of these challenges with 1,25-VD further enhances these changes in adipocytes. This suggests ROS production induced by glucose overload, and 1,25-VD promotes the prooxidative effect through similar pathways involving calcium influx and mitochondria uncoupling [98]. To determine if the prooxidative effects of 1,25-VD mediate cytokine induction, experiments to test if antioxidants can diminish 1,25-VD regulation of cytokine production are needed. In a microarray study, treatment with 1,25-VD induced the expression of NADPH oxidase 4 (NOX-4) and toll-like receptor-3 (TLR-3), suggesting two additional potential mechanisms by which 1,25-VD can promote oxidative stress and inflammatory response in adipose tissues [99]. Overall, 1,25-VD increases calcium influx, decreases mitochondria uncoupling, and increases NOX-4 expression to enhance the oxidative status in adipose tissues during nutrient overload. This prooxidative effect of 1,25-VD elicits cytokine production, which attracts inflammatory cell infiltration to further exaggerate oxidative stress level and cytokine production in adipose tissues.

5.4.2.2 Vascular Calcification

In the previous section, the evidence that vitamin D can protect vascular endothelial cells from oxidative injury, therefore improving endothelial function and preventing atherosclerosis, is discussed. Nonetheless, a high dose of vitamin D (300,000 IU/kg) is used in animals to create models of vascular calcification. Vascular calcification is found in several chronic diseases such as artherosclerosis, aorta stenosis, and diabetes. The relationship of oxidative injury level and vascular calcification was examined in high-dose VD2 and high-cholesterol diet (HCD)-fed rats that develop aortic vascular calcification [100]. VD2-fed groups had higher calcium deposition and systolic blood pressure in the aorta, compared with the control group. However, the oxidative stress level in the aorta, including increased lipid oxidation and superoxide levels and decreased SOD activity, was observed in the VD2+HCD fed group but not in the VD2-alone fed group. Therefore, the oxidative injury in this vascular calcification model was not the result of VD_2 uptake but from the HCD [100]. However, there was no HCD-alone-fed group to rule out whether VD_2 can participate in promoting HCD-induced oxidative injury. Another study applied VD_2 to induce aortic valve stenosis (AVS), which is a common aging-related valve degenerative heart disease [101]. The VD_2 feeding induced aortic valve structural, functional, and histological changes, including calcium and lipid deposition and macrophage infiltration, which are similar to AVS. Furthermore, the expression of VDUP1 (TXNIP) was increased in the aortic valves of the VD2-fed group [101]. This indicates that vitamin D–induced AVS development is associated with the prooxidative effect of VDUP1, which inactivates thioredoxin and increases oxidative stress in aortic valves.

5.4.3 Bone

Vitamin D is essential for bone health by maintaining mineral homeostasis and modulating cell growth and differentiation. Lipoxygenase (LOX) expression was found to be antiinflammatory, which can benefit degenerative bone disease involving inflammatory responses, such as periodontitis and arthritis [102]. Studies investigating the effect of vitamin D on LOX expression demonstrated that vitamin D metabolites (1,25-VD and 24,25-VD) and low calcemic analogs (JFK1624F2-2 and QW1624F2-2) can stimulate the expression of 12- and 15-LOX in human bone cells [103,104]. The activity of LOX measured by the formation of 12- and 15-hydroxyeicosatetraenoic acid (HETE), metabolites of 12- and 15-LOX, and ROS, correlates with the induction of LOX expression by vitamin D. However, it appears that the production of 15-HETE and ROS is a rapid response induced by vitamin D as it occurred 10 min after vitamin D treatment. The expression of LOX mRNA was the late response of vitamin D treatment measured at 3 days. It is therefore suggested that vitamin D regulates LOX activity and ROS formation in a nongenomic pathway. The vitamin D–induced ROS production and growth of bone cells was dependent on the activity of NOX. Although vitamin D–induced LOX expression can be beneficial in preventing inflammatory-related bone disease, the prooxidative effect of vitamin D displayed in these studies suggests that vitamin D can increase oxidative stress levels, resulting in oxidized fatty acids in bone, therefore favoring osteoclast differentiation and disturbing bone function [103,104].

5.5 CONCLUSION

The aforementioned evidence proves the regulatory role of vitamin D in redox balance in mediating both beneficial and harmful effects of vitamin D (summarized in Table 5.1). The antioxidative effect

TABLE 5.1

Effect and Mechanisms of Vitamin D in Regulating Redox Balance among Various Tissues and Potential Impact on Disease

Disease/System	Tissue/Cell, Model	Forms of Vitamin D	Route/Dosage	Assay	Mechanism	Reference
			Antioxidation			
Carcinogenesis	Colorectal mucosa	VD$_3$	800 IU/d VD 6 months	↓8-OH-dG	N/A	[44]
	Breast/MCF12F	25-VD	250 nM pre-treat 24 h	↓Cell death under H$_2$O$_2$ challenge	↓miR-182	[48]
	Prostate/ RWPE-1	1,25-VD	100 nM 6, 24, 48 h	Microarray	↑Nrf2, ↑GPX3, ↑TR1 ↑TR1, ↑MT1 ↑G6PDH	[51]
	Primary prostate epithelial cells		50 nM 6, 24 h			[52]
	BPH-1, RWPE-1	1,25-VD	1–100 nM 6–24 h	↑GSH, ↓ROS (DCF-DA), ↓lipid peroxidation		[54]
	Liver/rat DEN+PB		4 weeks before carcinogen 0.3 μg/100 μl twice/week for 20 weeks	Comet assay, ↓lipid peroxidation	↑SOD, ↑GSH,	[55]
	2-AAF+STZ		0.3 μg/100 μl twice/week for 8–13 weeks	↓Chromosomal aberrations, ↓DNA strand breaks, ↓DNA adducts	N/A	[57]
	DEN+STZ		0.3 μg/100 μl twice/week for 15 weeks	↓Lipid peroxidation	↑GSH, ↑GGT	[58]
Alcoholic myopathy	Right gastrocnemius muscle in rat	Serum VD level		↑Type IIa, b fiber area	↑GPX, ↑SOD	[60]
Cardiovascular complication in CKD patients	Heart in uremic rat	Paricalcitol	40 ng ip 3× per week for 4 weeks	↓Left ventricular hypertrophy	N/A	[38]
		Paricalcitol	200 ng 3× per week for 4 months	↓Alteration of Mn-SOD	↓NOX, ↑GSH	[37]
Atherosclerosis	Endothelial	25-VD	300,000 IU im monthly for 3 months	↓Lipid peroxidation	N/A	[61]
	ApoE KO mice	Paricalcitol	200 ng ip 3× per week for 16 weeks	↓Lipid peroxidation (MDA)	↑GSH, ↓cytokines, ↑Cu/Zn-SOD, ↓iNOS	[40]

Category	Model	Treatment	Dose	Effect	Mechanism	Ref
Diabetes	STZ rat	1,25-VD	23 days after the STZ, ip 0.4 µg/kg for 14 days	Erythrocytes CAT, SOD	N/A	[63]
	Cytokine-treated pancreatic islets		10 nM, 1 µM	↓Nitrite	N/A	[39]
Diabetic toxicity	Pancrease, liver, kidney in alloxan rat	1,25-VD	Gavage 5000 IU/kg/d 15 days b/a alloxan 3 weeks	↓Lipid peroxidation	↑SOD, CAT, GPX, nonenzymatic antioxidant	[67]
	Testes	1,25-VD		↓Lipid peroxidation	↑SOD, CAT, GPX	[68]
Skin	H_2O_2 in HaCaT keratinocytes	1,25-VD	0.1 nM >16 h 3–100 nM	↓Oxidative-stress-induced cell death	↑P38, JNK	[69,70]
	UV challenged keratinocytes and mice skin	1,25-VD	12 nM 24 h	↓Cell death, sunburn cell count	↑Metallothionein	[71]
Neurotoxicity	Rat primary cortical cells (Wy14643-> cyanide)	1,25-VD	1 nM 24 h	↓Thymine dimer, ↓NO	N/A	[72]
	Substantia nigra(Zn)	1,25-VD	1 nM 30′	↓ROS	↑GSH	[75]
	NG108-15 (H_2O_2)	1,25-VD, tacalcitol (PRI-2191) 1,25-VD	in vitro 8–24 nM, in vivo 6 pM/µl, 0.2 µl/min	↓Lipid peroxidation	N/A	[74]
			50–500 nM	↓LDH	N/A	[36]
	Locus coeruleus (Fe)	1,25-VD	ip 1 µg/kg/d 7 d before and after	↓Lipid peroxidation	N/A	[73]
Ischemia/ reperfusion (I/R) brain injury	Mesencephalic culture (H_2O_2)	1,25-VD	Pretreat 100 nM	↓ROS	N/A	[35]
	Focal cortical ischemia	1,25-VD + E_2	ip 1,25-VD 1µg/kg and E_2 7 µg/kg	↓HSP-27 ↓HSP32 (HMOX-1)	↓iNOS?	[77,78]
			Prooxidation			
Cancer/breast	MCF-7 +menadione	1,25-VD	100 nM 4 d	↓GSH, ↑antioxidant enzymes, ↑ROS	N/A	[80]
	+ doxorubicin		10 nM 2–4 d	↑Cytotoxicity	↓Cu/Zn SOD	[81]
	+ H_2O_2		100 nM 2–4 d	↑H_2O_2, ↓mitochondria membrane potential	N/A	[82]
	+TNFα		100 nM 1 d	↑Cytotoxicity	N/A	[83]
	+ IL-1, IL-6		10–100 nM 2 d	↑Cytotoxicity, ↓GSH	N/A	[84]

(continued)

TABLE 5.1 (Continued)

Effect and Mechanisms of Vitamin D in Regulating Redox Balance among Various Tissues and Potential Impact on Disease

Disease/System	Tissue/Cell, Model	Forms of Vitamin D	Route/Dosage	Assay	Mechanism	Reference
Leukemia	HL-60	1,25-VD/EB1089 1,25-VD	10 nM/1 nM 3 d 100 nM 48 h	↑ROS	N/A	[91]
		Ergocalciferol	60 μM 30′–12 h	↓TRx reducing activity ↑H$_2$O$_2$ (1 h), ↑superoxide (1 h), ↓GSH (30′)	↑VDUP1	[90]
					N/A	[93]
Obesity	Human adipocyte, 3T3-L1, RAW 264	1,25-VD	10 nM	↑MIF, ↑CD14, ↑IL-6	↑Calcium, mitochondria uncoupling	[97]
	3T3-L1, Zen-Bio human adipocytes sc adipocytes		10 nM, 40 pg/ml - >27 pg/ml	↑ROS, ↑TNFα, ↑IL6, ↑IL-8	↑Calcium, mitochondria uncoupling	[96,98]
Vascular calcification	Rat vessel	VD2	10 nM	microarray	↑NOX-4, ↑TLR-3	[99]
		VD2	300,000 IU/kg for 4 d (serum 1 μM 25-VD)	↑Lipid peroxidation, ↑oxLDL, ↑O2·⁻	SOD↓	[100]
AVS	Aortic valve in rabbits	VD2	25,000 IU/4 days weekly for 8 weeks	↑Macrophage infiltration ↓Endothelial function	↑VDUP1	[101]
Bone	SaOS2, hFOB, primary human bone cells	1,25-VD/24,25-VD/ JKF1624F(2)-2/ QW1624F(2)-2	25 nM/125 nM/1 nM/1 nM	↑ROS (1 h) -> DNA synthesis (24 h)	NOX dependent	[103,104]

of vitamin D participates in preventing carcinogenesis, impeding diabetes development, protecting neurons, and ameliorating cardiovascular complications. On the other hand, the prooxidative effect of vitamin D can reduce cancer growth but promote obesity development, trigger vascular calcification, and disturb osteoblast/osteoclast balance in bone.

The antioxidative effect of vitamin D involves reducing oxidative status or increasing survival signals when cells face oxidative stress challenges. The oxidative status can be reduced by vitamin D by increasing ROS scavenging pathways or decreasing the oxidative source. The ROS scavenging pathway regulated by vitamin D includes a proven target gene G6PDH [54] and many potential vitamin D target genes related to antioxidative pathways identified by gene profiling or other studies, such as TR1, Nrf2, SOD, CAT, and GPX [40,51,52,55,60,67]. Vitamin D decreases the oxidative source by inhibiting iNOS and NOX activity, and by its anti-inflammatory effect [40,61]. The survival signaling pathways regulated by vitamin D, including p38, JNK, and miR-182, protect cells from oxidative-stress-induced cytotoxicity [48,69,70].

The prooxidative effect of vitamin D can be beneficial by potentiating cancer cells' sensitivity to chemotherapy agents but can be harmful in promoting obesity, vascular calcification, and possibly bone function. Vitamin D can directly promote oxidative stress through a nongenomic mechanism that increases calcium influx to stimulate PKA inducing NOX [99,104] or through a genomic mechanism that upregulates VDUP1 to interfere with antioxidant thioredoxin [90,101]; or downregulates UCP2 to increase mitochondria potential [98]. Indirectly, the proinflammatory effect of vitamin D, inducing TLR-3 expression that increases cytokine production, can attract infiltrating immune cells secreting ROS or nitric oxide [99].

The antioxidative and prooxidative effect of vitamin D can occur simultaneously in the same tissue, such as mammary glands or the cardiovascular system, but in a cell-type specific manner. The antioxidative effect of vitamin D in a nonmalignant breast epithelial cell line (MCF12F) can protect against stress-induced cell death, whereas the prooxidative effect in a breast cancer cell line (MCF-7) promotes cytotoxicity from oxidative stress and chemotherapy agents [48,80,81]. In this scenario, two opposite effects are beneficial. Whether there are cell-type-specific (normal versus malignant) mechanisms mediating two different effects or this is just a cell-line-specific effect needs to be investigated. There is evidence suggesting that vitamin D indeed has differential effects in redox regulation among normal and malignant cell types such as the upregulation of G6PDH and metallothioneins specifically occurring in normal prostate but not cancer cells [52,54]. The antioxidative effect of vitamin D in preventing cardiovascular dysfunction in the model mimicking CKD [37,38] versus the prooxidative effect of vitamin D in stimulating the vascular calcification model [100,101] can be attributed to the different dosages of vitamin D applied in these animal models. Continuous treatment with high-dose vitamin D can result in vascular calcification and increase the susceptibility to cholesterol overload in endothelial cells. On the other hand, long-term treatment with safe dosages of vitamin D can protect endothelial cells from oxidative-stress-induced dysfunction. The vitamin D regulation of NOX and inflammatory cytokines displays tissue-specific effects with downregulation occurring in uremic rat aortas [37] but upregulation in adipocytes and bone cells [99,104]. Although this could be attributed to tissue specific mechanisms, the fundamental difference between these studies, such that the antioxidative effects were observed in an *in vivo* disease model, whereas the prooxidative effects were found in an *in vitro* normal cell line model, cannot be ignored.

In summary, the redox regulatory effect of vitamin D is proven significant in mediating several biological functions. This regulatory effect is complex and delicate since the treatment dosage and time, tissue, cell type, and environment can all modulate the regulatory mechanisms. Aside from cancer, diabetes, atherosclerosis, obesity, and neuron-degenerative diseases, there are other oxidative-stress-induced diseases, such as chronic lung disease (asthma and chronic obstructive pulmonary disease) and premature aging, which have been linked to vitamin D; therefore future research exploring the protective/destructive effect of vitamin D/VDR signaling is demanded.

REFERENCES

1. Feldman, D., J. W. Pike, and J. S. Adams. 2011. *Vitamin D*. 3rd edition, Amsterdam, Boston: Academic Press.
2. Garcion, E. et al. 2002. New clues about vitamin D functions in the nervous system. *Trends Endocrinol Metab* 13(3):100–105.
3. Di Rosa, M. et al. 2011. Vitamin D_3: A helpful immunomodulator. *Immunology* 134(2):123–139.
4. Takiishi, T. et al. 2010. Vitamin D and diabetes. *Endocrinol Metab Clin North Am* 39(2):419–446.
5. Plum, L. A. and H. F. De Luca. 2010. Vitamin D, disease and therapeutic opportunities. *Nat Rev Drug Discov* 9(12):941–955.
6. Trump, D. L., K. K. Deeb, and C. S. Johnson. 2010. Vitamin D: Considerations in the continued development as an agent for cancer prevention and therapy. *Cancer J* 16(1):1–9.
7. Norman, A. W. et al. 1992. 1,25(OH)2-vitamin D_3, a steroid hormone that produces biologic effects via both genomic and nongenomic pathways. *J Steroid Biochem Mol Biol* 41(3–8):231–240.
8. Casarett, L. J., J. Doull, and C. D. Klaassen. 2008. *Casarett and Doull's Toxicology: The Basic Science of Poisons*. 7th edition, New York: McGraw-Hill. xv, 1310 pp., 1 leaf of plates.
9. Newmeyer, D. D. and S. Ferguson-Miller. 2003. Mitochondria: Releasing power for life and unleashing the machineries of death. *Cell* 112(4):481–490.
10. Balaban, R. S., S. Nemoto, and T. Finkel. 2005. Mitochondria, oxidants, and aging. *Cell* 120(4):483–495.
11. Calabrese, E. J. 2001. Nitric oxide: Biphasic dose responses. *Crit Rev Toxicol* 31(4–5):489–501.
12. Goswami, S. K., N. Maulik, and D. K. Das. 2007. Ischemia-reperfusion and cardioprotection: A delicate balance between reactive oxygen species generation and redox homeostasis. *Ann Med* 39(4):275–289.
13. Mena, S., A. Ortega, and J. M. Estrela. 2009. Oxidative stress in environmental-induced carcinogenesis. *Mutat Res* 674(1–2):36–44.
14. Savolainen, K. M. et al. 1998. Interactions of excitatory neurotransmitters and xenobiotics in excitotoxicity and oxidative stress: Glutamate and lead. *Toxicol Lett* 102–103:363–367.
15. Montgomery, E. B. Jr. 1995. Heavy metals and the etiology of Parkinson's disease and other movement disorders. *Toxicology* 97(1–3):3–9.
16. Rock, E. and A. DeMichele. 2003. Nutritional approaches to late toxicities of adjuvant chemotherapy in breast cancer survivors. *J Nutr* 133(11 Suppl 1):3785S–3793S.
17. Mason, R. P. and C. F. Chignell. 1981. Free radicals in pharmacology and toxicology—Selected topics. *Pharmacol Rev* 33(4):189–211.
18. Riley, P. A. 1994. Free radicals in biology: Oxidative stress and the effects of ionizing radiation. *Int J Radiat Biol* 65(1):27–33.
19. Bickers, D. R. and M. Athar. 2006. Oxidative stress in the pathogenesis of skin disease. *J Invest Dermatol* 126(12):2565–2575.
20. Drummond, G. R. et al. 2011. Combating oxidative stress in vascular disease: NADPH oxidases as therapeutic targets. *Nat Rev Drug Discov* 10(6):453–471.
21. Stoll, G., S. Jander, and M. Schroeter. 2002. Detrimental and beneficial effects of injury-induced inflammation and cytokine expression in the nervous system. *Adv Exp Med Biol* 513:87–113.
22. Chung, J. W. et al. 2006. Vitamin D_3 upregulated protein 1 (VDUP1) is a regulator for redox signaling and stress-mediated diseases. *J Dermatol* 33(10):662–669.
23. Kinnula, V. L. and J. D. Crapo. 2003. Superoxide dismutases in the lung and human lung diseases. *Am J Respir Crit Care Med* 167(12):1600–1619.
24. Weiss, S. L. and R. A. Sunde. 1997. Selenium regulation of classical glutathione peroxidase expression requires the 3′ untranslated region in Chinese hamster ovary cells. *J Nutr* 127(7):1304–1310.
25. Nenoi, M. et al. 2001. Regulation of the catalase gene promoter by Sp1, CCAAT-recognizing factors, and a WT1/Egr-related factor in hydrogen peroxide-resistant HP100 cells. *Cancer Res* 61(15):5885–5894.
26. Fong, W. H. et al. 2010. Antiapoptotic actions of PPAR-gamma against ischemic stroke. *Mol Neurobiol* 41(2–3):180–186.
27. Dalton, T. P. et al. 2004. Genetically altered mice to evaluate glutathione homeostasis in health and disease. *Free Radic Biol Med* 37(10):1511–1526.
28. Rhee, S. G. et al. 2005. Controlled elimination of intracellular H(2)O(2): Regulation of peroxiredoxin, catalase, and glutathione peroxidase via post-translational modification. *Antioxid Redox Signal* 7(5–6):619–626.
29. Hussain, S. P. et al. 2004. p53-induced upregulation of MnSOD and GPx but not catalase increases oxidative stress and apoptosis. *Cancer Res* 64(7):2350–2356.

30. Pi, J., M. L. Freeman, and M. Yamamoto. 2010. Nrf2 in toxicology and pharmacology: The good, the bad and the ugly? *Toxicol Appl Pharmacol* 244(1):1–3.
31. Siegel, D. et al. 2004. NAD(P)H:quinone oxidoreductase 1: Role as a superoxide scavenger. *Mol Pharmacol* 65(5):1238–1247.
32. Turkseven, S. et al. 2005. Antioxidant mechanism of heme oxygenase-1 involves an increase in superoxide dismutase and catalase in experimental diabetes. *Am J Physiol Heart Circ Physiol* 289(2):H701–H707.
33. Arner, E. S. 2009. Focus on mammalian thioredoxin reductases—important selenoproteins with versatile functions. *Biochim Biophys Acta* 1790(6):495–526.
34. Immenschuh, S. and E. Baumgart-Vogt. 2005. Peroxiredoxins, oxidative stress, and cell proliferation. *Antioxid Redox Signal* 7(5–6):768–777.
35. Ibi, M. et al. 2001. Protective effects of 1 alpha,25-(OH)(2)D($_3$) against the neurotoxicity of glutamate and reactive oxygen species in mesencephalic culture. *Neuropharmacology* 40(6):761–771.
36. Tetich, M. et al. 2003. The third multidisciplinary conference on drug research, Pila 2002. Effects of 1alpha,25-dihydroxyvitamin D$_3$ and some putative steroid neuroprotective agents on the hydrogen peroxide-induced damage in neuroblastoma-glioma hybrid NG108-15 cells. *Acta Pol Pharm* 60(5):351–355.
37. Husain, K. et al. 2009. Combination therapy with paricalcitol and enalapril ameliorates cardiac oxidative injury in uremic rats. *Am J Nephrol* 29(5):465–472.
38. Mizobuchi, M. et al. 2010. Myocardial effects of VDR activators in renal failure. *J Steroid Biochem Mol Biol* 121(1–2):188–192.
39. Riachy, R. et al. 2001. Beneficial effect of 1,25 dihydroxyvitamin D$_3$ on cytokine-treated human pancreatic islets. *J Endocrinol* 169(1):161–168.
40. Husain, K. et al. 2010. Effects of paricalcitol and enalapril on atherosclerotic injury in mouse aortas. *Am J Nephrol* 32(4):296–304.
41. Wei, M. Y. et al. 2008. Vitamin D and prevention of colorectal adenoma: a meta-analysis. *Cancer Epidemiol Biomarkers Prev* 17(11):2958–2969.
42. Kallay, E. et al 2002. Vitamin D receptor activity and prevention of colonic hyperproliferation and oxidative stress. *Food Chem Toxicol* 40(8):1191–1196.
43. Kallay, E. et al. 2001. Characterization of a vitamin D receptor knockout mouse as a model of colorectal hyperproliferation and DNA damage. *Carcinogenesis* 22(9):1429–1435.
44. Fedirko, V. et al. 2010. Effects of supplemental vitamin D and calcium on oxidative DNA damage marker in normal colorectal mucosa: A randomized clinical trial. *Cancer Epidemiol Biomarkers Prev* 19(1):280–291.
45. Welsh, J. et al. 2003. Vitamin D-$_3$ receptor as a target for breast cancer prevention. *J Nutr* 133(7 Suppl):2425S–2433S.
46. Lou, Y. R. et al. 2010. 25-Hydroxyvitamin D($_3$) is an agonistic vitamin D receptor ligand. *J Steroid Biochem Mol Biol* 118(3):162–170.
47. Peng, X. et al. 2009. 25-Hydroxyvitamin D$_3$ is a natural chemopreventive agent against carcinogen induced precancerous lesions in mouse mammary gland organ culture. *Breast Cancer Res Treat* 113(1):31–41.
48. Peng, X. et al. 2010. Protection against cellular stress by 25-hydroxyvitamin D$_3$ in breast epithelial cells. *J Cell Biochem* 110(6):1324–1333.
49. Krishnan, A. V. and D. Feldman. 2010. Molecular pathways mediating the anti-inflammatory effects of calcitriol: Implications for prostate cancer chemoprevention and treatment. *Endocr Relat Cancer* 17(1):R19–R38.
50. Toyokuni, S. 2006. Novel aspects of oxidative stress-associated carcinogenesis. *Antioxid Redox Signal* 8(7–8):1373–1377.
51. Kovalenko, P. L. et al. 2010. 1,25 dihydroxyvitamin D–mediated orchestration of anticancer, transcript-level effects in the immortalized, nontransformed prostate epithelial cell line, RWPE1. *BMC Genomics* 11:26.
52. Peehl, D. M. et al. 2004. Molecular activity of 1,25-dihydroxyvitamin D$_3$ in primary cultures of human prostatic epithelial cells revealed by cDNA microarray analysis. *J Steroid Biochem Mol Biol* 92(3):131–141.
53. Krezel, A., Q. Hao, and W. Maret. 2007. The zinc/thiolate redox biochemistry of metallothionein and the control of zinc ion fluctuations in cell signaling. *Arch Biochem Biophys* 463(2):188–200.
54. Bao, B. Y. et al. 2008. Protective role of 1 alpha, 25-dihydroxyvitamin D$_3$ against oxidative stress in nonmalignant human prostate epithelial cells. *Int J Cancer* 122(12):2699–2706.

55. Banakar, M. C. et al. 2004. 1alpha, 25-dihydroxyvitamin D$_3$ prevents DNA damage and restores antioxidant enzymes in rat hepatocarcinogenesis induced by diethylnitrosamine and promoted by phenobarbital. *World J Gastroenterol* 10(9):1268–1275.
56. Karmakar, R., S. Banik, and M. Chatterjee. 2002. Inhibitory effect of vitamin D$_3$ on 3'methyl-4-dimethyl-amino-azobenzene-induced rat hepatocarcinogenesis: A study on antioxidant defense enzymes. *J Exp Ther Oncol* 2(4):193–199.
57. Saha, B. K. et al. 2001. 1Alpha,25-dihydroxyvitamin D$_3$ inhibits hepatic chromosomal aberrations, DNA strand breaks and specific DNA adducts during rat hepatocarcinogenesis. *Cell Mol Life Sci* 58(8):1141–1149.
58. Saha, B. K. et al. 2001. 1alpha, 25-Dihydroxyvitamin D$_3$ suppresses the effect of streptozotocin-induced diabetes during chemical rat liver carcinogenesis. *Cell Biol Int* 25(3):227–237.
59. Preedy, V. R. et al. 2002. Free radicals in alcoholic myopathy: Indices of damage and preventive studies. *Free Radic Biol Med* 32(8):683–637.
60. Gonzalez-Reimers, E. et al. 2010. Alcoholic myopathy: Vitamin D deficiency is related to muscle fibre atrophy in a murine model. *Alcohol Alcohol* 45(3):223–230.
61. Tarcin, O. et al. 2009. Effect of vitamin D deficiency and replacement on endothelial function in asymptomatic subjects. *J Clin Endocrinol Metab* 94(10):4023–4030.
62. Casteels, K. M. et al. 1998. Prevention of type I diabetes in nonobese diabetic mice by late intervention with nonhypercalcemic analogs of 1,25-dihydroxyvitamin D$_3$ in combination with a short induction course of cyclosporin A. *Endocrinology* 139(1):95–102.
63. Cetinkalp, S. et al. 2009. The effect of 1alpha,25(OH)2D$_3$ vitamin over oxidative stress and biochemical parameters in rats where type 1 diabetes is formed by streptozotocin. *J Diabetes Complications* 23(6):401–408.
64. Luca, G. et al. 2003. Multifunctional microcapsules for pancreatic islet cell entrapment: Design, preparation and *in vitro* characterization. *Biomaterials* 24(18):3101–3114.
65. Pop-Busui, R., A. Sima, and M. Stevens. 2006. Diabetic neuropathy and oxidative stress. *Diabetes Metab Res Rev* 22(4):257–273.
66. Greene, D. A. et al. 1999. Glucose-induced oxidative stress and programmed cell death in diabetic neuropathy. *Eur J Pharmacol* 375(1–3):217–223.
67. Hamden, K. et al. 2009. 1Alpha,25 dihydroxyvitamin D$_3$: Therapeutic and preventive effects against oxidative stress, hepatic, pancreatic and renal injury in alloxan-induced diabetes in rats. *J Nutr Sci Vitaminol (Tokyo)* 55(3):215–222.
68. Hamden, K. et al. 2008. Inhibitory effects of 1alpha, 25dihydroxyvitamin D$_3$ and Ajuga iva extract on oxidative stress, toxicity and hypo-fertility in diabetic rat testes. *J Physiol Biochem* 64(3):231–239.
69. Ravid, A. et al. 2002. Vitamin D inhibits the activation of stress-activated protein kinases by physiological and environmental stresses in keratinocytes. *J Endocrinol* 173(3):525–532.
70. Diker-Cohen, T. et al. 2003. Vitamin D protects keratinocytes from apoptosis induced by osmotic shock, oxidative stress, and tumor necrosis factor. *Ann N Y Acad Sci* 1010:350–353.
71. Lee, J. and J. I. Youn. 1998. The photoprotective effect of 1,25-dihydroxyvitamin D$_3$ on ultraviolet light B-induced damage in keratinocyte and its mechanism of action. *J Dermatol Sci* 18(1):11–18.
72. Gupta, R. et al. 2007. Photoprotection by 1,25 dihydroxyvitamin D$_3$ is associated with an increase in p53 and a decrease in nitric oxide products. *J Invest Dermatol* 127(3):707–715.
73. Chen, K. B., A. M. Lin, and T. H. Chiu. 2003. Systemic vitamin D$_3$ attenuated oxidative injuries in the locus coeruleus of rat brain. *Ann N Y Acad Sci* 993:313–324; Discussion 345–9.
74. Lin, A. M., K. B. Chen, and P. L. Chao. 2005. Antioxidative effect of vitamin D$_3$ on zinc-induced oxidative stress in CNS. *Ann N Y Acad Sci* 1053:319–329.
75. Li, L. et al. 2008. 1Alpha,25-dihydroxyvitamin D$_3$ attenuates cyanide-induced neurotoxicity by inhibiting uncoupling protein-2 upregulation. *J Neurosci Res* 86(6):1397–1408.
76. Ekici, F., B. Ozyurt, and H. Erdogan. 2009. The combination of vitamin D$_3$ and dehydroascorbic acid administration attenuates brain damage in focal ischemia. *Neurol Sci* 30(3):207–212.
77. Losem-Heinrichs, E. et al. 2005. 1alpha,25-dihydroxy-vitamin D$_3$ in combination with 17beta-estradiol lowers the cortical expression of heat shock protein-27 following experimentally induced focal cortical ischemia in rats. *Arch Biochem Biophys* 439(1):70–79.
78. Losem-Heinrichs, E. et al. 2004. A combined treatment with 1alpha,25-dihydroxy-vitamin D$_3$ and 17beta-estradiol reduces the expression of heat shock protein-32 (HSP-32) following cerebral cortical ischemia. *J Steroid Biochem Mol Biol* 89–90(1–5):371–374.
79. Byrne, B. M. and J. Welsh. 2005. Altered thioredoxin subcellular localization and redox status in MCF-7 cells following 1,25-dihydroxyvitamin D$_3$ treatment. *J Steroid Biochem Mol Biol* 97(1–2):57–64.

80. Marchionatti, A. M. et al. 2009. Antiproliferative action of menadione and 1,25(OH)2D3 on breast cancer cells. *J Steroid Biochem Mol Biol* 113(3–5):227–232.
81. Ravid, A. et al. 1999. 1,25-Dihydroxyvitamin D_3 enhances the susceptibility of breast cancer cells to doxorubicin-induced oxidative damage. *Cancer Res* 59(4):862–867.
82. Weitsman, G. E. et al. 2005. Vitamin D sensitizes breast cancer cells to the action of H2O2: Mitochondria as a convergence point in the death pathway. *Free Radic Biol Med* 39(2):266–278.
83. Weitsman, G. E. et al. 2003. Vitamin D enhances caspase-dependent and -independent TNFalpha-induced breast cancer cell death: The role of reactive oxygen species and mitochondria. *Int J Cancer* 106(2):178–186.
84. Koren, R. et al. 2000. Synergistic anticancer activity of 1,25-dihydroxyvitamin $D_{(3)}$ and immune cytokines: The involvement of reactive oxygen species. *J Steroid Biochem Mol Biol* 73(3–4):105–112.
85. Narvaez, C. J. and J. Welsh. 2001. Role of mitochondria and caspases in vitamin D-mediated apoptosis of MCF-7 breast cancer cells. *J Biol Chem* 276(12):9101–9107.
86. Biskobing, D. M. and J. Rubin. 1993. 1,25-Dihydroxyvitamin D_3 and phorbol myristate acetate produce divergent phenotypes in a monomyelocytic cell line. *Endocrinology* 132(2):862–866.
87. Chen, K. S. and H. F. De Luca. 1994. Isolation and characterization of a novel cDNA from HL-60 cells treated with 1,25-dihydroxyvitamin D_{-3}. *Biochim Biophys Acta* 1219(1):26–32.
88. Nishiyama, A. et al. 1999. Identification of thioredoxin-binding protein-2/vitamin $D_{(3)}$ upregulated protein 1 as a negative regulator of thioredoxin function and expression. *J Biol Chem* 274(31):21645–21650.
89. Yoshioka, J., E. R. Schreiter, and R. T. Lee. 2006. Role of thioredoxin in cell growth through interactions with signaling molecules. *Antioxid Redox Signal* 8(11–12):2143–2151.
90. Junn, E. et al. 2000. Vitamin D_3 upregulated protein 1 mediates oxidative stress via suppressing the thioredoxin function. *J Immunol* 164(12):6287–6295.
91. Bondza-Kibangou, P. et al. 2007. Antioxidants and doxorubicin supplementation to modulate CD14 expression and oxidative stress induced by vitamin D_3 and seocalcitol in HL60 cells. *Oncol Rep* 18(6):1513–159.
92. Bondza-Kibangou, P. et al. 2004. Modifications of cellular autofluorescence emission spectra under oxidative stress induced by 1 alpha,25dihydroxyvitamin $D_{(3)}$ and its analog EB1089. *Technol Cancer Res Treat* 3(4):383–391.
93. Chen, W. J. et al. 2008. Induction of apoptosis by vitamin D_3, ergocalciferol, via reactive oxygen species generation, glutathione depletion, and caspase activation in human leukemia cells. *J Agric Food Chem* 56(9):2996–3005.
94. Monteiro, R. and I. Azevedo. 2010. Chronic inflammation in obesity and the metabolic syndrome. *Mediators Inflamm*.
95. Sun, X. and M. B. Zemel. 2006. Dietary calcium regulates ROS production in aP2-agouti transgenic mice on high-fat/high-sucrose diets. *Int J Obes (Lond)* 30(9):1341–1346.
96. Sun, X. and M. B. Zemel. 2007. Calcium and 1,25-dihydroxyvitamin D_3 regulation of adipokine expression. *Obesity (Silver Spring)* 15(2):340–348.
97. Sun, X. and M. B. Zemel. 2008. Calcitriol and calcium regulate cytokine production and adipocyte-macrophage cross-talk. *J Nutr Biochem* 19(6):392–399.
98. Sun, X. and M. B. Zemel. 2007. 1Alpha,25-dihydroxyvitamin D_3 modulation of adipocyte reactive oxygen species production. *Obesity (Silver Spring)* 15(8):1944–1953.
99. Sun, X., K. L. Morris, and M. B. Zemel. 2008. Role of calcitriol and cortisol on human adipocyte proliferation and oxidative and inflammatory stress: A microarray study. *J Nutrigenet Nutrigenomics* 1(1–2):30–48.
100. Tang, F. T. et al. 2006. Hypercholesterolemia accelerates vascular calcification induced by excessive vitamin D via oxidative stress. *Calcif Tissue Int* 79(5):326–339.
101. Ngo, D. T. et al. 2008. Vitamin $D_{(3)}$ supplementation induces the development of aortic stenosis in rabbits: Interactions with endothelial function and thioredoxin-interacting protein. *Eur J Pharmacol* 590(1–3):290–296.
102. Serhan, C. N. et al. 2003. Reduced inflammation and tissue damage in transgenic rabbits overexpressing 15-lipoxygenase and endogenous anti-inflammatory lipid mediators. *J Immunol* 171(12):6856–6865.
103. Somjen, D. et al. 2010. Vitamin D analogs induce lipoxygenase mRNA expression and activity as well as reactive oxygen species (ROS) production in human bone cells. *J Steroid Biochem Mol Biol* 121(1–2):265–267.
104. Somjen, D. et al. 2011. Vitamin D metabolites and analogs induce lipoxygenase mRNA expression and activity as well as reactive oxygen species (ROS) production in human bone cell line. *J Steroid Biochem Mol Biol* 123(1–2):85–89.

6 Vitamin D₃ Upregulated Protein 1 (VDUP1): Roles in Redox Signaling and Stress-Mediated Diseases

Dong Oh Kim, Hyun-Woo Suh, Haiyoung Jung,
Young Jun Park, and Inpyo Choi

CONTENTS

6.1   VDUP1 NETWORK

Vitamin D₃ upregulated protein-1 (VDUP1), which is identical to TRX binding protein-2 (TBP-2) and thioredoxin interacting protein (TXNIP), was originally identified as a vitamin D–induced protein in HL-60 cells treated with 1,25-dihydroxyvitamin D (Chen and DeLuca 1994). Later, it was identified to bind to the reduced thioredoxin (Trx) by yeast two-hybrid screening and to inhibit the antioxidant function of Trx (Junn et al. 2000; Nishiyama et al. 1999). VDUP1 recognizes the catalytic active center of Trx via cysteine 247 residues. By binding the active center of Trx, VDUP1 can inhibit the reducing activity of Trx (Junn et al. 2000). VDUP1 has 2 beta-arrestin domains and 11 cysteine residues. This implies VDUP1 can bind many other proteins and interactions may be influenced by intracellular redox status. VDUP1 can interact with NLRP3 in a redox-dependent manner (Zhou et al. 2010). For binding with Trx, two cysteine residues, i.e., Cys 247 and Cys 63, are important, and mutation in these cysteines results in a failure of the binding of VDUP1 and Trx.

  Although biochemical studies demonstrated that VDUP1 functions as a redox regulator in cells, many reports have suggested unexpected functions of VDUP1 (Kim et al. 2007). VDUP1 expression is reduced in many tumor cells. Meanwhile, VDUP1 is induced by growth-arresting conditions such as serum deprivation and oxidative stress. It inhibits tumor growth and cell cycle progression (Han et al. 2003). It induces cell cycle arrest at the G_0/G_1 phase, and it increases p27 protein stability by regulating JAB. Antitumor reagents such as SAHA rapidly increase VDUP1 expression (Butler et al. 2002). VDUP1-deficient mice show hepatocarcinoma and hyperplasia (Sheth et al. 2006). These data indicate that VDUP1 is a novel tumor suppressor. In VDUP1-deficient mice, the level of free fatty acids is elevated, and Kreb cycle–mediated fatty acid utilization is impaired (Oka et al. 2006b). In addition,

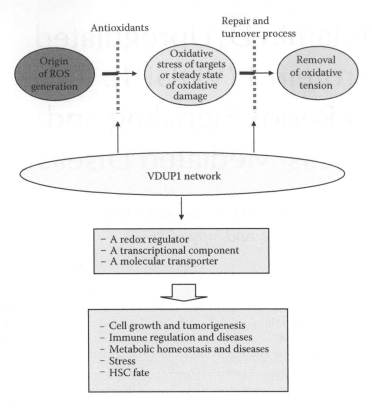

FIGURE 6.1 VDUP1 network in cells. VDUP1 is known to regulate the cellular redox state and oxidative stress. In addition, as a transcriptional regulator and a molecular transporter, VDUP1 regulates many cellular functions, such as tumorigenesis, immune regulation, metabolic diseases, and stem cell fate.

energy metabolism and glycolysis are affected by VDUP1 (Parikh et al. 2007). In this regard, it is suggested that VDUP1 is related to diabetes (van Greevenbroek et al. 2007). Recent results show that VDUP1 is a component of the inflammasome (Zhou et al. 2010). It binds to NLRP3 to regulate IL-1 production, which is a key molecule for inflammation. It seems that VDUP1 can control many cellular functions such as cell growth, immune responses, metabolism, and stress responses (Figure 6.1).

VDUP1 has diverse functions and many binding partners, which indicates that many functions may depend on the cellular environment. However, it is difficult to understand the mechanisms due to the lack of information on its *in vivo* roles and its protein structure. Overall, VDUP1 is like a key cellular molecule that binds to and regulates diverse partner molecules. To understand this interesting molecule in detail, some key roles of VDUP1 based on recent reports will be discussed.

6.2 VDUP1 AND REDOX SIGNALING

In normal status, cells are continuously encountering various stimuli, such as ultraviolet, chemicals, radiation, and virus infection, that induce intracellular generation of reactive oxygen species (ROSs). To overcome environmental stress or to eliminate intracellular stresses, cells have specialized intrinsic antioxidant systems. Oxidative stress caused by excessive amounts of ROS induces the formation of disulfide bonds in proteins, lipid peroxidation, and DNA cross linking and causes oxidative stress–related disorders such as atherosclerosis, diabetes, pulmonary fibrosis, neurodegenerative disorders, and arthritis. This is believed to be a major factor in aging (Watanabe et al. 2010; Yoshihara et al. 2010). Physiological redox regulation is an important function for the performance of biological events such as enzyme activation, gene expression, DNA synthesis, and cell cycle regulation in cells. Thus, the dosage of ROS is an important factor in the modulation of cellular function (Chung et al. 2006; Yoshihara et al. 2010).

The redox state of cysteinyl residues in proteins is mainly controlled by the Trx and the glutathione (GRX) systems (Holmgren 1989). Trx, which is a 12-kDa ubiquitous protein, was originally identified in *Escherichia coli* as a hydrogen donor to ribonucleotide reductase, an essential enzyme for DNA synthesis. Human Trx was cloned independently as an adult T-cell leukemia–derived factor (ADF) produced by human T-cell leukemia virus type-I (HTLV-I)–transformed T-cell line ALT-2 and as an IL-1-like growth factor produced by Epstein–Barr virus–transformed B–cell line 3B6 (Tagaya et al. 1989; Wollman et al. 1988). Two cysteine residues (Cys32 and Cys35) of the active site -Cys-Gly-Pro-Cys- are responsible for reducing the activity of Trx. Knockout of Trx mice is lethal at the early development, whereas Trx-transgenic mice are more resistant to oxidative stress with longer life spans, compared with wild-type mice (Matsui et al. 1996).

The involvement of VDUP1 in redox regulation was first suggested by the identification of VDUP1 as a thioredoxin (Trx)-binding protein-2 (TBP-2). VDUP1 is a member of the arrestin protein family, containing two characteristic arrestin-like domains with a PXXP sequence (a known binding motif for SH3-domains containing proteins) and a PPXY sequence (a known binding motif for the WW domain; Alvarez 2008; Oka et al. 2006a; Patwari et al. 2006; Patwari et al. 2009). VDUP1 is highly homologous to several other genes. One is TLIMP (TBP-2-like inducible membrane protein), a novel vitamin D3 or peroxisome proliferator–activated receptor (PPAR)-γ ligand-inducible membrane associated protein, which plays a regulatory role in cell proliferation and PPAR-γ activation. Another is DRH1, which is reported to be downregulated in hepatocellular carcinoma (Oka et al. 2006a; Yamamoto et al. 2001). Since the Trx system controls cellular redox by scavenging intracellular ROS, the function of VDUP1 is thought to be a negative regulator of Trx, thus increasing the intracellular level of ROS (Figure 6.2), although its inhibitory mechanism remains largely unknown. Functionally, VDUP1 has a potent growth-suppressive function and also is involved in lipid/glucose metabolism, as well as inhibition of the insulin-reducing activity. Thus, VDUP1 is an important molecule in connecting many pathways of various cellular functions and is a regulator in metabolic disorders, cancer, and inflammation. In some reports, it was found that VDUP1 inhibited the reducing activity and the expression of Trx, and the overexpression of VDUP1 increased the level of ROS in fibroblasts (Chung et al. 2006; Yoshioka et al. 2004). In addition, our study with VDUP1-deficient

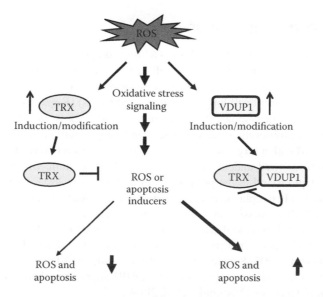

FIGURE 6.2 Regulation of redox signaling by VDUP1. VDUP1 as an inhibitor of Trx and regulates intracellular redox signaling and apoptosis.

mice showed that the absence of VDUP1 accompanied a decreased level of intracellular ROS. The function of VDUP1 as a negative regulator of Trx is associated with the fact that its binding with Trx requires a Trx active site, which makes VDUP1 compete with other proteins for binding to Trx (Chung et al. 2006; Lee et al. 2005). For example, overexpression of VDUP1 reduced the interaction of Trx with PAG or ASK-1, causing cells to become more sensitive to oxidative stress (Junn et al. 2000). PAG is a thiol-specific antioxidant that reduces hydrogen peroxide when Trx is present as an immediate electron donor, and the activity of ASK-1 is induced by ROS due to the dissociation with Trx (Saitoh et al. 1998). Thus, interference with the interaction between Trx and PAG or ASK-1 by VDUP1 makes the cells more susceptible to oxidative stress. In this regard, VDUP1-overexpressing cell lines, when treated with H_2O_2, show higher levels of IL-6 expression and JNK activity, which are known to be induced by ROS (Junn et al. 2000). Recently, one group reported that VDUP1 binds to Trx2 to reduce Trx2 binding to ASK1 and to activate ASK1 activity, resulting in induction of the mitochondrial pathway of apoptosis with cytochrome c release and caspase-3 cleavage in mitochondria. These data show VDUP1 shuttles between subcellular compartments in response to oxidative stress and a novel redox-sensitive mitochondrial VDUP1-Trx2-ASK1 signaling cascade (Saxena et al. 2010). VDUP1 regulates these ROS-mediated biological processes by direct interaction with antioxidant proteins. Downregulation of VDUP1 expression by antisense oligonucleotides in melanoma cells reduces ROS production and the expression of Fas ligands, which are related to tumor cell survival. Various tumor cells are known to generate ROS, contributing to the ability of tumors to transform and to advance metastasis (Song et al. 2003). This prooxidant state can promote tumor cells to neoplastic proliferation. In melanoma, endogenous ROS, the autocrinically stimulates constitutive activation of NF-κB, and it ultimately provokes tumor growth. In addition, antioxidants (such as NAC and catalase) or overexpression of MnSOD suppressed the tumor cell proliferation *in vitro* and *in vivo* (Borrello et al. 1993; Cerutti 1985; Szatrowski and Nathan 1991). On the other hand, positive effects of Trx in tumor growth have been reported in which Trx was proposed to have antiapoptotic properties (Brar et al. 2001; Soderberg et al. 2000; Zhao et al. 2001). The major role of ROS in melanoma cells seems to be to maintain a prooxidant state of the cells by reducing the expression of VDUP1. These reports show that VDUP1 can either positively or negatively regulate ROS signals depending on the stimuli and also suggest that VDUP1 plays an important role in a wide variety of biological functions such as the regulation of cell death, growth, differentiation, and energy metabolism.

6.3   VDUP1 AND TUMORIGENESIS

A number of studies have suggested that VDUP1 is a tumor suppressor and involved in the regulation of cell proliferation by associating with diverse cell cycle regulators (Kim et al. 2007). The expression of VDUP1 is downregulated in various types of cancer cells. For example, VDUP1 expression is decreased in human nonsmall cell lung carcinomas (NSCLCs) (Kopantzev et al. 2008). VDUP1 expression is also reduced in response to an increase in lung expansion that induces a marked acceleration in lung cell proliferation. On the other hand, lung deflation, which causes lung growth to cease, was associated with a large increase in VDUP1 expression (Filby et al. 2006). Most cases of renal cell carcinoma showed undetectable levels of VDUP1 protein, which was associated with the methylation of a CpG island in the promoter region. Moreover, loss of VDUP1 is essential for the proliferation of not only neoplastic but also nonneoplastic renal tubular cells (Dutta et al. 2005; Yang et al. 1998). VDUP1 expression in colorectal and gastric cancers was significantly lower than that in their adjacent normal tissues (Ikarashi et al. 2002). VDUP1 mRNA levels are also reduced in human breast, colon, and other tumor tissue, compared with matched samples of normal tissues (de Vos et al. 2003; Ohta et al. 2005; Solanas et al. 2009). In addition, VDUP1 is suppressed in HTLV-1–transformed T-cell lines, and overexpression of VDUP1 suppresses their cell growth, indicating its involvement in tumorigenesis (Nishinaka et al. 2004).

Myosin VI, an actin motor, is one of the top genes that demonstrated cancer-specific overexpression in human prostate and breast cancers (Dunn et al. 2006). The expression of VDUP1 was dramatically increased after myosin VI knockdown in both cell lines, resulting in reduced anchorage-independent growth (Dunn et al. 2006). VDUP1 deficiency may act in concert with other proliferative stimuli to induce tumor progression. The loss of VDUP1 contributes to the development and progression of malignancy in hepatocellular carcinoma (HCC) in VDUP1-deficient mice (Sheth et al. 2006). A recent report showed that VDUP1 expression is suppressed during human hepatic carcinogenesis, and mice lacking VDUP1 are much more susceptible to diethylnitrosamine-induced hepatocarcinogenesis compared with wild-type mice (Kwon et al. 2010). In addition, VDUP1 overexpression completely suppressed tumorigenesis in a lung cancer xenograft model (Kwon et al. 2010). A VDUP1 homologous gene, DRH1 is also downregulated in HCC, indicating that the familial members of VDUP1 also play a role in cancer suppression (Yamamoto et al. 2001). VDUP1-deficient fibroblasts were shown to exhibit increased proliferation, with reduced expression of p27[kip1], a cyclin-dependent kinase inhibitor. This appears to involve the interaction of VDUP1 with JAB1, blocking the JAB1-mediated translocation of p27[kip1] from the nucleus to the cytoplasm (Jeon et al. 2005).

VDUP1 is upregulated by carcinogens and various stress stimuli (Han et al. 2003; Junn et al. 2000). Suberoylanilide hydroxamic acid (SAHA) is a potent inhibitor of histone deacetylases (HDACs) that causes growth arrest (Butler et al. 2002). The expression of the VDUP1 gene was induced by SAHA in several human-transformed cell lines, including prostate carcinoma, bladder carcinoma, myeloma, breast carcinoma cell lines, and murine erythroleukemia cells. Induction of VDUP1 contributes to SAHA-mediated differentiation and antiproliferative effects (Butler et al. 2002; Huang and Pardee 2000). Interestingly, ectopic expression of hnRNP G in human cancer cells resulted in severe retardation of proliferation and reduction of *in vivo* tumorigenicity. The expression of VDUP1 is induced by hnRNP G, and this upregulation is associated with the tumor-suppressive function (Shin et al. 2008; Shin et al. 2006). VDUP1 is strongly upregulated by the gene 5-fluorouracil (5-FU) in the treatment of colon cancer (Takahashi et al. 2002). VDUP1 not only may be induced by stress response to 5-FU cytotoxicity but may also play a key role in 5-FU cytotoxicity in colon cancers (Takahashi et al. 2002). D-Allose is a novel antitumor monosaccharide that causes cell growth inhibition. D-Allose caused G_1 cell cycle arrest in HuH-7 cells and oral squamous carcinoma cell lines via VDUP1 induction (Hoshikawa et al. 2010; Yamaguchi et al. 2008a). The combination of D-Allose and 5-FU inhibited human HCC cell growth and enhanced the antitumor effect by inducing VDUP1 (Yamaguchi et al. 2008b). In addition, VDUP1 is upregulated during the senescence process. The overexpression of VDUP1 may block the action of Trx, eventually leading to the inhibition of cellular proliferation (Nishiyama et al. 1999; Yoshida et al. 2006). Introducing VDUP1 into breast cancer cell line MCF-7 significantly suppressed cellular proliferation. A subsequent upregulation of VDUP1 as cells undergo senescence was accompanied by a strong increase in levels of ROS (Cadenas et al. 2010). FOXO3A, whose activity increases in senescent cells, transcriptionally upregulates VDUP1 expression and miR-17-5p, which directly interacts with the 3′-untranslated region of VDUP1 transcripts and destabilizes VDUP1 mRNA in young cells (Zhuo et al. 2010). VDUP1 is highly induced in senescent normal human fibroblasts and overexpression of VDUP1 mRNA inhibits colony formation under aerobic conditions and sensitizes to paraquat-induced oxidative stress in HeLa cells (Joguchi et al. 2002).

VDUP1 has been shown to interact with transcription factors and chromatin remodeling complexes, which may play a role in cell cycle progression. VDUP1 expression is upregulated by serum or IL-2 deprivation, leading to growth arrest. VDUP1 interacts with various cellular targets such as JAB1 and FAZF and may be a component of a transcriptional repressor complex (Jeon et al. 2005; Nakamura et al. 2006). The transfection of VDUP1 in tumor cells reduced cell growth. VDUP1 expression was also increased when the cell cycle progression was arrested. Transfection of VDUP1-induced cell cycle arrest at the G_0/G_1 phase and interacted with transcriptional corepressors, indicating that VDUP1 possesses a tumor-suppressive activity (Han et al. 2003).

VDUP1 is currently thought to be a potent antitumor gene, and expression is downregulated in various tumors, as previously described. On the other hand, VDUP1 overexpression is observed in tumor cells of hypoxic perinecrotic areas in glioblastoma compared to nonhypoxic tumor cells (Le Jan et al. 2006). VDUP1 is also induced during hypoxia in pancreatic cancer cells in an HIF-1 α-dependent manner. The increase in VDUP1 expression is associated with the inhibition of Trx-1 redox activity and increased spontaneous apoptosis (Baker et al. 2008; Watanabe et al. 2010). In addition, CD437, a synthetic retinoid, induces endoplasmic reticulum stress-mediated apoptosis involving the upregulation of VDUP1 in ovarian adenocarcinoma cells (Matsuoka et al. 2008). Paclitaxel is a chemotherapy agent used in the treatment of breast cancer. Increases pacitaxel cytotoxicity is associated with an additive effect on hyperglycemia-mediated VDUP1 expression (Turturro et al. 2007a). Under diabetes-mediated oxidative stress in cancer, the increased level of VDUP1 leads to increased ROS levels through reduction of Trx activity that is reversed by p38 MAP kinase inhibition (Turturro et al. 2007b). Knockdown of HDAC10 significantly increases the mRNA expression levels of VDUP1 in SNU-620 human gastric cancer cells. Inhibition of HDAC10 activates apoptotic signaling molecules through accumulation of ROS (Lee et al. 2010). p21^{WAF1} acts as a tumor suppressor, but p21^{WAF1} can also promote tumor growth (Bachman et al. 2004). p21^{WAF1} can induce Trx secretion and angiogenesis in cancer cells by transcriptional repression of the VDUP1 promoter (Kuljaca et al. 2009). VDUP1 also suppresses the metastasis of melanoma in mice (Goldberg et al. 2003; Kaimul et al. 2007).

Collectively, the expression of VDUP1 is frequently lost in tumor tissue and cell lines, and ectopic expression of VDUP1 suppresses cellular proliferation along with increasing cell cycle arrest at the G_1 phase. Trx and VDUP1 are involved not only in cytoprotective functions against oxidative stress but also in the regulation of cellular proliferation and the aging process (Yoshida et al. 2005). VDUP1 is currently being evaluated as a potent tumor suppressor gene.

6.4   VDUP1 AND METABOLIC DISEASES

VDUP1 was first identified as a binding partner and inhibitor of the antioxidant protein Trx (Nishiyama et al. 1999). It has recently been recognized in vascular tissues that hyperglycemia promotes oxidative stress through the inhibition of the Trx function by increased expression and activity of VDUP1 (Schulze et al. 2004), leading to a functional inhibition of the antioxidative Trx function. This specific interaction results in a shift of the cellular redox balance that promotes increased intracellular oxidative stress (Kobayashi et al. 2003; Nishiyama et al. 1999). Recent studies show VDUP1 to be a novel metabolic regulator, with effects on glucose and lipids. VDUP1 expression is markedly glucose responsive (Schulze et al. 2004), and its expression is elevated in the skeletal muscle of diabetic and glucose-intolerant patients (Parikh et al. 2007). Mammals have three VDUP1 homologous proteins of unknown function, namely arrestin domain containing 2–4 (Arrdc2–4), which do not bind Trx (Patwari et al. 2006). Interestingly, VDUP1 is the only α-arrestin family member to bind Trx (Patwari et al. 2006). Another α-arrestin, Arrdc3, has been linked to the development of obesity in humans (Chutkow et al. 2010), and both Arrdc4 and a VDUP1 mutant that does not bind Trx have been shown to alter glucose metabolism (Patwari et al. 2009), supporting α-arrestins as a new class of metabolic regulators that operate independently of Trx (Chutkow et al. 2010). In the absence of VDUP1, the liver is intrinsically defective in maintaining blood glucose levels through glucose production and release. The effects of VDUP1 on glucose regulation are abolished by a single cysteine mutation that is required for the VDUP1-Trx interaction (Chutkow et al. 2008). These results implicate VDUP1 as a key regulator of hepatic glucose production and global glucose homeostasis (Chutkow et al. 2008).

VDUP1 also regulates glucose uptake in the periphery of the human body by acting as a glucose- and insulin-sensitive switch early in the disease process. VDUP1 expression would induce the loss of insulin-producing cells in the pancreas, thus further reducing insulin production and glucose uptake in the periphery and, ultimately, resulting in type-2 diabetes (Parikh et al. 2007). VDUP1 may lie upstream of the elevations in ROS and mitochondrial dysfunction that accompany insulin resistance. These findings focus on the pancreatic β-cells, which provide molecular insights into the pathogenesis of impaired insulin secretion and action that characterize the prediabetic state (De Fronzo 2004; Kahn 1994).

VDUP1 expression is strongly upregulated in human diabetes (Parikh et al. 2007) and diabetic complications (Kobayashi et al. 2003; Perrone et al. 2009), and increasing evidence suggests that VDUP1 is a potential diabetogenic signal. In diabetes mellitus patients, it was reported that genetic variation of the VDUP1 gene is associated with hyper-triglyceridemia and increased diastolic blood pressure (van Greevenbroek et al. 2007). Recent studies also found that VDUP1 inhibits glucose uptake in skeletal muscle (Parikh et al. 2007), mature adipocytes, and primary skin fibroblasts independently of Trx binding (Patwari et al. 2009), suggesting that VDUP1 is an important negative regulator of glucose uptake. In pancreatic β-cell function, overexpression of VDUP1 enhances β-cell death and impairs insulin secretion (Chen et al. 2006; Minn et al. 2005), which might further aggravate diabetes. In diabetic patients, increased VDUP1 expression is associated with chronic diabetic neuropathy and diabetic nephropathy (Parikh et al. 2007), which may be related to collagen synthesis. Most recently, it was reported that VDUP1 is a specific mediator of the mitochondrial death pathway in pancreatic β-cells under glucotoxic conditions (Chen et al. 2010). We have shown that coordinated regulation of PPAR α and insulin secretion by VDUP1 is crucial in feeding–fasting nutritional transition (Oka et al. 2009). VDUP1 is also reciprocally regulated by insulin and glucose (Shaked et al. 2009). These findings suggest that VDUP1 is a key regulator of glucose uptake, as well as insulin secretion in blood glucose homeostasis, and that VDUP1 functions as a homeostatic switch that integrates glucose sensing and insulin signaling to control cellular energy status (Figure 6.3). A nonsense mutation in the VDUP1 gene causes spontaneous hyperlipidemia and shares similar phenotypes with familial combined hyperlipidemia (FCHL) in HCB-19 strain mice (Bodnar et al. 2002), but the relationship between VDUP1 mutation and FCHL is controversial in humans (Pajukanta et al. 2004; van der Vleuten et al. 2004). The FCHL-associated gene is reported to be an upstream stimulation factor (USF); VDUP1 has a binding site for USF on its promoter

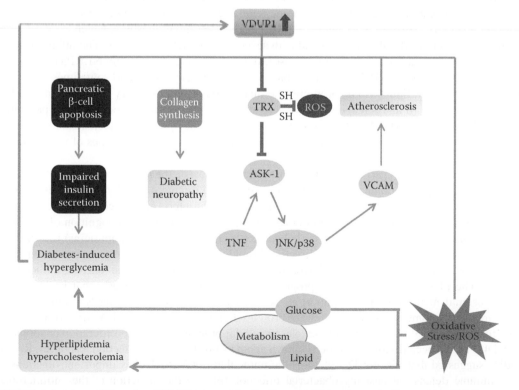

FIGURE 6.3 A possible role of VDUP1 in metabolic diseases. This schematic diagram shows the roles of VDUP1 in glucose/lipid metabolism and its related diseases in animal models and human diseases.

region (Ludwig et al. 2001; Pajukanta et al. 2004), suggesting that VDUP1 might be related to FCHL and that the mechanism should be elucidated.

Recently, Oka et al. showed that Krebs cycle–mediated fatty acid utilization was impaired in VDUP1 knockout mice, indicating the involvement of VDUP1 in lipid metabolism. In this study, they found reduced survival rates and concomitant with severe bleeding, dyslipidemia, fatty liver, hypoglycemica, and hepatic and renal dysfunction under fasting conditions in VDUP1-deficient mice although the mice were viable and fertile under normal housing conditions. Because these symptoms are similar to fatty acid utilization deficiency disorders, they argue that VDUP1-deficient mice might be a model of the Reyes syndrome, which is a metabolic syndrome due to disorders of mitochondrial fatty acid beta-oxidation as acute encephalopathy, hepatic dysfunction, and fatty infiltration of the visceral organs (Glasgow and Middleton 2001). In fact, the levels of free fatty acids were elevated in VDUP-deficient mice under fasting conditions, suggesting that the deposition of free fatty acids is correlated with the expression of VDUP1. In addition, they found that acetyl-CoA catabolism was dysregulated due to defective oxidation in VDUP1-deficient mice, defined as elevated serum levels of ketone bodies, which are converted from the excess amounts of acetyl-CoA (Oka et al. 2006a). Since carcinogenesis has been revealed to be associated with the dysregulation of basic energy metabolism such as augmentation of glycolysis and reduction of Krebs cycle function, it is suggested that the roles of VDUP1 as a tumor suppressor may involve the maintenance of an antitumorigenic metabolic phenotype (Glasgow and Middleton 2001).

6.5   VDUP1 AND INFLAMMATION

Since identification, a large number of studies have accumulated on the relationship among VDUP1, ROS, and cancer. However, the role of VDUP1 in immune cells has been largely unknown until recently. Our preliminary results showed that VDUP1 is expressed dominantly in immune cell populations, indicating VDUP1 may have some regulatory roles in the immune system. In a recent report (Zhou et al. 2010), VDUP1 interacted with the NLRP3 inflammasome, which is vital for the production of mature IL-1β and associated with several inflammatory diseases. The inflammasome is a multiprotein complex consisting of caspase1, the Nod-like receptor protein NLRP3 (also called cyopyrin or NALP3), CARDINAL, the adaptor protein ASC, caspase-1, and sometimes caspase 5 or caspase 11. The inflammasome promotes the maturation of inflammatory cytokines interleukin 1-β and interleukin 18 and is responsible for activating the inflammatory processes. The participation of VDUP1 in the NLRP3 inflammasome activating the may provide a mechanistic link to the observed involvement of proinflammatory cytokines such as IL-1β in many types of inflammation (Figure 6.4).

Phagocytosis is a vital part of the innate immune response, i.e., the process of uptake of particles such as bacteria, parasites, dead host cells, and cellular and foreign debris by phagocytes, such as macrophages, neutrophils, and dendritic cells. The binding of bacteria and phagocytes makes a available a rapid host mechanism for defending against microbial pathogens. Recognition of pathogen associated molecules (such as surface carbohydrates, peptidoglycans, or lipoproteins) is arbitrated by pattern recognition receptors, or indirectly, through mediation by opsonins. After recognition of pathogens, phagocytes create the phagosome through a plasma-membrane–derived intracellular vacuole. The phagosome undergoes a progression of fusion and fission events that is referred to as phagosomal maturation (Flannagan et al. 2009). Two of the most abundant lysosomal components are the lysosome-associated membrane proteins-1 and -2 (LAMP-1 and LAMP-2, respectively; Eskelinen et al. 2003). Although multiple roles in cellular processes such as proliferation or apoptosis for VDUP1 have been discovered, little is known about its involvement in infection responses. Recent studies suggested that vitamin D_3, which is the upregulator of VDUP1, has important functions in host immune defense against mycobacterial infection through the induction of the antimicrobial peptide cathelicidin. Activation of a toll-like receptor (TLR) induced the expression of the vitamin D receptor and vitamin D-1-hydroxylase genes; then, this stimulated the induction of the antimicrobial

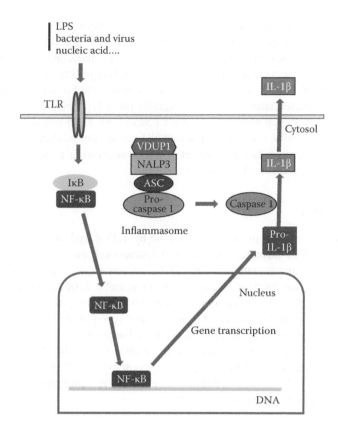

FIGURE 6.4 Regulation of inflammation and inflammasome by VDUP1. Inflammatory signals stimulate proinflammatory cytokines such as IL-1β. VDUP1 as a component of inflammasome regulates processing of IL-1β secretion.

peptide cathelicidin (Liu et al. 2006). Moreover, roles for vitamin D_3 reported that the antituberculosis activity of human monocytes (Rook et al. 1986) reduces the levels of matrix metalloproteinases.

Because the phagocytosis of apoptotic cells has distinctive morphologic features and unique downstream consequences, some scientist coined the unique term *efferocytosis* (taken from the Latin *effero*, meaning to take to the grave or to bury) (de Cathelineau and Henson 2003; Gardai et al. 2005). The phagocytosis of apoptotic cells by phagocytes suppresses the innate immune response and inflammatory mediator production through the action of transforming growth factor (TGF)-β1 and interleukin (IL)-10. The anti-inflammatory effect of efferocytosis can also be seen *in vivo* where the instillation of apoptotic cells suppresses endotoxin-induced lung inflammation in a TGF-β1– dependent manner. It is not clear yet that VDUP1 is involved in this process.

Bacterial lipopolysaccharide (LPS) is the major agent in the pathogenesis of endotoxin shock. Endotoxin shock is a severe systemic inflammatory response triggered by the interaction of LPS with host cells, in particular with monocytes and macrophages. Macrophages play a pivotal role in the immune response against bacteria through the production of cytokines and other inflammatory mediators, including tumor necrosis factor (TNF)-α, interleukin-1 (IL-1), IL-6, and nitric oxide (NO). LPS signals through TLR4, which involves downstream molecules MyD88, TIRAP/Mal, IRAK, and TRAF6. The transmitted signal triggers IκB kinase (IKK) and mitogen-activated protein kinase (MAPK) cascades, resulting in the activation of transcription factors NF-κB and AP-1. TNF-α and IL-1β can encourage NO production in macrophages through the action of inducible NO synthase (iNOS). The synthase functions as a dimer: one subunit transfers electrons from NADPH to FAD, then to FMN (flavin mononucleotide), and to the haem iron of the adjacent subunit to produce

NO• and citrulline from L-arginine and oxygen (Kuo et al. 2000; Lee et al. 1993). During inflammatory responses, macrophages become more active in order to more effectively clear offending pathogens. At these times, host cells are at increased risk of clearance, and newly arriving stem cells might require additional protection against phagocytosis. Although little research has addressed the relationship between VDUP1 and NO synthesis, it has been reported that the human promyelocytic cell line HL-60, when differentiated to a macrophagelike phenotype, acquires the ability to produce substantial amounts of NO upon stimulation with LPSs or with 1, 25-dihydroxyvitamin D3 (1,25-D_3; Wang et al. 2001). Recently, it was reported that NO production inhibited VDUP1 expression to facilitate Trx-mediated denitrosylation of target proteins, suggesting a close relation between NO production and VDUP1 regulation (Forrester et al. 2009).

6.6   CONCLUSIONS

Since VDUP1 was identified as an upregulated gene by 1,25-dihydroxyvitamin D treatment, the diverse functions of VDUP1 have been revealed. It controls cell proliferation, apoptosis, immune responses, inflammation, and metabolism. It can be assumed that controlling the expression of VDUP1 is an important issue for treating cancer, diabetes, and inflammatory diseases. Small molecule modulators of VDUP1 expression may be applied for controlling these diseases. In this regard, vitamin D and its analogs will be refocused as a regulator based on newly unrevealed functions of VDUP1. Vitamin D is known to be involved in cellular homeostasis, immune regulation, inflammation, and cancer prevention mostly in relation to vitamin D receptor. Now, we can consider its roles in disease control and cellular homeostasis in relation to VDUP1, as well as vitamin D receptor. Deciphering the crosstalk between vitamin D, vitamin D receptor, and VDUP1 will open new insights into controlling homeostasis in our body.

ACKNOWLEDGMENTS

This work was supported in part by grants from the Global Research Laboratory (GRL) Project and the New Drug Target Discovery Project (M10848000352-08N4800-35210) of the Ministry of Education, Science and Technology of Korea.

REFERENCES

Alvarez, C. E. 2008. On the origins of arrestin and rhodopsin. *BMC Evol Biol* 8:222.
Bachman, K. E., B. G. Blair, K. Brenner, A. Bardelli, S. Arena, S. Zhou, J. Hicks, A. M. De Marzo, P. Argani, and B. H. Park. 2004. p21(WAF1/CIP1) mediates the growth response to TGF-beta in human epithelial cells. *Cancer Biol Ther* 3:221–225.
Baker, A. F., M. Y. Koh, R. R. Williams, B. James, H. Wang, W. R. Tate, A. Gallegos, D. D. Von Hoff, H. Han, and G. Powis. 2008. Identification of thioredoxin-interacting protein 1 as a hypoxia-inducible factor 1alpha-induced gene in pancreatic cancer. *Pancreas* 36:178–186.
Bodnar, J. S. et al. 2002. Positional cloning of the combined hyperlipidemia gene Hyplip1. *Nat Genet* 30:110–116.
Borrello, S., M. E. De Leo, and T. Galeotti. 1993. Defective gene expression of MnSOD in cancer cells. *Mol Aspects Med* 14:253–258.
Brar, S. S., T. P. Kennedy, A. R. Whorton, A. B. Sturrock, T. P. Huecksteadt, A. J. Ghio, and J. R. Hoidal. 2001. Reactive oxygen species from NAD(P)H:quinone oxidoreductase constitutively activate NF-κB in malignant melanoma cells. *Am J Physiol Cell Physiol* 280:C659–676.
Butler, L. M., X. Zhou, W. S. Xu, H. I. Scher, R. A. Rifkind, P. A. Marks, and V. M. Richon. 2002. The histone deacetylase inhibitor SAHA arrests cancer cell growth, upregulates thioredoxin-binding protein-2, and downregulates thioredoxin. *Proc Natl Acad Sci USA* 99:11700–11705.
Cadenas, C. et al. 2010. Role of thioredoxin reductase 1 and thioredoxin interacting protein in prognosis of breast cancer. *Breast Cancer Res* 12:R44.
Cerutti, P. A. 1985. Prooxidant states and tumor promotion. *Science* 227:375–381.

Chen, J., F. M. Couto, A. H. Minn, and A. Shalev. 2006. Exenatide inhibits beta-cell apoptosis by decreasing thioredoxin-interacting protein. *Biochem Biophys Res Commun* 346:1067–1074.

Chen, J., G. Fontes, G. Saxena, V. Poitout, and A. Shalev. 2010. Lack of TXNIP protects against mitochondria-mediated apoptosis but not against fatty acid-induced ER stress-mediated beta-cell death. *Diabetes* 59:440–447.

Chen, K. S. and H. F. De Luca. 1994. Isolation and characterization of a novel cDNA from HL-60 cells treated with 1,25-dihydroxyvitamin D-$_3$. *Biochim Biophys Acta* 1219:26–32.

Chung, J. W., J. H. Jeon, S. R. Yoon, and I. Choi. 2006. Vitamin D3 upregulated protein 1 (VDUP1) is a regulator for redox signaling and stress-mediated diseases. *J Dermatol* 33:662–669.

Chutkow, W. A., P. Patwari, J. Yoshioka, and R. T. Lee. 2008. Thioredoxin-interacting protein (Txnip) is a critical regulator of hepatic glucose production. *J Biol Chem* 283:2397–2406.

Chutkow, W. A. et al. 2010. Deletion of the alpha-arrestin protein Txnip in mice promotes adiposity and adipogenesis while preserving insulin sensitivity. *Diabetes* 59:1424–1434.

de Vos, S., W. K. Hofmann, T. M. Grogan, U. Krug, M. Schrage, T. P. Miller, J. G. Braun, W. Wachsman, H. P. Koeffler, and J. W. Said. 2003. Gene expression profile of serial samples of transformed B-cell lymphomas. *Lab Invest* 83:271–285.

de Cathelineau, A. M. and P. M. Henson. 2003. The final step in programmed cell death: Phagocytes carry apoptotic cells to the grave. *Essays Biochem* 39:105–117.

De Fronzo, R. A. 2004. Pathogenesis of type 2 diabetes mellitus. *Med Clin North Am* 88:787–835, ix.

Dunn, T. A. et al. 2006. A novel role of myosin VI in human prostate cancer. *Am J Pathol* 169:1843–1854.

Dutta, K. K. et al. 2005. Two distinct mechanisms for loss of thioredoxin-binding protein-2 in oxidative stress-induced renal carcinogenesis. *Lab Invest* 85:798–807.

Eskelinen, E. L., Y. Tanaka, and P. Saftig. 2003. At the acidic edge: Emerging functions for lysosomal membrane proteins. *Trends Cell Biol* 13:137–145.

Filby, C. E., S. B. Hooper, F. Sozo, V. A. Zahra, S. J. Flecknoe, and M. J. Wallace. 2006. VDUP1: A potential mediator of expansion-induced lung growth and epithelial cell differentiation in the ovine fetus. *Am J Physiol Lung Cell Mol Physiol* 290:L250–258.

Flannagan, R. S., G. Cosio, and S. Grinstein. 2009. Antimicrobial mechanisms of phagocytes and bacterial evasion strategies. *Nat Rev Microbiol* 7:355–366.

Forrester, M. T., D. Seth, A. Hausladen, C. E. Eyler, M. W. Foster, A. Matsumoto, M. Benhar, H. E. Marshall, and J. S. Stamler. 2009. Thioredoxin-interacting protein (Txnip) is a feedback regulator of S-nitrosylation. *J Biol Chem* 284:36160–36166.

Gardai, S. J., K. A. McPhillips, S. C. Frasch, W. J. Janssen, A. Starefeldt, J. E. Murphy-Ullrich, D. L. Bratton, P. A. Oldenborg, M. Michalak, and P. M. Henson. 2005. Cell-surface calreticulin initiates clearance of viable or apoptotic cells through trans-activation of LRP on the phagocyte. *Cell* 123:321–334.

Glasgow, J. F. and B. Middleton. 2001. Reye syndrome: Insights on causation and prognosis. *Arch Dis Child* 85:351–353.

Goldberg, S. F., M. E. Miele, N. Hatta, M. Takata, C. Paquette-Straub, L. P. Freedman, and D. R. Welch. 2003. Melanoma metastasis suppression by chromosome 6: Evidence for a pathway regulated by CRSP3 and TXNIP. *Cancer Res* 63:432–440.

Han, S. H. et al. 2003. VDUP1 upregulated by TGF-beta1 and 1,25-dihydorxyvitamin D3 inhibits tumor cell growth by blocking cell-cycle progression. *Oncogene* 22:4035–4046.

Holmgren, A. 1989. Thioredoxin and glutaredoxin systems. *J Biol Chem* 264:13963–13966.

Hoshikawa, H., Mori, T., and N. Mori. 2010. *In vitro* and *in vivo* effects of D-allose: Upregulation of thioredoxin-interacting protein in head and neck cancer cells. *Ann Otol Rhinol Laryngol* 119:567–571.

Huang, L. and A. B. Pardee. 2000. Suberoylanilide hydroxamic acid as a potential therapeutic agent for human breast cancer treatment. *Mol Med* 6:849–866.

Ikarashi, M., Y. Takahashi, Y. Ishii, T. Nagata, S. Asai, and K. Ishikawa. 2002. Vitamin D3 upregulated protein 1 (VDUP1) expression in gastrointestinal cancer and its relation to stage of disease. *Anticancer Res* 22:4045–4048.

Jeon, J. H., K. N. Lee, C. Y. Hwang, K. S. Kwon, K. H. You, and I. Choi. 2005. Tumor suppressor VDUP1 increases p27(kip1) stability by inhibiting JAB1. *Cancer Res* 65:4485–4489.

Joguchi, A., I. Otsuka, S. Minagawa, T. Suzuki, M. Fujii, and D. Ayusawa. 2002. Overexpression of VDUP1 mRNA sensitizes HeLa cells to paraquat. *Biochem Biophys Res Commun* 293:293–297.

Junn, E. et al. 2000. Vitamin D3 upregulated protein 1 mediates oxidative stress via suppressing the thioredoxin function. *J Immunol* 164:6287–6295.

Kahn, C. R. 1994. Banting lecture: Insulin action, diabetogenes, and the cause of type II diabetes. *Diabetes* 43:1066–1084.

Kaimul, A. M., H. Nakamura, H. Masutani, and J. Yodoi. 2007. Thioredoxin and thioredoxin-binding protein-2 in cancer and metabolic syndrome. *Free Radic Biol Med* 43:861–868.

Kim, S. Y., H. W. Suh, J. W. Chung, S. R. Yoon, and I. Choi. 2007. Diverse functions of VDUP1 in cell proliferation, differentiation, and diseases. *Cell Mol Immunol* 4:345–351.

Kobayashi, T., S. Uehara, T. Ikeda, H. Itadani, and H. Kotani. 2003. Vitamin D3 upregulated protein-1 regulates collagen expression in mesangial cells. *Kidney Int* 64:1632–1642.

Kopantzev, E. P., G. S. Monastyrskaya, T. V. Vinogradova, M. V. Zinovyeva, M. B. Kostina, O. B. Filyukova, A. G. Tonevitsky, G. T. Sukhikh, and E. D. Sverdlov. 2008. Differences in gene expression levels between early and later stages of human lung development are opposite to those between normal lung tissue and non-small-lung-cell carcinoma. *Lung Cancer* 62:23–34.

Kuljaca, S., T. Liu, T. Dwarte, M. Kavallaris, M. Haber, M. D. Norris, J. Martin-Caballero, and G. M. Marshall. 2009. The cyclin-dependent kinase inhibitor, p21(WAF1), promotes angiogenesis by repressing gene transcription of thioredoxin-binding protein 2 in cancer cells. *Carcinogenesis* 30:1865–1871.

Kuo, H. P., C. H. Wang, Huang, K. S., H. C. Lin, C. T. Yu, C. Y. Liu, and L. C. Lu. 2000. Nitric oxide modulates interleukin-1beta and tumor necrosis factor-alpha synthesis by alveolar macrophages in pulmonary tuberculosis. *Am J Respir Crit Care Med* 161:192–199.

Kwon, H. J. et al. 2010. Vitamin D3 upregulated protein 1 suppresses TNF-alpha-induced NF-kappaB activation in hepatocarcinogenesis. *J Immunol* 185:3980–3989.

Le Jan, S., N. Le Meur, A. Cazes, J. Philippe, M. Le Cunff, J. Leger, P. Corvol, and S. Germain. 2006. Characterization of the expression of the hypoxia-induced genes neuritin, TXNIP and IGFBP3 in cancer. *FEBS Lett* 580:3395–3400.

Lee, J. H., E. G. Jeong, M. C. Choi, S. H. Kim, J. H. Park, S. H. Song, J. Park, Y. J. Bang, and T. Y. Kim. 2010. Inhibition of histone deacetylase 10 induces thioredoxin-interacting protein and causes accumulation of reactive oxygen species in SNU-620 human gastric cancer cells. *Mol Cells* 30:107–112.

Lee, K. N. et al. 2005. VDUP1 is required for the development of natural killer cells. *Immunity* 22:195–208.

Lee, S. C., D. W. Dickson, W. Liu, and C. F. Brosnan. 1993. Induction of nitric oxide synthase activity in human astrocytes by interleukin-1 beta and interferon-gamma. *J Neuroimmunol* 46:19–24.

Liu, P. T. et al. 2006. Toll-like receptor triggering of a vitamin D-mediated human antimicrobial response. *Science* 311:1770–1773.

Ludwig, D. L., H. Kotanides, T. Le, D. Chavkin, P. Bohlen, and L. Witte. 2001. Cloning, genetic characterization, and chromosomal mapping of the mouse VDUP1 gene. *Gene* 269:103–112.

Matsui, M., M. Oshima, H. Oshima, K. Takaku, T. Maruyama, J. Yodoi, and M. M. Taketo. 1996. Early embryonic lethality caused by targeted disruption of the mouse thioredoxin gene. *Dev Biol* 178: 179–185.

Matsuoka, S. et al. 2008. Involvement of thioredoxin-binding protein 2 in the antitumor activity of CD437. *Cancer Sci* 99:2485–2490.

Minn, A. H., C. A. Pise-Masison, M. Radonovich, J. N. Brady, P. Wang, C. Kendziorski, A. Shalev. 2005. Gene expression profiling in INS-1 cells overexpressing thioredoxin-interacting protein. *Biochem Biophys Res Commun* 336:770–778.

Nakamura, H., H. Masutani, and J. Yodoi. 2006. Extracellular thioredoxin and thioredoxin-binding protein 2 in control of cancer. *Semin Cancer Biol* 16:444–451.

Nishinaka, Y., A. Nishiyama, H. Masutani, S. Oka, K. M. Ahsan, Y. Nakayama, Y. Ishii, H. Nakamura, M. Maeda, and J. Yodoi. 2004. Loss of thioredoxin-binding protein-2/vitamin D3 upregulated protein 1 in human T-cell leukemia virus type I-dependent T-cell transformation: Implications for adult T-cell leukemia leukemogenesis. *Cancer Res* 64:1287–1292.

Nishiyama, A., M. Matsui, S. Iwata, K. Hirota, H. Masutani, H. Nakamura, Y. Takagi, H. Sono, Y. Gon, and J. Yodoi. 1999. Identification of thioredoxin-binding protein-2/vitamin D(3) upregulated protein 1 as a negative regulator of thioredoxin function and expression. *J Biol Chem* 274:21645–21650.

Ohta, S. et al. 2005. Downregulation of metastasis suppressor genes in malignant pheochromocytoma. *Int J Cancer* 114:139–143.

Oka, S., H. Masutani, W. Liu, H. Horita, D. Wang, S. Kizaka-Kondoh, and J. Yodoi. 2006a. Thioredoxin-binding protein-2-like inducible membrane protein is a novel vitamin D3 and peroxisome proliferator-activated receptor (PPAR)gamma ligand target protein that regulates PPARgamma signaling. *Endocrinology* 147:733–743.

Oka, S., W. Liu, H. Masutani, H. Hirata, Y. Shinkai, S. Yamada, T. Yoshida, H. Nakamura, and J. Yodoi. 2006b. Impaired fatty acid utilization in thioredoxin binding protein-2 (TBP-2)-deficient mice: A unique animal model of Reye syndrome. *FASEB J* 20:121–123.

Oka, S., E. Yoshihara, A. Bizen-Abe, W. Liu, M. Watanabe, J. Yodoi, and H. Masutani. 2009. Thioredoxin bind-ing protein-2/thioredoxin-interacting protein is a critical regulator of insulin secretion and peroxisome proliferator-activated receptor function. *Endocrinology* 150:1225–1234.

Pajukanta, P. et al. 2004. Familial combined hyperlipidemia is associated with upstream transcription factor 1 (USF1). *Nat Genet* 36:371–376.

Parikh, H. et al. 2007. TXNIP regulates peripheral glucose metabolism in humans. *PLoS Med* 4:e158.

Patwari, P., L. J. Higgins, W. A. Chutkow, J. Yoshioka, and R. T. Lee. 2006. The interaction of thioredoxin with Txnip. Evidence for formation of a mixed disulfide by disulfide exchange. *J Biol Chem* 281:21884–21891.

Patwari, P., W. A. Chutkow, K. Cummings, V. L. Verstraeten, J. Lammerding, E. R. Schreiter, and R. T. Lee. 2009. Thioredoxin-independent regulation of metabolism by the alpha-arrestin proteins. *J Biol Chem* 284:24996–25003.

Perrone, L., T. S. Devi, K. Hosoya, T. Terasaki, and L. P. Singh. 2009. Thioredoxin interacting protein (TXNIP) induces inflammation through chromatin modification in retinal capillary endothelial cells under diabetic conditions. *J Cell Physiol* 221:262–272.

Rook, G. A., J. Steele, L. Fraher, S. Barker, R. Karmali, J. O'Riordan, and J. Stanford. 1986. Vitamin D3, gamma interferon, and control of proliferation of Mycobacterium tuberculosis by human monocytes. *Immunology* 57:159–163.

Saitoh, M., H. Nishitoh, M. Fujii, K. Takeda, K. Tobiume, Y. Sawada, M. Kawabata, K. Miyazono, and H. Ichijo. 1998. Mammalian thioredoxin is a direct inhibitor of apoptosis signal–regulating kinase (ASK) 1. *EMBO J* 17:2596–2606.

Saxena, G., J. Chen, and A. Shalev. 2010. Intracellular shuttling and mitochondrial function of thioredoxin-interacting protein. *J Biol Chem* 285:3997–4005.

Schulze, P. C., J. Yoshioka, T. Takahashi, Z. He, G. L. King, and R. T. Lee. 2004. Hyperglycemia promotes oxi-dative stress through inhibition of thioredoxin function by thioredoxin-interacting protein. *J Biol Chem* 279:30369–30374.

Shaked, M., M. Ketzinel-Gilad, Y. Ariav, E. Cerasi, N. Kaiser, and G. Leibowitz. 2009. Insulin counteracts glucotoxic effects by suppressing thioredoxin-interacting protein production in INS-1E beta cells and in Psammomys obesus pancreatic islets. *Diabetologia* 52:636–644.

Sheth, S. S., J. S. Bodnar, A. Ghazalpour, C. K. Thipphavong, S. Tsutsumi, A. D. Tward, P. Demant, T. Kodama, H. Aburatani, and A. J. Lusis. 2006. Hepatocellular carcinoma in Txnip-deficient mice. *Oncogene* 25:3528–3536.

Shin, K. H., R. H. Kim, M. K. Kang, and N. H. Park. 2008. hnRNP G elicits tumor-suppressive activity in part by upregulating the expression of Txnip. *Biochem Biophys Res Commun* 372:880–885.

Shin, K. H., M. K. Kang, R. H. Kim, R. Christensen, N. H. Park. 2006. Heterogeneous nuclear ribonucleo-protein G shows tumor suppressive effect against oral squamous cell carcinoma cells. *Clin Cancer Res* 12:3222–3228.

Soderberg, A., B. Sahaf, and A. Rosen. 2000. Thioredoxin reductase, a redox-active selenoprotein, is secreted by normal and neoplastic cells: Presence in human plasma. *Cancer Res* 60:2281–2289.

Solanas, M., R. Moral, G. Garcia, L. Grau, E. Vela, R. Escrich, I. Costa, and E. Escrich. 2009. Differential expression of H19 and vitamin D3 upregulated protein 1 as a mechanism of the modulatory effects of high virgin olive oil and high corn oil diets on experimental mammary tumors. *Eur J Cancer Prev* 18:153–161.

Song, H., D. Cho, J. H. Jeon, S. H. Han, D. Y. Hur, Y. S. Kim, and I. Choi. 2003. Vitamin D(3) upregulating protein 1 (VDUP1) antisense DNA regulates tumorigenicity and melanogenesis of murine melanoma cells via regulating the expression of fas ligand and reactive oxygen species. *Immunol Lett* 86:235–247.

Szatrowski, T. P. and C. F. Nathan. 1991. Production of large amounts of hydrogen peroxide by human tumor cells. *Cancer Res* 51:794–798.

Tagaya, Y. et al. 1989. ATL-derived factor (ADF), an IL-2 receptor/Tac inducer homologous to thioredoxin; possible involvement of dithiol-reduction in the IL-2 receptor induction. *EMBO J* 8:757–764.

Takahashi, Y., T. Nagata, Y. Ishii, M. Ikarashi, K. Ishikawa, and S. Asai. 2002. Upregulation of vitamin D3 upregulated protein 1 gene in response to 5-fluorouracil in colon carcinoma SW620. *Oncol Rep* 9:75–79.

Turturro, F., G. Von Burton, and E. Friday. 2007a. Hyperglycemia-induced thioredoxin-interacting pro-tein expression differs in breast cancer-derived cells and regulates paclitaxel IC50. *Clin Cancer Res* 13:3724–3730.

Turturro, F., E. Friday, and T. Welbourne. 2007b. Hyperglycemia regulates thioredoxin-ROS activity through induction of thioredoxin-interacting protein (TXNIP) in metastatic breast cancer-derived cells MDA-MB-231. *BMC Cancer* 7:96.

van der Vleuten, G. M., A. Hijmans, L. A. Kluijtmans, H. J. Blom, A. F. Stalenhoef, and J. de Graaf. 2004. Thioredoxin-interacting protein in Dutch families with familial combined hyperlipidemia. *Am J Med Genet* A 130A:73–75.

van Greevenbroek, M. M., V. M. Vermeulen, E. J. Feskens, C. T. Evelo, M. Kruijshoop, B. Hoebee, C. J. van der Kallen, and T. W. de Bruin. 2007. Genetic variation in thioredoxin interacting protein (TXNIP) is associated with hypertriglyceridaemia and blood pressure in diabetes mellitus. *Diabet Med* 24:498–504.

Wang, C. H., H. C. Lin, C. Y. Liu, K. H. Huang, T. T. Huang, C. T. Yu, and H. P. Kuo. 2001. Upregulation of inducible nitric oxide synthase and cytokine secretion in peripheral blood monocytes from pulmonary tuberculosis patients. *Int J Tuberc Lung Dis* 5:283–291.

Watanabe, R., H. Nakamura, H. Masutani, and J. Yodoi. 2010. Antioxidative, anticancer and anti-inflammatory actions by thioredoxin 1 and thioredoxin-binding protein 2. *Pharmacol Ther* 127:261–270.

Wollman, E. E. et al. 1988. Cloning and expression of a cDNA for human thioredoxin. *J Biol Chem* 263:15506–15512.

Yamaguchi, F., M. Takata, K. Kamitori, M. Nonaka, Y. Dong, L. Sui, and M. Tokuda. 2008a. Rare sugar D-allose induces specific upregulation of TXNIP and subsequent G1 cell cycle arrest in hepatocellular carcinoma cells by stabilization of p27kip1. *Int J Oncol* 32:377–385.

Yamaguchi, F., K. Kamitori, K. Sanada, M. Horii, Y. Dong, L. Sui, and M. Tokuda. 2008b. Rare sugar D-allose enhances anti-tumor effect of 5-fluorouracil on the human hepatocellular carcinoma cell line HuH-7. *J Biosci Bioeng* 106:248–252.

Yamamoto, Y., M. Sakamoto, G. Fujii, K. Kanetaka, M. Asaka, and S. Hirohashi. 2001. Cloning and characterization of a novel gene, DRH1, downregulated in advanced human hepatocellular carcinoma. *Clin Cancer Res* 7:297–303.

Yang, X., L. H. Young, and J. M. Voigt. 1998. Expression of a vitamin D–regulated gene (VDUP-1) in untreated-and MNU-treated rat mammary tissue. *Breast Cancer Res Treat* 48:33–44.

Yoshida, T., H. Nakamura, H. Masutani, and J. Yodoi. 2005. The involvement of thioredoxin and thioredoxin binding protein-2 on cellular proliferation and aging process. *Ann N Y Acad Sci* 1055:1–12.

Yoshida, T., N. Kondo, S. Oka, M. K. Ahsan, T. Hara, H. Masutani, H. Nakamura, and J. Yodoi. 2006. Thioredoxin-binding protein-2 (TBP-2): Its potential roles in the aging process. *Biofactors* 27:47–51.

Yoshihara, E., Z. Chen, Y. Matsuo, H. Masutani, and J. Yodoi. 2010. Thiol redox transitions by thioredoxin and thioredoxin-binding protein-2 in cell signaling. *Methods Enzymol* 474:67–82.

Yoshioka, J., P. C. Schulze, M. Cupesi, J. D. Sylvan, C. MacGillivray, J. Gannon, H. Huang, and R. T. Lee. 2004. Thioredoxin-interacting protein controls cardiac hypertrophy through regulation of thioredoxin activity. *Circulation* 109:2581–2586.

Zhao, Y., Y. Xue, T. D. Oberley, K. K. Kiningham, S. M. Lin, H. C. Yen, H. Majima, J. Hines, and D. St. Clair. 2001. Overexpression of manganese superoxide dismutase suppresses tumor formation by modulation of activator protein-1 signaling in a multistage skin carcinogenesis model. *Cancer Res* 61:6082–6088.

Zhou, R., A. Tardivel, B. Thorens, I. Choi, and J. Tschopp. 2010. Thioredoxin-interacting protein links oxidative stress to inflammasome activation. *Nat Immunol* 11:136–140.

Zhuo, X., X. H. Niu, Y. C. Chen, D. Q. Xin, Y. L. Guo, and Z. B. Mao. 2010. Vitamin D3 upregulated protein 1(VDUP1) is regulated by FOXO3A and miR-17-5p at the transcriptional and post-transcriptional levels, respectively, in senescent fibroblasts. *J Biol Chem* 285:31491–31501.

7 Vitamin D and Its Role in Photoprotection of the Skin

Clare Gordon-Thomson, Wannit Tongkao-on,
and Rebecca S. Mason

CONTENTS

7.1 INTRODUCTION

The primary source of vitamin D is from the action of ultraviolet (UV) radiation on 7-dehydrocholesterol in skin. As noted in Chapter 1, high-energy UVB wavelengths (290–315 nm) convert 7-dehydrocholesterol to previtamin D, which then isomerizes to vitamin D at body temperature. Upon entering the circulation, it is transported by its binding protein and converted by 25-hydroxylase mainly in the liver to 25-dihydroxyvitamin D, the major circulating form of vitamin D. It is subsequently converted to the active hormone 1,25 dihydroxyvitamin D_3 ($1,25(OH)_2D_3$) principally in the kidney by the action of 1α-hydroxylase (Malloy et al. 1999; Reichrath 2007; Holick 2001, 2011). The local conversion of 7-dehydrocholesterol to the active hormone has also been demonstrated in the epidermis (Bikle et al. 1986; Lehmann et al. 1999, 2000, 2001). Continued irradiation of skin converts previtamin D and vitamin D into what are known as overirradiation products (Holick et al. 1981), which prevent excessive vitamin D from being produced by UV exposure, but have no other known function.

UV exposure, on the other hand, can also cause potentially harmful effects to skin cells. These include deoxyribonucleic acid (DNA) damage and the production of reactive oxygen (ROS) and nitrogen (RNS) species that interact with DNA and other cellular constituents and result in oxidative and nitrative damage. Failure of cells to respond, through inadequate DNA repair processes and deficiencies of antioxidant scavengers, leads to gene mutations, inflammation and immune suppression, predisposing skin to photoaging and carcinogenesis (Figure 7.1; Kochevar et al. 1999;

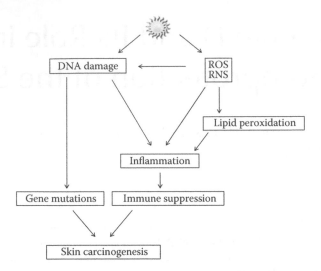

FIGURE 7.1 Sequence of events induced by UV radiation in the skin, which lead to skin carcinogenesis; refer to text for details. ROS: reactive oxygen species; and RNS: reactive nitrogen species.

Matsumura and Ananthaswamy 2002a; Halliday 2005; Mason et al. 2010; Dixon et al. 2011). There is now emerging evidence that local conversion of vitamin D to the active hormone $1,25(OH)_2D_3$ and other metabolites in skin cells acts as an adaptive photoprotective mechanism for cellular defense against further UV exposures. A number of *in vitro* and *in vivo* studies have shown that UV-induced DNA damage is reduced in the presence of vitamin D compounds in irradiated mice and humans and human skin cells (Wong et al. 2004; De Haes et al. 2005; Dixon et al. 2005; Gupta et al. 2007; Mason et al. 2010; Damian et al. 2010). Additionally, vitamin D compounds reduce UV-induced immune suppression and tumor formation in mice (Mason et al. 2010; Dixon et al. 2011). The focus of this chapter is on the photoprotective effects of vitamin D compounds, highlighting the possible roles of increased tumor suppressor p53 expression, the upregulation of the antioxidant metallothionein, and the reduction of RNS.

7.2 DNA DAMAGE AND THE DNA DAMAGE RESPONSE

7.2.1 UV-Induced DNA Damage

DNA damage induced by UV is an initiator and key mediator in the development of skin cancers. Various forms of DNA damage are produced by UV radiation at wavelengths ranging from 290 to 400 nm, the solar spectrum of UVB and UVA. Wavelengths below 290 nm (UVC), normally filtered out by stratospheric ozone, are potentially hazardous if the earth's ozone layer is depleted. DNA lesions can be produced directly by UV absorption or indirectly by the UV activation of other molecules that generate genotoxic free radicals. All are considered potential risks for genetic instability and infidelity. The most abundant DNA lesion produced directly by UV in human skin is the cyclobutane pyrimidine dimer (CPD). These are produced when photons strike two adjacent pyrimidines disrupting the double bonds. The subsequent formation of abnormal covalent bonds and a stable ring configuration between carbon 5 and carbon 6 of the two pyrimidines distorts the DNA helix (reviewed in the work of Ravanat et al. 2001 and Pattison and Davies 2006). Thymine–thymine dimers are the predominant CPDs found in irradiated DNA, including that in human skin cells, while thymine–cytosine, cytosine–cytosine bipyrimidines, and 6-4 photoproducts are detected with less frequency (Douki et al. 2000; Cooke et al. 2003;

Mouret et al. 2006). The current paradigm for CPD production involves high-energy UVB and UVC wavelengths, but there is evidence for UVA production of CPDs in skin cells (Mouret et al. 2006, 2009). There are also reports of chemical production of CPDs, which has been suggested to be through a triplet energy transfer mechanism (Lamola 1971; Lhiaubet-Vallet et al. 2007). Preliminary evidence for thymine dimer production in the absence of UV by the nitric oxide (NO) donor sodium nitroprusside (Mason et al. 2010) and their reduction by NO inhibitors has also been reported (Gupta et al. 2007).

Indirect production of DNA damage occurs when other cellular constituents or photoreceptors absorb UV and an outer orbiting electron is activated, releasing energy as a photon at a lower energy level than the original UV. The energy creates singlet oxygen, such as superoxide, which can cause oxidative damage to DNA (Kvam and Tyrrell 1997; Pattison and Davies 2006). The most prevalent UV-induced oxidative DNA damage arises from the conversion of the primary amine of guanine to 8-hydroxy-2′deoxyguanosine (8-oxoguanine) and is attributed to the UVA induction of ROS (Kvam and Tyrrell 1997; Agar et al. 2004).

High levels of NO are also produced in the skin from l-arginine by the UV activation of NO synthase (Deliconstantinos et al. 1996; Bruch-Gerharz et al. 1998; Cals-Grierson et al. 2004). Cellular NO levels can also increase by the UVA photo-decomposition of existing NO stores such as nitrosothiols and nitrite (Hess et al. 2001; Paunel et al. 2005; Mowbray et al. 2009). When produced in excess, NO loses its beneficial effects and acts as a reactive free radical. Furthermore, it can combine with superoxide or oxygen to form more toxic oxidating and nitrating intermediates, such as peroxynitrite and nitrous anhydride, respectively. Persistent high levels of these reactive species can modify DNA bases and attack the sugar–phosphate backbone. Peroxynitrite, a powerful oxidating and nitrating agent, converts guanine to either 8-oxoguanine or 8-nitroguanine (McAndrew et al. 1997; Burney et al. 1999; Niles et al. 2006; Pacher et al. 2007). The nitrated residue in the DNA has a very short half-life and undergoes rapid depurination to form an abasic site (Yermilov et al. 1995; Ohshima et al. 2006). Deleterious DNA strand breaks result from the nitrosation of primary amines, followed by deamination by nitrous anhydride (Wink 1991). All these lesions are considered promutagenic. Mutations can arise at unrepaired sites during DNA replication, where adenine is inserted as the default base to replace the damaged or non-coding site, thus altering the coding sequence (Matsumura and Ananthaswamy 2004). Sequence changes associated with 8-oxoguanine are mainly C:G to T:A transversions. These gene alterations correlate with mutations found in the tumor suppressor gene *p53* in human and mouse skin carcinomas (Nataraj et al. 1995; Berg et al. 1996; Brash et al. 1996; Soehnge et al. 1997; Agar et al. 2004) and in isolated DNA exposed to peroxynitrite (Murata et al. 2004).

7.2.2 DNA Damage Response

To maintain genome integrity, cells respond to DNA damage via signal transduction pathways that lead to cell cycle arrest. This facilitates either DNA repair before transcription and replication or the elimination of severely damaged cells by apoptosis (Ouhtit et al. 2000; Matsumura and Ananthaswamy 2004). The major DNA repair pathway for UV-induced CPDs is the nucleotide excision repair (NER) pathway with a half-life of 7–12 h (Mitchell et al. 1990; Katiyar et al. 2000; Matsumura and Ananthaswamy 2002a; Schul et al. 2002). Defects in this pathway cause a 1000-fold increase in the incidence of skin cancer in individuals with the inherited disorder xeroderma pigmentosum (McGregor et al. 1999). The alternate base excision repair (BER) pathway eliminates oxidative DNA damage such as 8-oxoguanine (Mitra et al. 2001; Nilsen and Krokan 2001; Zurer et al. 2004; David et al. 2007).

The key regulator of the DNA damage response is the tumor suppressor gene *p53*, which encodes a 53-kDa protein. Low levels of the protein are normally bound to its negative regulator (MDM2), which increases p53 susceptibility to proteolysis, resulting in a relatively short half-life (Hall et al. 1993; Berg et al. 1996; Momand et al. 2000). DNA damage activates p53 (Hall et al. 1993;

Eller et al. 1997, 2008), leading to its accumulation, which reaches maximum levels 12 h after UV radiation (Saito et al. 2003). The p53 protein is activated by various posttranslational modifications, including phosphorylation and acetylation on multiple sites in response to stress related protein kinases. Its activation promotes p53 accumulation by inhibiting its degradation and/or enhancing its transcription. As a transcription factor, p53 regulates genes implicated in growth control (Levine 1997; Matsumura and Ananthaswamy 2004), DNA repair (Smith et al. 1995; McKay et al. 1997), and proapoptotic or antiapoptotic pathways (Mullauer et al. 2001). Downstream gene products that are regulated by p53 contribute to NER (Hwang et al. 1999; Zhu et al. 2000; Amundson et al. 2002) and BER (Zurer et al. 2004) pathways.

Evidence that p53 plays a critical role in the DNA damage response has been demonstrated by the abrogation of UV-induced DNA repair in cells transfected with a dominant negative p53 gene (McKay et al. 1997). Its role as a tumor suppressor has been validated by the presence of inactivating p53 mutations in many cancers (Besaratinia et al. 2008), including skin cancers (Nataraj et al. 1995; Brash et al. 1996; Soehnge et al. 1997), and in premalignant actinic keratoses, the precursors for squamous-cell carcinomas (SCCs; Ziegler et al. 1994; Matsamura and Ananthaswamy 2002a). Mice with p53 knocked out show increased susceptibility to early onset tumor formation (Donehower et al. 1992), whereas heterozygotes age prematurely, associated with the depletion of adult stem cells, indicating that p53 also has antiaging activity (Liu et al. 2010). Phosphorylation of serine 15 of p53 by the strand break sensor protein kinase (ATM) in response to DNA damage (Siliciano et al. 1997; Watters et al. 1997; Banin et al. 1998; MacLaine and Hupp 2009) is reduced in the human genetic disorder ataxia telangiectasia and correlates with genome instability, immunodeficiency, and carcinogenesis in these individuals. Phosphorylation of serine 398 (the murine equivalent of human p53 serine 392) is a specific target of UV-induced DNA damage. A knock-in mutation in mice that blocked its phosphorylation promoted UV-induced skin cancer (Bruins et al. 2004). The p53 transcription factor also regulates GADD45, which binds to UV-damaged DNA, increasing accessibility to repair enzymes (Carrier et al. 1999), and has been identified as a $1,25(OH)_2D_3$-inducible gene associated with growth arrest in human prostate cancer cells (Flores and Burnsten 2010).

7.3 PHOTOPROTECTIVE EFFECTS OF VITAMIN D COMPOUNDS

7.3.1 VITAMIN D COMPOUNDS REDUCE UV-INDUCED DNA DAMAGE

There is substantial evidence that $1,25(OH)_2D_3$ protects irradiated skin cells from DNA damage. Thymine dimers are reduced in cells treated with $1,25(OH)_2D_3$ before or immediately after UV radiation. This has been substantiated by immunohistochemistry using a monoclonal antibody specific for thymine dimers. Image analysis showed reduced nuclear staining of $1,25(OH)_2D_3$-treated irradiated skin cells in culture (Wong et al. 2004; De Haes et al. 2005; Dixon et al. 2005, 2007; Gupta et al. 2007) and in mouse and human skin *in vivo* (Lee and Youn 1998; Dixon et al. 2005; Damian et al. 2010; Dixon et al. 2011). Significantly, thymine dimers continued to increase up to 6 h after UV exposure in irradiated skin cells (Gupta et al. 2007) but were suppressed by $1,25(OH)_2D_3$ within 30 min post-UV radiation, a result not entirely compatible with enhanced repair, given a half-life for NER of 7–12 h (Mitchell et al. 1990; Katiyar et al. 2000; Matsumura and Ananthaswamy 2002a; Schul et al. 2002). A similar reduction in thymine dimers 1 h after irradiation in Skh:hr1 hairless mice treated with silibinin has also been reported (Dhanalakshmi et al. 2004). The early reduction in thymine dimers by some agents, together with increasing detection of thymine dimers after irradiation, is consistent with a proposal that thymine dimers may be produced by metabolic processes as well as by direct DNA absorption of UV.

As discussed in Chapters 2 and 3, $1,25(OH)_2D_3$, similar to other steroid hormones, can function not only by a relative slow genomic pathway via the vitamin D receptor (VDR) in the nucleus and subsequent transactivation of target genes but also by a nongenomic pathway (Norman et al.

2000; Menegaz et al. 2011). Responses elicited by the genomic pathway normally emerge over hours or days, whereas the nongenomic pathway tends to be much faster acting, eliciting responses within seconds or minutes (Norman et al. 2004; Dixon and Mason 2009). It has been proposed that the photoprotective effects of $1,25(OH)_2D_3$ are mediated, at least in part, via the nongenomic pathway (Dixon and Mason 2009; Dixon et al. 2011). This is supported by studies with synthetic vitamin D analogs with different but fixed conformations. A 6-s-cis locked rapid-acting agonist 1,25-dihydroxylumisterol$_3$ (JN), which has been reported to have no gene-transactivating capacity (Norman et al. 2001) inhibited the formation of UV-induced thymine dimers in skin cells to the same extent as $1,25(OH)_2D_3$ (Wong et al. 2004; Dixon et al. 2005; Dixon and Mason 2009). By contrast, the photoprotective effects of $1,25(OH)_2D_3$ were completely abolished by a nongenomic antagonist 1β, 25-dihydroxyvitamin D$_3$ (HL; Wong et al. 2004; Dixon et al. 2005), whereas an antagonist of the genomic pathway, (23S)-25-dehydro-1α-hydroxyvitamin D$_3$-26,23-lactone (TEI-9647) did not alter photoprotection (Dixon et al. 2005). Nevertheless, the nongenomic analog 1,25-dihydroxylumisterol$_3$ (JN) was not as effective as $1,25(OH)_2D_3$ in reducing photocarcinogenesis in mice (Dixon et al. 2011). The molecular mechanisms of the nongenomic pathway in photoprotection are yet to be defined. It has been reported that nongenomic activation results in a rapid opening of voltage-gated chloride and calcium channels (Zanello and Norman 2004; Menegaz et al. 2011) and triggers extranuclear signal cascades probably via a cell surface receptor complex of the VDR translocated to the cell surface (Norman et al. 2001; Menegaz et al. 2011), possibly in association with the membrane-associated rapid response steroid binding protein, also known as ERp57/GRp58/ERp60 (Nemere et al. 2004). Further details on the actions of these two receptors in the nongenomic pathway are reviewed in the work of Khanal and Nemere (2007), Sequeira et al. (2009), and Menegaz et al. (2011).

7.3.2 Vitamin D Compounds Reduce Apoptosis and Sunburn Cell Formation

Cells bearing DNA lesions that are irreparable are removed from skin by apoptosis (Ouhtit et al. 2000; Matsumura and Ananthaswamy 2004). Protection against UV-induced cell loss by $1,25(OH)_2D_3$ treatment has been demonstrated in cultured human skin cells (Lee and Youn 1998; Mason and Holliday 2000; Manggau et al. 2001; Mason et al. 2002; De Haes et al. 2003, 2004; Wong et al. 2004; Gupta et al. 2007), in mouse (Gupta et al. 2007; Dixon et al. 2011), and human skin (Damian et al. 2010). This improved cell survival is almost certainly a consequence of the reduction in DNA damage (Gupta et al. 2007). A similar inhibition of apoptosis in irradiated skin cells in culture was shown with the vitamin D analogs, calcipotriol (Youn et al. 1997), 1α,25(OH)$_2$lumisterol$_3$ (JN), and 1α,25(OH)$_2$-7-dehydrocholesterol (JM; Wong et al. 2004). Inhibition of apoptosis by $1,25(OH)_2D_3$ and the rapid acting agonists JN and JM was reversed by the addition of the rapid response antagonist 1β,25(OH)$_2$D$_3$ (HL; Wong et al. 2004), suggesting that photoprotection from cell loss is also elicited by a nongenomic pathway. Apoptotic cells in skin, known as sunburn cells, are identifiable in irradiated skin by their pyknotic nuclei and eosinophilic cytoplasm in stained skin sections (Sheehan and Young 2002) and are likely eliminated from the skin before replication (Claerhout et al. 2006). $1,25(OH)_2D_3$ reduced sunburn cell formation in UV-irradiated mice (Hanada et al. 1995; Gupta et al. 2007; Dixon et al. 2007, 2011) after systemic or topical $1,25(OH)_2D_3$ treatment and in human subjects who were treated topically after irradiation (Damian et al. 2010).

One of the key mediators of apoptosis is the p53 upregulation of proapoptotic genes such as Bax and Fas/Apo-1 (Mullauer et al. 2001). UV-induced activation of the stress-activated c-Jun N-terminal kinase (JNK) was inhibited by $1,25(OH)_2D_3$ in human keratinocytes, leading to decreased apoptosis (De Haes et al. 2003). $1,25(OH)_2D_3$ has also been reported to protect keratinocytes from apoptosis induced by other stressors, e.g., oxidative stress and TNFα (Mason et al. 2002; Ravid et al. 2002; Diker-Cohen et al. 2003), possibly by increasing inherent antioxidant systems, e.g., metallothionein (Lee and Youn 1998). Cleavage of poly(ADP-ribose) polymerase by caspase 3, a hallmark of

UV-induced apoptosis (Kulms et al. 1999), is reduced in irradiated keratinocytes in the presence of $1,25(OH)_2D_3$ (De Haes et al. 2003). A relatively high dose of $1,25(OH)_2D_3$ ($10^{-7}M$) also improved cell survival in keratinocytes treated with UV, TNFα, and ceramides. This response appeared to be mediated by the production of sphingosine-1 phosphate, a breakdown product of membrane sphingolipid (Manggau et al. 2001).

Although there is compelling evidence that UV-induced DNA damage and apoptosis are significantly reduced in the presence of vitamin D compounds in human and mouse skin by the rapid nongenomic pathway (Mason et al. 2011), the definitive molecular mechanisms remain to be elucidated. A number of physiological responses to vitamin D treatment in UV-irradiated cells have been reported, which may contribute to its photoprotective effect against DNA damage, cell loss, inflammation, and immune suppression, which lead to skin carcinogenesis. These cellular responses include the upregulation of p53 and metallothionein, and a reduction in NO metabolites as measured by nitrite and 3-nitrotyrosine (Gupta et al. 2007; Dixon et al. 2011).

7.3.3 VITAMIN D COMPOUNDS INCREASE NUCLEAR P53

Levels of the expression of p53 are increased, which accumulates in the nuclei of skin cells following UV, indicating transactivating activity, which may facilitate DNA repair (Eller et al. 1997). This UV-induced accumulation of nuclear p53 is amplified in irradiated $1,25(OH)_2D_3$-treated cells (Dixon et al. 2005; Gupta et al. 2007; Dixon et al. 2011). However, as noted earlier, the reduction of thymine dimers by $1,25(OH)_2D_3$ occurs within 30 min (Gupta et al. 2007). This 30-min time period is inconsistent with the 7–12 h half-life of CPD repair by NER (Mitchell et al. 1990) and also the 12-h peak of the transactivating activity of p53 (Saito et al. 2003). It is probable that any increased repair of CPDs by $1,25(OH)_2D_3$ through upregulated NER pathway genes mediated by p53 is exerted via the genomic pathway. No differences were reported in levels of transcription of one of the key genes xeroderma pigmentosum complementation group G (XPG) in irradiated skin cells in the presence or absence of $1,25(OH)_2D_3$ (Mason et al. 2010). However, $1,25(OH)_2D_3$ interactions with the many other NER pathway genes still remain to be explored.

7.3.4 VITAMIN D COMPOUNDS INCREASE METALLOTHIONEIN

Another likely pathway for photoprotection against UV-induced DNA damage is an increase in antioxidant systems. UV radiation increases the rate of ROS and RNS production by the activation of photoreceptors in skin. Photoreceptor activation generates genotoxic and cytotoxic free radicals by electron transfer or hydrogen abstraction processes in other molecules or by energy transfer to molecular oxygen forming ROS (Pattison and Davies 2006). Low levels of ROS are produced during normal cellular metabolism but are converted to less damaging molecules by intrinsic antioxidant enzyme systems and scavengers that maintain the redox balance in cells. The antioxidant enzyme systems regulating ROS production are superoxide dismutase and catalase. Superoxide dismutase converts superoxide anions to hydrogen peroxide, which is then converted to water and oxygen by catalase. Insufficient enzyme activity increases the levels of superoxide or hydrogen peroxide, and the latter can form highly toxic hydroxyl ions. Free radical scavengers are also present in the skin and include glutathione, metallothionein, thioredoxin, vitamins C and E, and carotenoids. An imbalance between ROS and antioxidant production can cause genetic modifications and inflammation, which lead to skin cancer. Inflammatory cells that migrate into irradiated skin may also contribute to an increase in ROS (Halliday 2005).

UV induced a decrease in expression levels of glutathione peroxidase, superoxide dismutase, and catalase lasting for several days, although a late transient increase in mitochondrial superoxide dismutase was reported (Leccia et al. 2001). In addition, ROS and RNS induced by UV can inactivate endogenous free radical scavengers, thus enabling ROS and RNS to increase exponentially. A reduction in antioxidant enzymes and free radical scavengers by UV limits the capacity for cellular defense against the oxidation of DNA and protein residues and the peroxidation of membrane lipids (Leccia et al. 2001).

1,25(OH)$_2$D$_3$ protects keratinocytes from stress-activated protein kinases, JNK, TNF α, and hydrogen peroxide cytotoxicity induced by oxidative stress (Mason et al. 2002; Ravid et al. 2002; Diker-Cohen et al. 2003), possibly by increasing inherent antioxidant systems. Metallothionein, a cysteine-rich protein that is responsible for metal detoxification and an oxygen radical scavenger, has been observed in UVB-irradiated skin treated with cadmium chloride and reported to have a photoprotective effect by reducing sunburn cells, cell death, and photodamage (Hanada et al. 1991) and also reduced superoxide and hydroxyl radicals in irradiated mice (Hanada et al. 1992). Furthermore, a deficiency in the antioxidant action of metallothionein increased UV-induced immune suppression in a metallothionein knocked-out mouse (Reeve et al. 2000; Widyarini et al. 2006). Transcription of metallothionein is upregulated by 1,25(OH)$_2$D$_3$ (Karasawa et al. 1987), along with a reduction in UV-induced sunburn cells (Hanada et al. 1995; Lee and Youn 1998), which is some evidence for a genomic pathway involvement in photoprotection. Preliminary evidence has shown that, although 1,25(OH)$_2$D$_3$ is not a known antioxidant, it reduced UV-induced 8-oxoguanine by 48 ± 4% (SEM) within 30 min in UV-irradiated keratinocytes in culture (Gordon-Thomson et al. 2010), providing further evidence of photoprotective activities that are likely generated through the nongenomic pathway.

7.3.5 VITAMIN D COMPOUNDS REDUCE DERIVATIVES OF NO

As noted earlier, upregulated NO synthase activity by UV increases NO in skin (Deliconstntinos et al. 1996; Bruch-Gerhaz et al. 1998; Cals-Grierson and Ormerod 2004). NO combines with ROS to form potent genotoxic and cytotoxic derivatives such as peroxynitrite, causing DNA damage, nitrosylation of tyrosine residues in proteins, and initiating lipid peroxidation (Halliday 2005; Ohshima et al. 2006). Peroxynitrite also activates poly(ADP-ribose) polymerase, which converts NAD+ to nicotinamide and ADP-ribose, thus reducing NAD+ and ATP formation and resulting in energy depletion. This disrupts cellular functions and leads to cell death (Virag and Szabo 2002; Halliday 2005). NO overproduction induced by UV also inhibits CPD repair by preferentially inhibiting excision and ligation steps during NER (Jaiswal et al. 2000; Bau et al. 2001), and this can be reversed by inhibitors of NO synthase (Kuchel et al. 2005).

There is evidence to suggest that 1,25(OH)$_2$D$_3$ diminishes the incidence of oxidative and nitrative DNA damage by reducing the production of NO and other toxic RNS, which may well improve DNA repair mechanisms. Two relatively stable end products of the NO pathway, nitrite and 3-nitrotyrosine, which were used as measures for NO production, were significantly reduced in irradiated skin cells in the presence of 1,25(OH)$_2$D$_3$ when measured by the Griess assay (for nitrite) or a whole-cell enzyme-linked immunosorbent assay using a nitrotyrosine antibody (Gupta et al. 2007; Dixon et al. 2011). Similarly, NO synthase inhibitors such as aminoguanidine and L-N-monomethylarginine (L-NMMA) reduced nitrite and thymine dimer production in irradiated cells to an extent that was comparable to that produced by 1,25(OH)$_2$D$_3$ treatment (Gupta et al. 2007).

7.3.6 VITAMIN D AND IMMUNE PROTECTION

An innate inflammatory response initiated by UV is the familiar sunburn, which is induced by a combination of increased blood flow and infiltration of inflammatory cells into the injured site. The resulting inflammation is associated with the suppression of cell-mediated immune responses. The UVB component of sunlight has long been known to suppress immunity *in vivo* (Matsumura and Ananthaswamy 2002b; Ullrich 2005; Halliday 2005). Recent studies have shown that UVA also suppresses immunity, at least in humans (Halliday et al. 2004; Poon et al. 2005; Damian et al. 2011). However, there is conflicting evidence regarding the particular UVA wavelengths and its immunomodulatory effects from studies in mice and humans that are not easily explained. For further details on the various immune responses to UVA, refer to the work of Tyrrell and Reeve (2006), Ullrich et al. (2007), Matthews et al. (2010), and Damian et al. (2011).

Unrepaired DNA damage, particularly CPDs, have been demonstrated to be one of the key mediators of UV-induced inflammation and immune suppression (Kripke and Fisher 1976; Applegate et al. 1989; Kripke et al. 1992; Nishigori et al. 1996; Nghiem et al. 2002). Other important factors that initiate immune suppression include increased RNS and ROS production and proinflammatory cytokines released from resident skin cells (keratinocytes, Langerhans cells, macrophages, and mast cells) and infiltrating leucocytes. Additionally, the photoproducts cis-urocanic acid, membrane lipid peroxidation, prostaglandins, and the activation of an alternate complement pathway are also known to have inflammatory and immune-modulating activities (Matsumura and Ananthaswamy 2004; Halliday 2005; Ullrich 2005; Stapelberg et al. 2009; Halliday 2010; Sreevidya et al. 2010).

Lipid peroxidation by peroxynitrite is implicated in UV-induced immunosuppression (reviewed in Halliday 2005; Ullrich 2005), and antioxidant treatment has been shown to inhibit immune suppression mediated by the lipid peroxidation pathway in irradiated mice (Caceres-Dittmar et al. 1995). Lipid peroxidation leads to the production of prostaglandins and mediates cytokine induction and release. The transcription factors AP-1 and NF-κB, which regulate immune regulatory cytokines, are activated by signaling cascades initiated by the UV activation of Src located on the inner surface of the plasma membrane of keratinocytes. This can also be abolished by antioxidant treatment (Devary et al. 1992; Tobin et al. 1998; Ullrich 2005). The proinflammatory cytokine interleukin-10 (IL-10; Nishigori et al. 1996) is released by irradiated keratinocytes and, possibly, from mast cells (Grimbaldeston et al. 2007). IL-6 is also released by irradiated keratinocytes (Nishimura et al. 1999). Both of these interleukins suppress the release of immune-stimulating cytokines such as IL-12 (Schwarz et al. 1996; Halliday et al. 2008). Cytokines released in skin modulate cell-mediated T lymphocyte responses, which suppress immune responses at distant sites. Langerhans cell numbers are also depleted in the skin by UV, reducing their antigen presenting propensity to T helper-1 (Th-1) lymphocytes in regional lymph nodes (Schwarz et al. 2005; Gorman et al. 2007, 2010). This has been attributed to an increase in CPDs and/or oxidative stress (Vink et al. 1996; Kuchel et al. 2005).

Studies in human keratinocytes and in mice have demonstrated protection from UV-induced inflammation markers and immunosuppression by vitamin D compounds. This is supported by the reduction in proinflammatory IL-6 expression in cultured skin cells in the presence of 1,25(OH)$_2$D$_3$ (De Haes et al. 2003) and ergocalciferol (Mitani et al. 2004). Similarly, IL-6 (Mason et al. 2010) and skin fold thickness (Dixon et al. 2011), a measure of inflammation, were reduced by 1,25(OH)$_2$D$_3$ in irradiated skin of Skh:hr1 hairless mice.

Cell-mediated immune suppression by UV is demonstrated by a decrease in the immune response using contact hypersensitivity (CHS) or delayed-type hypersensitivity (DTH) protocols (Reeve 2002; Aubin 2003; Castano et al. 2006). CHS studies in UV-irradiated Skh:hr1 hairless mice have demonstrated that 1,25(OH)$_2$D$_3$ and the JN analog reduce UV-mediated immune suppression (Dixon et al. 2007, 2010, 2011; Mason et al. 2010). Because UV-induced immune suppression is associated with an increase in CPD and RNS (Kripke et al. 1992; Halliday et al. 2004), it is reasonable to propose that the reduction of CPD and NO derivatives by 1,25(OH)$_2$D$_3$ mediates, at least in part, its photoprotective effect against immune suppression (Mason et al. 2010). The upregulation of the antioxidant metallothionein by 1,25(OH)$_2$D$_3$ treatment may also reduce UV-induced ROS, including peroxynitrite. Evidence that cutaneous metallothionein acts as a scavenger of UV-induced free radicals (Hanada et al. 1992) and that its expression is enhanced by 1,25(OH)$_2$D$_3$ (Hanada et al. 1995) supports this view. A photoimmune protective role for metallothionein has also been demonstrated in metallothionein gene knockout mice, in which UV-induced immune suppression was exacerbated (Reeve et al. 2000; Widyarini et al. 2006). Increased antioxidant action against toxic free radicals would lead to a reduction in DNA damage, lipid peroxidation and proinflammatory cytokines, thus reducing UV-induced inflammation and immunosuppression.

Immune protection by $1,25(OH)_2D_3$ demonstrated in some mouse models is, however, not consistently found (as reviewed in the work of Kuritzky et al. 2008 and Dixon et al. 2010). In human subjects, a recall DTH Mantoux test showed basal immune suppression when treated with high doses of $1,25(OH)_2D_3$ (Damian et al. 2010). Immune suppression was at a similar level to that obtained with solar-simulated UV irradiation (Hanneman et al. 2006). Additionally, an increased suppressive activity of $CD4^+CD25^+$ regulatory T cells in the skin-draining lymph nodes of BALB/c mice following a single topical application of $1,25(OH)_2D_3$ has been reported (Gorman et al. 2007). Conversely, CHS studies in other mouse species apart from Skh:hr1 displayed no basal immunosuppressive response with $1,25(OH)_2D_3$ (unpublished observations). The vitamin D immunomodulatory effect in the skin appears to be elicited by a number of factors, including the inhibition of antigen-presenting Langerhans cell maturation and function by suppressing relB, a component of NF-κB, and modification of cytokine expression patterns (Singh et al. 1999; Griffin et al. 2001; Mathieu and Adorini 2002; Dong et al. 2005). Also, $1,25(OH)_2D_3$ and the analog 20-hydroxycholecalciferol were found to increase IkappaB alpha, an inhibitor to NF-κB activity in keratinocytes (Janjetovic et al. 2009, 2010). Recent reports have shown that $1,25(OH)_2D_3$ induced an increased production of the immune suppressor cytokine IL-10 in mast cells (Biggs et al. 2010) and activated regulatory T cells that suppress T-helper type 1 (Th1) responses (Gorman et al. 2007, 2010; Hart et al. 2011). In contrast to this, innate immunity is compromised in humans with low vitamin D status (Yang et al. 1993; Liu et al. 2006). Identification of the VDR in Langherhans cells, monocytes, macrophages, and activated T and B cells (Veldmanet al. 2000) supports the idea that $1,25(OH)_2D_3$ produced locally in skin cells may have a physiological role in modulating these cells and, therefore, skin immunity. The discrepancies in immune responses to $1,25(OH)_2D_3$ between humans and mouse strains and models may depend on a number of factors, such as genetic differences between species, dose, and the protocol used. It has also been suggested that the effects are due to the pathway used to elicit the response. Photoprotection is likely mediated, at least in part, via the nongenomic pathway, whereas the immunosuppressive effect of $1,25(OH)_2D_3$ could be mediated by the genomic pathway (as reviewed in the work of Kuritzky et al. 2008 and Dixon et al. 2010).

Although UV exposure results in reduced adaptive immunity, it does not result in increased infections. In fact, Nils Finsen won the Nobel Prize in 1903 for showing that UV exposure helps reduce lupus vulgaris, a cutaneous form of tuberculosis (Zasloff 2006). Recently, it has been reported that UV directly increases the expression of various antimicrobial peptides in skin cells (Glaser et al. 2009). The vitamin D hormone, $1,25(OH)_2D_3$, has also been shown to increase expression of antimicrobial peptides (Liu et al. 2006). These authors reported that stimulation of toll-like receptors on macrophages by pathogens resulted in increased expression of CYP27B1 (1-alpha-hydroxylase) in macrophages and increased expression of the VDR. If substrate 25OHD is adequate, conversion to $1,25(OH)_2D_3$, followed by binding of the vitamin D hormone to its receptor, results in increased expression of cathelicidins and other antimicrobial peptides (Liu et al. 2006). Ketoconazole, an inhibitor of vitamin D hydroxylases, as well as other enzymes, reduced the secretion of antimicrobial peptides in mice after UV (Hong et al. 2008). Whether this means that synthesis of vitamin D in skin in the presence of UV radiation and further local metabolism of this compound contributes to the improved innate immunity after UV remains to be examined.

7.3.7 VITAMIN D COMPOUNDS REDUCE PHOTOCARCINOGENESIS

UV-induced gene mutations and impaired immunity, which allows aberrant cells bearing genetic alterations to undergo clonal expansion, together promote skin carcinogenesis, as outlined in Figure 7.1 (Melnikova and Ananthaswamy 2005; Moloney et al. 2006; Halliday and Lyons 2008). Mutations arise in inadequately repaired DNA lesions and alter cellular actions. Furthermore, increased production of ROS and RNS at the sunburn site by inflammatory responses may also cause further promutagenic DNA damage (Halliday 2005). Inactivating p53 gene mutations have been identified in most skin cancers and also in actinic keratoses, the precursors of SCC, implicating p53 in cellular transformation to SCC (Ziegler et al. 1994; Brash et al. 1996; Leffell and Brash 1996). Immune suppression induced by UV was first shown to facilitate skin carcinogenesis in studies where UV-induced tumors were transplanted into immune-competent mice. The tumors continued to grow without rejection when transplanted into irradiated recipients, whereas they were rejected when transplanted into nonirradiated mice (Kripke 1974). Similarly, renal transplant patients and patients undergoing immunosuppressive chemotherapy are prone to UV-induced invasive SCC (Hill et al. 1976; Moloney et al. 2006).

Photocarcinogenesis studies in mice have demonstrated protection with vitamin D compounds against tumor formation induced by chronic and low doses of solar simulated UV. Topical treatment with $1,25(OH)_2D_3$ immediately after each UV exposure showed a significant reduction in size and the number of tumors (Dixon et al. 2011). Tumors were also reduced, but to a lesser extent, by the rapid acting cis-locked analog JN (Dixon et al. 2011), suggesting that the activation of the genomic pathway may also be important for photoprotection (Dixon et al. 2011). Reduction of UV-induced tumors by vitamin D compounds is in accord with other studies that demonstrated protection from chemically induced skin carcinogenesis by $1,25(OH)_2D_3$ (Wood et al. 1983) and related analogs (Kensler et al. 2000). These studies also complement reports showing increased susceptibility to photocarcinogenesis in VDR knock-out mice (Zinser et al. 2002; Ellison et al. 2008). Interestingly, human susceptibility to solar keratoses (Carless et al. 2008) and increased risk of SCC (Hans et al. 2007) are related to VDR polymorphisms.

7.4 CONCLUSIONS

UVB production of vitamin D in skin and its synthesis to the active hormone $1,25(OH)_2D_3$ locally may be an adaptive response to protect skin cells from further UV exposures (Mason et al. 2010, 2011), like melanogenesis and skin cornification (Kochevar et al. 1999; Matsumura and Ananthaswamy 2002b). The production of $1,25(OH)_2D_3$ and other metabolites after UV in skin takes several hours (Lehmann et al. 2001); therefore, like pigmentation and cornification, this process probably protects against the next UV response, not the current one. Interestingly, there is some evidence that the analog, $1\alpha,25(OH)_2$lumisterol$_3$ (JN), could be produced in skin from the overirradiation product, lumisterol$_3$ (Janjetovic et al. 2009; Slominski et al. 2009), providing a possible function for these compounds. These studies suggest that, because vitamin D and its various metabolites are produced in the skin by UVB, DNA damage is less than it would be otherwise after sun exposure. Because DNA damage is the first molecular event for tumor induction, less DNA damage produced in the presence of $1,25(OH)_2D_3$ would account for the reduction in the UV-induced downstream events of inflammation, immune suppression, and photocarcinogenesis. The action of $1,25(OH)_2D_3$ against DNA damage occurs within 30 min in keratinocytes after UV exposure, suggesting that it is mediated, at least in part, by the rapid nongenomic pathway, and this view is supported by evidence from functional studies with conformationally locked vitamin D analogs. However, the genomic pathway may also play a role in photoprotection, by mediating (1) the p53 transactivation of genes regulating cell survival and DNA repair and (2) the upregulation of the antioxidant metallothionein. It is therefore plausible that the photoprotective responses of $1,25(OH)_2D_3$ are driven by both pathways, but the precise molecular mechanisms still remain to be resolved.

We propose that photoprotection by $1,25(OH)_2D_3$ is brought about by the following factors as listed in Table 7.1 and illustrated in Figure 7.2. Increased p53 expression and its activation by UV leads to transcription of genes involved in DNA repair. Enhanced nuclear p53 by $1,25(OH)_2D_3$ accelerates DNA repair. $1,25(OH)_2D_3$ also reduces NO, an inhibitor of DNA repair, and thus also facilitates DNA repair and should reduce DNA damage by RNS. Increased levels of metallothionein

TABLE 7.1

Summary of the Cellular Responses of 1,25(OH)₂D₃ That Contribute to Its Photoprotective Effects and the Proposed Mediators of the Different Photoprotective Actions

UV-Induced Effects	1,25D Photoprotective Effects	Possible Mediators of 1,25D Photoprotection	References
DNA damage (direct and by ROS and RNS)	Thymine dimers are reduced oxidative DNA damage reduced	NO is reduced; therefore, there is less DNA damage. Increased p53 upregulates DNA repair genes Less NO, which inhibits DNA repair Increased antioxidants, e.g., metallothionein	Gupta et al. 2007 Dixon et al. 2011 De Haes et al. 2005 Gordon-Thomson et al. 2010
Cell death/sunburn cells	Cell survival Less sunburn cells	Less DNA damage Decreased stress-activated protein kinases, e.g., JNK	Mason et al. 2002 Dixon et al. 2007 Gupta et al. 2007 De Haes et al. 2003
Inflammation	Reduced eythema and edema in mice	NO reduction Less pro-inflammatory cytokines, eg., IL-6 Increased antioxidants, e.g., metallothionein	Gupta et al. 2007 Dixon et al. 2011 Mason et al. 2010 Hanada et al. 1995 De Haes et al. 2003
Immune suppression	Immune protection in mice	Less DNA damage Enhanced metallothionein	Dixon et al. 2007 Dixon et al. 2011 Hanada et al. 1995 Karasawa et al. 1987 Lee and Youn 1998
Photocarcinogenesis	Reduced tumors in mice	Immune protection Fewer mutations	Mason et al. 2010 Dixon et al. 2011 Zinser et al. 2002 Ellison et al. 2008

Notes: The publications in which the information was sourced are provided in the last column. Abbreviations: IL-6, interleukin-6; NO, nitric oxide; ROS, reactive oxygen species; and RNS, reactive nitrogen species.

in the presence of 1,25(OH)₂D₃ causes less DNA damage by scavenging UV-induced ROS. A reduction in NO by 1,25(OH)₂D₃ leads to a reduction in lipid peroxidation and proinflammatory cytokine release (IL6), thus reducing inflammation and immune suppression. Also, with less DNA damage and increased DNA repair, Langerhans cells remain active in instigating cell-mediated immune-protective responses. Subsequently, less DNA damage together with improved immune protection limits skin carcinogenesis.

Skin cancer incidence is the highest of all the cancers and is increasing. The risk of developing skin cancer also increases with age. Stable low-calcaemic and low-cost vitamin D–like compounds may be considered potential photoprotective agents in sunscreens and after-sun lotions. Their effects might be expected to reduce UV-induced DNA damage and other early molecular processes that trigger the cascade of events that lead to gene mutation and immunodeficiency. Positive outcomes, including reductions in the risk of premature aging and the development of skin cancers, therefore seem achievable.

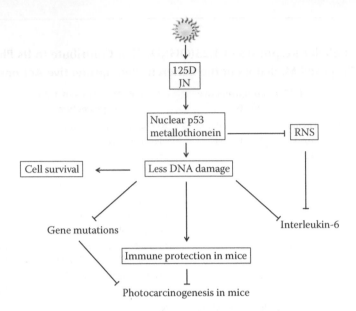

FIGURE 7.2   A proposed mechanism for the photoprotective effect of $1,25(OH)_2D_3$; refer to text for details. JN: 1,25-dihydroxylumisterol$_3$; NO: nitric oxide; and RNS: reactive nitrogen species.

REFERENCES

Agar, N. S., G. M. Halliday, R. S. Barnetson, H. N. Ananthaswamy, M. Wheeler, and A. M. Jones. 2004. The basal layer in human squamous tumors harbors more UVA than UVB fingerprint mutations: A role for UVA in human skin carcinogenesis. *Proceedings of the National Academy of Sciences of the United States of America* 101:4954–4959.

Amundson, S. A., A. Patterson, K. T. Do, and A. J. Fornace. 2002. A nucleotide excision repair master-switch: p53 regulated coordinate induction of global genomic repair genes. *Cancer Biology & Therapy* 1:145–149.

Applegate, L. A., R. D. Ley, J. Alcalay, and M. L. Kripke. 1989. Identification of the molecular target for the suppression of contact hypersensitivity by ultraviolet radiation. *Journal of Experimental Medicine* 170:1117–1131.

Aubin, F. 2003. Mechanisms involved in ultraviolet light–induced immunosuppression. *European Journal of Dermatology* 13:515–523.

Banin, S., L. Moyal, S. Y. Shieh, Y. Taya, C. W. Anderson, L. Chessa, N. I. Smorodinsky, C. Prives, Y. Reiss, Y. Shiloh, and Y. Ziv. 1998. Enhanced phosphorylation of p53 by ATM in response to DNA damage. *Science* 281:1674–1677.

Bau, D. T., J. R. Gurr, and K. Y. Jan. 2001. Nitric oxide is involved in arsenite inhibition of pyrimidine dimer excision. *Carcinogenesis* 22:709–716.

Berg, R. J. W., H. J. vanKranen, H. G. Rebel, A. deVries, W. A. vanVloten, C. F. vanKreijl, J. C. vander-Leun, and F. R. deGruijl. 1996. Early p53 alterations in mouse skin carcinogenesis by UVB radiation: Immunohistochemical detection of mutant p53 protein in clusters of preneoplastic epidermal cells. *Proceedings of the National Academy of Sciences of the United States of America* 93:274–278.

Besaratinia, A., S. Kim, and G. P. Pfeifer. 2008. Rapid repair of UVA-induced oxidized purines and persistence of UVB-induced dipyrimidine lesions determine the mutagenicity of sunlight in mouse cells. *FASEB Journal* 22:2379–2392.

Biggs, L., C. Yu, B. Fedoric, A. F. Lopez, S. J. Galli, and M. A. Grimbaldeston. 2010. Evidence that vitamin D(3) promotes mast cell–dependent reduction of chronic UVB-induced skin pathology in mice. *Journal of Experimental Medicine* 207:455–463.

Bikle, D. D., M. K. Nemanic, J. O. Whitney, and P. W. Elias. 1986. Neonatal human foreskin keratinocytes produce 1,25-dihydroxyvitamin D$_3$. *Biochemistry* 25:1545–1548.

Brash, D. E., A. Ziegler, A. S. Jonason, J. A. Simon, S. Kunala, and D. J. Leffell. 1996. Sunlight and sunburn in human skin cancer: p53, apoptosis, and tumor promotion. *Journal of Investigative Dermatology Symposium Proceedings* 1:136–142.

Bruch-Gerharz, D., T. Ruzicka, and V. Kolb-Bachofen. 1998. Nitric oxide in human skin: Current status and future prospects. *Journal of Investigative Dermatology* 110:1–7.

Bruins, W., E. Zwart, L. D. Attardi, T. Iwakuma, E. M. Hoogervorst, R. B. Beems, B. Miranda, C. T. van Oostrom, J. van den Berg, G. J. van den Aardweg, G. Lozano, H. van Steeg, T. Jacks, and A. de Vries. 2004. Increased sensitivity to UV radiation in mice with a p53 point mutation at Ser389. *Molecular and Cellular Biology* 24:8884–8894.

Burney, S., J. L. Caulfield, J. C. Niles, J. S. Wishnok, and S. R. Tannenbaum. 1999. The chemistry of DNA damage from nitric oxide and peroxynitrite. *Mutation Research—Fundamental and Molecular Mechanisms of Mutagenesis* 424:37–49.

Caceres-Dittmar, G., K. Ariizumi, S. Xu, F. J. Tapia, P. R. Bergstresser, and A. Takashima. 1995. Hydrogen-peroxide mediates UV-induced impairment of antigen presentation in a murine epidermal-derived dendritic cell-line. *Photochemistry and Photobiology* 62:176–183.

Cals-Grierson, M. M., and A. D. Ormerod. 2004. Nitric oxide function in the skin. *Nitric Oxide—Biology and Chemistry* 10:179–193.

Carless, M. A., T. Kraska, N. Lintell, R. E. Neale, A. C. Green, and L. R. Griffiths. 2008. Polymorphisms of the VDR gene are associated with presence of solar keratoses on the skin. *British Journal of Dermatology* 159:804–810.

Carrier, F., P. T. Georgel, P. Pourquier, M. Blake, H. U. Kontny, M. J. Antinore, M. Gariboldi, T. G. Myers, J. N. Weinstein, Y. Pommier, and A. J. Fornace Jr. 1999. Gadd45, a p53-Responsive stress protein, modifies DNA accessibility on damaged chromatin. *Molecular and Cellular Biology* 19:1673–1685.

Castano, A. P., P. Mroz, and M. R. Hamblin. 2006. Photodynamic therapy and antitumor immunity. *Nature Reviews Cancer* 6:535–545.

Claerhout, S., A. Van Laethem, P. Agostinis, and M. Garmyn. 2006. Pathways involved in sunburn cell formation: Deregulation in skin cancer. *Photochemical & Photobiological Sciences* 5:199–207.

Cooke, M. S., I. D. Podmore, N. Mistry, M. D. Evans, K. E. Herbert, H. R. Griffiths, and J. Lunec. 2003. Immunochemical detection of UV-induced DNA damage and repair. *Journal of Immunological Methods* 280:125–133.

Damian, D. L., Y. J. Kim, K. M. Dixon, G. M. Halliday, A. Javeri, and R. S. Mason. 2010. Topical calcitriol protects from UV-induced genetic damage but suppresses cutaneous immunity in humans. *Experimental Dermatology* 19:E23–E30.

Damian, D. L., Y. J. Matthews, T. A. Phan, and G. M. Halliday. 2011. An action spectrum for ultraviolet radiation-induced immunosuppression in humans. *British Journal of Dermatology* 164:657–659.

David, S. S., V. L. O'Shea, and S. Kundu. 2007. Base-excision repair of oxidative DNA damage. *Nature* 447:941–950.

De Haes, P., M. Garmyn, G. Carmeliet, H. Degreef, K. Vantieghem, R. Bouillon, and S. Segaert. 2004. Molecular pathways involved in the anti-apoptotic effect of 1,25-dihydroxyvitamin D_3 in primary human keratinocytes. *Journal of Cellular Biochemistry* 93:951–967.

De Haes, P., M. Garmyn, H. Degreef, K. Vantieghem, R. Bouillon, and S. Segaert. 2003. 1,25-Dihydroxyvitamin D_3 inhibits ultraviolet B-induced apoptosis, Jun kinase activation, and interleukin-6 production in primary human keratinocytes. *Journal of Cellular Biochemistry* 89:663–673.

De Haes, P., M. Garmyn, A. Verstuyf, P. De Clercq, M. Vandewalle, H. Degreef, K. Vantieghem, R. Bouillon, and S. Segaert. 2005. 1,25-Dihydroxyvitamin D_3 and analogues protect primary human keratinocytes against UVB-induced DNA damage. *Journal of Photochemistry & Photobiology. B Biology* 78:141–148.

Deliconstantinos, G., V. Villiotou, and J. C. Stavrides. 1996. Increase of particulate nitric oxide synthase activity and peroxynitrite synthesis in UVB-irradiated keratinocyte membranes. *Biochemical Journal* 320:997–1003.

Devary, Y., R. A. Gottlieb, T. Smeal, and M. Karin. 1992. The mammalian ultraviolet response is triggered by activation of src tyrosine kinases. *Cell* 71:1081–1091.

Dhanalakshmi, S., G. U. Mallikarjuna, and R. Agarwal. 2004. Silibinin prevents ultraviolet radiation-caused skin damages in Skh-1 hairless mice via a decrease in thymine dimer positive cells and an up-regulation of p53-p21/Cip1 in epidermis. *Carcinogenesis* 25:1459–1465.

Diker-Cohen, T., R. Koren, U. A. Liberman, and A. Ravid. 2003. Vitamin D protects keratinocytes from apoptosis induced by osmotic shock, oxidative stress, and tumor necrosis factor. *Annals of the New York Academy of Sciences* 1010:350–353.

Dixon, K. M., S. S. Deo, A. W. Norman, J. E. Bishop, G. M. Halliday, V. E. Reeve, and R. S. Mason. 2007. *In vivo* relevance for photoprotection by the vitamin D rapid response pathway. *The Journal of Steroid Biochemistry and Molecular Biology* 103:451–6.

Dixon, K. M., S. S. Deo, G. Wong, M. Slater, A. W. Norman, J. E. Bishop, G. H. Posner, S. Ishizuka, G. M. Halliday, V. E. Reeve, and R. S. Mason. 2005. Skin cancer prevention: A possible role of 1,25dihydroxyvitamin D_3 and its analogs. *The Journal of Steroid Biochemistry and Molecular Biology* 97:137–143.

Dixon, K. M., and R. S. Mason. 2009. Vitamin D. *International Journal of Biochemistry & Cell Biology* 41:982–985.

Dixon, K. M., A. W. Norman, V. B. Sequeira, R. Mohan, M. S. Rybchyn, V. E. Reeve, G. M. Halliday, and R. S. Mason. 2011. 1 alpha,25(OH)$_2$-vitamin D and a non-genomic vitamin D analog inhibit ultraviolet radiation-induced skin carcinogenesis. *Cancer Prevention Research* 4:1485–1494.

Dixon, K. M., V. B. Sequeira, A. J. Camp, and R. S. Mason. 2010. Vitamin D-fence. *Photochemical & Photobiological Sciences* 9:564–570.

Donehower, L., M. Harvey, B. Slagle, C. Jr Montgomery, J. Butel, and A. Bradley. 1992. p53-deficient mice develop normally but are susceptible to tumors. *Journal of Cellular Biochemistry Supplement* (16 Part B):114.

Dong, X. Y., W. Lutz, T. M. Schroeder, L. A. Bachman, J. J. Westendorf, R. Kumar, and M. D. Griffin. 2005. Regulation of relB in dendritic cells by means of modulated association of vitamin D receptor and histone deacetylase 3 with the promoter. *Proceedings of the National Academy of Sciences of the United States of America* 102:16007–16012.

Douki, T., M. Court, S. Sauvaigo, F. Odin, and J. Cadet. 2000. Formation of the main UV-induced thymine dimeric lesions within isolated and cellular DNA as measured by high performance liquid chromatography-tandem mass spectrometry. *Journal of Biological Chemistry* 275:11678–11685.

Eller, M. S., A. Asarch, and B. A. Gilchrest. 2008. Photoprotection in human skin—A multifaceted SOS response. *Photochemistry and Photobiology* 84:339–349.

Eller, M. S., T. Maeda, C. Magnoni, D. Atwal, and B. A. Gilchrest. 1997. Enhancement of DNA repair in human skin cells by thymidine dinucleotides: Evidence for a p53-mediated mammalian SOS response. *Proceedings of the National Academy of Sciences of the United States of America* 94:12627–12632.

Ellison, T. I., M. K. Smith, A. C. Gilliam, and P. N. MacDonald. 2008. Inactivation of the vitamin D receptor enhances susceptibility of murine skin to UV-induced tumorigenesis. *Journal of Investigative Dermatology* 128:2508–2517.

Flores, O., and K. L. Burnstein. 2010. GADD45 gamma: A new vitamin D–regulated gene that is antiproliferative in prostate cancer cells. *Endocrinology* 151:4654–4664.

Gläser, R., F. Navid, W. Schuller, C. Jantschitsch, J. Harder, J. M. Schröder, A. Schwarz, and T. Schwarz. 2009. UV-B radiation induces the expression of antimicrobial peptides in human keratinocytes *in vitro* and *in vivo*. *Journal of Allergy and Clinical Immunology* 123:1117–1123.

Gordon-Thomson, C., R. Gupta, V. E. Reeve, G.M. Halliday, and R. S. Mason. 2010. UV-induced oxidative and other forms of DNA damage are suppressed by 1,25 dihydroxyvitamin D_3 in human keratinocytes. *Proceedings of the Australian Health and Medical Research Congress*, Melbourne, Australia.

Gorman, S., L. A. Kuritzky, M. A. Judge, K. M. Dixon, J. P. McGlade, R. S. Mason, J. J. Finlay-Jones, and P. H. Hart. 2007. Topically applied 1,25-dihydroxyvitamin D-$_3$ enhances the suppressive activity of CD4(+) CD25(+), cells in the draining lymph nodes. *Journal of Immunology* 179:6273–6283.

Gorman, S., M. A. Judge, and P. H. Hart. 2010. Immune-modifying properties of topical vitamin D: Focus on dendritic cells and T cells. *The Journal of Steroid Biochemistry and Molecular Biology* 121:247–249.

Griffin, M. D., W. Lutz, V. A. Phan, L. A. Bachman, D. J. McKean, and R. Kumar. 2001. Dendritic cell modulation by 1alpha,25-dihydroxyvitamin D_3 and its analogs: A vitamin D receptor–dependent pathway that promotes a persistent state of immaturity *in vitro* and *in vivo*. *Journal of the American Society of Nephrology* 12(Program and Abstract Issue):857A.

Grimbaldeston, M. A., S. Nakae, J. Kalesnikoff, M. Tsai, and S. J. Galli. 2007. Mast cell-derived interleukin 10 limits skin pathology in contact dermatitis and chronic irradiation with ultraviolet B. *Nature Immunology* 8:1095–1104.

Gupta, R., K. M. Dixon, S. S. Deo, C. J. Holliday, M. Slater, G. M. Halliday, V. E. Reeve, and R. S. Mason. 2007. Photoprotection by 1,25 dihydroxyvitamin D_3 is associated with an increase in p53 and a decrease in nitric oxide products. *Journal of Investigative Dermatology* 127:707–715.

Hall, P. A., P. H. McKee, H. D. Menage, R. Dover, and D. P. Lane. 1993. High levels of p53 protein in UV-irradiated normal human skin. *Oncogene* 8:203–207.

Halliday, G. M. 2005. Inflammation, gene mutation and photoimmunosuppression in response to UVR-induced oxidative damage contributes to photocarcinogenesis. *Mutation Research* 571:107–120.

Halliday, G.M. 2010. Common links among the pathways leading to UV-induced immunosuppression. *Journal of Investigative Dermatology* 130:1209–1212.

Halliday, G. M., S. N. Byrne, J. M. Kuchel, T. S. C. Poon, and R. S. C. Barnetson. 2004. The suppression of immunity by ultraviolet radiation: UVA, nitric oxide and DNA damage. *Photochemical & Photobiological Sciences* 3:736–740.

Halliday, G. M., and J. G. Lyons. 2008. Inflammatory doses of UV may not be necessary for skin carcinogenesis. *Photochemistry and Photobiology* 84:272–283.

Halliday, G.M., M. Norval, Scott N. Byrne, X. Hwang, and P. Wolf. 2008. The effects of sunlight on the skin. In *Drug Discovery Today: Disease Mechanism* 52:e201–e209.

Hans, J. L., G. A. Colditz, and D. J. Hunter. 2007. Polymorphisms in the MTHFR and VDR genes and skin cancer risk. *Carcinogenesis* 28:390–397.

Hanada, K., T. Baba, I. Hashimoto, R. Fukui, and S. Watanabe. 1992. Possible role of cutaneous metallothionein in protection against photooxidative stress-epidermal localization and scavenging activity for superoxide and hydroxyl radicals. *Photodermatology, Photoimmunology & Photomedicine* 9:209–213.

Hanada, K., R. W. Gange, E. Siebert, and T. Hasan. 1991. Protective effects of cadmium chloride against UVB injury in mouse skin and in cultured human cells: A possible role of cadmium-induced metallothionein. *Photodermatology, Photoimmunology & Photomedicine* 8:111–115.

Hanada, K., D. Sawamura, H. Nakano, and I. Hashimoto. 1995. Possible role of 1,25-dihydroxyvitamin D_3-induced metallothionein in photoprotection against UVB injury in mouse skin and cultured rat keratinocytes. *Journal of Dermatological Science* 9:203–208.

Hanneman, K. K., H. M. Scull, K. D. Cooper, and E. D. Baron. 2006. Effect of topical vitamin D analogue on *in vivo* contact sensitization. *Archives of Dermatology* 142:1332–1334.

Hart, P. H., S. Gorman, and J. J. Finlay-Jones. 2011. Modulation of the immune system by UV radiation: More than just the effects of vitamin D? *Nature reviews. Immunology* 11:584–596.

Hess, D. T., A. Matsumoto, R. Nudelman, and J. S. Stamler. 2001. S-nitrosylation: Spectrum and specificity. *Nature Cell Biology* 3:E46–E49.

Hill, B. H. 1976. Immunosuppressive drug therapy as a potentiator of skin tumors in five patients with lymphoma. *Australasian Journal of Dermatology* 17:46–48.

Holick, M. F. 1981. The cutaneous photosynthesis of previtamin D_3: A unique photoendocrine system. *Journal of Investigative Dermatology* 77:51–58.

Holick, M. F. 2001. Calciotropic hormones and the skin: A millennium perspective. *Journal of Cosmetic Science* 52:146–148.

Holick, M. F. 2011. Vitamin D: Evolutionary, physiological and health perspectives. *Current Drug Targets* 12:4–18.

Hong, S. P., M. J. Kim, M. Y. Jung, H. Jeon, J. Goo, S. K. Ahn, S. H. Lee, P. M. Elias, and E. H. Choi. 2008. Biopositive effects of low-dose UVB on epidermis: Coordinate upregulation of antimicrobial peptides and permeability barrier reinforcement. *The Journal of Investigative Dermatology* 12:2880–2887.

Hwang, B. J., J. M. Ford, P. C. Hanawalt, and G. Chu. 1999. Expression of the p48 xeroderma pigmentosum gene is p53-dependent and is involved in global genomic repair. *Proceedings of the National Academy of Sciences of the United States of America* 96:424–428.

Jaiswal, M., N. F. LaRusso, L. J. Burgart, and G. J. Gores. 2000. Inflammatory cytokines induce DNA damage and inhibit DNA repair in cholangiocarcinoma cells by a nitric oxide-dependent mechanism. *Cancer Research* 60:184–190.

Janjetovic, Z., R. C. Tuckey, M. N. Nguyen, E. M. Thorpe, and A. T. Slominski. 2010. 20,23-Dihydroxyvitamin D_3, novel P450scc product, stimulates differentiation and inhibits proliferation and NF-kappa B activity in human keratinocytes. *Journal of Cellular Physiology* 223:36–48.

Janjetovic, Z., M. A. Zmijewski, R. C. Tuckey, D. A. DeLeon, M. N. Nguyen, L. M. Pfeffer, and A. T. Slominski. 2009. 20-Hydroxycholecalciferol, product of vitamin D_3 hydroxylation by P450scc, decreases NF-kappa B activity by increasing I kappa B alpha levels in human keratinocytes. *PLoS One* 4(6):e5988.

Karasawa, M., J. Hosoi, H. Hashiba, K. Nose, C. Tohyama, E. Abe, T. Suda, and T. Kuroki. 1987. Regulation of metallothionein gene-expression by 1-alpha,25-dihydroxyvitamin D_3 in cultured-cells and in mice. *Proceedings of the National Academy of Sciences of the United States of America* 84:8810–8813.

Katiyar, S. K. 2000. Kinetics of UV light-induced cyclobutane pyrimidine dimers in human skin *in vivo*: An immunohistochemical analysis of both epidermis and dermis. *Clinical Cancer Research* 6:3864–3869.

Kensler, T. W., P. M. Dolan, S. J. Gange, J. K. Lee, Q. Wang, and G. H. Posner. 2000. Conceptually new deltanoids (vitamin D analogs) inhibit multistage skin tumorigenesis. *Carcinogenesis* 21:1341–1345.

Khanal, R. C., and I. Nemere. 2007. The ERp57/GRp58/1,25D(3)-MARRS receptor: Multiple functional roles in diverse cell systems. *Current Medicinal Chemistry* 14:1087–1093.

Kochevar, I. E., M. A. Pathak, and J. A. Parrish. 1999. Photophysics, photochemistry and photobiology. In *Fitzpatrick's dermatology in general medicine*, ed. I. M. Freedberg, A. Eisen, K. Wolff, K. F. Austen, L. Goldsmith, S. Katz, and T. B. Fitzpatrick. New York: McGraw-Hill, pp. 220–229.

Kripke, M. L. 1974. Antigenicity of murine skin tumors induced by ultraviolet-light. *Journal of the National Cancer Institute* 53:1333–1336.

Kripke, M. L., P. A. Cox, L. G. Alas, and D. B. Yarosh. 1992. Pyrimidine dimers in DNA initiate systemic immunosuppression in UV-irradiated mice. *Proceedings of the National Academy of Sciences of the United States of America* 15:7516–7520.

Kripke, M. L., and M. S. Fisher. 1976. Immunological parameters of ultraviolet carcinogenesis. *Journal of the National Cancer Institute* 57:211–215.

Kuchel, J. M., R. S. Barnetson, and G. M. Halliday. 2005. Cyclobutane pyrimidine dimer formation is a molecular trigger for solar-simulated ultraviolet radiation-induced suppression of memory immunity in humans. *Photochemical & Photobiological Sciences* 4:577–582.

Kulms, D., B. Poppelmann, D. Yarosh, T. A. Luger, J. Krutmann, and T. Schwarz. 1999. Nuclear and cell membrane effects contribute independently to the induction of apoptosis in human cells exposed to UVB radiation. *Proceedings of the National Academy of Sciences of the United States of America* 96:7974–7979.

Kuritzky, L. A., J. J. Finlay-Jones, and P. H. Hart. 2008. The controversial role of vitamin D in the skin: Immunosuppression vs. photoprotection. *Clinical and Experimental Dermatology* 33:167–170.

Kvam, E., and R. M. Tyrrell. 1997. Induction of oxidative DNA base damage in human skin cells by UV and near visible radiation. *Carcinogenesis* 18:2379–2384.

Lamola, A. A. 1971. Production of pyrimidine dimers in DNA in the dark. *Biochemical & Biophysical Research Communications* 43:893–898.

Leccia, M. T., M. Yaar, N. Allen, M. Gleason, and B. A. Gilchrest. 2001. Solar simulated irradiation modulates gene expression and activity of antioxidant enzymes in cultured human dermal fibroblasts. *Experimental Dermatology* 10:272–279.

Lee, J., and J. I. Youn. 1998. The photoprotective effect of 1,25-dihydroxyvitamin D_3 on ultraviolet light B-induced damage in keratinocyte and its mechanism of action. *Journal of Dermatological Science* 18:11–18.

Leffell, D. J., and D. E. Brash. 1996. Sunlight and skin cancer. *Scientific American* 275:52–53, 56–59.

Lehmann, B., T. Genehr, P. Knuschke, J. Pietzsch, and M. Meurer. 2001. UVB-induced conversion of 7-dehydrocholesterol to 1alpha,25-dihydroxyvitamin D_3 in an *in vitro* human skin equivalent model. *Journal of Investigative Dermatology* 117:1179–1185.

Lehmann, B., T. Rudolph, J. Pietzsch, and M. Meurer. 2000. Conversion of vitamin D_3 to 1-alpha,25-dihydroxyvitamin D_3 in human skin equivalents. *Experimental Dermatology* 9:97–103.

Lehmann, B., O. Tiebel, and M. Meurer. 1999. Expression of vitamin D_3 25-hydroxylase (CYP27) mRNA after induction by vitamin D_3 or UVB radiation in keratinocytes of human skin equivalents—a preliminary study. *Archives of Dermatological Research* 291:507–510.

Levine, A. J. 1997. p53, the cellular gatekeeper for growth and division. *Cell* 88:323–331.

Lhiaubet-Vallet, V., M. Consuelo Cuquerella, J. V. Castell, F. Bosca, and M. A. Miranda. 2007. Triplet excited fluoroquinolones as mediators for thymine cyclobutane dimer formation in DNA. *Journal of Physical Chemistry B* 111:7409–7414.

Liu, D., L. Ou, G. D. Clemenson, Jr., C. Chao, M. E. Lutske, G. P. Zambetti, F. H. Gage, and Y. Xu. 2010. Puma is required for p53-induced depletion of adult stem cells. *Nature Cell Biology* 12:993–998.

Liu, P. T., S. Stenger, H. Y. Li, L. Wenzel, B. H. Tan, S. R. Krutzik, M. T. Ochoa, J. Schauber, K. Wu, C. Meinken, D. L. Kamen, M. Wagner, R. Bals, A. Steinmeyer, U. Zugel, R. L. Gallo, D. Eisenberg, M. Hewison, B. W. Hollis, J. S. Adams, B. R. Bloom, and R. L. Modlin. 2006. Toll-like receptor triggering of a vitamin D-mediated human antimicrobial response. *Science* 311:1770–1773.

Maclaine, N. J., and T. R. Hupp. 2009. The regulation of p53 by phosphorylation: A model for how distinct signals integrate into the p53 pathway. *Aging* 1:490–502.

Malloy, P. J., J. W. Pike, and D. Feldman. 1999. The vitamin D receptor and the syndrome of hereditary 1,25-dihydroxyvitamin D-resistant rickets. *Endocrine Reviews* 20:156–188.

Manggau, M., D. S. Kim, L. Ruwisch, R. Vogler, H. C. Korting, M. Schafer-Korting, and B. Kleuser. 2001. 1-alpha,25-dihydroxyvitamin D-$_3$ protects human keratinocytes from apoptosis by the formation of sphingosine-1-phosphate. *Journal of Investigative Dermatology* 117:1241–1249.

Mason, R. S., and C. J. Holliday. 2000. 1,25Dihydroxyvitamin D contributes to photoprotection in skin cells. In *Vitamin D endocrine system: Structural, biological, genetic and clinical aspects*, ed. A. W. Norman, R. Bouillon and M. Thomasset. Riverside: University of California, pp. 605–608.

Mason, R. S., C. J. Holliday, and R. Gupta. 2002. 1,25 Dihydroxyvitamin D and photoprotection in skin cells. In *Modern trends in skin pharmacology*, ed. D. Tsambos and H. Merk. Athens, Greece: Parissianos Medical Publications S.A. Athens, pp. 59–66.

Mason, R. S., K. M. Dixon, V. B. Sequeira, and C. Gordon-Thomson. 2011. Sunlight protection by vitamin D compounds. In *Vitamin D*, ed. D. Feldman, J. W. Pike and J. S. Adams: Academic Press, pp. 1943–1953.

Mason, R. S., V. B. Sequeira, K. M. Dixon, C. Gordon-Thomson, K. Pobre, A. Dilley, M. T. Mizwicki, A. W. Norman, D. Feldman, G. M. Halliday, and V. E. Reeve. 2010. Photoprotection by 1-alpha,25-dihydroxy-vitamin D and analogs: Further studies on mechanisms and implications for UV-damage. *Journal of Steroid Biochemistry and Molecular Biology* 121:164–168.

Mathieu, C., and L. Adorini. 2002. The coming of age of 1,25-dihydroxyvitamin D-3 analogs as immunomodulatory agents. *Trends in Molecular Medicine* 8:174–179.

Matsumura, Y., and H. N. Ananthaswamy. 2002a. Molecular mechanisms of photocarcinogenesis. *Frontiers in Bioscience* 7:d765–d783.

Matsumura, Y., and H. N. Ananthaswamy. 2002b. Short-term and long-term cellular and molecular events following UV irradiation of skin: Implications for molecular medicine. *Expert Reviews in Molecular Medicine* 4:1–22.

Matsumura, Y., and H. N. Ananthaswamy. 2004. Toxic effects of ultraviolet radiation on the skin. *Toxicology & Applied Pharmacology* 195:298–308.

Matthews, Y. J., G. M. Halliday, T. A. Phan, and D. L. Damian. 2010. Wavelength dependency for UVA-induced suppression of recall immunity in humans. *Journal of Dermatological Science* 59:192–197.

McAndrew, J., R. P. Patel, H. J. Jo, T. Cornwell, T. Lincoln, D. Moellering, C. R. White, S. Matalon, and V. DarleyUsmar. 1997. The interplay of nitric oxide and peroxynitrite with signal transduction pathways: Implications for disease. *Seminars in Perinatology* 21:351–366.

McGregor, J. M., and J. L. M. Hawk. 1999. Acute effects of ultraviolet radiation on the skin. In *Fitzpatrick's dermatology in general medicine*, ed. I. M. Freedberg, A. Eisen, K. Wolff, K. F. Austen, L. Goldsmith, S. Katz, and T. B. Fitzpatrick: McGraw-Hill, pp. 1555–1561.

McKay, B. C., M. A. Francis, and A. J. Rainbow. 1997. Wildtype p53 is required for heat shock and ultraviolet light enhanced repair of a UV-damaged reporter gene. *Carcinogenesis* 18:245–249.

Melnikova, V. O., and H. N. Ananthaswamy. 2005. Cellular and molecular events leading to the development of skin cancer. *Mutation Research-Fundamental and Molecular Mechanisms of Mutagenesis* 571:91–106.

Menegaz, D., M. T. Mizwicki, A. Barrientos-Duran, N. Chen, H. L. Henry, and A. W. Norman. 2011. Vitamin D receptor (VDR) regulation of voltage-gated chloride channels by ligands preferring a VDR-alternative pocket (VDR-AP). *Molecular Endocrinology* 25:1289–1300.

Mitani, H., E. Naru, M. Yamashita, K. Arakane, T. Suzuki, and T. Imanari. 2004. Ergocalciferol promotes *in vivo* differentiation of keratinocytes and reduces photodamage caused by ultraviolet irradiation in hairless mice. *Photodermatology, Photoimmunology & Photomedicine* 20:215–223.

Mitchell, D. L., J. E. Cleaver, and J. H. Epstein. 1990. Repair of pyrimidine(6-4)pyrimidone photoproducts in mouse skin. *Journal of Investigative Dermatology* 95:55–59.

Mitra, S., I. Boldogh, T. Izumi, and T. K. Hazra. 2001. Complexities of the DNA base excision repair pathway for repair of oxidative DNA damage. *Environmental and Molecular Mutagenesis* 38:180–190.

Moloney, F. J., H. Comber, P. O'Lorcain, P. O'Kelly, P. J. Conlon, and G. M. Murphy. 2006. A population-based study of skin cancer incidence and prevalence in renal transplant recipients. *The British Journal of Dermatology* 154:498–504.

Momand, J., H.-H. Wu, J. A. Thomas, L. Makmura, A. Areopagita, S. Furuta, and A. Munoz. 2000. p53 protein oxidation in cultured cells in response to PDTC: A novel method for relating the amount of p53 oxidation *in vivo* to the regulation of p53-responsive genes. *Free Radical Biology and Medicine* 29 (Supplement 1):S59.

Mouret, S., C. Baudouin, M. Charveron, A. Favier, J. Cadet, and T. Douki. 2006. Cyclobutane pyrimidine dimers are predominant DNA lesions in whole human skin exposed to UVA radiation. *Proceedings of the National Academy of Sciences of the United States of America* 103:13765–13770.

Mouret, S., C. Philippe, J. Gracia-Chantegrel, and T. Douki. 2009. Long UVA directly generate cyclobutane pyrimidine dimers in isolated and cellular DNA. *Journal of Investigative Dermatology* 129:S34.

Mowbray, M., S. McLintock, R. Weerakoon, N. Lomatschinsky, S. Jones, A. G. Rossi, and R. B. Weller. 2009. Enzyme-independent NO stores in human skin: Quantification and influence of UV radiation. *Journal of Investigative Dermatology* 129:834–842.

Mullauer, L., P. Gruber, D. Sebinger, J. Buch, S. Wohlfart, and A. Chott. 2001. Mutations in apoptosis genes: A pathogenetic factor for human disease. *Mutation Research* 488:211–231.

Murata, M., S. Ohnishi, K. Seike, K. Fukuhara, N. Miyata, and S. Kawanishi. 2004. Oxidative DNA damage induced by carcinogenic dinitropyrenes in the presence of p450 reductase. *Chemical Research in
Toxicology* 17:1750–1756.

Nataraj, A. J., J. C. Trent, and H. N. Ananthaswamy. 1995. p53 gene-mutations and photocarcinogenesis.
Photochemistry and Photobiology 62:218–230.

Nemere, I., S. E. Safford, B. Rohe, M. M. DeSouza, and M. C. Farach-Carson. 2004. Identification and characterization of 1,25D3 membrane–associated rapid response, steroid (1,25D3-MARRS) binding protein.
Journal of Steroid Biochemistry and Molecular Biology 89–90:281–285.

Nghiem, D. X., N. Kazimi, D. L. Mitchell, A. A. Vink, H. N. Ananthaswamy, M. L. Kripke, and S. E. Ullrich.
2002. Mechanisms underlying the suppression of established immune responses by ultraviolet radiation.
Journal of Investigative Dermatology 119:600–608.

Niles, J. C., J. S. Wishnok, and S. R. Tannenbaum. 2006. Peroxynitrite-induced oxidation and nitration products of guanine and 8-oxoguanine: Structures and mechanisms of product formation. *Nitric Oxide*
14:109–121.

Nilsen, H., and H. E. Krokan. 2001. Base excision repair in a network of defence and tolerance. *Carcinogenesis*
22:987–998.

Nishigori, C., D. B. Yarosh, S. E. Ullrich, A. A. Vink, C. D. Bucana, and L. Roza. 1996. Evidence that DNA
damage triggers interleukin 10 cytokine production in UV-irradiated murine keratinocytes. *Proceedings
of the National Academy of Sciences of the United States of America* 93:10354–10359.

Nishimura, N., C. Tohyama, M. Satoh, H. Nishimura, and V. E. Reeve. 1999. Defective immune response and
severe skin damage following UVB irradiation in interleukin 6–deficient mice. *Immunology* 97:77–83.

Norman, A. W., H. L. Henry, J. E. Bishop, X. D. Song, C. Bula, and W. H. Okamura. 2001. Different shapes
of the steroid hormone 1-alpha,25(OH)(2)-vitamin D($_3$) act as agonists for two different receptors in the
vitamin D endocrine system to mediate genomic and rapid responses. *Steroids* 66:147–158.

Norman, A. W., P. S. Manchand, M. R. Uskokovic, W. H. Okamura, J. A. Takeuchi, J. E. Bishop, J. I. Hisatake,
H. P. Koeffler, and S. Peleg. 2000. Characterization of a novel analogue of 1-alpha,25(OH)(2)-vitamin
D-$_3$ with two side chains: Interaction with its nuclear receptor and cellular actions. *Journal of Medicinal
Chemistry* 43:2719–2730.

Norman, A. W., M. T. Mizwicki, and D. P. Norman. 2004. Steroid-hormone rapid actions, membrane receptors
and a conformational ensemble model. *Nature Reviews* 3:27–41.

Ohshima, H., T. Sawa, and T. Akaike. 2006. 8-Nitroguanine, a product of nitrative DNA damage caused by
reactive nitrogen species: Formation, occurrence, and implications in inflammation and carcinogenesis.
Antioxidants & Redox Signaling 8:1033–1045.

Ouhtit, A., H. K. Muller, A. Gorny, and H. N. Ananthaswamy. 2000. UVB-induced experimental carcinogenesis: Dysregulation of apoptosis and p53 signaling pathway. *Redox Report* 5:128–129.

Pacher, P., J. S. Beckman, and L. Liaudet. 2007. Nitric oxide and peroxynitrite in health and disease.
Physiological Reviews 87:315–424.

Pattison, D. I., and M. J. Davies. 2006. Actions of ultraviolet light on cellular structures. *EXS* (96):131–157.

Paunel, A., A. Dejam, S. Thelen, M. Kirsch, M. Horstjann, P. Gharini, M. Muertz, M. Kelm, H. de Groot, V.
Kolb-Bachofen, and C. Suschek. 2005. UVA induces immediate and enzyme-independent nitric oxide
formation in healthy human skin leading to NO-specific signalling. *Journal of Investigative Dermatology*
125:A3.

Poon, T. S., R. S. Barnetson, and G. M. Halliday. 2005. Sunlight-induced immunosuppression in humans is initially because of UVB, then UVA, followed by interactive effects. *Journal of Investigative Dermatology*
125:840–846.

Ravanat, J. L., T. Douki, and J. Cadet. 2001. Direct and indirect effects of UV radiation on DNA and its components. *Journal of photochemistry and photobiology. B, Biology* 63:88–102.

Ravid, A., E. Rubinstein, A. Gamady, C. Rotem, U. A. Liberman, and R. Koren. 2002. Vitamin D inhibits the
activation of stress-activated protein kinases by physiological and environmental stresses in keratinocytes. *Journal of Endocrinology* 173:525–532.

Reeve, V. E. 2002. Ultraviolet radiation and the contact hypersensitivity reaction in mice. *Methods* 28:20–4.

Reeve, V. E., N. Nishimura, M. Bosnic, A. E. Michalska, and K. H. A. Choo. 2000. Lack of metallothionein-I
and -II exacerbates the immunosuppressive effect of ultraviolet B radiation and cis-urocanic acid in mice.
Immunology 100:399–404.

Reichrath, J. 2007. Vitamin D and the skin: An ancient friend, revisited. *Experimental Dermatology* 16:618–625.

Saito, S., H. Yamaguchi, Y. Higashimoto, C. Chao, Y. Xu, A. J. Fornace, E. Appella, and C. W. Anderson. 2003.
Phosphorylation site interdependence of human p53 posttranslational modifications in response to stress.
Journal of Biological Chemistry 278:37536–37544.

Schul, W., J. Jans, Y. M. A. Rijksen, K. H. M. Klemann, A. P. M. Eker, J. de Wit, O. Nikaido, S. Nakajima, A. Yasui, J. H. J. Hoeijmakers, and G. T. J. van der Horst. 2002. Enhanced repair of cyclobutane pyrimidine dimers and improved UV resistance in photolyase transgenic mice. *EMBO Journal* 21:4719–4729.

Schwarz, A., S. Grabbe, Y. Aragane, K. Sandkuhl, H. Riemann, and T. A. Luger. 1996. Interleukin-12 prevents ultraviolet B-induced local immunosuppression and overcomes UVB-induced tolerance. *Journal of Investigative Dermatology* 106:1187–1191.

Schwarz, A., A. Maeda, and T. Schwarz. 2005. Induction of regulatory T cells by UV-radiation is an active process driven by UV-damaged Langerhans cells. *Journal of Investigative Dermatology* 125:A54.

Sequeira, V. B., K. M. Dixon, and R. S. Mason. 2009. Resisting the sun with vitamin D. *Immunology Endocrine & Metabolic Agents in Medicinal Chemistry* 9:129–136.

Sheehan, J. M., and J. R. Young. 2002. The sunburn cell revisited: An update on mechanistic aspects. *Photochemical & Photobiological Sciences* 1:365–377.

Siliciano, J. D., C. E. Canman, Y. Taya, K. Sakaguchi, E. Appella, and M. B. Kastan. 1997. DNA damage induces phosphorylation of the amino terminus of p53. *Genes & Development* 11:3471–3481.

Singh, S., S. Aiba, H. Manome, and H. Tagami. 1999. The effects of dexamethasone, cyclosporine, and vitamin D-$_3$ on the activation of dendritic cells stimulated by haptens. *Archives of Dermatological Research* 291:548–554.

Slominski, A. T., M. A. Zmijewski, I. Semak, T. Sweatman, Z. Janjetovic, W. Li, J. K. Zjawiony, and R. C. Tuckey. 2009. Sequential metabolism of 7-dehydrocholesterol to steroidal 5,7-dienes in adrenal glands and its biological implication in the skin. *PLoS One* 4 (2):e4309.

Smith, M. L., I. T. Chen, Q. Zhan, P. M. O'Connor, and A. J. Fornace, Jr. 1995. Involvement of the p53 tumor suppressor in repair of UV-type DNA damage. *Oncogene* 10:1053–1059.

Soehnge, H., A. Ouhtit, and O. N. Ananthaswamy. 1997. Mechanisms of induction of skin cancer by UV radiation. *Frontiers in Bioscience* 2:d538–d5351.

Sreevidya, C. S., A. Fukunaga, N. M. Khaskhely, T. Masaki, R. Ono, C. Nishigori, and S. E. Ullrich. 2010. Agents that reverse UV-induced immune suppression and photocarcinogenesis affect DNA repair. *Journal of Investigative Dermatology* 130:1428–1437.

Stapelberg, M. P. F., R. B. H. Williams, S. N. Byrne, and G. M. Halliday. 2009. The alternative complement pathway seems to be a UVA sensor that leads to systemic immunosuppression. *Journal of Investigative Dermatology* 129:2694–2701.

Tobin, D., M. van Hogerlinden, and R. Toftgard. 1998. UVB-induced association of tumor necrosis factor (TNF) receptor 1 TNF receptor-associated factor-2 mediates activation of Rel proteins. *Proceedings of the National Academy of Sciences of the United States of America* 95:565–569.

Tyrrell, R. M., and V. E. Reeve. 2006. Potential protection of skin by acute UVA irradiation—From cellular to animal models. *Progress in Biophysics & Molecular Biology* 92:86–91.

Ullrich, S. E. 2005. Mechanisms underlying UV-induced immune suppression. *Mutation Research* 571:185–205.

Ullrich, S. E., D. X. Nghiem, and P. Khaskina. 2007. Suppression of an established immune response by UVA—A critical role for mast cells. *Photochemistry and Photobiology* 83:1095–1100.

Veldman, C. M., M. T. Cantorna, and H. F. DeLuca. 2000. Expression of 1,25-dihydroxyvitamin D-$_3$ receptor in the immune system. *Archives of Biochemistry and Biophysics* 374:334–338.

Vink, A. A., F. M. Strickland, C. Bucana, P. A. Cox, L. Roza, D. B. Yarosh, and M. L. Kripke. 1996. Localization of DNA damage and its role in altered antigen-presenting cell function in ultraviolet-irradiated mice. *Journal of Experimental Medicine* 183:1491–1500.

Virag, L., and C. Szabo. 2002. The therapeutic potential of poly(ADP-ribose) polymerase inhibitors. *Pharmacological Reviews* 54:375–429.

Watters, D., K. K. Khanna, H. Beamish, G. Birrell, K. Spring, P. Kedar, M. Gatei, D. Stenzel, K. Hobson, S. Kozlov, N. Zhang, A. Farrell, J. Ramsay, R. Gatti, and M. Lavin. 1997. Cellular localisation of the ataxia-telangiectasia (ATM) gene product and discrimination between mutated and normal forms. *Oncogene* 14:1911–1921.

Widyarini, S., M. Allanson, N. L. Gallagher, J. Pedley, G. M. Boyle, P. G. Parsons, D. C. Whiteman, C. Walker, and V. E. Reeve. 2006. Isoflavonoid photoprotection in mouse and human skin is dependent on metallothionein. *The Journal of Investigative Dermatology* 126:198–204.

Wink, D. A., K. S. Kasprzak, C. M. Maragos, R. K. Elespuru, M. Misra, T. M. Dunams, T. A. Cebula, W. H. Koch, A. W. Andrews, and J. S. Allen. 1991. DNA deaminating ability and genotoxicity of nitric oxide and its progenitors. *Science* 254:1001–1003.

Wong, G., R. Gupta, K. M. Dixon, S. S. Deo, S. M. Choong, G. M. Halliday, J. E. Bishop, S. Ishizuka, A. W. Norman, G. H. Posner, and R. S. Mason. 2004. 1,25-Dihydroxyvitamin D and three low-calcemic analogs decrease UV-induced DNA damage via the rapid response pathway. *Journal of Steroid Biochemistry and Molecular Biology* 89–90:567–570.

Wood, A. W., R. L. Chang, M. T. Huang, M. Uskokovic, and A. H. Conney. 1983. 1 alpha, 25-Dihydroxyvitamin D_3 inhibits phorbol ester-dependent chemical carcinogenesis in mouse skin. *Biochemical & Biophysical Research Communications* 116:605–611.

Yang, S. L., C. Smith, J. M. Prahl, X. L. Luo, and H. F. Deluca. 1993. Vitamin D deficiency suppresses cell-mediated-immunity *in vivo*. *Archives of Biochemistry and Biophysics* 303:98–106.

Yermilov, V., J. Rubio, M. Becchi, M. D. Friesen, B. Pignatelli, and H. Ohshima. 1995. Formation of 8-nitro-guanine by the reaction of guanine with peroxynitrite *in vitro*. *Carcinogenesis* 16:2045–2050.

Youn, J. I., B. S. Park, J. H. Chung, and J. H. Lee. 1997. Photoprotective effect of calcipotriol upon skin photo-reaction to UVA and UVB. *Photodermatology, Photoimmunology & Photomedicine* 13:109–114.

Zanello, L. P., and A. W. Norman. 2004. Rapid modulation of osteoblast ion channel responses by 1α, 25(OH)2-vitamin D_3 requires the presence of a functional vitamin D nuclear receptor. *Proceedings of the National Academy of Sciences of the United States of America* 101:1589–1594.

Zasloff, M. 2006. Fighting infections with vitamin D. *Nature Medicine* 12:388–390.

Zhu, Q., M. A. Wani, M. El-Mahdy, and A. A. Wani. 2000. Decreased DNA repair efficiency by loss or disruption of p53 function preferentially affects removal of cyclobutane pyrimidine dimers from non-transcribed strand and slow repair sites in transcribed strand. *The Journal of Biological Chemistry* 275:11492–11497.

Ziegler, A., A. S. Jonason, D. J. Leffell, J. A. Simon, H. W. Sharma, J. Kimmelman, L. Remington, T. Jacks, and D. E. Brash. 1994. Sunburn and p53 in the onset of skin cancer. *Nature* 372:773–776.

Zinser, G. M., J. P. Sundberg, and J. Welsh. 2002. Vitamin D$(_3)$ receptor ablation sensitizes skin to chemically induced tumorigenesis. *Carcinogenesis* 23:2103–2109.

Zurer, I., L. J. Hofseth, Y. Cohen, M. Xu-Welliver, S. P. Hussain, C. C. Harris, and V. Rotter. 2004. The role of p53 in base excision repair following genotoxic stress. *Carcinogenesis* 25:11–19.

8 Vitamin D and Adipose Tissue: 1α, 25-Dihydroxyvitamin D₃ (Calcitriol) Modulation of Energy Metabolism, Reactive Oxygen Species, and Inflammatory Stress

Antje Bruckbauer and Michael B. Zemel

CONTENTS

8.1   INTRODUCTION

The concept of calcium and 1α, 25-dihydroxy $(OH)_2$-vitamin D_3 (calcitriol), which is the active form of vitamin D, modulation of adipocyte and energy metabolism was originally derived from an apparent "antiobesity" effect found in clinical trials, which investigated the effects of high calcium intake for other endpoints (e.g., skeletal and cardiovascular) [1–3]. Subsequently, a number of observational and epidemiological studies [3–8] demonstrated a strong inverse association between calcium intake, vitamin D levels, and relative risk of obesity, suggesting that vitamin D and calcium play a key regulatory role in adipocyte lipid metabolism. These observations are supported by a large number of experimental *in vitro* and *in vivo* studies that provide a clear mechanistic framework for both genomic and nongenomic actions of calcitriol in regulation of adipocyte function.

Alterations of the vitamin D–endocrine system are associated with obesity [9], with increased circulating levels of the active form of the vitamin (1α, 25-dihydroxyvitamin D) and decreased plasma levels of the precursor (25-hydroxyvitamin D) [10,11]. In addition, because of its fat solubility, a greater fraction of vitamin D is stored in the fat tissue in obese people, possibly resulting in decreased levels and bioavailability of circulatory vitamin D [10]. These findings, coupled with the direct effects of calcitriol on adipocyte lipid metabolism, implicate the increase in calcitriol found on low-calcium diets as a contributory factor to excess adiposity.

Finally, calcitriol modulation of adipocyte reactive oxygen species (ROSs) and inflammatory cytokine production will be discussed as contributory factors in the development of obesity comorbidities, such as cardiovascular disease, insulin resistance, and diabetes.

8.2 CALCITRIOL REGULATION OF ENERGY METABOLISM

8.2.1 RELATIONSHIP BETWEEN DIETARY CALCIUM, VITAMIN D, AND CALCITRIOL

A key function of vitamin D is tight regulation of serum calcium homeostasis. Transient decreases of circulating calcium plasma levels stimulate parathyroid hormone (PTH) secretion, which stimulates 25-hydroxyvitamin D-1α-hydrolase (1α-OHase), which, in turn, is the enzyme producing the active form 1α, 25-dihydroxy $(OH)_2$-vitamin D_3 (calcitriol) from its precursor 25-hydroxyvitamin D_3 (25-OH-D) [12]. Anderson et al. demonstrated that low calcium and vitamin D intake in rats increased renal 1α-OHase mRNA expression and contributed to 12–18-fold increased levels of serum 1α, 25-dihydroxy $(OH)_2$-vitamin D_3 [13]. Although 1α-OHase is predominately expressed in the kidney and regulated by feedback inhibition by calcitriol, its activity has also been detected in numerous other tissues including adipose tissue and macrophages [14,15]. Moreover, extrarenal 1α-OHase appears to be differently regulated than the kidney enzyme, resulting in substrate (25-OH-D) being the primary determinant of 1α, 25-dihydroxy $(OH)_2$-vitamin D_3 in extrarenal tissues [16]. This local production of 1α, 25-dihydroxy $(OH)_2$-vitamin D_3 in extrarenal tissues is believed not to flow into the circulatory system and therefore not to measurably increase the plasma concentration of 1α, 25-dihydroxy $(OH)_2$-vitamin D_3; however, it generates local biological responses and therefore plays an important role for paracrine and autocrine functions [14,15].

8.2.2 VITAMIN D STATUS AND OBESITY

A growing body of evidence indicates that the vitamin D endocrine system is related to obesity. Population-based and case-control studies indicate lower levels of serum 25-hydroxyvitamin D and higher levels of PTH and 1α, 25-dihydroxy $(OH)_2$-vitamin D_3 in obese, compared to leaner, subjects [5,17,18]. Although it is not clear whether adiposity results from low vitamin D levels or whether low vitamin D levels are a consequence of obesity, it has been suggested that causes are less sun exposure due to limited mobility, as well as higher storage of vitamin D in adipose tissue [10,18]. Nonetheless, weight loss of 10% increased 25-hydroxyvitamin D levels in obese women [19], whereas both increased dairy calcium intake and increased serum 25-hydroxyvitamin D levels during a weight loss intervention period were independently significantly associated with successful weight loss at 24 mo [5].

8.2.3 CALCIUM PARADOX

Low dietary calcium levels generally result in a paradoxical increase in intracellular calcium. This "calcium paradox" results from PTH and calcitriol stimulation of rapid Ca^{2+} influx in many cell types, thereby increasing the intracellular calcium concentration [20–22]. This rapid response (within seconds to minutes) results from nongenomic action on calcium channels, as discussed later in this chapter, and results predominately from Ca^{2+} influx from extracellular calcium sources [20].

8.2.4 Ca^{2+} Signaling in Adipose Tissue

Ca^{2+} signaling is a universal intracellular messenger system and controls many cell processes, such as proliferation, secretion, and contraction. Earlier work on the agouti gene demonstrated that Ca^{2+} signaling also plays an important role in adipocyte lipid metabolism. In mice, the agouti gene is primarily expressed in melanocytes of the skin and hair follicle and plays a role in pigmentation patterns of the hair by antagonizing the binding of α-melanocyte-stimulating hormone to its receptor, which produces a predominately black hair shaft with a subapical yellow band. However, dominant mutations of the agouti gene cause ectopic and ubiquitous expression of agouti characterized by pleiotropic effects including yellow coat color, obesity, and insulin resistance [23–24]. The human agouti homolog is normally expressed in white adipose tissue suggesting a role in lipid metabolism. Indeed, agouti protein–stimulated lipogenesis by activation of fatty acid synthase (FAS), which is the key regulatory enzyme of lipogenesis, and an agouti/Ca^{2+} response sequence in the FAS promoter region has been demonstrated [25–27]. Further, agouti-protein-inhibited lipolysis and increasing Ca^{2+} influx through either voltage-or receptor-operated Ca^{2+} channels produced the same effects, whereas Ca^{2+} channel antagonists blocked these effects, indicating a Ca^{2+}-dependent mechanism [26–30]. This antilipolytic effect was further demonstrated to be mediated via Ca^{2+}-induced activation of phosphodiesterase 3B, resulting in reduced cAMP levels and, consequently, in inhibition of hormone sensitive lipase activity [30]. In addition, intracellular Ca^{2+} plays a role in adipocyte differentiation in a biphasic manner with inhibitory effects in the early stages of adipogenesis and stimulatory effects in the late stages of adipocyte differentiation and lipid filling [31]. Thus, a sustained increase in intracellular calcium favors a smaller number of hypertrophic adipocytes.

These studies demonstrated a key role for intracellular Ca^{2+} and calcitrophic hormones in the regulation of adipocyte metabolism, which also provides the primary mechanistic basis for the antiobesity effect of dietary calcium. We have found calcitriol to stimulate rapid increases in human adipocyte intracellular Ca^{2+} via a nongenomic membrane vitamin D receptor (VDR, also referred to as 1,25-D3-MARRS, i.e., membrane-associated rapid-response steroid binding protein), which later was found to be identical with the multifunctional protein ERp57/GRp58 [3,22,32–34]. Accordingly, the increase in calcitriol in response to low-calcium diets stimulates adipocyte Ca^{2+}-influx and consequently promotes adiposity by inhibiting lipolysis and stimulating lipogenesis [35]; conversely, suppression of calcitriol by high-calcium diets exert the opposite effects by decreasing the intracellular Ca^{2+} concentration [22]. This "calcium paradox" of increasing dietary calcium resulting in decreases in intracellular Ca^{2+} is mediated by suppression of calcitriol and has been confirmed in multiple cellular, animal, and clinical studies, which also demonstrate that an increase in dietary calcium intake is associated with significant reductions in adipose tissue mass in the absence of caloric restriction and augments weight and body fat loss under caloric restriction [3,35–38].

8.2.5 Role of Adipocyte nVDR

In addition to regulating adipocyte metabolism via a nongenomic membrane receptor, calcitriol also acts via the "classical" nuclear vitamin D receptor (nVDR) in adipocytes. The nVDR belongs to the nuclear receptor superfamily expressed in most tissues and is responsible for the transcriptional regulation of many cell processes by binding directly either as a homodimer or heterodimer to specific response elements (VDREs) or by interacting with other DNA-binding transcription factors [39]. Interestingly, polymorphisms in the nVDR gene are associated with the susceptibility to obesity in humans, suggesting a role for nVDR in adipose tissue [40,41].

The nVDR is responsible for mediating the effects of calcitriol on preadipocyte differentiation. Interestingly, the expression of the nVDR in preadipocytes is very low but is transiently stimulated during early adipogenesis with a maximum achieved at 4–8 h and returning back to baseline levels over the next two days [42]. This appears to be a window for the inhibitory actions of calcitriol on the early adipogenic processes since calcitriol added to the differentiation medium after 48 h of initiation of the adipogenic program was not able to block adipocyte differentiation and instead stimulated adipocyte differentiation and lipid filling [43]. In addition, the unliganded nVDR protein itself exerts inhibitory effects on adipogenesis, as overexpression of hVDR by adenovirus in 3T3-L1 adipocytes blocked adipocyte differentiation completely even in the absence of calcitriol. However, in the presence of calcitriol, the VDR protein appeared more stabilized particularly during the late stages of adipogenesis, thus preventing the decline of VDR [44].

In mature adipocytes, the expression of nVDR is low but inducible by multiple compounds, including calcitriol and glucocorticoids [45], as discussed later in this chapter. The nVDR is important for calcitriol effects on adipocyte function by inhibiting the expression of uncoupling protein 2 (UCP2) [46]. The uncoupling proteins belong to a family of mitochondrial carrier proteins present in the inner membrane of mitochondria. UCP1, which is the first UCP family member identified, is mainly expressed in brown adipose tissue and has been shown to stimulate mitochondrial proton leak during oxidative ATP generation, resulting in uncoupling of ATP synthesis from the mitochondrial chain and dissipation of energy as heat [47]. The role of the other homologues to UCP1 (UCP2-5) is still not fully understood. However, in addition to its possible role in thermogenesis, UCP2, which is ubiquitously expressed at high levels in macrophages and white adipose tissue, and UCP3, which is the main form in skeletal muscle, appear to play a role in fatty acid metabolism, ROS production, and inflammation [48]. Calcitriol acts via the nVDR to inhibit UCP2 expression and to suppress UCP2 responses to stimulants such as isoproterenol and fatty acids [46]. These effects were prevented by antisense oligonucleotide-mediated nVDR knock-out in adipocytes but not by 1,25(OH)$_2$D-MARRS antagonists. Consistent with these observations, VDR-null mutant mice accumulated less body fat mass than wild-type mice when fed a high-fat diet and exhibited higher rates of adipose fatty acid β-oxidation and higher expression of UCP1, UCP2, and UCP3 suggesting higher energy expenditure [49]. Similarly, suppression of calcitriol by feeding high calcium diets to mice results in increased adipose tissue UCP2 and skeletal muscle UCP3 expression and attenuation of the decline in thermogenesis, which otherwise occurs with energy restriction [35,38]. Thus, despite the undetermined role of UCP2 in thermogenesis, UCP2 serves also to mediate mitochondrial fatty acid transport and oxidation, suggesting that calcitriol suppression of UCP2 expression contributes to decreased fat oxidation and increased energetic efficiency, thereby increasing lipid accumulation and obesity risk on low-calcium diets.

8.2.6 REGULATION OF ADIPOCYTE APOPTOSIS

Calcitriol may also modulate energy metabolism by inhibiting adipocyte apoptosis [50]. This effect is mediated in part via inhibition of UCP2 expression and a consequent increase in mitochondrial potential, which is a key regulator of apoptosis, and in part via calcitriol regulation of cytosolic Ca^{2+} and of Ca^{2+} flux between endoplasmic reticulum and mitochondria [51]. The effects of calcitriol are dose dependent. While high concentrations (≥100 nM) of calcitriol are proapoptotic, which is consistent with the effects of calcitriol in several other tissues [52–54], lower levels in the physiological range (0.1–10 nM) exert inhibitory effects on apoptosis by suppressing proapoptotic genes such as caspase-1 and caspase-3, and stimulating antiapoptotic genes such as BCL-2 and increasing the BCL-2/BAX ratio [55]. Furthermore, calcitriol dose-dependently induced mitochondrial potential (Δψ) and ATP production, whereas overexpression of UCP2 in adipocytes exerted the opposite effects, indicating that suppression of UCP2 expression and consequent increases in mitochondrial potential and ATP production may contribute to its antiapoptotic effects [55]. In contrast, the apoptosis induced by pharmacological doses of calcitriol (≥100 nM) is associated with an increase in mitochondrial Ca^{2+}-levels [55]. Since mitochondrial are often located in close spatial relationship to the endoplasmatic reticulum (ER), they are often exposed to higher concentration of Ca^{2+} than the remaining cytosol

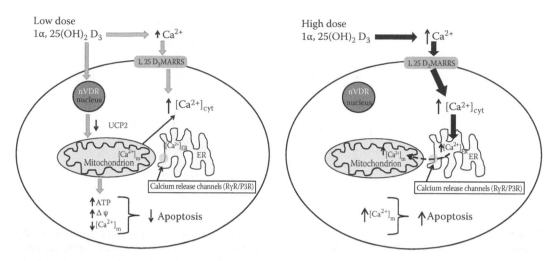

FIGURE 8.1 1,25-(OH)$_2$-D modulation of apoptosis. Low (physiological) doses of 1,25-(OH)$_2$-D inhibit apoptosis by inhibiting UCP2 and thereby restore UCP2-induced loss of mitochondrial potential and ATP reduction, protecting adipocytes from apoptotic death. The reduction in mitochondrial Ca^{2+} levels found with lower doses of 1,25-(OH)$_2$-D$_3$ may further contribute to this antiapoptotic effect. In contrast, high levels of 1,25-(OH)$_2$-D$_3$ cause markedly greater increases in cytosolic [Ca^{2+}]$_{cyt}$, probably resulting in increased [Ca^{2+}]$_{ER}$. Since ER can open their Ca^{2+} release channels in response to elevations in [Ca^{2+}]$_{cyt}$ and contribute to Ca^{2+}-induced Ca^{2+} release (CICR), the high Ca^{2+} levels achieved at these contact sites favors Ca^{2+} uptake into mitochondria. Calcium overload in mitochondria, in turn, triggers apoptosis.

through the Ca^{2+} release by the inositol-1,4,5-triphosphate receptor (IP3R) and ryonidine receptor (RyR) of the ER. These contact sites favor Ca^{2+} uptake into mitochondria, thus making them highly susceptible to abnormalities in Ca^{2+} signaling [56]. Therefore, the proapoptotic effects of supra-physiologic high calcitriol doses are mainly mediated by apoptosis due to mitochondrial Ca^{2+} overload, whereas the antiapoptotic effects of calcitriol at physiological concentrations are primarily caused by suppression of UCP2 expression (Figure 8.1).

Supporting evidence for the effects of calcitriol on adipocyte apoptosis comes also from a microarray study of human adipocytes showing that physiological concentrations of calcitriol suppressed the proapoptotic gene stanniocalcin (STC2) but stimulated the antiapoptotic gene STC1 [57]. In addition, *in vivo* data demonstrate that suppression of calcitriol by high-calcium diets resulted in significant increases in white adipose tissue apoptosis in diet-induced obesity, thus providing further evidence for this concept [55].

8.2.7 MODULATION OF ADIPOCYTE GLUCOCORTICOID PRODUCTION

Calcitriol participates also in regulation of energy metabolism by altering adipose tissue fat deposition and expansion. Although obesity is not associated with increased glucocorticoid levels in general, excessive visceral adiposity may result from local increases in glucocorticoid production. This is under the control of 11β-hydroxysteroid dehydrogenase (11β-HSD), which is the enzyme that generates active cortisol from inactive cortisone, leading to increased intracellular levels of glucocorticoids [58]. Human adipose tissue expresses significant amounts of 11β-hydroxysteroid dehydrogenase-1 (11β-HSD-1) [59,60]. Moreover, the expression of 11β-HSD-1 is greater in visceral than in subcutaneous adipose tissue, resulting in greater glucocorticoid production in visceral than in subcutaneous adipose tissue [61]. This may lead to greater visceral fat depot expansion and altered metabolic response since glucocorticoids influence many adipocyte functions, including preadipocyte differentiation, fat accumulation, insulin sensitivity, and cytokine production [62]. Accordingly, selective overexpression of 11-β-HSD-1 in white adipose tissue of mice results in

Regulation of calcitriol on nVDR in adipocytes

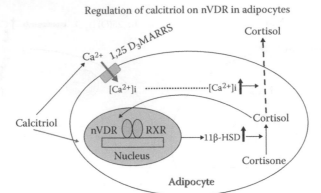

FIGURE 8.2 Calcitriol regulation of adipocyte nVDR expression and glucocorticoid production [45]. Positive feedback between calcitriol stimulated glucocorticoid production, and nVDR leads to augmented glucocorticoid production; suppression of calcitriol reduces adipose tissue cortisol production and limits expansion of visceral adipose tissue.

central obesity with features of the metabolic syndrome [63,64], whereas homozygous 11β-HSD-1 knockout mice exhibit protection from features of the metabolic syndrome [65].

Calcitriol has been shown to exert both short- and long-term effects on 11β-HSD-1 and cortisol release in human adipocytes, resulting in approximately twofold increases in 11β-HSD-1 expression and up to sixfold increases in net cortisol production [66]. In addition, data analysis from a microarray study in human adipocytes demonstrated an additional upregulation of 11β-HSD expression by cortisol when combined with calcitriol [67], which may result from an indirect positive feedback loop of cortisol on its own production. Cortisol increases the expression of the nVDR in adipocytes, leading to augmented action of calcitriol on 11β-HSD expressions and, thereby, to increased cortisol production and release. Furthermore, it stimulates cortisol release independent from the nVDR but instead mediated by intracellular calcium concentration (Figure 8.2) [45]. Consequently, suppressing calcitriol levels via high-calcium diets results in a corresponding decrease in adipose tissue cortisol production and thereby contributes to reduced visceral adipose tissue mass.

8.2.8 REGULATION OF ENERGY PARTITIONING BETWEEN ADIPOCYTE AND SKELETAL MUSCLE

Calcium and calcitriol play a role in energy metabolism by regulating crosstalk between adipose tissue and skeletal muscle via modulation of adipocytokine expression and production. Adipose tissue not only functions as a large energy reserve that provides fuel in the form of fatty acids for metabolic processes in other tissues such as muscle, but it also serves as an active endocrine organ that secretes a variety of signaling molecules. These serve to regulate interactions between adipocytes and other tissues and therefore play an important role in the regulation of glucose and lipid metabolism, as well as in the regulation of energy balance [68].

Suppression of calcitriol by dietary calcium, particularly when derived from dairy products, has been demonstrated to reduce adiposity and to promote lean tissue mass in multiple studies [35,37,69], indicating a role of calcium and calcitriol on energy partitioning between adipose tissue and skeletal muscle. Indeed, calcitriol significantly decreased fatty acid oxidation and UCP3 expression in muscle cells and concurrently increased FAS and PPAR-gamma gene expression in differentiated 3T3-L1 adipocytes, whereas the calcium-channel-blocker nifedipine partially inhibited these effects [70]. In addition, calcitriol decreased mitochondrial biogenesis and associated regulatory gene expression in both myocytes and adipocytes [71]. Comparable effects were also found in myocytes treated with conditioned medium derived from adipocytes, indicating that one or more factors derived from adipocytes regulated skeletal muscle energy metabolism [70,71]. Interestingly,

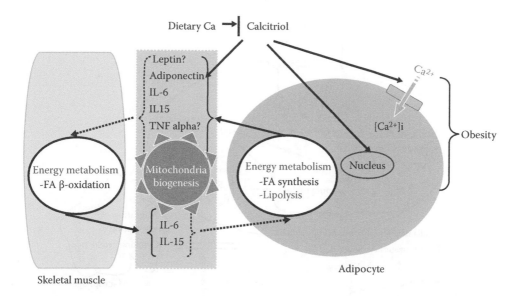

FIGURE 8.3 Calcitriol regulation of adipocyte-muscle crosstalk: Calcitriol coordinately controls adipocyte and skeletal muscle lipid metabolism by modulating production and secretion of adipocytokines. Accordingly, suppression of calcitriol results in increased mitochondrial biogenesis in adipose and muscle tissue, and increased lipolysis in adipose tissue provides the fatty acids for the increased fatty acid β-oxidation in muscle.

it was found that calcitriol inhibited adiponectin production in adipocytes, which was reversed by addition of nifedipine, whereas adiponectin treatment, in turn, increased interleukin (IL)-15 and IL-6 release by myotubes and partially reversed the inhibitory effects of calcitriol [70]. Thus, the calcitriol modulation of adipocyte-skeletal muscle crosstalk may be mediated, at least in part, by alterations in adipocyte adiponectin and skeletal muscle IL-15 and IL-6 response; however, additional studies are necessary to identify further specific adipokines (Figure 8.3).

8.3 CALCITRIOL REGULATION OF ADIPOCYTE OXIDATIVE STRESS AND INFLAMMATORY STRESS

8.3.1 REGULATION OF ROS PRODUCTION

Adipose tissue is a major site of ROS production, and oxidative stress is increased in obesity and associated disorders [72]. Although most of the ROS in cells is produced in the mitochondria, ROS-generating enzymes such as NADPH oxidase and myeloperoxidase also play an important role. Calcium signaling is critical for ROS production, as increases in intracellular Ca^{2+} concentration ($[Ca^{2+}]_i$) leads to activation of ROS-generating enzymes. However, there is also a bidirectional interaction between ROS and $[Ca^{2+}]_i$, wherein increased ROS production may increase $[Ca^{2+}]_i$ by activating calcium channels of the plasma membrane and of the ER while manipulation of Ca^{2+} signaling regulates cellular ROS production [73].

It has been demonstrated that calcitriol stimulates ROS production in adipose tissue via both genomic and nongenomic actions [74]. The calcitriol-induced increase in intracellular Ca^{2+} resulted in an increased expression of NADPH oxidase in human and murine adipocytes [75], whereas suppression of calcitriol by dietary calcium inhibited adipocyte NADPH oxidase expression and ROS production in aP2-transgenic mice [76]. Interestingly, greater visceral adiposity predisposes to enhanced ROS production since NADPH oxidase is markedly higher expressed in visceral than in subcutaneous fat [61]. Accordingly, suppression of calcitriol by dietary calcium resulted in greater reduction in ROS in visceral fat than in subcutaneous fat.

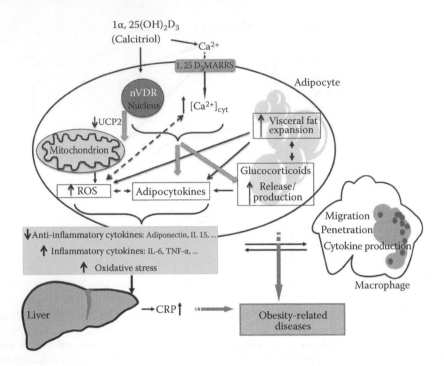

FIGURE 8.4 Effects of calcitriol on adipocyte ROS and inflammatory cytokine production. Calcitriol decreases UCP2 and increases cytosolic Ca^{2+} in adipocytes, resulting in increased ROS and altered adipocytokine production. Glucocorticoid production and release is also augmented, causing increased visceral fat expansion, resulting in further dysregulation of adipocytokines. The additional bidirectional interactions between ROS, adipocytokines, and Ca^{2+} further enhance oxidative and inflammatory stress. Together, these effects result in obesity-related diseases.

Mitochondria are the major site of cellular ROS production. Mitochondrial ROS production is dependent on mitochondrial potential, which is determined by the backflow of H^+ across the mitochondrial membrane. Thus, uncoupling of the electron transport chain by uncoupling proteins leads to reduced mitochondrial potential and, consequently, to diminished ROS production. Accordingly, calcitriol suppression of adipocyte UCP2 expression and the resultant increase in mitochondrial potential enhances mitochondrial ROS production [75]. Calcitriol stimulation of adipocyte mitochondrial ROS is augmented by addition of a mitochondrial uncoupler inhibitor and reversed by overexpression of UCP2 [65]. The effects of calcitriol on ROS production in adipocytes are illustrated in Figure 8.4.

8.3.2 REGULATION OF ADIPOCYTE INFLAMMATORY CYTOKINE PRODUCTION

In addition to stimulating ROS production in adipose tissue, calcitriol also increases the production of adipocyte-derived pro inflammatory cytokine production. Adipose tissue is a key source of bioactive molecules, which includes anti-inflammatory adipokines, such as adiponectin and IL-15, as well as pro-inflammatory cytokines such as tumor necrosis factor alpha (TNF-α), Interleukin (IL)-6, and IL-8 [77–79]. Dysregulation of these cytokines play a central role in the pathogenesis of inflammatory and metabolic diseases associated with obesity [80]. In addition, TNF-α appears to be an important activator of ROS production. Most of this effect is mediated by inducing NADPH oxidase gene expression and activity [81–83]. Increased ROS, in turn, stimulates the NF-κB-dependent activation of pro-inflammatory cytokine expression including TNF-α [84]. Therefore, this bidirectional interaction may act as a vicious cycle under conditions of either enhanced ROS or cytokine production as seen in obesity, particularly in combination with low-calcium diets.

Calcitriol has been shown to stimulate the expression of TNF-α, IL-6, and Il-8 in differentiated adipocytes; this effect is blocked by a calcium channel antagonist, suggesting that the calcitriol-induced Ca^{2+} influx is the key mediator of this effect [85]. Conversely, suppression of calcitriol by dietary-calcium-attenuated obesity-induced inflammatory stress in mouse models of obesity by inhibiting the expression of pro-inflammatory factors such as TNF-α and IL-6 in visceral fat and stimulating the expression of the anti-inflammatory factors such as IL-15 and adiponectin [76,85]. Furthermore, clinical data demonstrate that high-calcium diets result in suppression of key inflammatory stress markers (circulating C-reactive protein, TNF-α, IL-6, and monocyte chemotactic protein-1) and in an increase of adiponectin levels during both weight loss and weight maintenance in obese subjects [76,86]. These effects were found within seven days of diet initiation and increased with the duration of the diet. Moreover, recent results of high-calcium diet feeding in the form of dairy in metabolic syndrome patients with elevated oxidative and inflammatory stress markers showed significant improvements of their metabolic state (25%–35% decrease in biomarkers of oxidative stress and 35%–55% decrease in biomarkers of inflammatory stress), which were independent of dairy-induced adiposity changes [87]. These findings indicate an important role of calcitriol in inducing both oxidative and inflammatory stress. Thus, suppression of calcitriol by dietary calcium may attenuate obesity-induced oxidative and inflammatory stress and consequently may suppress key mediators of obesity-induced disease and morbidity (Figure 8.4).

8.3.3 Calcitriol Regulation of Adipocyte-Macrophage Crosstalk in Inflammation

In addition to adipocytes, adipose tissue consists of a variety of other cell types from the stromal/vascular fraction such as blood cells, endothelial cells, and macrophages [88,89]. The recruited macrophages from the stromal/vascular fraction, in particular, play a major role in adipose-derived cytokine production. First, they provide an additional source of cytokine production, and second, there is a bidirectional interaction between adipocytes and macrophages. While infiltration and differentiation of adipose-tissue resident macrophages are under the local control of chemokines, which are, in large part, produced by adipocytes, macrophage-produced cytokines also stimulate adipogenesis and adipocyte inflammatory response, thus aggravating inflammatory changes in adipose tissue [89,90].

Calcitriol's effects on the immune response are multiplex because they are dose and time dependent, as well as cell-type specific. Its importance as a regulator of the immune system is further suggested by the fact that activated macrophages not only express the VDR but also possess the necessary enzyme (1α-OHase) to synthesize and secret calcitriol [91]. Most literature reports an overall immune-suppressive effect of calcitriol in higher concentrations (40–100 nM) [92,93]. In monocytes, calcitriol induces cell growth inhibition and differentiation to a monocyte-macrophage phenotype at near-physiological concentrations [94,95]. This was associated with an increase in phagocytic activity and activation of the oxidase system for production of superoxide [91]. This response may be important for the immune defense in the presence of an infection; however, since obesity is already associated with a low-grade inflammatory state, activation of adipose-tissue-derived macrophages may lead to an inappropriate increase in oxidative stress and inflammatory cytokine production. In fact, our data show that calcitriol at physiological concentrations (10 nM) induces the production of an array of inflammatory cytokines, including macrophage surface-specific protein CD14, macrophage inhibitory factor, macrophage colony-stimulating factor, macrophage inflammatory protein, and monocyte chemoattractant protein-1 in both adipocytes and adipose-tissue-derived macrophages. In addition, calcitriol augmented the crosstalk between adipocytes and macrophages, resulting in an augmentation of inflammatory cytokine production [96]. Both the calcium-channel antagonist nifedipine and the mitochondrial uncoupler dinitrophenol were able to block these effects, suggesting that the calcitriol-induced inflammatory cytokine production involves both calcium-dependent and mitochondrial uncoupling-dependent mechanisms (Figure 8.4).

Finally, there is evidence that calcitriol not only augments adipocyte/macrophage crosstalk and, consecutively, inflammatory response in adipose tissue but also increases adherence and infiltration of monocytes across vascular endothelium, which is an early step in atherosclerosis. When monolayers of human endothelial cells were perfused with fluorescently labeled human monocytic cells, which were incubated with conditioned media from adipocytes treated either with vehicle or calcitriol for 48 h, calcitriol increased monocyte adherence by approximately twofold (~200%) [97]. These data further indicate that food rich in calcium may be beneficial in reducing obesity-induced inflammatory response and, consequently, obesity-associated co-morbidities.

8.4 VITAMIN D SUPPLEMENTATION AND OBESITY

Since obesity is associated with lower serum vitamin D levels, supplementation has been proposed to be beneficial for improving weight, immunity, and bone function, as well as overall health in overweight and obese people. However, there is a paucity of randomized clinical trials of vitamin D with weight, cardiovascular, and/or immune function as primary outcomes. Therefore, recommendations for other health benefits than bone function are more based on observational studies.

Weight loss is positively correlated with increased serum vitamin D levels, whereas baseline concentrations of serum vitamin D do not appear to be predictive of subsequent weight loss [19,98]. Additionally, supplemental vitamin D (20000 IU/twice a week) in the absence of caloric restriction exerted no effect in vitamin D–replete individuals in a one-year intervention study [99]. In contrast, secondary analysis of food consumption for vitamin D and calcium in a two-year weight loss intervention suggested that higher consumption of both calcium and vitamin D was associated with improved weight loss over the two-year period [5]. Zittermann et al. reported no effects of vitamin D supplementation (83 ug/d) on weight loss in a 12-month study of overweight subjects; however, it did result in improvements in both triglycerides and serum TNF-α [101]. These effects appear to be dependent on baseline vitamin D status, as no effect of vitamin D (20,000 or 40,000 IU/ week) on inflammatory markers was found in overweight and obese subjects with adequate serum 25-hydroxy-vitamin D levels at baseline [100]. Furthermore, since higher adiposity is generally associated with lower vitamin D levels, it is possible that higher doses or longer duration of supplementation will be required in obese individuals [102].

8.5 CONCLUSION

Calcitriol has been shown to regulate energy metabolism by modulating adipocyte function via both nongenomic and genomic actions. Nongenomic actions are mediated by a rapid increase in intracellular calcium concentration that results in stimulated lipogenesis and inhibited lipolysis. Genomic action involves calcitriol interaction with the nVDR, resulting in suppression of UCP2, which causes an increase in energy efficiency. In addition, calcitriol regulates adipocyte apoptosis in a dose-dependent manner and increases visceral adipose tissue fat depot location and expansion by promoting glucocorticoid production and release. It also mediates adipocyte and skeletal muscle interactions that influences metabolic flexibility. Lastly, calcitriol plays an important role in production of inflammatory and oxidative stress by augmenting adipocyte/macrophage crosstalk, resulting in increased adipose-derived pro-inflammatory cytokine production, which consecutively may play a role for the development of obesity-co-morbidities.

REFERENCES

1. Heaney, R. P., K. M. Davies, and M. J. Barger-Lux. 2002. Calcium and weight: Clinical studies. *J Am Coll Nutr* 21:152S–155S.
2. Davies, K. M., R. P. Heaney, R. R. Recker, J. M. Lappe, M. J. Barger-Lux, K. Rafferty, and S. Hinders. 2000. Calcium intake and body weight. *J Clin Endocrinol Metab* 85:4635–4638.

3. Zemel, M., H. Shi, B. Greer, D. Dirienzo, and P. Zemel. 2000. Regulation of adiposity by dietary calcium. *FASEB J* 14:1132–1138.
4. Loos, R. J., T. Rankinen, A. S. Leon, J. S. Skinner, J. H. Wilmore, D. C. Rao, and C. Bouchard. 2004. Calcium intake is associated with adiposity in Black and White men and White women of the HERITAGE Family Study. *J Nutr* 134:1772–1778.
5. Shahar, D. R., D. Schwarzfuchs, D. Fraser, H. Vardi, J. Thiery, G. M. Fiedler, M. Bluher, M. Stumvoll, M.J. Stampfer, and I. Shai. 2010. Dairy calcium intake, serum vitamin D, and successful weight loss. *Am J Clin Nutr* 92:1017–1022.
6. Martini, L. A. and R. J. Wood. 2006. Vitamin D status and the metabolic syndrome. *Nutr Rev* 64:479–486.
7. Cheng, S., J. M. Massaro, C. S. Fox, M. G. Larson, M. J. Keyes, E. L. McCabe, S. J. Robins, C. J. O'Donnell, U. Hoffmann, et al. 2010. Adiposity, cardiometabolic risk, and vitamin D status: The Framingham Heart Study. *Diabetes* 59:242–248.
8. Albertson, A. M., C. K. Good, N. M. Holschuh, and E. L. Eldridge. 2004. The relationship between dietary calcium intake and body mass index in adult women from three National dietary intake databases. *FASEB J* 18:Abstract 6259.
9. Bell, N. H., S. Epstein, A. Greene, J. Shary, M. J. Oexmann, and S. Shaw. 1985. Evidence for alteration of the vitamin D–endocrine system in obese subjects. *J Clin Invest* 76:370–373.
10. Wortsman, J., L. Y. Matsuoka, T. C. Chen, Z. Lu, and M. F. Holick. 2000. Decreased bioavailability of vitamin D in obesity. *Am J Clin Nutr* 72:690–693.
11. Bortolotti, M., S. Rudelle, P. Schneiter, H. Vidal, E. Loizon, L. Tappy, and K. J. Acheson. 2008. Dairy calcium supplementation in overweight or obese persons. Its effect on markers of fat metabolism. *Am J Clin Nutr* 88:877–885.
12. Henry, H. L., C. Dutta, N. Cunningham, R. Blanchard, R. Penny, C. Tang, G. Marchetto, and S. Y. Chou. 1992. The cellular and molecular regulation of 1,25(OH)2D3 production. *J Steroid Biochem Mol Biol* 41:401–407.
13. Anderson, P. H., A. M. Lee, S. M. Anderson, R. K. Sawyer, P. D. O'Loughlin, and H. A. Morris. 2010. The effect of dietary calcium on 1,25(OH)2D3 synthesis and sparing of serum 25(OH)D3 levels. *J Steroid Biochem Mol Biol* 121:288–292.
14. Norman, A. W. 2008. From vitamin D to hormone D: Fundamentals of the vitamin D endocrine system essential for good health. *Am J Clin Nutr* 88:491S–499S.
15. Li, J., M. E. Byrne, E. Chang, Y. Jiang, S. S. Donkin, K. K. Buhman, J. R. Burgess, and D. Teegarden. 2008. 1alpha,25-Dihydroxyvitamin D hydroxylase in adipocytes. *J Steroid Biochem Mol Biol* 112:122–126.
16. Hewison, M., D. Zehnder, R. Bland, and P. M. Stewart. 2000. 1alpha-Hydroxylase and the action of vitamin D. *J Mol Endocrinol* 25:141–148.
17. Grethen, E., R. McClintock, C. E. Gupta, R. Jones, B. M. Cacucci, D. Diaz, A. D. Fulford, S. M. Perkins, R. V. Considine, and M. Peacock. 2011. Vitamin D and hyperparathyroidism in obesity. *J Clin Endocrinol Metab* 96:1320–1326.
18. Snijder, M. B., R. M. van Dam, M. Visser, D. J. Deeg, J. M. Dekker, L. M. Bouter, J. C. Seidell, and P. Lips. 2005. Adiposity in relation to vitamin D status and parathyroid hormone levels: A population-based study in older men and women. *J Clin Endocrinol Metab* 90:4119–4123.
19. Tzotzas, T., F. G. Papadopoulou, K. Tziomalos, S. Karras, K. Gastaris, P. Perros, and G. E. Krassas. 2010. Rising serum 25-hydroxy-vitamin D levels after weight loss in obese women correlate with improvement in insulin resistance. *J Clin Endocrinol Metab* 95:4251–4257.
20. Uchida, Y., T. Endoh, Y. Shibukawa, M. Tazaki, and K. Sueishi. 2010. 1alpha,25-dihydroxyvitamin D rapidly modulates Ca(2+) influx in osteoblasts mediated by Ca(2+) channels. *Bull Tokyo Dent Coll* 51:221–226.
21. Picotto, G. 2001. Rapid effects of calciotropic hormones on female rat enterocytes: Combined actions of 1,25(OH)2-vitamin D3, PTH and 17beta-estradiol on intracellular Ca2+ regulation. *Horm Metab Res* 33:733–738.
22. Shi, H., A. Norman, W. Okamura, A. Sen, and M. Zemel. 2001. 1alpha,25-Dihydroxyvitamin D3 modulates human adipocyte metabolism via nongenomic action. *FASEB J* 15:2751–2753.
23. Michaud, E., R. Mynatt, R. Miltenberger, M. Klebig, J. Wilkinson, M. Zemel, W. Wilkison, and R. Woychik. 1997. Role of the agouti gene in obesity. *J Endocrinol* 155:207–209.
24. Zemel, M. B., J. H. Kim, R. P. Woychik, E. J. Michaud, S. H. Kadwell, I. R. Patel, and W. O. Wilkison. 1995. Agouti regulation of intracellular calcium: role in the insulin resistance of viable yellow mice. *Proc Natl Acad Sci USA* 92:4733–4737.
25. Moustaid, N., K. Sakamoto, S. Clarke, R. S. Beyer, and H. S. Sul. 1993. Regulation of fatty acid synthase gene transcription: Sequences that confer a positive insulin effect and differentiation-dependent expression in 3T3-L1 preadipocytes are present in the 332 bp promoter. *Biochem J* 292 (Pt 3):767–772.

26. Claycombe, K., Y. Wang, B. Jones, S. Kim, W. Wilkison, M. Zemel, J. Chun, and N. Moustaid-Moussa. 2000. Transcriptional regulation of the adipocyte fatty acid synthase gene by agouti: interaction with insulin. *Physiol Genomics* 3:157–162.

27. Xue, B. and M. Zemel. 2000. Relationship between human adipose tissue agouti and fatty acid synthase (FAS). *J Nutr* 130:2478–2481.

28. Kim, J., L. Kiefer, R. Woychik, W. Wilkison, A. Truesdale, O. Ittoop, D. Willard, J. Nichols, and M. Zemel. 1997. Agouti regulation of intracellular calcium: Role of melanocortin receptors. *Am J Physiol* 272:E379–E384.

29. Xue, B. 1998. The agouti gene product inhibits lipolysis in human adipocytes via a Ca2+-dependent mechanism. *FASEB J* 12:1391–1396.

30. Xue, B., A. Greenberg, F. Kraemer, and M. Zemel. 2001. Mechanism of intracellular calcium ([Ca2+]i) inhibition of lipolysis in human adipocytes. *FASEB J* 15:2527–2529.

31. Shi, H., Y. Halvorsen, P. Ellis, W. Wilkison, and M. Zemel. 2000. Role of intracellular calcium in human adipocyte differentiation. *Physiol Genomics* 3:75–82.

32. Nemere, I., S. E. Safford, B. Rohe, M. M. De Souza, and M. C. Farach-Carson. 2004. Identification and characterization of 1,25D3-membrane-associated rapid response, steroid (1,25D3-MARRS) binding protein. *J Steroid Biochem Mol Biol* 89–90:281–285.

33. Nemere, I., M. C. Farach-Carson, B. Rohe, T. M. Sterling, A. W. Norman, B. D. Boyan, and S. E. Safford. 2004. Ribozyme knockdown functionally links a 1,25(OH)2D3 membrane binding protein (1,25D3-MARRS) and phosphate uptake in intestinal cells. *Proc Natl Acad Sci USA* 101:7392–7397.

34. Khanal, R. C. and I. Nemere. 2007. The ERp57/GRp58/1,25D3-MARRS receptor: Multiple functional roles in diverse cell systems. *Curr Med Chem* 14:1087–1093.

35. Shi, H., D. Dirienzo, and M. Zemel. 2001. Effects of dietary calcium on adipocyte lipid metabolism and body weight regulation in energy-restricted aP2-agouti transgenic mice. *FASEB J* 15:291–293.

36. Zemel, M., J. Richards, A. Milstead, and P. Campbell. 2005. Effects of calcium and dairy on body composition and weight loss in African-American adults. *Obes Res* 13:1218–1225.

37. Zemel, M., W. Thompson, A. Milstead, K. Morris, and P. Campbell. 2004. Calcium and dairy acceleration of weight and fat loss during energy restriction in obese adults. *Obes Res* 12:582–590.

38. Sun, X. and M. Zemel. 2004. Calcium and dairy products inhibit weight and fat regain during ad libitum consumption following energy restriction in Ap2-agouti transgenic mice. *J Nutr* 134:3054–3060.

39. Carlberg, C. and S. Seuter. 2007. The vitamin D receptor. *Dermatol Clin* 25:515–523, viii.

40. Ye, W. Z., A. F. Reis, D. Dubois-Laforgue, C. Bellanne-Chantelot, J. Timsit, and G. Velho. 2001. Vitamin D receptor gene polymorphisms are associated with obesity in type 2 diabetic subjects with early age of onset. *Eur J Endocrinol* 145:181–186.

41. Barger-Lux, M. J., R. P. Heaney, J. Hayes, H. F. De Luca, M. L. Johnson, and G. Gong. 1995. Vitamin D receptor gene polymorphism, bone mass, body size, and vitamin D receptor density. *Calcif Tissue Int* 57:161–162.

42. Wood, R. J. 2008. Vitamin D and adipogenesis: New molecular insights. *Nutr Rev* 66:40–46.

43. Bellows, C. G., Y. H. Wang, J. N. Heersche, and J. E. Aubin. 1994. 1,25-dihydroxyvitamin D3 stimulates adipocyte differentiation in cultures of fetal rat calvaria cells: Comparison with the effects of dexamethasone. *Endocrinology* 134:2221–2229.

44. Kong, J. and Y. C. Li. 2006. Molecular mechanism of 1,25-dihydroxyvitamin D3 inhibition of adipogenesis in 3T3-L1 cells. *Am J Physiol Endocrinol Metab* 290:E916–E924.

45. Sun, X. and M. B. Zemel. 2008. 1Alpha, 25-dihydroxyvitamin D and corticosteroid regulate adipocyte nuclear vitamin D receptor. *Int J Obes (Lond)* 32:1305–1311.

46. Shi, H., A. Norman, W. Okamura, A. Sen, and M. Zemel. 2002. 1alpha,25-dihydroxyvitamin D3 inhibits uncoupling protein 2 expression in human adipocytes. *FASEB J* 16:1808–1810.

47. Echtay, K. S. 2007. Mitochondrial uncoupling proteins: What is their physiological role? *Free Radic Biol Med* 43:1351–1371.

48. Nubel, T. and D. Ricquier. 2006. Respiration under control of uncoupling proteins: Clinical perspective. *Horm Res* 65:300–310.

49. Wong, K. E., F. L. Szeto, W. Zhang, H. Ye, J. Kong, Z. Zhang, X. J. Sun, and Y. C. Li. 2009. Involvement of the vitamin D receptor in energy metabolism: Regulation of uncoupling proteins. *Am J Physiol Endocrinol Metab* 296:E820–E828.

50. Soares, M. J. and W. L. Chan She-Ping Delfos. 2010. Postprandial energy metabolism in the regulation of body weight: Is there a mechanistic role for dietary calcium? *Nutrients* 2:586–898.

51. Sun, X. and M. Zemel. 2004. Role of uncoupling protein 2 (UCP2) expression and 1alpha, 25-dihydroxyvitamin D3 in modulating adipocyte apoptosis. *FASEB J* 18:1430–1432.

52. Sergeev, I. N. 2009. 1,25-Dihydroxyvitamin D_3 induces Ca2+-mediated apoptosis in adipocytes via activation of calpain and caspase-12. *Biochem Biophys Res Commun* 384:18–21.
53. Elias, J., B. Marian, C. Edling, B. Lachmann, C. R. Noe, S. H. Rolf, and I. Schuster. 2003. Induction of apoptosis by vitamin D metabolites and analogs in a glioma cell line. *Recent Results Cancer Res* 164:319–332.
54. Wagner, N., K. D. Wagner, G. Schley, L. Badiali, H. Theres, and H. Scholz. 2003. 1,25-dihydroxyvitamin D_3-induced apoptosis of retinoblastoma cells is associated with reciprocal changes of Bcl-2 and bax. *Exp Eye Res* 77:1–9.
55. Sun, X. and M. B. Zemel. 2004. Role of uncoupling protein 2 (UCP2) expression and 1alpha, 25-dihydroxyvitamin D_3 in modulating adipocyte apoptosis. *FASEB J* 18:1430–1432.
56. Rizzuto, R., P. Pinton, W. Carrington, F. S. Fay, K. E. Fogarty, L. M. Lifshitz, R. A. Tuft, and T. Pozzan. 1998. Close contacts with the endoplasmic reticulum as determinants of mitochondrial Ca2+ responses. *Science* 280:1763–1766.
57. Sun, X., K. L. Morris, and M. B. Zemel. 2008. Role of calcitriol and cortisol on human adipocyte proliferation and oxidative and inflammatory stress: A microarray study. *J Nutrigenet Nutrigenomics* 1:30–48.
58. Wang, M. 2005. The role of glucocorticoid action in the pathophysiology of the metabolic syndrome. *Nutr Metab (Lond)* 2:3.
59. Seckl, J. R. and B. R. Walker. 2001. Minireview: 11beta-hydroxysteroid dehydrogenase type 1—A tissue-specific amplifier of glucocorticoid action. *Endocrinology* 142:1371–1376.
60. Rask, E., T. Olsson, S. Soderberg, R. Andrew, D. E. Livingstone, O. Johnson, and B. R. Walker. 2001. Tissue-specific dysregulation of cortisol metabolism in human obesity. *J Clin Endocrinol Metab* 86: 1418–1421.
61. Sun, X. and M. Zemel. 2006. Dietary calcium regulates ROS production in aP2-agouti transgenic mice on high-fat/high-sucrose diets. *Int J Obes (Lond)* 30:1341–1346.
62. Morton, N. M. 2010. Obesity and corticosteroids: 11beta-hydroxysteroid type 1 as a cause and therapeutic target in metabolic disease. *Mol Cell Endocrinol* 316:154–164.
63. Masuzaki, H., J. Paterson, H. Shinyama, N. M. Morton, J. J. Mullins, J. R. Seckl, and J. S. Flier. 2001. A transgenic model of visceral obesity and the metabolic syndrome. *Science* 294:2166–170.
64. Masuzaki, H., H. Yamamoto, C. J. Kenyon, J. K. Elmquist, N. M. Morton, J. M. Paterson, H. Shinyama, M. G. Sharp, S. Fleming, et al. 2003. Transgenic amplification of glucocorticoid action in adipose tissue causes high blood pressure in mice. *J Clin Invest* 112:83–90.
65. Kotelevtsev, Y., M. C. Holmes, A. Burchell, P. M. Houston, D. Schmoll, P. Jamieson, R. Best, R. Brown, C. R. Edwards, et al. 1997. 11beta-hydroxysteroid dehydrogenase type 1 knockout mice show attenuated glucocorticoid-inducible responses and resist hyperglycemia on obesity or stress. *Proc Natl Acad Sci USA* 94:14924–14929.
66. Morris, K. and M. Zemel. 2005. 1,25-dihydroxyvitamin D^3 modulation of adipocyte glucocorticoid function. *Obes Res* 13:670–677.
67. Sun, X., K. L. Morris, and M. B. Zemel. 2008. Role of calcitriol and cortisol on human adipocyte proliferation and oxidative and inflammatory stress: A microarray study. *J Nutrigenet Nutrigenomics* 1:30–48.
68. Kokta, T. A., M. V. Dodson, A. Gertler, and R. A. Hill. 2004. Intercellular signaling between adipose tissue and muscle tissue. *Domest Anim Endocrinol* 27:303–331.
69. Zemel, M. B., J. Richards, S. Mathis, A. Milstead, L. Gebhardt, and E. Silva. 2005. Dairy augmentation of total and central fat loss in obese subjects. *Int J Obes (Lond)* 29:391–397.
70. Sun, X. and M. Zemel. 2007. Leucine and calcium regulate fat metabolism and energy partitioning in murine adipocytes and muscle cells. *Lipids* 42:297–305.
71. Sun, X. and M. B. Zemel. 2009. Leucine modulation of mitochondrial mass and oxygen consumption in skeletal muscle cells and adipocytes. *Nutr Metab (Lond)* 6:26.
72. Furukawa, S., T. Fujita, M. Shimabukuro, M. Iwaki, Y. Yamada, Y. Nakajima, O. Nakayama, M. Makishima, M. Matsuda, and I. Shimomura. 2004. Increased oxidative stress in obesity and its impact on metabolic syndrome. *J Clin Invest* 114:1752–61.
73. Gordeeva, A. V., R. A. Zvyagilskaya, and Y. A. Labas. 2003. Cross talk between reactive oxygen species and calcium in living cells. *Biochemistry (Mosc)* 68:1077–1080.
74. Zemel, M. and B. X. Sun. 2008. Calcitriol and energy metabolism. *Nutr Rev* 66:S139–S146.
75. Sun, X. and M. B. Zemel. 2007. 1Alpha,25-dihydroxyvitamin D_3 modulation of adipocyte reactive oxygen species production. *Obesity (Silver Spring)* 15:1944–1953.
76. Zemel, M. B. and X. Sun. 2008. Dietary calcium and dairy products modulate oxidative and inflammatory stress in mice and humans. *J Nutr* 138:1047–1052.

77. Whitehead, J. P., A. A. Richards, I. J. Hickman, G. A. Macdonald, and J. B. Prins. 2006. Adiponectin: A key adipokine in the metabolic syndrome. *Diabetes Obes Metab* 8:264–280.

78. Mohamed-Ali, V., J. H. Pinkney, and S. W. Coppack. 1998. Adipose tissue as an endocrine and paracrine organ. *Int J Obes Relat Metab Disord* 22:1145–1158.

79. Coppack, S. W. 2001. Proinflammatory cytokines and adipose tissue. *Proc Nutr Soc* 60:349–356.

80. Rabe, K., M. Lehrke, K. G. Parhofer, and U. C. Broedl. 2008. Adipokines and insulin resistance. *Mol Med* 14:741–751.

81. Manea, A., S. A. Manea, A. V. Gafencu, M. Raicu, and M. Simionescu. 2008. AP 1–dependent transcriptional regulation of NADPH oxidase in human aortic smooth muscle cells: Role of p22phox subunit. *Arterioscler Thromb Vasc Biol* 28:878–885.

82. Basuroy, S., S. Bhattacharya, C. W. Leffler, and H. Parfenova. 2009. Nox4 NADPH oxidase mediates oxidative stress and apoptosis caused by TNF-alpha in cerebral vascular endothelial cells. *Am J Physiol Cell Physiol* 296:C422–C432.

83. Kuwano, Y., K. Tominaga, T. Kawahara, H. Sasaki, K. Takeo, K. Nishida, K. Masuda, T. Kawai, S. Teshima-Kondo, and K. Rokutan. 2008. Tumor necrosis factor alpha activates transcription of the NADPH oxidase organizer 1 (NOXO1) gene and upregulates superoxide production in colon epithelial cells. *Free Radic Biol Med* 45:1642–1652.

84. Janssen-Heininger, Y. M., I. Macara, and B. T. Mossman. 1999. Cooperativity between oxidants and tumor necrosis factor in the activation of nuclear factor (NF)-kappaB: requirement of Ras/mitogen-activated protein kinases in the activation of NF-kappaB by oxidants. *Am J Respir Cell Mol Biol* 20:942–952.

85. Sun, X. and M. Zemel. 2007. Calcium and 1,25-dihydroxyvitamin D_3 regulation of adipokine expression. *Obesity (Silver Spring)* 15:340–348.

86. Zemel, M. B., X. Sun, T. Sobhani, and B. Wilson. 2011. Effects of dairy compared with soy on oxidative and inflammatory stress in overweight and obese subjects. *Am J Clin Nutr* 91:16–22.

87. Zemel, M. B. and R. Stancliffe. 2010. Dairy attenuates oxidative and inflammatory stress in metabolic syndrome. *FASEB J* 24:Abstract 105.3

88. Weisberg, S. P., D. McCann, M. Desai, M. Rosenbaum, R. L. Leibel, and A. W. Ferrante Jr. 2003. Obesity is associated with macrophage accumulation in adipose tissue. *J Clin Invest* 112:1796–808.

89. Curat, C. A., A. Miranville, C. Sengenes, M. Diehl, C. Tonus, R. Busse, and A. Bouloumie. 2004. From blood monocytes to adipose tissue-resident macrophages: induction of diapedesis by human mature adipocytes. *Diabetes* 53:1285–92.

90. Nishimura, S., I. Manabe, M. Nagasaki, Y. Hosoya, H. Yamashita, H. Fujita, M. Ohsugi, K. Tobe, T. Kadowaki, et al. 2007. Adipogenesis in obesity requires close interplay between differentiating adipocytes, stromal cells, and blood vessels. *Diabetes* 56:1517–26.

91. Overbergh, L., B. Decallonne, D. Valckx, A. Verstuyf, J. Depovere, J. Laureys, O. Rutgeerts, R. Saint-Arnaud, R. Bouillon, and C. Mathieu. 2000. Identification and immune regulation of 25-hydroxyvitamin D-1-alpha-hydroxylase in murine macrophages. *Clin Exp Immunol* 120:139–146.

92. Helming, L., J. Bose, J. Ehrchen, S. Schiebe, T. Frahm, R. Geffers, M. Probst-Kepper, R. Balling, and A. Lengeling. 2005. 1alpha,25-Dihydroxyvitamin D_3 is a potent suppressor of interferon gamma-mediated macrophage activation. *Blood* 106:4351–4358.

93. Guillot, X., L. Semerano, N. Saidenberg-Kermanac'h, G. Falgarone, and M. C. Boissier. 2010. Vitamin D and inflammation. *Joint Bone Spine* 77:552–557.

94. Rigby, W. F., L. Shen, E. D. Ball, P. M. Guyre, and M. W. Fanger. 1984. Differentiation of a human monocytic cell line by 1,25-dihydroxyvitamin D_3 (calcitriol): A morphologic, phenotypic, and functional analysis. *Blood* 64:1110–1115.

95. Olsson, I., U. Gullberg, I. Ivhed, and K. Nilsson. 1983. Induction of differentiation of the human histiocytic lymphoma cell line U-937 by 1 alpha,25-dihydroxycholecalciferol. *Cancer Res* 43:5862–867.

96. Sun, X. and M. Zemel. 2007. Calcitriol and calcium regulate cytokine production and adipocyte-macrophage cross-talk. *J Nutr Biochem* 19(6):392–9. Epub 2007 Sep 14.

97. Curry, B., J. Biggerstaff, and M. B. Zemel. 2010. Effects of leucine and calcitriol on monocyte-vascular endothelial cell adhesion. *FASEB J* 24:230.5 (Abstract).

98. Mason, C., L. Xiao, I. Imayama, C. R. Duggan, C. Bain, K. E. Foster-Schubert, A. Kong, K. L. Campbell, C. Y. Wang, et al. 2010. Effects of weight loss on serum vitamin D in postmenopausal women. *Am J Clin Nutr* 94(1):95–103. Epub 2011 May 25.

99. Sneve, M., Y. Figenschau, and R. Jorde. 2008. Supplementation with cholecalciferol does not result in weight reduction in overweight and obese subjects. *Eur J Endocrinol* 159:675–684.

100. Jorde, R., M. Sneve, P. A. Torjesen, Y. Figenschau, L. G. Goransson, and R. Omdal. 2010. No effect of supplementation with cholecalciferol on cytokines and markers of inflammation in overweight and obese subjects. *Cytokine* 50:175–180.
101. Zittermann, A., S. Frisch, H. K. Berthold, C. Gotting, J. Kuhn, K. Kleesiek, P. Stehle, H. Koertke, and R. Koerfer. 2009. Vitamin D supplementation enhances the beneficial effects of weight loss on cardiovascular disease risk markers. *Am J Clin Nutr* 89:1321–1327.
102. Lee, P., J. R. Greenfield, M. J. Seibel, J. A. Eisman, and J. R. Center. 2009. Adequacy of vitamin D replacement in severe deficiency is dependent on body mass index. *Am J Med* 122:1056–1060.

9 Membrane Receptors for Vitamin D Metabolites and the Role of Reactive Oxygen Species

Ramesh C. Khanal and Ilka Nemere

CONTENTS

9.1 INTRODUCTION

Vitamin D plays a central role in modulating calcium and phosphate homeostasis in the body. However, its role in many other physiological and disease states, primarily those of a chronic nature, such as diabetes, cancer, and Alzheimer's, is also widely known. While the effects of its principal stimulatory metabolite, 1,25-dihydroxyvitamin D$_3$[1,25(OH)$_2$D$_3$], is studied extensively in many physiological and disease states, research results have now emerged suggesting the possible biological role for another metabolite, i.e., 24,25-dihydroxyvitamin D$_3$[24,25(OH)$_2$D$_3$]. Actions of 1,25(OH)$_2$D$_3$ on Ca^{2+} and Pi homeostasis are believed to be mediated by both pregenomic and genomic pathways involving the classical vitamin D receptor and the more recently identified 1,25D3-MARRS (membrane-associated rapid-response steroid binding) receptor, which is also known as PDIA3/ERp57/GRp58/ERp60/ERp61. The metabolite 24,25(OH)$_2$D$_3$ is made when an animal is vitamin D replete and provides an endogenous feedback loop for intestinal phosphate absorption. It acts to decrease the specific activity of the antioxidative enzyme catalase, resulting in higher levels of H$_2$O$_2$. Hydrogen peroxide, in turn, acts at two points: oxidation of the 1,25D$_3$-MARRS receptor and inhibition of PKC. These effects can be overcome with diets supplemented with antioxidants. Here, we discuss these two recently identified putative receptors for vitamin D metabolites, their role in relation to Ca^{2+} and Pi transport, and the role of reactive oxygen species in relation to oxidative stress that is usually associated with chronic diseases, along with a brief overview of 25-hydroxy-vitamin D3 [25(OH)D$_3$] as a functional metabolite.

9.2   PDIA3/ERp57/GRp58 AS A MEMBRANE ASSOCIATED RECEPTOR FOR 1,25(OH)$_2$D$_3$

It is now well accepted that the steroid hormone 1,25(OH)$_2$D$_3$ is capable of inducing membrane-initiated signaling phenomena [see the work of Farach-Carson and Nemere (2003) for review] independent of the classical vitamin D receptor (VDR; Nemere 2005; Nemere et al. 2010), and its pregenomic actions have been described in detail elsewhere in the book. In intestinal epithelial cells and perfused duodena, the 1,25D$_3$-MARRS receptor mediates the rapid (seconds to minutes) stimulation of phosphate (Nemere 1996; Nemere et al. 2012a,b; Ferraro et al. 1999) and calcium (Nemere et al. 1994; Nemere 2005; Nemere et al. 2010) uptake. Both minerals are required for the maintenance of health but are available for duodenal absorption from the diet for only the few minutes that it takes for transition to more distal regions of the small intestine (for review of mechanism see Nemere and Norman 1982). Thus, rapid hormonal stimulation of mineral transport is vital.

Based on increased fluid and ion permeability in the uterus following estrogen administration in estrogen-withdrawn rats (reviewed by Walters 1985), Szego and Roberts (1953) hypothesized that the increase in permeability was the primary mechanism of hormonal action in general. Thus, the search for membrane-initiated steroid hormone effects began. In the case of vitamin D, Toffolon et al. (1975) reported an enhanced calcium transport in the everted rat intestine *in vitro* 30 min after injection of 1,25(OH)$_2$D$_3$ *in vivo*. Nemere and Szego (1981) demonstrated a direct rapid effect of 1,25(OH)$_2$D$_3$ on calcium uptake in isolated rat intestinal epithelial cells, within 5 min, and proposed for the first time the existence of a membrane receptor for the steroid hormone. In addition, using rat enterocytes, Lieberherr et al. (1989) reiterated the proposition for the requirement of a functioning cell membrane-type receptor when 1,25(OH)$_2$D$_3$ treatment showed immediate effects (5–20 s) on phosphoinositide metabolism. In mice with a targeted knock-out of the gene for the 1,25D$_3$-MARRS receptor, we have recently demonstrated that this protein—not the VDR—is required for rapid uptake of calcium from the intestine (Nemere et al. 2010; Nemere et al. 2012b).

An early indicator that the membrane-associated receptor is distinct from the VDR came from studies with analogs (Norman et al. 1994). Figure 9.1 depicts naturally occurring steroid hormones and some analogs that were tested for their ability to initiate the rapid transport of calcium. The compound 1,25(OH)$_2$-lumisterol (which is also referred to by the two-letter designation of JN) is conformationally locked in the 6-s-cis configuration, whereas the naturally occurring 1,25(OH)$_2$D$_3$ is free to rotate at the double bond. Such analogs do not compete with [^{3}H]1,25(OH)$_2$D$_3$ for binding to the VDR but compete quite successfully for binding to the 1,25D$_3$-MARRS receptor (Figure 9.2). The analog JM is also somewhat able to compete for binding to the cell surface moiety but less so than JN. Of particular interest is the additional observation that JN stimulates the rapid transport of calcium in perfused duodenal loops to a larger extent than JM (Nemere et al. 1994; Norman et al. 1994), thus linking receptor affinity to physiological performance.

Currently, there are two main candidates for the 1,25D$_3$ membrane receptor: the 1,25D$_3$–MARRS protein/ERp57/GRp58 and the classical VDR. Since we have provided further details as to why the VDR is probably not the right candidate for membrane-associated rapid responses in recent publications (Khanal and Nemere 2007a,b; 2009; Nemere et al. 2010; Nemere et al. 2012b), only the 1,25D$_3$-MARRS protein will be discussed here.

Using the N-terminal sequence, a highly specific antibody, Ab099, was generated using the multiple antigenic peptide format and used to demonstrate the cell surface localization of the receptor in the intestinal cells (Nemere et al. 2000). The newly characterized antisera were used to further distinguish the membrane receptor from the nuclear receptor. In Western analyses of basaolateral membrane proteins, Ab099 was monospecific for a 64.5 kDa band, which was also labeled by the affinity ligand [^{14}C]1,25(OH)$_2$D$_3$-bromoacetate; the label was diminished in the presence of excess secosterol (Nemere et al. 2000). The monoclonal antibody 9A7 against the VDR, however, failed to detect an appropriate band in the BLM fractions (Nemere et al. 2000). Preincubation of isolated intestinal cells with Ab099, on the other hand, acted as an agonist for intracellular calcium fluxes as

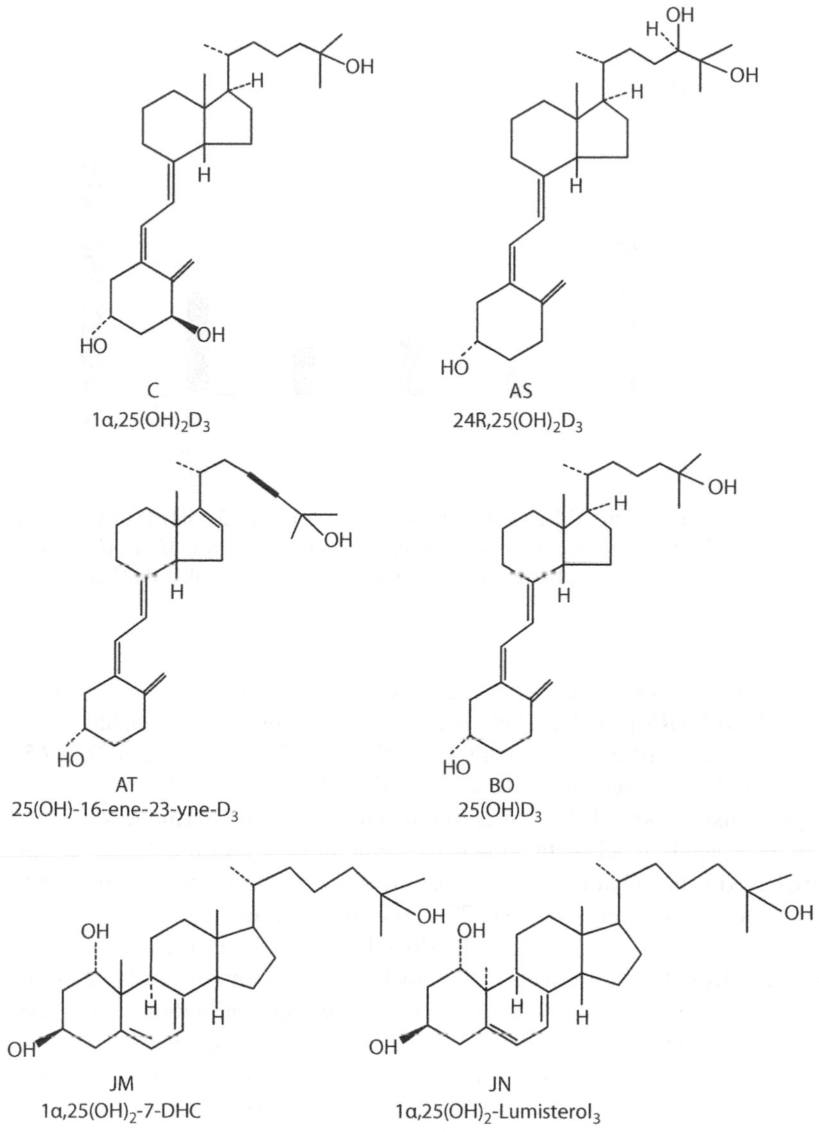

FIGURE 9.1   Structures of vitamin D metabolites and analogs.

detected by fura-2, suggesting a close coupling of the 1,25D$_3$-MARRS receptor and a basal lateral calcium channel. Both of these signal transduction pathways have been shown to be stimulated by 1,25(OH)$_2$D$_3$ in rats and chicken (de Boland et al. 1990; Morelli et al. 1993; Boyan et al. 1999; Facchinetti et al. 1999).

The 1,25D$_3$-MARRS protein as a separate membrane receptor was then isolated from chick intestinal basal lateral membranes (Nemere et al. 2004), characterized (Rohe et al. 2005), and its functional link to rapid uptake of phosphate in the intestinal cells established using ribozyme knockdown (Nemere et al. 2004). The same Ab099 was then used to demonstrate the cell surface localization of the receptor protein in osteoblasts and odontoblasts, and its upregulation was observed in differentiated dental cells *in vivo* (Teillaud et al. 2005). Tissue formation and biomineralization process as a model system was used to study in detail the pregenomic pathways of the secosteroid by immunolocalization of the receptor during prenatal human development (Berdal et

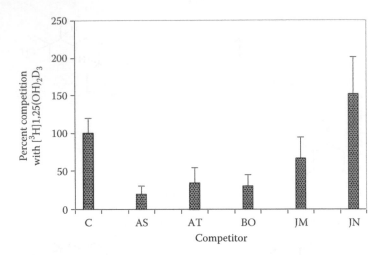

FIGURE 9.2   Competitive binding of 1,25 $(OH)_2D_3$-analogs with $[^3H]1,25$ $(OH)_2D_3$. (Adapted from Norman, A. W. et al., in *Vitamin D, A Pluripotent Steroid Hormone: Structural Studies, Molecular Endocrinology and Clinical Applications* (Norman, A. W., Bouillon, R., Thomasset, M., eds.), Walter de Gruyter, Berlin, 324–331, 1994.)

al. 2003). Odontogenesis coincided with a sequence of upregulation and downregulation of immunoreactive $1,25D_3$-MARRS protein, whereas biomineralization was associated with upregulation of the receptor protein in the adjoining cells in all tissues (Mesbah et al. 2002). A 65-kDa protein recognized by Ab099 was detected by Western blotting (Mesbah et al. 2002).

We have previously reported that, in cultured intestinal cells, hormone-stimulated phosphate uptake can be completely blocked by either a specific antibody to the $1,25D_3$-MARRS receptor (Ab099) or by RNAi using a hammerhead ribozyme directed against the receptor mRNA (Nemere et al. 2004). Recently, we have crossed ERp57$^{flx/flx}$ mice with commercially available mice expressing villin-driven cre-recombinase (Nemere et al. 2010). Using confocal microscopy, we demonstrated that intestinal cells from both litter mates and knock-out mice contain similar staining patterns for cell surface VDR and the $1,25D_3$-MARRS receptor; however, antibodies against the VDR failed to inhibit $1,25D_3$-mediated rapid stimulation of calcium uptake. We have previously reported that chick intestinal cells rely solely on the $1,25D_3$-MARRS receptor to mediate steroid hormone effects on enhanced calcium uptake (Nemere 2005; Khanal et al. 2008), whereas $1,25D_3$-stimulated calcium uptake in isolated rat enterocytes is inhibited by either antibodies to the $1,25D_3$-MARRS receptor or the VDR (Nemere 2005). This has led us to postulate that the evolution of mammals may have led to the dual cell surface receptor mechanism. In view of our recent findings with mice enterocytes (Nemere et al. 2010; Nemere et al. 2012a,b), this appears not to be the case. It remains to be determined what the function of cell surface VDR is in intestinal cells of mice. One possibility might be to mediate the actions of $25(OH)D_3$, which is a metabolite that has been shown to increase calcium uptake in chick intestinal cells at physiological concentrations (Sterling and Nemere 2007) and which has been reported to bind to the VDR in prostate cancer cell lines (Lambert et al. 2007), as well as thyroparathyroid explants (Ritter and Brown 2011).

Every endocrine stimulatory system is regulated by a feedback loop. Under conditions of calcium, phosphate, and vitamin D repletion, the metabolite $24,25(OH)_2D_3$ is produced. The $24,25(OH)_2D_3$ metabolite has been shown to be part of the feedback loop that inhibits stimulated phosphate and calcium transport (Nemere 1996; Zhao and Nemere 2002; Nemere et al. 2002). On a molecular level, the inhibitory action is caused by decreased catalase activity with increased H_2O_2 production (Nemere et al. 2006). Similar results have been observed in human dermal fibroblasts, where major targets for oxidative stress induced by H_2O_2 are the ER resident proteins, including ERp57 (van der Vlies et al. 2002). This, in turn, oxidizes the thioredoxin domains, leading to reduced $1,25(OH)_2D_3$ binding (Nemere et al. 2006). Indeed, we have tested this model *in vivo*: Chicks were raised on grower diet alone or diet supplemented with antioxidants for three weeks and then tested for net phosphate absorption. The birds in the antioxidant supplemented groups exhibited phosphate absorption that was twice that of chicks on the control diet. Furthermore, $1,25(OH)_2D_3$ binding to the $1,25D_3$-MARRS receptor was dramatically increased in mucosal fractions prepared from chicks on the antioxidant diets, relative to equivalent preparations from controls. Alkylation of cysteines was also found to be inhibitory for calcium transport (Nemere and Norman 1987)

9.3 STRUCTURE OF PDIA3/ERp57/1,25D₃-MARRS RECEPTOR

The $1,25D_3$-MARRS receptor, which is also known as PDIA3/ERp57/GRp58, is a glycoprotein specific thiol-oxidoreductase and an isoform of protein disulphide isomerase (PDI). It promotes the intra- and intermolecular disulphide bonds during glycoprotein folding (Molinari and Helenius 1999) and interacts specifically with N-glucosylated integral membrane proteins (Elliott et al. 1997; Tamura et al. 2002; Silvennoin et al. 2005). Its generalized linear structure showing different sites is given in Figure 9.3 and contains two thioredoxin like domains, six protein kinase C phosphorylation sites, seven casein kinase II phosphorylation sites, three tyrosine phosphorylation sites, two sulfation sites, and one N-myrisoylation site. It is composed of at least four domains named a, b, b′, and a′, of which a and a′ are two thioredoxin like domains in positions corresponding to those of domains a and a′ in the PDI polypeptide (Urade et al. 2004). The PDI and ERp57 also show significant sequence similarities in regions corresponding to the b and b′ domains and have an overall identity of 33% (Ferrari and Soling 1999). Although ERp57 has the modular structure of active and inactive domains of PDI, it does not have the acidic C-terminus but instead a basic C-terminus that is found in ERp57 orthologs (Pollock et al. 2004). The redox activity of thioredoxin-like domains is provided by two cysteine residues present in a characteristic CXXC sequence motif (Alticri et al. 1993; Urade et al. 1997), which are thought to be responsible for catalytic activity of the enzyme (Urade et al. 1997; Hirano et al. 1995). One motif is presented near the N-terminus, whereas the other is presented near the C-terminus (Li and Camacho 2004). The mutation of the first motif resulted in 30% activity; the second, 70%. Both mutations abolished the activity completely (Li and Camacho 2004). This may provide an opportunity to study and design the functionalities and importance of the molecule that may prove to be of tremendous medicinal implications.

Human ERp57 has three major protease-sensitive regions, the first of which is located between amino acid residues 120 and 150, the second is located between 201 and 215, and the third is located between 313 and 341 (Silvennoinen et al. 2004), with the data thus being consistent with a four-domain structure aba′b′. The cDNA encoding human ERp57/GRp58 was cloned independently

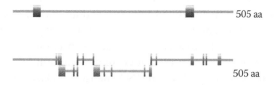

FIGURE 9.3   (See color insert.) Linear structure of $1,25D_3$ MARRS/ERp57/GRp58/PDIA3 showing different sites. Figure was generated by Prosite using the amino acid sequence of ERp57. (i) Two thioredoxin sites, aa 49–67 and 398–416 (upper panel). (ii) Six protein kinase C phosphorylation sites: Thin green vertical bars, aa 102–14, aa 150–152, aa 212–214, aa 229–231, aa 303–305, and aa 330–332. (iii) Three tyrosine phosphorylation sites: Red vertical bars (yellow on top), aa 1107–115, aa 410–416, and aa 471–479. (iv) Two sulfation sites: Thick vertical green bars, aa 108–122 aa 189–203. (v) Seven casein kinase II phosphorylation sites: Blue vertical bars, aa 141–144, aa 150–153, aa 163–166, aa 187–190, aa 319–322, aa 437–440, and aa 443–446. (vi) One N-myristoylation site: Medium think vertical green bar, aa 315–320.

in three different laboratories (Bourdi et al. 1995; Hirano et al. 1995; Koivunen et al. 1997). All of them reported that the gene encodes a 505-amino acid polypeptide with significant homology to human PDI. The sequence includes a putative nuclear localization motif and an endoplasmic reticulum (ER)-retention/retrieval motif (Bourdi et al. 1995). Genomic organization of the ERp57 gene revealed that it is encoded in 13 exons spanning 18 kb with no similarity between the genomic structures of the ERp57, PDI, and thioredoxin genes. Northern blotting showed that ERp57 is expressed as a 2-kb message most abundantly in liver, placenta, and lung, and, at lower levels, in many other tissues. Upon mapping by fluorescence *in situ* hybridization, the ERp57 was assigned to 15q15, and the processed gene was assigned to 1q21, so that neither was located on the same chromosome as the human PDI and thioredoxin genes (Bourdi et al. 1995).

9.4   1,25D$_3$-MARRS RECEPTOR IN TISSUES OTHER THAN INTESTINE

Given the name ER protein, it is ubiquitous in all tissues, including the intestine, the major organ for dietary Ca^{2+}, and Pi absorption. Since the protein is involved in posttranslational modification and given its involvement as a chaperone protein to clear out the improperly formed proteins, global knock-out causes death *in utero*. Using Ab099, cell surface localization of the 1,25D$_3$-MARRS receptor protein was demonstrated in osteoblasts and odontoblasts, and its upregulation was observed in differentiated dental cells *in vivo* (Teillaud et al. 2005). Odontogenesis coincided with a sequence of upregulation and downregulation of immunoreactive 1,25D$_3$-MARRS protein, whereas biomineralization was associated with upregulation of the receptor protein in the adjoining cells in all tissues with Ab099 recognizing a 65 kDa protein during Western blotting (Mesbah et al. 2002). Its presence in the liver, placenta, lung, and many other tissues has been determined (Bourdi et al. 1995). More recently, osteoblast-like MC3T3-E1 cells showed that silencing of PDIA3/1,25(OH)$_2$D$_3$-MARRS receptor abolished 1,25(OH)$_2$D$_3$-induced rapid activation of PKC while overexpression of the receptor resulted in augmentation of PKC activity by the secosteroid (Wang et al. 2010). These data further proved that 1,25D$_3$-MARRS receptor plays a crucial role in 1,25(OH)$_2$D$_3$-regulated bone formation and the 1,25(OH)$_2$D$_3$-PKC signaling pathway is involved in this process.

9.5   1,25D$_3$-MARRS RECEPTOR AND DISEASES

Vitamin D has been extensively investigated for its critical role in modulating different diseases. However, only the role in relation to its membrane receptor will be discussed here. The ability of proteins to fold into a predefined and functional conformation is one of the fundamental processes in biological systems. Certain conditions, be they genetic or environmental, however, initiate misfolding or unfolding of proteins, resulting in the loss of functional protein. Defective protein folding has now been recognized as the basis of various human diseases (Thomas et al. 1995; Winklhofer et al. 2006; Scheibel and Buchner 2006). Given that ERp57 is critical for posttranslational modification of proteins, its work as a molecular chaperone, and its critical need in major histocompatibility complex (MHC) class-I assembly and regulation of gene expression, it is highly likely that the protein is involved in some of the disease states, such as prion neurotoxicity, Alzheimer's, and cancer. This has led to new approaches aimed at correcting protein folding defects, including the potential of molecular chaperones (e.g., through RNA interference) as novel therapeutic targets, and the usage of chemical or pharmacological chaperones, including ERp57, as new therapeutic agents (Amaral 2006).

Association of ERp57 in the formation of MHC class-I molecules and STAT3, both of which are associated with immune system and cancer, have been well documented (Coe and Michalak 2010; Garbi et al. 2006; Antoniou et al. 2002). Moreover, impaired assembly of the MHC class-I peptide-loading complex in the mice deficient in ERp57 resulted in death *in utero* (Garbi et al. 2006). Earlier, high correlation was observed between intracellular amounts of ERp57 and immunoglobulin production in hybridoma cells, indicating a possible role of ERp57 in disulphide bond formation (Kozaki et al. 1994). Involvement of ERp57 in the oncogenic transformation by controlling both extra- and intracellular redox activities through its thiol-dependent reductase activity has been postulated previously (Hirano et al. 1995). In human hepatoma Hep3B cells, ERp57 may regulate signaling by sequestering inactive and activated STAT3 (Guo et al. 2002). When C57B6 mice were exposed to the carcinogen diethylhexyl phthalate for seven days, downregulation of ERp57 gene in the liver, but not in kidney or testis, was observed (Muhlenkamp and Gill 1998). A reduced ERp57 protein level in mouse treated with diethylhexyl phthalate *in vivo* was also observed (Muhlenkamp et al. 1995). Similarly, expression levels of several proteins, including ERp57, were downregulated in human dendritic cells inhibited by tumor-derived gangliosides (Tourkova et al. 2005). More recently, it has been suggested that $1,25D_3$-MARRS receptor expression interferes with the growth inhibitory activity of $1,25(OH)_2D_3$ in breast cancer cells, possibly through the nuclear VDR. Further research should examine the potential for pharmacological or natural agents that modify $1,25D_3$-MARRS expression or activity as anticancer agents (Richard et al. 2010). Moreover, nuclear translocation of the $1,25D_3$-MARRS receptor protein and NFkappaB in differentiating NB4 leukemia cells has been observed (Wu et al. 2010).

When human breast, lung, uterus, and stomach tumors were treated with mitomycin (an anticancer drug used in bioreductive chemotherapy that requires reductive activation to exert its effect) and compared with normal tissues of similar origin, a unique cytosolic activity that activated mitomycin, leading to DNA cross-linking, was found (Celli and Jaiswal 2003). The unique cytosolic activity that catalyzed the mitomycn-induced DNA cross-linking was identified as ERp57, and a higher expression of ERp57 in all tumors of interest was observed (Celli and Jaiswal 2003). In murine tumor cell lines, downregulation of MHC class-I surface expression has been frequently detected. In a study to understand the underlying mechanisms of such deficiencies, dysregulation of MHC class-I antigen processing machinery, of which ERp57 is a component, was observed (Seliger et al. 2000). Similarly, a nuclear protein complex containing ERp57 as one of its components has been found to be involved in the cytotoxic response to DNA modified by incorporation of anticancer nucleoside analogs (Krynetski et al. 2003).

In diabetes, changes in assisted protein folding are largely unexplored. A reductive shift in the redox status of rat liver microsomes after four weeks of streptozotocin-induced diabetes has been identified (Nardai et al. 2005). This change was reflected by a significant increase in the total- and protein-sulfhydryl content, as well as in the free sulfhydryl groups of ERp57 but not other chaperones. A parallel decrease of the protein-disulfide oxidoreductase activity was detected in the microsomal fraction of diabetic livers. The oxidant of PDI, Ero1-Lalpha showed a more oxidized status in diabetic rats. Their results revealed major changes in the redox status of the ER and its redox chaperones in diabetic rats, which may contribute to the defective protein secretion of the diabetic liver.

A deficiency in Perk (EIF2AK3) causes multiple neonatal defects in humans known as the Wolcott Rallison syndrome. Perk KO mice exhibit the same array of defects including permanent neonatal diabetes, which in mice is due to a decrease in beta cell proliferation and insulin secretion. Acute ablation of Perk in INS 832/13 beta cells exhibited all of the major defects seen in Perk KO mice and revealed abnormal expression and redox state of key ER chaperone proteins, including ERp57 (Feng et al. 2009). The oxidized state of ERp57 was increased, suggesting an imbalance in the redox state of the ER, leading to the dysfunction of pancreatic beta cells and impaired insulin secretion (Feng et al. 2009). Similarly, ERp57 has been shown to be essential for insulin biosynthesis, and enhancing chaperone capacity can improve beta-cell function in the presence of prolonged hyperglycemia (Zhang et al. 2009).

ERp57 should be mentioned briefly in relation to a few other diseases. Prion disorders are fatal transmissible neurodegenerative diseases that are characterized by extensive neuronal apoptosis and accumulation of misfolded prion protein. It has been indicated that expression of ERp57 is an early cellular response to prion replication and acts as a neuroprotective factor against prion neurotoxicity (Hetz et al. 2005). A direct interaction between ERp57 and prion protein was observed, and the data suggested that targeting ERp57 interaction may have applications for developing novel strategies for early diagnosis of the disease and its consequent treatment (Hetz et al. 2005). Creation of knockout and transgenic animals overexpressing ERp57 to analyze the progression of the disease may be beneficial in this regard. Another disease of major importance that has been associated with ERp57 is Alzheimer's, in which β-amyloids are the major components of the plaque observed in the brains of its patients. Although β-amyloids are produced in everyone during posttranslational processing in the ER of its precursor protein, deposits are observed only in elderly (Dobson 2006; Thomas et al. 1995). The reports of Ericksson and co-workers suggested that normal people have a carrier protein(s), keeping them in solution (Ericksson et al. 2005). One such carrier protein is ERp57, which prevents aggregation of β-amyloids and that the deposits are due to faulty posttranslational processing of its precursor protein with the failure to form this complex with ERp57 (Ericksson et al. 2005). An *in vitro* study suggested that newly synthesized amyloid precursor protein is subject to amyloidogenic processing during the initial phases of the secretory pathway, where ERp57 is involved (Selivanova et al. 2007).

9.5.1 Biological Activity of $24,25(OH)_2D_3$: A Historical Perspective

Although $1,25(OH)_2D_3$ is well known for its activity in diverse cell types and its ability to enhance intestinal calcium and phosphate absorption, $24,25(OH)_2D_3$ has not been as widely studied, perhaps because of the perception that it is an inactivation product. A biological role for $24,25(OH)_2D_3$ was initially observed when hens raised to sexual maturity from hatching with $1,25(OH)_2D_3$ as their only source of vitamin D_3 produced viable fertile eggs that appeared to develop well but failed to hatch (Henry and Norman 1978). However, when hens received a combination of $1,25(OH)_2D_3$ and $24,25(OH)_2D_3$ in their diet, hatchability equivalent to that with hens given vitamin D_3 was achieved (Henry and Norman 1978). Similar results were obtained in Japanese quail (*Coturnix coturnix japonica*) (Norman et al. 1983). In the case of bone, not only was biological activity unique to $24,25(OH)_2D_3$ observed in the maintenance of the integrity of certain parameters of its structure (Malluche et al. 1980; Seo et al. 1997) but also the requirements of $24,25(OH)_2D_3$ in bone formation (Dekel et al. 1983) and repair process were demonstrated (Lidor et al. 1987; Schwartz et al. 1988; Yukihiro et al. 1994; Seo and Norman 1997; Seo et al. 1997; Kato et al. 1998) with a concomitant increase in renal 24-hydroxylase activity and serum $1,25(OH)_2D_3$ concentration (Seo and Norman 1997; Norman et al. 2002). Earlier results also suggested that, when $24,25(OH)_2D_3$ is present at normal physiological concentrations, it is an essential vitamin D_3 metabolite for both normal bone integrity and healing of fracture in chicks (Seo et al. 1997). Effective stimulation of both epiphysis and diaphysis mineralization was observed with $24,25(OH)_2D_3$ at a dose of 10 IU, and the effect was maintained after a tenfold increase of the dose (Alekseeva et al. 1982). Although such an effect on epiphyseal growth was proposed to be regulated through an increased ornithine decarboxylase activity (Sömjen et al. 1983), no consistent effects of $24,25(OH)_2D_3$ on ornithine decarboxylase or creatine kinase activity was observed in other works (Ittel et al. 1986). Specific roles for $24,25(OH)_2D_3$ were also observed in the growth and differentiation of cartilage cells (Binderman and Sömjen 1984), which appeared to be effected through acid and alkaline posphatases (Lieberherr et al. 1977; Lieberherr et al. 1979; Hale et al. 1986). Formation of $24,25(OH)_2D_3$ from $25(OH)D_3$ in rat calvaria and cartilage further indicated its possible involvement in overall bone mineralization process (Lieberherr et al. 1979; Garabedian et al. 1978; Garabedian et al. 1981; Grosse et al. 1993). Effects of $24,25(OH)_2D_3$ on the formation and function of osteoclasts, osteoblasts, chondrocytes, osteosarcoma, and odontoblasts, both *in vitro* and *in vivo* has been observed (Sylvia et al. 1993; Dean et al. 2001; Kamao et al. 2001; Yamato et al. 1993; Hale et al. 1986; Schwartz et al. 2000).

Such biological effects of $24,25(OH)_2D_3$ have recently been extended to dietary calcium and phosphate transport across the intestine, both *in vivo* and *in vitro* (Nemere 1996; Nemere et al. 2002; Larsson et al. 2006; Nemere et al. 2006). In chick intestine, $24,25(OH)_2D_3$ inhibits the rapid stimulation of calcium and phosphate transport mediated by $1,25(OH)_2D_3$ (Nemere 1996; Nemere et al. 2002). In Atlantic cod fish, physiologically relevant concentrations of $24,25(OH)_2D_3$ have been found to decrease the Ca^{2+} transport in the perfused intestine (Sundell and Bjornsson 1990; Larsson et al. 1995). This ability of $24,25(OH)_2D_3$ to block the intestinal calcium transport appears to be an important survival mechanism in marine fish (Sundell and Bjornsson 1990; Larsson et al. 1995; Larsson 1999). The ability of $24,25(OH)_2D_3$ to inhibit $1,25(OH)_2D_3$-mediated protein kinases A and C activities in chick intestine and kidney (Nemere 1999; Khanal et al. 2007b), as well as matrix vesicles (Sylvia et al. 1997) and chondrocytes (Schwartz et al. 2000; Boyan et al. 2002), suggests a possibility for the existence of a specific binding protein for $24,25(OH)_2D_3$ that has receptor like properties. Moreover, Sylvia et al. (2001) showed that $24,25(OH)_2D_3$ regulation of PKC activity through phospholipase D required a separate functional membrane receptor for the seco-steroid. In concert with these findings, putative membrane receptor proteins for $24,25(OH)_2D_3$ in plasma membranes and evidence for physiological actions were demonstrated in several other animal species and tissues (Seo et al. 1996; Kato et al. 1998; Pedrozo et al. 1999; Larsson et al. 2001, 2003). These putative receptors for $24,25(OH)_2D_3$ show allosteric binding to $24,25(OH)_2D_3$ and mediate rapid membrane-initiated responses regulating calcium homeostasis, cell differentiation, and maturation, and trigger second messenger systems in chondrocytes, osteoblasts, and intestinal cells (Lieberherr 1987; Yukihiro et al. 1994; Boyan et al. 2002; Zhao and Nemere 2002; Larsson and Nemere 2003; Takeuchi and Guggino 1996).

9.5.1.1 Membrane Receptors for $24,25(OH)_2D_3$

Once the possibility for the existence of a specific binding protein that has receptor like properties for $24,25(OH)_2D_3$ was suggested, efforts were directed to isolate and characterize the protein. Percoll-gradient resolution of differential centrifugation fractions from chick intestinal mucosal homogenates revealed the highest levels of specific $[^3H]24,25(OH)_2D_3$ binding to be in lysosomes (Nemere et al. 2002). Protein dependence studies demonstrated linear binding between 0.05 and 0.155 mg of lysosomal protein, and chromatographic studies revealed a single protein band (upon SDS-PAGE and silver staining) of 66 kDa $24,25(OH)_2D_3$ (Nemere et al. 2002). This endogenous protein binding to $24,25(OH)_2D_3$ was identified as catalase (Larsson et al. 2006). Figure 9.4 provides the 527 amino acid sequence with tryptic peptides indicated in bold and underlined. When Western analysis was performed, Ab365 (as primary antibody against samples of chick intestinal catalase) directed against a protein equivalent to that which was sequenced, recognized both commercially available catalase and protein in P_2 fractions (Larsson et al. 2006). Its binding parameters have been presented in Table 9.1. The fact that catalase is tightly bound to the cell surface (Watanabe et al. 2003) gives further credence as the most viable candidate for the membrane receptor of $24,25(OH)_2D_3$.

In endocrinology, every stimulatory system is regulated by a feedback loop (see Figure 9.5). Under conditions of calcium, phosphate, and vitamin D repletion, the metabolite $24,25(OH)_2D_3$ is produced. The $24,25(OH)_2D_3$ metabolite as part of that feedback loop that inhibits $1,25(OH)_2D_3$

1 msdsrdpasd qmkqwkeqra sqrpdvlttg ggnpigdkln imtagsrgpl lvqdvvftde

61 mahfdrerip ervvhakgag afgyfevthd itryskakvf ehig**rtpia vr**fstvages

121 gsadtvrdpr gfavkfyted gnwdlvgnnt piffirdail fpsfihsqkr npqthlkdpd

181 mvwdfwslrp eslhqvsfls sdrgipdghr hmngygshtf klvnadgeav yckfhyktdq

241 giknlpvgea **grlaqedpdy gl**rdlfnaia ngnypswtfy iqvmtfkeae t**fpfnpfdlt**

301 **kvwphkdypl ipvgklvlnk npvnyfaeve qmafdpsnmp pgiepspdkm lggrlfaypd**

361 **thr**hrlgpny lqipvncpyr arvanyqrdg pmcmhdnqgg apnyypnsfs apeqqrsale

421 hsvqcavdvk rfnsanedsv tqvrtfytkv lneeerkrlc enaiaghlkda qlfigkkav**k**

481 **nftdvhpdyg ar**igalldky naekpknaih tytqagshma akgkanl

FIGURE 9.4 Amino acid sequence of catalase, which is the putative receptor of $24,25(OH)_2D_3$. (From Larsson, D. et al., *J. Cell. Biochem.*, 97, 1259–1266, 2006. With permission.)

TABLE 9.1

Binding Parameters of a Putative Membrane Receptor for 24,25(OH)$_2$D$_3$

Tissue	Organelle	Bmax (fmol/mg)	Kd (nM)	Reference
Small intestine	Lysosome	142 ± 16	7.4 ± 1.8	Nemere et al. 2002
-do-	BLM	149 ± 25	8.5	-do-
-do-	BLM	209 ± 34	5.6 ± 2.7	Larsson et al. 2006
Tibia (callus)	Plasma membrane	43.9 ± 6.0	18.3 ± 1.9	Kato et al. 1998
Resting zone cells	Matrix vesicles	52.6	69.2	Pedrozo et al. 1999

stimulated phosphate and calcium transport has been well documented (Nemere 1996; Zhao and Nemere 2002; Nemere et al. 2002). On a molecular level, the inhibitory action is caused by decreased catalase activity in both intestine (Nemere et al. 2006; Peery 2006) and kidney (Khanal et al. 2007) models. In the case of kidney, 24,25(OH)$_2$D$_3$ affects phosphate uptake and catalase activity not only in the short term but also in the long term (Khanal et al. 2007). The 24,25(OH)$_2$D$_3$-mediated inhibition of catalase activity was reversed with commercially available anticatalase antibody (Larsson et al. 2006), which was caused by reduced PKC activity (Khanal et al. 2007).

Because 24,25(OH)$_2$D$_3$ resulted in a time-dependent decrease in the ability of the 1,25D$_3$-MARRS receptor to bind its ligand and decreased binding was not evident after 5 min of incubation (Nemere et al. 2006), two things became evident: 24,25(OH)$_2$D$_3$ does not compete with 1,25(OH)$_2$D$_3$ for binding to the 1,25D$_3$-MARRS receptor, as previously reported (Nemere et al. 1994), and there must be another rapid mechanism to account for the inhibitory effects of 24,25(OH)$_2$D$_3$. Since 24,25(OH)$_2$D$_3$-inhibited PKC activity in various models (Boyan et al. 2002; Khanal et al. 2007), experiments were carried out to assess the involvement of H$_2$O$_2$ in the process. Moreover, an increased H$_2$O$_2$ production was observed in relation to reduced catalase activity (Nemere et al. 2006). Increased H$_2$O$_2$ production, in turn, oxidizes the thioredoxin domains of the 1,25D$_3$-MARRS receptor, leading to reduced 1,25(OH)$_2$D$_3$ binding (Nemere et al. 2006). Indeed, we have tested this model *in vivo*: Chicks were raised on starter diet alone or diet supplemented with antioxidants for three weeks and then tested for net phosphate absorption (Nemere et al. 2006). The birds in the antioxidant

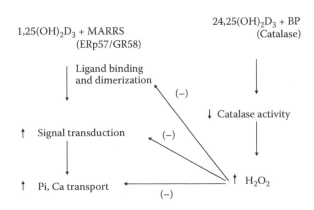

FIGURE 9.5 Model for the interaction between the 1,25(OH)$_2$D$_3$-MARRS receptor and catalase. Upon binding 1,25(OH)$_2$D$_3$, the 1,25(OH)$_2$D$_3$-MARRS receptor dimerizes and initiates signal transduction through PKA for enhanced calcium uptake and PKC for enhanced phosphate uptake. Feedback inhibition of this stimulatory pathway is initiated when 24,25(OH)$_2$D$_3$ binds to catalase, thereby decreasing enzymatic activity. The increased H$_2$O$_2$ levels are known to inhibit stimulated ion uptake. Mechanistically, this occurs through oxidation of the 1,25(OH)$_2$D$_3$-MARRS receptor to decrease 1,25(OH)$_2$D$_3$ binding, and subsequent signal transduction, as well as directly inhibiting hormone-stimulated PKC activity.

supplemented groups exhibited phosphate absorption that was twice that of chicks on the control diet (Nemere et al. 2006). Furthermore, $1,25(OH)_2D_3$ binding to the $1,25D_3$-MARRS receptor was dramatically increased in mucosal fractions prepared from chicks on the antioxidant diets, relative to equivalent preparations from controls (Nemere et al. 2006).

Involvement of H_2O_2 in downregulation of PKC activity by $24,25(OH)_2D_3$ was tested *in vitro*. It was demonstrated that phorbol ester stimulates phosphate uptake in isolated enterocytes (Peery 2006; Zhao and Nemere 2002) and cultured kidney cells (Khanal et al. 2008), and enhanced uptake is susceptible to inhibition by H_2O_2. By measuring PKC activity in response to phorbol, it was found that H_2O_2 did directly abolish PKC stimulation (Peery and Nemere 2006).

9.5.1.2 Physiological Importance of Catalase

Reactive oxygen species, including hydrogen peroxide, are physiologically generated in biological systems and are involved in many pathological processes and aging (Halliwell and Gutteridge 1984; Ames and Shigenaga 1992; Wick et al. 2000; Junqueria et al. 2004; Ames et al. 1992). Hydrogen peroxide is a normal metabolite produced by the enzyme superoxide dismutase during the dismutation of superoxide anion (Halliwell and Gutteridge 1984) and can also be produced by tumor necrosis factor α and cytokines to enhance their own activities (Fiers 1991). It can freely cross membranes and can generate hydroxyl radical, which is the most reactive free radical (Halliwell and Gutteridge 1984). Moreover, H_2O_2 produced by the glucose oxidation system can damage nuclear and mitochondrial DNAs (Salazar and Van Houten 1997). Although generation of H_2O_2 is a constant process, so is its removal by antioxidant enzymes, such as catalase. Predominantly located in cellular peroxisomes and found in nearly all living beings relying on oxygen, catalase is involved in detoxification by converting H_2O_2 to less reactive water and oxygen molecules (Evans et al. 1907; Derevianko et al. 1997; Chelikani et al. 2004).

Catalases are bifunctional and work in two different ways. If the concentration of H_2O_2 is high, the enzyme works catalytically, i.e., removes H_2O_2 by forming water and oxygen molecules; however, at low H_2O_2 concentrations and in the presence of a suitable hydrogen donor, such as phenols and alcohol, it acts peroxidically, removing H_2O_2 but oxidizing its substrate (peroxide reaction) (Kirkman and Gaetani 2007; Scibior and Czeczet 2006). More recently, catalase peroxidases have also been shown to possess oxidase activity (Singh et al. 2004). In addition to bacterial bifunctional catalases, mammalian catalases also have limited ability to act in a manner and condition similar to their bacterial or fungal counterparts (Kirkman and Gaetani 2007).

The mechanism of catalysis has not been fully elucidated, but the overall reaction is as follows:

$$2H_2O_2 \rightarrow 2\,H_2O + O_2$$

The protective function of the enzyme in preventing H_2O_2 accumulation in the cell formed during the oxidation process connected with cell respiration was first proposed more than a hundred years ago (Loew 1902), which was later substantiated through further experimental evidence in bacteria (McLeod and Gordon 1922), as well as in animal tissues (Dixon 1925).

It has been demonstrated that the reactive oxygen species H_2O_2 is increased through the binding of $24,25(OH)_2D_3$ to catalase, and this mediates the inhibition of stimulated phosphate uptake as previously described (Nemere et al. 2006). Furthermore, $24,25(OH)_2D_3$ binds to catalase in both intestinal and kidney epithelia with similar effects (Khanal et al. 2008; Nemere et al. 2006; Peery and Nemere 2007).

9.5.1.3 25(OH)D₃

The current paradigm suggests that 25-hydroxyvitamin D_3, $25(OH)D_3$, is a prehormone activated through the action of 25-hydroxyvitamin D_3 1α-hydroxylase, with no biological activity of its own.

Limited research on $25(OH)D_3$ may, however, suggest otherwise. For instance, one of the early reports suggested that 24-hydroxylation of $25(OH)D_3$ was not required for the embryonic development of chicks (Ameenuddin et al. 1982), whereas 25(OH)D$_3$ was suggested to be essential for chick hatchability. Similarly, normal embryonic development was found in eggs from hens given 25(OH)D$_3$ or 24,24-difluoro-25-hydroxyvitamin D$_3$, in contrast to embryos from hens given 1,25(OH)$_2$D$_3$, 24,25(OH)$_2$D$_3$ or in combination: the latter were abnormal and failed to hatch (Hart and DeLuca 1985). In another report, maximum embryonic survival with lowest embryonic mortality was observed when 0.20 µg/egg of 1,25(OH)$_2$D$_3$ or 0.60 µg/egg of 25(OH)D$_3$ was injected (Ameenuddin et al. 1983). Embryos from hens fed 1,25(OH)$_2$D$_3$ and/or 24,25(OH)$_2$D$_3$ had vitamin D deficiency: low bone ash, low plasma calcium, low total body calcium, and extremely high plasma phosphorus (Hart and DeLuca 1985). More recently, 25(OH)D$_3$ has been shown to suppress parathyroid hormone synthesis by parathyroid cells (Ritter et al. 2006; Ritter and Brown 2011). Lou et al. (2004) suggested that the metabolite at a high but physiological concentration acts as an active hormone with respect to vitamin D$_3$–responsive gene regulation and suppression of cell proliferation. Quantitative real-time RT-PCR analysis revealed that 25(OH)D$_3$-induced 25-hydroxyvitamin D$_3$ 24-hydroxylase (24-hydroxylase) mRNA in a dose- and time-dependent manner (Lou et al. 2004). In prostate cell cancer studies, 25(OH)D$_3$ has been shown to inhibit the proliferation of primary prostatic epithelial cells (Barreto et al. 2000). Moreover, Lambert et al. (2007) demonstrated that the antiproliferative effect of 25(OH)D$_3$-3-bromoacetate is predominantly mediated by VDR in ALVA-31 prostate cancer cells. In another study related to oxidative stress, Peng et al. (2010) used multiple *in vitro* stress models including serum starvation, hypoxia, oxidative stress, and apoptosis induction in breast epithelial cells using an established breast epithelial cell line MCF12F. 25(OH)D$_3$ (250 nmol/L) significantly protected cells against cell death under all situations, demonstrating that the prehormone was involved in the stress process. Microarray analysis further demonstrated that stress induced by serum starvation caused significant alteration in the expression of multiple miRNAs, but the presence of 25(OH)D$_3$ effectively reversed this alteration. These data suggested that there is a significant protective role for 25(OH)D$_3$ against cellular stress in the breast epithelial cells, and these effects may be mediated by altered miRNA expression.

Rapid calcium transport was observed (within 2 min) after perfusion of isolated duodenal loops of normal chickens with 100 nM 25(OH)D$_3$ relative to the vehicle controls (Nemere et al. 1984; Yoshimoto and Norman 1986; Phadnis and Nemere 2003). Although physiologically relevant 100 nM 25(OH)D$_3$ did not stimulate ^{33}P movement from the lumen to venous effluent, 20-nM steroid did from 6 to 40 min (Nemere 1996). In fetal rat calvaria cultures, a significant increase in alkaline phosphatase and tartrate-resistant acid phosphatase activities was observed when treated with the metabolite compared with the untreated controls at 10 and 30 min (Municio and Traba 2004). These results indicate the biological significance of not only the metabolite but also the possible pregenomic effects. Furthermore, 25(OH)D$_3$ had no effect on PKC activity, but PKA activity in isolated enterocytes was stimulated by 200%, relative to vehicle controls (Phadnis and Nemere 2003). We are not aware of the work directed toward identification and characterization of a cell surface 25(OH)D$_3$ binding protein, if it exists or what role it would play in oxidative stress under physiological or disease states.

These results not only suggest the physiological significance of the metabolite 25(OH)D$_3$ but also of the possible genomic and pregenomic pathways for eliciting the effect. The pregenomic pathway may involve direct activation of the VDR (Ritter et al. 2006). Further investigation is needed in at least a few primary areas 1) to provide conclusive evidence on its physiological roles, 2) to elucidate the complete signal transduction pathways originating from actions at the level of plasma membrane, 3) to identify and characterize any cell surface binding proteins, and 4) to fully characterize its binding to the VDR.

In conclusion, the natures of the receptors for 1,25(OH)$_2$D$_3$, 24,25(OH)$_2$D$_3$, and 25(OH)D$_3$ have not been fully elucidated; there may be multiple receptors for each of these metabolites. In addition, each receptor may be responsible for mediating its action through specific signal transduction pathways, a possibility that has yet to be investigated. Given the extensive role that vitamin D plays in many different disease states, it is likely that some of the mechanisms may be directly linked to its membrane receptor. This is an area that is in its early infancy and needs extensive research in the future.

REFERENCES

Alekseeva, I. A., V. B. Spiricheva, N. V. Blazhevich, S. V. Sokolova, and O. G. Pereverzeva. 1982. Comparative evaluation of biological activity of 1 α,25-dihydroxycholecalciferol and 24,25-dihydroxycholecalciferol in rats. *Vopr. Med. Khim.* 28:71–78.

Altieri, F., B. Maras, M. Eufemi, A. Ferraro, and C. Turano. 1993. Purification of a 57kDa nuclear matrix protein associated with thiol:protein-disulfide oxidoreductase and phospholipase C activities. *Biochem. Biophys. Res. Commun.* 194:992–1000.

Amaral, M. D. 2006. Therapy through chaperones: Sense or antisense? Cystic fibrosis as a model disease. *J. Inherit. Metab. Dis.* 29:477–487.

Ameenuddin, S., M. Sunde, H. F. De Luca, N. Ikekawa, and Y. Kobayashi. 1982. 24-hydroxylation of 25-hydroxyvitamin D_3: Is it required for embryonic development in chicks? *Science* 217:451–452.

Ameenuddin, S., M. L. Sunde, H. F. DeLuca, N. Ikekawa, and Y. Kobayashi. 1983. Support of embryonic chick survival by vitamin D metabolites. *Arch. Biochem. Biophys.* 226:666–670.

Ames, B. N. and M. K. Shigenaga. 1992. Oxidants are a major contributor to aging. *Ann. N. Y. Acad. Sci.* 663: 85–96.

Ames, B. N., M. K. Shigenaga, and T. M. Hagen. 1992. Oxidants, antioxidants, and the degenerative diseases of aging. *Proc. Natl. Acad. Sci. USA* 90:7915–7922.

Antoniou, A. N., S. Ford, M. Alphey, A. Osborne, T. Elliott, and S. J. Powis. 2002. The oxidoreductase ERp57 efficiently reduces partially folded in preference to fully folded MHC class I molecules. *EMBO J.* 21:2655–2663.

Barreto, A. M., G. G. Schwartz, R. Woodruff, and S. D. Cramer. 2000. 25-Hydroxyvitamin D_3, the prohormone of 1,25-dihydroxyvitamin D_3, inhibits the proliferation of primary prostatic epithelial cells. *Cancer Epidemiol. Biomarkers Prev.* 9:265–270.

Berdal, A., M. Mesbah, P. Papagerakis, and I. Nemere. 2003. Putative membrane receptor for 1,25(OH)$_2$ vitamin D_3 in human mineralized tissues during prenatal development. *Connect. Tissue Res.* 44 (Suppl 1):136–140.

Binderman, I. and D. Somjen. 1984. 24,25-Dihydroxycholecalciferol induces the growth of chick cartilage *in vitro*. *Endocrinol.* 115:430–432.

Bourdi, M., D. Demady, J. L. Martin, S. K. Jabbour, B. M.. Martin, J. W. George, and L. R. Pohl. 1995. cDNA cloning and baculovirus expression of the human liver endoplasmic reticulum P58: Characterization as a protein disulfide isomerase isoform, but not as a protease or a carnitine acyltransferase. *Arch. Biochem. Biophys.* 323:397–403.

Boyan, B. D., L. F. Bonewald, V. L. Sylvia, I. Nemere, D. Larsson, A. W. Norman, J. Rosser, D. D. Dean, and Z. Schwartz. 2002. Evidence for distinct membrane receptors for 1 α,25-(OH)$_2$D$_3$ and 24R,25-(OH)$_2$D$_3$ in osteoblasts. *Steroids* 67:235–246.

Boyan, B. D., V. L. Sylvia, D. D. Dean, F. Del Toro, and Z. Schwartz. 2002. Differential regulation of growth plate chondrocytes by 1 α,25-(OH)$_2$D$_3$ and 24R,25-(OH)$_2$D$_3$ involves cell-maturation-specific membrane-receptor-activated phospholipid metabolism. *Crit. Rev. Oral. Biol. Med.* 13:143–154.

Boyan, B. D., V. L. Sylvia, D. D. Dean, H. Pedrozo, F. Del Toro, I. Nemere, G. H. Posner, and Z. Schwartz. 1999. 1,25-(OH)$_2$D$_3$ modulates growth plate chondrocytes via membrane receptor-mediated protein kinase C by a mechanism that involves changes in phospholipid metabolism and the action of arachidonic acid and PGE2. *Steroids* 64:129–136.

Boyan, B. D., V. L. Sylvia, D. D. Dean, and Z. Schwartz. 2002. Membrane mediated signaling mechanisms are used differentially by metabolites of vitamin D(3) in musculoskeletal cells. *Steroids* 67:421–427.

Celli, C. M. and A. K. Jaiswal. 2003. Role of GRP58 in mitomycin C-induced DNA cross-linking. *Cancer Res.* 63:6016–6025.

Chelikani, P., I. Fita, and P. C. Loewen. 2004. Diversity of structures and properties among catalases. *Cell. Mol. Life Sci.* 61:192–208.

Coe, H. and M. Michalak. 2010. ERp57, a multifunctional endoplasmic reticulum resident oxidoreductase. *Int. J. Biochem. Cell Biol.* 42:796–799.

De Boland, A. R., I. Nemere, and A. W. Norman. 1990. Ca^{2+}-channel agonist bay K8644 mimics 1,25(OH)$_2$-vitamin D_3 rapid enhancement of Ca^{2+} transport, in chick perfused duodenum. *Biochem. Biophys. Res. Commun.* 166:217–220.

Dean, D. D., B. D. Boyan, Z. Schwart, O. E. Muniz, M. R. Carreno, S. Maeda, and D. S. Howell. 2001. Effect of 1alpha,25-dihydroxyvitamin D_3 and 24R,25-dihydroxyvitamin D_3 on metalloproteinase activity and cell maturation in growth plate cartilage *in vivo*. *Endocrine* 14:311–323.

Dekel, S., R. Salama, and S. Edelstein. 1983. The effect of vitamin D and its metabolites on fracture repair in chicks. *Clin. Sci. (Lond).* 65:429–436.

Derevianko, A., R. D'Amico, T. Graeber, H. Keeping, and H. H. Simms. 1997. Endogenous PMN-derived reactive oxygen intermediates provide feedback regulation on respiratory burst signal transduction. *J. Leukoc. Biol.* 62:268–276.

Dixon, M. 1925. Studies on xanthine oxidase—Part V: The function of catalase. *Biochem. J.* 19:507–512.

Dobson, C. 2006. Protein aggregation and its consequences for human disease. *Protein Pept. Lett.* 13:219–227.

Elliott, J. G., J. D. Oliver, and S. High. 1997. The thiol-dependent reductase ERp57 interacts specifically with N-glycosylated integral membrane proteins. *J. Biol. Chem.* 272:13849–13855.

Erickson, R. R., L. M. Dunning, D. A. Olson, S. J. Cohen, A. T. Davis, W. G. Wood, R. A. Kratzke, and J. L. Holtzman. 2005. In cerebrospinal fluid ER chaperones ERp57 and calreticulin bind beta-amyloid. *Biochem. Biophys. Res. Commun.* 332:50–57.

Evans, C. A. L. 1907. On the catalytic decomposition of hydrogen peroxide by the catalase of blood. *Biochem. J.* 2:133–155.

Facchinetti, M. M. and A. R. de Boland. 1999. Effect of ageing on the expression of protein kinase C and its activation by 1,25(OH)$_2$-vitamin D$_3$ in rat skeletal muscle. *Cell Signal* 1139–1144.

Farach-Carson, M. C. and I. Nemere. 2003. Membrane receptors for vitamin D steroid hormones: Potential new drug targets. *Curr. Drug. Targets* 4:67–76.

Feng, D., J. Wei, S. Gupta, B. C. McGrath, and D. R. Cavener. 2009. Acute ablation of PERK results in ER dysfunctions followed by reduced insulin secretion and cell proliferation. *BMC Cell Biol.* 10:61.

Ferrari, D. M. and H. D. Söling. 1999. The protein disulphide-isomerase family: Unraveling a string of folds. *Biochem J.* 339(Part 1):1–10.

Ferraro, A., F. Altieri, S. Coppari, M. Eufemi, S. Chichiarelli, and C. Turano. 1999. Binding of the protein disulphide isomerase isoform ERp60 to the nuclear matrix-associated regions. *J. Cell. Biochem.* 72:528–539.

Fiers, W. 1991. Tumor necrosis factor: Characterization at the molecular, cellular and *in vivo* level. *FEBS Lett.* 285:199–212.

Garabedian, M., M. Lieberherr, T. M. N'Guyen, M. T. Corvol, M. B. Du Bois, and S. Balsan. 1978. The *in vitro* production and activity of 24, 25-dihydroxycholecalciferol in cartilage and calvarium. *Clin. Orthop. Relat. Res.* 135:241–248.

Garabedian, M., M. Lieberherr, C. L. Thil, H. Guillozo, B. Grosse, and S. Balsan. 1981. Vitamin D metabolism in rat calvarium *in vitro*. *Ann. Endocrinol. (Paris)* 42:488–491.

Garbi, N., S. Tanaka, F. Momburg, and G. J. Hammerling. 2006. Impaired assembly of the major histocompatibility complex class I peptide-loading complex in mice deficient in the oxidoreductase ERp57. *Nat. Immunol.* 7:93–102.

Grosse, B., A. Bourdeau, and M. Lieberherr. 1993. Oscillations in inositol 1,4,5-trisphosphate and diacyglycerol induced by vitamin D$_3$ metabolites in confluent mouse osteoblasts. *J. Bone Miner. Res.* 8:1059–1069.

Guo, G. G., K. Patel, V. Kumar, M. Shah, V. A. Fried, J. D. Etlinger, and P. B. Sehgal. 2002. Association of the chaperone glucose-regulated protein 58 (GRP58/ER-60/ERp57) with Stat3 in cytosol and plasma membrane complexes. *J. Interferon. Cytokine Res.* 22:555–563.

Hale, L. V., M. L. Kemick, and R. E. Wuthier. 1986. Effect of vitamin D metabolites on the expression of alkaline phosphatase activity by epiphyseal hypertrophic chondrocytes in primary cell culture. *J. Bone Miner. Res.* 1:489–495.

Halliwell, B. and J. M. Gutteridge. 1984. Oxygen toxicity, oxygen radicals, transition metals and disease. *Biochem. J.* 219:1–14

Hart, L. E. and H. F. De Luca. 1985. Effect of vitamin D$_3$ metabolites on calcium and phosphorus metabolism in chick embryos. *Am. J. Physiol.* 248:E281–285.

Henry, H. L. and A. W. Norman. 1978. Vitamin D: Two dihydroxylated metabolites are required for normal chicken egg hatchability. *Science* 201:835–837.

Hetz, C., M. Russelakis-Carneiro, S. Wälchli, S. Carboni, E. Vial-Knecht, K. Maundrell, J. Castilla, and C. Soto. 2005. The disulfide isomerase Grp58 is a protective factor against prion neurotoxicity. *J. Neurosci.* 25:2793–2802.

Hirano, N., F. Shibakashi, R. Sakai, T. Tanaka, J. Nishida, Y. Yazaki, T. Takenawa, and H. Hirai. 1995. Molecular cloning of the human glucose–regulated protein ERp57/GRP58, a thiol-dependent reductase. Identification of its secretory form and inducible expression by the oncogenic transformation. *Eur. J. Biochem.* 234:336–342.

Ittel, T. H., F. P. Ross, and A. W. Norman. 1986. Activity of ornithine decarboxylase and creatine kinase in soft and hard tissue of vitamin D-deficient chicks following parenteral application of 1,25-dihydroxyvitamin D$_3$ or 24R,25-dihydroxyvitamin D$_3$. *J. Bone Miner. Res.* 1:23–31.

Junqueira, V. B., S. B. Barros, S. S. Chan, L. Rodrigues, L. Giavarotti, R. L. Abud, and G. P. Deucher. 2004. Aging and oxidative stress. *Mol. Aspects Med.* 25:5–16

Kamao, M., S. Tatematsu, G. S. Reddy, S. Hatakeyama, M. Sugiura, N. Ohashi, N. Kubodera, and T. Okano. 2001. Isolation, identification and biological activity of 24R,25-dihydroxy-3-epi-vitamin D3: A novel metabolite of 24R,25-dihydroxyvitamin D3 produced in rat osteosarcoma cells (UMR 106). *J. Nutr. Sci. Vitaminol (Tokyo)* 47:108–115.

Kato, A., E. G. Seo, T. A. Einhorn, J. E. Bishop, and A. W. Norman. 1998. Studies on 24R,25-dihydroxyvitamin D_3: Evidence for a nonnuclear membrane receptor in the chick tibial fracture-healing callus. *Bone* 23:141–146.

Khanal, R. C. and I. Nemere. 2007a. Membrane receptors for vitamin D metabolites. *Crit. Rev. Eukaryot. Gene. Expr.* 17:31–47.

Khanal, R. C. and I. Nemere. 2007b. The ERp57/GRp58/1,25D3-MARRS receptor: Multiple functional roles in diverse cell systems. *Curr. Med. Chem.* 14:1087–1093.

Khanal, R. C. and I. Nemere. 2008. Regulation of intestinal calcium transport. *Annu. Rev. Nutr.* 28:179–196.

Khanal, R. C. and I. Nemere. 2009. Newly recognized receptors for vitamin D metabolites. *Immunol. Endocrine Metabol. Agents Med. Chem.* 9:143–152.

Khanal, R. C., T. M. Peters, N. M. Smith, and I. Nemere. 2008. Membrane receptor–initiated signaling in 1,25(OH)$_2$D$_3$-stimulated calcium uptake in intestinal epithelial cells. *J. Cell. Biochem.* 105:1109–1116.

Khanal, R. C., N. M. Smith, and I. Nemere. 2007. Phosphate uptake in chick kidney cells: Effects of 1,25(OH)$_2$D$_3$ and 24,25(OH)$_2$D$_3$. *Steroids* 72:158–164.

Kirkman, H. N. and G. F. Gaetani. 1984. Catalase: A tetrameric enzyme with four tightly bound molecules of NADPH. *Proc. Natl. Acad. Sci. USA* 81:4343–4347.

Kirkman, H. N. and G. F. Gaetani. 2007. Mammalian catalase: A venerable enzyme with new mysteries. *Trends Biochem. Sci.* 32:44–50.

Koivunen, P., N. Horelli-Kuitunen, T. Helaakoski, P. Karvonen, M. Jaakkola, A. Palotie, and K. I. Kivirikko. 1997. Structures of the human gene for the protein disulfide isomerase-related polypeptide ERp60 and a processed gene and assignment of these genes to 15q15 and 1q21. *Genomics* 42:397–404.

Kozaki, K., O. Miyaishi, N. Asai, K. Iida, K. Sakata, M. Hayashi, T. Nishida, M. Matsuyama, S. Shimizu, T. Kaneda, et al. 1994. Tissue distribution of ERp61 and association of its increased expression with IgG production in hybridoma cells. *Exp. Cell Res.* 213:348–358.

Krynetski, E. Y., N. F. Krynetskaia, M. E. Bianchi, and W. E. Evans. 2003. A nuclear protein complex containing high mobility group proteins B1 and B2, heat shock cognate protein 70, ERp60, and glyceraldehyde-3-phosphate dehydrogenase is involved in the cytotoxic response to DNA modified by incorporation of anticancer nucleoside analogues. *Cancer Res.* 63:100–106.

Lambert, J. R., C. D. Young, K. S. Persons, and R. Ray. 2007. Mechanistic and pharmacodynamic studies of a 25-hydroxyvitamin D$_3$ derivative in prostate cancer cells. *Biochem. Biophys. Res. Commun.* 361:189–195.

Larsson, B. and I. Nemere. 2003. Effect of growth and maturation on membrane-initiated actions of 1,25-dihydroxyvitamin D$_3$—Part I: Calcium transport, receptor kinetics, and signal transduction in intestine of male chickens. *Endocrinology* 144:1726–1735.

Larsson, B. and I. Nemere. 2003. Effect of growth and maturation on membrane-initiated actions of 1,25-dihydroxyvitamin D$_3$—Part II: Calcium transport, receptor kinetics, and signal transduction in intestine of female chickens. *J. Cell. Biochem.* 90:901–913.

Larsson, D., D. Anderson, N. M. Smith, and I. Nemere. 2006. 24,25-dihydroxyvitamin D$_3$ binds to catalase. *J. Cell. Biochem.* 97:1259–1266.

Larsson, D., B. T. Bjornsson, and K. Sundell. 1995. Physiological concentrations of 24,25-dihydroxyvitamin D$_3$ rapidly decrease the in vitro intestinal calcium uptake in the Atlantic cod, Gadus morhua. *Gen. Comp. Endocrinol.* 100:211–217.

Larsson, D., I. Nemere, L. Aksnes, and K. Sundell. 2003. Environmental salinity regulates receptor expression, cellular effects, and circulating levels of two antagonizing hormones, 1,25-dihydroxyvitamin D$_3$ and 24,25-dihydroxyvitamin D$_3$, in rainbow trout. *Endocrinol.* 144:559–566.

Larsson, D., I. Nemere, and K. Sundell. 2001. Putative basal lateral membrane receptors for 24,25-dihydroxyvitamin D$_3$ in carp and Atlantic cod enterocytes: Characterization of binding and effects on intracellular calcium regulation. *J. Cell. Biochem.* 83:171–186.

Larsson, D. 1999. Vitamin D in teleost fish: Nongenomic regulation of intestinal calcium transport. PhD Dissertation. ISBN 91-628-3681-1. Göteborg University, Gothenburg.

Loew, O. 1902. Spielt Wasserstoffsuperoxyd eine rolle in der lebenden zelle? *Ber. Dtsch. Chem.* 35:2487–2488.

Li, Y. and P. Camacho. 2004. Ca2+-dependent redox modulation of SERCA 2b by ERp57. *J. Cell Biol.* 164:35–46.

Lidor, C., S. Dekel, and S. Edelstein. 1987. The metabolism of vitamin D$_3$ during fracture healing in chicks. *Endocrinol.* 120:389–393.

Lieberherr, M., M. Garabedian, H. Guillozo, M. Bailly du Bois, and S. Balsan. 1979. Interaction of 24,25-dihy-droxyvitamin D3 and parathyroid hormone on bone enzymes *in vitro*. *Calcif. Tissue Int.* 27:47–52.

Lieberherr, M., B. Grosse, P. Duchambon, and T. Drueke. 1989. A functional cell surface type receptor is required for the early action of 1,25-dihydroxyvitamin D_3 on the phosphoinositide metabolism in rat enterocytes. *J. Biol. Chem.* 264:20403–20406.

Lieberherr, M., E. Pezant, M. Garabedian, and S. Balsan. 1977. Phosphatase content of rat calvaria after *in vivo* administration of vitamin D3 metabolites. *Calcif. Tissue Res.* 23:235–239.

Lieberherr, M. 1987. Effects of vitamin D_3 metabolites on cytosolic free calcium in confluent mouse osteo-blasts. *J. Biol. Chem.* 262:13168–13173.

Lou, Y. R., I. Laaksi, H. Syvälä, M. Bläuer, T. L. Tammela, T. Ylikomi, and P. Tuohimaa. 2004. 25-hydroxy-vitamin D_3 is an active hormone in human primary prostatic stromal cells. *FASEB J.* 18:332–334.

Malluche, H. H., H. Henry, W. Meyer-Sabellak, D. Sherman, S. G. Massry, and A. W. Norman. 1980. Effects and interactions of 24R,25(OH)$_2$D$_3$ and 1,25(OH)$_2$D$_3$ on bone. *Am. J. Physiol.* 238:E494–498.

McLeod, J. W. and J. Gordon. 1922. Production of hydrogen peroxide by bacteria. *Biochem. J.* 16:499–506.

Mesbah, M., I. Nemere, P. Papagerakis, J. R. Nefussi, S. Orestes-Cardoso, C. Nessmann, and A. Berdal. 2002. Expression of a 1,25-dihydroxyvitamin D_3 membrane-associated rapid-response steroid-binding protein during human tooth and bone development and biomineralization. *J. Bone Miner. Res.* 17:1588–1596.

Molinari, M. and A. Helenius. 1999. Glycoproteins form mixed disulphides with oxidoreductases during fold-ing in living cells. *Nature* 402(6757):90–93.

Morelli, S., A. R. de Boland, and R. L. Boland. 1993. Generation of inositol phosphates, diacylglycerol and calcium fluxes in myoblasts treated with 1,25-dihydroxyvitamin D3. *Biochem. J.* 289:675–679.

Muhlenkamp, C. R. and S. S. Gill. 1998. A glucose-regulated protein, GRP58, is downregulated in C57B6 mouse liver after diethylhexyl phthalate exposure. *Toxicol. Appl. Pharmacol.* 148:101–108.

Municio, M. J. and M. L. Traba. 2004. Effects of 24,25(OH)$_2$D$_3$, 1,25(OH)$_2$D$_3$ and 25(OH)D$_3$ on alkaline and tartrate-resistant acid phosphatase activities in fetal rat calvaria. *J. Physiol. Biochem.* 60:219–224.

Nardai, G., K. Stadler, E. Papp, T. Korcsmáros, J. Jakus, and P. Csermely. 2005. Diabetic changes in the redox status of the microsomal protein folding machinery. *Biochem. Biophys. Res. Commun.* 334:787–795.

Nemere, I., M. C. Dormanen, M. W. Hammond, W. H. Okamura, and A. W. Norman. 1994. Identification of a specific binding protein for 1 alpha,25-dihydroxyvitamin D_3 in basal-lateral membranes of chick intesti-nal epithelium and relationship to transcaltachia. *J. Biol. Chem.* 269:23750–23756.

Nemere, I., M. C. Farach-Carson, B. Rohe, T. M. Sterling, A. W. Norman, B. D. Boyan, and S. E. Safford. 2004. Ribozyme knockdown functionally links a 1,25(OH)$_2$D$_3$ membrane binding protein (1,25D$_3$-MARRS) and phosphate uptake in intestinal cells. *Proc. Natl. Acad. Sci. USA.* 101:7392–7397.

Nemere, I., N. Garcia-Garbi, G. J. Hämmerling, and R. C. Khanal. 2010a. Intestinal cell calcium uptake and the targeted knockout of the 1,25D$_3$-MARRS receptor/PDIA3/Erp57. *J. Biol. Chem.* 285:31859–31866.

Nemere, I., N. Garcia-Garbi, G. J. Hämmerling, and Q. Winger. 2012. Intestinal cell phosphate uptake and the targeted knockout of the 1,25D$_3$-MARRS receptor/PDIA3/Erp57. *Endocrinology* 153:1609–1615.

Nemere, I., N. Garbi, G. J. Hämmerling, and K. J. Hintze. 2012b. Role of the 1,25D$_3$-MARRS receptor in the 1,25(OH)$_2$D$_3$-stimulated uptake of calcium and phosphate in intestinal cells. *Steroids* 77:897–902.

Nemere, I., R. Ray, and W. McManus. 2000. Immunochemical studies on the putative plasmalemmal receptor for 1, 25(OH)$_2$D$_3$—Part I: Chick intestine. *Am. J. Physiol. Endocrinol. Metab.* 278:E1104–1114.

Nemere, I. and A.W. Norman. 1982. Vitamin D and intestinal cell membranes. *Biochim. Biophys. Acta* 694:307–327.

Nemere, I. and A.W. Norman. 1987. Rapid actions of 1,25-dihydroxyvitamin D_3 on calcium transport in pre-fused duodena: Effect of inhibitors. *J. Bone Mineral Res.* 2:99–107.

Nemere, I., S. E. Safford, B. Rohe, M. M. De Souza, and M. C. Farach-Carson. 2004. Identification and char-acterization of 1,25D$_3$-membrane-associated rapid response, steroid (1,25D3-MARRS) binding protein. *J. Steroid Biochem. Mol. Biol.* 89–90:281–285.

Nemere, I., C. Wilson, W. Jensen, M. Steinbeck, B. Rohe, and M. C. Farach-Carson. 2006. Mechanism of 24,25-dihydroxyvitamin D_3-mediated inhibition of rapid, 1,25-dihydroxyvitamin D_3-induced responses: Role of reactive oxygen species. *J. Cell. Biochem.* 99:1572–1581.

Nemere, I., D. Yazzie-Atkinson, D. O. Johns, and D. Larsson. 2002. Biochemical characterization and puri-fication of a binding protein for 24,25-dihydroxyvitamin D_3 from chick intestine. *J. Endocrinol.* 172: 211–219.

Nemere, I., Y. Yoshimoto, and A. W. Norman. 1984. Calcium transport in perfused duodena from normal chicks: Enhancement within 14 minutes of exposure to 1,25-dihydroxyvitamin D_3. *Endocrinology* 115:1476–1483.

Nemere, I. and C. M. Szego. 1981. Early actions of parathyroid hormone and 1,25-dihydroxycholecalciferol on isolated epithelial cells from rat intestine—Part I: Limited lysosomal enzyme release and calcium uptake. *Endocrinology* 108:1450–1462.

Nemere, I. 1996. Apparent nonnuclear regulation of intestinal phosphate transport: Effects of 1,25-dihydroxy-vitamin D_3, 24,25-dihydroxyvitamin D_3, and 25-hydroxyvitamin D_3. *Endocrinology* 137:2254–2261.

Nemere, I. 1999. 24,25-dihydroxyvitamin D_3 suppresses the rapid actions of 1,25-dihydroxyvitamin D_3 and parathyroid hormone on calcium transport in chick intestine. *J. Bone Miner. Res.* 14:1543–1549.

Nemere, I. 2005. The $1,25D_3$-MARRS protein: Contribution to steroid stimulated calcium uptake in chicks and rats. *Steroids* 70:455–457.

Norman, A. W., M. Dormanen, W. H. Okamura, M. Hammond, and I. Nemere. 1994. Nonnuclear actions of $1\alpha,25(OH)_2D_3$ and $24R,25(OH)_2D_3$ in mediating intestinal calcium transport: The use of analogs to study membrane receptors for vitamind D metabolites and to determine receptor ligand conformational preferences. In *Vitamin D, a Pluripotent steroid hormone: Structural Studies, Molecular Endocrinology and Clinical Applications* (Norman, A. W., Bouillon, R., Thomasset, M., eds.), Walter de Gruyter, Berlin, 324–331.

Norman, A. W., V. Leathers, and J. E. Bishop. 1983. Normal egg hatchability requires the simultaneous administration to the hen of 1 alpha,25-dihydroxycholecalciferol and 24R,25-dihydroxycholecalciferol. *J. Nutr.* 113:2505–2515.

Norman, A. W., W. H. Okamura, J. E. Bishop, and H. L. Henry. 2002. Update on biological actions of 1alpha,25(OH)2-vitamin D3 (rapid effects) and $24R,25(OH)_2$-vitamin D_3. *Mol. Cell. Endocrinol.* 197: 1–13.

Norman, A. W., C. J. Olivera, F. R. Barreto Silva, and J. E. Bishop. 2002. A specific binding protein/receptor for 1alpha,25-dihydroxyvitamin D_3 is present in an intestinal caveolae membrane fraction. *Biochem. Biophys. Res. Commun.* 298:414–419.

Pedrozo, H. A., Z. Schwartz, S. Rimes, V. L. Sylvia, I. Nemere, G. H. Posner, D. D. Dean, and B. D. Boyan. 1999. Physiological importance of the $1,25(OH)_2D_3$ membrane receptor and evidence for a membrane receptor specific for $24,25(OH)_2D_3$. *J. Bone Miner. Res.* 14:856–867.

Peery, S. L. 2006. Catalase activity mediates the inhibitory actions of 24,25 dihydroxyvitamin D_3. Master's Thesis. Department of Nutrition and Food Science, Utah State University, Logan, USA.

Peery, S. L. and I. Nemere. 2007. Contributions of pro-oxidant and anti-oxidant conditions to the actions of 24,25-dihydroxyvitamin D_3 and 1,25-dihydroxyvitamin D_3 on phosphate uptake in intestinal cells. *J. Cell. Biochem.* 101:1176–1184.

Peng, X., A. Vaishnav, G. Murillo, F. Alimirah, K. E. Torres, and R. G. Mehta. 2010. Protection against cellular stress by 25-hydroxyvitamin D3 in breast epithelial cells. *J. Cell. Biochem.* 110:1324–1333.

Phadnis, R. and I. Nemere. 2003. Direct, rapid effects of 25-hydroxyvitamin D_3 on isolated intestinal cells. *J. Cell. Biochem.* 90:287–293.

Pollock, S., G. Kozlov, M. F. Pelletier, J. F. Trempe, G. Jansen, D. Sitnikov, J. J. Bergeron, K. Gehring, I. Ekiel, and D. Y. Thomas. 2004. Specific interaction of ERp57 and calnexin determined by NMR spectroscopy and an ER two-hybrid system. *EMBO J.* 23:1020–1029.

Richard, C. L., M. C. Farach-Carson, B. Rohe, I. Nemere, and K. A. Meckling. 2010. Involvement of 1,25D3-MARRS (membrane associated, rapid response steroid-binding), a novel vitamin D receptor, in growth inhibition of breast cancer cells. *Exp. Cell. Res.* 316:695–703.

Ritter, C. S. and A. J. Brown. 2011. Direct suppression of Pth gene expression by the vitamin D prohormones doxercalciferol and calcidiol requires the vitamin D receptor. *J. Mol. Endocrinology* 46:63–66.

Ritter, C. S., H. J. Armbrecht, E. Slatopolsky, and A. J. Brown. 2006. 25-Hydroxyvitamin D_3 suppresses PTH synthesis and secretion by bovine parathyroid cells. *Kidney Int.* 70:654–659.

Rohe, B., S. E. Safford, I. Nemere, and M. C. Farach-Carson. 2005. Identification and characterization of $1,25D_3$-membrane-associated rapid response, steroid ($1,25D_3$-MARRS)-binding protein in rat IEC-6 cells. *Steroids* 70:458–463.

Salazar, J. J. and B. Van Houten. 1997. Preferential mitochondrial DNA injury caused by glucose oxidase as a steady generator of hydrogen peroxide in human fibroblasts. *Mutat. Res.* 385:139–149.

Scheibel, T. and J. Buchner. 2006. Protein aggregation as a cause for disease. *Handb. Exp. Pharmacol.* 172:199–220.

Schwartz, Z., D. L. Schlader, L. D. Swain, and B. D. Boyan. 1988. Direct effects of 1,25-dihydroxyvitamin D_3 and 24,25-dihydroxyvitamin D_3 on growth zone and resting zone chondrocyte membrane alkaline phosphatase and phospholipase-A2 specific activities. *Endocrinology* 123:2878–2884.

Schwartz, Z., V. L. Sylvia, F. Del Toro, R. R. Hardin, D. D. Dean, and B. D. Boyan. 2000. 24R,25-(OH)$_2$D$_3$ mediates its membrane receptor-dependent effects on protein kinase C and alkaline phosphatase via phospholipase A$_2$ and cyclooxygenase-1 but not cyclooxygenase-2 in growth plate chondrocytes. *J. Cell. Physiol.* 182:390–401.

Scibior, D. and H. Czeczot. 2006. Catalase: Structure, properties, functions. *Postepy. Hig. Med. Dosw.* (Online) 60:170–180.

Seliger, B., U. Wollscheid, F. Momburg, T. Blankenstein, and C. Huber. 2000. Coordinate downregulation of multiple MHC class I antigen processing genes in chemical-induced murine tumor cell lines of distinct origin. *Tissue Antigens* 56:327–336.

Selivanova, A., B. Winblad, N. P. Dantuma, and M. R. Farmery. 2007. Biogenesis and processing of the amyloid precursor protein in the early secretory pathway. *Biochem. Biophys. Res. Commun.* 357:1034–1039.

Seo, E. G., T. A. Einhorn, and A. W. Norman. 1997. 24R,25-dihydroxyvitamin D$_3$: An essential vitamin D$_3$ metabolite for both normal bone integrity and healing of tibial fracture in chicks. *Endocrinology* 138:3864–3872.

Seo, E. G., A. Kato, and A. W. Norman. 1996. Evidence for a 24R,25(OH)$_2$-vitamin D$_3$ receptor/binding protein in a membrane fraction isolated from a chick tibial fracture-healing callus. *Biochem. Biophys. Res. Commun.* 225:203–208.

Seo, E. G. and A. W. Norman. 1997. Threefold induction of renal 25-hydroxyvitamin D$_3$-24-hydroxylase activity and increased serum 24,25-dihydroxyvitamin D$_3$ levels are correlated with the healing process after chick tibial fracture. *J. Bone Miner. Res.* 12:598–606.

Silvennoinen, L., P. Koivunen, J. Myllyharju, I. Kilpeläinen, and P. Permi. 2005. NMR assignment of the N-terminal domain a of the glycoprotein chaperone ERp57. *J. Biomol. NMR.* 33:136.

Silvennoinen, L., J. Myllyharju, M. Ruoppolo, S. Orrù, M. Caterino, K. I. Kivirikko, and P. Koivunen. 2004. Identification and characterization of structural domains of human ERp57: Association with calreticulin requires several domains. *J. Biol. Chem.* 279:13607–13615.

Singh, R., B. Wiseman, T. Deemagarn, L. J. Donald, H. W. Duckworth, X. Carpena, I. Fita, and P. C. Loewen. 2004. Catalase-peroxidases (KatG) exhibit NADH oxidase activity. *J. Biol. Chem.* 279:43098–43106.

Sömjen, D., I. Binderman, and Y. Weisman. 1983. The effects of 24R,25-dihydroxycholecalciferol and of 1 α,25-dihydroxycholecalciferol on ornithine decarboxylase activity and on DNA synthesis in the epiphysis and diaphysis of rat bone and in the duodenum. *Biochem. J.* 214:293–298.

Sterling, T. M. and I. Nemere. 2005. 1,25-dihydroxyvitamin D$_3$ stimulates vesicular transport within 5 s in polarized intestinal epithelial cells. *J. Endocrinol.* 185:81–91.

Sundell, K. and B. T. Bjornsson. 1990. Effects of vitamin D$_3$, 25(OH) vitamin D$_3$, 24,25(OH)$_2$ vitamin D$_3$, and 1,25(OH)$_2$ vitamin D$_3$ on the in vitro intestinal calcium absorption in the marine teleost, Atlantic cod (Gadus morhua). *Gen. Comp. Endocrinol.* 78:74–79.

Sylvia, V. L., Z. Schwartz, F. Del Toro, P. De Veau, R. Whetstone, R. R. Hardin, D. D. Dean, and B. D. Boyan. 2001. Regulation of phospholipase D (PLD) in growth plate chondrocytes by 24R,25-(OH)$_2$D$_3$ is dependent on cell maturation state (resting zone cells) and is specific to the PLD$_2$ isoform. *Biochim. Biophys. Acta* 1499:209–221.

Sylvia, V. L., Z. Schwartz, S. C. Holmes, D. D. Dean, and B. D. Boyan. 1997. 24,25-(OH)$_2$D$_3$ regulation of matrix vesicle protein kinase C occurs both during biosynthesis and in the extracellular matrix. *Calcif. Tissue Int.* 61:313–321.

Sylvia, V. L., Z. Schwartz, L. Schuman, R. T. Morgan, S. Mackey, R. Gomez, and B. D. Boyan. 1993. Maturation-dependent regulation of protein kinase C activity by vitamin D$_3$ metabolites in chondrocyte cultures. *J. Cell. Physiol.* 157:271–278.

Szego, C. M. and S. Roberts. 1953. Steroid action and interaction in uterine metabolism. *Rec. Prog. Horm. Res.* 8:419–459.

Takeuchi, K. and S. E. Guggino. 1996. 24R,25-(OH)$_2$ vitamin D$_3$ inhibits 1alpha,25-(OH)$_2$ vitamin D$_3$ and testosterone potentiation of calcium channels in osteosarcoma cells. *J. Biol. Chem.* 271:33335–33343.

Tamura, T., T. Yamashita, H. Segawa, and H. Taira. 2002. N-linked oligosaccharide chains of Sendai virus fusion protein determine the interaction with endoplasmic reticulum molecular chaperones. *FEBS Lett.* 513:153–158.

Teillaud, C., I. Nemere, F. Boukhobza, C. Mathiot, N. Conan, M. Oboeuf, D. Hotton, M. Macdougall, and A. Berdal. 2005. Modulation of 1α,25-dihydroxyvitamin D$_3$-membrane associated, rapid response steroid binding protein expression in mouse odontoblasts by 1α,25-(OH)$_2$D$_3$. *J. Cell. Biochem.* 94:139–152.

Thomas, P. J., B. H. Qu, and P. L. Pedersen. 1995. Defective protein folding as a basis of human disease. *Trends Biochem. Sci.* 20:456–459.

Toffolon, E. P., M. M. Pechett, and K. Isselbacher. 1975. Demonstration of the rapid action of pure crystalline 1 alpha-hydroxy vitamin D_3 and 1 alpha,25-dihydroxy vitamin D_3 on intestinal calcium uptake. *Proc. Natl. Acad. Sci. USA* 72:229–230.

Tourkova, I. L., G. V. Shurin, G. S. Chatta, and L. Perez. 2005. Restoration by IL-15 of MHC class I antigen-processing machinery in human dendritic cells inhibited by tumor-derived gangliosides. *J. Immunol.* 175:3045–3052.

Urade, R., T. Oda, H. Ito, T. Moriyama, S. Utshumi, and M. Kito. 1997. Functions of characteristic Cys-Gly-His-Cys (CGHC) and Gln-Glu-Asp-Leu (QEDL) motifs of microsomal ER-60 protease. *J. Biochem. (Tokyo)* 122:834–842.

Urade, R., H. Okudo, H. Kato, T. Moriyama, and Y. Arakaki. 2004. ER-60 domains responsible for interaction with calnexin and calreticulin. *Biochem.* 43:8858–8868.

van der Vlies, D., E. H. W. Pap, J. A. Post, J. E. Celis, and K. W. Wirtz. 2002. Endoplasmic reticulum resident proteins of normal human dermal fibroblasts are the major targets for oxidative stress-induced by hydrogen peroxide. *Biochem J.* 366:825–830.

Walters, M. R. 1985. Steroid hormone receptors and the nucleus. *Endocr. Rev.* 6:512–543.

Wang, Y., J. Chen, C. S. D. Lee, A. Nizkorodov, K. Riemenschneider, D. Martin, S. Hyzy, Z. Schwartz, and B. D. Boyan. 2010. Disruption of Pdia3 gene results in bone abnormality and affects 1α,25-dihydroxy-vitamin D_3–induced rapid activation of PKC. *J. Steroid Biochem. Mol. Biol.* 121:257–260.

Watanabe, N., T. Iwamoto, K. D. Bowen, D. A. Dickinson, M. Torres, and H. J. Forman. 2003. Bioeffectiveness of Tat-catalase conjugate: A potential tool for the identification of H_2O_2-dependent cellular signal transduction pathways. *Biochem. Biophys. Res. Commun.* 303.287–293.

Wick, G., P. Jansen-Dürr, P. Berger, I. Blasko, and B. Grubeck-Loebenstein. 2000. Diseases of aging. *Vaccine* 18:1567–1583.

Winklhofer, K. F. and J. Tatzelt. 2006. The role of chaperones in Parkinson's disease and prion diseases. *Handb. Exp. Pharmacol.* 172:221–258.

Wu, W., G. Beilhartz, Y. Roy, C. L. Richard, M. Curtin, L. Brown, D. Cadieux, M. Coppolino, M. C. Farach-Carson, I. Nemere, and K. A. Meckling. 2010. Nuclear translocation of the 1,25D3-MARRS (membrane-associated rapid response to steroids) receptor protein and NFkappaB in differentiating NB4 leukemia cells. *Exp. Cell. Res.* 31:1101–1118.

Yamato, H., R. Okazaki, T. Ishii, E. Ogata, M. Sato, M. Kumegawa, K. Akaogi, N. Taniguchi, and T. Matsumoto. 1993. Effect of 24R,25-dihydroxyvitamin D_3 on the formation and function of osteoclastic cells. *Calcif. Tissue Int.* 52:255–260.

Yoshimoto, Y. and A. W. Norman. 1986. Biological activity of vitamin D metabolites and analogs: Dose-response study of ^{45}Ca transport in an isolated chick duodenum perfusion system. *J. Steroid Biochem.* 25:905–909.

Yukihiro, S., G. H. Posner, and S. E. Guggino. 1994. Vitamin D_3 analogs stimulate calcium currents in rat osteosarcoma cells. *J. Biol. Chem.* 269:23889–23893.

Zhang, L., E. Lai, T. Teodoro, and A. Volchuk. 2009. GRP78, but not protein-disulfide isomerase, partially reverses hyperglycemia-induced inhibition of insulin synthesis and secretion in pancreatic β-cells. *J. Biol. Chem.* 284:5289–5298.

Zhao, B. and I. Nemere. 2002. 1,25$(OH)_2D_3$-mediated phosphate uptake in isolated chick intestinal cells: Effect of 24,25$(OH)_2D_3$, signal transduction activators, and age. *J. Cell. Biochem.* 86:497–508.

Section III

Immunity and Disease

10 Vitamin D and Human Innate Immunity

Eun-Kyeong Jo, Dong-Min Shin, and Robert L. Modlin

CONTENTS

It has become increasingly evident that vitamin D has a functional role in innate immunity and inflammation. First, vitamin D–induced antimicrobial activity against pathogens is a key component of the monocyte/macrophage response in infectious disease (Liu et al. 2006; Yuk et al. 2009; Gombart 2009; Jo 2010). Second, it is now recognized that vitamin D insufficiency/deficiency, defined by serum 25D levels, is a clinical problem of global proportions (Oliveri et al. 2004; Rosen 2011; O'Sullivan et al. 2008; Holick 2007; Hollis et al. 2007; Adams and Hewison 2008) and potentially compromises the intracrine metabolism of precursor 25-hydroxyvitamin D_3 (25D; Hewison 2010b) to sufficient levels of active 1,25-dihydroxyvitamin D_3 (1,25D) in monocytes/macrophages (Adams and Hewison 2008). Consequently, the intracellular or local levels of 1,25D achieved during an immune response are suboptimal and, therefore, not sufficient to activate the necessary downstream immune effector functions. Third, investigations into the vitamin D receptor (VDR) have revealed exaggerated inflammatory responses through defects in VDR signaling, suggesting a potential link between decreased circulating vitamin D levels and VDR signaling with chronic inflammatory diseases (Sun et al. 2006; Szeto et al. 2007; Zehnder et al. 2008; Proal et al. 2009; Wu and Sun 2011; Liu et al. 2011).

In this chapter, we detail the observations and discoveries that provide new insight into the role of vitamin D in innate immunity with relevance to human health. This scientific progress has contributed to a more complete appreciation of how vitamin D regulates immune responses, including cellular homeostatic and host defense pathways.

10.1 INTRODUCTION

Vitamin D functions as a fat-soluble hormone and is essential for a large number of physiological processes. Conventionally, it is clear that vitamin D regulates bone mineralization and calcium homeostasis. However, accumulating evidence has revealed a key function of vitamin D/VDR signaling in the regulation of innate immune function (Lemire 1992; Deluca and Cantorna 2001; Mora et al. 2008; Hewison 2011a; Wu and Sun 2011; Sun 2010). The cells of the innate immune system express VDR, which is a ligand-activated transcription factor that interacts with vitamin D response elements in vitamin D–regulated genes, either as homodimers or heterodimers with the retinoid X receptor. Innate immune cells, by converting 25D to bioactive 1,25D, can activate the VDR and downstream gene programs.

The innate immune system detects and rapidly responds to a diverse range of microbial pathogens and a variety of alterations to the host. Detection is mediated by pattern recognition receptors (PRRs), which directly bind to pathogen-associated molecular patterns (PAMPs) and danger-associated molecular patterns (DAMPs; Aderem 2001; Medzhitov 2007; Takeuchi and Akira 2010; Kono and Rock 2008). Because DAMPs are expressed by host cells during noninfectious inflammation (e.g., as a result of tissue injury), the innate immune system, by recognizing PAMPs and DAMPs, can respond to a vast array of inflammatory stimuli. The cells of the innate immune system, including phagocytes [macrophages, neutrophils, and dendritic cells (DCs)], mast cells, basophils, eosinophils, and natural killer cells, are recruited to sites of infection through the secretion of chemical effectors and mediators called cytokines (Medzhitov 2007; Steinke and Borish 2006). Epithelial cells, including epidermal keratinocytes, are resident cells that are not only simple physical barriers but also play an active role as part of the active innate immune system (Ryu et al. 2010). As a response to infection or harmful stimuli, the innate immune cells secrete antimicrobial peptides (AMPs), crucial effector molecules that directly target microbes and also have immunomodulatory properties (Brown and Hancock 2006; Mookherjee and Hancock 2007). Reactive oxygen species and nitrogen species are critically involved in the direct killing of microbes and the induction of secondary messengers to control multiple biological responses (Bogdan et al. 2000; Fang 2004). Activation of the complement cascade leads to the phagocytosis of bacteria, cell activation, and the clearance of antibody complexes. Cells of the innate immune system, through antigen presentation and the simultaneous release of cytokines, instruct and activate the adaptive immune response.

Vitamin D has long been known to be a crucial component of calcium homeostasis and the regulation of multiple physiological functions, including immune responses (Heaney 2008; Hewison 2011a). In the past few decades, it has been shown that the active form of vitamin D (1,25D) has an immunomodulatory (Lemire 1992; Deluca and Cantorna 2001) and antiproliferative function (Johnson et al. 2006). However, it is not the level of 1,25D that defines clinical insufficiency or deficiency, because the circulating levels of 1,25D, measured as the total of $1,25D_3$ (cholecalciferol) and 1,25D2 (ergocalciferol), are maintained in a relatively constant range (Holick 2007; Hollis et al. 2007; Adams and Hewison 2008). Instead, the levels of the 1,25D precursor, 25D, are found to correlate with bone and other diseases (Tsugawa and Okano 2006; Cranney et al. 2007; Haussler et al. 1997; Takiishi et al. 2010; de Borst et al. 2011). Therefore, clinicians measure 25D levels to determine vitamin D deficiency. In recent years, there has been considerable new insight into the mechanisms by which vitamin D and the VDRs regulate innate immune function and contribute to the pathophysiology of disease (Hewison 2010a). These discoveries underscore the importance of conducting large-scale clinical trials of vitamin D supplementation for the prevention and for the treatment of human infectious and inflammatory diseases.

10.2   INNATE IMMUNE SYSTEM AND VITAMIN D

Macrophages and DCs are the key cell types of the innate immune system, expressing germ line–encoded receptors that function as crucial sentinels of microbial and noninfectious antigens and mediate host innate immunity. Among a variety of innate receptors, Toll-like receptors (TLRs) were the first family of PRRs discovered in mammals (Medzhitov et al. 1997; Akira 2003). Extensive studies have shown that TLRs can respond to PAMPs or endogenous molecules released in response to stress, trauma, and cell damage, called DAMPs (Akira et al. 2001; Takeda et al. 2003; Takeda and

Akira 2007; Essakalli et al. 2009; Kumar et al. 2011). The interaction between PAMPs, DAMPs, and innate receptors leads to the selective induction of inflammatory responses against dangerous stimuli or tolerance to avoid harmful damage to self (Aderem 2001; Nakayama et al. 2004; Takeuchi and Akira 2010; Piccinini and Midwood 2010). Through this interplay, the innate immune system, the first line of immune defense, is integrally involved in maintaining tissue homeostasis and integrity. Dysregulated innate responses result in sustaining proinflammatory responses and in perpetuating chronic inflammation and tissue damage (Ospelt and Gay 2010; Drexler and Foxwell 2010).

For many years, adequate levels of vitamin D have been known to be required for maintaining healthy bone and preventing several types of bone diseases, whereas there have been relatively few reports exploring the mechanism by which vitamin D contributes to innate immunity (Lemire 1992; Deluca and Cantorna 2001). However, during the last 5 years, there has been considerable interest in the function of vitamin D in the regulation of innate immune defense responses (Liu et al. 2006; Mora et al. 2008; Yuk et al. 2009; Jo 2010; Hewison 2011a). Studies of innate immune responses against intracellular pathogens such as *Mycobacterium tuberculosis* have revealed PRR-induced activation of the localized conversion of 25D to 1,25D and subsequent induction of the human cationic antimicrobial peptide, cathelicidin hCAP18/LL-37, as a key defense strategy against infection (Liu et al. 2006; Yuk et al. 2009; Jo 2010). The ability to upregulate cathelicidin is highly dependent on vitamin D status and the inducible expression of the enzyme 25-hydroxyvitamin D-1 α-hydroxylase (CYP27B1; Hewison 2011b). There is accumulating evidence that supports the connection between optimal levels of circulating 25D and a lower risk of multiple human diseases, including those caused by infectious agents (Hewison 2011b). Given the clinical relevance of these observations, there is renewed interest in understanding the mechanisms by which vitamin D contributes to human health.

10.2.1 ROLE OF VITAMIN D IN REGULATING THE MONOCYTE/MACROPHAGE FUNCTION

It has become increasingly clear that the locally produced 1,25D can act on immune cells in an autocrine or a paracrine manner. The regulatory function of vitamin D on innate immunity has been shown from the discovery of specific high-affinity intracellular receptors (VDR) in activated inflammatory cells (Provvedini et al. 1983). The VDR is expressed in at least 30 different target tissues and a variety of cell types, including activated monocytes/macrophages, DCs, NK cells, T cells, and B cells of the immune system, and even in malignant cells (Deluca and Cantorna 2001; Bouillon et al. 1995; James et al. 1999; Gombart 2009). It was also found that VDR expression is upregulated in response to 1,25D, probably because it increases the lifetime or stability of the VDR mRNA rather than the VDR mRNA transcription levels (Wiese et al. 1992). Binding of 1,25D to the VDR increases the gene expression of CYP24A1 but downregulates CYP27B1 (Avila et al. 2007). This negative feedback control mechanism might contribute to the homeostatic regulation of excessive production of 1,25D. However, macrophages can synthesize an alternative splice variant of CYP24 that leads to the production of a dominant negative-acting protein that is a catalytically inactive, amino-terminally truncated 24-OHase protein that prevents the catabolism of 1,25D (Ren et al. 2005). Thus, macrophages do not limit their production of 1,25D, resulting in enhanced immune defense against infectious disease in humans (Ren et al. 2005). Together, these mechanisms that regulate vitamin D levels play an important role in regulating innate immune pathways.

Initially, 1,25D was found to suppress the proliferation of mouse myeloid leukemia cells and to promote their differentiation into macrophages (Miyaura et al. 1981). Additionally, incubating human monocytic cells with 1,25D significantly increased the expression of interleukin-1β (IL-1β; Adorini and Penna 2009) and nucleotide-binding oligomerization domain 2 (NOD2; Wang et al. 2010), as well as activating phagocytosis and superoxide synthesis (Adams et al. 2007). A molecular basis for the role of vitamin D in host defense is based on two major sets of findings: (1) 1,25D induces the expression of genes encoding AMPs (Wang et al. 2004; Gombart et al. 2005; Weber et al. 2005; Jo 2010; Schauber et al. 2007) and (2) TLR signaling increases the conversion of 25D to

1,25D in monocytes/macrophages upregulating antimicrobial responses (Liu et al. 2006; Shin et al. 2010). These will be discussed in detail in the latter part of this chapter.

10.2.2 ROLE OF VITAMIN D IN REGULATING THE DC FUNCTION

DCs are the principal antigen-presenting cells in cell-mediated immune responses. This function was attenuated by physiological levels of 1,25D and a related $1,25D_3$ analog. Conditioning of bone marrow cultures with a vitamin D_3 analog inhibited the expression of major histocompatibility complex class II and costimulatory receptors, which were not reversed by maturing stimuli, including CD40 ligation, macrophage products, or lipopolysaccharides (Griffin and Kumar 2003; Griffin et al. 2001, 2004; Dong et al. 2003). Additionally, 1,25D can affect DC phenotype and function by inducing tolerogenic DC through the downregulation of costimulatory molecules (CD40, CD80, and CD86) with increased IL-10 but decreased IL-12 levels (Adorini and Penna 2009; Sochorová et al. 2009). For the vitamin D regulation of chemokine expression in DCs, 1,25D showed a similar and additive effect on the expression of RANTES, CCR5, and CCR7, whereas it had an inducing effect on the expression of MIP-1 β and MCP-1 (Xing et al. 2002).

10.2.3 ROLE OF VITAMIN D IN REGULATING OTHER HEMATOPOIETIC CELL FUNCTIONS

The addition of exogenous 1,25D can inhibit the maturation of mast cell progenitors, suggesting a physiological role for vitamin D in mast cell development (Baroni et al. 2007). Moreover, mast cell VDR signaling was shown to be important in modulating ultraviolet-B irradiation–induced skin pathology (Biggs et al. 2010). Thus, the "negative" or "positive" regulation of vitamin D on immune cell maturation/differentiation may contribute to the beneficial activity of this hormone in maintaining the overall homeostasis of the immune system, preventing autoimmune disease and graft rejection, as well as chronic inflammatory conditions.

10.2.4 ROLE OF VITAMIN D IN EPITHELIAL SYSTEMS

Airway epithelial cells serve not only as a physical barrier to the outside environment but also represent part of an efficient innate immune system to prevent infections from continuous exposure to inhaled pathogens. The synthesis of 1,25D by any tissue or cell is dependent on the level of the enzyme CYP27B1. These airway epithelial cells express high baseline levels of CYP27B1 but low levels of CYP24 (Yim et al. 2007; Hansdottir et al. 2008). Thus, they can constitutively convert inactive 25D to the active 1,25D form, but they cannot readily inactivate both through 24 hydroxylation. The net effect is to increase the expression of those vitamin D–regulated genes with crucial innate immune functions, including cathelicidin and CD14. In human bronchial epithelial cells, 1,25D induced cathelicidin expression and antimicrobial activity against the pathogens *Bordetella bronchiseptica* and *Pseudomonas aeruginosa* (Yim et al. 2007). However, 1,25D attenuates the expression of proinflammatory chemokine IL-8 in response to dsRNA, thus influencing host defense in response to viruses (Hansdottir et al. 2008). Because bronchial epithelial cells are common targets of respiratory pathogenic bacteria and viruses, a seasonal variation of serum 25OHD levels likely contribute to increased frequencies of lower respiratory tract infection in the wintertime (Herr et al. 2011).

Evidence has accumulated to reveal important roles for vitamin D in a variety of epithelial cells, including those lining the oral cavity, lung, gastrointestinal system, genitourinary system, skin, and even the surface of the eye (Schwalfenberg 2011). The innate immune system in epithelial cells provides protection against a broad spectrum of pathogenic insults through a variety of effector factors and mechanisms. The release of AMPs plays an important role in this host defense system (Beisswenger and Bals 2005). Of particular interest, vitamin D has been reported to be a major regulator of AMPs in the epithelial system, contributing to host defense through direct antimicrobial activity and indirect immunomodulatory activities (Cole and Waring 2002; Schutte and McCray

2002; Bals and Hiemstra 2004; Hiemstra 2007). Novel insight into the mechanisms of vitamin D–mediated AMP expression and function will likely lead to innovative approaches for the treatment of infectious and inflammatory disorders in which epithelial barriers are critical for host defense.

As in the case with macrophages, keratinocytes express CYP27B1 via TLR2-induced mechanisms. The keratinocyte production of transforming growth factor-β1 (TGF-β1) can stimulate the induction of CYP27B1 and 1,25D, subsequently resulting in the upregulation of TLR2 and cathelicidin production, such as in macrophages (Schauber et al. 2007). Because TGF-β1 is associated with wound repair, TGF-β1-mediated functional VDR signaling and cathelicidin production provides a link between innate immune surveillance and wound repair (Schauber et al. 2007). In the epidermis, 1,25D is also known to promote differentiation and mediate antiproliferative effects on keratinocytes (Luderer and Demay 2010). The diverse roles of vitamin D in hematopoietic and nonhematopoietic cells have been summarized in Figure 10.1. The mechanisms by which vitamin D

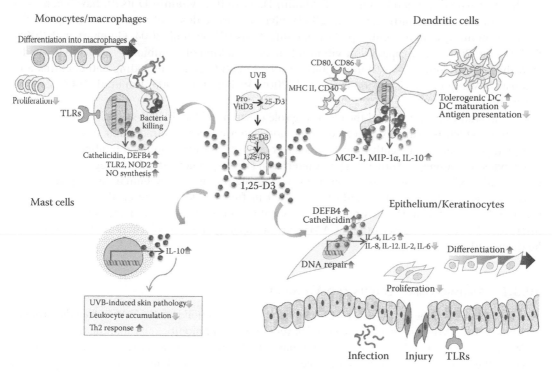

FIGURE 10.1 **(See color insert.)** Diverse roles of vitamin D in different cell types. Vitamin D_3 is synthesized in the skin and hydroxylated in the liver to $25D_3$, the main circulating form. In kidney, 25D is hydroxylated to 1,25D by the CYP27A1. The physiologically active form, 1,25D, circulates in the blood, where it has multiple systemic effects on different cell types. Systemically or locally produced 1,25D exerts its effects on several different cell types (e.g., monocytes/macrophages, DCs, airway epithelial cells/keratinocytes, and mast cells). Notably, these cells can express VDR and induce CYP27B1, which is responsible for the hydroxylation of 25D to 1,25D. In monocytes/macrophages, 1,25D can induce the differentiation of monocytes into macrophages and the expression of IL-1, NO synthesis and cathelicidin/DEFB4, thereby promoting innate immune responses against bacterial infection. In DCs, 1,25D decreases DC maturation and antigen presentation by suppressing the expression of MHC class II, CD40, CD80, and CD86, whereas it increases tolerogenic DCs. In addition, 1,25D reduces IL-12 production while inducing IL-10, MCP-1, and MIP-1α production. In epithelial cells/keratinocytes, 1,25D induces the differentiation of keratinocytes, DNA repair, and wound healing, whereas it reduces keratinocyte proliferation. 1,25D can also induce the production of IL-4 and IL-5, whereas it decreases IL-8, IL-12, IL-2, and IL-6. In mast cells, 1,25D induces the expression of IL-10 to limit skin pathology. Abbreviations: DEFB4, defensin-β4; MHC, major histocompatibility complex; CD, clusters of differentiation; MCP-1, monocyte chemotactic protein-1; MIP-1α, macrophage inflammatory protein-1α; NO, nitric oxide; UVB, Ultraviolet B.

regulates the immune function and contributes to treatment efficacy in psoriasis and other autoimmune/inflammatory disorders will be discussed in the latter part of this chapter.

10.3 MECHANISMS BY WHICH VITAMIN D REGULATES INNATE IMMUNE EFFECTOR FUNCTIONS

It is apparent that vitamin D and VDR signaling are involved in the enhancement of the innate immune function and the eradication of intracellular microbes (Liu et al. 2006; Yuk et al. 2009; Shin et al. 2010). The AMPs, including hCAP18/LL-37, are key players in vitamin D–mediated control of innate immune pathways in human monocytes/macrophages (Liu et al. 2006; Yuk et al. 2009; Shin et al. 2010). Prior to the development of antibiotics, sunlight and cod liver oil, both of which are good sources of vitamin D, as well as vitamin D itself, have been used to treat tuberculosis (Martineau et al. 2007). Pharmacologic doses of vitamin D can boost the innate immune system to combat pathogenic infections (Wang et al. 2004, Gombart et al. 2005; Weber et al. 2005). The capacity of vitamin D to act as a kind of "antibiotic" seems to be mediated through the induction of human antimicrobial peptide genes, including cathelicidins and defensins (Wang et al. 2004; Gombart et al. 2005; Weber et al. 2005; Liu et al. 2006; Zasloff 2006a,b).

Recent studies have revealed the emerging role of vitamin D and VDR signaling in autophagy activation, which is critically involved in the host defense against intracellular infection and in autophagy-mediated cell death for certain cancer cells. Moreover, combined treatment of human monocytes/macrophages with 1,25D and interferon (IFN)-γ synergistically upregulate inducible nitric oxide (NO) synthase expression. However, nitrite production by human macrophages was low, and the function of NO should be clarified in future studies (Lee et al. 2009). Through this regulation, vitamin D can strengthen the local response of innate immune cells to invading microbes. The detailed description of AMPs and autophagy in terms of vitamin D signaling will be discussed.

10.3.1 ANTIMICROBIAL PEPTIDES AND VITAMIN D

As aforementioned, the synthesis of 1,25D in response to infection is thought to induce the production of several AMPs, the crucial effectors/players in innate immunity and inflammation.

We now know that VDR signaling regulates both innate and adaptive immune responses through the induction of a variety of genes, including AMPs. In human innate immune cells, including monocytes and keratinocytes, there is a feedback regulatory system in concert with the induction of hCAP18/LL-37. Pathogenic or injurious stimuli can trigger the local conversion of 25D to 1,25D by the upregulated expression of CYP27B1 in macrophages (Liu et al. 2006) and keratinocytes (Miller and Gallo 2010). This conversion of 25D to 1,25D results in a local increase in the antimicrobial response through the induction of AMPs and the expression of PRRs such as TLR2 and CD14 (Schauber et al. 2007). In terms of innate response activation, 1,25D can stimulate the expression of the cytosolic PRR NOD2/CARD15/IBD1 in primary human monocytic and epithelial cells (Wang et al. 2010). Pretreatment with 1,25D, followed by the NOD2 ligand, muramyl dipeptide, resulted in the synergistic effect of VDR and nuclear factor-κB (NF-κB) signaling to induce the expression of the antimicrobial peptide defensin β2. This combined effect was significantly decreased in macrophages from patients with Crohn disease, homozygous for nonfunctional NOD2 variants, suggesting a link between VDR signaling and NOD function (Wang et al. 2010).

The addition of 1,25D to oral epithelial cells induced LL-37 expression and antimicrobial activity against the periodontal pathogen *Aggregatibacter actinomycetescomitans*, thereby activating antibacterial innate defense in the oral cavity (McMahon et al. 2011). Recent studies have reported the antimicrobial activity of LL-37 on *Candida albicans* through targeting of the cell wall carbohydrates (Tsai et al. 2011; den Hertog et al. 2005). In addition to numerous antimicrobial functions, the multifunctional protein cathelicidin LL-37 has several other activities related to host defense, as a principal mediator between innate and adaptive immunity (Zanetti 2005; Davidson et al. 2004), including chemotactic migration (Yang et al. 2000; Tjabringa et al. 2003, 2006; Kai-Larsen and Agerberth 2008), endotoxin neutralization (Rosenfeld et al. 2006; Mookherjee et al. 2006; Kai-Larsen and Agerberth 2008), histamine release (Niyonsaba et al. 2001; Kai-Larsen and Agerberth 2008), angiogenesis (Koczulla et al. 2003; Kai-Larsen and Agerberth 2008), reepithelialization (Heilborn et al. 2003; Shaykhiev et al. 2005; Kai-Larsen and Agerberth 2008), wound healing (Schauber et al. 2007), and autophagy processes (Yuk et al. 2009; Jo 2010).

10.3.2 AUTOPHAGY AND VITAMIN D

Autophagy, an intracellular quality and quantity control pathway, has recently been recognized as an immune mechanism (Deretic 2006). Currently, our understanding of the role of autophagy has extended into the cooperation of traditional innate immunity systems with autophagy pathways identified as participating in defense against intracellular pathogens in macrophages (Amano et al. 2006; Deretic 2011). In addition, the innate immune system has multiple connections with autophagy, acting as one of the immune effectors of PRR signaling pathways (Delgado and Deretic 2009; Delgado et al. 2009). Autophagy is a catabolic process and can be assigned a cellular protective function through the removal of damaged proteins and organelles and in facilitating microbicidal responses against intracellular pathogens such as mycobacteria (Amano et al. 2006; Deretic 2011).

Recent studies have shown that vitamin D links autophagy to innate immunity, including antimicrobial responses and the induction of cancer cell death (Høyer-Hansen et al. 2010; Wu and Sun 2011). Our previous studies have shown that autophagy and vitamin D–mediated antimicrobial activities against mycobacteria-infected macrophages are dependent on the regulatory function of hCAP-18/LL-37 (Yuk et al. 2009). The addition of 1,25D at physiological concentrations was sufficient for the induction of direct antimicrobial activity in human macrophages infected with mycobacteria (Yuk et al. 2009). The human cationic antimicrobial peptide hCAP-18/LL-37 was found to be essential for the fusion of autophagosomes with lysosomes and antimycobacterial activity in 1,25D-treated human monocytes/macrophages (Yuk et al. 2009). The cathelicidin induced by 1,25D treatment further induced the maturation of autophagic processes, thereby activating host innate immune responses (Yuk et al. 2009; Jo 2010). In separate studies, activation of TLR2/1 by mycobacterial lipoproteins triggered the vitamin D–dependent induction of cathelicidin expression and antimicrobial activity against intracellular *M. tuberculosis* (Liu et al. 2006), as well as the autophagic process through VDR signaling (Shin et al. 2010). Cathelicidin expression was also required for TLR2/1-induced autophagy in human monocytes/macrophages (Shin et al. 2010). Liu et al. (2006) and Shin et al. (2010) showed that the upregulation of cathelicidin was dependent on the conversion of $25D_3$ to $1,25D_3$ through CYP27B1. We have found that Th1 cytokine IFN-γ also triggers the vitamin D antimicrobial pathway, which includes the vitamin D–dependent induction of autophagy activation (Fabri et al. 2011). IFN-γ induction of autophagy was also dependent on cathelicidin expression (unpublished data). These data suggest that the innate and adaptive immune responses share a common pathway involving the vitamin D–dependent induction of antimicrobial peptides and autophagy. Furthermore, combined treatment of 1,25D and IFN-γ enhances antimicrobial function of human macrophages in response to *M. tuberculosis*, compared with the results by treatment with either one alone (Lee et al. 2009). These data strongly suggest that vitamin D–mediated innate immunity cooperates with adaptive immunity to provide maximal advantages to the host antimycobacterial immune responses through the autophagy activation and antimycobacterial effector function.

10.4   CROSS TALK BETWEEN VDR SIGNALING AND INNATE IMMUNITY

Initial evidence for the interaction of vitamin D with the innate immune system was provided from studies of the disease sarcoidosis, a chronic granulomatous disorder, often manifested by hypercalcemia and high levels of circulating calcitriol (Papapoulos et al. 1979). It was observed that, in alveolar macrophages from patients with sarcoidosis, the ectopic synthesis of 1,25D was significantly increased (Adams et al. 1983). Now, it is clear that normal macrophages can synthesize 1,25D when activated by various agents, including IFN-γ, lipopolysaccharide, and TLR2/1 stimulation (Stoffels et al. 2006).

Earlier studies showed that VDR ligation with exogenous 1,25D induces a superoxide burst (Sly et al. 2001) and reverses the inhibition of phagolysosome fusion (Hmama et al. 2004) in *M. tuberculosis*–infected macrophages; both phenomena are mediated through a Phosphoinositide 3-kinase (PI3K) signaling pathway (Hmama et al. 2004; Rebsamen et al. 2002). It is also relevant to note that vitamin D enhances antimicrobial activity against mycobacteria, principally through genomic regulatory mechanisms. In response to 1,25D, VDR heterodimerizes with nuclear receptors of the retinoic X receptor family, binds to vitamin D response elements in the promoters of a variety of vitamin D–responsive genes, and transactivates the expression of target genes (Mora et al. 2008). Through this genomic regulation, 1,25D enhances host defense by the upregulation of protective innate host responses, including the induction of NO synthase, NOS2A (Rockett et al. 1998). In addition, NF-κB and VDR signaling pathways were synergized in the induction of defensin-β4 expression, which contributed to the TLR2/1-mediated antimicrobial responses against mycobacteria (Liu et al. 2009).

Cross talk between innate immune signaling and the VDR pathway may synergize for the optimal induction of antimicrobial activities through the expression of AMPs. Previous studies have shown that T-cell cytokines differentially regulated TLR2/1-induced expression of AMPs, cathelicidin, and defensin-β4. The Th1 cytokine IFN-γ upregulated TLR2/1-induced CYP27B1 expression in human monocytes, whereas the Th2 cytokine IL-4 led to the catabolism of 25D to the inactive metabolite 24,25D through CYP24A1 expression (Edfeldt et al. 2010). The immune activation of macrophages with IFN-γ derived from Th1 cells results in the induction of CYP27B1 and the conversion of 25D to 1,25D (Stoffels et al. 2006, 2007). IFN-γ also potentiates this effect by inhibiting the induction of 24-hydroxylase, a key enzyme for the inactivation of 1,25D (Vidal et al. 2002). Direct protein–protein interactions between activated STAT1 and the DNA binding domains of the VDR prevents VDR–retinoid X receptor binding to the vitamin D–response element, thus diverting the VDR from its normal genomic target on the 24-hydroxylase promoter and antagonizing 1,25D-VDR transactivation of the 24-hydroxylase promoter. This functional relevance between the signaling pathways for the 1,25D and the cytokine IFN-γ may enhance the immunomodulatory activity of 1,25D on monocytes/macrophages and may be associated with the pathogenesis of abnormal 1,25D homeostasis in granulomatous diseases (Vidal et al. 2002).

The activation of human monocytes with IL-15 was found to be sufficient to induce the macrophage differentiation and upregulated expression of CYP27B1, VDR, and cathelicidin (Krutzik et al. 2008). The addition of $25D_3$ to IL-15–differentiated macrophages resulted in the conversion of $25D_3$ to bioactive $1,25D_3$ and the subsequent induction of antimicrobial activity against intracellular *M. tuberculosis*. The IL-15–induced macrophage vitamin D–dependent antimicrobial program may also contribute to host defense in leprosy (Montoya et al. 2009). These advances identify potential therapeutic applications of vitamin D in the control of infectious diseases and other chronic inflammatory diseases (Sun et al. 2006; Szeto et al. 2007; Zehnder et al. 2008; Proal et al. 2009; Wu and Sun 2011; Liu et al. 2011). The cross talk between vitamin D and innate immune signaling pathways have been described in Figure 10.2.

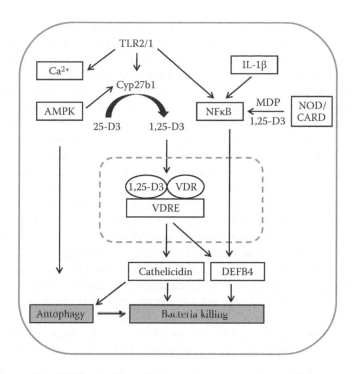

FIGURE 10.2 Cross talk of VDR signaling with innate immune pathways. TLR engagement with specific ligands triggers a series of intracellular signaling pathways, culminating in the activation of NF-κB pathways, which are important for proinflammatory cytokine production and the upregulation of antimicrobial peptide human β-defensin (DEFB4). TLR signaling also induces AMPK activation, which is required for CYP27B1 expression. Activation of the intracellular calcium–AMPK pathway is known to enhance autophagy in many reports. CYP27B1 is essentially required for the hydroxylation of the inactive form of vitamin D (25D) into the active form (1,25D). The resulting functional VDR activation is responsible for the upregulation of antimicrobial peptides including cathelicidin and DEFB4, leading to bacteria killing. Cathelicidins, but not DEFB4, play a critical role in the induction of 1,25D-induced autophagy activation. The cytosolic sensor NOD/CARD signals can also activate the NF-κB pathway, thereby promoting DEFB4 expression. Abbreviations: AMPK, AMP-activated protein kinase; CARD, caspase recruitment domain.

10.5 VITAMIN D LEVELS LINKED TO HUMAN INFECTIOUS/ INFLAMMATORY DISEASES

The importance of vitamin D and 1,25D, the bioactive metabolite of vitamin D, in innate immune function was initially recognized by clinicians by identifying that vitamin D deficiency is highly associated with rickets (associated with respiratory, infectious problems). However, it has recently been revealed that inadequate serum vitamin D concentrations are involved in various health problems, including infectious diseases such as tuberculosis (Walker and Modlin 2009). The prospective cohort study has shown that severe deficiency in serum 25D is associated with higher mortality in patients with acute hospital admission (<30 days) with community-acquired pneumonia (Leow et al. 2011).

Psoriasis is a common, persistent, chronic inflammatory disorder, thought to be of autoimmune etiology, affecting primarily the skin and the joints. In psoriasis, AMPs are strongly expressed in lesional skin and act as inflammatory "alarmins" in this chronic inflammatory skin disease. Topical treatment of psoriasis with vitamin D analogs has been effective as monotherapy or adjunctive therapy of this disease, with relatively minor side effects (Kurian and Barankin 2011; Mitra and Wu 2010). In addition to psoriasis, vitamin D and various synthetic analogs are used in the treatment of other inflammatory and hyperproliferative dermatoses (Pinette et al. 2003; Masuda and Jones 2006). The upregulated expression of CYP27B1 has been reported in psoriasis and sarcoidosis. Dysregulated increased expression of CYP27B1 and the resulting hypercalcemia are also associated with skin lesions in patients with sarcoidosis (Karakelides et al. 2006).

Several lines of evidence have suggested that 1,25D and its derivatives attenuate epidermal inflammation. One possible mechanism is suggested by the finding that vitamin D reduces NF-κB activity by increasing IκB α protein levels, thus inhibiting IL-8 production in normal human keratinocytes (Riis et al. 2004). Similarly, vitamin D treatment of macrophages resulted in NF-κB downregulation (Cohen-Lahav et al. 2006, 2007). The negative modulating effects of 1,25D on NF-κB signaling pathways through an increase of IκBα have been observed in keratinocytes, and this effect was mediated through VDR expression (Janjetovic et al. 2009). Because the NF-κB pathway is principally involved in chronic inflammation and proliferation, the activity of vitamin D in NF-κB modulation suggests that 1,25D and its analogs provide an attractive therapeutic strategy for inflammatory and autoimmune disorders. Although the mechanism by which topical vitamin D therapy is effective in the treatment of psoriasis is not clear, it is intriguing that narrowband ultraviolet-B phototherapy is linked to the inhibition of Th17 and type I and type II IFN signaling pathways. Recent studies have revealed that 1,25D levels increased after ultraviolet-B irradiation of the skin contributed to the mast cell production of IL-10, which can limit the skin pathology (Biggs et al. 2010). Moreover, serum 25D levels are inversely correlated with TNF-α levels in healthy women (Peterson and Heffernan 2008). Thus, the tight regulation of vitamin D production is important in the homeostatic control of cellular function, especially in innate and inflammatory responses.

It is now apparent that 1,25D dampens inflammatory responses and that vitamin D levels are linked to the occurrence and severity of autoimmune diseases. Vitamin D deficiency (as measured by circulating 25D levels) increases the risk of inflammatory diseases and pathogenesis of adverse skeletal effects in autoimmune diseases (Kamen and Alele 2010). Previous studies have shown that 1,25D inhibits IL-12 production by activated macrophages and DCs through the downregulation of NF-κB activation (D'Ambrosio et al. 1998). In addition, 1,25D treatment resulted in a reduction of proinflammatory cytokines IL-12, IL-23, and IL-6, whereas it promoted regulatory T-cell profiles with an increase of IL-10, TGF-β, FoxP3, and CTLA4 (Daniel et al. 2008). These modulatory effects of 1,25D on regulatory T-cell function and anti-inflammatory cytokine production may be important for therapeutic treatment. Detailed discussions on the role of vitamin D in modulating inflammation is covered in Chapter 12, Vitamin D: Anti-Inflammatory Actions to Prevent and Treat Disease.

Because 1,25D has some adverse effects, such as hypercalcemia, its direct clinical applications are limited. Combined treatment of 1,25D with other drugs may be one approach to overcome these side effects. Vitamin D analogs with low calcemic activity will provide another approach. Further development of tissue-selective or function-selective therapeutics of VDR ligands without inducing systemically adverse effects will be necessary for the treatment of infectious/inflammatory diseases.

10.6   CONCLUDING REMARKS AND FUTURE DIRECTIONS

Approaches that provide insight into vitamin D physiology in innate immunity are critical for understanding the potential link between innate immune signaling and VDR pathways contributing to host defense. Because of its key roles in immune cell differentiation and functional regulation, vitamin D is an important mediator regulating essential biological processes such as host defense, inflammation, tissue injury, and immunity. Through genomic mechanisms involving VDR signaling, vitamin D contributes to the maintenance of immune homeostasis under steady state and influences a complex pathologic microenvironment during the development of infectious/inflammatory diseases. The action of vitamin D conducting specialized tasks is dependent on the individual cell type and is associated with the key mechanisms that involve the production of CYP27B1 and VDR signaling. AMPs, including cathelicidins and defensins, are key effector molecules of vitamin

D–mediated innate immune responses. Autophagy induced by vitamin D and VDR signaling activation opens a new era for the application of vitamin D not only in the therapeutic approach to infectious diseases but also in a variety of human diseases involving a defect in autophagy and vitamin D insufficiency.

Deciphering the underlying mechanisms by which VDR signaling cross talks with innate immune pathways will remain a rich area of research. Based on the recent intriguing studies for vitamin D–mediated innate immune functions, clinical trials of vitamin D and its analogs in infectious diseases, as well as other diseases in which vitamin D insufficiency is thought to be involved, should be very exciting. Vitamin D supplementation to maintain normal levels of circulating 25D could be a helpful option for effective innate immune responses in humans. However, the hypercalcemic activity of 1,25D provides significant potential for toxicity, which might limit therapeutic applications (Haschek et al. 1978; Bernardi et al. 2001). Hence, the vitamin D analogs and investigations into VDR signaling and effector molecules such as cathelicidins could provide a useful strategy against infectious and inflammatory diseases.

REFERENCES

Adams, J. S. and M. Hewison. 2008. Unexpected actions of vitamin D: New perspectives on the regulation of innate and adaptive immunity. *Nature Clinical Practice. Endocrinology & Metabolism* 4(2):80–90.

Adams, J. S., O. P. Sharma, M. A. Gacad, and F. R. Singer. 1983. Metabolism of 25-hydroxyvitamin D3 by cultured pulmonary alveolar macrophages in sarcoidosis. *Journal of Clinical Investigation* 72(5):1856–1860.

Adams, J. S., P. T. Liu, R. Chun, R. L. Modlin, and M. Hewison. 2007. Vitamin D in defense of the human immune response. *Annals of the New York Academy of Sciences* 1117:94–105.

Aderem, A. 2001. Role of Toll-like receptors in inflammatory response in macrophages. *Critical Care Medicine* 29(7 Suppl):S16–S18.

Adorini, L. and G. Penna. 2009. Dendritic cell tolerogenicity: A key mechanism in immunomodulation by vitamin D receptor agonists. *Human Immunology* 70(5):345–352.

Akira, S. 2003. Mammalian Toll-like receptors. *Current Opinion in Immunology* 15(1):5–11.

Akira, S., K. Takeda, and T. Kaisho. 2001. Toll-like receptors: Critical proteins linking innate and acquired immunity. *Nature Immunology* 2(8):675–680.

Amano, A., I. Nakagawa, and T. Yoshimori. 2006. Autophagy in innate immunity against intracellular bacteria. *Journal of Biochemistry* 140(2):161–166.

Avila, E., L. Díaz, D. Barrera et al. 2007. Regulation of Vitamin D hydroxylases gene expression by 1,25-dihydroxyvitamin D3 and cyclic AMP in cultured human syncytiotrophoblasts. *The Journal of Steroid Biochemistry and Molecular Biology* 103(1):90–96.

Bals R. and P. S. Hiemstra. 2004. Innate immunity in the lung: How epithelial cells fight against respiratory pathogens. *The European Respiratory Journal* 23(2):327–333.

Baroni, E., M. Biffi, F. Benigni, A. Monno, D. Carlucci, G. Carmeliet, R. Bouillon, and D. D'Ambrosio. 2007. VDR-dependent regulation of mast cell maturation mediated by 1,25-dihydroxyvitamin D_3. *Journal of Leukocyte Biology* 81(1):250–262.

Beisswenger, C. and R. Bals. 2005. Functions of antimicrobial peptides in host defense and immunity. *Current Protein & Peptide Science* 6(3):255–264.

Bernardi, R. J., D. L. Trump, W. D. Yu, T. F. McGuire, P. A. Hershberger, and C. S. Johnson. 2001. Combination of 1alpha,25-dihydroxyvitamin D(3) with dexamethasone enhances cell cycle arrest and apoptosis: role of nuclear receptor cross-talk and Erk/Akt signaling. *Clinical Cancer Research* 7(12):4164–4173.

Biggs, L., C. Yu, B. Fedoric, A. F. Lopez, S. J. Galli, and M. A. Grimbaldeston. 2010. Evidence that vitamin D(3) promotes mast cell–dependent reduction of chronic UVB-induced skin pathology in mice. *Journal of Experimental Medicine* 207(3):455–463.

Bogdan, C., M. Röllinghoff, and A. Diefenbach. 2000. Reactive oxygen and reactive nitrogen intermediates in innate and specific immunity. *Current Opinion in Immunology* 12(1):64–76.

Bouillon, R., M. Garmyn, A. Verstuyf, S. Segaert, K. Casteels, and C. Mathieu. 1995. Paracrine role for calcitriol in the immune system and skin creates new therapeutic possibilities for vitamin D analogs. *European Journal of Endocrinology* 133(1):7–16.

Brown, K. L., and R. E. Hancock. 2006. Cationic host defense (antimicrobial) peptides. *Current Opinion in Immunology* 18(1):24–30.

Cohen-Lahav, M., S. Shany, D. Tobvin, C. Chaimovitz, and A. Douvdevani. 2006. Vitamin D decreases NFkappaB activity by increasing IkappaBalpha levels. *Nephrology, Dialysis, Transplantation* 21(4): 889–897.

Cohen-Lahav, M., A. Douvdevani, C. Chaimovitz, and S. Shany. 2007. The anti-inflammatory activity of 1,25-dihydroxyvitamin D3 in macrophages. *The Journal of Steroid Biochemistry and Molecular Biology* 103(3–5):558–562.

Cole, A. M. and A. J. Waring. 2002. The role of defensins in lung biology and therapy. *American Journal of Respiratory Medicine* 1(4):249–259.

Cranney, A., T. Horsley, S. O'Donnell et al. 2007. Effectiveness and safety of vitamin D in relation to bone health. *Evidence Report/Technology Assessment* (158):1–235.

D'Ambrosio, D., M. Cippitelli, M. G. Cocciolo et al. 1998. Inhibition of IL-12 production by 1,25-dihydroxyvitamin D3: Involvement of NF-kappaB downregulation in transcriptional repression of the p40 gene. *Journal of Clinical Investigation* 101(1):252–262.

Daniel, C., N. A. Sartory, N. Zahn, H. H. Radeke, and J. M. Stein. 2008. Immune modulatory treatment of trinitrobenzene sulfonic acid colitis with calcitriol is associated with a change of a T helper (Th) 1/Th17 to a Th2 and regulatory T cell profile. *Journal of Pharmacology and Experimental Therapeutics* 324(1):23–33.

Davidson, D. J., A. J. Currie, G. S. Reid et al. 2004. The cationic antimicrobial peptide LL-37 modulates dendritic cell differentiation and dendritic cell–induced T cell polarization. *Journal of Immunology* 172(2):1146–1156.

de Borst, M. H., R. A. de Boer, R. P. Stolk, J. P. Slaets, B. H. Wolffenbuttel, and G. Navis. 2011. Vitamin D deficiency: Universal risk factor for multifactorial diseases? *Current Drug Targets* 12(1):97–106.

Delgado, M. A. and V. Deretic. 2009. Toll-like receptors in control of immunological autophagy. *Cell Death and Differentiation* 16(7):976–983.

Delgado, M., S. Singh, S. De Haro et al. 2009. Autophagy and pattern recognition receptors in innate immunity. *Immunological Reviews* 227(1):189–202.

Deluca, H. F. and M. T. Cantorna. 2001. Vitamin D: Its role and uses in immunology. *FASEB Journal* 15(14):2579–2585.

den Hertog, A. L., J. van Marle, H. A. van Veen et al. 2005. Candidacidal effects of two antimicrobial peptides: Histatin 5 causes small membrane defects, but LL-37 causes massive disruption of the cell membrane. *The Biochemical Journal* 388(Pt 2):689–695.

Deretic, V. 2006. Autophagy as an immune defense mechanism. *Current Opinion in Immunology* 18(4):375–382 (Epub 2006 Jun 19).

Deretic, V. 2011. Autophagy in immunity and cell-autonomous defense against intracellular microbes. *Immunological Reviews* 240(1):92–104.

Dong, X., T. Craig, N. Xing et al. 2003. Direct transcriptional regulation of RelB by 1alpha,25-dihydroxyvitamin D3 and its analogs: Physiologic and therapeutic implications for dendritic cell function. *Journal of Biological Chemistry* 278(49):49378–49385.

Drexler, S. K. and B. M. Foxwell. 2010. The role of toll-like receptors in chronic inflammation. *The International Journal of Biochemistry & Cell Biology* 42(4):506–518.

Edfeldt, K., P. T. Liu, R. Chun et al. 2010. T-cell cytokines differentially control human monocyte antimicrobial responses by regulating vitamin D metabolism. *Proceedings of the National Academy of Sciences of the United States of America* 107(52):22593–22598.

Essakalli, M., O. Atouf, N. Bennani, N. Benseffaj, S. Ouadghiri, and C. Brick. 2009. [Toll-like receptors]. *Pathologie-Biologie* 57(5):430–438.

Fang, F. C. 2004. Antimicrobial reactive oxygen and nitrogen species: Concepts and controversies. *Nature Reviews. Microbiology* 2(10):820–832.

Fabri, M., S. Stenger, D. M. Shin et al. 2011. Vitamin D is required for IFN-gamma-mediated antimicrobial activity of human macrophages. *Sci Transl Med* 3(104):104ra102.

Gombart, A. F. 2009. The vitamin D–antimicrobial peptide pathway and its role in protection against infection. *Future Microbiology* 4(9):1151–1165.

Gombart, A. F., N. Borregaard, and H. P. Koeffler. 2005. Human cathelicidin antimicrobial peptide (CAMP) gene is a direct target of the vitamin D receptor and is strongly up-regulated in myeloid cells by 1,25-dihydroxyvitamin D3. *FASEB Journal* 19(9):1067–1077.

Griffin, M. D. and R. Kumar. 2003. Effects of 1alpha,25(OH)2D3 and its analogs on dendritic cell function. *Journal of Cellular Biochemistry* 88(2):323–326.

Griffin, M. D., W. Lutz, V. A. Phan, L. A. Bachman, D. J. McKean, and R. Kumar. 2001. Dendritic cell modulation by 1alpha,25 dihydroxyvitamin D3 and its analogs: A vitamin D receptor–dependent pathway that promotes a persistent state of immaturity *in vitro* and *in vivo*. *Proceedings of the National Academy of Sciences of the United States of America* 98(12):6800–6805.

Griffin, M. D., N. Xing, and R. Kumar. 2004. Gene expression profiles in dendritic cells conditioned by 1alpha,25-dihydroxyvitamin D3 analog. *The Journal of Steroid Biochemistry and Molecular Biology* 89–90(1–5):443–448.

Hansdottir, S., M. M. Monick, S. L. Hinde, N. Lovan, D. C. Look, and G. W. Hunninghake. 2008. Respiratory epithelial cells convert inactive vitamin D to its active form: Potential effects on host defense. *Journal of Immunology* 181(10):7090–7099.

Haschek, W. M., L. Krook, F. A. Kallfelz, and W. G. Pond. 1978. Vitamin D toxicity: Initial site and mode of action. *The Cornell Veterinarian* 68(3):324–364.

Haussler, M. R., C. A. Haussler, P. W. Jurutka et al. 1997. The vitamin D hormone and its nuclear receptor: Molecular actions and disease states. *Journal of Endocrinology* 154 Suppl:S57–S73.

Heaney, R. P. 2008. Vitamin D in health and disease. *Clinical Journal of the American Society of Nephrology* 3(5):1535–1541.

Heilborn, J. D., M. F. Nilsson, and G. Kratz. 2003. The cathelicidin antimicrobial peptide LL-37 is involved in reepithelialization of human skin wounds and is lacking in chronic ulcer epithelium. *The Journal of Investigative Dermatology* 120(3):379–389.

Herr, C., T. Greulich, R. A. Koczulla et al. 2011. The role of vitamin D in pulmonary disease: COPD, asthma, infection, and cancer. *Respiratory Research* 12:31.

Hewison, M. 2010a. Vitamin D and the immune system: new perspectives on an old theme. *Endocrinology and Metabolism Clinics of North America* 39(2):365–379.

Hewison, M. 2010b. Vitamin D and the intracrinology of innate immunity. *Molecular and Cellular Endocrinology* 321(2):103–111.

Hewison, M. 2011a. Vitamin D and innate and adaptive immunity. *Vitamins and Hormones* 86:23–62.

Hewison, M. 2011b. Antibacterial effects of vitamin D. *Nature Reviews. Endocrinology* (Epub ahead of print).

Hiemstra, P. S. 2007. The role of epithelial beta-defensins and cathelicidins in host defense of the lung. *Experimental Lung Research* 33(10):537–542.

Hmama, Z., K. Sendide, A. Talal, R. Garcia, K. Dobos, and N. E. Reiner. 2004. Quantitative analysis of phagolysosome fusion in intact cells: Inhibition by mycobacterial lipoarabinomannan and rescue by an 1alpha,25-dihydroxyvitamin D3-phosphoinositide 3-kinase pathway. *Journal of Cell Science* 117(Pt 10):2131–2140.

Holick, M. F. 2007. Vitamin D deficiency. *New England Journal of Medicine* 357(3):266–281.

Hollis, B. W., C. L. Wagner, M. K. Drezner, and N. C. Binkley. 2007. Circulating vitamin D3 and 25-hydroxyvitamin D in humans: An important tool to define adequate nutritional vitamin D status. *The Journal of Steroid Biochemistry and Molecular Biology* 103(3–5):631–634.

Høyer-Hansen, M., S. P. Nordbrandt, and M. Jäättelä. 2010. Autophagy as a basis for the health-promoting effects of vitamin D. *Trends in Molecular Medicine* 16(7):295–302.

Janjetovic, Z., M. A. Zmijewski, R. C. Tuckey et al. 2009. 20-Hydroxycholecalciferol, product of vitamin D3 hydroxylation by P450scc, decreases NF-kappaB activity by increasing IkappaB alpha levels in human keratinocytes. *PLoS One* 4(6):e5988.

James, S. Y., M. A. Williams, A. C. Newland, and K. W. Colston. 1999. Leukemia cell differentiation: Cellular and molecular interactions of retinoids and vitamin D. *General Pharmacology* 32(1):143–154.

Jo, E. K. 2010. Innate immunity to mycobacteria: Vitamin D and autophagy. *Cellular Microbiology* 12(8):1026–1035.

Johnson, C. S., J. R. Muindi, P. A. Hershberger, and D. L. Trump. 2006. The antitumor efficacy of calcitriol: Preclinical studies. *Anticancer Research* 26(4A):2543–2549.

Kai-Larsen, Y. and B. Agerberth. 2008. The role of the multifunctional peptide LL-37 in host defense. *Frontiers in Bioscience* 13:3760–3767.

Kamen, D. L. and J. D. Alele. 2010. Skeletal manifestations of systemic autoimmune diseases. *Current Opinion in Endocrinology, Diabetes, and Obesity* 17(6):540–545.

Karakelides, H., J. L. Geller, A. L. Schroeter et al. 2006. Vitamin D–mediated hypercalcemia in slack skin disease: Evidence for involvement of extrarenal 25-hydroxyvitamin D 1alpha-hydroxylase. *Journal of Bone and Mineral Research* 21(9):1496–1499.

Koczulla, R., G. von Degenfeld, C. Kupatt et al. 2003. An angiogenic role for the human peptide antibiotic LL-37/hCAP-18. *Journal of Clinical Investigation* 111(11):1665–1672.

Kono, H. and K. L. Rock. 2008. How dying cells alert the immune system to danger. *Nature Reviews. Immunology* 8:279–289.

Krutzik, S. R., M. Hewison, P. T. Liu et al. 2008. IL-15 links TLR2/1-induced macrophage differentiation to the vitamin D–dependent antimicrobial pathway. *Journal of Immunology* 181(10):7115–7120.

Kumar, H., T. Kawai, and S. Akira. 2011. Pathogen recognition by the innate immune system. *International Reviews of Immunology* 30(1):16–34.

Kurian, A. and B. Barankin. 2011. Current effective topical therapies in the management of psoriasis. *Skin Therapy Letter* 16(1):4–7.

Lee, J. S., C. S. Yang, D. M. Shin, J. M. Yuk, J. W. Son, and E. K. Jo. 2009. Nitric oxide synthesis is modulated by 1,25-dihydroxyvitamin D3 and interferon-gamma in human macrophages after mycobacterial infection. *Immune Network* 9(5):192–202.

Lemire, J. M. 1992. Immunomodulatory role of 1,25-dihydroxyvitamin D3. *Journal of Cellular Biochemistry* 49(1):26–31.

Leow, L., Simpson, T. Cursons, R. Karalus, N., and R. J. Hancox. 2011. Vitamin D, innate immunity and outcomes in community acquired pneumonia. *Respirology* 16(4):611–616.

Liu, P. T., S. Stenger, H. Li et al. 2006. Toll-like receptor triggering of a vitamin D–mediated human antimicrobial response. *Science* 311(5768):1770–1773.

Liu, P. T., M. Schenk, V. P. Walker et al. 2009. Convergence of IL-1beta and VDR activation pathways in human TLR2/1-induced antimicrobial responses. *PLoS One* 4(6):e5810.

Liu, N. Q., A. T. Kaplan, V. Lagishetty et al. 2011. Vitamin D and the regulation of placental inflammation. *Journal of Immunology* 186(10):5968–5974.

Luderer, H. F. and M. B. Demay. 2010. The vitamin D receptor, the skin and stem cells. *The Journal of Steroid Biochemistry and Molecular Biology* 121(1–2):314–316.

Martineau, A. R., F. U. Honecker, R. J. Wilkinson, and C. J. Griffiths. 2007. Vitamin D in the treatment of pulmonary tuberculosis. *The Journal of Steroid Biochemistry and Molecular Biology* 103(3–5):793–798.

Masuda, S. and G. Jones. 2006. Promise of vitamin D analogues in the treatment of hyperproliferative conditions. *Molecular Cancer Therapeutics* 5(4):797–808.

McMahon, L., K. Schwartz, O. Yilmaz, E. Brown, L. K. Ryan, and G. Diamond. 2011. Vitamin D–mediated induction of innate immunity in gingival epithelial cells. *Infection and Immunity* 79(6):2250–2256.

Medzhitov, R. 2007. Recognition of microorganisms and activation of the immune response. *Nature* 449(7164):819–826.

Medzhitov, R., P. Preston-Hurlburt, and C. A. Janeway, Jr. 1997. A human homologue of the Drosophila Toll protein signals activation of adaptive immunity. *Nature* 388(6640):394–397.

Miller, J. and R. L. Gallo. 2010. Vitamin D and innate immunity. *Dermatologic Therapy* 23(1):13–22.

Mitra, A. and Y. Wu. 2010. Topical delivery for the treatment of psoriasis. *Expert Opinion on Drug Delivery* 7(8):977–992.

Miyaura, C., E. Abe, T. Kuribayashi et al. 1981. 1 alpha,25-Dihydroxyvitamin D3 induces differentiation of human myeloid leukemia cells. *Biochemical and Biophysical Research Communications* 102(3):937–943.

Montoya, D., D. Cruz, and R. M. Teles. 2009. Divergence of macrophage phagocytic and antimicrobial programs in leprosy. *Cell Host & Microbe* 6(4):343–353.

Mookherjee, N. and R. E. Hancock. 2007. Cationic host defence peptides: Innate immune regulatory peptides as a novel approach for treating infections. *Cellular and Molecular Life Sciences* 64(7–8):922–933.

Mookherjee N., K. L. Brown, D. M. Bowdish et al. 2006. Modulation of the TLR-mediated inflammatory response by the endogenous human host defense peptide LL-37. *Journal of Immunology* 176(4): 2455–2464.

Mora, J. R., M. Iwata, and U. H. von Andrian. 2008. Vitamin effects on the immune system: Vitamins A and D take center stage. *Nature Reviews. Immunology* 8(9):685–698.

Nakayama, K., S. Okugawa, S. Yanagimoto et al. 2004. Involvement of IRAK-M in peptidoglycan-induced tolerance in macrophages. *Journal of Biological Chemistry* 279(8):6629–6634.

Niyonsaba, F., A. Someya, M. Hirata, H. Ogawa, and I. Nagaoka. 2001. Evaluation of the effects of peptide antibiotics human beta-defensins-1/-2 and LL-37 on histamine release and prostaglandin D(2) production from mast cells. *European Journal of Immunology* 31(4):1066–1075.

O'Sullivan, M., T. Nic Suibhne, G. Cox, M. Healy, and C. O'Morain. 2008. High prevalence of vitamin D insufficiency in healthy Irish adults. *Irish Journal of Medical Science* 177(2):131–134.

Oliveri, B., L. Plantalech, A. Bagur et al. 2004. High prevalence of vitamin D insufficiency in healthy elderly people living at home in Argentina. *European Journal of Clinical Nutrition* 58(2):337–342.

Ospelt, C. and S. Gay. 2010. TLRs and chronic inflammation. *The International Journal of Biochemistry & Cell Biology* 42(4):495–505.

Papapoulos, S. E., T. L. Clemens, L. J. Fraher, I. G. Lewin, L. M. Sandler, and J. L. O'Riordan. 1979. 1,25-Dihydroxycholecalciferol in the pathogenesis of the hypercalcaemia of sarcoidosis. *Lancet* 1(8117):627–630.

Peterson, C. A. and M. E. Heffernan. 2008. Serum tumor necrosis factor-alpha concentrations are negatively correlated with serum 25(OH)D concentrations in healthy women. *Journal of Inflammation* 5:10.

Piccinini, A. M. and K. S. Midwood. 2010. DAMPening inflammation by modulating TLR signaling. *Mediators of Inflammation* pii: 672395 (Epub 2010 Jul 13).

Pinette, K. V., Y. K. Yee, B. Y. Amegadzie, and S. Nagpal. 2003. Vitamin D receptor as a drug discovery target. *Mini Reviews in Medicinal Chemistry* 3(3):193–204.

Proal, A. D., P. J. Albert, and T. G. Marshall. 2009. Dysregulation of the vitamin D nuclear receptor may contribute to the higher prevalence of some autoimmune diseases in women. *Annals of the New York Academy of Sciences* 1173:252–259.

Provvedini, D. M., C. D. Tsoukas, L. J. Deftos, and S. C. Manolagas. 1983. 1,25-Dihydroxyvitamin D3 receptors in human leukocytes. *Science* 221(4616):1181–1183.

Rebsamen, M. C., J. Sun, A. W. Norman, and J. K. Liao. 2002. 1alpha,25-Dihydroxyvitamin D3 induces vascular smooth muscle cell migration via activation of phosphatidylinositol 3-kinase. *Circulation Research* 91(1):17–24.

Ren, S., L. Nguyen, S. Wu, C. Encinas, J. S. Adams, and M. Hewison. 2005. Alternative splicing of vitamin D-24-hydroxylase: A novel mechanism for the regulation of extrarenal 1,25-dihydroxyvitamin D synthesis. *Journal of Biological Chemistry* 280(21):20604–20611.

Riis, J. L., C. Johansen, B. Gesser et al. 2004. 1alpha,25(OH)(2)D(3) regulates NF-kappaB DNA binding activity in cultured normal human keratinocytes through an increase in IkappaBalpha expression. *Archives of Dermatological Research* 296(5):195–202.

Rockett, K. A., R. Brookes, I. Udalova, V. Vidal, A. V. Hill, and D. Kwiatkowski. 1998. 1,25-Dihydroxyvitamin D3 induces nitric oxide synthase and suppresses growth of *Mycobacterium tuberculosis* in a human macrophage–like cell line. *Infection and Immunity* 66(11):5314–5321.

Rosen, C. J. 2011. Clinical practice. Vitamin D insufficiency. *New England Journal of Medicine* 364(3):248–254.

Rosenfeld, Y., N. Papo, and Y. Shai. 2006. Endotoxin (lipopolysaccharide) neutralization by innate immunity host-defense peptides. Peptide properties and plausible modes of action. *Journal of Biological Chemistry* 281(3):1636–1643.

Ryu, J. H., C. H. Kim, and J. H. Yoon. 2010. Innate immune responses of the airway epithelium. *Molecules and Cells* 30(3):173–183.

Schauber, J., R. A. Dorschner, and A. B. Coda. 2007. Injury enhances TLR2 function and antimicrobial peptide expression through a vitamin D–dependent mechanism. *Journal of Clinical Investigation* 117(3):803–811.

Schutte, B. C. and P. B. McCray, Jr. 2002. [beta]-Defensins in lung host defense. *Annual Review of Physiology* 64:709–748.

Schwalfenberg, G. K. 2011. A review of the critical role of vitamin D in the functioning of the immune system and the clinical implications of vitamin D deficiency. *Molecular Nutrition & Food Research* 55(1):96–108.

Shaykhiev, R., C. Beisswenger, K. Kandler et al. 2005. Human endogenous antibiotic LL-37 stimulates airway epithelial cell proliferation and wound closure. *American Journal of Physiology. Lung Cellular and Molecular Physiology* 289(5):L842–L848.

Shin, D. M., J. M. Yuk, and H. M. Lee. 2010. Mycobacterial lipoprotein activates autophagy via TLR2/1/CD14 and a functional vitamin D receptor signaling. *Cellular Microbiology* 12(11):1648–1665.

Sly, L. M., M. Lopez, W. M. Nauseef, and N. E. Reiner. 2001. 1alpha,25-Dihydroxyvitamin D3–induced monocyte antimycobacterial activity is regulated by phosphatidylinositol 3-kinase and mediated by the NADPH-dependent phagocyte oxidase. *Journal of Biological Chemistry* 276(38):35482–35493.

Sochorová, K., V. Budinský, D. Rozková et al. 2009. Paricalcitol (19-nor-1,25-dihydroxyvitamin D2) and calcitriol (1,25-dihydroxyvitamin D3) exert potent immunomodulatory effects on dendritic cells and inhibit induction of antigen-specific T cells. *Clinical Immunology* 133(1):69–77.

Steinke, J. W. and L. Borish. 2006. Cytokines and chemokines. *The Journal of Allergy and Clinical Immunology* 117(2 Suppl Mini-Primer):S441–S445.

Stoffels, K., L. Overbergh, A. Giulietti, L. Verlinden, R. Bouillon, and C. Mathieu. 2006. Immune regulation of 25-hydroxyvitamin-D3-1alpha-hydroxylase in human monocytes. *Journal of Bone and Mineral Research* 21(1):37–47.

Stoffels, K., L. Overbergh, R. Bouillon, and C. Mathieu. 2007. Immune regulation of 1alpha-hydroxylase in murine peritoneal macrophages: Unraveling the IFNgamma pathway. *The Journal of Steroid Biochemistry and Molecular Biology* 103(3–5):567–571.

Sun, J. 2010. Vitamin D and mucosal immune function. *Current Opinion in Gastroenterology* 26(6):591–595.

Sun, J., J. Kong, Y. Duan et al. 2006. Increased NF-kappaB activity in fibroblasts lacking the vitamin D receptor. *American Journal of Physiology. Endocrinology and Metabolism* 291(2):E315–E322.

Szeto, F. L., J. Sun, J. Kong et al. 2007. Involvement of the vitamin D receptor in the regulation of NF-kappaB activity in fibroblasts. *The Journal of Steroid Biochemistry and Molecular Biology* 103(3–5):563–566.

Takeda, K. and S. Akira. 2007. Toll-like receptors. *Current Protocols in Immunology* Chapter 14:Unit 14.12.

Takeda, K., T. Kaisho, and S. Akira. 2003. Toll-like receptors. *Annual Review of Immunology* 21:335–376.

Takeuchi, O. and S. Akira. 2010. Pattern recognition receptors and inflammation. *Cell* 140(6):805–820.

Takiishi, T., C. Gysemans, R. Bouillon and C. Mathieu. 2010. Vitamin D and diabetes. *Endocrinol Metab Clin North Am* 39(2):419–446.

Tjabringa, G. S., J. Aarbiou, D. K. Ninaber et al. 2003. The antimicrobial peptide LL-37 activates innate immunity at the airway epithelial surface by transactivation of the epidermal growth factor receptor. *Journal of Immunology* 171(12):6690–6696.

Tjabringa, G. S., D. K. Ninaber, J. W. Drijfhout, K. F. Rabe, and P. S. Hiemstra. 2006. Human cathelicidin LL-37 is a chemoattractant for eosinophils and neutrophils that acts via formyl-peptide receptors. *International Archives of Allergy and Immunology* 140(2):103–112.

Tsai, P. W., C. Y. Yang, H. T. Chang, and C. Y. Lan. 2011. Human antimicrobial peptide LL-37 inhibits adhesion of *Candida albicans* by interacting with yeast cell-wall carbohydrates. *PLoS One* 6(3):e17755.

Tsugawa, N. and T. Okano. 2006. Bone and bone related biochemical examinations. Hormone and hormone-related substances. Vitamin D (25D, 1,25D); measurements and clinical significances. *Clinical Calcium* 16(6):920–926.

Vidal, M., C. V. Ramana, and A. S. Dusso. 2002. Stat1–vitamin D receptor interactions antagonize 1,25-dihydroxyvitamin D transcriptional activity and enhance stat1-mediated transcription. *Molecular and Cellular Biology* 22(8):2777–2787.

Walker, V. P. and R. L. Modlin. 2009. The vitamin D connection to pediatric infections and immune function. *Pediatric Research* 65(5 Pt 2):106R–113R.

Wang, T. T., F. P. Nestel, V. Bourdeau et al. 2004. Cutting edge: 1,25-dihydroxyvitamin D3 is a direct inducer of antimicrobial peptide gene expression. *Journal of Immunology* 173:2909–2911.

Wang, T. T., B. Dabbas, D. Laperriere et al. 2010. Direct and indirect induction by 1,25-dihydroxyvitamin D3 of the NOD2/CARD15-defensin beta2 innate immune pathway defective in Crohn disease. *Journal of Biological Chemistry* 285(4):2227–2231.

Weber, G., J. D. Heilborn, C. I. Chamorro Jimenez, A. Hammarsjo, H. Törmä, and M. Stahle. 2005. Vitamin D induces the antimicrobial protein hCAP18 in human skin. *The Journal of Investigative Dermatology* 124(5):1080–1082.

Wiese, R. J., A. Uhland-Smith, T. K. Ross, J. M. Prahl, and H. F. DeLuca. 1992. Up-regulation of the vitamin D receptor in response to 1,25-dihydroxyvitamin D3 results from ligand-induced stabilization. *Journal of Biological Chemistry* 267(28):20082–20086.

Wu, S. and J. Sun. 2011. Vitamin D, vitamin D receptor, and macroautophagy in inflammation and infection. *Discovery Medicine* 11(59):325–335.

Xing, N., M. L. Maldonado, L. A. Bachman, D. J. McKean, R. Kumar, and M. D. Griffin. 2002. Distinctive dendritic cell modulation by vitamin D(3) and glucocorticoid pathways. *Biochemical and Biophysical Research Communications* 297(3):645–652.

Yang, D., Q. Chen, A. P. Schmidt et al. 2000. LL-37, the neutrophil granule– and epithelial cell–derived cathelicidin, utilizes formyl peptide receptor–like 1 (FPRL1) as a receptor to chemoattract human peripheral blood neutrophils, monocytes, and T cells. *Journal of Experimental Medicine* 192(7):1069–1074.

Yim, S., P. Dhawan, C. Ragunath, S. Christakos, and G. Diamond. 2007. Induction of cathelicidin in normal and CF bronchial epithelial cells by 1,25-dihydroxyvitamin D(3). *Journal of Cystic Fibrosis* 6(6):403–410.

Yuk, J. M., D. M. Shin, H. M. Lee et al. 2009. Vitamin D3 induces autophagy in human monocytes/macrophages via cathelicidin. *Cell Host & Microbe* 6(3):231–243.

Zanetti, M. 2005. The role of cathelicidins in the innate host defenses of mammals. *Current Issues in Molecular Biology* 7(2):179–196.

Zasloff, M. 2006a. Fighting infections with vitamin D. *Nature Medicine* 12(4):388–390.

Zasloff, M. 2006b. Inducing endogenous antimicrobial peptides to battle infections. *Proceedings of the National Academy of Sciences of the United States of America* 103(24):8913–8914.

Zehnder, D., M. Quinkler, K. S. Eardley et al. 2008. Reduction of the vitamin D hormonal system in kidney disease is associated with increased renal inflammation. *Kidney International* 74(10):1343–1353.

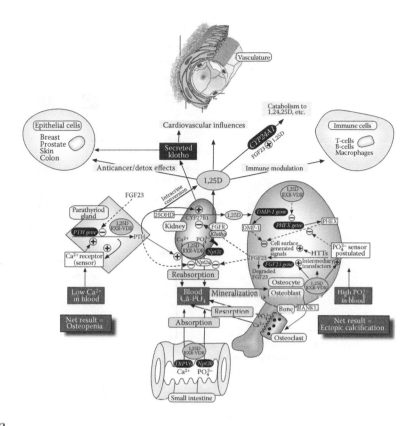

FIGURE 1.2

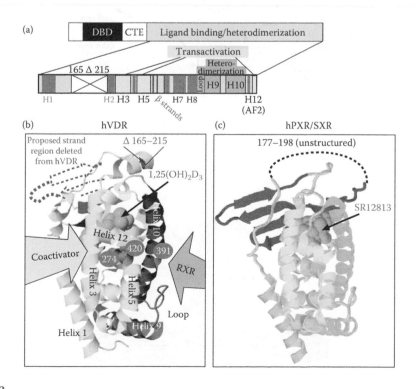

FIGURE 1.3

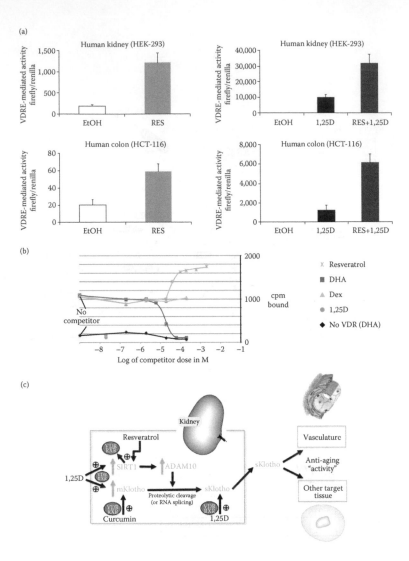

FIGURE 1.8

FIGURE 1.10

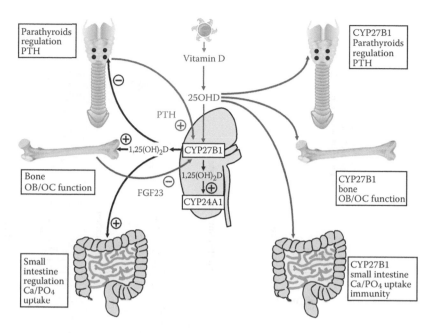

FIGURE 4.1

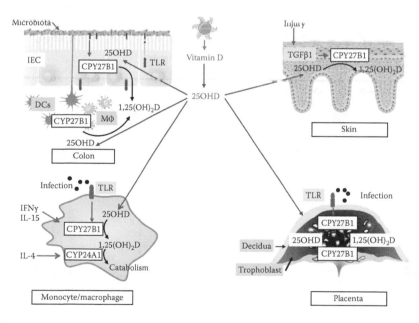

FIGURE 4.2

FIGURE 9.3

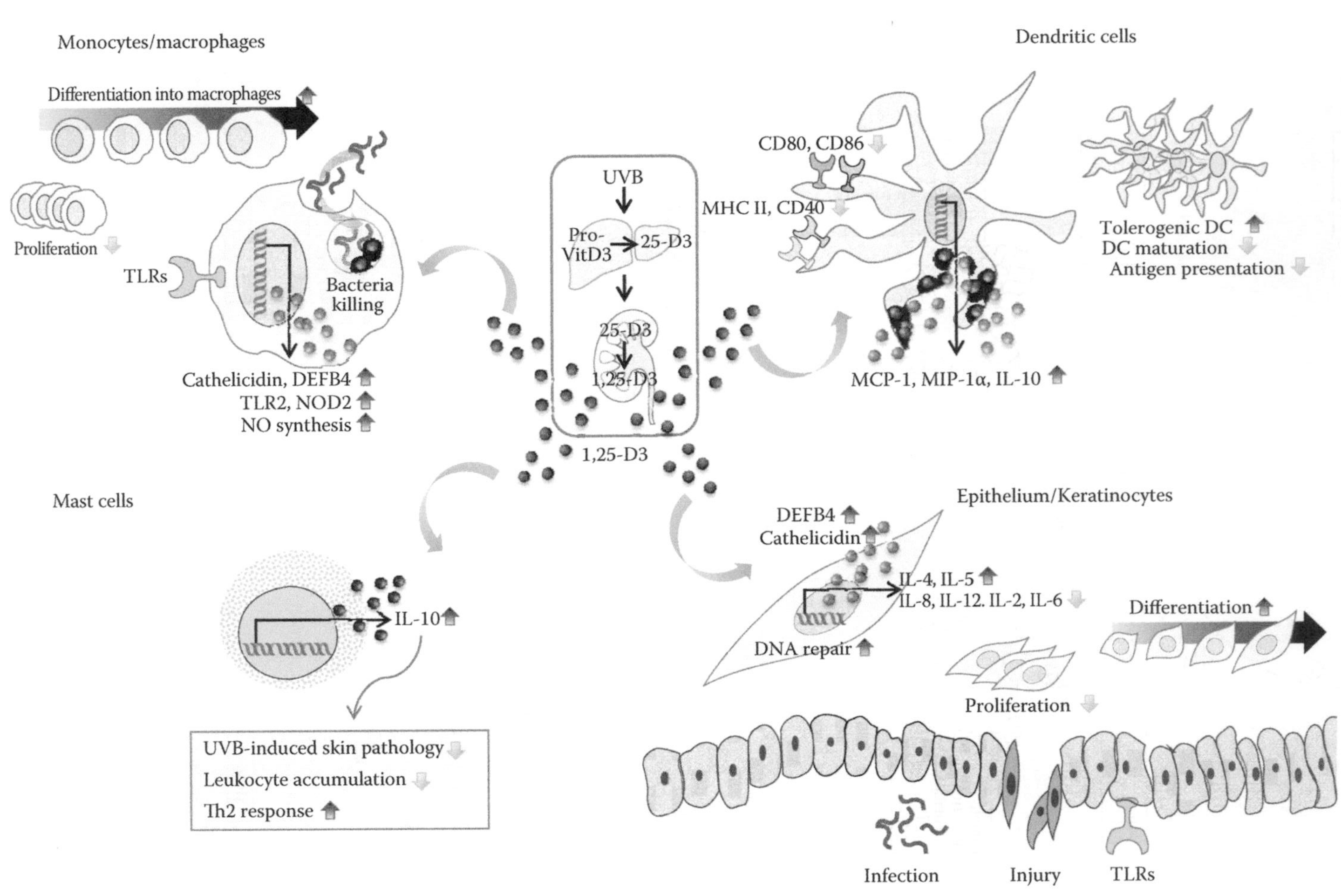

FIGURE 10.1

11 Vitamin D and Autoimmune Disease

Colleen E. Hayes, Corwin D. Nelson, and Justin A. Spanier

CONTENTS

11.1 INTRODUCTION

Autoimmune disease syndromes afflict an estimated 50 million Americans, compared with 81 million with heart disease and 11 million with cancer [1]. Among the autoimmune diseases, rheumatoid arthritis (RA), type 1 diabetes mellitus (T1D), multiple sclerosis (MS), inflammatory bowel diseases (IBD), systemic lupus erythematosus (SLE), psoriasis, and scleroderma are the most common. According to estimates from the U.S. National Institute of Allergy and Infectious Diseases, autoimmune diseases contribute more than $100 billion to health care costs in the United States. These costs are rising as the prevalence of autoimmune diseases increases. The effect of autoimmune diseases in terms of human and health care costs emphasizes the urgency of defining the complex etiology and pathogenic mechanisms of autoimmune diseases, and translating these insights into effective autoimmune disease prevention and therapeutic strategies. Remarkable new data support the view that the vitamin D endocrine system performs a variety of critical biological functions that decrease the risk of autoimmune diseases, in particular MS, T1D, IBD, RA, and SLE (and possibly many more). By evaluating experimental data across several different human autoimmune diseases and their respective animal models from a vitamin D perspective, we aim at identifying both highly specific and common mechanisms of vitamin D hormone action that enable us to better understand the function of this important secosteroid hormone *in vivo* and harness its potential to both prevent and treat autoimmunity.

11.2 VITAMIN D METABOLISM, ANALYSIS, AND RECEPTOR FUNCTION

11.2.1 VITAMIN D METABOLISM

Among hormone systems, the vitamin D_3 system is uniquely responsive to sunlight because ultraviolet (UV) light catalyzes vitamin D_3 synthesis in skin [2,3]. Few foods naturally contain significant amounts of vitamin D (e.g., wild-caught oily fish), so humans derive approximately 90% of their vitamin D requirements from exposure to sunlight [4]. Vitamin D and its metabolites are not water soluble, so they circulate bound to the vitamin D binding protein (DBP). The DBP delivers vitamin D_3 to the liver, where it is hydroxylated to 25-hydroxyvitamin D_3 [25(OH)D_3] (Table 11.1). This metabolite is an inactive prohormone that is widely used as an indicator of vitamin D_3 supplies [5]. New data show that circulating 25(OH)D_3 levels are approximately 115 nmol/L in populations living in a traditional manner in East Africa, whereas these levels are considerably lower in westernized populations [6]. The enzyme 25-hydroxyvitamin D-1α -hydroxylase (1α-OHase) catalyzes the rate-limiting step in the synthesis of the active hormone 1,25-dihydroxyvitamin D_3 [1,25(OH)$_2$D$_3$ or calcitriol]. The circulating 1,25(OH)$_2$D$_3$ levels are strictly controlled. If 1,25(OH)$_2$D$_3$ rises above the homeostatic set point, it induces the 1,25-dihydroxyvitamin D_3-24-hydroxylase (24-OHase), which inactivates the hormone. Thus, the circulating 1,25(OH)$_2$D$_3$ level depends on the relative activities of the activating enzyme, 1α-OHase, and its catabolic counterpart 24-OHase in the kidney [7].

The source and availability of 1,25(OH)$_2$D$_3$ in tissues such as the central nervous system (CNS), pancreas, and gastrointestinal (GI) tract, which are targeted by autoimmunity, is debated. Data reviewed here inversely correlate the risk of MS, T1D, and IBD with plasma 25(OH)D levels, not with 1,25(OH)$_2$D$_3$ levels. Moreover, in an animal model of MS, the level of 1,25(OH)$_2$D$_3$ in the inflamed CNS varied directly with vitamin D_3 intake and plasma 25(OH)D_3 levels, whereas plasma

TABLE 11.1

Genes, Enzymes, and Proteins of Vitamin D Metabolism

Gene (Human Chromosome)	Catalyst, Enzyme, or Protein (Abbreviation)	Substrate or Ligand	Product or Function (Abbreviation)	Relevant Tissues	References
	UV light (290–315 nm)	7-Dehydro-cholesterol	Vitamin D_3 (cholecalciferol)	Skin	[21]
CYP2R1 (11p15.2)	Vitamin D_3-25-hydroxylase [25(OH)ase]	Vitamin D_3	25-hydroxyvitamin D_3 [25(OH)D_3 or calcidiol]	Liver	
CYP27B1 (12q13)	25-Hydroxyvitamin D_3-1α-hydroxylase [1α(OH)ase]	25-Hydroxy-vitamin D_3	1,25-Dihydroxy-vitamin D_3 [1,25(OH)$_2D_3$ or calcitriol]	Kidney, macrophages DCs, T lymphocytes, B lymphocytes	[407]
CYP24A1 (20q13)	1,25-Dihydroxy-vitamin D_3-24-hydroxylase [24(OH)ase]	1,25-Dihydroxy-vitamin D_3	1,24,25-Trihydroxy-vitamin D_3 [1,24,25(OH)$_3D_3$]	Kidney, macrophages DCs, T lymphocytes, B lymphocytes	[407]
VDR (12q13.1)	VDR	1,25-Dihydroxy-vitamin D_3	Ligand-dependent transcriptional regulation	Immune system, Th1, Th2, Th17, CD8 T cells, B cells, monocytes, CNS, pancreatic β-cells, GI tract	[544,545] [28] [407] [546]
GC (4q12–13)	DBP	Vitamin D_3, 25(OH)D_3, 1,25(OH)$_2D_3$	Carrier protein for vitamin D metabolites		

1,25(OH)$_2D_3$ levels did not fluctuate [8]. These data, together with data showing 1α-OHase expression in immune cells, strongly suggest that immune cell 1,25(OH)$_2D_3$ synthesis is supported by high circulating levels of the 25(OH)D–DBP complex and probably determines hormone availability in autoimmune target tissues [7,9,10]. Importantly, recent data show that circulating 25(OH)D levels are strongly linked to polymorphisms in the *GC* gene encoding the DBP (D432E; $p = 1.8 \times 10^{-49}$; [11]). Additional new data also show that more than 75% of Caucasians and more than 90% of blacks, Hispanics, and Asians in the United States now suffer from vitamin D insufficiency [25(OH)D < 30 ng/mL], double the prevalence of vitamin D insufficiency seen just a decade ago [12]. Similar data are available for other nations [13]. If the correlation between low vitamin D_3 supplies and high risk of autoimmune disease reflects a causal relationship in autoimmune disease etiology, as many scientists now believe [14], then reversing the vitamin D insufficiency epidemic must become an urgent global priority.

11.2.2 VITAMIN D ANALYSIS

Before summarizing the evidence, it is important to mention the complex considerations that apply to vitamin D_3 studies and their interpretation. The vitamin D_3 metabolite that best indicates near-term UV light exposure and vitamin D_3 consumption is 25(OH)D_3 [5]. Most methods for quantifying circulating $25(OH)D_3$ do not distinguish $25(OH)D_3$ from $25(OH)D_2$, so the analytical results are reported simply as $25(OH)D$. The analytical methods have gradually evolved, and variations in the results have been noted between assay methods and between laboratories performing similar assays [15,16]. Consequently, comparing $25(OH)D$ levels between studies can be problematic. The season of sampling is an important consideration because UV light availability varies seasonally at high latitudes but not at low latitudes [17]. The population studied is also an important consideration because skin pigmentation and genes independent of pigmentation influence circulating $25(OH)D$ levels [11,18–23].

11.2.3 Vitamin D Receptor

The vitamin D receptor (VDR) is a nuclear protein that dimerizes with the retinoid X receptor (RXR) to regulate gene expression through vitamin D–responsive elements (VDRE) in the promoters of $1,25-(OH)_2D_3$–responsive genes [24]. The binding of the biologically active hormone $1,25-(OH)_2D_3$ to the VDR activates the transcriptional regulatory functions of the RXR–VDR complex [2]. The VDRE is composed of two hexameric half-sites, arranged as direct repeats separated by three random base pairs, for example, GGTTCACGAGGTTCA [25]. The VDR is expressed in the cells that maintain mineral ion homeostasis and skeletal health [2].

The VDR is also expressed in immune system cells (reviewed in [14]). Early data showed few VDR molecules in human nonactivated T and B cells (<1000 VDR/cell) and a 10-fold increase on activation [26]. Human macrophages had no detectable VDR, but dendritic cells (DC) had approximately 6000 VDR/cell [26]. In rodents, nonactivated T and B cells had few VDR molecules, whereas VDR protein was more abundant in activated $CD8^+$ T cells than in activated $CD4^+$ T cells and was present at only trace levels in macrophages and LPS-activated B cells [27]. Within $CD4^+$ T-cell subsets, the interleukin-4 (IL-4)–producing Th2 cells had more abundant VDR mRNA than interferon-γ (IFN-γ)–producing Th1 cells [28]. Very recent data showed high VDR mRNA expression in IL-17–producing Th17 cells, but very low VDR mRNA expression in natural and induced $Foxp3^+CD4^+$ T regulatory (Treg) cells [29,30]. High-level VDR expression is a prerequisite for high $1,25-(OH)_2D_3$ responsiveness; therefore, these results favor immunological mechanisms involving direct $1,25-(OH)_2D_3$ actions in activated $CD8^+$ T cells, Th1 cell, Th17 cells, and DC, and possibly B cells (human), with direct actions on macrophages and Treg cells being less likely. It will be important to determine the VDR phenotype of other innate immune system cells such as natural killer (NK) cells, invariant NK T-cells (iNKT), neutrophils, mast cells, and others for a more complete understanding of VDR and immunomodulation.

11.3 MULTIPLE SCLEROSIS

11.3.1 Pathogenesis and Etiology

MS, a complicated and enigmatic neurodegenerative disease, afflicts approximately 2.5 million individuals worldwide and is the leading cause of nontraumatic neurological disability in young adults [31,32]. In the 85% of patients with MS (mainly women) who have relapsing–remitting disease (RRMS), periods of acute neurological dysfunction (e.g., sensory, optic, motor impairment, cognitive dysfunction, incontinence) are followed by periods of remission [33]. Remissions eventually cease (age ~54 in women, age ~51 in men), and the disease progresses relentlessly (secondary progressive MS) [34]. Approximately 15% of MS patients (mainly men) experience unrelenting neurological dysfunction from the outset (primary progressive MS) [35]. The MS disease course varies considerably between individuals, but most patients with MS become wheelchair-bound and bed-bound within 18 and 28 years of diagnosis, respectively [33].

Neurological dysfunction in MS is attributed to focal lesions in the CNS, which show inflammation, demyelination, oligodendrocyte loss, reactive astrocyte formation, and axonal injury and loss [31,32]. Data linking T lymphocyte responses to neural antigens with MS strongly implicate a pathogenic autoimmune reaction directed to components of the myelin sheath in MS etiology. Specifically, nascent MS lesions harbor activated T-cells and other immune cells [31], the HLA class II region genes encoding proteins that present peptide antigens to CD4+ T lymphocytes exert the strongest genetic influence on MS risk [36], particular T-cell receptor (TCR) gene polymorphisms have recently been linked with MS [37], and inflammatory T cells specific for myelin sheath antigens cause an MS-like disease in mice [38,39]. Accordingly, efforts to treat the disease have focused on eliminating or containing these autoreactive T lymphocyte responses.

The MS disease process likely begins years before neurological dysfunction becomes clinically evident, possibly as early as thymic development [31,32,40]. Myelin-reactive T cells that are not deleted in the thymus may later become activated in response to unknown antigens, migrate through the blood–brain barrier into perivascular spaces, and ignite the MS disease process [41]. Interactions between these T cells and CNS microglia are thought to amplify the autoimmune reaction, recruiting additional inflammatory cells, penetrating the parenchyma, and releasing harmful inflammatory mediators, oxidizing free radicals, proapoptotic factors, and matrix-degrading proteases. In addition, there seem to be neurodegenerative processes in MS that are independent of the immune system [32]. The origins of the immune-mediated and non–immune-mediated pathological processes, whether they occur simultaneously or sequentially, and their relative contributions to distinct MS disease phenotypes (relapsing versus progressive) are actively debated subjects.

Intensive research into MS etiology has yielded the paradigm that the disease occurs when there is a confluence of genetic and environmental risk factors. The high frequency (~1–2 in 1000 people) of MS in some ethnic populations (e.g., those of northern European origin) and low frequency in others (e.g., those of African, Asian, southern or eastern European origin), the familial aggregation of MS, the 15- to 20-fold increased relative risk of MS among siblings, and the approximately 25% disease concordance between monozygotic twins, all suggest a heritable component in MS etiology [42]. Epistatic interactions among human leukocyte antigen (HLA) class II haplotypes largely determine MS genetic risk (odds ratio, 5.4; [43]), whereas genes outside this region exert extremely small influences (odds ratio, < 1.2). No gene or combination of genes successfully predicts this disease [44,45]. Moreover, no genomic, epigenomic, or transcriptomic differences correlated with MS disease phenotype in disease-discordant monozygotic twins [46]. This study included only one twin pair and did not examine T lymphocyte repertoire differences; therefore, epigenetic differences cannot yet be ruled out as an explanation for monozygotic twin disease discordance. Consequently, interactions between genetic and environmental risk factors are currently suspected to underlie the risk of MS [47].

The environmental risk factors for MS seem to be acting at a population level, and to determine disease development in genetically at-risk populations [48,49]. The evidence in support of a dominant role for environmental risk factors includes the approximately 75% MS discordance between monozygotic twins [50], geographical gradients in MS prevalence [51], seasonality in MS patients' birth and disease onset months, and alteration of MS risk in genetically at-risk populations by migration [52,53]. Importantly, the migration studies suggest that protective environmental factors may exert their greatest effect during childhood [52]. These data demand a close investigation of the environmental factors of pathophysiological processes suspected in MS disease etiology, and raise the possibility that modifying environmental risk factors in the young may significantly increase the threshold for MS development and prevent a majority of MS cases.

11.3.2 MS Disease Association with UV Light Exposure and Vitamin D

The possibility that UV light–catalyzed vitamin D_3 synthesis might influence MS susceptibility was proposed in 1974 [54], 1978 [55], and 1992 [56], but without a plausible mechanistic explanation, the idea languished. With the proposal that UV light–catalyzed vitamin D_3 synthesis might inhibit MS through a selective action of the hormonal form of vitamin D on VDR-expressing T lymphocytes [57], and supporting data from an experimental animal model of MS [58], investigation of the vitamin D–MS hypothesis quickly intensified. Abundant and compelling evidence in support of this hypothesis has since emerged and is summarized in a recent chapter [14] and in several excellent reviews [40,48,59–62]. Here, we update this evidence and note similarities between MS and T1D regarding the protective influence of UV light and vitamin D_3.

11.3.2.1 Latitude

A geographical gradient in worldwide MS prevalence has been reported, with a low of one to two cases/10^5 population near the equator to a high of more than 200 cases/10^5 population at latitudes above 55° [63,64]. A comprehensive review and meta-analysis of 321 geographical gradient studies confirmed a direct association between latitude and MS prevalence, and computed an odds ratio of approximately 20 for the latitude-linked factor [51]. Adjustment for potential sources of bias had no effect on the gradient. Adjustment for *HLA-DRB1* susceptibility allele frequencies reversed an inverse gradient in the Italian region, but did not alter the European gradient. Similarly, the *HLA-DRB1* susceptibility allele frequencies did not explain the association between latitude and MS prevalence in Ireland [65]. The global incidence of MS has increased sharply and there are now more than 400 cases/10^5 population residing at latitudes above 55° [51].

11.3.2.2 Season

Seasonal influences in MS birth month, onset, and disease severity have been noted [14]. New data from France [66], Sweden [67], Scotland (22% excess of spring births) [68], and Finland (9% excess of spring births) confirmed the excess of April MS births and a paucity of November MS births that was previously reported for Canada [69] and Sardinia [70]. The Finnish data found no influence of *HLA-DRB1* genotype on the birth month data [71]. Consistent with the inverted pattern of UV light availability in the southern hemisphere, Australian data showed an inverted birth month pattern, with an excess of November MS births and a paucity of April MS births [72]. This birth month effect disappeared after adjusting for ambient UV light. Another study in the United States found that winter-born MS patients were, on average, 2.6 years younger at symptom onset than others [73]. In Israel, where ambient UV light is plentiful, there was no birth month effect [74]. Together, the data suggest that a seasonally fluctuating environmental factor influences MS risk beginning with gestation [75].

The seasonally fluctuating factor also influences disease activity in individuals with MS [14]. Striking new data from Boston (42°N) showed a seasonally skewed risk of new magnetic resonance imaging (MRI) lesions in patients with untreated RRMS [76]. The new lesion frequency was threefold higher between mid-March and early September than the rest of the year. Patients with primary progressive MS showed less elevation in seasonal new lesion risk, and the peak occurred about 2 months earlier. These data are consistent with much of the previous data on seasonality of disease activity in northern regions. In Scotland (50–55°N), hospital admissions for MS attacks varied seasonally, peaking in April [77]. Data from Israel (31°N) [78] and Brazil (32°S) [79] did not show strong seasonal skewing of MS attacks.

11.3.2.3 Ambient UV Radiation

There are large seasonal fluctuations in ambient UV light at high latitudes but not at low latitudes. Accordingly, UV light is a prime candidate for an environmental factor capable of influencing MS risk at the population level [14,48,63,64,80]. New data from Australia showed an inverse correlation between actinic damage (an objective measure of long-term UV light exposure) and the risk of a first demyelinating event [81], complementing previous Australian data on UV exposure and the risk of MS [82–84]. Two studies investigated the age at which ambient UV light may be influencing MS risk. One study highlighted sun exposure at 6 to 15 years of age [85]; RRMS onset was 2.1 years earlier in patients who lived in low-to-medium solar radiation areas during childhood than in patients who lived in high solar radiation areas. Another study highlighted sun exposure at 0 to 14 years of age [52]; MS disease risk was lower for young United Kingdom and Irish immigrants to Australia (ages 0–14; 22/10^5 population) than for older immigrants (ages 15–39; 54/10^5 population). These new data confirm previous reports correlating youthful sun exposure with reduced MS risk [83,86–88].

There was a striking sex-based difference in the strength of the association between MS prevalence and ambient UV light [89]. Plotting MS prevalence as a function of annual mean UVB irradiation for 22 regions in France yielded a correlation coefficient of -0.76 ($p < 0.001$) for women and -0.46 ($p < 0.001$) for men. Thus, females showed a significantly higher rate of MS prevalence increase with decreasing UVB than males.

11.3.2.4 Vitamin D

UV light–catalyzed vitamin D synthesis is the most likely environmental factor influencing MS risk [14]. Circulating 25(OH)D varies according to latitude and seasonal ambient UV radiation, reaching a nadir two months after the winter solstice, and a zenith two months after the summer solstice [90]. A 1978 study found a strong ($r = 0.78$; $p < 0.001$) direct correlation between the frequency of decayed, missing, and filled teeth among school-aged children and MS prevalence in 45 countries [55]. This close correlation applied in regions that are exceptions to the latitude gradient in MS prevalence (e.g., among Alaskan Eskimos), among ethnic groups of similar skin pigmentation but discrete disease prevalence (e.g., Bantus compared with dark South Africans), among genetically similar populations living in rural versus urban settings (e.g., on the Scottish Island of Lewis), between individuals of African and Caucasian descent whose MS rates differed sharply (e.g., military veterans in the United States), in analyses of populations whose disease rates changed on migration (e.g., Asian immigrants to the United Kingdom), in comparisons between women and men (e.g., higher disease rates in women), and among different occupations (e.g., agricultural workers compared with office managers). Thus, the direct correlation between childhood dental disease and MS prevalence withstood the many challenges posed by anomalies in MS epidemiology. The data correlating childhood dental disease and MS prevalence are significant because they may be the first data to implicate vitamin D deficiency as an etiological factor in MS [55]. Dental rickets is the term given to the dental disease that develops in individuals with nutritional or hereditary skeletal rickets [91–94]. Features of dental rickets include enamel hypoplasia, attrition and fracture, dentin hypoplasia, poor calcification, loss of lamina dura and periodontal ligament, and spontaneous oral abscesses.

Later studies actually analyzed circulating 25(OH)D and reported that vitamin D deficiency was common in patients with MS [95–97]. The patients had low bone mineral density, rapid bone loss, and high fracture rates. Many similar studies have now been published [98–100], some of which found no correlation [101]. However, all such correlative studies are subject to the concern of reverse causation. The MS patients had low vitamin D intakes, little or no weekly sun exposure, mobility problems, and corticosteroid treatments, all of which would negatively influence vitamin D nutritional status [97,102–104]. These correlations do not provide evidence for vitamin D deficiency in MS etiology, but they do suggest that most MS patients have vitamin D insufficiency of long duration and would benefit from comprehensive efforts to detect and address this problem.

Longitudinal studies evaluating circulating 25(OH)D levels prior to an MS diagnosis have associated low vitamin D supplies with high MS risk and vice versa [105,106]. These and similar studies have been reviewed [14]. Further work has now correlated a mother's gestational vitamin D intake with the risk of MS in her offspring [107]. The longitudinal studies eliminate the reverse causality concern and support a possible causal relationship between high baseline circulating 25(OH)D levels, particularly from gestation through adolescence, and a subsequently reduced risk of MS.

Data correlating circulating 25(OH)D levels with MS disease activity came from Finland [108,109] and have now been replicated in further studies [110,111]. A prospective longitudinal study revealed a close correspondence between decreasing serum 25(OH)D levels, hypocalcemia, a blunted parathyroid hormone (PTH) response, and increasingly frequent MS relapses. There was a lag of some months between the decline in 25(OH)D and the increase in relapses. Two prospective longitudinal studies have confirmed this groundbreaking study. The new data showed that each 10 nM increase in serum 25(OH)D correlated with an approximately 13.6% decrease in attacks for pediatric MS patients [110], and a 12% decrease in attacks for adult MS patients [111]. Both studies enrolled newly diagnosed patients with very mild disability and used a blinded study design. Research on vitamin D and MS attacks has now been extended to include patients with inflammatory diseases of the spinal cord [112]. The results showed significantly lower vitamin D levels in patients with recurrent transverse myelitis than in controls (adjusting for season, age, sex, and race). Additional new longitudinal data showed a seasonally skewed risk of new MRI lesions in untreated RRMS patients residing in Boston [76]. Most striking in the MRI lesion data were the rapid decline (~67%) in lesions/scan over 2 months beginning at the end of August, which is the peak of circulating 25(OH)D levels in Boston [113], and the rapid increase (~67%) in lesions/scan over 2 months beginning at the end of February [76], which is the nadir of circulating 25(OH)D levels.

In summary, there is now substantial evidence showing a strong, consistent, and specific correlation between poor vitamin D nutrition and high MS risk and disease severity. A biological gradient is evident, with the lowest 25(OH)D levels corresponding to the highest MS risk and disability. Temporal trends have been documented, with low gestational and childhood vitamin D levels correlating with later risk of MS, and low seasonal vitamin D levels correlating with subsequent increases in disease activity. Thus, the data have begun to support the assertion of a causal role for vitamin D insufficiency in MS etiology [114]. Moreover, the data provide a strong scientific rationale for investigating whether increasing vitamin D supplies would prevent, delay, or otherwise limit demyelinating diseases [14,112,115].

11.3.3 MS DISEASE ASSOCIATION WITH *HLA, VDR, CYP27B1,* AND *GC* GENES

The polygenic hypothesis to explain the modest heritable component of MS risk (15-fold increased risk observed in biological first-degree relatives of MS patients) posits that genetic risk derives from the cumulative effects of several risk alleles, each of which is common in the population and exerts a small phenotypic effect [42,116]. Genomewide association (GWA) studies have attempted to identify these risk alleles. The most recent global study examined approximately 0.5 million single nucleotide polymorphisms in each of approximately 10,000 MS cases and approximately 17,000 controls [45]. Polymorphisms in approximately 16 genes had statistically significant associations with MS, but with an odds ratio of less than 1.3, they had very weak effects (with the exception of HLA region genes). To date, it has not been possible to compute an MS genetic burden score that is predictive of MS based on the combinations of these common genetic variants [44,116]. Close attention is now focused on gene–environment interactions [117] and epigenetic mechanisms [47]. In the context of gene–environment interactions, it is important to note that GWA studies may fail to identify environmentally sensitive genetic variants relevant to MS disease because they combine patients with heritable and spontaneous forms of disease living in high- and low-risk environments. In contrast, studies of MS multiplex families with heritable disease or case and control subjects residing in high-risk environments have the potential to reveal environmentally sensitive genetic risk factors. Recent studies of this type have begun to reveal important associations between vitamin D pathway genes and MS disease.

11.3.3.1 *HLA*

The association between MS susceptibility and particular *HLA* regions was discovered nearly 30 years ago [118] and has been reviewed [36]. The *HLA-DRB1*1501* allele is the strongest genetic determinant of MS risk (odds ratio, 5.4); it is associated with earlier disease onset and a more severe disease course, particularly in women [119]. A crucial recent advance was the discovery of a potential VDRE within the *HLA-DRB1*1501* promoter region immediately 5' to the transcriptional start site [120]. This sequence bound VDR protein and drove a modest 1,25(OH)$_2$D$_3$-mediated increase in HLA-DR2b protein expression in lymphoblastoid cell lines, and was absent in the *DRB1*04*, *DRB1*07*, and *DRB1*09* alleles associated with MS resistance. Whether this element functions as a VDRE *in vivo* and how vitamin D regulation of *HLA-DRB1*1501* gene expression may contribute to MS disease development awaits further experimentation.

11.3.3.2 *VDR*

Associations between alleles of the *VDR* locus on chromosome 12q13.1 and MS in particular populations have been reviewed [117,121,122]. Two concerns regarding *VDR* and MS association data are the inconsistencies between populations [123] and a lack of evidence that the polymorphic variants have altered transcriptional functions. An exception is the *VDRFokI* polymorphism. The *VDR*[f] variant (*FokI* site in the first ATG codon) encodes a 427 amino acid VDR protein, whereas the *VDR*[F] allele (without the *FokI* site) encodes a 424 amino acid VDR protein [124]. The *VDR*[F]-encoded short protein had elevated transcriptional activity compared with the *VDR*[f]-encoded long protein [125]. The *FokI*[f]-encoded, less active VDR protein was associated with higher circulating 25(OH)D levels [22,126], lower MS risk in males but not in females [127], and lower MS disability a decade after disease onset [128]. Moreover, the association between vitamin D supplement use and reduced MS risk was evident in women who were homozygous for the *VDRFokI*[f] variant, but not in women who were homozygous for the *VDRFokI*[F] variant [129]. A second exception is the *VDR* gene *Cdx-2*[G] variant. An A > G substitution within the *Cdx-2* factor binding element in the 5′UTR of the *VDR* gene reduced the promoter activity by 70% [130]. The *Cdx-2*[G] variant correlated with an increased risk of MS in children (ages 6–10) who had ≤2 h/day of winter sun exposure [88]. One interpretation of this observation is that high childhood sun exposure overcame the effect of reduced *VDR* gene expression.

11.3.3.3 *CYP27B1*

The chromosome 12q13.1-q13.3 *CYP27B1* gene encodes the rate-limiting 1α-hydroxylase in $1,25(OH)_2D_3$ synthesis. The syntenic region in rodents has been implicated in animal models of MS, diabetes, arthritis, and lupus. Homozygosity for deleterious *CYP27B1* mutations causes vitamin D–dependent rickets type I (VDDRI; [131,132]). An association between familial VDDRI and MS was reported in 2008 [133] and confirmed in 2010 [134]. The chance occurrence of MS and VDDRI is highly improbable because both diseases are relatively rare (MS prevalence, ~1 in 1000; VDDRI prevalence, ~1 in 5×10^5). Several GWA studies have not found this association, despite large sample sizes and comprehensive gene coverage [123,129,135]. The explanation for the discrepant findings between the familial studies and the GWA studies derives from the different assumptions that underlie these two gene-searching techniques. Familial studies assume that uncommon genetic risk alleles with large phenotypic effects are segregating within a multiplex family, whereas GWA studies assume that common genetic risk alleles with small phenotypic effects are combining to precipitate disease in the population. A critically important new MS multiplex family study performed in Canada, a region with low ambient UV radiation, has unambiguously associated rare but highly penetrant loss-of-function alleles of the *CYP27B1* gene with MS [136]. Exon sequencing within candidate MS susceptibility genes was performed on one representative member of each of the 43 MS multiplex families. A known null *CYP27B1* variant was identified and showed significant heterozygous association in more than 3000 parent–affected child trios. Three additional null *CYP27B1* variants also conferred a significant risk of MS (odds ratio, 4.7). Remarkably, heterozygous parents who carried one null *CYP27B1* allele transmitted it to their MS offspring in 35 of 35 cases ($p = 3 \times 10^{-9}$). These data unambiguously implicate $1,25(OH)_2D_3$ synthesis as a factor in MS etiology, independently of UV light, although the relevant disease mechanisms are still undetermined. The data also reinforce the concept that searching for uncommon genetic risk alleles with large phenotypic effects within multiplex families will prove fruitful in penetrating this enigmatic disease.

11.3.3.4 GC

The *GC* gene (group-specific component) on human chromosome 4q12-q13 encodes the DBP, the plasma carrier protein for vitamin D metabolites [137]. DBP circulates in approximately 100-fold molar excess of 25(OH)D_3, its primary bound metabolite, and has approximately 10-fold higher affinity for 25(OH)D_3 than for 1,25-(OH)$_2D_3$. *GC* is one of the most polymorphic human genes. The need to prolong vitamin D supplies in light-skinned populations living in regions with limited winter UV light availability may have driven the evolution of *GC* diversity [137]. Consistent with this suggestion, *GC* polymorphisms were strongly associated with plasma 25(OH)D levels in a recent GWA study [11]. A *GC* polymorphism was weakly associated with MS in an Italian population [138], but not in other nationalities [129,135,139–141].

11.3.4 VITAMIN D TREATMENT OF MS

Vitamin D_3 supplementation has been evaluated in small pilot studies of patients with RRMS. A randomized vitamin D_3 dose escalation study of 1 year duration [142] followed a small phase I safety study [143]. In these clinical studies, vitamin D_3 supplementation decreased the number of MRI lesions, lowered the relapse rate, and slowed the progression of disability. Furthermore, T lymphocyte responses to test antigens (including myelin and glial antigens) decreased significantly in the vitamin D_3 supplemented group ($p = 0.002$), but not in the controls, although the two groups did not differ in their cytokine profiles [IL-1β, IL-2, IL-4, IL-5, IL-6, IL-10, IL-12p40, IL-13, IFN-γ, and tumor necrosis factor-α (TNF-α)]. These studies were not blinded, so the results are considered exploratory. It is important to note that vitamin D_3 supplementation increases dietary calcium absorption, so calcium supplementation (e.g., >2 g/day) is unnecessary and possibly harmful [144].

The first double-blind, randomized, placebo controlled, trial of vitamin D_3 supplementation in MS patients was recently completed [145]. The patients ($n = 66$) were receiving disease-modifying therapy. Once weekly vitamin D_3 supplement (20,000 IU) increased the mean circulating 25(OH)D from 54 to 110 nM, and decreased new gadolinium-enhancing T1 lesions on brain MRI by a remarkable 85% ($p = 0.004$) compared with controls. Moreover, despite its short duration and small number of patients, the study showed trends toward decreased total lesion burden (vitamin D_3, 83 mm^3; control, 287 mm^3; $p = 0.105$), lower disability ($p = 0.071$), and improved ambulation ($p = 0.076$) in the vitamin D_3 group. These data show remarkable vitamin D_3 benefits in MS. It will be important to determine whether vitamin D_3 provides benefits to MS patients and patients with a first demyelinating event (clinically isolated syndrome) independently of disease-modifying therapy.

Vitamin D_2 supplementation has also been evaluated in patients with RRMS. The vitamin D_2 intakes of those receiving high-dose vitamin D_2 were adjusted individually to achieve serum 25(OH)D levels of 130 to 175 nM. Low-dose controls received 1000 IU/day of vitamin D_2. Unfortunately, placebo and vitamin D_3 controls were not included. Six months later, the high vitamin D_2 dose group had more relapses and higher levels of disability than the low dose group. The interpretation of these data is confounded by the lack of reference data. However, it seems that high vitamin D_2 doses exacerbated RRMS disease, similar to the deterioration in glycemic control observed in T1D patients given high vitamin D_2 doses [146].

Two pilot studies have tested oral vitamin D_3 metabolites in patients with RRMS. The first study tested 1-α-hydroxyvitamin D_3, which is metabolized to 1,25-(OH)$_2D_3$ [147]. The second study tested escalating doses of 1,25-(OH)$_2D_3$ to a target dose of 2.5 μg/day in patients with RRMS [148]. From a baseline relapse rate of 1.0, the composite on study relapse rate declined to 0.27, then increased to 0.8 in the post-study period. The median disability score was unchanged from the baseline (2.2) during the study, then increased to 3.1 in the post-study period. Patients who adhered to the calcium intake guidelines had no adverse events, but two diet-noncompliant patients withdrew with symptomatic hypercalcemia.

11.3.5 Experimental Autoimmune Encephalomyelitis Model of MS

Experimental autoimmune encephalomyelitis (EAE) is an induced paralytic autoimmune disease that has been intensively studied as an MS model [39]. The disease is induced by immunizing rodents with CNS tissue, proteins, or peptides, or by transferring encephalitogenic T lymphocytes isolated from an EAE-diseased animal into a nondiseased normal animal. EAE severity and disease course depend on the strain used and is determined by examining animals for signs of ascending paralysis (weakness, ataxia, hind and foreleg paralysis), performing histopathological analysis of spinal cord sections for focal lesions with inflammatory cell infiltration, demyelination, and axonal damage, and assessing lymphocyte responses and cytokine production to neural antigen stimulation. EAE is thoroughly characterized neurologically, immunologically, and genetically, and has proven valuable in addressing complex questions of immunological self-tolerance and immunological mechanisms in autoimmune disease.

Autoimmune T cell–mediated mechanisms in EAE are well understood [41]. Normally, tight junctions between the endothelial cells of the blood–brain barrier and the epithelial cells of the blood–cerebrospinal fluid barrier limit blood cell access to CNS tissue. Activated CD4+ T cells express adhesion molecules, chemokine receptors, and integrins that allow them to migrate into the CNS for immune surveillance. If these T cells are reactivated by antigen-presenting cells (APC) presenting peptides derived from myelin sheath components in the parenchymal space, they induce and sustain an inflammation. Chemokines attract additional inflammatory cells, and proinflammatory cytokines activate the parenchymal microglia. The activated macrophages and microglia produce cytokines that sustain T cell activation (e.g., IL-12 and IL-23), and neurotoxic chemicals (e.g., NO and oxygen radicals) that cause demyelination and axonal damage.

MS and EAE have strong similarities. They are similar in their relapsing–remitting or unremitting disease phenotype, depending on the individual patient or rodent strain, the progressive accumulation of neuromuscular dysfunction, visual impairments, and other physical manifestations of neurologic dysfunction. They are also similar histologically with focal perivascular lesions in white matter, gray matter, brain stem, and optic nerves, gliosis, myelin sheath destruction (and partial remyelination), and axonal damage. Finally, they show common immunological features, such as activated, neural antigen–specific T lymphocytes populating newly formed lesions, activated macrophages and microglia, inflammatory cytokines, chemokines, and reactive oxygen species, and the presence of oligoclonal immunoglobulins in the CNS and cerebrospinal fluid. The strong parallels between EAE and MS support the use of EAE as a model to study MS from the perspective of immune system dysfunction(s), genetic susceptibility and resistance, and nutritional influences. Moreover, EAE researchers have developed three approved MS therapeutics, and examined the mechanisms of action of a fourth MS therapeutic, the antiviral cytokine IFN-β [39]. There are differences that require caution in extrapolating data from the EAE model to human MS. Whereas EAE is an induced disease with a defined autoimmune origin, MS is a spontaneous disease of uncertain etiology, so EAE is not a useful MS model from the perspective of possible microbial contributions to MS etiology.

11.3.6 Vitamin D and EAE Prevention and Treatment

Early studies found that moderate daily UV light exposure for 1 week prior to EAE induction significantly reduced the incidence, severity, and neuropathology of EAE, but exposure at the time of immunization or later had no effect [149]. EAE inhibition by moderate daily UV light exposure was recently confirmed [150]. Another study found that a single high UV light dose followed 1 h later by neural antigen immunization exacerbated EAE, and this was attributed to antibody-mediated CNS pathology [151]. Early studies also evaluated the effects of vitamin D_3 deficiency and vitamin D_3 sufficiency on EAE induction. Female vitamin D-deficient mice had more rapid disease onset than vitamin D-sufficient mice [58]. Furthermore, vitamin D-sufficient females had reduced EAE disease incidence, mortality, peak disease severity, cumulative index, and spinal cord pathology compared with vitamin D-deficient females, but vitamin D-sufficient and vitamin D-deficient males were not significantly different [8]. These unexpected sex-based differences have been confirmed [152]. Recent work investigated 25(OH)D$_3$ pretreatments and found that the lowest dose tested, 0.2 µg/day, had no effect on EAE disease, whereas the next dose tested, 10 µg/day, yielded toxic serum 25(OH)D$_3$ levels (~1050 nM; [150]).

With regard to 1,25-(OH)$_2$D$_3$, early studies showed that initiating treatments before EAE induction delayed the disease onset and decreased disease morbidity and mortality in male and female SJL/J and B10.PL/J mice (modeling RRMS), and C57BL/6 mice (modeling progressive MS, as reviewed in [14]). In contrast, 1,25-(OH)$_2$D$_3$ treatments initiated before transferring pathogenic IFN-γ–producing T helper type 1 (Th1) cells into nonimmunized animals did not inhibit EAE disease [28]. In the transfer model, the 1,25-(OH)$_2$D$_3$–treated mice had activated CD4+ T cells in the CNS, whereas in the induced EAE model, the 1,25-(OH)$_2$D$_3$–treated mice had only nonactivated CD4+ T cells in the CNS. This difference suggests that in the induced EAE model, the presence of 1,25-(OH)$_2$D$_3$ during neural antigen immunization may have promoted the development of regulatory cells capable of preventing CD4+ T cells from expressing an activated phenotype in the CNS. Further evidence in favor of regulatory cell development in the induced EAE model was the durability of inhibition after 1,25-(OH)$_2$D$_3$ was discontinued, and the requirement for *Rag-1*–expressing cells (lymphocytes) other than encephalitogenic CD4+ T cells for 1,25-(OH)$_2$D$_3$ inhibition.

Administering 1,25-(OH)$_2$D$_3$ to rodents continuously beginning at the first clinical signs of EAE prevented disease progression [58,153–156]. Administering 1,25-(OH)$_2$D$_3$ at the peak of acute EAE disability reversed the EAE disease signs [155,157]. Treatment of established EAE disease with a 1,25-(OH)$_2$D$_3$ analog (Ro 63-2023) was also reported [158]. Dramatic reductions in histopathology (inflammatory infiltrate, demyelination) accompanied the reversal of EAE [155]. Importantly, when 1,25-(OH)$_2$D$_3$ treatments were initiated in mice with established EAE disease, progression resumed when the 1,25-(OH)$_2$D$_3$ treatments were stopped [58].

11.3.7 Vitamin D Hormone Modulation of Immune Function in Demyelinating Disease

Mechanistic studies aimed at defining how vitamin D may modulate MS risk and severity have focused on its immunomodulating functions in adults [14], because strong evidence has implicated a pathogenic T lymphocyte–mediated attack directed to components of the myelin sheath [31]. There is still debate as to the sequence and relative importance of the autoimmune and neurodegenerative phases of the disease [32,40,159]. Some recent reports have addressed the role of vitamin D in neuroprotection. Here, we update information on immunomodulation and neuroprotection.

11.3.7.1 Calcium

The suppressive effects of 1,25-(OH)$_2$D$_3$ in the EAE model correlate with moderate elevations in serum Ca [58,160,161] and vary inversely with dietary Ca [160]. These data support a role for extracellular Ca in protective mechanisms, although it is clear that hypercalcemia alone is not sufficient to inhibit EAE [161]. The peptide hormone calcitonin (CT), combined with high dietary Ca, reduced the amount of exogenously supplied 1,25-(OH)$_2$D$_3$ needed to prevent EAE in female mice and reduced the hypercalcemic risk [162]. Direct participation of PTH [163] and CT [164] in the mechanism of 1,25-(OH)$_2$D$_3$ protection have been ruled out.

It is noteworthy that Finnish investigators found dysregulated Ca homeostasis in patients with MS [109]. As serum 25(OH)D levels declined in winter, patients with MS showed a blunted PTH response and developed hypocalcemia, whereas healthy controls showed increased PTH and maintained serum Ca homeostasis. There was a striking correlation between low 25(OH)D, high intact PTH, low Ca and MS attacks. All attacks occurred at 25(OH)D < 85 nM, intact PTH > 20 ng/L, and Ca ≤ 2.24 mM. If these data reflect a causal relationship, they suggest that MS attacks might be avoidable by maintaining 25(OH)D > 85 nmol/L, PTH < 20 ng/L, and Ca ≥ 2.24 mmol/L. It is critically important to determine if this relationship holds in additional populations, and whether it reflects causation or reverse causation.

11.3.7.2   1,25-(OH)$_2$D$_3$ Synthesis in the CNS

Renal calcitriol synthesis and calcitriol function in mineral ion homeostasis are well established [2]. Immune cell 1,25-(OH)$_2$D$_3$ synthesis seems to have important intracrine, autocrine, and paracrine functions in the regulation of immune responses (reviewed in [165]). The known *CYP27B1* inducers in innate immune cells are IFN-γ, Toll-like receptor (TLR) 4 ligands, and TLR 2/1 ligands using IL-15. Also, human B cells activated through the immunoglobulin receptor and CD40 [166], and CD4$^+$ T cells activated through the TCR and CD28 [100] both produced and responded to 1,25-(OH)$_2$D$_3$. Langerhans cells produced 1,25-(OH)$_2$D$_3$ and programmed T cells to migrate to the epidermis [167], and programmed DC cells to migrate to lymph nodes [168].

CNS 1,25-(OH)$_2$D$_3$ synthesis is controversial because there is a paucity of direct evidence [169]. Weak 1α-hydroxylase immunohistochemical staining was noted in human neuronal and glial cells [170,171]. Furthermore, rat microglia produced calcitriol *in vitro* when stimulated with IFN-γ [169]. Some direct evidence for 1,25-(OH)$_2$D$_3$ synthesis *in vivo* in the CNS has been reported [8]. The 1,25-(OH)$_2$D$_3$ was extracted and quantified in spinal cord tissue of vitamin D–deficient and vitamin D–sufficient mice with EAE [8]. Spinal cord 1,25-(OH)$_2$D$_3$ correlated with serum 25(OH)D$_3$ but not with 1,25-(OH)$_2$D$_3$. Spinal cord 1,25-(OH)$_2$D$_3$ was significantly higher in vitamin D–sufficient females than males, and correlated directly with EAE inhibition in females but not in males.

11.3.7.3   Dendritic Cells

Evidence for a 1,25-(OH)$_2$D$_3$ action directly on myeloid DC to establish a tolerogenic phenotype, and subsequent tolerogenic DC induction of CD4$^+$Foxp3$^+$ Treg cells, has been suggested as a mechanism for autoimmune disease prevention and has been reviewed [172]. The myeloid DC express VDR and respond to 1,25-(OH)$_2$D$_3$ *in vitro* by increasing phagocytic function and decreasing major histocompatibility complex (MHC) class II molecules, costimulatory molecules, and IL-12 production. Plasmacytoid DC do not show these responses. Additional experiments showed that human DC responded to 1,25-(OH)$_2$D$_3$ *in vitro* with increased expression of indoleamine 2,3-dioxygenase, the enzyme that degrades tryptophan to kynurenine [100]. Microenvironments with a high kynurenine to tryptophan ratio are believed to favor the development of CD4$^+$CD25$^+$Foxp3$^+$ Treg cells [173,174]. In diabetes [175] and colitis models [176], the appearance of tolerogenic DC correlated with 1,25-(OH)$_2$D$_3$–mediated disease prevention. However, in EAE [28,30] and in experimental autoimmune uveitis [177], direct 1,25-(OH)$_2$D$_3$ action on myeloid DC was not sufficient to prevent disease. Collectively, the data suggest that 1,25-(OH)$_2$D$_3$ acts on myeloid DC to induce a tolerogenic phenotype that was sufficient to prevent autoimmune diseases affecting peripheral tissue (diabetes and colitis), but was insufficient to prevent autoimmune diseases from affecting immune privileged sites in the eye and the CNS.

11.3.7.4   NK Cells and iNKT Cells

NK cells are innate immune system cells with unique cytotoxic capability that are believed to perform an immunoregulatory role in preventing autoimmune-mediated tissue pathology, although they have the potential to perform a pathogenic role as well. Evidence that NK cell numbers and functions are decreased in MS has been summarized [40,178]. To our knowledge, the effects of vitamin D$_3$ on NK cells have not been reported.

iNKT cells are distinct from NK cells [178]. The iNKT cells are T cells that express invariant TCR α-chains and bridge the innate and adaptive immune systems. The iNKT cells have the dual potential of acting as either protective or pathogenic lymphocytes, confounding the interpretation of data obtained from human and rodent studies. Recent studies noted reduced numbers of iNKT cells (defined as TCR β$^+$αGalCer-CD1d tetramer$^+$) in homozygous *VDR*-null mice [179], homozygous *CYP27B1*-null mice, and wild-type mice made vitamin D–deficient from birth [180]. Vitamin D$_3$

repletion reportedly did not restore iNKT cells, but the vitamin D_3 used in repletion was not specified and the circulating $25(OH)D_3$ in the repleted mice was not reported [180]. Some restoration of iNKT cells was achieved with $1,25\text{-}(OH)_2D_3$ treatments in animals with circulating $25(OH)D_3$ < 5 nmol/L. Consequently, additional research will be needed to determine whether vitamin D_3 repletion to a higher level, with or without $1,25\text{-}(OH)_2D_3$ treatments, will restore the iNKT cells.

11.3.7.5 Treg Lymphocytes

Treg cells are $CD4^+$ T lymphocytes defined functionally by their ability to limit autoimmune T cell expansion and prevent autoimmune-mediated pathology [181]. Two Treg subsets are recognized, natural Treg that arise during thymic development [182] and adaptive Treg that arise during TCR engagement in the periphery [183].

The data concerning Treg cells an MS are contradictory, with Treg cell numbers and functions being reported variably as normal or decreased (reviewed in [184]). The use of different phenotypic markers to enumerate these cells [185], and possible differences between cells in the CNS and in the peripheral blood [186] are two confounding factors. Correlations between circulating $25(OH)D$ levels and Treg proportions in the peripheral blood of MS patients have been reported [187–189]. It is difficult to extrapolate from these data a cause–effect relationship because a change in cell proportions could be due to a gain in one compartment or a loss in another, and it is not clear that peripheral blood is an acceptable surrogate for cell populations in the CNS or even in peripheral lymphoid tissue.

In EAE, Treg cells have been characterized as $CD25^+CD44^{high}CTLA\text{-}4^{high}GITR^{high}$ cells expressing Foxp3 as a lineage-specific transcription factor [190], and suppressing disease by an IL-10–dependent mechanism [191–195]. Data showing residual protection from EAE disease after $1,25\text{-}(OH)_2D_3$ treatments were discontinued [196], and both *Rag-1* and *IL-10* gene expression requirements for protection [28,161] are consistent with the induction of a regulatory lymphocyte population. However, reports on the contributions of $1,25\text{-}(OH)_2D_3$ to Treg development in EAE are conflicting. Two studies found very low *VDR* expression in natural Treg and induced Treg cells [30,197]. Consistent with the low *VDR* expression data, one group found no changes in the Treg cells in the CNS or spleen of mice lacking VDR expression only in T cells, and no effect of $1,25\text{-}(OH)_2D_3$ on the Treg cells in the CNS or spleen of wild-type mice in the EAE model [30]. Another group found that $1,25\text{-}(OH)_2D_3$ reduced $Foxp3^+$ Treg cells in the spleen *in vivo*, and downregulated *Foxp3* transcription *in vitro* in a VDR-dependent manner [197]. A third group reported that $1,25\text{-}(OH)_2D_3$ enhanced $Foxp3^+$ T cells in the spleen and CNS in the EAE model, and upregulated *Foxp3* transcription *in vitro* through VDR-dependent recruitment of histone deacetylases [198]. These discrepant results remain unexplained. However, it seems unlikely that $1,25\text{-}(OH)_2D_3$–mediated increases in $Foxp3^+$ Treg cells will be both necessary and sufficient for EAE inhibition.

An especially interesting proposal is $1,25\text{-}(OH)_2D_3$–mediated induction of Treg cells by exposure of the skin to UV light [199]. In this system, UV light–induced immunosuppression depended on a functional *VDR* gene [200]. This should be a productive area of investigation in the future.

11.3.7.6 T Effector Lymphocytes

There is a growing consensus that $1,25\text{-}(OH)_2D_3$ acts directly on encephalitogenic $CD4^+$ Th1 and Th17 cells to inhibit autoimmune disease (reviewed in [14]). Important new insights came from research investigating the need for *VDR* gene expression in $CD4^+$ T cells [30]. The $1,25\text{-}(OH)_2D_3$ inhibited EAE induction in bone marrow chimeric mice with a disrupted *VDR* gene in nonhematopoietic cells, but completely failed to inhibit disease in chimeric mice with a disrupted *VDR* gene only in hematopoietic cells. Targeting the *VDR* gene specifically in T cells completely abrogated $1,25\text{-}(OH)_2D_3$–mediated EAE inhibition. However, neither the *VDR* targeting in T cells nor $1,25\text{-}(OH)_2D_3$ treatment in targeted or wild-type mice influenced Th1 and Th17 cell priming and cytokine production in the spleen and lymph nodes in the early stages of EAE. Thus, no influence of *VDR* or $1,25\text{-}(OH)_2D_3$ on pathogenic T cell development in the periphery was observed. Previous data ruled out 1,25-$(OH)_2D_3$–mediated restriction of pathogenic T cell access to the CNS as a protective mechanism [28]. In the later stages of EAE, others found fewer splenic Th17 cells and lower IL-17 production in 1,25-$(OH)_2D_3$–treated mice compared with placebo controls [29,197]. They reported that 1,25-$(OH)_2D_3$ suppressed IL-17 protein synthesis by a posttranscriptional mechanism [29]. Another group reported that 1,25-$(OH)_2D_3$ suppressed *IL-17A* gene transcription through blocking of nuclear factor for activated T cells and sequestration of Runt-related transcription factor 1 [198]. In summary, rodent data showed no 1,25-$(OH)_2D_3$-mediated effect on Th17 cell development in the periphery *in vivo* [30], but *in vitro* data showed 1,25-$(OH)_2D_3$-mediated transcriptional repression of the *IL-17A* gene [198] and post-transcriptional suppression of IL-17 protein synthesis [29]. It is not clear how to reconcile the discrepant *in vivo* and *in vitro* findings. In humans, adding 1,25-$(OH)_2D_3$ to purified and activated CD4+ T cells from controls and patients with RRMS significantly reduced the frequency of CD4+ Th17 cells [100].

Analyses of Th1 cells *in vivo* have yielded similarly inconsistent findings. Administering 1,25-$(OH)_2D_3$ to Biozzi AB/H mice before EAE induction decreased the Th1 cell frequency in peripheral lymphoid tissues correlating with disease inhibition [158]. In contrast, administering 1,25-$(OH)_2D_3$ to B10.PL mice before EAE induction had no effect on the Th1 cell response (frequency or cytokine production), although disease was inhibited [28,58,157]. In humans, vitamin D_3 or 1,25-$(OH)_2D_3$ had no effect on Th1 cell IFN-γ responses [142,201,202]. This finding was confirmed in a recent vitamin D_3 repletion study [142].

Analyses of Th2 cells *in vivo* have also yielded inconsistent results. Administering 1,25-$(OH)_2D_3$ to mice before EAE induction increased the IL-4 transcripts in the lymph nodes and in the CNS compared with the placebo controls [157]. Also, targeted disruption of the *IL-4* gene moderately decreased the protective function of 1,25-$(OH)_2D_3$ in EAE [203]. A subsequent report found no significant differences between 1,25-$(OH)_2D_3$–pretreated and placebo-pretreated B10.PL mice regarding IL-4 mRNA in the lymph nodes or in the CNS after immunization with myelin basic protein [28]. There was also no effect of 1,25-$(OH)_2D_3$ on the IL-4 protein synthesis per Th2 cell. Another report found that administering 1,25-$(OH)_2D_3$ to Biozzi AB/H mice before EAE induction had no effect on the IL-4–producing Th2 cell frequency [158]. Thus, there are some inconsistencies regarding IL-4 that remain to be resolved. Because 1,25-$(OH)_2D_3$ efficacy against EAE decreased slightly in *IL-4*–null mice, there may be some IL-4 contribution to the mechanism [203], but it is possible that Th2 cells are not the source of the protective IL-4 [28,158].

The mechanisms involved in the amelioration of established EAE center on the eradication of the T cell–mediated inflammatory cascade in the CNS (reviewed in [14]). Administering 1,25-$(OH)_2D_3$ to rodents with established EAE results in very rapid disease amelioration that correlates with a significant decrease in CNS pathology and fewer CD4+ T cells in the CNS [153,155]. A gene expression analysis performed 6 h after 1,25-$(OH)_2D_3$ treatment found 1,25-$(OH)_2D_3$–driven transcriptional changes favoring apoptosis, suggesting the hypothesis that the 1,25-$(OH)_2D_3$ may have sensitized CNS-resident pathogenic T cells to programmed cell death signals [204]. Additional evidence for this mechanism came from flow cytometric analyses showing a decrease in living CD4+ T cells accompanied by an increase in apoptotic CD4+CD44^{high} T cells, immunohistochemical analyses showing dying cells with fragmented DNA in the meningeal lesions, and PCR analyses showing less IFN-γ production, all of which occurred in the spinal cord within 6 to 12 h of the 1,25-$(OH)_2D_3$ treatment [204,205]. Thereafter, decreases in chemokine synthesis, nitric oxide synthesis, and immune cell recruitment into the CNS were documented [205]. None of these changes were observed in the lymph nodes of the animals or in cell cultures, suggesting that the T-cell depletion mechanism was specific to the CNS [205].

Collectively, the EAE disease prevention data suggest a model wherein vitamin D_3 supports 1,25-$(OH)_2D_3$ synthesis in the CNS, which in turn initiates a disease prevention pathway that constrains the autoimmune T cell–driven inflammatory cascade in the CNS by a mechanism that is not fully understood. The Th1, Th2, Th17, and Treg cells in the periphery may be unchanged by vitamin D_3 and 1,25-$(OH)_2D_3$, but there is a growing consensus that the pathogenic subsets, Th1 and Th17 cells, are eliminated in the CNS by mechanisms that require them to express a functional VDR, and may depend on extracellular Ca. The $1,25\text{-}(OH)_2D_3$–mediated protection also requires *Rag-1*–dependent regulatory lymphocytes, although there is no agreement on the phenotype of these lymphocytes, and possibly increases in TGF-β 1, IL-4, and IL-10, and decreases in IL-17 transcripts and proteins. Together, the mechanisms lead to the inhibition of activated neural antigen–specific pathogenic T cell accumulation in the CNS. The disease treatment data suggest a model wherein $1,25\text{-}(OH)_2D_3$ initiates a disease resolution pathway that centers on activation-induced cell death of CNS-resident pathogenic T cells responding to signals *in situ*.

11.3.8 Vitamin D and MS Summary

The evidence supporting an inverse correlation between vitamin D supplies and MS risk is now sufficiently compelling to assert that it probably reflects a causal relationship. The association has been documented globally, it holds where ambient UV light varies due to climate and altitude rather than latitude, and is consistently strongest where seasonal fluctuations in vitamin D supplies are greatest. Very recent genetic data unambiguously implicated the *CYP27B1* gene and $1,25\text{-}(OH)_2D_3$ synthesis in MS risk. Additionally, functionally significant polymorphisms in the *VDR* gene have been consistently correlated with MS risk when the sex, MHC genes, and latitude of residence of the subjects are considered. Temporal studies also support an assertion of causality because UV light and vitamin D supplies fluctuate before changes in MS disease activity. Dose–response relationships have also been documented, establishing that there is a biological gradient linking UV light and vitamin D supplies with MS risk and activity. Vitamin D supplementation studies in limited numbers of MS patients have provided the strongest evidence to date, and larger studies are in progress. The vitamin D–MS hypothesis also provides possible explanations for puzzling facts of MS disease, namely, the female sex bias and the relapsing–remitting disease phenotype. Collectively, the data nearly fulfill the Hill criteria for causality in the study of environmental factors and disease, and warrant a strong effort on the part of the global MS research community to prevent MS using vitamin D–based intervention strategies.

11.4 TYPE 1 DIABETES

11.4.1 Pathogenesis and Etiology

T1D results from the selective destruction of the insulin-producing β cells in the pancreatic islets of Langerhans by $CD4^+$ and $CD8^+$ T lymphocytes specific for peptides derived from β-cell proteins (insulin, glutamic acid decarboxylase, islet-specific glucose 6-phosphatase; [206,207]). Upon losing β cells, insulin production, and blood glucose control, T1D patients experience acute (ketoacidosis and severe hypoglycemia) and secondary complications (heart disease, blindness, or kidney failure) of their disease, despite insulin replacement therapy.

The first step in T1D pathogenesis may be the escape of high-avidity, islet cell antigen-specific T cells from the thymus, where they would normally be deleted [206]. Subsequently, unknown insults cause β cells to release islet antigens that are acquired by APCs and presented to the islet cell antigen-specific T lymphocytes in pancreatic lymph nodes. On activation, these diabetogenic T cells migrate into the islets and begin the inflammatory cascade that is lethal to β cells. Natural killer T cells (NKT) and $Foxp3^+CD4^+CD25^+$ Treg cells oppose and regulate the activities of the destructive T lymphocytes [208,209]. Thus, immune dysregulation (functional and numerical defects) in the NKT and Treg cell subsets may contribute to T1D etiology.

11.4.2 T1D Disease Association with UV Light Exposure and Vitamin D

Globally, T1D incidence is increasing by approximately 2% to 5%/year [210]. At the same time, the proportion of T1D patients with the high-risk *HLA-DR3/4* genotype is decreasing [211].

Consequently, increasing exposure to environmental risks such as declining UV light exposure and vitamin D supplies rather than genetic risk may be driving this trend. Data linking T1D with insufficient UV light exposure and low vitamin D supplies has been reviewed [212–220]. Here, we update the evidence linking vitamin D and T1D and note parallels with other autoimmune diseases.

11.4.2.1 Latitude

Latitude has been correlated with T1D incidence in Europe, Scandinavia, China, Australia, and Canada [221]. A new comprehensive global study of 51 regions varying in latitude from 55°S (Tierra del Fuego, Argentina) to 60°N (Finland) found a U-shaped incidence curve centered at 0° latitude [222], very similar to that reported for MS [63]. The T1D incidence varied over a 350-fold range from <0.5/10^5 population in equatorial regions to a high in Scotland and in the Scandinavian nations. The Mediterranean island of Sardinia, at latitude 38°N and with a T1D incidence of approximately 44/10^5 population, is a notable exception to the strong correlation between latitude and T1D incidence, as it is for latitude and MS [223].

11.4.2.2 Season

The seasonality of T1D patient birth months and onset suggest a seasonally variable environmental factor. Children born in the spring in northern latitudes had an increased risk of T1D [224]. Furthermore, the T1D diagnosis frequency in young children conformed to a sinusoidal model with a winter peak [225,226]. Recent research from Canada, the United States, the United Kingdom, Australia, and Italy has confirmed this seasonality [227–231]. A global analysis of T1D incidence data (105 centers in 53 countries) showed peaks from October to January and troughs from June to August in the northern hemisphere, and a reverse pattern in the southern hemisphere [232]. The correlation disappeared after adjustment for latitude. The seasonality in birth month and disease onset for T1D-affected individuals signals a contribution of seasonal environmental variables to prenatal and early postnatal developmental events relevant to the T1D disease process.

11.4.2.3 Ambient UV Radiation

Ambient UV radiation is the candidate environmental variable for T1D risk because it varies inversely with latitude and sinusoidally with season [228,233]. In Australia, ambient winter UV radiation correlated inversely with T1D ($r = -0.80$; [234]), as it did with MS and RA [235]. In Newfoundland, T1D incidence correlated better with UV radiation than with latitude [236]. In Australia, ambient UV radiation correlated inversely with T1D in rural but not in urban environments because personal UV exposure patterns differed between these populations [230]. The new global study also found a close association between low ambient UV radiation and high incidence of T1D [222]. Importantly, from a disease prevention viewpoint, the T1D incidence approached zero in regions with high ambient UV radiation [222].

11.4.2.4 Vitamin D

Interest in vitamin D$_3$ as a variable relevant to T1D began with the discoveries (in rodents and humans) that pancreatic β-cells expressed the *VDR* [237–239] and *CYP27B1* genes [240], vitamin D deficiency impaired β-cell insulin release [241], and vitamin D replenishment [242–246] or 1,25(OH)$_2$D$_3$ administration enhanced β-cell insulin release [247,248]. No VDRE has been identified in the *INS* gene promoter [249], so 1,25(OH)$_2$D$_3$ likely does not transcriptionally regulate insulin production [250]. These studies established the vitamin D dependence of β-cell insulin release, and set the stage for probing UV light–linked vitamin D$_3$ synthesis as a variable relevant to T1D risk [213,214,216–220].

A decade ago, the EURODIAB study correlated vitamin D$_3$ supplementation in infancy with a decreased risk of T1D (odds ratio, 0.67) [251]. A recent study confirmed those data (odds ratio, 0.71) [252]. Likewise, recent studies have confirmed Finnish data correlating supplementary vitamin D$_3$ (2000 IU/day) with an 88% lower risk of childhood T1D compared with unsupplemented children

(T1D incidence 200/10^5 population; [253]). Finland, at 60° to 70°N latitude, has insufficient UV light to catalyze cutaneous vitamin D$_3$ synthesis approximately 6 months of the year [254], and has the highest incidence of T1D in the world. Analyzing approximately 50 years of Finnish T1D incidence data revealed a startling inverse relationship between supplementary vitamin D$_3$ and T1D risk [217]. Before 1964, health officials recommended 4500 IU/day of supplementary vitamin D$_3$ for children, and the T1D incidence was approximately 17/10^5 population (age < 15). Health officials reduced the vitamin D$_3$ recommendation to 2000 IU/day in 1964, to 1000 IU/day in 1975, and to 400 IU/day in 1992. As those reductions were implemented, the T1D incidence quadrupled to approximately 65/10^5 population by 2005. These time–trend data support a cause–effect relationship between declining vitamin D$_3$ intake and increasing T1D incidence. If this is correct, then the sequential recommendations to drastically reduce vitamin D$_3$ intake in this UV light–deprived country may have imposed T1D on many thousands of Finnish children who might otherwise have been healthy.

Recent data have shown an inverse dose–response relationship between circulating 25(OH)D and T1D risk. For vitamin D–insufficient subjects from the United Kingdom [50 nmol/L ≤ 25(OH)D < 75 nmol/L], the T1D odds ratio was 3.3; for deficient subjects [25 nmol/L ≤ 25(OH)D < 50 nmol/L], it was 5.5; and for severely deficient subjects [25(OH)D < 25 nmol/L], it was 8.4 [231]. This inverse correlation applied *in utero*, where risk of islet cell autoimmunity in children varied inversely with maternal vitamin D intake [255]. It also applied in older adults, where fasting blood glucose varied inversely with circulating 25(OH)D [256]. Studies from Italy [227], Sweden [257], Australia [258], the United States [259,260], Switzerland [261], and Egypt [262] have all documented low circulating 25(OH)D levels in newly diagnosed T1D youth. In fact, more than 80% of newly diagnosed youth met the criteria for vitamin D insufficiency or deficiency [259,261]. The severity of cardiovascular problems [260,263] and retinopathy [264] in T1D individuals also varied inversely with circulating 25(OH)D. Together, the data suggest that vitamin D deficiency may contribute to T1D risk and to the severity of vascular damage in T1D individuals.

In summary, the observational data show a strong, consistent, and specific association between low circulating 25(OH)D levels and high T1D risk. There is also a clear biological gradient with the lowest 25(OH)D levels corresponding to the highest T1D risk, and a temporal trend, with low maternal vitamin D intake correlating with high risk of T1D in her offspring. Thus, the data have begun to support the assertion that vitamin D insufficiency plays a causal role in T1D, according to the Hill criteria for causation in the study of environmental factors and disease [114].

11.4.3 T1D Disease Association with the *VDR, CYP2R1, CYP27B1,* and *GC* Genes

Recent genetic data are suggestive of a role for vitamin D in T1D risk [211]. The approximately 25% to 65% T1D concordance rate for monozygotic twins (depending on length of follow-up), with 4% to 10% for dizygotic twins, indicates some contribution to T1D risk from heritable genetic factors [265,266]. However, more than 85% of T1D patients have no affected family member [211]. This high incidence of apparently spontaneous disease, and the lack of complete concordance in monozygotic twins, supports a causal role for environmental factors. UV light exposure and vitamin D are uniquely capable of multiple environment–gene interactions relevant to T1D disease due to the UV light responsiveness of the vitamin D endocrine system and the transcription activation function of the VDR in immune cells [24]. Consequently, new studies have probed for associations between the *VDR, CYP2R1, CYP27B1,* and *GC* genes and T1D. An important caveat regarding the genetic associations that have been found is that with rare exceptions, the genetic polymorphisms associated with T1D have no demonstrated effect on gene expression or protein function.

11.4.3.1 *VDR*

Associations between the *VDR* (Chr 12q13.1) and T1D were first detected in linkage studies involving Indian Asian [267] and German families [268]. Family linkage studies use genetic data from familial clusters of T1D individuals to detect alleles with strong phenotypic effects [269].

Interestingly, the association between a *VDR* allele and T1D was strongest in Finnish individuals carrying the *HLA-DQB1*0302* high-risk allele [270]. In an Irish study, a rare *VDR* allele defined by polymorphisms at the *VDR* gene 3′ end was associated with protection from diabetic nephropathy in T1D patients [271]. However, follow-up case-control and family studies have reported inconsistent results [272–279].

The GWA studies of more than 30,000 individuals with T1D identified 41 genomic regions associated with T1D, but genes like *VDR* that are involved in vitamin D metabolism and function were not among them [269]. Apart from the *HLA* region, the *INS* locus (insulin), and *PTPN22* (lymphoid-specific protein tyrosine phosphatase), the genomic regions associated with T1D have an odds ratio of less than 1.3, indicating very weak phenotypic effects. As discussed previously, global GWA studies on populations that include both heritable and spontaneous disease cases are unlikely to identify as genetic risk factors those genes that are sensitive to an environmental variable for their influence on disease phenotype.

A recent review of *VDR* and T1D associations revealed a novel environment–gene interaction that may explain the inconsistent results obtained in case-control analyses performed in different world regions [280]. The investigators reasoned that the *VDR* might be associated with T1D only where circulating $25(OH)D_3$ was sufficient to support $1,25\text{-}(OH)_2D_3$ synthesis in cells relevant to T1D. Consistent with this reasoning, the association between the *VDRFokI*[F] allele and T1D risk increased as the ambient winter UV radiation increased in the region where the study was performed. The *VDRFokI*[F] allele, also implicated in MS risk [127], encodes a shorter (424 amino acid), more transcriptionally active VDR protein than the *FokI*[f] allele (427 amino acids) due to a start codon polymorphism [281,282]. The *VDRFokI*[F] allele was also associated with lower circulating $25(OH)D_3$ levels [123]. Thus, where winter ambient UV radiation was plentiful, the functional differences between the long and short VDR protein variants became evident and relevant to T1D risk. These insightful studies implicate ambient UV radiation, vitamin D, and the VDR pathway in T1D etiology, and emphasize the importance of analyzing genetic data in the context of environmental variables.

There is a feed-forward loop wherein $1,25\text{-}(OH)_2D_3$ upregulates *VDR* gene transcription fourfold through an action of the VDR–RXR complex on several intronic enhancers [283]. Given sufficient $25(OH)D_3$ to support $1,25\text{-}(OH)_2D_3$ synthesis, this feed-forward loop could amplify functional differences between the short and long VDR protein variants in immune cells, providing a biochemical explanation for the genetic epidemiological data.

In T1D, as in MS, there was an epistatic interaction between *VDR* alleles and the *HLA-DR* region. An association between the *VDRFokI*[F] alleles and T1D was only evident in the 86% of North Indian T1D patients who had the high-risk *HLA-DRB1*0301* allele [284]. Sequencing of the *HLA-DRB1*0301* promoter unexpectedly revealed a putative VDRE identical in sequence to the VDRE in the high-risk MS allele, *HLA-DRB1*1501* [120]. This exciting parallel suggests that UV light and the vitamin D endocrine system may regulate *HLA-DRB1*0301* and *HLA-DRB1*1501* expression and presentation of antigens relevant to T1D and MS etiology, respectively. The nature of the antigens and the timing and outcome of antigen presentation are unknown. Among many possibilities, low *HLA-DRB1* expression during thymic development could result in the failure to purge self-reactive T lymphocytes from the repertoire, or low expression during peripheral immune responses could result in the failure to present a microbial or viral antigen for a protective immune response.

11.4.3.2 *CYP2R1*

Four genes, *GC, DHCR7, CYP2R1,* and *CYP24A1,* profoundly influence circulating $25(OH)D_3$ levels, according to recent data. The *CYP2R1* gene encoding the vitamin D-25-hydroxylase is one of these four genes [11,231,285,286]. Associations between a *CYP2R1* allele (rs10741657 G variant), low circulating $25(OH)D_3$ levels, and T1D were first uncovered in German families [287]. The two *CYP2R1* variants did not differ with regard to mRNA abundance, but may differ with regard to enzyme activity. Another *CYP2R1* allele (rs12794714, A variant) was associated with low circulating

$25(OH)D_3$ levels and T1D in British families [231]. These associations between *CYP2R1* and circulating $25(OH)D_3$ levels represent a second type of gene–environment interaction that is relevant to T1D etiology.

11.4.3.3 *CYP27B1*

Additional recent genetic data have associated T1D with variants of the *CYP27B1* gene [231,288–291], but not the *CYP24A1* gene [231,290]. There is a promoter polymorphism in the *CYP27B1* gene (–1260; C/A) that may affect transcription. The *CYP27B1* C/C homozygotes were overrepresented among T1D individuals in German case-control and family linkage studies [288,289]. This association was confirmed in large British T1D case-control and family linkage studies [231,290]. However, this association was not evident in Polish T1D patients [292] or in the exhaustive GWA studies [269]. These data suggest a role for $1,25(OH)_2D_3$ synthesis in T1D etiology, but the effect (if any) of the promoter polymorphism on *CYP27B1* gene transcription in cells relevant to T1D remains to be determined.

11.4.3.4 *GC*

Not surprisingly, the *GC* gene encoding the DBP was also one of four genes that influenced circulating $25(OH)D_3$ levels in the recent analyses [11,231,286]. Case-control studies performed in French [293,294] and British populations [231] found that the *GC* alleles associated with lower circulating $25(OH)D_3$ levels were also associated with T1D.

In summary, recent genetic evidence has highlighted a possible contribution of particular variants of the *VDR*, *CYP2R1*, *CYP27B1*, and *GC* genes, ambient UV radiation, and circulating $25(OH)D_3$ levels in the development of spontaneous and heritable forms of T1D. The precise mechanisms by which these genetic polymorphisms contribute to the disease are unknown.

11.4.4 VITAMIN D PREVENTION OF T1D

A Finnish birth cohort study reported that 2000 IU/day of supplementary vitamin D_3 reduced the T1D risk by 88% compared with nonsupplemented children [253]. However, recent T1D prevention trials using 400 IU/day of supplementary vitamin D_3 have not significantly reduced the T1D risk. Swedish investigators found that infants of mothers who reported taking multivitamins (400 IU/day of vitamin D_3) had lower levels of T1D-related autoantibodies at age 1 but not at age 2.5 compared with nonsupplementing mothers [295]. Furthermore, infants who received 400 IU/day of vitamin D_3 had the same risk of T1D-related autoantibodies as nonsupplemented infants. Similar results were reported from Finland [296]. Neither study actually measured circulating 25(OH)D levels. It is well known that 400 IU/day (10 µg/day) of vitamin D_3 (without UV light exposure) does not increase the circulating 25(OH)D levels to more than 50 nmol/L, the cutoff for vitamin D deficiency [297–300]. Thus, the negative outcomes in the two T1D prevention trials undoubtedly reflect the use of a supplementary vitamin D_3 dose that was insufficient to increase circulating 25(OH)D, and thus could not have been expected to influence T1D etiology [215]. The Finnish data, wherein the T1D incidence quadrupled as the recommended supplementary vitamin D_3 dropped from 4500 to 400 IU/day, should inform future T1D prevention efforts [217].

11.4.5 VITAMIN D TREATMENT OF T1D

Importantly, humans with vitamin D deficiency and osteomalacia had a 30% impairment in β-cell insulin release that was ameliorated with 6 months of vitamin D_3 supplementation (2000 IU/day; [301]). A more recent study found that 4000 IU/day of supplementary vitamin D_3 for 12 weeks significantly improved glycemic control in young adults with T1D (disease duration 7.1 ± 6.7 years) and vitamin D deficiency [302]. Increasing 25(OH)D from less than 50 nmol/L to more than 75 ± 17 nmol/L decreased the glycosylated hemoglobin fraction from 9.4 ± 2.3% at baseline to 8.2 ±

2.3% ($p < 0.02$). These data support further investigation of 2000 to 4000 IU/day of supplementary vitamin D_3 as a possible intervention to improve glycemic control in T1D patients.

One vitamin D supplementation study with a negative outcome involved vitamin D_2 (ergocalciferol) rather than vitamin D_3 (cholecalciferol; [146]). Three patients with vitamin D deficiency and type 2 diabetes were given 300,000 IU of ergocalciferol by intramuscular injection. Unlike the previous studies in which vitamin D_3 improved glycemic control [301,302], the vitamin D_2–supplemented patients showed deterioration in their glycemic control [146]. Clinical assays for circulating 25(OH)D do not distinguish 25(OH)D_2 and 25(OH)D_3, but these compounds are not biologically equivalent [303,304]. Vitamin D_3 supplementation (50,000 IU) increased the circulating 25(OH)D by 15 nmol/L 1 month later, but vitamin D_2 supplementation (50,000 IU) actually decreased circulating 25(OH)D by 5 nmol/L [303]. Moreover, vitamin D_2 was only about half as effective as vitamin D_3 in an *in vitro* assay of vitamin D metabolism and T lymphocyte function [167]. These data caution against using vitamin D_2 to support immune function *in vivo*.

The hormone 1,25(OH)$_2$D$_3$ has been evaluated as a T1D treatment. Adults with recent onset T1D were randomly assigned to receive placebo or 1,25(OH)$_2$D$_3$, and circulating pro-insulin C-peptide was analyzed 9 months later as a measure of β-cell function [305]. The placebo and 1,25(OH)$_2$D$_3$ treatment groups did not differ significantly for serum 1,25(OH)$_2$D$_3$, T1D-related autoantibodies, fasting C-peptide, peak C-peptide, or daily insulin dose needed for glycemic control. Both groups lost 40% of β-cell function during the study. Similar data were reported by others [306,307]. Neither study reported circulating 25(OH)D. It is not surprising that the 1,25(OH)$_2$D$_3$ treatment failed to increase serum 1,25(OH)$_2$D$_3$ levels [305] because 1,25(OH)$_2$D$_3$ induction of the catabolic 24-OHase maintains strict homeostatic control over circulating 1,25(OH)$_2$D$_3$ [2]. Because the 1,25(OH)$_2$D$_3$ treatment failed to increase serum 1,25(OH)$_2$D$_3$ levels, it is not surprising that it failed to influence T1D disease. To date, the T1D data do not support daily 1,25(OH)$_2$D$_3$ treatments for disease prevention or treatment. The data solidly support further investigation of 2000 to 4000 IU/day of supplementary vitamin D_3 (not vitamin D_2) for T1D disease prevention and early treatment.

11.4.6 NONOBESE DIABETIC MOUSE

Research into the pathogenesis of spontaneous diabetes in the inbred nonobese diabetic (NOD) mouse strain has contributed greatly to current knowledge on T1D etiology. The nonobese diabetes-resistant mice are used for comparisons. In common with human T1D, diabetes in NOD mice begins with a breakdown in immune regulation (e.g., defects in thymic T-cell negative selection) and autoreactive T and B lymphocyte infiltration of pancreatic islets at an early age (insulitis; [209,308]). Activation of the innate immune system occurs, and the adaptive and innate immune systems collaborate to destroy the pancreatic β-cells (destructive insulitis). Also in common with human T1D, MHC class II genes encoding molecules involved in antigen presentation to CD4+ T cells are the major genetic determinants of T1D risk in rodents [309]. A thorough discussion of immune dysregulation in T1D is available [209]. Here, we note important discrepancies between human and NOD mouse T1D, and focus on mechanisms that may explain a causal role for vitamin D insufficiency in T1D.

11.4.7 VITAMIN D AND T1D PREVENTION IN NOD MICE

Human T1D correlated inversely with circulating 25(OH)D [213–220]. In sharp contrast, T1D in the NOD mouse did not correlate inversely with circulating 25(OH)D. Vitamin D deficiency increased glucose intolerance (particularly in males) and the T1D incidence in NOD mice [310–312]. However, supplementary vitamin D_3 did not reduce the T1D incidence in male or female NOD mice [310,313]. Thus, T1D correlated inversely with circulating 25(OH)D in humans but not in NOD mice.

Why supplementary vitamin D_3 did not inhibit T1D in NOD mice is unknown, but may involve a defect in NOD immune cell 1,25(OH)$_2$D$_3$ synthesis. The 1,25(OH)$_2$D$_3$ effects on immune cells occur at more than 0.1 nmol/L, 100-fold higher than concentrations needed to influence bone homeostasis. Because immune cells like IFN-γ–activated macrophages produce $1,25(OH)_2D_3$, it is believed that concentrations of more than 0.1 nmol/L can be attained locally in inflamed tissues, given sufficient circulating 25(OH)D to support immune cell $1,25(OH)_2D_3$ synthesis [165]. In this context, it is highly significant that NOD macrophages had a defect in immune regulation of *CYP27B1* transcription [314]. For unknown reasons, IFN-γ induced *CYP27B1* transcription approximately sixfold in nonobese diabetes-resistant and C57BL/6 macrophages, but less than twofold in NOD macrophages. Thus, inflammatory stimuli like IFN-γ may not trigger sufficient immune cell $1,25(OH)_2D_3$ synthesis to elicit anti-inflammatory mechanisms in the NOD strain. Intervening downstream of this potential block with $1,25(OH)_2D_3$ treatment at an early age (ages 3–14 weeks) inhibited T1D development, but intervening after insulitis was present (age >14 weeks) did not [310,312,315,316].

11.4.8 VITAMIN D DEPENDENCE OF β CELL INSULIN RELEASE

One relevant mechanism for vitamin D prevention of T1D undoubtedly involves the vitamin D dependence of β-cell insulin release [317]. The *VDR* [237–239] and *CYP27B1* genes [240] were both expressed in β cells, so an intracrine role for $1,25(OH)_2D_3$ in β-cell insulin release was suspected. In humans, the *VDR* genotype determined insulin secretion phenotype [318]. Moreover, in rodents and humans, vitamin D deficiency decreased β-cell insulin release [241], and vitamin D replenishment restored it [242–246]. Further evidence supporting the vitamin D dependence of β-cell insulin release came from a study of glucose tolerance in mice expressing a functionally inactive mutant VDR protein [319]. The mice were fed a rescue diet to normalize mineral homeostasis and weight. The fasting, VDR mutant mice had normal baseline blood glucose levels, β cell mass, islet architecture, and islet neogenesis. However, after glucose loading, their blood glucose was elevated and both insulin mRNA and circulating insulin levels were below normal. These data implicate the VDR in control of insulin synthesis and release. Conversely, rats with streptozotocin-induced diabetes had 88% lower circulating $1,25(OH)_2D_3$ levels than controls, and insulin treatment restored their circulating $1,25(OH)_2D_3$ to normal levels [320]. These data implicate insulin in control of circulating $1,25(OH)_2D_3$ homeostasis. If the VDR controls insulin synthesis and release, and insulin controls circulating levels of the VDR ligand, $1,25(OH)_2D_3$, then one can postulate integration of the insulin and vitamin D endocrine systems in metabolic control. Because no VDRE has been identified in the *INS* gene [249], VDR control of insulin may be indirect, for example through control of β-cell calcium homeostasis [321]. Understanding whether and how the insulin and vitamin D endocrine systems collaborate in metabolic control would be an important future research area.

11.4.9 VITAMIN D HORMONE MODULATION OF IMMUNE FUNCTION IN DIABETES

Viewed from an immune system perspective, T1D is a very complex autoimmune disease. The details of this complexity have been reviewed [207] and are beyond the scope of this article. Here, we summarize the immune abnormalities in T1D patients that are thought to play causal roles in T1D pathogenesis, and discuss how the vitamin D pathway may modulate these abnormalities. There is no doubt that diabetogenic CD4+ and CD8+ T lymphocytes perform a predominant causal role in human T1D pathogenesis [207, Figure 1]. Data supporting this conclusion came from a study of T1D patients who received transplanted pancreatic tissue from HLA-identical donors as a diabetes treatment [322]; T1D diabetes recurred unless the patients received immunosuppressive therapy. It is also clear that innate APC, like DCs and activated macrophages (MΦ), contribute directly to pathogenicity by producing proinflammatory mediators such as IL-1β, IL-12, TNF, IFN-γ, and reactive oxygen species, and indirectly by stimulating β cell–specific, diabetogenic CD4+ and CD8+ T cells with β cell–derived peptides. The β-cell destruction is thought to occur through direct cytotoxic actions of diabetogenic CD8+ T cells and NK cells, and through apoptosis induction by diabetogenic CD4$^+$ T cells and the proinflammatory cytokines IL-1β, IL-12, TNF, and IFN-γ. What remains puzzling is the genesis of the autoimmune reaction to β cell–derived proteins.

11.4.9.1 Myeloid Cells

One hypothesis is that some type of β-cell damage, possibly of viral origin, causes the release of β-cell proteins into a proinflammatory microenvironment. The evidence on links between infection and T1D has been reviewed [207, Table 1]. Consistent with the infection hypothesis, investigators detected proinflammatory DC and MΦ producing IL-1β and TNF in pancreatic tissue from patients with recent T1D onset [323]. Moreover, monocytes from T1D patients produced excessive IL-1β in response to TLR ligation [260,324]. The germ line–encoded TLR recognize molecular patterns on microbial pathogens, and initiate innate immune responses to those pathogens [325]. Additional data showed that preincubating monocytes from T1D patients with 1,25(OH)$_2$D$_3$ decreased TLR4 expression, TLR4 responsiveness, and IL-1β production [260,324].

It is highly significant that TLR signaling induces 1,25(OH)$_2$D$_3$ synthesis in myeloid cells [165], and the hormone acts through an intracrine pathway to induce expression of the cathelicidin antimicrobial peptide (*CAMP*) gene [326]. Very recent data suggest that vitamin D also has direct antiviral effects (particularly against enveloped viruses) that may be linked to cathelicidin and β defensin 2 upregulation [327]. Thus, one can envision a natural pathway protecting β cells from a viral pathogen, whereby the pathogen engages a TLR, and TLR signaling induces 1,25(OH)$_2$D$_3$ synthesis. Thereafter, the hormone upregulates defensins to eliminate the pathogen, and downregulates the TLR and the proinflammatory cytokines to restore tissue homeostasis. If vitamin D supplies were insufficient to support 1,25(OH)$_2$D$_3$ synthesis in inflamed pancreatic tissue under pathogen attack, this innate immune pathway could fail, and β-cell damage and release of proteins into an inflammatory microenvironment could result.

Monocytes from vitamin D–deficient NOD mice, like those from T1D patients, produced excessive amounts of IL-1 compared with controls [311], indicating a strong proinflammatory phenotype. However, NOD monocytes did not increase *CYP27B1* transcription and 1,25(OH)$_2$D$_3$ synthesis in response to TLR stimulation [314]. Whether this is a significant mechanism predisposing NOD mice to autoimmunity remains to be studied. Unlike humans, induction of the *CAMP* gene and cathelicidin synthesis are not under VDR control in the mouse. Therefore, the possible role of a natural pathway involving TLR signaling, 1,25(OH)$_2$D$_3$ synthesis, and cathelicidin action protecting against β-cell damage cannot be studied in NOD mice.

The DCs contribute to diabetes by producing IL-12 and presenting β-cell–specific peptides to autoreactive CD4$^+$ and CD8$^+$ T cells [207]. The DC and IL-12 synthesis are thought to be directly targeted by 1,25(OH)$_2$D$_3$ as a mechanism of diabetes prevention in the NOD strain [328,329]. Early data showed that the presence of CD11c$^+$ DC with an immature, tolerogenic phenotype correlated with inhibition of Th1 cell accumulation in pancreatic islets and a lower T1D incidence in NOD mice treated with a 1,25(OH)$_2$D$_3$ analog [175]. Moreover, when added to *in vitro* bone marrow cell cultures, 1,25(OH)$_2$D$_3$ promoted the outgrowth of immature, tolerogenic CD11c$^+$ DC with lower surface expression of MHC II and CD86 [310,330,331]. These and other data have been reviewed [219]. The data have been interpreted as evidence that 1,25(OH)$_2$D$_3$ prevention of full DC maturation is an essential step in the mechanism for preventing diabetes development [332].

Recent data from NOD mice carrying a *VDR* loss-of-function allele brought new questions regarding the tolerogenic DC hypothesis [333]. This hypothesis predicts that NOD mice lacking VDR function, like vitamin D–deficient NOD mice, would have mature DC and aggravated diabetes presentation. Surprisingly, the VDR$^{-/-}$ NOD mice had immature DC, yet identical diabetes presentation to the VDR$^{+/+}$ NOD mice. This puzzling result indicates that VDR is not necessary to maintain the immature DC phenotype, and that immature DC alone are not sufficient to protect NOD mice from developing T1D. Recent data from the rodent model of MS showed that 1,25(OH)$_2$D$_3$ action on DC cells alone was not sufficient to inhibit autoimmune disease [30], so a direct 1,25(OH)$_2$D$_3$ action on DC is not a common mechanism between the rodent T1D and MS models.

11.4.9.2 NK Cells

The role of NK cells in diabetes is not entirely clear. On the one hand, highly activated and destructive NK cells have been detected infiltrating pancreatic tissue in T1D patients and NOD mice [207]. Furthermore, NK cell depletion prevented rodent T1D, supporting a pathogenic role for NK cells. In this context, it may be significant that $1,25(OH)_2D_3$ inhibited human NK cell cytotoxicity and IFN-γ secretion in a dose-dependent manner [334]. On the other hand, T1D subjects had fewer NK cells in the peripheral blood than healthy controls, and their NK cells responded poorly to IL-2 and IL-15, had faulty survival signaling and cytotoxic function, and low IFN-γ secretion [335]. These data support the hypothesis that NK cell dysfunction might contribute to T1D autoimmune pathogenesis. In this context, it may be significant that $1,25(OH)_2D_3$ enhanced human NK cell IL-10 synthesis [336]. It is not clear why the NK cell phenotype differed significantly between pancreatic tissue and the circulation. Clarification of the role of NK cells in T1D, and of $1,25(OH)_2D_3$ as a modifier of NK cell function, awaits further experimentation.

11.4.9.3 Treg Lymphocytes

Diabetogenic T lymphocytes are known to have a causal role in human T1D pathogenesis, so mechanistic studies have focused on T cells. Early rodent data suggested $1,25(OH)_2D_3$ might enrich a regulatory cell population because cotransfer of splenocytes from $1,25(OH)_2D_3$-treated NOD mice prevented diabetic NOD splenocytes from transferring T1D into young NOD mice [315]. Treg cells were more frequent in the pancreatic lymph nodes of NOD mice treated with a $1,25(OH)_2D_3$ analog than placebo controls [175], and less frequent in NOD mice carrying the VDR^- allele [333]. These data and the NOD DC data have been interpreted as evidence that $1,25(OH)_2D_3$ increases Treg development either directly, or through actions that maintain DC in an immature state. It is noteworthy that rodent pathogenic Th1 and Th17 cells had a high level VDR mRNA expression, whereas Foxp3$^+$ Treg cells had a low VDR mRNA expression [28,30], suggesting the pathogenic T cell may be the preferred target for $1,25(OH)_2D_3$ action. Moreover, effects of $1,25(OH)_2D_3$ on Foxp3$^+$ Treg cells in the rodent model of MS have been sought but not found [30]. Thus, it seems that a direct $1,25(OH)_2D_3$ action on Treg cells is not a common mechanism between the rodent T1D and MS models.

The Treg hypothesis has also been investigated in humans. The $1,25(OH)_2D_3$ and IL-2 combined to diminish human peripheral blood T cell IFN-γ and IL-17 cytokine production and enhance CTLA-4 and Foxp3$^+$ expression *in vitro* without substantially influencing cell division [337]. The mechanism for this apparent CD4$^+$ T cell subset interconversion was not detailed. Similarly, the $1,25(OH)_2D_3$ analog, TX527, added to human peripheral blood T cells, diminished effector cytokine mRNA (IFN-γ, IL-4, IL-17), and promoted the emergence of IL-10–producing Treg cells marked by CTLA-4 and OX40 expression, and the capacity to suppress effector T cell functions [338]. Moreover, the analog decreased the lymphoid organ homing receptor (CD62L) and two chemokine receptors (CCR7, CXCR4), while increasing the chemokine receptors (CCR5, CXCR3, CXCR6) that would enable homing to inflamed sites. These data suggest promotion of a phenotypic change that would enable CD4$^+$ T cells to home to inflamed sites and suppress inflammation.

Interconversion between pathogenic and regulatory CD4$^+$ T cell phenotypes has been observed in rodents and humans, promoting an emerging concept that there is plasticity within the CD4$^+$ T cell lineage [339,340]. T1D patients and controls had comparable frequencies of natural and induced (adaptive) Foxp3$^+$ Treg cells in the peripheral blood [341], but the T1D patients had a higher frequency of adaptive Treg cells that produced IFN-γ and may represent the subset that can be converted into pathogenic Th1 cells [342]. The extent to which peripheral blood T cell subsets reflect T cell subsets in the pancreatic islets is not known. However, the plasticity within the CD4$^+$ T-cell lineage poses a substantial challenge to the strategy of increasing the frequency of induced (adaptive) Treg cells as a means of treating autoimmune diseases, due to the potential of these cells to become pathogenic. Whether vitamin D$_3$ might support stabilization of the adaptive Foxp3$^+$ Treg cell lineage *in vivo* remains to be determined. In summary, there is evidence supporting the concept that $1,25(OH)_2D_3$ promotes Treg cells, but significant questions remain as to whether Treg cells express VDR and are themselves $1,25(OH)_2D_3$ responsive, or whether the reported effects reflect an action of $1,25(OH)_2D_3$ on VDR-expressing pathogenic $CD4^+$ T cells to eliminate them or promote their conversion to a regulatory phenotype.

11.4.9.4 T Effector Lymphocytes

Activated effector T lymphocytes (IFN-γ–producing Th1 cells; IL-17–producing Th17 cells) express the VDR and respond to $1,25(OH)_2D_3$ [343,344], but the details of the T-cell response to the hormone are still under investigation (reviewed in [27–30]). The effects of $1,25(OH)_2D_3$ on activated T cells have been reviewed recently [219]. Important new research showed that upregulation of VDR protein in naive human T cells was required for full activation [343]. Compared with antigen-primed T cells, naive T cells had low VDR expression and weak coupling of TCR signals to the phospholipase C-γ1 (PLCγ1) pathway that regulates intracellular calcium mobilization and cell cycle entry. Initial TCR signaling rapidly induced VDR protein using the mitogen-activated protein kinase p38 pathway. Thereafter, PLCγ1 was induced 75-fold in a VDR-dependent manner, whereupon the T cells showed greater TCR signaling, calcium mobilization, and cell cycle entry. Additional new evidence showed coordinated upregulation of *VDR* and *CYP27B1* transcripts during T cell activation, and $1,25(OH)_2D_3$-mediated increases in *CYP24A1* transcripts and CCR10, a skin-homing receptor, as well as decreases in cytokine transcripts (IFN-γ, IL-10) in long-term T cell cultures devoid of APC [344]. In this report, downregulation of IL-10 was not consistent with other reports on $1,25(OH)_2D_3$-mediated upregulation of IL-10 in T-cells [338,345], and may signal the outgrowth of distinct T-cell subpopulations in the long-term cultures, rather than direct effects on IL-10 production.

Because UV light–catalyzed vitamin D_3 synthesis seems to be a common environmental factor contributing to the risk of several T cell-mediated autoimmune diseases, it is reasonable to expect some common biological mechanisms to underpin vitamin D's protective effects. The $1,25(OH)_2D_3$-mediated sensitization of autoreactive T cells to TCR activation-induced cell death (AICD) may be one common mechanism. The NOD mice have defects in central and peripheral tolerance that are related to the failure of AICD. The NOD thymocytes had a T-cell intrinsic defect in AICD correlating with strong TCR-mediated upregulation of caspase 8-homologous FLICE inhibitory protein [308,346]. Similarly, NOD peripheral T cells had a defect in AICD that was linked to the *Idd5* region encompassing the *Ctla-4* gene [347], and correlated with lower caspase 8 levels [348–350]. Caspase 8 transduces death signals in AICD pathways. Importantly, $1,25-(OH)_2D_3$ treatment restored AICD sensitivity in NOD peripheral T cells [348,350]. Similarly, resistance to AICD was noted in peripheral T cells in the EAE model of MS, and $1,25-(OH)_2D_3$ treatment restored AICD sensitivity [204,205], possibly through upregulation of FLASH, a protein involved in Fas-mediated activation of caspase-8 [351], and downregulation of cIAP-2 (cellular inhibitor of apoptosis protein 2; [352]). Thus, a direct $1,25(OH)_2D_3$ action on pathogenic effector T lymphocytes to increase sensitivity to AICD may be a common mechanism between the rodent T1D and MS models. AICD resistance could allow autoreactive thymocytes to escape censure in the thymus and seed the periphery with potentially pathogenic T cells, and in the periphery, it could trigger and potentiate autoimmunity.

11.4.10   Vitamin D and T1D Summary

The evidence correlating reduced vitamin D supplies with high T1D risk is strong and diverse, suggesting that this link may also reflect a causal relationship. Latitudinal, seasonal, climatological, and nutritional data have been collected globally, and these data show a consistent inverse relationship that is strongest where fluctuations in vitamin D supplies are greatest. Noteworthy in the recent genetic studies are data showing that functionally significant polymorphisms in the *VDR* gene are correlated with T1D risk, when MHC genes and latitude of residence are taken into account. The finding that the high T1D risk allele, *HLA-DRB1*0301*, may be under VDR control provides a strong parallel with the MS risk allele, *HLA-DRB1*1501*, both findings implying VDR control of an important antigen presentation engagement. Data linking *CYP27B1* gene variants with T1D are not yet strong, but may be forthcoming when multiplex families are studied. Compelling temporal data in favor of a causal relationship derived from Finnish birth cohort studies, and from noting that T1D incidence quadrupled over several decades as the recommended vitamin D intake was decreased from 4500 to 400 IU/day. These data also suggest a biological gradient linking higher vitamin D supplies with lower T1D risk and vice versa. There are plausible biological mechanisms of action, in that β-cell insulin release shows a dependence on vitamin D, and $1,25(OH)_2D_3$ modulates the function of immune cells (DC, NK cells, Treg cells, and T effector cells) known to contribute to β-cell destruction. Finally, vitamin D_3 supplementation has increased glycemic control in T1D patients. Together, the data linking vitamin D and T1D nearly fulfill the Hill criteria for asserting that vitamin D is a protective environmental factor in this disease.

11.5 INFLAMMATORY BOWEL DISEASE

11.5.1 PATHOGENESIS AND ETIOLOGY

The IBDs, mainly ulcerative colitis (UC) and Crohn disease (CD), are chronic inflammatory disorders of the GI tract affecting 1.4 million North Americans [353–355]. UC pathology is limited to the mucosa and submucosa with cryptitis and crypt abscesses, beginning in the rectum and spreading proximally. CD pathology shows thickened submucosa, transmural inflammation, fissuring ulceration, and granuloma formation affecting any part of the GI tract, most often the terminal ileum, in a discontinuous fashion. Both diseases are characterized by periods of active disease (flares) alternating with periods of little or no disease activity (remission; [355]). Both diseases result from disruption of intestinal homeostasis, breaching of the intestinal barrier, and an inappropriately aggressive and sustained mucosal immune response to commensal microbes of the gut in the genetically susceptible individual [356,357].

The GI microbiota exists in a symbiotic relationship with the human host that benefits both organisms [358,359]. Coordinated crosstalk between the columnar intestinal epithelial cells (IEC), commensal bacteria, and localized immune cells maintain the delicate symbiotic relationship between microbes and host known as intestinal homeostasis. The IEC form a physical barrier that is impenetrable to microbes, yet allows nutrient absorption. The IEC sample the intestinal microenvironment, sense beneficial and harmful microbes, and secrete compounds such as mucins (produced by goblet cells) and antimicrobial peptides (produced by Paneth cells) that restrict microbial colonization [354].

Normally, the mucosal immune system is hyporesponsive toward the commensal bacteria because the cells and molecular profiles of this system have developed under the influence of these microbes to sustain symbiotic relationships [354,357,360]. The microbial pattern recognition receptors (PRR), for example, the Nod-like receptors and the TLRs expressed in myeloid cells and IEC [361], are crucial for microbial sensing by innate immune cells and maintenance of intestinal homeostasis. TLR2 senses lipopeptides from Gram-positive bacteria, whereas TLR4 senses lipopolysaccharides from Gram-negative bacteria. In the healthy intestine, the IEC produce cytokines that limit DC production of IL-12 and IL-23 and promote IL-10 production, thereby dampening innate immune responses and favoring anti-inflammatory adaptive immune responses rather than proinflammatory innate and effector CD4+ T cell responses to microbial antigens. Moreover, the IEC produce cytokines that promote B lymphocyte IgA production and IgA transport into the mucus layer to defend against microbial invasion of the epithelium [362]. Thus, the IEC, the microbiota, and the innate and adaptive mucosal immune systems collaborate to insure healthy GI homeostasis.

The observed changes in diseased GI tissue suggest that many factors contribute to IBD pathogenesis [356]. Among them are goblet cell destruction and reduced mucin secretion, as well as Paneth cell dysfunction and impaired defensin secretion, which reduce epithelial barrier function [354]. Other contributors seem to be defects in the handling of bacteria (induction of autophagy, antigen presentation, and microbial clearance) or introduction of pathogenic bacteria (dysbiosis). Finally, sustained innate immune system activation through excessive PRR signaling [363], excessive inflammatory actions of Th1 and Th17 cells (in CD) and Th2 cells (in UC [356,357]), and improper immune regulation due to regulatory cell dysfunction [355] seems to promote IBD pathogenesis. The antigenic targets of the Th1 and Th17 cells are uncertain. If these T cells are responding to microbial antigens, such as bacterial flagellin as they seem to be [364], then IBD may be an unchecked antimicrobial response more than an autoimmune disease [356]. Underlying the pathogenic changes are contributions from genetic susceptibility factors [353] and environmental factors [355].

11.5.2 IBD Association with UV Light Exposure and Vitamin D

Like MS and T1D, global IBD incidence has tripled in the last half century [365]. The rate at which IBD incidence has risen suggests increasing environmental risk factors may be responsible. Candidate factors include changes in sunlight exposure, hygiene, the colonic microbiota and infection, nutritional factors, smoking, and environmental pollutants [366]. A clue to environmental risk factors relevant to IBD came from older studies reporting that IBD incidence and morbidity varied inversely with latitude of residence in Europe and the United States [367]. This north to south gradient has weakened, because IBD rates have risen in southern regions [365]. Nevertheless, similarities in the geographic distributions of MS, T1D, and IBD suggested the hypothesis that high ambient UV light and vitamin D supplies might be protective in these three diseases due to selective regulation of immune responses by the vitamin D hormone [221]. Recent reviews have summarized the data linking ambient UV light, vitamin D, and IBD [368–371]. Here, we focus on the most current evidence for these links, the mechanisms that apply, and parallels with MS and T1D.

11.5.2.1 Season

Unlike strong seasonality trend data in T1D and MS, data on seasonality trends in IBD are conflicting. Slovakian data showed a peak of IBD individuals born in June [372], whereas Israeli and Belgian data showed a winter peak [373,374], and Italian [375], Dutch [376], and United Kingdom [377] data showed no birth month seasonality. Canadian data showed a peak of symptom flares in winter [378]. Spring was identified as the season of most frequent IBD disease onset in an Italian study [379], and most frequent symptom flares in Chinese [380] and American studies (CD only; [381]). Thus, it seems that birth month, onset, and symptom flares were mainly associated with winter and spring, although the seasonal influences do not seem to be as strong in IBD as they seem to be in T1D and MS.

11.5.2.2 Ambient UV Radiation

Scientific interest in UV light, vitamin D and colon health began with the 1980 observation that colon cancer varied inversely with ambient UV light [382]. Additional data linked outdoor occupations with a lower IBD risk [383]. Recent data confirmed a possible protective effect of sunlight in CD [384]. The relative risk of CD was highest in the regions of France receiving the least ambient UV radiation. The data showed a significantly decreased relative risk in high or medium sunlight exposure regions. A very recent study performed in the United States found a similar latitude gradient. In the Nurses' Health Study I and II cohorts, the incidence of CD and UC increased significantly with increasing latitude of residence at age 30 years [385]. Further study is needed to confirm this association in other global regions and to determine if it reflects a causal relationship.

11.5.2.3 Vitamin D

Interest in vitamin D and colon health was stimulated by data showing that human IEC expressed the *VDR* and *CYP27B1* genes [386], VDR-null mice showed ulcerative damage to the colon [387], and $1,25(OH)_2D_3$ prevented disease in a rodent IBD model [388]. These data focused new research on the relationships between circulating 25(OH)D and IBD risk and severity.

Hypovitaminosis D is common in IBD patients, but interpretation of these data may be confounded by reverse causation (nutrient malabsorption in the inflamed GI tract; [389]). Nearly 80% of newly diagnosed Canadian IBD patients [390], 58% of Australian pediatric IBD patients [391], and 80% of American pediatric IBD patients [392] had less than 75 nmol/L of serum 25(OH)D. In an Irish study, 59% of CD patients had less than 50 nmol/L of serum 25(OH)D [393]. Similar results were reported for British children [394], and American [395–397], Dutch [398], and Indian [399] adults with IBD. Many of these studies also noted hypovitaminosis D in the control populations, confounding data interpretation.

Curiously, studies have inversely correlated circulating 25(OH)D with severity of CD, but not UC [397,399,400]. However, this correlation did not apply in British children [394]. The inconsistencies and confounding factors make it difficult to draw firm conclusions. Altogether, the epidemiological evidence suggests a possible inverse association between vitamin D and IBD risk and severity, but the evidence is not as strong as it is for MS and T1D. Noteworthy is the inverse association between ambient UV light and circulating 25(OH)D with CD, but not UC [384,397,399,400]. These data suggest there may be distinct etiologies and roles for the vitamin D endocrine system in these two diseases. More rigorous epidemiological studies, particularly longitudinal studies exploring the interrelatedness of genetic and environmental risk factors in determining IBD risk and severity are needed to confirm the suggested effects of vitamin D.

11.5.3 IBD ASSOCIATION WITH THE *VDR* AND *GC* GENES

Heritable genetic factors contribute to IBD risk, as evidenced by the low disease concordance rates among monozygotic twins for CD (36%) and UC (16%) [401]. GWA studies have suggested IBD associations with 99 nonoverlapping genetic risk loci, 28 of them shared between CD and UC [353]. Nucleotide-binding oligomerization domain 2 (*NOD2*) gene polymorphisms are prominent in CD; 30% to 40% of Caucasian CD patients have at least one polymorphic *NOD2* allele [353,356]. Notably, the *NOD2* gene encodes an intracellular PRR that recognizes *N*-acetyl muramyl dipeptide derived from bacterial peptidoglycan [402].

The genetic contribution to UC risk seems to be considerably weaker, but polymorphisms in the *IL-10* gene have been associated with UC and CD [353], and also with MS [403]. IL-10 dampens innate and adaptive inflammatory processes, thereby limiting immune-mediated tissue damage, particularly at mucosal interfaces in the GI tract and the lung [404,405]. The CD association with *NOD2*, and the CD and UC associations with *IL10*, are highly significant from a vitamin D perspective because *NOD2* [406] and *IL10* [407,408] gene expression are reportedly under $1,25(OH)_2D_3$ control. The mechanistic implications of $1,25(OH)_2D_3$ control of *NOD2* and *IL10* are discussed later. Interestingly, GWA studies have identified *IL10* alleles that protect against CD but confer risk of T1D [409].

11.5.3.1 *VDR*

Some recent data have re-enforced a previously reported association between VDR gene polymorphisms and CD risk [410]. Curiously, the *f* allele of the *Fok*I polymorphism was implicated in CD and UC risk [411], whereas the *F* allele was associated with MS [127] and T1D risk [280]. In a Han Chinese cohort [412] and a Jewish Ashkenazi population [413], the *B* allele of the *Bsm*I polymorphism was associated with an approximately twofold increased risk of UC. However, an Irish study found no associations between *VDR* polymorphisms and IBD. The relative paucity of data and the inconsistencies between populations preclude using these data to support or refute a vitamin D hypothesis in IBD etiology.

11.5.3.2   GC

The GC gene encoding the DBP, implicated in control of circulating $25(OH)D_3$ and associated with MS and T1D (see above), has also been associated with IBD [414]. The DBP double variant with 416 Asp and 420 Thr was more frequent in IBD cases than controls (odds ratio, 4.4). These residues are not near the vitamin D metabolite binding pocket. However, they may influence other DBP functions, for example its role as a precursor of a macrophage activating factor, or enhancer of C5a-mediated neutrophil chemotaxis [415].

11.5.4   VITAMIN D TREATMENT OF IBD

Although the monozygotic twin concordance data [401] and the rapidly increasing incidence data [365] argue for strong environmental factors in determining IBD risk, the existing evidence showing inconsistent associations of IBD with season, ambient UV light, hypovitaminosis D, and VDR gene polymorphisms does not fulfill the Hill criteria for asserting a causal relationship [114]. The most compelling evidence for causation comes from experimental intervention to prevent disease, but due to the uncertainties, no studies have examined vitamin D effects on IBD risk. Prospective studies are needed to determine if vitamin D_3 supplementation might mitigate IBD risk.

Two studies have examined vitamin D_3 for IBD treatment. In one study, vitamin D_3 supplementation (1000 IU/day) did not result in measurable changes in disease parameters over 1 year compared with baseline in a small group of CD patients with inactive disease [416]. The omission of 25(OH)D measurements and the lack of placebo controls confound the interpretation of this study. A second randomized, double-blind, placebo-controlled study yielded more convincing data. Danish CD patients in remission were randomized to receive 1200 IU/day of vitamin D_3 plus Ca or Ca alone [417]. After 3 months, the circulating 25(OH)D increased to 96 ± 27 nmol/L in the supplemented group, but remained unchanged at 69 ± 31 nmol/L in the controls. After 1 year, the relapse rate trended substantially lower in the supplemented group (13%) compared with the placebo group (29%; $p < 0.06$). These results warrant further investigation of vitamin D_3 as a treatment for CD, preferably using vitamin D_3 supplements that elevate circulating 25(OH)D well above the 80 nmol/L threshold for insufficiency and into the natural range found in high ambient UV light environments [418].

Alfacalcidol (1α-hydroxyvitamin D_3), has also been evaluated for IBD treatment. The liver readily converts alfacalcidol into $1,25(OH)_2D_3$; alfacalcidol's prolonged pharmacokinetics and lower calcemic function are favorable compared with $1,25(OH)_2D_3$. In a small study lacking placebo controls, CD patients in remission received alfacalcidol (0.5 µg/day) [416]. Six weeks later, their CD activity index scores and C-reactive protein (CRP) levels were lower than baseline, and their quality of life scores were improved. There was, however, high variability in these measurements. Nonetheless, these results warrant further investigation of alfacalcidol as a short-term CD treatment.

11.5.5   ANIMAL MODELS OF IBD

Animal models of colitis have proven useful in dissecting IBD pathogenesis and potential beneficial contributions of the vitamin D endocrine system. The most important models are dextran sodium sulfate (DSS) colitis, adoptive transfer $CD4^+CD45^+RB^{hi}T$ cell colitis (transfer of these cells into the immunodeficient recipients), microbial colitis (e.g., induced by *H. hepaticus* or segmented filamentous bacteria), and spontaneous colitis in various gene-targeted mouse strains such as *IL-10*$^{-/-}$ or *IL-10R*$^{-/-}$ strains [419]. DSS, administered in drinking water, is cytotoxic to the enterocytes of the basal crypts. In the DSS model, there is loss of crypts and barrier function, microbial translocation, and strong innate and adaptive immune responses to microbial antigens. The role of the NOD2 pathway in mucosal defense and regeneration has been studied extensively in the DSS model.

In contrast to the DSS model, 2,4,6-trinitrobenzene sulfonic acid (TNBS), instilled into the colon, causes acute oxidative injury, exposure of the lamina propria to bacteria, and a massive TNBS-specific Th1-mediated inflammatory response. This model and adoptive transfer CD4+CD45+RB[hi] T cell colitis have been valuable for elucidating pathogenic effector Th1 and Th17 cell mechanisms, and the regulatory circuitry that opposes these pathogenic effector cells. Strains with targeted deletions in the genes encoding TGF-β, Foxp3, and IL-10 have been instructive in the analysis of Treg development and function, respectively, because Treg function is impaired in these animals allowing spontaneous colitis to develop [419]. Treg cells seem to be much more effective in colitis prevention than in colitis treatment. Although none of the models of intestinal inflammation fully resemble CD or UC pathogenesis, many of them mimic the uncontained Th1/Th17 mediated inflammation observed in human IBD [420].

Initial evidence that vitamin D_3 or $1,25(OH)_2D_3$ suppressed IBD came from studies with *IL10*[-/-] mice [388]. The mice fed vitamin D_3 or $1,25(OH)_2D_3$ showed improved weight gain and survival and decreased intestinal inflammation. Similar data have been reported in the TNBS and DSS colitis models [176,421] and for an analog of $1,25(OH)_2D_3$ in the DSS colitis model [422]. In DSS-induced colitis, the *CYP27B1* transcripts increased fourfold in the proximal colon, and mice lacking the *CYP27B1* gene were more susceptible [423]. Similarly, mice with a disrupted *VDR* gene (*VDR*[-/-]) were more susceptible to spontaneous colitis in the CD4+CD45+RB[hi] T-cell transfer and *IL10*[-/-] colitis models [424], and in DSS-induced colitis [421]. Together, these data support a need for $1,25(OH)_2D_3$ synthesis in the inflamed colon, and a protective role for vitamin D signaling in three murine colitis models.

11.5.6 VITAMIN D AND INTESTINAL BARRIER MAINTENANCE

Some evidence from the DSS colitis model suggests that vitamin D deficiency may compromise the mucosal barrier, increasing susceptibility to mucosal damage and IBD [425]. Whereas Wt mice were mostly resistant to DSS colitis, *VDR*[-/-] mice lacking a functional VDR protein developed severe diarrhea, rectal bleeding and weight loss, leading rapidly to death. In the *VDR*[-/-] mice compared with Wt mice, microscopic examination of the colonic epithelium showed severe disruption in epithelial junctions, a decrease in the intestinal transepithelial electric resistance, extensive ulceration, and impaired wound healing. Additional data showed that $1,25(OH)_2D_3$ markedly enhanced junction protein expression, tight junction formation and structural integrity, and transepithelial electric resistance in Caco-2 cells cultured in the presence of DSS. VDR knockdown with small interfering RNA eliminated these benefits. $1,25(OH)_2D_3$ can also stimulate epithelial cell migration *in vitro*. These data suggest a critical role for $1,25(OH)_2D_3$ and the VDR to preserve the integrity of junction complexes and the healing capacity of the colonic epithelium.

11.5.7 VITAMIN D AND INNATE IMMUNE DEFENSE OF THE INTESTINAL BARRIER

A key beneficial mechanism for vitamin D in IBD seems to be defense of the intestinal barrier to microbial invasion [426]. Loss of barrier function is a very early pathological change observed in IBD [354,356]. Moreover, *NOD2* loss-of-function mutations are strongly and probably causally associated with CD and UC [427], reinforcing the importance of pathogen recognition and killing for intestinal barrier function in the healthy intestine [353]. In this context, the antibacterial effects of vitamin D are highly significant [426]. Recent seminal experiments demonstrated that $1,25(OH)_2D_3$ potently induced *NOD2* transcripts and protein in human monocytes and epithelial cells [406]. Moreover, in macrophages from healthy individuals, *N*-acetyl muramyl dipeptide ligation of NOD2 stimulated NF-κB activation, which synergized with $1,25(OH)_2D_3$ to induce *DEFB4* (also termed *DEFB2*) transcription and synthesis of the antimicrobial peptide defensin β4. The *N*-acetyl muramyl dipeptide did not trigger this defensin pathway in macrophages from homozygous *NOD2* mutant CD patients. These studies linked CD genetics and vitamin D to an antimicrobial response at the molecular level, illustrating a gene–environment interaction that supports an essential intestinal barrier function.

Paneth cells may be particularly important in the proposed vitamin D–NOD2–defensin mechanism because they are the primary defensin producers in the healthy GI tract, and Paneth cell dysfunction occurs early in IBD [354]. It is conceivable that vitamin D insufficiency contributes causally to IBD etiology by exacerbating defects in barrier function attributable to *NOD2* mutations. Vitamin D–mediated IBD prevention studies in individuals carrying *NOD2* mutations could test this concept and detect possible direct effects of $1,25(OH)_2D_3$ on Paneth cell antimicrobial defenses.

A second key beneficial mechanism for vitamin D in IBD seems to be the regulation of innate immune defense to intracellular microbes [426]. An inappropriately aggressive innate immune response to the GI microbiota is observed very early in IBD, likely triggered by microbial invasion of the epithelium [354,356]. Innate immune cells in the lamina propria sense macromolecules in invading microbes using PRR such as NOD2 and the TLR [354,419]. These innate cells gather microbial antigens, migrate to the gut-associated lymphoid tissue (Peyer's patches and mesenteric lymph nodes), and present the antigens to T lymphocytes for an adaptive immune response. Importantly, monocyte recognition of microbial lipopolysaccharide by TLR1/2 induced *CYP27B1* and *VDR* transcripts, thereby establishing an intracrine system for producing and using $1,25(OH)_2D_3$ [428]. This vitamin D intracrine system depended on TLR1/2 stimulation of IL-15 synthesis to link microbial pattern recognition with *CYP27B1* gene expression and $1,25(OH)_2D_3$ synthesis [429].

In human monocytes, the $1,25(OH)_2D_3$ potently induced the *CAMP* gene encoding cathelicidin, which killed bacteria taken into the autophagosome [430–432]. Additionally, in human monocytes, $1,25(OH)_2D_3$ suppressed TLR2 and TLR4 transcripts and proteins in a dose-dependent manner [433]. Accompanying the TLR2 and TLR4 downregulation, there was impaired translocation of NF-kB/RelA complexes to the nucleus and reduced inflammatory signaling. These data suggest a model wherein innate immune cells sense microbial breaching of the intestinal barrier through TLR2 and TLR4, upregulate $1,25(OH)_2D_3$ and VDR to support *NOD2*, *CAMP*, and *DEFB4* transcription and microbicidal functions, and subsequently downregulate the TLR2 and TLR4 to re-establish a hyporesponsive state [426,434]. According to this model, vitamin D insufficiency could contribute causally to IBD etiology by undermining local $1,25(OH)_2D_3$ synthesis, weakening antimicrobial mechanisms, lengthening the lifespan of intracellular bacteria, and prolonging inflammatory TLR signaling.

Regrettably, the relationship between vitamin D signaling in the colonic innate immune system, induction of the *CAMP*, *DEFB4*, and *NOD2* genes, and protection from IBD pathogenesis cannot be studied in rodent IBD models because $1,25(OH)_2D_3$ regulation of *CAMP* is primate-specific [435]. Moreover, $1,25(OH)_2D_3$ regulation of *NOD2* and *DEFB4* have not been reported in species other than humans. However, development of mice expressing a human *CAMP* transgene under $1,25(OH)_2D_3$ control may alleviate this problem and allow testing of the vitamin D–antimicrobial defense model [326].

In support of this model, the magnitude of *CAMP* induction in innate cells cultured with human serum varied according to the 25(OH)D level in the serum (within the physiological range; [428]). Additionally, a study of DSS-induced colitis in vitamin D–deficient and vitamin D–sufficient mice demonstrated greater weight loss and bacterial loads in the vitamin D–deficient group [436]. Higher colonic expression of angiogenin-4, an antimicrobial protein, correlated with the reduced bacterial load in the vitamin D–sufficient animals.

11.5.8 VITAMIN D AND iNKT CELLS OF THE INTESTINE

The iNKT cells have a role in immune tolerance and protection from autoimmune inflammation [437]. Development and function of iNKT cells is impaired in $VDR^{-/-}$ mice, predisposing them to the development of colitis [438]. Vitamin D deficiency *in utero* also impaired the development of iNKT cells, which could not be overcome later in life [180].

11.5.9 Vitamin D and Adaptive Immune Defense of the Intestinal Barrier

The normal hyporesponsive state of the mucosal immune system toward the commensal bacteria is disrupted in IBD [354,363]. Enrichment of CD4+ T cells at mucosal sites, and excessive levels of proinflammatory cytokines in the chronically inflamed GI tract have supported the view that these cells are pathogenic in IBD. Immune-mediated pathology in the gut of CD patients has been associated with Th1 cells producing IFN-γ, or IL-23–driven Th17 cells producing IL-17, whereas in UC pathology has been correlated with Th2 cells producing IL-4 [439–441].

11.5.9.1 T Effector Lymphocytes

In the animal IBD models, there is evidence for direct $1,25(OH)_2D_3$ effects on VDR-expressing pathogenic T cells. In the $IL-10^{-/-}$ colitis model, mice also lacking VDR expression developed more severe and fatal colitis than mice with a functional VDR gene [424]. Further experiments in CD4+CD45RB[hi] T-cell transfer colitis showed that $VDR^{-/-}$ effector T-cell transfer yielded more severe colitis than Wt effector T-cell transfer. These data indicate a need for VDR expression in the transferred effector T cells to limit autoimmune-mediated colon tissue damage.

Recent reports have focused specifically on Th17 cells. Higher levels of colonic IL-17 correlated with severe DSS-induced colitis in mice lacking the $CYP27B1$ gene and hence $1,25(OH)_2D_3$ synthesis [423]. Conversely, administering $1,25(OH)_2D_3$ in the TNBS-induced colitis model reduced the colonic expression of Th1-related molecules (IL-12p70, T-bet, and IFN-γ) and Th17-related molecules (IL-6, IL17, and IL-23p19), and also increased the expression of Th2-related molecules (GATA3 and IL-4) and Treg molecules (Foxp3, CTLA4, IL-10, and TGF-β; [176]). Furthermore, $1,25(OH)_2D_3$ inhibited Th17 cell development in cell cultures [442]. However, it is unclear in these studies whether the observed changes resulted from direct $1,25(OH)_2D_3$ effects on antigen-presenting cells like DC, or the T cells (or both). The importance of VDR expression in non-T cells was demonstrated in CD4+CD45RB[hi] T-cell transfer colitis [442]. Transferring Wt effector T cells into $VDR^{-/-}$ recipients yielded increased proportions of Th17 cells (but not Th1 cells) and more severe colitis compared with Wt recipients, possibly due to fewer tolerogenic CD103+ DC in the $VDR^{-/-}$ recipients.

11.5.9.2 IL-10

Genetic epidemiology has indicated that IL-10 signaling plays a crucial role in intestinal homeostasis. Polymorphisms in the $IL-10$ and $IL-10R$ genes have been implicated in early onset IBD [353], and impaired IL-10 signaling has been documented in UC [443]. Sources of IL-10 include macrophages, DC, T cells, B cells, neutrophils, eosinophils, and mast cells [405]. In the healthy intestine, IL-10 performs many anti-inflammatory functions and also supports activated B-cell isotype switching from IgM to IgA in the gut-associated lymphoid tissues, such that microbial antigen-specific IgA is secreted and transported into the mucus layer to prevent microbial invasion [362].

Importantly, $1,25(OH)_2D_3$ enhancement of IL-10 production was reported in human peripheral blood CD4+ T lymphocytes from CD patients and healthy controls [444]. When cultured with $1,25(OH)_2D_3$, the purified and activated CD4+ T cells from both groups decreased their proliferation and their IFN-γ production, but doubled their IL-10 production compared with placebo. A more recent study confirmed these data and showed a clear dose–response relationship between $1,25(OH)_2D_3$ and decreased IFN-γ and increased IL-10 transcripts *in vitro* [445]. The later study also found that IL-10 transcripts increased approximately fourfold in the peripheral blood CD4+ T lymphocytes from human volunteers 3 days after they ingested $1,25(OH)_2D_3$. It is not yet clear whether these data reflect a direct effect of $1,25(OH)_2D_3$ on $IL-10$ gene expression, or an effect on T cell subset proportions. Enhancement of IL-10 production has also been reported for mouse mast cells [446] and human B lymphocytes [408]. In the human B cells, $1,25(OH)_2D_3$ enhanced VDR recruitment to the $IL10$ gene promoter, suggesting a direct effect on transcription.

Experimental colitis models have shown that IL-10–secreting CD4+ T cells (Foxp3+ and Foxp3−) are particularly important for intestinal homeostasis [354]. Ablating IL-10 production in Foxp3+

CD4+ T cells resulted in immune-mediated tissue damage at mucosal interfaces (colon and lungs), but no deleterious effect on systemic autoimmunity [404]. These data suggest that IL-10–secreting CD4+ T-cell suppressive functions may be nonredundant and tissue specific. Development, migration, or function of induced CD4+ Treg cells in the gut may be VDR dependent because there were fewer of them in $VDR^{-/-}$ mice than Wt mice, and the CD4+ Treg cells from the $VDR^{-/-}$ mice were refractory to IL-6 inhibition [442].

Rodent research also demonstrated a requirement for VDR expression to support the development and recruitment to the gut of the CD8αα intraepithelial lymphocytes (IEL) that produce IL-10 in response to commensal microbial antigens [447,448]. There were fewer CD8αα IEL in $VDR^{-/-}$ mice than in Wt mice, and in $IL10^{-/-}$ recipients of transferred $VDR^{-/-}$ T cells than transferred Wt T cells. The CD8αα IEL deficit was cell intrinsic and due to decreased thymic development, CCR9 expression, and homing to the small intestine. In summary, genetic epidemiology and immunological studies in humans and mice support an essential role for IL-10 in intestinal homeostasis and suggest that $1,25(OH)_2D_3$-mediated enhancement of IL-10 production by several types of cells could contribute substantially to the maintenance of homeostasis in the healthy intestine.

11.5.10 VITAMIN D AND ESTROGEN MAY COOPERATE IN FEMALE INTESTINAL HOMEOSTASIS

A recent study in postmenopausal women detected a link between the estrogen and vitamin D endocrine systems that is relevant to protection from IBD [449]. Rectal mucosal biopsies were obtained from the women before and after a four week estrogen intervention that raised circulating estradiol from 17 ± 10 to 47 ± 19 pg/mL. Transcript analysis showed VDR mRNA increased significantly, as did multiple downstream VDR target transcripts. Notably, transcripts associated with antigen presentation, proinflammatory chemokines and cytokines, and complement components were all downregulated in this study. These changes are consistent with the $1,25(OH)_2D_3$-mediated decreases in DC antigen-presenting function that result in tolerogenic DC [172]. Thus, there may be a beneficial role for estrogen replacement therapy in post-menopausal women to support VDR-mediated homeostatic and anti-inflammatory functions in the GI tract.

11.5.11 VITAMIN D AND IBD SUMMARY

In summary, there are now detailed and plausible molecular mechanisms linking CD and UC with vitamin D enhancement of antimicrobial defenses of the intestinal barrier. Those molecular mechanisms include vitamin D hormone signaling contributions to epithelial junction maintenance, potentiation of innate antimicrobial defenses at the mucosal surface, and maintenance of innate and adaptive immune system hyporesponsiveness to commensal bacteria. Mechanisms involved in this hyporesponsiveness include actions on APC to induce a tolerogenic state, actions on pathogenic effector T cells to limit their function, and actions on iNKT cells, CD8αα T cells, and other IL-10–producing regulatory cells to foster homeostasis in the healthy intestine. Despite the weaknesses in the epidemiological data, the strength of the genetic and mechanistic data clearly supports further study of vitamin D supplementation (or alphacalcidol) treatment for individuals with active IBD. In that context, subject genotyping and biopsy sample analysis would allow testing of the proposed molecular mechanisms and might reveal further gene–environment interactions relevant to IBD.

11.6 SYSTEMIC LUPUS ERYTHEMATOSUS

11.6.1 PATHOGENESIS AND ETIOLOGY

SLE is a genetically and immunologically complex autoimmune disorder that predominantly strikes young women, especially non-Caucasian women, and is not well understood mechanistically [450]. The term *lupus erythematosus* originally referred to skin lesions, but it is now clear that SLE is a systemic autoimmune disease that involves many organ systems (skin, kidney, heart, lungs, brain, blood, and joints). The SLE diagnosis is based on the presence of any 4 of 11 criteria, specifically rashes, photosensitivity, oral ulcers, arthritis, serositis, kidney disorders, neurologic disorders, blood abnormalities (hemolytic anemia, leukopenia, lymphopenia), and circulating autoantibodies to chromatin components [450]. Genetic predisposition is believed to contribute to SLE disease etiology, with hormonal and environmental factors determining disease development and phenotype. Two rare single gene deficiencies are known to cause SLE, *C4* deficiency (required for clonal elimination of self-reactive transitional B lymphocytes) and *C1q* deficiency (required for elimination of necrotic tissue). These results directly implicate defective elimination of self-reactive B lymphocytes and of necrotic tissue in SLE pathogenesis. The GWA studies to date have identified 29 loci as possible determinants of SLE risk and accounted for approximately 15% of SLE heritability [451]. As in MS, the SLE concordance rate in monozygotic twin pairs was approximately 25% [452]. Together, these data suggest that environmental factors seem to dominate SLE risk.

11.6.2 SLE DISEASE ASSOCIATION WITH UV LIGHT EXPOSURE AND VITAMIN D

In MS, T1D, and IBD, ambient UV radiation correlates inversely with disease risk. In sharp contrast, UV light–induced skin injury may have a direct causal role in SLE pathogenesis [453–456]. Strong UVB radiation exposure causes skin tissue damage (necrosis and keratinocyte apoptosis). Normally, erythemal damage is repaired within a few days without inducing inflammation. The repair pathway involves phagocytocytic cell engulfment of necrotic tissue and apoptotic keratinocytes without inflammation. However, in lupus, the repair pathway is somehow disrupted such that erythemal damage elicits inflammation (chemokine production, inflammatory cell recruitment) and skin lesions. The likelihood that erythemal damage contributes to SLE pathogenesis confounds the interpretation of data attempting to determine whether vitamin D is linked in some way to SLE risk or disease activity. Photosensitivity, sunlight avoidance, and sunscreen use put SLE patients at high risk of vitamin D deficiency [457] because vitamin D supplies derive mainly from ambient UV radiation [458]. Individuals with photosensitivity may avoid sun exposure even before a SLE diagnosis, thereby limiting their vitamin D supplies. Recent reviews have summarized the data on vitamin D and lupus [457,459–461].

11.6.2.1 Latitude

There is very limited information on geographical gradients in SLE risk and disease activity. The SLE incidence was higher ($7.62/10^5$ population) in Northern Ireland (54°N) than in the United Kingdom West Midlands ($3.56/10^5$ population; 52°N) [462]. These data are similar to Irish data documenting a twofold higher MS disease prevalence in Northern Ireland than in Southern Ireland [65]. The SLE study did not address whether the populations in question differed genetically. In sharp contrast to the direct relationship between SLE incidence and latitude, there was an inverse latitudinal gradient in SLE mortality in the United States [463–465]. Data from the U.S. National Center for Health Statistics showed a high SLE standardized mortality ratio (1.54) in the Southwest (34°N) and a low mortality ratio (0.65) in the upper Midwest (49°N), particularly for Caucasian women. High ambient UV radiation, poverty, and Hispanic ethnicity each correlated directly with SLE mortality. Although the SLE data are very limited, the contrasting direct relationship between latitude and risk and inverse relationship between latitude and mortality are striking. Additional research is needed to determine if these contrasting gradients for disease risk and disease mortality hold independently of ethnicity and socioeconomic status. The results could be informative with regard to the role of ambient UV radiation and vitamin D in the biological processes implicated in SLE risk and severity.

11.6.2.2 Season and Ambient UV Radiation

No studies have investigated possible seasonality in SLE patient birth or onset months. Such studies would be valuable because they might reveal a seasonally varying environmental factor involved in SLE risk and the developmental window within which the factor might act. One longitudinal study examined seasonal variations in SLE disease activity at latitude 70°N [466]. Photosensitive rashes accumulated in the summer, when sunlight was available 24 h/day, whereas no SLE flares occurred in January, when there was less than 6 h of sunlight in the entire month. Other clinical and laboratory measures of SLE activity did not fluctuate during the year-long study. These data are consistent with data showing a direct correlation between ambient UV radiation and SLE mortality and support the proposition that erythemal damage contributes to SLE pathogenesis.

11.6.2.3 Vitamin D

Efforts are underway to investigate a possible correlation between vitamin D supplies and SLE risk. These data are subject to the reverse causation concern because photosensitive individuals may have low vitamin D supplies as a consequence of avoiding sun exposure. Analyzing data from the Nurses' Health Study and Nurses' Health Study II cohorts, investigators reported no correlation between vitamin D intake and SLE risk [467], unlike MS risk, which was correlated with vitamin D intake in this cohort [105]. However, further research will be needed to thoroughly address the vitamin D–SLE risk question because vitamin D intake provides only a small fraction of the total vitamin D supplies.

Additional studies have attempted to correlate circulating 25(OH)D levels and SLE. In single time point studies, it is essential to adjust 25(OH)D for season of sampling, particularly in regions with large seasonal fluctuations in ambient UV radiation. Not all studies have performed this adjustment. One study of seasonally adjusted 25(OH)D found newly diagnosed SLE cases had significantly lower levels than matched controls [468]. The individuals of African ancestry had significantly lower levels than Caucasians (15.9 versus 31.3 ng/mL), adding to the large body of evidence showing reduced vitamin D_3 synthesis in highly pigmented skin [4,18,19]. The 25(OH)D levels were critically low in 36.8% of pediatric SLE patients compared with 9.2% of controls; 60% of darkly pigmented SLE children were severely deficient [469]. The SLE disease activity index increased as the 25(OH)D levels decreased, and the 25(OH)D levels decreased as the body mass index increased. Together, these data suggest that dark pigmentation and obesity may be exacerbating vitamin D insufficiency and SLE disease activity in children of African ancestry. At least two studies have reported that circulating $1,25(OH)_2D_3$ was significantly lower in SLE patients than in controls (e.g., 32.2 vs. 54.4 pg/mL; [469,470]). Adjusting for the 25(OH)D concentration in a multivariable linear regression model did not affect the $1,25(OH)_2D$ deficit in SLE, indicating that the $1,25(OH)_2D$ deficit did not depend on 25(OH)D and probably represented an impairment in $1,25(OH)_2D$ synthesis [469]. Despite reduced 25(OH)D and $1,25(OH)_2D$, patients with SLE had lower intact PTH levels than controls [469]. These data are reminiscent of hypoparathyroidism noted in a Finnish MS patient cohort [108], and justify further investigation into possible abnormalities in the vitamin D–PTH axis in autoimmune disease.

A large number of studies of adult SLE patients have now reported critically low 25(OH)D levels (<30 ng/mL) in the vast majority of patients [470–476]. Moreover, 11 independent studies have inversely correlated serum 25(OH)D levels with SLE disease activity [469,475,477–485]. After adjustment for age, seasonal variation, and race/ethnicity, one study correlated lower 25(OH)D levels with higher fasting serum glucose [475], whereas another correlated lower 25(OH)D levels with insulin resistance [480]. These data are consistent with data showing vitamin D dependence of insulin production and function [317]. Another study correlated lower 25(OH)D levels with higher osteocalcin, bone-specific alkaline phosphatase, and circulating IL-6 [470]. Two studies reported equivalent serum 25(OH)D levels in patients with SLE and in controls, and an inverse correlation between serum 25(OH)D levels and SLE disease activity. It is notable that these two studies were performed at lower latitudes: Brisbane, Australia [480], and Texas [482]. A few studies have not observed a significant relationship between serum 25(OH)D and SLE disease activity [486–489]. This discrepancy remains unexplained.

A second discrepancy concerning circulating Ab to dsDNA also remains unexplained. Several recent studies reported that Ab to dsDNA (and other measures of disease activity) varied inversely with 25(OH)D [481,485,490], but one earlier report correlated Ab to dsDNA directly with circulating 25(OH)D [471]. The data showing an inverse relationship between vitamin D and Ab to dsDNA are consistent with *in vitro* data showing that $1,25(OH)_2D_3$ treatment inhibited cell proliferation and stimulated the death of activated B cells from SLE patients [407]. Further study is clearly warranted regarding vitamin D and B cell biology in patients with SLE.

11.6.3 SLE Disease Association with the *VDR* Gene

Recent data have correlated the B allele of the *VDR* gene *Bsm*I polymorphism with SLE risk, antinucleosome antibodies, and lupus nephritis in Han Chinese patients [491,492]. Significantly, SLE patients with the *VDR* B allele had markedly less VDR mRNA than patients without the B allele. These data suggest that reduced VDR expression may contribute to increased SLE risk in Asians. However, the *VDR* gene *Bsm*I polymorphism was not associated with SLE risk in an Iranian population [493], and the *VDR* gene *Fok*I polymorphism was not associated with SLE risk in a Brazilian population [493]. Other genes in the vitamin D pathway have not been investigated.

Particular *HLA-DRB1* genes have been linked with SLE risk. Studying Caucasian SLE trios, investigators from the United Kingdom determined that the *HLA-DRB1*0301* allele conferred an increased risk of SLE (odds ratio, 2.3; [495]). Other investigators reported that the *HLA-DRB1*1501* allele conferred an increased risk of SLE in a Japanese population [496,497]. Remarkably, *HLA-DRB1*0301* and *HLA-DRB1*1501* are the same *HLA-DRB1* alleles that were associated with T1D risk in a North Indian population [284] and MS risk in individuals of Northern European ancestry [36], respectively, and reportedly had identical putative VDRE sequences in their promoters. An association between the VDR gene *Fok*I^F polymorphism and T1D was only evident in the North Indian population carrying the *HLA-DRB1*0301* allele [284].

Given the complexities of investigating whether vitamin D is linked in some way to SLE risk or disease activity through epidemiological studies where photosensitivity and erythemal damage are confounders, it may be more fruitful to address the question of vitamin D and SLE risk through genetic studies. For example, evaluating possible contributions of rare null *CYP27B1* variants to disease risk in SLE multiplex families, as in MS [136], could unambiguously implicate the vitamin D pathway independently of UV light. Furthermore, investigating *VDR* gene variants in the context of *HLA-DRB1*0301* and *HLA-DRB1*1501* in populations with limited vitamin D supplies (due to pigmentation or low ambient UV radiation), as in T1D [284], could clarify discrepancies regarding *VDR* gene contributions to disease risk. Finally, searching for a SLE patient birth month effect in those who carry *HLA-DRB1*0301* and *HLA-DRB1*1501* genes [498] could reveal whether these genes interact with a seasonally fluctuating environmental variable *in utero*.

11.6.4 Animal Models of SLE and $1,25(OH)_2D_3$

Studies of spontaneous and induced SLE-like syndromes in diverse mouse strains have identified several candidate genetic risk factors, provided insight into environmental risk factors, yielded a better understanding of the cellular mechanisms involved in the disease, and served to screen potential therapeutics [499,500]. Strains that spontaneously develop SLE-like syndromes include (NZB × NZW)F1, NZM2410, BXSB, and MRL/lpr [501]. These strains display uniformly high penetrance of lupus phenotypes, lymphoid hyperplasia, and uniformly progressive disease. In contrast, human SLE patients display variable and incompletely penetrant lupus phenotypes, lymphopenia, and a fluctuating disease course with flares and remissions. Thus, as useful as the spontaneous rodent SLE models are, they do not faithfully mimic human SLE disease.

The MRL/lpr strain develops a lymphoproliferative disorder and lupus-like signs because of a null mutation at the *lpr* locus encoding the FAS death receptor [502]. Increasing production of IL-17 correlated with progressively worsening nephritis in these mice [503]. Administering $1,25(OH)_2D_3$

to MRL/lpr mice, beginning at age 4 weeks, had no effect on lymphoid hyperplasia, but it completely inhibited dermatologic lesions (alopecia and skin necrosis), and reduced the production of Ab to dsDNA and glomerular nephritis [504]. These results have been confirmed, and the importance of dietary calcium demonstrated [505]. The mechanisms of $1,25(OH)_2D_3$-mediated SLE disease inhibition in MRL/lpr mice have not been studied.

11.6.5 VITAMIN D DEPENDENCE OF ERYTHEMAL DAMAGE PREVENTION

One vitamin D mechanism that may be highly relevant to SLE involves preventing the erythemal damage that is induced by UVB irradiation. Chronic UVB irradiation exposure damages skin tissue. Erythemal damage is normally repaired without inducing inflammation, but in patients with SLE, erythemal damage leads to leukocyte infiltration, epidermal hyperplasia, and epidermal necrosis. Mast cells in the epidermis play an unexpected role as suppressors of erythemal damage through their production of IL-10 [506]. New research has established that this mast cell function is under direct VDR control [446]. The UVB irradiation that induced erythemal damage also increased the synthesis of $1,25(OH)_2D_3$. Adding $1,25(OH)_2D_3$ to *in vitro* cultures of mouse mast cells increased IL-10 transcripts and IL-10 protein secretion. In an elegant set of experiments, genetically mast cell–deficient WBB6F$_1$-Kit$^{W/W-v}$ mice were engrafted with mast cells harboring a functional or nonfunctional VDR, and were later examined for mast cell–mediated suppression of the inflammatory response to chronic UVB irradiation. Mice whose mast cells harbored a functional VDR produced IL-10 and suppressed the local production of proinflammatory cytokines, epidermal hyperplasia, and epidermal ulceration. However, mice whose mast cells harbored a nonfunctional VDR did not suppress inflammation and developed skin lesions. The data suggest that suppression of inflammation and skin pathology at sites of chronic UVB irradiation required $1,25(OH)_2D_3$ and VDR-dependent induction of IL-10. This is reminiscent of data showing IL-10 pathway dependence of $1,25(OH)_2D_3$ action in the EAE model of MS [161].

11.6.6 VITAMIN D HORMONE MODULATION OF IMMUNE FUNCTION IN LUPUS

Very little is known about the influences of $1,25(OH)_2D_3$ on immunological processes of relevance to SLE. A study of activated B cells from SLE patients found that $1,25(OH)_2D_3$ inhibited proliferation and stimulated the death of these cells, but the mechanisms were not explored [407]. Another report described an activating effect of SLE patient sera on human APC that was inhibited by $1,25(OH)_2D_3$, but no mechanistic details were described [507]. Recent data have correlated elevated Th17 cell numbers, and higher circulating IL-12, IL-17, and CXCL10 with disease activity in SLE patients, implicating these inflammatory cells and mediators in SLE pathogenesis [508]. T cells lacking the CD4 and CD8 surface molecules and producing IL-17 and IFN-γ were identified in kidney biopsies of patients with lupus nephritis [509]. Additional studies have confirmed the association between Th17 cells and SLE disease [510–514]. Although no experiments have evaluated the possible inhibitory effects of vitamin D$_3$ repletion on CD4$^+$ Th1 and Th17 cells in SLE patients, such effects might be expected, based on the growing consensus that the hormonal form, $1,25\text{-}(OH)_2D_3$, inhibits these cells in MS and T1D animal models.

11.6.7 VITAMIN D AND SLE SUMMARY

There are some data suggesting protective effects of the vitamin D system in SLE, but the evidence is incomplete, sometimes inconsistent, and often confounded by the fact that UVB irradiation, the main source of vitamin D$_3$ synthesis, is also suspected to contribute to SLE pathogenesis. There is clearly an inverse latitudinal gradient in SLE mortality, and a seasonality in SLE flares that are consistent with a pathogenic role for UVB-induced erythemal damage. Whether there is a direct latitudinal gradient in SLE risk is less clear. Limited evaluations of circulating vitamin D supplies and SLE risk to date are not conclusive. However, compelling evidence has linked criti-cally low serum 25(OH)D levels with high SLE disease activity, particularly in obese individuals of African ancestry. These data argue strongly for vitamin D_3 repletion efforts in patients with SLE. Very interesting genetic data have associated the T1D-linked *HLA-DRB1*0301* allele and the MS-linked *HLA-DRB1*1501* allele with SLE risk. It will be critical to determine if the identi-cal putative VDRE sequences in the promoters of these two genes fluctuate in response to vitamin D supplies *in vivo*, which could implicate vitamin D in SLE risk. Very significant new data from the genetically mast cell–deficient $WBB6F_1$-$Kit^{W/W-v}$ mouse strain show that cutaneous mast cell IL-10 synthesis and protection from UVB-induced erythemal damage is under VDR gene control. Whether vitamin D repletion will improve protection from erythemal damage in patients with SLE remains to be determined.

11.7 RHEUMATOID ARTHRITIS

11.7.1 Pathogenesis and Etiology

RA is another autoimmune disease that has been associated with vitamin D [515–517]. RA is marked by chronic inflammation in the synovial fluid, eventually leading to destruction of the cartilage and bone of the affected joint [518]. The cellular and secretory interactions of synovial fibroblasts, pro-fessional APCs (B cells, DC, MΦ), osteoclasts, and autoreactive Th1 and Th17 cells, coupled with a breakdown of regulatory mechanisms, are ultimately what drive the chronic inflammation seen in RA [519,520]. Thus, like MS and T1D, RA represents an uncontrolled immune response in the synovial joint that is perpetuated by autoreactive Th1 and Th17 cells.

11.7.2 RA Disease Association with Vitamin D

The few studies that have evaluated a possible link between vitamin D status and the risk of devel-oping RA have produced conflicting results. Data from the Iowa Women's Health Study correlated increased vitamin D intake with a significant decrease in the risk of developing RA (RR, 0.67; [521]). However, data from the Nurses' Health Study and the Nurses' Health Study II did not show this cor-relation [467]. Likewise, serum 25(OH)D < 20 nM at 1, 2, and 5 years before onset of RA was not associated with increased RA risk in a Dutch population [522]. Thus, there is insufficient evidence to suggest a casual relationship between vitamin D status and RA risk.

Multiple observational studies have examined vitamin D status in patients with RA. Several studies involving African Americans, veterans, and others have found that vitamin D deficiency is common in patients with RA [523–526]. Several other investigators have found that the prevalence of vitamin D deficiency or insufficiency in patients with RA is not different from controls [527–529]. Thus, the widespread occurrence of vitamin D insufficiency and deficiency in the population pre-cludes drawing conclusions about vitamin D status and RA prevalence.

In contrast with RA prevalence, there is evidence to suggest that vitamin D may modulate RA disease activity. An inverse relationship between serum 25(OH)D and disease activity score (DAS28, a measure of disease activity using 28 joint counts) was noted in several different studies of Americans and Europeans [528,530,531]. Antibodies to cyclic citrullinated peptide were more common in RA patients with serum 25(OH)D < 75 nM, and serum 25(OH)D < 50 nM was associ-ated with higher tender joint counts and CRP levels [525]. An inverse correlation between serum 25(OH)D and the Health Assessment Questionnaire of Disability Index and the DAS28 scores was observed [528]. Others noted an inverse correlation between serum 25(OH)D and CRP levels [529]. Furthermore, in a study of patients with early inflammatory polyarthritis, CRP levels decreased 25% for each 10 ng/mL increase in serum 25(OH)D [532]. So, although data on vitamin D status and RA prevalence are inconclusive, multiple studies indicate that vitamin D status is inversely cor-related with RA disease activity. The association between vitamin D status and RA disease activity may reflect a causal relationship, or a reverse causal relationship if RA patients with active disease spend less time outdoors.

11.7.3   RA Disease Association with the *VDR* Gene

The relationship between *VDR* polymorphisms and RA risk has been evaluated in several studies [491,533–537]. These studies have identified that *Fok*I and *Bsm*I polymorphisms of the *VDR* are associated with RA risk. No associations have been identified with the *CYP27B1* or *CYP24A1* polymorphisms.

11.7.4   Vitamin D Treatment of RA

A few intervention studies evaluating the effectiveness of vitamin D metabolites for the treatment of RA have been conducted to date. The first of these evaluated 2 µg/day of alfacalcidiol (1α-hydroxyvitamin D_3) as a treatment for RA, and found that 89% of the patients had a reduced incidence of disease 3 months later [538]. However, the study did not include a control group, and no follow-up studies have been conducted. More recently, the effects of a disease-modifying antirheumatic drug plus calcium were compared with disease-modifying antirheumatic drug, calcium, and calcitriol combination therapy in patients with early onset RA [539]. The group receiving calcitriol reported higher pain relief than the noncalcitriol group after 3 months of therapy. A study evaluating the effectiveness of 12-week calcidiol [25(OH)D_3] therapy in patients with RA receiving methotrexate found that calcidiol did not have an effect on DAS28 scores compared with patients receiving methotrexate therapy only [540].

11.7.5   Collagen-Induced Arthritis Model of RA

Animal models of RA have provided evidence for a role for vitamin D signaling in the prevention and treatment in RA. The first evidence for a protective role for vitamin D came from a study using collagen-induced arthritis (CIA) in rats [541]. Oral administration of alfacalcidiol inhibited the onset of arthritis symptoms and joint swelling. Likewise, dietary 1,25(OH)$_2$$D_3$ treatment prevented CIA and stopped the progression of established CIA in mice [542]. Most recently, feeding a vitamin D–replete diet as compared with a vitamin D–deficient diet inhibited adoptive transfer of adjuvant-induced arthritis in rats [543]. Resolution of arthritis symptoms was also delayed in the rats fed a vitamin D–deficient diet.

11.7.6   Vitamin D and RA Summary

RA prevalence does not correlate well with vitamin D status and little evidence exists to suggest that vitamin D status is related to RA risk. Disease activity in patients with RA, on the other hand, seems to be inversely correlated with serum 25(OH)D levels, although this may be due to decreased sunlight exposure in those individuals with the highest disease activity. Evidence from intervention studies and animal models of RA, however, suggests that further investigation into the role of the vitamin D signaling pathway in RA pathogenesis is warranted. One can envision that the vitamin D signaling pathway functions in a manner similar to its mechanisms in Th1/Th17-driven autoimmune diseases to suppress inflammation.

11.8   CONCLUSIONS AND FUTURE RESEARCH DIRECTIONS

Research into the relationship between vitamin D deficiency and autoimmune disease risk intensified in the late 1990s and has reached a point in which one can suggest that a cause–effect relationship may exist between low vitamin D supplies and high autoimmune disease risk. The concept of a cause–effect relationship was founded on robust data showing ambient UV light gradients in MS, T1D, and IBD disease prevalence, and equally robust data on seasonal fluctuations in birth month, disease onset, and disease activity in regions with wide fluctuations in ambient UV light. Indeed, the MS data support a computed odds ratio of approximately 20 for the UV light–linked factor, far greater than the odds ratio for any combination of genetic risk factors for any auto-immune disease.

The articulation of a plausible biological mechanism, UV light–catalyzed vitamin D_3 synthesis inhibiting autoimmune disease through a selective action of the hormonal form of vitamin D on VDR-expressing, autoreactive T lymphocytes, together with supporting data in mouse models of autoimmunity, solidified the concept of a cause–effect relationship and stimulated clinical research. Birth cohort studies and longitudinal population studies contributed the insight that ambient UV exposure, vitamin D_3 intake, and circulating 25(OH)D levels in early life correlated strongly with autoimmune disease risk and disease activity years later.

Genetic studies have also advanced the concept of a cause–effect relationship between low vitamin D supplies and high autoimmune disease risk, with the discovery of putative VDRE within the promoter regions of *HLA-DRB1* alleles implicated in MS, T1D, and SLE, and the unambiguous association of deleterious *CYP27B1* mutations with risk of MS and T1D.

Research into biological mechanisms has strengthened the case for a causal link. Disease-specific mechanisms have emerged from animal model studies: for example, vitamin D support for neuro-protection in EAE, pancreatic islet β-cell insulin release in T1D, defense of the intestinal barrier in IBD, and repair of UVB-induced erythemal damage in SLE. Common biological mechanisms have also emerged from animal modeling, specifically, control of autoreactive T-cell pathogenic-ity through downregulation of pathogenic molecules, T-cell elimination, or T-cell conversion to a nonpathogenic, possibly protective Treg cell phenotype.

Clinical applications of the vitamin D–autoimmune disease hypothesis have also emerged. Efforts are underway to modify autoimmune disease activity through vitamin D_3 supplementation. Successful pilot study data in the MS and T1D fields serve as a harbinger of what may be learned in future treatment studies. In this way, vitamin D and autoimmune disease research exemplifies the value of basic research in linkage with human clinical research.

A promising translational effect originating in the present research in vitamin D and autoimmune disease relates to autoimmune disease prevention. Based on the approximately 75% discordance rates for autoimmune diseases in monozygotic twin pairs, up to approximately 75% of the current autoimmune disease burden may be preventable through modification of environmental factors. Using the twin data as a guide and assuming that vitamin D accounts for approximately 50% of the environmental risk, a theoretical estimate might be the prevention of approximately 40% of autoim-mune disease cases using a vitamin D–based strategy. This may be an underestimate. Autoimmune disease prevention efforts could conceivably spare millions of people and billions of dollars in health care expenditures.

What future research will be needed if we are to harness the potential of vitamin D_3 to reach these goals? Environmental risk together with a foundation of genetic predisposition is a common theme in autoimmune disease causality. For IBD and SLE, in which evidence for a vitamin D–disease link is not as compelling as it is for MS and T1D, additional research modeled on research performed in the MS and T1D fields would be valuable to strengthen the evidence. For example, exomic sequenc-ing of vitamin D pathway genes (or genes believed to be under vitamin D control) in T1D, IBD, and SLE multiplex families living in regions with low ambient UV radiation, modeled on the studies of CYP27B1 in MS families, may be able to test the hypothesis that vitamin D is a regulator of dis-ease risk. Longitudinal studies to support a temporal sequence of vitamin D insufficiency followed by disease development would also be valuable. Additional interventional studies in autoimmune disease–affected individuals are also urgently needed, particularly to define the vitamin D_3 intake required to obtain health benefits. Such evidence is urgently needed to inform public health policy about the recommended daily intakes for particular groups of people.

Presently, it is not possible to identify with certainty the individuals who are at high risk of autoimmune disease and who might benefit from higher vitamin D_3 intakes than are currently recommended. Screening procedures for at-risk individuals must be based on a molecular understanding of disease pathogenesis. Accordingly, research is needed into vitamin D's influences on specific genetic pathways associated with autoimmune disease risk, and conversely, on genetic contributions to biological processes that are relevant to autoimmunity and responsive to vitamin D. A more detailed understanding of maternal vitamin D influences on autoimmune disease risk in genetically susceptible offspring, and of hormonal interactions with the vitamin D system will also be needed. These types of gene–environment interaction studies will serve to identify genetic and hormonal signatures of individuals who may be at high risk of autoimmune disease, given exposure to vitamin D insufficiency.

Research is also needed to discover windows of opportunity and points of no return, if they exist, regarding vitamin D–based interventions. For example, a better understanding of how the vitamin D system regulates the development and function of both innate and adaptive immune cells may reveal whether developmental defects imposed by early vitamin D insufficiency can be overcome later in life by compensatory mechanisms. Viewed from the vitamin D perspective, the many strong parallels between MS, T1D, IBD, and possibly SLE suggest that a collaboration between the autoimmune disease research communities to report their research results, refine a joint research agenda, and envision and implement collaborative research aimed at reducing the human and financial effects of these diseases on a global scale could be very rewarding.

ACKNOWLEDGMENTS

We are grateful to the National Multiple Sclerosis Society (Research Grants RG4076A5 and RG3107D6), a HATCH McIntyre Stennis Award (MSN119798), and the Multiple Sclerosis Research Fund in Biochemistry (UW Foundation) for supporting our research and the writing of this book chapter.

REFERENCES

1. Blumberg, R. S., B. Dittel, D. Hafler, M. von Herrath, and F. O. Nestle. 2012. Unraveling the autoimmune translational research process layer by layer. *Nature Medicine* 18:35–41.
2. Prosser, D. E., and G. Jones. 2004. Enzymes involved in the activation and inactivation of vitamin D. *Trends in Biochemical Sciences* 29:664–673.
3. Verstuyf, A., G. Carmeliet, R. Bouillon, and C. Mathieu. 2010. Vitamin D: A pleiotropic hormone. *Kidney International* 78:140–145.
4. Holick, M. F., T. C. Chen, Z. Lu, and E. Sauter. 2007. Vitamin D and skin physiology: A D-lightful story. *Journal of Bone and Mineral Research* 22 Suppl 2:V28–33.
5. Zerwekh, J. E. 2008. Blood biomarkers of vitamin D status. *The American Journal of Clinical Nutrition* 87:1087S–1091S.
6. Holick, M. F. 2010. Vitamin D: Extraskeletal health. *Endocrinology and Metabolism Clinics of North America* 39:381–400.
7. Adams, J. S., H. Chen, R. Chun, S. Ren, S. Wu, M. Gacad, L. Nguyen, J. Ride, P. Liu, R. Modlin, and M. Hewison. 2007. Substrate and enzyme trafficking as a means of regulating 1,25-dihydroxyvitamin D synthesis and action: The human innate immune response. *Journal of Bone and Mineral Research* 22 Suppl 2:V20–24.
8. Spach, K. M., and C. E. Hayes. 2005. Vitamin D3 confers protection from autoimmune encephalomyelitis only in female mice. *Journal of Immunology* 175:4119–4126.
9. Hewison, M., F. Burke, K. N. Evans, D. A. Lammas, D. M. Sansom, P. Liu, R. L. Modlin, and J. S. Adams. 2007. Extra-renal 25-hydroxyvitamin D3-1alpha-hydroxylase in human health and disease. *The Journal of Steroid Biochemistry and Molecular Biology* 103:316–321.
10. Chun, R. F., A. L. Lauridsen, L. Suon, L. A. Zella, J. W. Pike, R. L. Modlin, A. R. Martineau, R. J. Wilkinson, J. Adams, and M. Hewison. 2010. Vitamin D–binding protein directs monocyte responses to 25-hydroxy- and 1,25-dihydroxyvitamin D. *The Journal of Clinical Endocrinology and Metabolism* 95:3368–3376.

11. Ahn, J., K. Yu, R. Stolzenberg-Solomon, K. C. Simon, M. L. McCullough, L. Gallicchio, E. J. Jacobs, A. Ascherio, K. Helzlsouer, K. B. Jacobs, Q. Li, S. J. Weinstein, M. Purdue, J. Virtamo, R. Horst, W. Wheeler, S. Chanock, D. J. Hunter, R. B. Hayes, P. Kraft, and D. Albanes. 2010. Genome-wide association study of circulating vitamin D levels. *Human Molecular Genetics* 19:2739–2745.

12. Adams, J. S., and M. Hewison. 2010. Update in vitamin D. *The Journal of Clinical Endocrinology and Metabolism* 95:471–478.

13. Holick, M. F. 2008. The vitamin D deficiency pandemic and consequences for nonskeletal health: Mechanisms of action. *Molecular Aspects of Medicine* 29:361–368.

14. Hayes, C. E., F. E. Nashold, C. G. Mayne, J. A. Spanier, and C. D. Nelson. 2011. Vitamin D and multiple sclerosis. In *Vitamin D*, edited by D. Feldman, J. W. Pike, and J. S. Adams, 3rd ed., 1843–1877. Elsevier, San Diego, CA.

15. Lips, P., M. C. Chapuy, B. Dawson-Hughes, H. A. Pols, and M. F. Holick. 1999. An international comparison of serum 25-hydroxyvitamin D measurements. *Osteoporosis International* 9:394–397.

16. Hollis, B. W. 2008. Measuring 25-hydroxyvitamin D in a clinical environment: Challenges and needs. *The American Journal of Clinical Nutrition* 88:507S–510S.

17. Holick, M. F. 2003. Vitamin D: A millennium perspective. *Journal of Cellular Biochemistry* 88:296–307.

18. Clemens, T. L., J. S. Adams, S. L. Henderson, and M. F. Holick. 1982. Increased skin pigment reduces the capacity of skin to synthesise vitamin D3. *Lancet* 1:74–76.

19. Bell, N. H., A. Greene, S. Epstein, M. J. Oexmann, S. Shaw, and J. Shary. 1985. Evidence for alteration of the vitamin D–endocrine system in blacks. *Journal of Clinical Investigation* 76:470–473.

20. Bell, N. H. 1997. Bone and mineral metabolism in African Americans. *Trends in Endocrinology and Metabolism* 8:240–245.

21. Chen, T. C., F. Chimeh, Z. Lu, J. Mathieu, K. S. Person, A. Zhang, N. Kohn, S. Martinello, R. Berkowitz, and M. F. Holick. 2007. Factors that influence the cutaneous synthesis and dietary sources of vitamin D. *Archives of Biochemistry and Biophysics* 460:213–217.

22. Orton, S. M., A. P. Morris, B. M. Herrera, S. V. Ramagopalan, M. R. Lincoln, M. J. Chao, R. Vieth, A. D. Sadovnick, and G. C. Ebers. 2008. Evidence for genetic regulation of vitamin D status in twins with multiple sclerosis. *The American Journal of Clinical Nutrition* 88:441–447.

23. Fu, L., F. Yun, M. Oczak, B. Y. Wong, R. Vieth, and D. E. Cole. 2009. Common genetic variants of the vitamin D binding protein (DBP) predict differences in response of serum 25-hydroxyvitamin D [25(OH)D] to vitamin D supplementation. *Clinical Biochemistry* 42:1174–1177.

24. Pike, J. W., and M. B. Meyer. 2010. The vitamin D receptor: New paradigms for the regulation of gene expression by 1,25-dihydroxyvitamin D(3). *Endocrinology and Metabolism Clinics of North America* 39:255–269.

25. Toell, A., P. Polly, and C. Carlberg. 2000. All natural DR3-type vitamin D response elements show a similar functionality *in vitro*. *The Biochemical Journal* 352 (Pt 2):301–309.

26. Brennan, A., D. R. Katz, J. D. Nunn, S. Barker, M. Hewison, L. J. Fraher, and J. L. O'Riordan. 1987. Dendritic cells from human tissues express receptors for the immunoregulatory vitamin D3 metabolite, dihydroxycholecalciferol. *Immunology* 61:457–461.

27. Veldman, C. M., M. T. Cantorna, and H. F. DeLuca. 2000. Expression of 1,25-dihydroxyvitamin D(3) receptor in the immune system. *Archives of Biochemistry and Biophysics* 374:334–338.

28. Nashold, F. E., K. A. Hoag, J. Goverman, and C. E. Hayes. 2001. Rag-1–dependent cells are necessary for 1,25-dihydroxyvitamin D(3) prevention of experimental autoimmune encephalomyelitis. *Journal of Neuroimmunology* 119:16–29.

29. Chang, S. H., Y. Chung, and C. Dong. 2010. Vitamin D suppresses Th17 cytokine production by inducing C/EBP homologous protein (CHOP) expression. *The Journal of Biological Chemistry* 285:38751–38755.

30. Mayne, C. G., J. A. Spanier, L. M. Relland, C. B. Williams, and C. E. Hayes. 2011. 1,25-Dihydroxyvitamin D3 acts directly on the T lymphocyte vitamin D receptor to inhibit experimental autoimmune encephalomyelitis. *European Journal of Immunology* 41:822–832.

31. Hauser, S. L., and J. R. Oksenberg. 2006. The neurobiology of multiple sclerosis: Genes, inflammation, and neurodegeneration. *Neuron* 52:61–76.

32. Dutta, R., and B. D. Trapp. 2011. Mechanisms of neuronal dysfunction and degeneration in multiple sclerosis. *Progress in Neurobiology* 93:1–12.

33. Scalfari, A., A. Neuhaus, A. Degenhardt, G. P. Rice, P. A. Muraro, M. Daumer, and G. C. Ebers. 2010. The natural history of multiple sclerosis—A geographically based study 10: Relapses and long-term disability. *Brain* 133:1914–1929.

34. Koch, M., E. Kingwell, P. Rieckmann, and H. Tremlett. 2010. The natural history of secondary progressive multiple sclerosis. *Journal of Neurology, Neurosurgery, and Psychiatry* 81:1039–1043.

35. Ramagopalan, S. V., J. K. Byrnes, S. M. Orton, D. A. Dyment, C. Guimond, I. M. Yee, G. C. Ebers, and A. D. Sadovnick. 2010. Sex ratio of multiple sclerosis and clinical phenotype. *European Journal of Neurology* 17:634–637.
36. Ramagopalan, S. V., and G. C. Ebers. 2009. Multiple sclerosis: Major histocompatibility complexity and antigen presentation. *Genome Medicine* 1:105.
37. Watson, C. T., A. E. Para, M. R. Lincoln, S. V. Ramagopalan, S. M. Orton, K. M. Morrison, L. Handunnetthi, A. E. Handel, M. J. Chao, J. Morahan, A. D. Sadovnick, F. Breden, and G. C. Ebers. 2011. Revisiting the T-cell receptor alpha/delta locus and possible associations with multiple sclerosis. *Genes and Immunity* 12:59–66.
38. Steinman, L., and S. S. Zamvil. 2006. How to successfully apply animal studies in experimental allergic encephalomyelitis to research on multiple sclerosis. *Annals of Neurology* 60:12–21.
39. Baxter, A. G. 2007. The origin and application of experimental autoimmune encephalomyelitis. *Nature Reviews. Immunology* 7:904–912.
40. Ramagopalan, S. V., R. Dobson, U. C. Meier, and G. Giovannoni. 2010. Multiple sclerosis: Risk factors, prodromes, and potential causal pathways. *Lancet Neurology* 9:727–739.
41. Goverman, J. 2009. Autoimmune T cell responses in the central nervous system. *Nature Reviews. Immunology* 9:393–407.
42. Oksenberg, J. R., and S. E. Baranzini. 2010. Multiple sclerosis genetics—is the glass half full, or half empty? *Nature Reviews. Neurology* 6:429–437.
43. Lincoln, M. R., S. V. Ramagopalan, M. J. Chao, B. M. Herrera, G. C. Deluca, S. M. Orton, D. A. Dyment, A. D. Sadovnick, and G. C. Ebers. 2009. Epistasis among HLA-DRB1, HLA-DQA1, and HLA-DQB1 loci determines multiple sclerosis susceptibility. *Proceedings of the National Academy of Sciences of the United States of America* 106:7542–7547.
44. De Jager, P. L., L. B. Chibnik, J. Cui, J. Reischl, S. Lehr, K. C. Simon, C. Aubin, D. Bauer, J. F. Heubach, R. Sandbrink, M. Tyblova, P. Lelkova, E. Havrdova, C. Pohl, D. Horakova, A. Ascherio, D. A. Hafler, and E. W. Karlson. 2009. Integration of genetic risk factors into a clinical algorithm for multiple sclerosis susceptibility: A weighted genetic risk score. *Lancet Neurology* 8:1111–1119.
45. Sawcer, S., G. Hellenthal, M. Pirinen, C. C. Spencer, N. A. Patsopoulos, L. Moutsianas, A. Dilthey, Z. Su, C. Freeman, S. E. Hunt, S. Edkins, E. Gray, D. R. Booth, S. C. Potter, A. Goris, G. Band, A. B. Oturai, A. Strange, J. Saarela, C. Bellenguez, B. Fontaine, M. Gillman, B. Hemmer, R. Gwilliam, F. Zipp, A. Jayakumar, R. Martin, S. Leslie, S. Hawkins, E. Giannoulatou, S. D'Alfonso, H. Blackburn, F. M. Boneschi, J. Liddle, H. F. Harbo, M. L. Perez, A. Spurkland, M. J. Waller, M. P. Mycko, M. Ricketts, M. Comabella, N. Hammond, I. Kockum, O. T. McCann, M. Ban, P. Whittaker, A. Kemppinen, P. Weston, C. Hawkins, S. Widaa, J. Zajicek, S. Dronov, N. Robertson, S. J. Bumpstead, L. F. Barcellos, R. Ravindrarajah, R. Abraham, L. Alfredsson, K. Ardlie, C. Aubin, A. Baker, K. Baker, S. E. Baranzini, L. Bergamaschi, R. Bergamaschi, A. Bernstein, A. Berthele, M. Boggild, J. P. Bradfield, D. Brassat, S. A. Broadley, D. Buck, H. Butzkueven, R. Capra, W. M. Carroll, P. Cavalla, E. G. Celius, S. Cepok, R. Chiavacci, F. Clerget-Darpoux, K. Clysters, G. Comi, M. Cossburn, I. Cournu-Rebeix, M. B. Cox, W. Cozen, B. A. Cree, A. H. Cross, D. Cusi, M. J. Daly, E. Davis, P. I. de Bakker, M. Debouverie, M. B. D'Hooghe, K. Dixon, R. Dobosi, B. Dubois, D. Ellinghaus, I. Elovaara, F. Esposito, C. Fontenille, S. Foote, A. Franke, D. Galimberti, A. Ghezzi, J. Glessner, R. Gomez, O. Gout, C. Graham, S. F. Grant, F. R. Guerini, H. Hakonarson, P. Hall, A. Hamsten, H. P. Hartung, R. N. Heard, S. Heath, J. Hobart, M. Hoshi, C. Infante-Duarte, G. Ingram, W. Ingram, T. Islam, M. Jagodic, M. Kabesch, A. G. Kermode, T. J. Kilpatrick, C. Kim, N. Klopp, K. Koivisto, M. Larsson, M. Lathrop, J. S. Lechner-Scott, M. A. Leone, V. Leppa, U. Liljedahl, I. L. Bomfim, R. R. Lincoln, J. Link, J. Liu, A. R. Lorentzen, S. Lupoli, F. Macciardi, T. Mack, M. Marriott, V. Martinelli, D. Mason, J. L. McCauley, F. Mentch, I. L. Mero, T. Mihalova, X. Montalban, J. Mottershead, K. M. Myhr, P. Naldi, W. Ollier, A. Page, A. Palotie, J. Pelletier, L. Piccio, T. Pickersgill, F. Piehl, S. Pobywajlo, H. L. Quach, P. P. Ramsay, M. Reunanen, R. Reynolds, J. D. Rioux, M. Rodegher, S. Roesner, J. P. Rubio, I. M. Ruckert, M. Salvetti, E. Salvi, A. Santaniello, C. A. Schaefer, S. Schreiber, C. Schulze, R. J. Scott, F. Sellebjerg, K. W. Selmaj, D. Sexton, L. Shen, B. Simms-Acuna, S. Skidmore, P. M. Sleiman, C. Smestad, P. S. Sorensen, H. B. Sondergaard, J. Stankovich, R. C. Strange, A. M. Sulonen, E. Sundqvist, A. C. Syvanen, F. Taddeo, B. Taylor, J. M. Blackwell, P. Tienari, E. Bramon, A. Tourbah, M. A. Brown, E. Tronczynska, J. P. Casas, N. Tubridy, A. Corvin, J. Vickery, J. Jankowski, P. Villoslada, H. S. Markus, K. Wang, C. G. Mathew, J. Wason, C. N. Palmer, H. E. Wichmann, R. Plomin, E. Willoughby, A. Rautanen, J. Winkelmann, M. Wittig, R. C. Trembath, J. Yaouanq, A. C. Viswanathan, H. Zhang, N. W. Wood, R. Zuvich, P. Deloukas, C. Langford, A. Duncanson, J. R. Oksenberg, M. A. Pericak-Vance, J. L. Haines, T. Olsson, J. Hillert, A. J. Ivinson, P. L. De Jager, L. Peltonen, G. J. Stewart, D. A. Hafler, S. L. Hauser, G. McVean, P. Donnelly, and A. Compston. 2011. Genetic risk and a primary role for cell-mediated immune mechanisms in multiple sclerosis. *Nature* 476:214–219.

46. Baranzini, S. E., J. Mudge, J. C. van Velkinburgh, P. Khankhanian, I. Khrebtukova, N. A. Miller, L. Zhang, A. D. Farmer, C. J. Bell, R. W. Kim, G. D. May, J. E. Woodward, S. J. Caillier, J. P. McElroy, R. Gomez, M. J. Pando, L. E. Clendenen, E. E. Ganusova, F. D. Schilkey, T. Ramaraj, O. A. Khan, J. J. Huntley, S. Luo, P. Y. Kwok, T. D. Wu, G. P. Schroth, J. R. Oksenberg, S. L. Hauser, and S. F. Kingsmore. 2010. Genome, epigenome and RNA sequences of monozygotic twins discordant for multiple sclerosis. *Nature* 464:1351–1356.
47. Burrell, A. M., A. E. Handel, S. V. Ramagopalan, G. C. Ebers, and J. M. Morahan. 2011. Epigenetic mechanisms in multiple sclerosis and the major histocompatibility complex (MHC). *Discovery Medicine* 11:187–196.
48. Ebers, G. C. 2008. Environmental factors and multiple sclerosis. *Lancet Neurology* 7:268–277.
49. Lauer, K. 2010. Environmental risk factors in multiple sclerosis. *Expert Review of Neurotherapeutics* 10:421–440.
50. Ebers, G. C., D. E. Bulman, A. D. Sadovnick, D. W. Paty, S. Warren, W. Hader, T. J. Murray, T. P. Seland, P. Duquette, T. Grey, et al. 1986. A population-based study of multiple sclerosis in twins. *New England Journal of Medicine* 315:1638–1642.
51. Simpson, S., Jr., L. Blizzard, P. Otahal, I. Van der Mei, and B. Taylor. 2011. Latitude is significantly associated with the prevalence of multiple sclerosis: A meta-analysis. *Journal of Neurology, Neurosurgery, and Psychiatry* 82:1132–1141.
52. McLeod, J. G., S. R. Hammond, and J. F. Kurtzke. 2011. Migration and multiple sclerosis in immigrants to Australia from United Kingdom and Ireland: A reassessment—Part I: Risk of MS by age at immigration. *Journal of Neurology* 258:1140–1149.
53. Orton, S. M., S. V. Ramagopalan, D. Brocklebank, B. M. Herrera, D. A. Dyment, I. M. Yee, A. D. Sadovnick, and G. C. Ebers. 2010. Effect of immigration on multiple sclerosis sex ratio in Canada: The Canadian Collaborative Study. *Journal of Neurology, Neurosurgery, and Psychiatry* 81:31–36.
54. Goldberg, P. 1974. Multiple sclerosis: Vitamin D and calcium as environmental determinants of prevalence (a viewpoint)—Part 1: Sunlight, dietary factors and epidemiology. *International Journal of Environmental Studies* 6:19–27.
55. Craelius, W. 1978. Comparative epidemiology of multiple sclerosis and dental caries. *Journal of Epidemiology and Community Health* 32:155–165.
56. Schwartz, G. G. 1992. Multiple sclerosis and prostate cancer: What do their similar geographies suggest? *Neuroepidemiology* 11:244–254.
57. Hayes, C. E., M. T. Cantorna, and H. F. DeLuca. 1997. Vitamin D and multiple sclerosis. *Proceedings of the Society for Experimental Biology and Medicine* 216:21–27.
58. Cantorna, M. T., C. E. Hayes, and H. F. DeLuca. 1996. 1,25-Dihydroxyvitamin D3 reversibly blocks the progression of relapsing encephalomyelitis, a model of multiple sclerosis. *Proceedings of the National Academy of Sciences of the United States of America* 93:7861–7864.
59. Smolders, J., J. Damoiseaux, P. Menheere, and R. Hupperts. 2008. Vitamin D as an immune modulator in multiple sclerosis, a review. *Journal of Neuroimmunology* 194:7–17.
60. Ascherio, A., K. L. Munger, and K. C. Simon. 2010. Vitamin D and multiple sclerosis. *Lancet Neurology* 9:599–612.
61. Hanwell, H. E., and B. Banwell. 2011. Assessment of evidence for a protective role of vitamin D in multiple sclerosis. *Biochimica et Biophysica Acta* 1812:202–212.
62. Correale, J., M. C. Ysrraelit, and M. I. Gaitan. 2011. Vitamin D–mediated immune regulation in multiple sclerosis. *Journal of the Neurological Sciences* 311:23–31.
63. Acheson, E. D., C. A. Bachrach, and F. M. Wright. 1960. Some comments on the relationship of the distribution of multiple sclerosis to latitude, solar radiation and other variables. *Acta Psychiatrica Scandinavica* 35 (Supplement 147):132–147.
64. Ebers, G. C., and A. D. Sadovnick. 1993. The geographic distribution of multiple sclerosis: A review. *Neuroepidemiology* 12:1–5.
65. Lonergan, R., K. Kinsella, P. Fitzpatrick, J. Brady, B. Murray, C. Dunne, R. Hagan, M. Duggan, S. Jordan, M. McKenna, M. Hutchinson, and N. Tubridy. 2011. Multiple sclerosis prevalence in Ireland: Relationship to vitamin D status and HLA genotype. *Journal of Neurology, Neurosurgery, and Psychiatry* 82:317–322.
66. Fernandes de Abreu, D. A., M. C. Babron, I. Rebeix, C. Fontenille, J. Yaouanq, D. Brassat, B. Fontaine, F. Clerget-Darpoux, F. Jehan, and F. Feron. 2009. Season of birth and not vitamin D receptor promoter polymorphisms is a risk factor for multiple sclerosis. *Multiple Sclerosis (Houndmills, Basingstoke, England)* 15:1146–1152.
67. Salzer, J., A. Svenningsson, and P. Sundstrom. 2010. Season of birth and multiple sclerosis in Sweden. *Acta Neurologica Scandinavica* 121:20–23.

68. Bayes, H. K., C. J. Weir, and C. O'Leary. 2010. Timing of birth and risk of multiple sclerosis in the Scottish population. *European Neurology* 63:36–40.
69. Willer, C. J., D. A. Dyment, A. D. Sadovnick, P. M. Rothwell, T. J. Murray, and G. C. Ebers. 2005. Timing of birth and risk of multiple sclerosis: Population-based study. *British Medical Journal* 330:120.
70. Sotgiu, S., M. Pugliatti, M. A. Sotgiu, M. L. Fois, G. Arru, A. Sanna, and G. Rosati. 2006. Seasonal fluctuation of multiple sclerosis births in Sardinia. *Journal of Neurology* 253:38–44.
71. Saastamoinen, K. P., M. K. Auvinen, and P. J. Tienari. 2012. Month of birth is associated with multiple sclerosis but not with HLA-DR15 in Finland. *Multiple Sclerosis (Houndmills, Basingstoke, England)* 18:563–568.
72. Staples, J., A. L. Ponsonby, and L. Lim. 2010. Low maternal exposure to ultraviolet radiation in pregnancy, month of birth, and risk of multiple sclerosis in offspring: Longitudinal analysis. *British Medical Journal* 340:c1640.
73. McDowell, T. Y., S. Amr, P. Langenberg, W. Royal, C. Bever, W. J. Culpepper, and D. D. Bradham. 2010. Time of birth, residential solar radiation and age at onset of multiple sclerosis. *Neuroepidemiology* 34:238–244.
74. Givon, U., G. Zeilig, M. Dolev, and A. Achiron. 2012. The month of birth and the incidence of multiple sclerosis in the israeli population. *Neuroepidemiology* 38:64–68.
75. Ebers, G. 2010. Month of birth and multiple sclerosis risk in Scotland. *European Neurology* 63:41–42.
76. Meier, D. S., K. E. Balashov, B. Healy, H. L. Weiner, and C. R. Guttmann. 2010. Seasonal prevalence of MS disease activity. *Neurology* 75:799–806.
77. Handel, A. E., G. Disanto, L. Jarvis, R. McLaughlin, A. Fries, G. C. Ebers, and S. V. Ramagopalan. 2011. Seasonality of admissions with multiple sclerosis in Scotland. *European Journal of Neurology: The Official Journal of the European Federation of Neurological Societies* 18:1109–1111.
78. Saaroni, H., A. Sigal, I. Lejbkowicz, and A. Miller. 2010. Mediterranean weather conditions and exacerbations of multiple sclerosis. *Neuroepidemiology* 35:142–151.
79. Damasceno, A., F. Von Glehn, L. de Deus-Silva, and B. P. Damasceno. 2012. Monthly variation of multiple sclerosis activity in the southern hemisphere: Analysis from 996 relapses in Brazil. *European Journal of Neurology: The Official Journal of the European Federation of Neurological Societies* 19:660–662.
80. Ascherio, A., and K. L. Munger. 2007. Environmental risk factors for multiple sclerosis—Part II: Noninfectious factors. *Annals of Neurology* 61:504–513.
81. Lucas, R. M., A. L. Ponsonby, K. Dear, P. C. Valery, M. P. Pender, B. V. Taylor, T. J. Kilpatrick, T. Dwyer, A. Coulthard, C. Chapman, I. van der Mei, D. Williams, and A. J. McMichael. 2011. Sun exposure and vitamin D are independent risk factors for CNS demyelination. *Neurology* 76:540–548.
82. van der Mei, I. A., A. L. Ponsonby, L. Blizzard, and T. Dwyer. 2001. Regional variation in multiple sclerosis prevalence in Australia and its association with ambient ultraviolet radiation. *Neuroepidemiology* 20:168–174.
83. van der Mei, I. A., A. L. Ponsonby, T. Dwyer, L. Blizzard, R. Simmons, B. V. Taylor, H. Butzkueven, and T. Kilpatrick. 2003. Past exposure to sun, skin phenotype, and risk of multiple sclerosis: Case-control study. *British Medical Journal* 327:316.
84. Dwyer, T., I. van der Mei, A. L. Ponsonby, B. V. Taylor, J. Stankovich, J. D. McKay, R. J. Thomson, A. M. Polanowski, and J. L. Dickinson. 2008. Melanocortin 1 receptor genotype, past environmental sun exposure, and risk of multiple sclerosis. *Neurology* 71:583–589.
85. McDowell, T. Y., S. Amr, W. J. Culpepper, P. Langenberg, W. Royal, C. Bever, and D. D. Bradham. 2011. Sun exposure, vitamin D and age at disease onset in relapsing multiple sclerosis. *Neuroepidemiology* 36:39–45.
86. Islam, T., W. J. Gauderman, W. Cozen, and T. M. Mack. 2007. Childhood sun exposure influences risk of multiple sclerosis in monozygotic twins. *Neurology* 69:381–388.
87. Kampman, M. T., T. Wilsgaard, and S. I. Mellgren. 2007. Outdoor activities and diet in childhood and adolescence relate to MS risk above the Arctic Circle. *Journal of Neurology* 254:471–477.
88. Dickinson, J., D. Perera, A. van der Mei, A. L. Ponsonby, A. Polanowski, R. Thomson, B. Taylor, J. McKay, J. Stankovich, and T. Dwyer. 2009. Past environmental sun exposure and risk of multiple sclerosis: A role for the Cdx-2 Vitamin D receptor variant in this interaction. *Multiple Sclerosis (Houndmills, Basingstoke, England)* 15:563–570.
89. Orton, S. M., L. Wald, C. Confavreux, S. Vukusic, J. P. Krohn, S. V. Ramagopalan, B. M. Herrera, A. D. Sadovnick, and G. C. Ebers. 2011. Association of UV radiation with multiple sclerosis prevalence and sex ratio in France. *Neurology* 76:425–431.
90. Hine, T. J., and N. B. Roberts. 1994. Seasonal variation in serum 25-hydroxy vitamin D3 does not affect 1,25-dihydroxy vitamin D. *Annals of Clinical Biochemistry* 31 (Pt 1):31–34.

91. Archard, H. O. 1971. The dental defects of vitamin D–resistant rickets. *Birth Defects Original Article Series* 7:196–199.

92. Purvis, R. J., W. J. Barrie, G. S. MacKay, E. M. Wilkinson, F. Cockburn, and N. R. Belton. 1973. Enamel hypoplasia of the teeth associated with neonatal tetany: A manifestation of maternal vitamin-D deficiency. *Lancet* 2:811–814.

93. Berdal, A., N. Balmain, P. Cuisinier-Gleizes, and H. Mathieu. 1987. Histology and microradiography of early post-natal molar tooth development in vitamin-D deficient rats. *Archives of Oral Biology* 32:493–498.

94. Nishino, M., K. Kamada, K. Arita, and T. Takarada. 1990. [Dentofacial manifestations in children with vitamin D–dependent Rickets type II]. *Shoni Shikagaku Zasshi* 28:346–358.

95. Cosman, F., J. Nieves, J. Herbert, V. Shen, and R. Lindsay. 1994. High-dose glucocorticoids in multiple sclerosis patients exert direct effects on the kidney and skeleton. *Journal of Bone and Mineral Research* 9:1097–1105.

96. Nieves, J., F. Cosman, J. Herbert, V. Shen, and R. Lindsay. 1994. High prevalence of vitamin D deficiency and reduced bone mass in multiple sclerosis. *Neurology* 44:1687–1692.

97. Cosman, F., J. Nieves, L. Komar, G. Ferrer, J. Herbert, C. Formica, V. Shen, and R. Lindsay. 1998. Fracture history and bone loss in patients with MS. *Neurology* 51:1161–1165.

98. Ascherio, A., and K. Munger. 2008. Epidemiology of multiple sclerosis: From risk factors to prevention. *Seminars in Neurology* 28:17–28.

99. Smolders, J., P. Menheere, A. Kessels, J. Damoiseaux, and R. Hupperts. 2008. Association of vitamin D metabolite levels with relapse rate and disability in multiple sclerosis. *Multiple Sclerosis (Houndmills, Basingstoke, England)* 14:1220–1224.

100. Correale, J., M. C. Ysrraelit, and M. I. Gaitan. 2009. Immunomodulatory effects of vitamin D in multiple sclerosis. *Brain* 132:1146–1160.

101. Holmoy, T., S. M. Moen, T. A. Gundersen, M. F. Holick, E. Fainardi, M. Castellazzi, and I. Casetta. 2009. 25-hydroxyvitamin D in cerebrospinal fluid during relapse and remission of multiple sclerosis. *Multiple Sclerosis (Houndmills, Basingstoke, England)* 15:1280–1285.

102. van der Mei, I. A., A. L. Ponsonby, T. Dwyer, L. Blizzard, B. V. Taylor, T. Kilpatrick, H. Butzkueven, and A. J. McMichael. 2007. Vitamin D levels in people with multiple sclerosis and community controls in Tasmania, Australia. *Journal of Neurology* 254:581–590.

103. Terzi, T., M. Terzi, B. Tander, F. Canturk, and M. Onar. 2010. Changes in bone mineral density and bone metabolism markers in premenopausal women with multiple sclerosis and the relationship to clinical variables. *Journal of Clinical Neuroscience* 17:1260–1264.

104. Neau, J. P., M. S. Artaud-Uriot, V. Lhomme, J. Y. Bounaud, F. Lebras, L. Boissonnot, N. Moinot, J. Ciron, D. Larrieu, S. Mathis, G. Godeneche, and P. Ingrand. 2011. [Vitamin D and multiple sclerosis. A prospective survey of patients of Poitou–Charentes area]. *Revue Neurologique* 167:317–323.

105. Munger, K. L., S. M. Zhang, E. O'Reilly, M. A. Hernan, M. J. Olek, W. C. Willett, and A. Ascherio. 2004. Vitamin D intake and incidence of multiple sclerosis. *Neurology* 62:60–65.

106. Munger, K. L., L. I. Levin, B. W. Hollis, N. S. Howard, and A. Ascherio. 2006. Serum 25-hydroxyvitamin D levels and risk of multiple sclerosis. *Journal of the American Medical Association* 296:2832–2838.

107. Mirzaei, F., K. B. Michels, K. Munger, E. O'Reilly, T. Chitnis, M. R. Forman, E. Giovannucci, B. Rosner, and A. Ascherio. 2011. Gestational vitamin D and the risk of multiple sclerosis in offspring. *Annals of Neurology* 70:30–40.

108. Soilu-Hanninen, M., L. Airas, I. Mononen, A. Heikkila, M. Viljanen, and A. Hanninen. 2005. 25-Hydroxyvitamin D levels in serum at the onset of multiple sclerosis. *Multiple Sclerosis (Houndmills, Basingstoke, England)* 11:266–271.

109. Soilu-Hanninen, M., M. Laaksonen, I. Laitinen, J. P. Eralinna, E. M. Lilius, and I. Mononen. 2008. A longitudinal study of serum 25-hydroxyvitamin D and intact parathyroid hormone levels indicate the importance of vitamin D and calcium homeostasis regulation in multiple sclerosis. *Journal of Neurology, Neurosurgery, and Psychiatry* 79:152–157.

110. Mowry, E. M., L. B. Krupp, M. Milazzo, D. Chabas, J. B. Strober, A. L. Belman, J. C. McDonald, J. R. Oksenberg, P. Bacchetti, and E. Waubant. 2010. Vitamin D status is associated with relapse rate in pediatric-onset multiple sclerosis. *Annals of Neurology* 67:618–624.

111. Simpson, S., B. Taylor, L. Blizzard, A.-L. Ponsonby, F. Pittas, H. Tremlett, T. Dwyer, P. Gies, and I. A. van der Mei. 2010. Higher 25-hydroxyvitamin D is associated with lower relapse risk in multiple sclerosis. *Annals of Neurology* 68:193–203.

112. Mealy, M. A., S. Newsome, B. M. Greenberg, D. Wingerchuk, P. Calabresi, and M. Levy. 2012. Low serum vitamin D levels and recurrent inflammatory spinal cord disease. *Archives of Neurology* 69:352–356.

113. Webb, A. R., L. Kline, and M. F. Holick. 1988. Influence of season and latitude on the cutaneous synthesis of vitamin D3: Exposure to winter sunlight in Boston and Edmonton will not promote vitamin D3 synthesis in human skin. *The Journal of Clinical Endocrinology and Metabolism* 67: 373–378.

114. Hill, A. B. 1965. The environment and disease: Association or causation? *Proceedings of the Royal Society of Medicine* 58:295–300.

115. Mowry, E. M. 2011. Vitamin D: Evidence for its role as a prognostic factor in multiple sclerosis. *Journal of the Neurological Sciences* 311:19–22.

116. Gourraud, P. A., J. P. McElroy, S. J. Caillier, B. A. Johnson, A. Santaniello, S. L. Hauser, and J. R. Oksenberg. 2011. Aggregation of multiple sclerosis genetic risk variants in multiple and single case families. *Annals of Neurology* 69:65–74.

117. Berlanga-Taylor, A. J., G. Disanto, G. C. Ebers, and S. V. Ramagopalan. 2011. Vitamin D–gene interactions in multiple sclerosis. *Journal of the Neurological Sciences* 311:32–36.

118. Jersild, C., T. Fog, G. S. Hansen, M. Thomsen, A. Svejgaard, and B. Dupont. 1973. Histocompatibility determinants in multiple sclerosis, with special reference to clinical course. *Lancet* 2:1221–1225.

119. Okuda, D. T., R. Srinivasan, J. R. Oksenberg, D. S. Goodin, S. E. Baranzini, A. Beheshtian, E. Waubant, S. S. Zamvil, D. Leppert, P. Qualley, R. Lincoln, R. Gomez, S. Caillier, M. George, J. Wang, S. J. Nelson, B. A. Cree, S. L. Hauser, and D. Pelletier. 2009. Genotype–phenotype correlations in multiple sclerosis: HLA genes influence disease severity inferred by 1HMR spectroscopy and MRI measures. *Brain* 132:250–259.

120. Ramagopalan, S. V., N. J. Maugeri, L. Handunnetthi, M. R. Lincoln, S. M. Orton, D. A. Dyment, G. C. Deluca, B. M. Herrera, M. J. Chao, A. D. Sadovnick, G. C. Ebers, and J. C. Knight. 2009. Expression of the multiple sclerosis–associated MHC class II allele HLA-DRB1*1501 is regulated by vitamin D. *PLoS Genetics* 5:e1000369.

121. Smolders, J., E. Peelen, M. Thewissen, P. Menheere, J. W. Cohen Tervaert, R. Hupperts, and J. Damoiseaux. 2009. The relevance of vitamin D receptor gene polymorphisms for vitamin D research in multiple sclerosis. *Autoimmunity Reviews* 8:621–626.

122. Huang, J., and Z. F. Xie. 2012. Polymorphisms in the vitamin D receptor gene and multiple sclerosis risk: A meta-analysis of case-control studies. *Journal of the Neurological Sciences* 313:79–85.

123. Orton, S. M., S. V. Ramagopalan, A. E. Para, M. R. Lincoln, L. Handunnetthi, M. J. Chao, J. Morahan, K. M. Morrison, A. D. Sadovnick, and G. C. Ebers. 2011. Vitamin D metabolic pathway genes and risk of multiple sclerosis in Canadians. *Journal of the Neurological Sciences* 305:116–120.

124. Gross, C., A. V. Krishnan, P. J. Malloy, T. R. Eccleshall, X. Y. Zhao, and D. Feldman. 1998. The vitamin D receptor gene start codon polymorphism: A functional analysis of FokI variants. *Journal of Bone and Mineral Research* 13:1691–1699.

125. Jurutka, P. W., L. S. Remus, G. K. Whitfield, P. D. Thompson, J. C. Hsieh, H. Zitzer, P. Tavakkoli, M. A. Galligan, H. T. Dang, C. A. Haussler, and M. R. Haussler. 2000. The polymorphic N terminus in human vitamin D receptor isoforms influences transcriptional activity by modulating interaction with transcription factor IIB. *Molecular Endocrinology* 14:401–420.

126. Smolders, J., J. Damoiseaux, P. Menheere, J. W. Tervaert, and R. Hupperts. 2009. Fok-I vitamin D receptor gene polymorphism (rs10735810) and vitamin D metabolism in multiple sclerosis. *Journal of Neuroimmunology* 207:117–121.

127. Partridge, J. M., S. J. Weatherby, J. A. Woolmore, D. J. Highland, A. A. Fryer, C. L. Mann, M. D. Boggild, W. E. Ollier, R. C. Strange, and C. P. Hawkins. 2004. Susceptibility and outcome in MS: Associations with polymorphisms in pigmentation-related genes. *Neurology* 62:2323–2325.

128. Mamutse, G., J. Woolmore, E. Pye, J. Partridge, M. Boggild, C. Young, A. Fryer, P. R. Hoban, N. Rukin, J. Alldersea, R. C. Strange, and C. P. Hawkins. 2008. Vitamin D receptor gene polymorphism is associated with reduced disability in multiple sclerosis. *Multiple Sclerosis (Houndmills, Basingstoke, England)* 14:1280–1283.

129. Simon, K. C., K. L. Munger, Y. Xing, and A. Ascherio. 2010. Polymorphisms in vitamin D metabolism related genes and risk of multiple sclerosis. *Multiple Sclerosis (Houndmills, Basingstoke, England)* 16:133–138.

130. Arai, H., K. I. Miyamoto, M. Yoshida, H. Yamamoto, Y. Taketani, K. Morita, M. Kubota, S. Yoshida, M. Ikeda, F. Watabe, Y. Kanemasa, and E. Takeda. 2001. The polymorphism in the caudal-related homeodomain protein Cdx-2 binding element in the human vitamin D receptor gene. *Journal of Bone and Mineral Research* 16:1256–1264.

131. Sakaki, T., N. Kagawa, K. Yamamoto, and K. Inouye. 2005. Metabolism of vitamin D3 by cytochromes P450. *Frontiers in Bioscience* 10:119–134.

132. Yamamoto, K., E. Uchida, N. Urushino, T. Sakaki, N. Kagawa, N. Sawada, M. Kamakura, S. Kato, K. Inouye, and S. Yamada. 2005. Identification of the amino acid residue of CYP27B1 responsible for binding of 25-hydroxyvitamin D3 whose mutation causes vitamin D–dependent rickets type 1. *Journal of Biological Chemistry* 280:30511–30516.

133. Torkildsen, O., P. M. Knappskog, H. I. Nyland, and K. M. Myhr. 2008. Vitamin D–dependent rickets as a possible risk factor for multiple sclerosis. *Archives of Neurology* 65:809–811.

134. Ramagopalan, S. V., H. E. Hanwell, G. Giovannoni, P. M. Knappskog, H. I. Nyland, K. M. Myhr, G. C. Ebers, and O. Torkildsen. 2010. Vitamin D–dependent rickets, HLA-DRB1, and the risk of multiple sclerosis. *Archives of Neurology* 67:1034–1035.

135. Steckley, J. L., D. A. Dyment, A. D. Sadovnick, N. Risch, C. Hayes, and G. C. Ebers. 2000. Genetic analysis of vitamin D related genes in Canadian multiple sclerosis patients. Canadian Collaborative Study Group. *Neurology* 54:729–732.

136. Ramagopalan, S. V., D. A. Dyment, M. Z. Cader, K. M. Morrison, G. Disanto, J. M. Morahan, A. J. Berlanga-Taylor, A. Handel, G. C. De Luca, A. D. Sadovnick, P. Lepage, A. Montpetit, and G. C. Ebers. 2011. Rare variants in the CYP27B1 gene are associated with multiple sclerosis. *Annals of Neurology* 70:881–886.

137. Speeckaert, M., G. Huang, J. R. Delanghe, and Y. E. Taes. 2006. Biological and clinical aspects of the vitamin D binding protein (Gc-globulin) and its polymorphism. *Clinica Chimica Acta* 372:33–42.

138. Di Bacco, M., D. Luiselli, M. L. Manca, and G. Siciliano. 2002. Bayesian approach to searching for susceptibility genes: Gc2 and EsD1 alleles and multiple sclerosis. *Collegium Antropologicum* 26:77–84.

139. Hollsberg, P., S. Haahr, P. M. Larsen, and S. J. Fey. 1988. MS and the group specific component. *Acta Neurologica Scandinavica* 78:158–160.

140. Lindblom, B., G. Wetterling, and H. Link. 1988. Distribution of group-specific component subtypes in multiple sclerosis. *Acta Neurologica Scandinavica* 78:443–444.

141. Niino, M., S. Kikuchi, T. Fukazawa, I. Yabe, and K. Tashiro. 2002. No association of vitamin D–binding protein gene polymorphisms in Japanese patients with MS. *Journal of Neuroimmunology* 127:177–179.

142. Burton, J. M., S. Kimball, R. Vieth, A. Bar-Or, H. M. Dosch, R. Cheung, D. Gagne, C. D'Souza, M. Ursell, and P. O'Connor. 2010. A phase I/II dose-escalation trial of vitamin D3 and calcium in multiple sclerosis. *Neurology* 74:1852–1859.

143. Kimball, S. M., M. R. Ursell, P. O'Connor, and R. Vieth. 2007. Safety of vitamin D3 in adults with multiple sclerosis. *The American Journal of Clinical Nutrition* 86:645–651.

144. Marcus, J. F., S. M. Shalev, C. A. Harris, D. S. Goodin, and S. A. Josephson. 2012. Severe hypercalcemia following vitamin D supplementation in a patient with multiple sclerosis: A note of caution. *Archives of Neurology* 69:129–132.

145. Soilu-Hanninen, M., J. Aivo, B. M. Lindstrom, I. Elovaara, M. L. Sumelahti, M. Farkkila, P. Tienari, S. Atula, T. Sarasoja, L. Herrala, I. Keskinarkaus, J. Kruger, T. Kallio, M. A. Rocca, and M. Filippi. 2012. A randomized, double blind, placebo controlled trial with vitamin D3 as an add on treatment to interferon beta-1b in patients with multiple sclerosis. *Journal of Neurology, Neurosurgery, and Psychiatry* 83:565–571.

146. Taylor, A. V., and P. H. Wise. 1998. Vitamin D replacement in Asians with diabetes may increase insulin resistance. *Postgraduate Medical Journal* 74:365–366.

147. Achiron, A., Y. Barak, S. Miron, Y. Izhak, M. Faibel, and S. Edelstein. 2003. Alfacalcidol treatment in multiple sclerosis. *Clinical Neuropharmacology* 26:53.

148. Wingerchuk, D. M., J. Lesaux, G. P. Rice, M. Kremenchutzky, and G. C. Ebers. 2005. A pilot study of oral calcitriol (1,25-dihydroxyvitamin D3) for relapsing–remitting multiple sclerosis. *Journal of Neurology, Neurosurgery, and Psychiatry* 76:1294–1296.

149. Hauser, S. L., H. L. Weiner, M. Che, M. E. Shapiro, F. Gilles, and N. L. Letvin. 1984. Prevention of experimental allergic encephalomyelitis (EAE) in the SJL/J mouse by whole body ultraviolet irradiation. *Journal of Immunology* 132:1276–1281.

150. Becklund, B. R., K. S. Severson, S. V. Vang, and H. F. DeLuca. 2010. UV radiation suppresses experimental autoimmune encephalomyelitis independent of vitamin D production. *Proceedings of the National Academy of Sciences of the United States of America* 107:6418–6423.

151. Tsunoda, I., L. Q. Kuang, I. Z. Igenge, and R. S. Fujinami. 2005. Converting relapsing remitting to secondary progressive experimental allergic encephalomyelitis (EAE) by ultraviolet B irradiation. *Journal of Neuroimmunology* 160:122–134.

152. Nashold, F. E., K. M. Spach, J. A. Spanier, and C. E. Hayes. 2009. Estrogen controls vitamin D3-mediated resistance to experimental autoimmune encephalomyelitis by controlling vitamin D3 metabolism and receptor expression. *Journal of Immunology* 183:3672–3681.

153. Nataf, S., E. Garcion, F. Darcy, D. Chabannes, J. Y. Muller, and P. Brachet. 1996. 1,25 Dihydroxyvitamin D3 exerts regional effects in the central nervous system during experimental allergic encephalomyelitis. *Journal of Neuropathology and Experimental Neurology* 55:904–914.

154. Garcion, E., S. Nataf, A. Berod, F. Darcy, and P. Brachet. 1997. 1,25-Dihydroxyvitamin D3 inhibits the expression of inducible nitric oxide synthase in rat central nervous system during experimental allergic encephalomyelitis. *Brain Research. Molecular Brain Research* 45:255–267.

155. Nashold, F. E., D. J. Miller, and C. E. Hayes. 2000. 1,25-Dihydroxyvitamin D3 treatment decreases macrophage accumulation in the CNS of mice with experimental autoimmune encephalomyelitis. *Journal of Neuroimmunology* 103:171–179.

156. Garcion, E., L. Sindji, S. Nataf, P. Brachet, F. Darcy, and C. N. Montero-Menei. 2003. Treatment of experimental autoimmune encephalomyelitis in rat by 1,25-dihydroxyvitamin D3 leads to early effects within the central nervous system. *Acta Neuropathologica* 105:438–448.

157. Cantorna, M. T., W. D. Woodward, C. E. Hayes, and H. F. DeLuca. 1998. 1,25-Dihydroxyvitamin D3 is a positive regulator for the two anti-encephalitogenic cytokines TGF-beta 1 and IL-4. *Journal of Immunology* 160:5314–5319.

158. Mattner, F., S. Smiroldo, F. Galbiati, M. Muller, P. Di Lucia, P. L. Poliani, G. Martino, P. Panina-Bordignon, and L. Adorini. 2000. Inhibition of Th1 development and treatment of chronic-relapsing experimental allergic encephalomyelitis by a non-hypercalcemic analogue of 1,25-dihydroxyvitamin D(3). *European Journal of Immunology* 30:498–508.

159. Goodin, D. S. 2009. The causal cascade to multiple sclerosis: A model for MS pathogenesis. *PloS One* 4:e4565.

160. Cantorna, M. T., J. Humpal-Winter, and H. F. DeLuca. 1999. Dietary calcium is a major factor in 1,25-dihydroxycholecalciferol suppression of experimental autoimmune encephalomyelitis in mice. *Journal of Nutrition* 129:1966–1971.

161. Spach, K. M., F. E. Nashold, B. N. Dittel, and C. E. Hayes. 2006. IL-10 signaling is essential for 1,25-dihydroxyvitamin D3–mediated inhibition of experimental autoimmune encephalomyelitis. *Journal of Immunology* 177:6030–6037.

162. Becklund, B. R., D. W. Hansen, Jr., and H. F. Deluca. 2009. Enhancement of 1,25-dihydroxyvitamin D3–mediated suppression of experimental autoimmune encephalomyelitis by calcitonin. *Proceedings of the National Academy of Sciences of the United States of America* 106:5276–5281.

163. Meehan, T. F., J. Vanhooke, J. Prahl, and H. F. Deluca. 2005. Hypercalcemia produced by parathyroid hormone suppresses experimental autoimmune encephalomyelitis in female but not male mice. *Archives of Biochemistry and Biophysics* 442:214–221.

164. Becklund, B. R., B. J. James, R. F. Gagel, and H. F. DeLuca. 2009. The calcitonin/calcitonin gene related peptide-alpha gene is not required for 1alpha,25-dihydroxyvitamin D3–mediated suppression of experimental autoimmune encephalomyelitis. *Archives of Biochemistry and Biophysics* 488:105–108.

165. Hewison, M. 2010. Vitamin D and the intracrinology of innate immunity. *Molecular and Cellular Endocrinology* 321:103–111.

166. Heine, G., U. Niesner, H. D. Chang, A. Steinmeyer, U. Zugel, T. Zuberbier, A. Radbruch, and M. Worm. 2008. 1,25-dihydroxyvitamin D(3) promotes IL-10 production in human B cells. *European Journal of Immunology* 38:2210–2218.

167. Sigmundsdottir, H., J. Pan, G. F. Debes, C. Alt, A. Habtezion, D. Soler, and E. C. Butcher. 2007. DCs metabolize sunlight-induced vitamin D3 to 'program' T cell attraction to the epidermal chemokine CCL27. *Nature Immunology* 8:285–293.

168. Enioutina, E. Y., D. Bareyan, and R. A. Daynes. 2009. TLR-induced local metabolism of vitamin D3 plays an important role in the diversification of adaptive immune responses. *Journal of Immunology* 182:4296–4305.

169. Neveu, I., P. Naveilhan, C. Menaa, D. Wion, P. Brachet, and M. Garabedian. 1994. Synthesis of 1,25-dihydroxyvitamin D3 by rat brain macrophages *in vitro. Journal of Neuroscience Research* 38:214–220.

170. Zehnder, D., R. Bland, M. C. Williams, R. W. McNinch, A. J. Howie, P. M. Stewart, and M. Hewison. 2001. Extrarenal expression of 25-hydroxyvitamin D(3)-1 alpha-hydroxylase. *The Journal of Clinical Endocrinology and Metabolism* 86:888–894.

171. Eyles, D. W., S. Smith, R. Kinobe, M. Hewison, and J. J. McGrath. 2005. Distribution of the vitamin D receptor and 1 alpha-hydroxylase in human brain. *Journal of Chemical Neuroanatomy* 29:21–30.

172. Adorini, L., and G. Penna. 2009. Dendritic cell tolerogenicity: A key mechanism in immunomodulation by vitamin D receptor agonists. *Human Immunology* 70:345–352.

173. Fallarino, F., U. Grohmann, S. You, B. C. McGrath, D. R. Cavener, C. Vacca, C. Orabona, R. Bianchi, M. L. Belladonna, C. Volpi, P. Santamaria, M. C. Fioretti, and P. Puccetti. 2006. The combined effects of tryptophan starvation and tryptophan catabolites down-regulate T cell receptor zeta-chain and induce a regulatory phenotype in naive T cells. *Journal of Immunology* 176:6752–6761.

174. Curti, A., S. Pandolfi, B. Valzasina, M. Aluigi, A. Isidori, E. Ferri, V. Salvestrini, G. Bonanno, S. Rutella, I. Durelli, A. L. Horenstein, F. Fiore, M. Massaia, M. P. Colombo, M. Baccarani, and R. M. Lemoli. 2007. Modulation of tryptophan catabolism by human leukemic cells results in the conversion of CD25– into CD25+ T regulatory cells. *Blood* 109:2871–2877.

175. Gregori, S., N. Giarratana, S. Smiroldo, M. Uskokovic, and L. Adorini. 2002. A 1alpha,25-dihydroxy-vitamin D(3) analog enhances regulatory T-cells and arrests autoimmune diabetes in NOD mice. *Diabetes* 51:1367–1374.

176. Daniel, C., N. A. Sartory, N. Zahn, H. H. Radeke, and J. M. Stein. 2008. Immune modulatory treatment of trinitrobenzene sulfonic acid colitis with calcitriol is associated with a change of a T helper (Th) 1/Th17 to a Th2 and regulatory T cell profile. *Journal of Pharmacology and Experimental Therapeutics* 324:23–33.

177. Tang, J., R. Zhou, D. Luger, W. Zhu, P. B. Silver, R. S. Grajewski, S. B. Su, C. C. Chan, L. Adorini, and R. R. Caspi. 2009. Calcitriol suppresses antiretinal autoimmunity through inhibitory effects on the Th17 effector response. *Journal of Immunology* 182:4624–4632.

178. Sakuishi, K., S. Miyake, and T. Yamamura. 2010. Role of NK cells and invariant NKT cells in multiple sclerosis. *Results and Problems in Cell Differentiation* 51:127–147.

179. Yu, S., and M. T. Cantorna. 2008. The vitamin D receptor is required for iNKT cell development. *Proceedings of the National Academy of Sciences of the United States of America* 105:5207–5212.

180. Yu, S., and M. T. Cantorna. 2011. Epigenetic reduction in invariant NKT cells following *in utero* vitamin D deficiency in mice. *Journal of Immunology* 186:1384–1390.

181. Wing, K., and S. Sakaguchi. 2010. Regulatory T cells exert checks and balances on self tolerance and autoimmunity. *Nature Immunology* 11:7–13.

182. Sakaguchi, S., M. Miyara, C. M. Costantino, and D. A. Hafler. 2010. FOXP3+ regulatory T cells in the human immune system. *Nature Reviews. Immunology* 10:490–500.

183. Bailey-Bucktrout, S. L., and J. A. Bluestone. 2011. Regulatory T cells: Stability revisited. *Trends in Immunology* 32:301–306.

184. Venken, K., N. Hellings, R. Liblau, and P. Stinissen. 2010. Disturbed regulatory T cell homeostasis in multiple sclerosis. *Trends in Molecular Medicine* 16:58–68.

185. Libera, D. D., D. D. Mitri, A. Bergami, D. Centonze, C. Gasperini, M. G. Grasso, S. Galgani, V. Martinelli, G. Comi, C. Avolio, G. Martino, G. Borsellino, F. Sallusto, L. Battistini, and R. Furlan. 2011. T regulatory cells are markers of disease activity in multiple sclerosis patients. *PloS One* 6:e21386.

186. Edwards, L. J., B. Sharrack, A. Ismail, H. Tumani, and C. S. Constantinescu. 2011. Central inflammation versus peripheral regulation in multiple sclerosis. *Journal of Neurology* 258:1518–1527.

187. Royal, W., 3rd, Y. Mia, H. Li, and K. Naunton. 2009. Peripheral blood regulatory T cell measurements correlate with serum vitamin D levels in patients with multiple sclerosis. *Journal of Neuroimmunology* 213:135–141.

188. Smolders, J., M. Thewissen, E. Peelen, P. Menheere, J. W. Cohen Tervaert, J. Damoiseaux, and R. Hupperts. 2009. Vitamin D status is positively correlated with regulatory T cell function in patients with multiple sclerosis. *PLoS One* 4:e6635.

189. Smolders, J., P. Menheere, M. Thewissen, E. Peelen, J. W. Cohen Tervaert, R. Hupperts, and J. Damoiseaux. 2010. Regulatory T cell function correlates with serum 25-hydroxyvitamin D, but not with 1,25-dihydroxyvitamin D, parathyroid hormone and calcium levels in patients with relapsing remitting multiple sclerosis. *The Journal of Steroid Biochemistry and Molecular Biology* 121:243–246.

190. McGeachy, M. J., L. A. Stephens, and S. M. Anderton. 2005. Natural recovery and protection from autoimmune encephalomyelitis: Contribution of CD4+CD25+ regulatory cells within the central nervous system. *Journal of Immunology* 175:3025–3032.

191. Bettelli, E., M. P. Das, E. D. Howard, H. L. Weiner, R. A. Sobel, and V. K. Kuchroo. 1998. IL-10 is critical in the regulation of autoimmune encephalomyelitis as demonstrated by studies of IL-10- and IL-4–deficient and transgenic mice. *Journal of Immunology* 161:3299–3306.

192. Stohlman, S. A., L. Pei, D. J. Cua, Z. Li, and D. R. Hinton. 1999. Activation of regulatory cells suppresses experimental allergic encephalomyelitis via secretion of IL-10. *Journal of Immunology* 163:6338–6344.

193. Burkhart, C., G. Y. Liu, S. M. Anderton, B. Metzler, and D. C. Wraith. 1999. Peptide-induced T cell regulation of experimental autoimmune encephalomyelitis: A role for IL-10. *International Immunology* 11:1625–1634.

194. Sundstedt, A., E. J. O'Neill, K. S. Nicolson, and D. C. Wraith. 2003. Role for IL-10 in suppression mediated by peptide-induced regulatory T cells *in vivo*. *Journal of Immunology* 170:1240–1248.
195. Zhang, X., D. N. Koldzic, L. Izikson, J. Reddy, R. F. Nazareno, S. Sakaguchi, V. K. Kuchroo, and H. L. Weiner. 2004. IL-10 is involved in the suppression of experimental autoimmune encephalomyelitis by CD25+CD4+ regulatory T cells. *International Immunology* 16:249–256.
196. Lemire, J. M., and D. C. Archer. 1991. 1,25-dihydroxyvitamin D3 prevents the *in vivo* induction of murine experimental autoimmune encephalomyelitis. *Journal of Clinical Investigation* 87:1103–1107.
197. Chang, J. H., H. R. Cha, D. S. Lee, K. Y. Seo, and M. N. Kweon. 2010. 1,25-Dihydroxyvitamin D3 inhibits the differentiation and migration of T(H)17 cells to protect against experimental autoimmune encephalomyelitis. *PLoS One* 5:e12925.
198. Joshi, S., L. C. Pantalena, X. K. Liu, S. L. Gaffen, H. Liu, C. Rohowsky-Kochan, K. Ichiyama, A. Yoshimura, L. Steinman, S. Christakos, and S. Youssef. 2011. 1,25-Dihydroxyvitamin D(3) ameliorates Th17 autoimmunity via transcriptional modulation of interleukin-17A. *Molecular and Cellular Biology* 31:3653–3669.
199. Loser, K., and S. Beissert. 2009. Regulation of cutaneous immunity by the environment: An important role for UV irradiation and vitamin D. *International Immunopharmacology* 9:587–589.
200. Ghoreishi, M., P. Bach, J. Obst, M. Komba, J. C. Fleet, and J. P. Dutz. 2009. Expansion of antigen-specific regulatory T cells with the topical vitamin D analog calcipotriol. *Journal of Immunology* 182:6071–6078.
201. Muller, K., J. Gram, J. Bollerslev, M. Diamant, T. Barington, M. B. Hansen, and K. Bendtzen. 1991. Down-regulation of monocyte functions by treatment of healthy adults with 1 alpha,25 dihydroxyvitamin D3. *International Journal of Immunopharmacology* 13:525–530.
202. Mahon, B. D., S. A. Gordon, J. Cruz, F. Cosman, and M. T. Cantorna. 2003. Cytokine profile in patients with multiple sclerosis following vitamin D supplementation. *Journal of Neuroimmunology* 134:128–132.
203. Cantorna, M. T., J. Humpal-Winter, and H. F. DeLuca. 2000. *In vivo* upregulation of interleukin-4 is one mechanism underlying the immunoregulatory effects of 1,25-dihydroxyvitamin D(3). *Archives of Biochemistry and Biophysics* 377:135–138.
204. Spach, K. M., L. B. Pedersen, F. E. Nashold, T. Kayo, B. S. Yandell, T. A. Prolla, and C. E. Hayes. 2004. Gene expression analysis suggests that 1,25-dihydroxyvitamin D3 reverses experimental autoimmune encephalomyelitis by stimulating inflammatory cell apoptosis. *Physiological Genomics* 18:141–151.
205. Pedersen, L. B., F. E. Nashold, K. M. Spach, and C. E. Hayes. 2007. 1,25-Dihydroxyvitamin D3 reverses experimental autoimmune encephalomyelitis by inhibiting chemokine synthesis and monocyte trafficking. *Journal of Neuroscience Research* 85:2480–2490.
206. Santamaria, P. 2010. The long and winding road to understanding and conquering type 1 diabetes. *Immunity* 32:437–445.
207. Lehuen, A., J. Diana, P. Zaccone, and A. Cooke. 2010. Immune cell crosstalk in type 1 diabetes. *Nature Reviews. Immunology* 10:501–513.
208. Todd, J. A. 2010. Etiology of type 1 diabetes. *Immunity* 32:457–467.
209. Bluestone, J. A., K. Herold, and G. Eisenbarth. 2010. Genetics, pathogenesis and clinical interventions in type 1 diabetes. *Nature* 464:1293–1300.
210. Maahs, D. M., N. A. West, J. M. Lawrence, and E. J. Mayer-Davis. 2010. Epidemiology of type 1 diabetes. *Endocrinology and Metabolism Clinics of North America* 39:481–497.
211. Steck, A. K., and M. J. Rewers. 2011. Genetics of type 1 diabetes. *Clinical Chemistry* 57:176–185.
212. Baz-Hecht, M., and A. B. Goldfine. 2010. The impact of vitamin D deficiency on diabetes and cardiovascular risk. *Current Opinion in Endocrinology, Diabetes, and Obesity* 17:113–119.
213. Hypponen, E. 2010. Vitamin D and increasing incidence of type 1 diabetes—evidence for an association? *Diabetes, Obesity & Metabolism* 12:737–743.
214. Balay, J. L. 2010. Vitamin D. What is its role in diabetes? *Diabetes Self-Management* 27:11–14.
215. Mathieu, C. 2010. Vitamin D and diabetes: The devil is in the D-tails. *Diabetologia* 53:1545–1548.
216. Mitri, J., and A. G. Pittas. 2010. Diabetes—Shining a light: The role of vitamin D in diabetes mellitus. *Nature Reviews. Endocrinology* 6:478–480.
217. Mohr, S. B., F. C. Garland, C. F. Garland, E. D. Gorham, and C. Ricordi. 2010. Is there a role of vitamin D deficiency in type 1 diabetes of children? *American Journal of Preventive Medicine* 39:189–190.
218. Pittas, A. G., and B. Dawson-Hughes. 2010. Vitamin D and diabetes. *The Journal of Steroid Biochemistry and Molecular Biology* 121:425–429.
219. Takiishi, T., C. Gysemans, R. Bouillon, and C. Mathieu. 2010. Vitamin D and diabetes. *Endocrinology and Metabolism Clinics of North America* 39:419–446.
220. Boucher, B. J. 2011. Vitamin D insufficiency and diabetes risks. *Current Drug Targets* 12:61–87.
221. Hayes, C. E., F. E. Nashold, K. M. Spach, and L. B. Pedersen. 2003. The immunological functions of the vitamin D endocrine system. *Cellular and Molecular Biology (Noisy-le-Grand, France)* 49:277–300.

222. Mohr, S. B., C. F. Garland, E. D. Gorham, and F. C. Garland. 2008. The association between ultraviolet B irradiance, vitamin D status and incidence rates of type 1 diabetes in 51 regions worldwide. *Diabetologia* 51:1391–1398.

223. Pugliatti, M., S. Sotgiu, G. Solinas, P. Castiglia, M. I. Pirastru, B. Murgia, L. Mannu, G. Sanna, and G. Rosati. 2001. Multiple sclerosis epidemiology in Sardinia: Evidence for a true increasing risk. *Acta Neurologica Scandinavica* 103:20–26.

224. Kahn, H. S., T. M. Morgan, L. D. Case, D. Dabelea, E. J. Mayer-Davis, J. M. Lawrence, S. M. Marcovina, and G. Imperatore. 2009. Association of type 1 diabetes with month of birth among U.S. youth: The SEARCH for Diabetes in Youth Study. *Diabetes Care* 32:2010–2015.

225. Levy-Marchal, C., C. Patterson, and A. Green. 1995. Variation by age group and seasonality at diagnosis of childhood IDDM in Europe. The EURODIAB ACE Study Group. *Diabetologia* 38:823–830.

226. Karvonen, M., V. Jantti, S. Muntoni, M. Stabilini, L. Stabilini, and J. Tuomilehto. 1998. Comparison of the seasonal pattern in the clinical onset of IDDM in Finland and Sardinia. *Diabetes Care* 21:1101–1109.

227. Pozzilli, P., S. Manfrini, A. Crino, A. Picardi, C. Leomanni, V. Cherubini, L. Valente, M. Khazrai, and N. Visalli. 2005. Low levels of 25-hydroxyvitamin D3 and 1,25-dihydroxyvitamin D3 in patients with newly diagnosed type 1 diabetes. *Hormone and Metabolic Research* 37:680–683.

228. Sloka, S., M. Grant, and L. A. Newhook. 2008. Time series analysis of ultraviolet B radiation and type 1 diabetes in Newfoundland. *Pediatric Diabetes* 9:81–86.

229. Gorham, E. D., E. Barrett-Connor, R. M. Highfill-McRoy, S. B. Mohr, C. F. Garland, F. C. Garland, and C. Ricordi. 2009. Incidence of insulin-requiring diabetes in the US military. *Diabetologia* 52:2087–2091.

230. Elliott, J. C., R. M. Lucas, M. S. Clements, and H. J. Bambrick. 2010. Population density determines the direction of the association between ambient ultraviolet radiation and type 1 diabetes incidence. *Pediatric Diabetes* 11:394–402.

231. Cooper, J. D., D. J. Smyth, N. M. Walker, H. Stevens, O. S. Burren, C. Wallace, C. Greissl, E. Ramos-Lopez, E. Hypponen, D. B. Dunger, T. D. Spector, W. H. Ouwehand, T. J. Wang, K. Badenhoop, and J. A. Todd. 2011. Inherited variation in vitamin D genes is associated with predisposition to autoimmune disease type 1 diabetes. *Diabetes* 60:1624–1631.

232. Moltchanova, E. V., N. Schreier, N. Lammi, and M. Karvonen. 2009. Seasonal variation of diagnosis of type 1 diabetes mellitus in children worldwide. *Diabetic Medicine* 26:673–678.

233. Godar, D. E. 2005. UV doses worldwide. *Photochemistry and Photobiology* 81:736–749.

234. Staples, J. A., A. L. Ponsonby, L. L. Lim, and A. J. McMichael. 2003. Ecologic analysis of some immune-related disorders, including type 1 diabetes, in Australia: Latitude, regional ultraviolet radiation, and disease prevalence. *Environmental Health Perspectives* 111:518–523.

235. Ponsonby, A. L., R. M. Lucas, and I. A. van der Mei. 2005. UVR, vitamin D and three autoimmune diseases—multiple sclerosis, type 1 diabetes, rheumatoid arthritis. *Photochemistry and Photobiology* 81:1267–1275.

236. Sloka, S., M. Grant, and L. A. Newhook. 2010. The geospatial relation between UV solar radiation and type 1 diabetes in Newfoundland. *Acta Diabetologica* 47:73–78.

237. Christakos, S., and A. W. Norman. 1979. Studies on the mode of action of calciferol. XVIII. Evidence for a specific high affinity binding protein for 1,25 dihydroxyvitamin D3 in chick kidney and pancreas. *Biochemical and Biophysical Research Communications* 89:56–63.

238. Clark, S. A., W. E. Stumpf, M. Sar, H. F. DeLuca, and Y. Tanaka. 1980. Target cells for 1,25 dihydroxyvitamin D3 in the pancreas. *Cell and Tissue Research* 209:515–520.

239. Pike, J. W. 1982. Receptors for 1,25-dihydroxyvitamin D3 in chick pancreas: A partial physical and functional characterization. *Journal of Steroid Biochemistry* 16:385–395.

240. Bland, R., D. Markovic, C. E. Hills, S. V. Hughes, S. L. Chan, P. E. Squires, and M. Hewison. 2004. Expression of 25-hydroxyvitamin D3-1alpha-hydroxylase in pancreatic islets. *The Journal of Steroid Biochemistry and Molecular Biology* 89–90:121–125.

241. Norman, A. W., J. B. Frankel, A. M. Heldt, and G. M. Grodsky. 1980. Vitamin D deficiency inhibits pancreatic secretion of insulin. *Science* 209:823–825.

242. Chertow, B. S., W. I. Sivitz, N. G. Baranetsky, S. A. Clark, A. Waite, and H. F. Deluca. 1983. Cellular mechanisms of insulin release: The effects of vitamin D deficiency and repletion on rat insulin secretion. *Endocrinology* 113:1511–1518.

243. Tanaka, Y., Y. Seino, M. Ishida, K. Yamaoka, H. Yabuuchi, H. Ishida, S. Seino, and H. Imura. 1984. Effect of vitamin D3 on the pancreatic secretion of insulin and somatostatin. *Acta Endocrinologica* 105:528–533.

244. Kadowaki, S., and A. W. Norman. 1984. Dietary vitamin D is essential for normal insulin secretion from the perfused rat pancreas. *The Journal of Clinical Investigation* 73:759–766.

245. Cade, C., and A. W. Norman. 1986. Vitamin D3 improves impaired glucose tolerance and insulin secretion in the vitamin D–deficient rat *in vivo*. *Endocrinology* 119:84–90.

246. Kumar, S., M. Davies, Y. Zakaria, E. B. Mawer, C. Gordon, A. O. Olukoga, and A. J. Boulton. 1994. Improvement in glucose tolerance and beta-cell function in a patient with vitamin D deficiency during treatment with vitamin D. *Postgraduate Medical Journal* 70:440–443.

247. Clark, S. A., W. E. Stumpf, and M. Sar. 1981. Effect of 1,25 dihydroxyvitamin D3 on insulin secretion. *Diabetes* 30:382–386.

248. Cade, C., and A. W. Norman. 1987. Rapid normalization/stimulation by 1,25-dihydroxyvitamin D3 of insulin secretion and glucose tolerance in the vitamin D–deficient rat. *Endocrinology* 120:1490–1497.

249. Hay, C. W., and K. Docherty. 2006. Comparative analysis of insulin gene promoters: Implications for diabetes research. *Diabetes* 55:3201–3213.

250. Mathieu, C., and K. Badenhoop. 2005. Vitamin D and type 1 diabetes mellitus: State of the art. *Trends in Endocrinology and Metabolism* 16:261–266.

251. 1999. Vitamin D supplement in early childhood and risk for type I (insulin-dependent) diabetes mellitus. The EURODIAB Substudy 2 Study Group. *Diabetologia* 42:51–54.

252. Zipitis, C. S., and A. K. Akobeng. 2008. Vitamin D supplementation in early childhood and risk of type 1 diabetes: a systematic review and meta-analysis. *Archives of Disease in Childhood* 93:512–517.

253. Hypponen, E., E. Laara, A. Reunanen, M. R. Jarvelin, and S. M. Virtanen. 2001. Intake of vitamin D and risk of type 1 diabetes: A birth-cohort study. *Lancet* 358:1500–1503.

254. Holick, M. F. 1995. Environmental factors that influence the cutaneous production of vitamin D. *The American Journal of Clinical Nutrition* 61:638S–645S.

255. Fronczak, C. M., A. E. Baron, H. P. Chase, C. Ross, H. L. Brady, M. Hoffman, G. S. Eisenbarth, M. Rewers, and J. M. Norris. 2003. *In utero* dietary exposures and risk of islet autoimmunity in children. *Diabetes Care* 26:3237–3242.

256. Need, A. G., P. D. O'Loughlin, M. Horowitz, and B. E. Nordin. 2005. Relationship between fasting serum glucose, age, body mass index and serum 25 hydroxyvitamin D in postmenopausal women. *Clinical Endocrinology* 62:738–741.

257. Littorin, B., P. Blom, A. Scholin, H. J. Arnqvist, G. Blohme, J. Bolinder, A. Ekbom-Schnell, J. W. Eriksson, S. Gudbjornsdottir, L. Nystrom, J. Ostman, and G. Sundkvist. 2006. Lower levels of plasma 25-hydroxyvitamin D among young adults at diagnosis of autoimmune type 1 diabetes compared with control subjects: Results from the nationwide Diabetes Incidence Study in Sweden (DISS). *Diabetologia* 49:2847–2852.

258. Greer, R. M., M. A. Rogers, F. G. Bowling, H. M. Buntain, M. Harris, G. M. Leong, and A. M. Cotterill. 2007. Australian children and adolescents with type 1 diabetes have low vitamin D levels. *The Medical Journal of Australia* 187:59–60.

259. Svoren, B. M., L. K. Volkening, J. R. Wood, and L. M. Laffel. 2009. Significant vitamin D deficiency in youth with type 1 diabetes mellitus. *Journal of Pediatrics* 154:132–134.

260. Devaraj, S., J. M. Yun, C. R. Duncan-Staley, and I. Jialal. 2011. Low vitamin D levels correlate with the proinflammatory state in type 1 diabetic subjects with and without microvascular complications. *American Journal of Clinical Pathology* 135:429–433.

261. Janner, M., P. Ballinari, P. E. Mullis, and C. E. Fluck. 2010. High prevalence of vitamin D deficiency in children and adolescents with type 1 diabetes. *Swiss Medical Weekly* 140:w13091.

262. Hamed, E. A., N. H. Abu Faddan, H. A. Adb Elhafeez, and D. Sayed. 2011. Parathormone–25(OH)–vitamin D axis and bone status in children and adolescents with type 1 diabetes mellitus. *Pediatric Diabetes* 12:536–546.

263. Young, K. A., J. K. Snell-Bergeon, R. G. Naik, J. E. Hokanson, D. Tarullo, P. A. Gottlieb, S. K. Garg, and M. Rewers. 2011. Vitamin D deficiency and coronary artery calcification in subjects with type 1 diabetes. *Diabetes Care* 34:454–458.

264. Kaur, H., K. C. Donaghue, A. K. Chan, P. Benitez-Aguirre, S. Hing, M. Lloyd, J. Cusumano, A. Pryke, and M. E. Craig. 2011. Vitamin D deficiency is associated with retinopathy in children and adolescents with type 1 diabetes. *Diabetes Care* 34:1400–1402.

265. Hyttinen, V., J. Kaprio, L. Kinnunen, M. Koskenvuo, and J. Tuomilehto. 2003. Genetic liability of type 1 diabetes and the onset age among 22,650 young Finnish twin pairs: A nationwide follow-up study. *Diabetes* 52:1052–1055.

266. Redondo, M. J., J. Jeffrey, P. R. Fain, G. S. Eisenbarth, and T. Orban. 2008. Concordance for islet autoimmunity among monozygotic twins. *The New England Journal of Medicine* 359:2849–2850.

267. McDermott, M. F., A. Ramachandran, B. W. Ogunkolade, E. Aganna, D. Curtis, B. J. Boucher, C. Snehalatha, and G. A. Hitman. 1997. Allelic variation in the vitamin D receptor influences susceptibility to IDDM in Indian Asians. *Diabetologia* 40:971–975.

268. Pani, M. A., M. Knapp, H. Donner, J. Braun, M. P. Baur, K. H. Usadel, and K. Badenhoop. 2000. Vitamin D receptor allele combinations influence genetic susceptibility to type 1 diabetes in Germans. *Diabetes* 49:504–507.

269. Pociot, F., B. Akolkar, P. Concannon, H. A. Erlich, C. Julier, G. Morahan, C. R. Nierras, J. A. Todd, S. S. Rich, and J. Nerup. 2010. Genetics of type 1 diabetes: What's next? *Diabetes* 59:1561–1571.

270. Turpeinen, H., R. Hermann, S. Vaara, A. P. Laine, O. Simell, M. Knip, R. Veijola, and J. Ilonen. 2003. Vitamin D receptor polymorphisms: No association with type 1 diabetes in the Finnish population. *European Journal of Endocrinology/European Federation of Endocrine Societies* 149:591–596.

271. Martin, R. J., A. J. McKnight, C. C. Patterson, D. M. Sadlier, and A. P. Maxwell. 2010. A rare haplotype of the vitamin D receptor gene is protective against diabetic nephropathy. *Nephrology, Dialysis, Transplantation: Official Publication of the European Dialysis and Transplant Association—European Renal Association* 25:497–503.

272. Chang, T. J., H. H. Lei, J. I. Yeh, K. C. Chiu, K. C. Lee, M. C. Chen, T. Y. Tai, and L. M. Chuang. 2000. Vitamin D receptor gene polymorphisms influence susceptibility to type 1 diabetes mellitus in the Taiwanese population. *Clinical Endocrinology* 52:575–580.

273. Nguyen, M., A. d'Alesio, J. M. Pascussi, R. Kumar, M. D. Griffin, X. Dong, H. Guillozo, M. Rizk-Rabin, C. Sinding, P. Bougneres, F. Jehan, and M. Garabedian. 2006. Vitamin D–resistant rickets and type 1 diabetes in a child with compound heterozygous mutations of the vitamin D receptor (L263R and R391S): Dissociated responses of the CYP-24 and rel-B promoters to 1,25-dihydroxyvitamin D3. *Journal of Bone and Mineral Research* 21:886–894.

274. Guo, S. W., V. L. Magnuson, J. J. Schiller, X. Wang, Y. Wu, and S. Ghosh. 2006. Meta analysis of vitamin D receptor polymorphisms and type 1 diabetes: A HuGE review of genetic association studies. *American Journal of Epidemiology* 164:711–724.

275. Mimbacas, A., J. Trujillo, C. Gascue, G. Javiel, and H. Cardoso. 2007. Prevalence of vitamin D receptor gene polymorphism in a Uruguayan population and its relation to type 1 diabetes mellitus. *Genetics and Molecular Research* 6:534–542.

276. Lemos, M. C., A. Fagulha, E. Coutinho, L. Gomes, M. Bastos, L. Barros, F. Carrilho, E. Geraldes, F. J. Regateiro, and M. Carvalheiro. 2008. Lack of association of vitamin D receptor gene polymorphisms with susceptibility to type 1 diabetes mellitus in the Portuguese population. *Human Immunology* 69:134–138.

277. Kahles, H., G. Morahan, J. A. Todd, and K. Badenhoop. 2009. Association analyses of the vitamin D receptor gene in 1654 families with type I diabetes. *Genes and Immunity* 10 Suppl 1:S60–63.

278. Mory, D. B., E. R. Rocco, W. L. Miranda, T. Kasamatsu, F. Crispim, and S. A. Dib. 2009. Prevalence of vitamin D receptor gene polymorphisms FokI and BsmI in Brazilian individuals with type 1 diabetes and their relation to beta-cell autoimmunity and to remaining beta-cell function. *Human Immunology* 70:447–451.

279. Mohammadnejad, Z., M. Ghanbari, R. Ganjali, J. T. Afshari, M. Heydarpour, S. M. Taghavi, S. Fatemi, and H. Rafatpanah. 2011. Association between vitamin D receptor gene polymorphisms and type 1 diabetes mellitus in Iranian population. *Molecular Biology Reports* 39:831–837.

280. Ponsonby, A. L., A. Pezic, J. Ellis, R. Morley, F. Cameron, J. Carlin, and T. Dwyer. 2008. Variation in associations between allelic variants of the vitamin D receptor gene and onset of type 1 diabetes mellitus by ambient winter ultraviolet radiation levels: A meta-regression analysis. *American Journal of Epidemiology* 168:358–365.

281. Gross, C., A. V. Krishnan, P. J. Malloy, T. R. Eccleshall, X. Y. Zhao, and D. Feldman. 1998. The vitamin D receptor gene start codon polymorphism: A functional analysis of FokI variants. *Journal of Bone and Mineral Research: The Official Journal of the American Society for Bone and Mineral Research* 13:1691–1699.

282. van Etten, E., L. Verlinden, A. Giulietti, E. Ramos-Lopez, D. D. Branisteanu, G. B. Ferreira, L. Overbergh, A. Verstuyf, R. Bouillon, B. O. Roep, K. Badenhoop, and C. Mathieu. 2007. The vitamin D receptor gene FokI polymorphism: Functional impact on the immune system. *European Journal of Immunology* 37:395–405.

283. Zella, L. A., M. B. Meyer, R. D. Nerenz, S. M. Lee, M. L. Martowicz, and J. W. Pike. 2010. Multifunctional enhancers regulate mouse and human vitamin D receptor gene transcription. *Molecular Endocrinology* 24:128–147.

284. Israni, N., R. Goswami, A. Kumar, and R. Rani. 2009. Interaction of vitamin D receptor with HLA DRB1 0301 in type 1 diabetes patients from North India. *PLoS One* 4:e8023.

285. Cheng, J. B., M. A. Levine, N. H. Bell, D. J. Mangelsdorf, and D. W. Russell. 2004. Genetic evidence that the human CYP2R1 enzyme is a key vitamin D 25-hydroxylase. *Proceedings of the National Academy of Sciences of the United States of America* 101:7711–7715.

286. Wang, T. J., F. Zhang, J. B. Richards, B. Kestenbaum, J. B. van Meurs, D. Berry, D. P. Kiel, E. A. Streeten, C. Ohlsson, D. L. Koller, L. Peltonen, J. D. Cooper, P. F. O'Reilly, D. K. Houston, N. L. Glazer, L. Vandenput, M. Peacock, J. Shi, F. Rivadeneira, M. I. McCarthy, P. Anneli, I. H. de Boer, M. Mangino, B. Kato, D. J. Smyth, S. L. Booth, P. F. Jacques, G. L. Burke, M. Goodarzi, C. L. Cheung, M. Wolf, K. Rice, D. Goltzman, N. Hidiroglou, M. Ladouceur, N. J. Wareham, L. J. Hocking, D. Hart, N. K. Arden, C. Cooper, S. Malik, W. D. Fraser, A. L. Hartikainen, G. Zhai, H. M. Macdonald, N. G. Forouhi, R. J. Loos, D. M. Reid, A. Hakim, E. Dennison, Y. Liu, C. Power, H. E. Stevens, L. Jaana, R. S. Vasan, N. Soranzo, J. Bojunga, B. M. Psaty, M. Lorentzon, T. Foroud, T. B. Harris, A. Hofman, J. O. Jansson, J. A. Cauley, A. G. Uitterlinden, Q. Gibson, M. R. Jarvelin, D. Karasik, D. S. Siscovick, M. J. Econs, S. B. Kritchevsky, J. C. Florez, J. A. Todd, J. Dupuis, E. Hypponen, and T. D. Spector. 2010. Common genetic determinants of vitamin D insufficiency: A genome-wide association study. *Lancet* 376:180–188.

287. Ramos-Lopez, E., P. Bruck, T. Jansen, J. Herwig, and K. Badenhoop. 2007. CYP2R1 (vitamin D 25-hydroxylase) gene is associated with susceptibility to type 1 diabetes and vitamin D levels in Germans. *Diabetes Metab Res Rev* 23:631–636.

288. Lopez, E. R., O. Zwermann, M. Segni, G. Meyer, M. Reincke, J. Seissler, J. Herwig, K. H. Usadel, and K. Badenhoop. 2004. A promoter polymorphism of the CYP27B1 gene is associated with Addison's disease, Hashimoto's thyroiditis, Graves' disease and type 1 diabetes mellitus in Germans. *European Journal of Endocrinology/European Federation of Endocrine Societies* 151:193–197.

289. Lopez, E. R., K. Regulla, M. A. Pani, M. Krause, K. H. Usadel, and K. Badenhoop. 2004. CYP27B1 polymorphisms variants are associated with type 1 diabetes mellitus in Germans. *The Journal of Steroid Biochemistry and Molecular Biology* 89–90:155–157.

290. Bailey, R., J. D. Cooper, L. Zeitels, D. J. Smyth, J. H. Yang, N. M. Walker, E. Hypponen, D. B. Dunger, E. Ramos-Lopez, K. Badenhoop, S. Nejentsev, and J. A. Todd. 2007. Association of the vitamin D metabolism gene CYP27B1 with type 1 diabetes. *Diabetes* 56:2616–2621.

291. Pani, M. A., K. Regulla, M. Segni, M. Krause, S. Hofmann, M. Hufner, J. Herwig, A. M. Pasquino, K. H. Usadel, and K. Badenhoop. 2002. Vitamin D 1alpha-hydroxylase (CYP1alpha) polymorphism in Graves' disease, Hashimoto's thyroiditis and type 1 diabetes mellitus. *European Journal of Endocrinology/ European Federation of Endocrine Societies* 146:777–781.

292. Fichna, M., M. Zurawek, D. Januszkiewicz-Lewandowska, P. Fichna, and J. Nowak. 2010. PTPN22, PDCD1 and CYP27B1 polymorphisms and susceptibility to type 1 diabetes in Polish patients. *International Journal of Immunogenetics* 37:367–372.

293. Ongagna, J. C., M. C. Kaltenbacher, R. Sapin, M. Pinget, and A. Belcourt. 2001. The HLA-DQB alleles and amino acid variants of the vitamin D–binding protein in diabetic patients in Alsace. *Clinical Biochemistry* 34:59–63.

294. Ongagna, J. C., M. Pinget, and A. Belcourt. 2005. Vitamin D–binding protein gene polymorphism association with IA-2 autoantibodies in type 1 diabetes. *Clinical Biochemistry* 38:415–419.

295. Brekke, H. K., and J. Ludvigsson. 2007. Vitamin D supplementation and diabetes-related autoimmunity in the ABIS study. *Pediatric Diabetes* 8:11–14.

296. Marjamaki, L., S. Niinisto, M. G. Kenward, L. Uusitalo, U. Uusitalo, M. L. Ovaskainen, C. Kronberg-Kippila, O. Simell, R. Veijola, J. Ilonen, M. Knip, and S. M. Virtanen. 2010. Maternal intake of vitamin D during pregnancy and risk of advanced beta cell autoimmunity and type 1 diabetes in offspring. *Diabetologia* 53:1599–1607.

297. Barger-Lux, M. J., R. P. Heaney, S. Dowell, T. C. Chen, and M. F. Holick. 1998. Vitamin D and its major metabolites: Serum levels after graded oral dosing in healthy men. *Osteoporosis International* 8:222–230.

298. Vieth, R. 2004. Why the optimal requirement for Vitamin D3 is probably much higher than what is officially recommended for adults. *The Journal of Steroid Biochemistry and Molecular Biology* 89–90:575–579.

299. Heaney, R. P., L. A. Armas, J. R. Shary, N. H. Bell, N. Binkley, and B. W. Hollis. 2008. 25-Hydroxylation of vitamin D3: Relation to circulating vitamin D3 under various input conditions. *The American Journal of Clinical Nutrition* 87:1738–1742.

300. Bischoff-Ferrari, H. A., A. Shao, B. Dawson-Hughes, J. Hathcock, E. Giovannucci, and W. C. Willett. 2010. Benefit–risk assessment of vitamin D supplementation. *Osteoporosis International* 21:1121–1132.

301. Gedik, O., and S. Akalin. 1986. Effects of vitamin D deficiency and repletion on insulin and glucagon secretion in man. *Diabetologia* 29:142–145.

302. Aljabri, K. S., S. A. Bokhari, and M. J. Khan. 2010. Glycemic changes after vitamin D supplementation in patients with type 1 diabetes mellitus and vitamin D deficiency. *Annals of Saudi Medicine* 30:454–458.

303. Armas, L. A., B. W. Hollis, and R. P. Heaney. 2004. Vitamin D2 is much less effective than vitamin D3 in humans. *The Journal of Clinical Endocrinology and Metabolism* 89:5387–5391.

304. Houghton, L. A., and R. Vieth. 2006. The case against ergocalciferol (vitamin D2) as a vitamin supplement. *The American Journal of Clinical Nutrition* 84:694–697.
305. Walter, M., T. Kaupper, K. Adler, J. Foersch, E. Bonifacio, and A. G. Ziegler. 2010. No effect of the 1alpha,25-dihydroxyvitamin D3 on beta-cell residual function and insulin requirement in adults with new-onset type 1 diabetes. *Diabetes Care* 33:1443–1448.
306. Orwoll, E., M. Riddle, and M. Prince. 1994. Effects of vitamin D on insulin and glucagon secretion in non–insulin-dependent diabetes mellitus. *The American Journal of Clinical Nutrition* 59:1083–1087.
307. Bizzarri, C., D. Pitocco, N. Napoli, E. Di Stasio, D. Maggi, S. Manfrini, C. Suraci, M. G. Cavallo, M. Cappa, G. Ghirlanda, and P. Pozzilli. 2010. No protective effect of calcitriol on beta-cell function in recent-onset type 1 diabetes: The IMDIAB XIII trial. *Diabetes Care* 33:1962–1963.
308. Kishimoto, H., and J. Sprent. 2001. A defect in central tolerance in NOD mice. *Nature Immunology* 2:1025–1031.
309. Driver, J. P., D. V. Serreze, and Y. G. Chen. 2011. Mouse models for the study of autoimmune type 1 diabetes: A NOD to similarities and differences to human disease. *Seminars in Immunopathology* 33:67–87.
310. Mathieu, C., E. van Etten, B. Decallonne, A. Guilietti, C. Gysemans, R. Bouillon, and L. Overbergh. 2004. Vitamin D and 1,25-dihydroxyvitamin D3 as modulators in the immune system. *The Journal of Steroid Biochemistry and Molecular Biology* 89–90:449–452.
311. Giulietti, A., C. Gysemans, K. Stoffels, E. van Etten, B. Decallonne, L. Overbergh, R. Bouillon, and C. Mathieu. 2004. Vitamin D deficiency in early life accelerates type 1 diabetes in non-obese diabetic mice. *Diabetologia* 47:451–462.
312. Zella, J. B., L. C. McCary, and H. F. DeLuca. 2003. Oral administration of 1,25 dihydroxyvitamin D3 completely protects NOD mice from insulin-dependent diabetes mellitus. *Archives of Biochemistry and Biophysics* 417:77–80.
313. Hawa, M. I., M. G. Valorani, L. R. Buckley, P. E. Beales, A. Afeltra, F. Cacciapaglia, R. D. Leslie, and P. Pozzilli. 2004. Lack of effect of vitamin D administration during pregnancy and early life on diabetes incidence in the non-obese diabetic mouse. *Hormone and Metabolic Research* 36:620–624.
314. Overbergh, L., B. Decallonne, D. Valckx, A. Verstuyf, J. Depovere, J. Laureys, O. Rutgeerts, R. Saint-Arnaud, R. Bouillon, and C. Mathieu. 2000. Identification and immune regulation of 25-hydroxyvitamin D–1-alpha-hydroxylase in murine macrophages. *Clinical and Experimental Immunology* 120:139–146.
315. Mathieu, C., M. Waer, J. Laureys, O. Rutgeerts, and R. Bouillon. 1994. Prevention of autoimmune diabetes in NOD mice by 1,25 dihydroxyvitamin D3. *Diabetologia* 37:552–558.
316. Gysemans, C. A., A. K. Cardozo, H. Callewaert, A. Giulietti, L. Hulshagen, R. Bouillon, D. L. Eizirik, and C. Mathieu. 2005. 1,25-Dihydroxyvitamin D3 modulates expression of chemokines and cytokines in pancreatic islets: Implications for prevention of diabetes in nonobese diabetic mice. *Endocrinology* 146:1956–1964.
317. Wolden-Kirk, H., L. Overbergh, H. T. Christesen, K. Brusgaard, and C. Mathieu. 2011. Vitamin D and diabetes: Its importance for beta cell and immune function. *Molecular and Cellular Endocrinology* 347:106–120.
318. Ogunkolade, B. W., B. J. Boucher, J. M. Prahl, S. A. Bustin, J. M. Burrin, K. Noonan, B. V. North, N. Mannan, M. F. McDermott, H. F. DeLuca, and G. A. Hitman. 2002. Vitamin D receptor (VDR) mRNA and VDR protein levels in relation to vitamin D status, insulin secretory capacity, and VDR genotype in Bangladeshi Asians. *Diabetes* 51:2294–2300.
319. Zeitz, U., K. Weber, D. W. Soegiarto, E. Wolf, R. Balling, and R. G. Erben. 2003. Impaired insulin secretory capacity in mice lacking a functional vitamin D receptor. *FASEB Journal* 17:509–511.
320. Schneider, L. E., H. P. Schedl, T. McCain, and M. R. Haussler. 1977. Experimental diabetes reduces circulating 1,25-dihydroxyvitamin D in the rat. *Science* 196:1452–1454.
321. Bikle, D. D. 2011. Vitamin D: An ancient hormone. *Experimental Dermatology* 20:7–13.
322. Sibley, R. K., D. E. Sutherland, F. Goetz, and A. F. Michael. 1985. Recurrent diabetes mellitus in the pancreas iso- and allograft. A light and electron microscopic and immunohistochemical analysis of four cases. *Laboratory Investigation* 53:132–144.
323. Uno, S., A. Imagawa, K. Okita, K. Sayama, M. Moriwaki, H. Iwahashi, K. Yamagata, S. Tamura, Y. Matsuzawa, T. Hanafusa, J. Miyagawa, and I. Shimomura. 2007. Macrophages and dendritic cells infiltrating islets with or without beta cells produce tumour necrosis factor-alpha in patients with recent-onset type 1 diabetes. *Diabetologia* 50:596–601.
324. Du, T., Z. G. Zhou, S. You, J. Lin, L. Yang, W. D. Zhou, G. Huang, and C. Chao. 2009. Regulation by 1, 25-dihydroxy-vitamin D3 on altered TLRs expression and response to ligands of monocyte from autoimmune diabetes. *Clinica Chimica Acta; International Journal of Clinical Chemistry* 402:133–138.
325. Kawai, T., and S. Akira. 2011. Toll-like receptors and their crosstalk with other innate receptors in infection and immunity. *Immunity* 34:637–650.

326. Gombart, A. F. 2009. The vitamin D–antimicrobial peptide pathway and its role in protection against infection. *Future Microbiology* 4:1151–1165.
327. Beard, J. A., A. Bearden, and R. Striker. 2011. Vitamin D and the anti-viral state. *Journal of Clinical Virology* 50:194–200.
328. Baeke, F., T. Takiishi, H. Korf, C. Gysemans, and C. Mathieu. 2010. Vitamin D: Modulator of the immune system. *Current Opinion in Pharmacology* 10:482–496.
329. Bikle, D. D. 2011. Vitamin D regulation of immune function. *Vitamins and Hormones* 86:1–21.
330. Adorini, L. 2003. Tolerogenic dendritic cells induced by vitamin D receptor ligands enhance regulatory T cells inhibiting autoimmune diabetes. *Annals of the New York Academy of Sciences* 987:258–261.
331. van Etten, E., O. Dardenne, C. Gysemans, L. Overbergh, and C. Mathieu. 2004. 1,25-Dihydroxyvitamin D3 alters the profile of bone marrow–derived dendritic cells of NOD mice. *Annals of the New York Academy of Sciences* 1037:186–192.
332. Adorini, L., and G. Penna. 2009. Dendritic cell tolerogenicity: A key mechanism in immunomodulation by vitamin D receptor agonists. *Human Immunology* 70:345–352.
333. Gysemans, C., E. van Etten, L. Overbergh, A. Giulietti, G. Eelen, M. Waer, A. Verstuyf, R. Bouillon, and C. Mathieu. 2008. Unaltered diabetes presentation in NOD mice lacking the vitamin D receptor. *Diabetes* 57:269–275.
334. Leung, K. H. 1989. Inhibition of human natural killer cell and lymphokine-activated killer cell cytotoxicity and differentiation by vitamin D3. *Scandinavian Journal of Immunology* 30:199–208.
335. Qin, H., I. F. Lee, C. Panagiotopoulos, X. Wang, A. D. Chu, P. J. Utz, J. J. Priatel, and R. Tan. 2011. Natural killer cells from children with type 1 diabetes have defects in NKG2D-dependent function and signaling. *Diabetes* 60:857–866.
336. Deniz, G., G. Erten, U. C. Kucuksezer, D. Kocacik, C. Karagiannidis, E. Aktas, C. A. Akdis, and M. Akdis. 2008. Regulatory NK cells suppress antigen-specific T cell responses. *Journal of Immunology* 180:850–857.
337. Jeffery, L. E., F. Burke, M. Mura, Y. Zheng, O. S. Qureshi, M. Hewison, L. S. Walker, D. A. Lammas, K. Raza, and D. M. Sansom. 2009. 1,25-Dihydroxyvitamin D3 and IL-2 combine to inhibit T cell production of inflammatory cytokines and promote development of regulatory T cells expressing CTLA-4 and FoxP3. *Journal of Immunology* 183:5458–5467.
338. Baeke, F., H. Korf, L. Overbergh, A. Verstuyf, L. Thorrez, L. Van Lommel, M. Waer, F. Schuit, C. Gysemans, and C. Mathieu. 2011. The vitamin D analog, TX527, promotes a human CD4+CD25highCD127low regulatory T cell profile and induces a migratory signature specific for homing to sites of inflammation. *Journal of Immunology* 186:132–142.
339. Zhou, X., S. L. Bailey-Bucktrout, L. T. Jeker, C. Penaranda, M. Martinez-Llordella, M. Ashby, M. Nakayama, W. Rosenthal, and J. A. Bluestone. 2009. Instability of the transcription factor Foxp3 leads to the generation of pathogenic memory T cells *in vivo*. *Nature Immunology* 10:1000–1007.
340. O'Shea, J. J., and W. E. Paul. 2010. Mechanisms underlying lineage commitment and plasticity of helper CD4+ T cells. *Science* 327:1098–1102.
341. Brusko, T., C. Wasserfall, K. McGrail, R. Schatz, H. L. Viener, D. Schatz, M. Haller, J. Rockell, P. Gottlieb, M. Clare-Salzler, and M. Atkinson. 2007. No alterations in the frequency of FOXP3+ regulatory T-cells in type 1 diabetes. *Diabetes* 56:604–612.
342. McClymont, S. A., A. L. Putnam, M. R. Lee, J. H. Esensten, W. Liu, M. A. Hulme, U. Hoffmuller, U. Baron, S. Olek, J. A. Bluestone, and T. M. Brusko. 2011. Plasticity of human regulatory T cells in healthy subjects and patients with type 1 diabetes. *Journal of Immunology* 186:3918–3926.
343. von Essen, M. R., M. Kongsbak, P. Schjerling, K. Olgaard, N. Odum, and C. Geisler. 2010. Vitamin D controls T cell antigen receptor signaling and activation of human T cells. *Nature Immunology* 11:344–349.
344. Baeke, F., H. Korf, L. Overbergh, E. van Etten, A. Verstuyf, C. Gysemans, and C. Mathieu. 2010. Human T lymphocytes are direct targets of 1,25-dihydroxyvitamin D3 in the immune system. *The Journal of Steroid Biochemistry and Molecular Biology* 121:221–227.
345. Correale, J., M. C. Ysrraelit, and M. I. Gaitan. 2010. Gender differences in 1,25 dihydroxyvitamin D3 immunomodulatory effects in multiple sclerosis patients and healthy subjects. *Journal of Immunology* 185:4948–4958.
346. Lesage, S., S. B. Hartley, S. Akkaraju, J. Wilson, M. Townsend, and C. C. Goodnow. 2002. Failure to censor forbidden clones of CD4 T cells in autoimmune diabetes. *Journal of Experimental Medicine* 196:1175–1188.
347. Colucci, F., M. L. Bergman, C. Penha-Goncalves, C. M. Cilio, and D. Holmberg. 1997. Apoptosis resistance of nonobese diabetic peripheral lymphocytes linked to the Idd5 diabetes susceptibility region. *Proceedings of the National Academy of Sciences of the United States of America* 94:8670–8674.

348. Decallonne, B., and C. Mathieu. 2003. Defective activation-induced cell death in NOD T lymphocytes: 1,25-dihydroxyvitamin D3 restores defect. *Annals of the New York Academy of Sciences* 1005:176–177.

349. Decallonne, B., E. van Etten, A. Giulietti, K. Casteels, L. Overbergh, R. Bouillon, and C. Mathieu. 2003. Defect in activation-induced cell death in non-obese diabetic (NOD) T lymphocytes. *Journal of Autoimmunity* 20:219–226.

350. Decallonne, B., E. van Etten, L. Overbergh, D. Valckx, R. Bouillon, and C. Mathieu. 2005. 1Alpha,25-dihydroxyvitamin D3 restores thymocyte apoptosis sensitivity in non-obese diabetic (NOD) mice through dendritic cells. *Journal of Autoimmunity* 24:281–289.

351. Imai, Y., T. Kimura, A. Murakami, N. Yajima, K. Sakamaki, and S. Yonehara. 1999. The CED-4-homologous protein FLASH is involved in Fas-mediated activation of caspase-8 during apoptosis. *Nature* 398:777–785.

352. Zheng, C., V. Kabaleeswaran, Y. Wang, G. Cheng, and H. Wu. 2010. Crystal structures of the TRAF2:cIAP2 and the TRAF1:TRAF2:cIAP2 complexes: Affinity, specificity, and regulation. *Molecular Cell* 38:101–113.

353. Khor, B., A. Gardet, and R. J. Xavier. 2011. Genetics and pathogenesis of inflammatory bowel disease. *Nature* 474:307–317.

354. Maloy, K. J., and F. Powrie. 2011. Intestinal homeostasis and its breakdown in inflammatory bowel disease. *Nature* 474:298–306.

355. Hardenberg, G., T. S. Steiner, and M. K. Levings. 2011. Environmental influences on T regulatory cells in inflammatory bowel disease. *Seminars in Immunology* 23:130–138.

356. Kaser, A., S. Zeissig, and R. S. Blumberg. 2010. Inflammatory bowel disease. *Annual Review of Immunology* 28:573–621.

357. Koboziev, I., F. Karlsson, and M. B. Grisham. 2010. Gut-associated lymphoid tissue, T cell trafficking, and chronic intestinal inflammation. *Annals of the New York Academy of Sciences* 1207 Suppl 1:E86–93.

358. Weaver, C. T., and R. D. Hatton. 2009. Interplay between the TH17 and TReg cell lineages: A (co-)evolutionary perspective. *Nature Reviews. Immunology* 9:883–889.

359. Lee, Y. K., and S. K. Mazmanian. 2010. Has the microbiota played a critical role in the evolution of the adaptive immune system? *Science* 330:1768–1773.

360. Round, J. L., R. M. O'Connell, and S. K. Mazmanian. 2010. Coordination of tolerogenic immune responses by the commensal microbiota. *Journal of Autoimmunity* 34:J220–225.

361. Moresco, E. M., D. LaVine, and B. Beutler. 2011. Toll-like receptors. *Current Biology* 21:R488–493.

362. Fagarasan, S., S. Kawamoto, O. Kanagawa, and K. Suzuki. 2010. Adaptive immune regulation in the gut: T cell–dependent and T cell–independent IgA synthesis. *Annual Review of Immunology* 28:243–273.

363. Kaser, A., and R. S. Blumberg. 2011. Autophagy, microbial sensing, endoplasmic reticulum stress, and epithelial function in inflammatory bowel disease. *Gastroenterology* 140:1738–1747.

364. Lodes, M. J., Y. Cong, C. O. Elson, R. Mohamath, C. J. Landers, S. R. Targan, M. Fort, and R. M. Hershberg. 2004. Bacterial flagellin is a dominant antigen in Crohn disease. *Journal of Clinical Investigation* 113:1296–1306.

365. Loftus, E. V., Jr. 2004. Clinical epidemiology of inflammatory bowel disease: Incidence, prevalence, and environmental influences. *Gastroenterology* 126:1504–1517.

366. Shapira, Y., N. Agmon-Levin, and Y. Shoenfeld. 2010. Defining and analyzing geoepidemiology and human autoimmunity. *Journal of Autoimmunity* 34:J168–177.

367. Sonneberg, A., and I. H. Wasseman. 1991. Epidemiology of inflammatory bowel diseases among U.S. military veterans. *Gastroenterology* 101:122–130.

368. Ardizzone, S., A. Cassinotti, M. Bevilacqua, M. Clerici, and G. B. Porro. 2011. Vitamin D and inflammatory bowel disease. *Vitamins and Hormones* 86:367–377.

369. Cross, H. S., T. Nittke, and E. Kallay. 2011. Colonic vitamin D metabolism: Implications for the pathogenesis of inflammatory bowel disease and colorectal cancer. *Molecular and Cellular Endocrinology* 347:70–79

370. Mascitelli, L., F. Pezzetta, and M. R. Goldstein. 2011. Inflammatory bowel disease and the vitamin D endocrine system. *Internal Medicine Journal* 41:369–370.

371. Raman, M., A. N. Milestone, J. R. Walters, A. L. Hart, and S. Ghosh. 2011. Vitamin D and gastrointestinal diseases: Inflammatory bowel disease and colorectal cancer. *Therapeutic Advances in Gastroenterology* 4:49–62.

372. Mikulecky, M., and I. Cierna. 2005. Seasonality of births and childhood inflammatory bowel disease. *Wiener Klinische Wochenschrift* 117:554–557.

373. Chowers, Y., S. Odes, Y. Bujanover, R. Eliakim, S. Bar Meir, and B. Avidan. 2004. The month of birth is linked to the risk of Crohn's disease in the Israeli population. *The American Journal of Gastroenterology* 99:1974–1976.

374. Van Ranst, M., M. Joossens, S. Joossens, K. Van Steen, M. Pierik, S. Vermeire, and P. Rutgeerts. 2005. Crohn's disease and month of birth. *Inflammatory Bowel Diseases* 11:597–599.
375. Angelucci, E., A. Cocco, M. Cesarini, A. Crudeli, S. Necozione, R. Caprilli, and G. Latella. 2009. Monthly and seasonal birth patterns and the occurrence of Crohn's disease. *The American Journal of Gastroenterology* 104:1608–1609.
376. Romberg-Camps, M. J., M. A. Hesselink-van de Kruijs, L. J. Schouten, P. C. Dagnelie, C. B. Limonard, A. D. Kester, L. P. Bos, J. Goedhard, W. H. Hameeteman, F. L. Wolters, M. G. Russel, and R. W. Stockbrugger. 2009. Inflammatory bowel disease in south limburg (the Netherlands) 1991–2002: Incidence, diagnostic delay, and seasonal variations in onset of symptoms. *Journal of Crohn's & Colitis* 3:115–124.
377. Sonnenberg, A. 2009. Date of birth in the occurrence of inflammatory bowel disease. *Inflammatory Bowel Diseases* 15:206–211.
378. Zeng, L., and F. H. Anderson. 1996. Seasonal change in the exacerbations of Crohn's disease. *Scandinavian Journal of Gastroenterology* 31:79–82.
379. Aratari, A., C. Papi, B. Galletti, E. Angelucci, A. Viscido, V. D'Ovidio, A. Ciaco, M. Abdullahi, and R. Caprilli. 2006. Seasonal variations in onset of symptoms in Crohn's disease. *Digestive and Liver Disease* 38:319–323.
380. Bai, A., Y. Guo, Y. Shen, Y. Xie, X. Zhu, and N. Lu. 2009. Seasonality in flares and months of births of patients with ulcerative colitis in a Chinese population. *Digestive Diseases and Sciences* 54:1094–1098.
381. Lewis, J. D., F. N. Aberra, G. R. Lichtenstein, W. B. Bilker, C. Brensinger, and B. L. Strom. 2004. Seasonal variation in flares of inflammatory bowel disease. *Gastroenterology* 126:665–673.
382. Garland, C. F., and F. C. Garland. 1980. Do sunlight and vitamin D reduce the likelihood of colon cancer? *International Journal of Epidemiology* 9:227–231.
383. Sonnenberg, A. 1990. Occupational distribution of inflammatory bowel disease among German employees. *Gut* 31:1037–1040.
384. Nerich, V., P. Jantchou, M. C. Boutron-Ruault, E. Monnet, A. Weill, V. Vanbockstael, G. R. Auleley, C. Balaire, P. Dubost, S. Rican, H. Allemand, and F. Carbonnel. 2011. Low exposure to sunlight is a risk factor for Crohn's disease. *Alimentary Pharmacology & Therapeutics* 33:940–945.
385. Khalili, H., E. S. Huang, A. N. Ananthakrishnan, L. Higuchi, J. M. Richter, C. S. Fuchs, and A. T. Chan. 2012. Geographical variation and incidence of inflammatory bowel disease among US women. *Gut* 10.1136/gutjnl-2011-301574.
386. Cross, H. S., P. Bareis, H. Hofer, M. G. Bischof, E. Bajna, S. Kriwanek, E. Bonner, and M. Peterlik. 2001. 25-Hydroxyvitamin D(3)-1alpha-hydroxylase and vitamin D receptor gene expression in human colonic mucosa is elevated during early cancerogenesis. *Steroids* 66:287–292.
387. Kallay, E., P. Pietschmann, S. Toyokuni, E. Bajna, P. Hahn, K. Mazzucco, C. Bieglmayer, S. Kato, and H. S. Cross. 2001. Characterization of a vitamin D receptor knockout mouse as a model of colorectal hyperproliferation and DNA damage. *Carcinogenesis* 22:1429–1435.
388. Cantorna, M. T., C. Munsick, C. Bemiss, and B. D. Mahon. 2000. 1,25-Dihydroxycholecalciferol prevents and ameliorates symptoms of experimental murine inflammatory bowel disease. *Journal of Nutrition* 130:2648–2652.
389. Pappa, H. M., R. J. Grand, and C. M. Gordon. 2006. Report on the vitamin D status of adult and pediatric patients with inflammatory bowel disease and its significance for bone health and disease. *Inflammatory Bowel Diseases* 12:1162–1174.
390. Leslie, W. D., N. Miller, L. Rogala, and C. N. Bernstein. 2008. Vitamin D status and bone density in recently diagnosed inflammatory bowel disease: The Manitoba IBD Cohort Study. *The American Journal of Gastroenterology* 103:1451–1459.
391. Levin, A. D., V. Wadhera, S. T. Leach, H. J. Woodhead, D. A. Lemberg, A. C. Mendoza-Cruz, and A. S. Day. 2011. Vitamin D deficiency in children with inflammatory bowel disease. *Digestive Diseases and Sciences* 56:830–836.
392. Bowden, S. A., R. F. Robinson, R. Carr, and J. D. Mahan. 2008. Prevalence of vitamin D deficiency and insufficiency in children with osteopenia or osteoporosis referred to a pediatric metabolic bone clinic. *Pediatrics* 121:e1585–1590.
393. Gilman, J., F. Shanahan, and K. D. Cashman. 2006. Determinants of vitamin D status in adult Crohn's disease patients, with particular emphasis on supplemental vitamin D use. *European Journal of Clinical Nutrition* 60:889–896.
394. El-Matary, W., S. Sikora, and D. Spady. 2011. Bone mineral density, vitamin D, and disease activity in children newly diagnosed with inflammatory bowel disease. *Digestive Diseases and Sciences* 56:825–829.

395. Pappa, H. M., E. J. Langereis, R. J. Grand, and C. M. Gordon. 2011. Prevalence and risk factors for hypovitaminosis D in young patients with inflammatory bowel disease: A retrospective study. *Journal of Pediatric Gastroenterology and Nutrition* 53:361–364.

396. Atia, A., R. Murthy, B. A. Bailey, T. Manning, L. L. Garrett, D. Youssef, and A. N. Peiris. 2011. Vitamin D status in veterans with inflammatory bowel disease: Relationship to health care costs and services. *Military Medicine* 176:711–714.

397. Ulitsky, A., A. N. Ananthakrishnan, A. Naik, S. Skaros, Y. Zadvornova, D. G. Binion, and M. Issa. 2011. Vitamin D deficiency in patients with inflammatory bowel disease: Association with disease activity and quality of life. *JPEN. Journal of Parenteral and Enteral Nutrition* 35:308–316.

398. Bours, P. H., J. P. Wielders, J. R. Vermeijden, and A. van de Wiel. 2010. Seasonal variation of serum 25-hydroxyvitamin D levels in adult patients with inflammatory bowel disease. *Osteoporosis International* 22:2857–2867.

399. Joseph, A. J., B. George, A. B. Pulimood, M. S. Seshadri, and A. Chacko. 2009. 25 (OH) vitamin D level in Crohn's disease: Association with sun exposure and disease activity. *The Indian Journal of Medical Research* 130:133–137.

400. Tajika, M., A. Matsuura, T. Nakamura, T. Suzuki, A. Sawaki, T. Kato, K. Hara, K. Ookubo, K. Yamao, M. Kato, and Y. Muto. 2004. Risk factors for vitamin D deficiency in patients with Crohn's disease. *Journal of Gastroenterology* 39:527–533.

401. Van Limbergen, J., R. K. Russell, E. R. Nimmo, and J. Satsangi. 2007. The genetics of inflammatory bowel disease. *The American Journal of Gastroenterology* 102:2820–2831.

402. Kanneganti, T. D., M. Lamkanfi, and G. Nunez. 2007. Intracellular NOD-like receptors in host defense and disease. *Immunity* 27:549–559.

403. Baranzini, S. E., N. W. Galwey, J. Wang, P. Khankhanian, R. Lindberg, D. Pelletier, W. Wu, B. M. Uitdehaag, L. Kappos, C. H. Polman, P. M. Matthews, S. L. Hauser, R. A. Gibson, J. R. Oksenberg, and M. R. Barnes. 2009. Pathway and network-based analysis of genome-wide association studies in multiple sclerosis. *Human Molecular Genetics* 18:2078–2090.

404. Rubtsov, Y. P., J. P. Rasmussen, E. Y. Chi, J. Fontenot, L. Castelli, X. Ye, P. Treuting, L. Siewe, A. Roers, W. R. Henderson, Jr., W. Muller, and A. Y. Rudensky. 2008. Regulatory T cell–derived interleukin-10 limits inflammation at environmental interfaces. *Immunity* 28:546–558.

405. Cyktor, J. C., and J. Turner. 2011. Interleukin-10 and immunity against prokaryotic and eukaryotic intracellular pathogens. *Infection and Immunity* 79:2964–2973.

406. Wang, T. T., B. Dabbas, D. Laperriere, A. J. Bitton, H. Soualhine, L. E. Tavera-Mendoza, S. Dionne, M. J. Servant, A. Bitton, E. G. Seidman, S. Mader, M. A. Behr, and J. H. White. 2010. Direct and indirect induction by 1,25-dihydroxyvitamin D3 of the NOD2/CARD15-defensin beta2 innate immune pathway defective in Crohn disease. *Journal of Biological Chemistry* 285:2227–2231.

407. Chen, S., G. P. Sims, X. X. Chen, Y. Y. Gu, and P. E. Lipsky. 2007. Modulatory effects of 1,25-Dihydroxyvitamin D3 on human B cell differentiation. *Journal of Immunology* 179:1634–1647.

408. Heine, G., U. Niesner, H. D. Chang, A. Steinmeyer, U. Zugel, T. Zuberbier, A. Radbruch, and M. Worm. 2008. 1,25-Dihydroxyvitamin D(3) promotes IL-10 production in human B cells. *European Journal of Immunology* 38:2210–2218.

409. Wang, K., R. Baldassano, H. Zhang, H. Q. Qu, M. Imielinski, S. Kugathasan, V. Annese, M. Dubinsky, J. I. Rotter, R. K. Russell, J. P. Bradfield, P. M. Sleiman, J. T. Glessner, T. Walters, C. Hou, C. Kim, E. C. Frackelton, M. Garris, J. Doran, C. Romano, C. Catassi, J. Van Limbergen, S. L. Guthery, L. Denson, D. Piccoli, M. S. Silverberg, C. A. Stanley, D. Monos, D. C. Wilson, A. Griffiths, S. F. Grant, J. Satsangi, C. Polychronakos, and H. Hakonarson. 2010. Comparative genetic analysis of inflammatory bowel disease and type 1 diabetes implicates multiple loci with opposite effects. *Human Molecular Genetics* 19:2059–2067.

410. Simmons, J. D., C. Mullighan, K. I. Welsh, and D. P. Jewell. 2000. Vitamin D receptor gene polymorphism: Association with Crohn's disease susceptibility. *Gut* 47:211–214.

411. Naderi, N., A. Farnood, M. Habibi, F. Derakhshan, H. Balaii, Z. Motahari, M. R. Agah, F. Firouzi, M. G. Rad, R. Aghazadeh, H. Zojaji, and M. R. Zali. 2008. Association of vitamin D receptor gene polymorphisms in Iranian patients with inflammatory bowel disease. *Journal of Gastroenterology and Hepatology* 23:1816–1822.

412. Pei, F. H., Y. J. Wang, S. L. Gao, B. R. Liu, Y. J. Du, W. Liu, H. Y. Yu, L. X. Zhao, and B. R. Chi. 2011. Vitamin D receptor gene polymorphism and ulcerative colitis susceptibility in Han Chinese. *Journal of Digestive Diseases* 12:90–98.

413. Dresner-Pollak, R., Z. Ackerman, R. Eliakim, A. Karban, Y. Chowers, and H. H. Fidder. 2004. The BsmI vitamin D receptor gene polymorphism is associated with ulcerative colitis in Jewish Ashkenazi patients. *Genetic Testing* 8:417–420.

414. Eloranta, J. J., C. Wenger, J. Mwinyi, C. Hiller, C. Gubler, S. R. Vavricka, M. Fried, and G. A. Kullak-Ublick. 2011. Association of a common vitamin D–binding protein polymorphism with inflammatory bowel disease. *Pharmacogenetics and Genomics* 21:559–564.

415. Nagasawa, H., Y. Uto, H. Sasaki, N. Okamura, A. Murakami, S. Kubo, K. L. Kirk, and H. Hori. 2005. Gc protein (vitamin D–binding protein): Gc genotyping and GcMAF precursor activity. *Anticancer Research* 25:3689–3695.

416. Miheller, P., G. Muzes, I. Hritz, G. Lakatos, I. Pregun, P. L. Lakatos, L. Herszenyi, and Z. Tulassay. 2009. Comparison of the effects of 1,25 dihydroxyvitamin D and 25 hydroxyvitamin D on bone pathology and disease activity in Crohn's disease patients. *Inflammatory Bowel Diseases* 15:1656–1662.

417. Jorgensen, S. P., J. Agnholt, H. Glerup, S. Lyhne, G. E. Villadsen, C. L. Hvas, L. E. Bartels, J. Kelsen, L. A. Christensen, and J. F. Dahlerup. 2010. Clinical trial: Vitamin D3 treatment in Crohn's disease—a randomized double-blind placebo-controlled study. *Alimentary Pharmacology & Therapeutics* 32:377–383.

418. Hollis, B. W. 2005. Circulating 25-hydroxyvitamin D levels indicative of vitamin D sufficiency: Implications for establishing a new effective dietary intake recommendation for vitamin D. *Journal of Nutrition* 135:317–322.

419. Saleh, M., and C. O. Elson. 2011. Experimental inflammatory bowel disease: Insights into the host-microbiota dialog. *Immunity* 34:293–302.

420. Mizoguchi, A., and E. Mizoguchi. 2010. Animal models of IBD: Linkage to human disease. *Current Opinion in Pharmacology* 10:578–587.

421. Froicu, M., and M. T. Cantorna. 2007. Vitamin D and the vitamin D receptor are critical for control of the innate immune response to colonic injury. *BMC Immunology* 8:5.

422. Laverny, G., G. Penna, S. Vetrano, C. Correale, M. Nebuloni, S. Danese, and L. Adorini. 2010. Efficacy of a potent and safe vitamin D receptor agonist for the treatment of inflammatory bowel disease. *Immunology Letters* 131:49–58.

423. Liu, N., L. Nguyen, R. F. Chun, V. Lagishetty, S. Ren, S. Wu, B. Hollis, H. F. DeLuca, J. S. Adams, and M. Hewison. 2008. Altered endocrine and autocrine metabolism of vitamin D in a mouse model of gastrointestinal inflammation. *Endocrinology* 149:4799–4808.

424. Froicu, M., V. Weaver, T. A. Wynn, M. A. McDowell, J. E. Welsh, and M. T. Cantorna. 2003. A crucial role for the vitamin D receptor in experimental inflammatory bowel diseases. *Molecular Endocrinology* 17:2386–2392.

425. Kong, J., Z. Zhang, M. W. Musch, G. Ning, J. Sun, J. Hart, M. Bissonnette, and Y. C. Li. 2008. Novel role of the vitamin D receptor in maintaining the integrity of the intestinal mucosal barrier. *American Journal of Physiology. Gastrointestinal and Liver Physiology* 294:G208–216.

426. Hewison, M. 2011. Antibacterial effects of vitamin D. *Nature Reviews. Endocrinology* 7:337–345.

427. van Heel, D. A., S. Ghosh, M. Butler, K. A. Hunt, A. M. Lundberg, T. Ahmad, D. P. McGovern, C. Onnie, K. Negoro, S. Goldthorpe, B. M. Foxwell, C. G. Mathew, A. Forbes, D. P. Jewell, and R. J. Playford. 2005. Muramyl dipeptide and toll-like receptor sensitivity in NOD2-associated Crohn's disease. *Lancet* 365:1794–1796.

428. Liu, P. T., S. Stenger, H. Li, L. Wenzel, B. H. Tan, S. R. Krutzik, M. T. Ochoa, J. Schauber, K. Wu, C. Meinken, D. L. Kamen, M. Wagner, R. Bals, A. Steinmeyer, U. Zugel, R. L. Gallo, D. Eisenberg, M. Hewison, B. W. Hollis, J. S. Adams, B. R. Bloom, and R. L. Modlin. 2006. Toll-like receptor triggering of a vitamin D–mediated human antimicrobial response. *Science* 311:1770–1773.

429. Krutzik, S. R., M. Hewison, P. T. Liu, J. A. Robles, S. Stenger, J. S. Adams, and R. L. Modlin. 2008. IL-15 links TLR2/1-induced macrophage differentiation to the vitamin D–dependent antimicrobial pathway. *Journal of Immunology* 181:7115–7120.

430. Wang, T. T., F. P. Nestel, V. Bourdeau, Y. Nagai, Q. Wang, J. Liao, L. Tavera-Mendoza, R. Lin, J. W. Hanrahan, S. Mader, and J. H. White. 2004. Cutting edge: 1,25-dihydroxyvitamin D3 is a direct inducer of antimicrobial peptide gene expression. *Journal of Immunology* 173:2909–2912.

431. Weber, G., J. D. Heilborn, C. I. Chamorro Jimenez, A. Hammarsjo, H. Torma, and M. Stahle. 2005. Vitamin D induces the antimicrobial protein hCAP18 in human skin. *The Journal of Investigative Dermatology* 124:1080–1082.

432. Gombart, A. F., N. Borregaard, and H. P. Koeffler. 2005. Human cathelicidin antimicrobial peptide (CAMP) gene is a direct target of the vitamin D receptor and is strongly up-regulated in myeloid cells by 1,25-dihydroxyvitamin D3. *FASEB Journal* 19:1067–1077.

433. Sadeghi, K., B. Wessner, U. Laggner, M. Ploder, D. Tamandl, J. Friedl, U. Zugel, A. Steinmeyer, A. Pollak, E. Roth, G. Boltz-Nitulescu, and A. Spittler. 2006. Vitamin D3 down-regulates monocyte TLR expression and triggers hyporesponsiveness to pathogen-associated molecular patterns. *European Journal of Immunology* 36:361–370.

434. Verway, M., M. A. Behr, and J. H. White. 2010. Vitamin D, NOD2, autophagy and Crohn's disease. *Expert Review of Clinical Immunology* 6:505–508.

435. Gombart, A. F., T. Saito, and H. P. Koeffler. 2009. Exapation of an ancient Alu short interspersed element provides a highly conserved vitamin D–mediated innate immune response in humans and primates. *BMC Genomics* 10:321.

436. Lagishetty, V., A. V. Misharin, N. Q. Liu, T. S. Lisse, R. F. Chun, Y. Ouyang, S. M. McLachlan, J. S. Adams, and M. Hewison. 2010. Vitamin D deficiency in mice impairs colonic antibacterial activity and predisposes to colitis. *Endocrinology* 151:2423–2432.

437. Taniguchi, M., K. Seino, and T. Nakayama. 2003. The NKT cell system: Bridging innate and acquired immunity. *Nature Immunology* 4:1164–1165.

438. Yu, S., and M. T. Cantorna. 2008. The vitamin D receptor is required for iNKT cell development. *Proceedings of the National Academy of Sciences of the United States of America* 105:5207–5212.

439. Strober, W., I. Fuss, and P. Mannon. 2007. The fundamental basis of inflammatory bowel disease. *Journal of Clinical Investigation* 117:514–521.

440. Sarra, M., F. Pallone, T. T. Macdonald, and G. Monteleone. 2010. IL-23/IL-17 axis in IBD. *Inflammatory Bowel Diseases* 16:1808–1813.

441. Shen, W., and S. K. Durum. 2010. Synergy of IL-23 and Th17 cytokines: New light on inflammatory bowel disease. *Neurochemical Research* 35:940–946.

442. Bruce, D., S. Yu, J. H. Ooi, and M. T. Cantorna. 2011. Converging pathways lead to overproduction of IL-17 in the absence of vitamin D signaling. *International Immunology* 23:519–528.

443. Thompson, A. I., and C. W. Lees. 2011. Genetics of ulcerative colitis. *Inflammatory Bowel Diseases* 17:831–848.

444. Bartels, L. E., S. P. Jorgensen, J. Agnholt, J. Kelsen, C. L. Hvas, and J. F. Dahlerup. 2007. 1,25-dihydroxyvitamin D3 and dexamethasone increase interleukin-10 production in CD4+ T cells from patients with Crohn's disease. *International Immunopharmacology* 7:1755–1764.

445. Urry, Z., E. Xystrakis, D. F. Richards, J. McDonald, Z. Sattar, D. J. Cousins, C. J. Corrigan, E. Hickman, Z. Brown, and C. M. Hawrylowicz. 2009. Ligation of TLR9 induced on human IL-10–secreting Tregs by 1alpha,25-dihydroxyvitamin D3 abrogates regulatory function. *Journal of Clinical Investigation* 119:387–398.

446. Biggs, L., C. Yu, B. Fedoric, A. F. Lopez, S. J. Galli, and M. A. Grimbaldeston. 2010. Evidence that vitamin D(3) promotes mast cell–dependent reduction of chronic UVB-induced skin pathology in mice. *Journal of Experimental Medicine* 207:455–463.

447. Yu, S., D. Bruce, M. Froicu, V. Weaver, and M. T. Cantorna. 2008. Failure of T cell homing, reduced CD4/CD8alphaalpha intraepithelial lymphocytes, and inflammation in the gut of vitamin D receptor KO mice. *Proceedings of the National Academy of Sciences of the United States of America* 105:20834–20839.

448. Bruce, D., and M. T. Cantorna. 2011. Intrinsic requirement for the vitamin D receptor in the development of CD8alphaalpha-expressing T cells. *Journal of Immunology* 186:2819–2825.

449. Protiva, P., H. S. Cross, M. E. Hopkins, E. Kallay, G. Bises, E. Dreyhaupt, L. Augenlicht, M. Lipkin, M. Lesser, E. Livote, and P. R. Holt. 2009. Chemoprevention of colorectal neoplasia by estrogen: Potential role of vitamin D activity. *Cancer Prevention Research (Philadelphia, Pa.)* 2:43–51.

450. Tsokos, G. C. 2011. Systemic lupus erythematosus. *New England Journal of Medicine* 365:2110–2121.

451. Flesher, D. L., X. Sun, T. W. Behrens, R. R. Graham, and L. A. Criswell. 2010. Recent advances in the genetics of systemic lupus erythematosus. *Expert Review of Clinical Immunology* 6:461–479.

452. Grennan, D. M., A. Parfitt, N. Manolios, Q. Huang, V. Hyland, H. Dunckley, T. Doran, P. Gatenby, and C. Badcock. 1997. Family and twin studies in systemic lupus erythematosus. *Disease Markers* 13:93–98.

453. Sanders, C. J., H. Van Weelden, G. A. Kazzaz, V. Sigurdsson, J. Toonstra, and C. A. Bruijnzeel-Koomen. 2003. Photosensitivity in patients with lupus erythematosus: A clinical and photobiological study of 100 patients using a prolonged phototest protocol. *The British Journal of Dermatology* 149:131–137.

454. Meller, S., F. Winterberg, M. Gilliet, A. Muller, I. Lauceviciute, J. Rieker, N. J. Neumann, R. Kubitza, M. Gombert, E. Bunemann, U. Wiesner, P. Franken-Kunkel, H. Kanzler, M. C. Dieu-Nosjean, A. Amara, T. Ruzicka, P. Lehmann, A. Zlotnik, and B. Homey. 2005. Ultraviolet radiation-induced injury, chemokines, and leukocyte recruitment: An amplification cycle triggering cutaneous lupus erythematosus. *Arthritis and Rheumatism* 52:1504–1516.

455. Bijl, M., and C. G. Kallenberg. 2006. Ultraviolet light and cutaneous lupus. *Lupus* 15:724–727.

456. Kuhn, A., M. Herrmann, S. Kleber, M. Beckmann-Welle, K. Fehsel, A. Martin-Villalba, P. Lehmann, T. Ruzicka, P. H. Krammer, and V. Kolb-Bachofen. 2006. Accumulation of apoptotic cells in the epidermis of patients with cutaneous lupus erythematosus after ultraviolet irradiation. *Arthritis and Rheumatism* 54:939–950.

457. Breslin, L. C., P. J. Magee, J. M. Wallace, and E. M. McSorley. 2011. An evaluation of vitamin D status in individuals with systemic lupus erythematosus. *The Proceedings of the Nutrition Society* 70:399–407.
458. Holick, M. F. 2008. Sunlight, UV-radiation, vitamin D and skin cancer: How much sunlight do we need? *Advances in Experimental Medicine and Biology* 624:1–15.
459. Kamen, D. L. 2010. Vitamin D in lupus—new kid on the block? *Bulletin of the NYU Hospital for Joint Diseases* 68:218–222.
460. Cutolo, M., M. Plebani, Y. Shoenfeld, L. Adorini, and A. Tincani. 2011. Vitamin D endocrine system and the immune response in rheumatic diseases. *Vitamins and Hormones* 86:327–351.
461. Zandman-Goddard, G., M. Solomon, Z. Rosman, E. Peeva, and Y. Shoenfeld. 2012. Environment and lupus-related diseases. *Lupus* 21:241–250.
462. Somers, E. C., S. L. Thomas, L. Smeeth, W. M. Schoonen, and A. J. Hall. 2007. Incidence of systemic lupus erythematosus in the United Kingdom, 1990–1999. *Arthritis and Rheumatism* 57:612–618.
463. Walsh, S. J., and L. M. DeChello. 2001. Geographical variation in mortality from systemic lupus erythematosus in the United States. *Lupus* 10:637–646.
464. Grant, W. B. 2004. Solar UV-B radiation is linked to the geographic variation of mortality from systemic lupus erythematosus in the USA. *Lupus* 13:281–282.
465. Walsh, S. J., and A. Gilchrist. 2006. Geographical clustering of mortality from systemic lupus erythematosus in the United States: Contributions of poverty, Hispanic ethnicity and solar radiation. *Lupus* 15:662–670.
466. Haga, H. J., J. G. Brun, O. P. Rekvig, and L. Wetterberg. 1999. Seasonal variations in activity of systemic lupus erythematosus in a subarctic region. *Lupus* 8:269–273.
467. Costenbader, K. H., D. Feskanich, M. Holmes, E. W. Karlson, and E. Benito-Garcia. 2008. Vitamin D intake and risks of systemic lupus erythematosus and rheumatoid arthritis in women. *Annals of the Rheumatic Diseases* 67:530–535.
468. Kamen, D. L., G. S. Cooper, H. Bouali, S. R. Shaftman, B. W. Hollis, and G. S. Gilkeson. 2006. Vitamin D deficiency in systemic lupus erythematosus. *Autoimmunity Reviews* 5:114–117.
469. Wright, T. B., J. Shults, M. B. Leonard, B. S. Zemel, and J. M. Burnham. 2009. Hypovitaminosis D is associated with greater body mass index and disease activity in pediatric systemic lupus erythematosus. *Journal of Pediatrics* 155:260–265.
470. Borba, V. Z., J. G. Vieira, T. Kasamatsu, S. C. Radominski, E. I. Sato, and M. Lazaretti-Castro. 2009. Vitamin D deficiency in patients with active systemic lupus erythematosus. *Osteoporosis International* 20:427–433.
471. Thudi, A., S. Yin, A. E. Wandstrat, Q. Z. Li, and N. J. Olsen. 2008. Vitamin D levels and disease status in Texas patients with systemic lupus erythematosus. *The American Journal of the Medical Sciences* 335:99–104.
472. Ruiz-Irastorza, G., M. V. Egurbide, N. Olivares, A. Martinez-Berriotxoa, and C. Aguirre. 2008. Vitamin D deficiency in systemic lupus erythematosus: Prevalence, predictors and clinical consequences. *Rheumatology (Oxford)* 47:920–923.
473. Chen, S., G. P. Sims, X. X. Chen, Y. Y. Gu, and P. E. Lipsky. 2007. Modulatory effects of 1,25-dihydroxyvitamin D3 on human B cell differentiation. *Journal of Immunology* 179:1634–1647.
474. Damanhouri, L. H. 2009. Vitamin D deficiency in Saudi patients with systemic lupus erythematosus. *Saudi Medical Journal* 30:1291–1295.
475. Wu, P. W., E. Y. Rhew, A. R. Dyer, D. D. Dunlop, C. B. Langman, H. Price, K. Sutton-Tyrrell, D. D. McPherson, D. Edmundowicz, G. T. Kondos, and R. Ramsey-Goldman. 2009. 25-Hydroxyvitamin D and cardiovascular risk factors in women with systemic lupus erythematosus. *Arthritis and Rheumatism* 61:1387–1395.
476. Bogaczewicz, J., A. Sysa-Jedrzejowska, C. Arkuszewska, J. Zabek, E. Kontny, D. P. McCauliffe, and A. Wozniacka. 2011. Vitamin D status in systemic lupus erythematosus patients and its association with selected clinical and laboratory parameters. *Lupus* 21:477–484.
477. Amital, H., Z. Szekanecz, G. Szucs, K. Danko, E. Nagy, T. Csepany, E. Kiss, J. Rovensky, A. Tuchynova, D. Kozakova, A. Doria, N. Corocher, N. Agmon-Levin, V. Barak, H. Orbach, G. Zandman-Goddard, and Y. Shoenfeld. 2010. Serum concentrations of 25-OH vitamin D in patients with systemic lupus erythematosus (SLE) are inversely related to disease activity: Is it time to routinely supplement patients with SLE with vitamin D? *Annals of the Rheumatic Diseases* 69:1155–1157.
478. Ben-Zvi, I., C. Aranow, M. Mackay, A. Stanevsky, D. L. Kamen, L. M. Marinescu, C. E. Collins, G. S. Gilkeson, B. Diamond, and J. A. Hardin. 2010. The impact of vitamin D on dendritic cell function in patients with systemic lupus erythematosus. *PLoS One* 5:e9193.
479. Hamza, R. T., K. S. Awwad, M. K. Ali, and A. I. Hamed. 2011. Reduced serum concentrations of 25-hydroxy vitamin D in Egyptian patients with systemic lupus erythematosus: Relation to disease activity. *Medical Science Monitor* 17:CR711–718.

480. Reynolds, J. A., S. Haque, J. L. Berry, P. Pemberton, L. S. Teh, P. Ho, R. Gorodkin, and I. N. Bruce. 2011. 25-Hydroxyvitamin D deficiency is associated with increased aortic stiffness in patients with systemic lupus erythematosus. *Rheumatology (Oxford)* 51:544–551.
481. Szodoray, P., T. Tarr, A. Bazso, G. Poor, G. Szegedi, and E. Kiss. 2011. The immunopathological role of vitamin D in patients with SLE: Data from a single centre registry in Hungary. *Scandinavian Journal of Rheumatology* 40:122–126.
482. Word, A. P., F. Perese, L. C. Tseng, B. Adams-Huet, N. J. Olsen, and B. F. Chong. 2012. 25-Hydroxyvitamin D levels in African-American and Caucasian/Hispanic subjects with cutaneous lupus erythematosus. *The British Journal of Dermatology* 166:372–379.
483. Mok, C. C., D. J. Birmingham, H. W. Leung, L. A. Hebert, H. Song, and B. H. Rovin. 2011. Vitamin D levels in Chinese patients with systemic lupus erythematosus: Relationship with disease activity, vascular risk factors and atherosclerosis. *Rheumatology (Oxford)* 51:644–652.
484. Mok, C. C., D. J. Birmingham, L. Y. Ho, L. A. Hebert, H. Song, and B. H. Rovin. 2012. Vitamin D deficiency as marker for disease activity and damage in systemic lupus erythematosus: A comparison with anti-dsDNA and anti-C1q. *Lupus* 21:36–42.
485. Bonakdar, Z. S., L. Jahanshahifar, F. Jahanshahifar, and A. Gholamrezaei. 2011. Vitamin D deficiency and its association with disease activity in new cases of systemic lupus erythematosus. *Lupus* 20:1155–1160.
486. Lopez-Robles, C., R. Rios-Fernandez, J. L. Callejas-Rubio, and N. Ortego-Centeno. 2011. Vitamin D deficiency in a cohort of patients with systemic lupus erythematous from the South of Spain. *Lupus* 20:330–331.
487. Ruiz-Irastorza, G., S. Gordo, N. Olivares, M. V. Egurbide, and C. Aguirre. 2010. Changes in vitamin D levels in patients with systemic lupus erythematosus: Effects on fatigue, disease activity, and damage. *Arthritis Care & Research* 62:1160–1165.
488. Toloza, S. M., D. E. Cole, D. D. Gladman, D. Ibanez, and M. B. Urowitz. 2010. Vitamin D insufficiency in a large female SLE cohort. *Lupus* 19:13–19.
489. Robinson, A. B., M. Thierry-Palmer, K. L. Gibson, and C. E. Rabinovich. 2012. Disease activity, proteinuria, and vitamin D status in children with systemic lupus erythematosus and juvenile dermatomyositis. *Journal of Pediatrics* 160:297–302.
490. Ritterhouse, L. L., S. R. Crowe, T. B. Niewold, D. L. Kamen, S. R. Macwana, V. C. Roberts, A. B. Dedeke, J. B. Harley, R. H. Scofield, J. M. Guthridge, and J. A. James. 2011. Vitamin D deficiency is associated with an increased autoimmune response in healthy individuals and in patients with systemic lupus erythematosus. *Annals of the Rheumatic Diseases* 70:1569–1574.
491. Lee, Y. H., S. C. Bae, S. J. Choi, J. D. Ji, and G. G. Song. 2011. Associations between vitamin D receptor polymorphisms and susceptibility to rheumatoid arthritis and systemic lupus erythematosus: A meta-analysis. *Molecular Biology Reports* 38:3643–3651.
492. Luo, X. Y., M. H. Yang, F. X. Wu, L. J. Wu, L. Chen, Z. Tang, N. T. Liu, X. F. Zeng, J. L. Guan, and G. H. Yuan. 2012. Vitamin D receptor gene BsmI polymorphism B allele, but not BB genotype, is associated with systemic lupus erythematosus in a Han Chinese population. *Lupus* 21:53–59.
493. Abbasi, M., Z. Rezaieyazdi, J. T. Afshari, M. Hatef, M. Sahebari, and N. Saadati. 2010. Lack of association of vitamin D receptor gene BsmI polymorphisms in patients with systemic lupus erythematosus. *Rheumatology International* 30:1537–1539.
494. Monticielo, O. A., J. C. Brenol, J. A. Chies, M. G. Longo, G. G. Rucatti, R. Scalco, and R. M. Xavier. 2012. The role of BsmI and FokI vitamin D receptor gene polymorphisms and serum 25-hydroxyvitamin D in Brazilian patients with systemic lupus erythematosus. *Lupus* 21:43–52.
495. Fernando, M. M., C. R. Stevens, P. C. Sabeti, E. C. Walsh, A. J. McWhinnie, A. Shah, T. Green, J. D. Rioux, and T. J. Vyse. 2007. Identification of two independent risk factors for lupus within the MHC in United Kingdom families. *PLoS Genetics* 3:e192.
496. Furukawa, F., Y. Yamamoto, N. Kanazawa, and M. Muto. 2009. Race differences in immunogenetic features and photosensitivity of cutaneous lupus erythematosus from the aspect of Japanese studies. *Annals of the New York Academy of Sciences* 1173:552–556.
497. Furukawa, F., and M. Muto. 2009. Ethnic differences in immunogenetic features and photosensitivity of cutaneous lupus erythematosus. *Archives of Dermatological Research* 301:111–115.
498. Ramagopalan, S. V., J. Link, J. K. Byrnes, D. A. Dyment, G. Giovannoni, R. Q. Hintzen, E. Sundqvist, I. Kockum, C. Smestad, B. A. Lie, H. F. Harbo, L. Padyukov, L. Alfredsson, T. Olsson, A. D. Sadovnick, J. Hillert, and G. C. Ebers. 2009. HLA-DRB1 and month of birth in multiple sclerosis. *Neurology* 73:2107–2111.
499. Fairhurst, A. M., A. E. Wandstrat, and E. K. Wakeland. 2006. Systemic lupus erythematosus: Multiple immunological phenotypes in a complex genetic disease. *Advances in Immunology* 92:1–69.

500. Perry, D., A. Sang, Y. Yin, Y. Y. Zheng, and L. Morel. 2011. Murine models of systemic lupus erythematosus. *Journal of Biomedicine & Biotechnology* 2011:271694.
501. Gulinello, M., and C. Putterman. 2011. The MRL/lpr mouse strain as a model for neuropsychiatric systemic lupus erythematosus. *Journal of Biomedicine & Biotechnology* 2011:207504.
502. Cohen, P. L., and R. A. Eisenberg. 1991. Lpr and gld: Single gene models of systemic autoimmunity and lymphoproliferative disease. *Annual Review of Immunology* 9:243–269.
503. Zhang, Z., V. C. Kyttaris, and G. C. Tsokos. 2009. The role of IL-23/IL-17 axis in lupus nephritis. *Journal of Immunology* 183:3160–3169.
504. Lemire, J. M., A. Ince, and M. Takashima. 1992. 1,25-Dihydroxyvitamin D3 attenuates the expression of experimental murine lupus of MRL/l mice. *Autoimmunity* 12:143–148.
505. Deluca, H. F., and M. T. Cantorna. 2001. Vitamin D: Its role and uses in immunology. *FASEB Journal* 15:2579–2585.
506. Grimbaldeston, M. A., S. Nakae, J. Kalesnikoff, M. Tsai, and S. J. Galli. 2007. Mast cell–derived interleukin 10 limits skin pathology in contact dermatitis and chronic irradiation with ultraviolet B. *Nature Immunology* 8:1095–1104.
507. Lerman, M., J. Burnham, and E. Behrens. 2011. 1,25 Dihydroxyvitamin D3 limits monocyte maturation in lupus sera. *Lupus* 20:749–753.
508. Wong, C. K., L. C. Lit, L. S. Tam, E. K. Li, P. T. Wong, and C. W. Lam. 2008. Hyperproduction of IL-23 and IL-17 in patients with systemic lupus erythematosus: Implications for Th17-mediated inflammation in auto-immunity. *Clinical Immunology* 127:385–393.
509. Crispin, J. C., M. Oukka, G. Bayliss, R. A. Cohen, C. A. Van Beek, I. E. Stillman, V. C. Kyttaris, Y. T. Juang, and G. C. Tsokos. 2008. Expanded double negative T cells in patients with systemic lupus erythematosus produce IL-17 and infiltrate the kidneys. *Journal of Immunology* 181:8761–8766.
510. Mok, M. Y., H. J. Wu, Y. Lo, and C. S. Lau. 2010. The relation of interleukin 17 (IL-17) and IL-23 to Th1/Th2 cytokines and disease activity in systemic lupus erythematosus. *The Journal of Rheumatology* 37:2046–2052.
511. Tanasescu, C., E. Balanescu, P. Balanescu, R. Olteanu, C. Badea, C. Grancea, C. Vagu, C. Bleotu, C. Ardeleanu, and A. Georgescu. 2010. IL-17 in cutaneous lupus erythematosus. *European Journal of Internal Medicine* 21:202–207.
512. Crispin, J. C., and G. C. Tsokos. 2010. IL-17 in systemic lupus erythematosus. *Journal of Biomedicine & Biotechnology* 2010:943254.
513. Apostolidis, S. A., J. C. Crispin, and G. C. Tsokos. 2011. IL-17–producing T cells in lupus nephritis. *Lupus* 20:120–124.
514. Perry, D., A. B. Peck, W. C. Carcamo, L. Morel, and C. Q. Nguyen. 2011. The current concept of T(h)17 cells and their expanding role in systemic lupus erythematosus. *Arthritis* 2011:810649.
515. Cutolo, M., C. Pizzorni, and A. Sulli. 2011. Vitamin D endocrine system involvement in autoimmune rheumatic diseases. *Autoimmunity Reviews* 11:84–87.
516. Marques, C. D., A. T. Dantas, T. S. Fragoso, and A. L. Duarte. 2010. The importance of vitamin D levels in autoimmune diseases. *Revista Brasileira de Reumatologia* 50:67–80.
517. Wen, H., and J. F. Baker. 2011. Vitamin D, immunoregulation, and rheumatoid arthritis. *Journal of Clinical Rheumatology* 17:102–107.
518. Tran, C. N., S. K. Lundy, and D. A. Fox. 2005. Synovial biology and T cells in rheumatoid arthritis. *Pathophysiology* 12:183–189.
519. Karouzakis, E., M. Neidhart, R. E. Gay, and S. Gay. 2006. Molecular and cellular basis of rheumatoid joint destruction. *Immunology Letters* 106:8–13.
520. Lundy, S. K., S. Sarkar, L. A. Tesmer, and D. A. Fox. 2007. Cells of the synovium in rheumatoid arthritis. T lymphocytes. *Arthritis Research & Therapy* 9:202.
521. Merlino, L. A., J. Curtis, T. R. Mikuls, J. R. Cerhan, L. A. Criswell, and K. G. Saag. 2004. Vitamin D intake is inversely associated with rheumatoid arthritis: Results from the Iowa Women's Health Study. *Arthritis and Rheumatism* 50:72–77.
522. Nielen, M. M., D. van Schaardenburg, W. F. Lems, R. J. van de Stadt, M. H. de Koning, H. W. Reesink, M. R. Habibuw, I. E. van der Horst-Bruinsma, J. W. Twisk, and B. A. Dijkmans. 2006. Vitamin D deficiency does not increase the risk of rheumatoid arthritis: Comment on the article by Merlino et al. *Arthritis and Rheumatism* 54:3719–3720.
523. Als, O. S., B. Riis, and C. Christiansen. 1987. Serum concentration of vitamin D metabolites in rheumatoid arthritis. *Clinical Rheumatology* 6:238–243.

524. Craig, S. M., F. Yu, J. R. Curtis, G. S. Alarcon, D. L. Conn, B. Jonas, L. F. Callahan, E. A. Smith, L. W. Moreland, S. L. Bridges, Jr., and T. R. Mikuls. 2010. Vitamin D status and its associations with disease activity and severity in African Americans with recent-onset rheumatoid arthritis. *The Journal of Rheumatology* 37:275–281.

525. Kerr, G. S., I. Sabahi, J. S. Richards, L. Caplan, G. W. Cannon, A. Reimold, G. M. Thiele, D. Johnson, and T. R. Mikuls. 2011. Prevalence of vitamin D insufficiency/deficiency in rheumatoid arthritis and associations with disease severity and activity. *The Journal of Rheumatology* 38:53–59.

526. Kroger, H., I. M. Penttila, and E. M. Alhava. 1993. Low serum vitamin D metabolites in women with rheumatoid arthritis. *Scandinavian Journal of Rheumatology* 22:172–177.

527. Heidari, B., K. Hajian-Tilaki, and P. Heidari. 2012. The status of serum vitamin D in patients with rheumatoid arthritis and undifferentiated inflammatory arthritis compared with controls. *Rheumatology International* 32:991–995.

528. Rossini, M., S. Maddali Bongi, G. La Montagna, G. Minisola, N. Malavolta, L. Bernini, E. Cacace, L. Sinigaglia, O. Di Munno, and S. Adami. 2010. Vitamin D deficiency in rheumatoid arthritis: Prevalence, determinants and associations with disease activity and disability. *Arthritis Research & Therapy* 12:R216.

529. Turhanoglu, A. D., H. Guler, Z. Yonden, F. Aslan, A. Mansuroglu, and C. Ozer. 2010. The relationship between vitamin D and disease activity and functional health status in rheumatoid arthritis. *Rheumatology International* 31:911–914.

530. Cutolo, M., K. Otsa, K. Laas, M. Yprus, R. Lehtme, M. E. Secchi, A. Sulli, S. Paolino, and B. Seriolo. 2006. Circannual vitamin D serum levels and disease activity in rheumatoid arthritis: Northern versus southern Europe. *Clinical and Experimental Rheumatology* 24:702–704.

531. Haque, U. J., and S. J. Bartlett. 2010. Relationships among vitamin D, disease activity, pain and disability in rheumatoid arthritis. *Clinical and Experimental Rheumatology* 28:745–747.

532. Patel, S., T. Farragher, J. Berry, D. Bunn, A. Silman, and D. Symmons. 2007. Association between serum vitamin D metabolite levels and disease activity in patients with early inflammatory polyarthritis. *Arthritis and Rheumatism* 56:2143–2149.

533. Gonzalez, M. P., Z. Gandara, Y. Fall, and G. Gomez. 2008. Radial distribution function descriptors for predicting affinity for vitamin D receptor. *European Journal of Medicinal Chemistry* 43:1360–1365.

534. Karray, E. F., I. Ben Dhifallah, K. Ben Abdelghani, I. Ben Ghorbel, M. Khanfir, H. Houman, K. Hamzaoui, and L. Zakraoui. 2012. Associations of vitamin D receptor gene polymorphisms FokI and BsmI with susceptibility to rheumatoid arthritis and Behcet's disease in Tunisians. *Joint Bone Spine* 79:144–148.

535. Maalej, A., E. Petit-Teixeira, L. Michou, A. Rebai, F. Cornelis, and H. Ayadi. 2005. Association study of VDR gene with rheumatoid arthritis in the French population. *Genes and Immunity* 6:707–711.

536. Milchert, M. 2010. [Association between BsmI vitamin D receptor gene polymorphism and serum concentration of vitamin D with progression of rheumatoid arthritis]. *Annales Academiae Medicae Stetinensis* 56:45–56.

537. Rass, P., A. Pakozdi, P. Lakatos, E. Zilahi, S. Sipka, G. Szegedi, and Z. Szekanecz. 2006. Vitamin D receptor gene polymorphism in rheumatoid arthritis and associated osteoporosis. *Rheumatology International* 26:964–971.

538. Andjelkovic, Z., J. Vojinovic, N. Pejnovic, M. Popovic, A. Dujic, D. Mitrovic, L. Pavlica, and D. Stefanovic. 1999. Disease modifying and immunomodulatory effects of high dose 1 alpha(OH)D3 in rheumatoid arthritis patients. *Clinical and Experimental Rheumatology* 17:453–456.

539. Gopinath, K., and D. Danda. 2011. Supplementation of 1,25 dihydroxy vitamin D3 in patients with treatment naive early rheumatoid arthritis: A randomized controlled trial. *International Journal of Rheumatic Diseases* 14:332–339.

540. Salesi, M., and Z. Farajzadegan. 2012. Efficacy of vitamin D in patients with active rheumatoid arthritis receiving methotrexate therapy. *Rheumatology International* 32:2129–2133.

541. Tsuji, M., K. Fujii, T. Nakano, and Y. Nishii. 1994. 1 alpha-Hydroxyvitamin D3 inhibits type II collagen-induced arthritis in rats. *FEBS Letters* 337:248–250.

542. Cantorna, M. T., C. E. Hayes, and H. F. DeLuca. 1998. 1,25-Dihydroxycholecalciferol inhibits the progression of arthritis in murine models of human arthritis. *Journal of Nutrition* 128:68–72.

543. Moghaddami, M., G. Mayrhofer, P. H. Anderson, H. A. Morris, M. Van Der Hoek, and L. G. Cleland. 2012. Efficacy and mechanisms of action of vitamin D in experimental polyarthritis. *Immunology and Cell Biology* 90:168–177.

544. Provvedini, D. M., C. D. Tsoukas, L. J. Deftos, and S. C. Manolagas. 1983. 1,25-Dihydroxyvitamin D3 receptors in human leukocytes. *Science* 221:1181–1183.

545. Bhalla, A. K., E. P. Amento, T. L. Clemens, M. F. Holick, and S. M. Krane. 1983. Specific high-affinity receptors for 1,25-dihydroxyvitamin D3 in human peripheral blood mononuclear cells: Presence in monocytes and induction in T lymphocytes following activation. *The Journal of Clinical Endocrinology and Metabolism* 57:1308–1310.

546. Nyomba, B. L., J. Auwerx, and V. Bormans. 1986. Pancreatic secretion in man with subclinical vitamin D deficiency. *Diabetologia* 29:34–38.

12 Vitamin D: Anti-Inflammatory Actions to Prevent and Treat Diseases

Jun Sun

CONTENTS

12.1 CHRONIC INFLAMMATION AND HUMAN DISEASES

12.1.1 INFLAMMATION

Inflammation (*inflammare* in Latin, which means "to set on fire") is the body's immediate response to damage to its tissues and cells by pathogens, chemicals, or physical injury (Weiss 2008; Medzhitov 2008). Inflammation was first recorded in ca. 1650 BC. In the first century AD, Cornelius Celsus defined inflammation as *Rubor et tumor cum clore et dolore*. The brief history of inflammation research is shown in Table 12.1.

Inflammatory responses cause localized redness, swelling, heat, and pain. Changes in the capillary wall structure allow interstitial fluid and white blood cells to leak out into the tissue and promote macrophage activity. This response is mediated by the microvasculature, in which fluid and electrolytes accumulate at the site of injury. It can involve both nonimmune and immune mechanisms. Based on the duration and the character of the inflammatory reactions, inflammation can be classified as acute, chronic, or granulomatous. Acute and chronic inflammation is of longer duration and is associated with the presence of mononuclear inflammatory cells, as well as the proliferation of blood vessels and connective tissue. Granulomatous inflammation is a distinct circumscribed locus of chronic inflammation (Rieber et al. 2012; Ben Dror et al. 1993; Burke et al. 2010).

Controlled inflammation is beneficial. Acute inflammation lasts from a few minutes to a few days and is characterized by fluid exudation (edema) and infiltration by leukocytes, predominantly neutrophils. Acute inflammation usually results in healing: leukocytes infiltrate the damaged region,

TABLE 12.1
History of Inflammation Research

Time	Investigator(s)	Hallmark
ca. 1650 BC	Smith Papyrus	Treatise on the treatment of wounds
460–380 BC	Hippocrates	Suppuration
First century AD	Cornelius Celsus	Defined inflammation as *Rubor et tumor cum clore et dolore*
1793	John Hunter	Considered inflammation a "salutary reaction"
1858	Rudolf Virchow	Functio laesa
1867	Julius Cohnheim	Studied inflammation *in vivo* vascular dilation, increased permeability, diapedesis
1882	Elie Metchnikoff	Phagocytosis
1924	Sir Thomas Lewis	Triple response and histamine

removing the stimulus and repairing the tissue (Weiss 2008). Without inflammation, wounds and infections would never heal. Hence, acute inflammation is considered a host defense strategy to remove injurious stimuli. It is "good" inflammation.

A cascade of biochemical events propagates and matures the inflammatory response, involving various cells within the injured tissue, local vascular system, and systemic immune response. Inflammation is not a synonym of infection. Infection is caused by an exogenous pathogen such as bacteria, virus, and parasites, whereas inflammation is one of the host responses to the pathogen. Although a successful inflammatory response is normally closely regulated by the body, inflammation could be pathologic and out of control. If the acute inflammation fails to eliminate the pathogen, the inflammatory process persists and acquires new characteristics (Medzhitov 2008).

Chronic inflammation is a prolonged, dysregulated, and maladaptive response that involves active inflammation, tissue destruction, and attempts at tissue repair (Weiss 2008). Compelling evidence demonstrates that persistent inflammation is associated with many chronic human conditions and diseases (Handschin and Spiegelman 2008; Galli et al. 2008; Mantovani et al. 2008; Medzhitov 2008). Therefore, chronic inflammation is considered "bad" inflammation.

12.1.2 INFLAMMATION AND HUMAN DISEASES

Based on its sites, inflammation could be local or systemic. Exogenous and endogenous inducers lead to chronic inflammation, thus resulting in a progressive shift in the type of cells present at the site of inflammation and may involve various organs or tissues (Handschin and Spiegelman 2008; Galli et al. 2008; Mantovani et al. 2008; Medzhitov 2008). Hence, inflammation is associated with various human diseases, including atherosclerosis, asthma, allergy, arthritis, cancer, type 2 diabetes, and brain diseases (Figure 12.1).

The link between inflammation and cancer is now generally accepted (Grivennikov et al. 2010). Inflammatory diseases increase the risk of developing many types of cancer, including bladder, cervical, gastric, intestinal, esophageal, ovarian, prostate, and thyroid cancer. It has become evident that an inflammatory microenvironment is an essential component of all tumors (Mantovani et al. 2008). Nonsteroidal anti-inflammatory drugs reduce the risk of developing certain cancers (such as colon and breast cancer) and reduce the mortality caused by these cancers.

Aging is closely associated with inflammation. Low-level chronic inflammation is associated with all the classic diseases of old age, including Alzheimer's. High levels of inflammatory cytokines are also associated with muscle weakness, known as sarcopenia, which often accompanies old age (Visser et al. 2002). Recently, Franceschi and Bonafe have developed a theory called "inflamm-ageing," pointing out that many of the conditions that afflict the elderly are associated with inflammatory responses (Franceschi and Bonafe 2003).

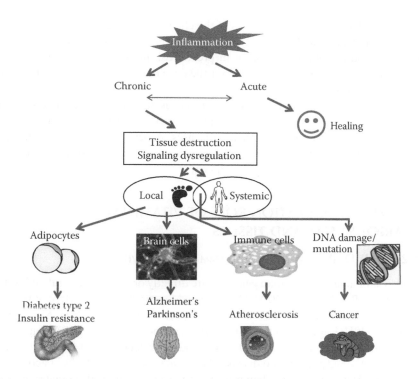

FIGURE 12.1 Inflammation and human diseases. Inflammation includes acute and chronic inflammation. Acute inflammation leads to tissue healing and the elimination of the pathogen. Chronic inflammation leads tissue destruction and signaling dysregulation, which involves local and systemic inflammation. Chronic inflammation is also associated with various human diseases, including type 2 diabetes, brain diseases, atherosclerosis, asthma, and cancer.

Overall, inflammation is closely associated with various human diseases. For prevention and therapy, the identification of key inflammatory factors and target molecules could lead to improved diagnosis and treatment.

12.1.3 MOLECULAR MECHANISMS OF INFLAMMATION

The process by which acute inflammation is initiated and develops are well defined, but much less is known about the causes of chronic inflammation and the associated molecular and cellular pathways (Medzhitov 2008). Recent studies advance our knowledge of the inducers of chronic inflammation, as well as the inflammatory mediators and cells that are involved (Weiss 2008).

Both exogenous and endogenous inducers are associated with chronic inflammation (Medzhitov 2008). The molecular pathways of inflammation are now being unraveled. Key signaling pathways modulating inflammation include nuclear factor-κB (NF-κB), p38 mitogen-activated protein kinase (MAPK), signal transducers and activators of transcription (STAT), and the activator protein-1 (AP-1). NF-κB, STAT, and AP-1 are key transcription factors that regulate the expression and activity of chemokines and cytokines, the inflammatory mediators. For example, in cancer-related inflammation, the inflammatory signaling pathways operate downstream of oncogenic mutations (such as mutations in the genes encoding RAS, MYC, and RET).

Chemokines are key players of inflammation-related diseases. Most chemokines belong to one of two major subfamilies—the CXC and CC subfamilies—based on the distribution of the first two of four cysteine residues. The two cysteines are separated by a single amino acid in CXC chemokines, whereas the two cysteines are adjacent in CC chemokines (Laing and Secombes 2004a,b). The chemokines induce immune cell migration and are involved in a wide variety of processes, including acute and chronic inflammation. Components of the chemokine system affect multiple pathways of tumor progression, including leukocyte recruitment, neoangiogenesis, tumor cell proliferation and survival, invasion, and metastasis. Evidence in preclinical and clinical settings suggests that the chemokine system represents a valuable target for the development of innovative therapeutic strategies (Allavena et al. 2011).

Therefore, targeting inflammatory mediators [such as tumor necrosis factor-α (TNF-α) and interleukin-1β (IL-1β)], key transcription factors (NF-κB and STAT3), or inflammatory cells decrease the incidence of inflammation-related diseases. A better understanding of inflammatory responses will ultimately result in the design of more effective therapies for the numerous debilitating diseases with a chronic inflammatory component.

12.2 THE ANTI-INFLAMMATION ROLE OF VITAMIN D IN VARIOUS CELLS AND TISSUES

12.2.1   1,25(OH)$_2$D$_3$ Possesses Anti-Inflammatory Activity

Approximately 3% of mouse and human genomes are directly or indirectly regulated by the vitamin D endocrine system (Bouillon et al. 2008; Carlberg and Seuter 2009). Therefore, it is not surprising that vitamin D is associated with a wide variety of diseases (Wu and Sun 2011; Motiwala and Wang 2011; Manson et al. 2011; Lagishetty et al. 2011; Krishnan and Feldman 2011). Vitamin D deficiency is a critical factor in the pathology of at least 17 varieties of cancer, as well as autoimmune diseases, diabetes, osteoarthritis, and periodontal disease. The tissue-specific synthesis of active 1,25(OH)$_2$D$_3$ from its precursor 25OHD has been shown to be important for both the innate and the adaptive immune systems (Adams and Hewison 2008; Ghoreishi et al. 2009).

Increasing evidence shows that 1,25(OH)$_2$D$_3$ possesses anti-inflammatory activity. Associations between vitamin D status and inflammation have been suggested not only in rheumatoid arthritis, lupus, inflammatory bowel disease (IBD), and type 1 diabetes but also in infections, malignancies, transplant rejection, and cardiovascular disease (Wu and Sun 2011; Motiwala and Wang 2011; Manson et al. 2011; Lagishetty et al. 2011; Krishnan and Feldman 2011; Stein and Shane 2011). In this chapter, we focus on the very recent reports on vitamin D and inflammation in various cells and tissues.

Vitamin D deficiency is associated with inflammation-linked vascular endothelial dysfunction in middle-aged and older adults (Jablonski et al. 2011). A recent study showed that vitamin D deficiency is associated with specific subtypes of anemia in the elderly, especially in those with anemia of inflammation (Perlstein et al. 2011). Vitamin D status is also associated with diseases at different ages. Experimental data indicate that fetal (trophoblastic) vitamin D plays a pivotal role in controlling placental inflammation. In humans, this may be a key factor in placental responses to infection and associated adverse outcomes of pregnancy (Liu et al. 2011). Higher vitamin D intake in pregnant women has also been linked to reduced rates of wheezing and asthma in their offspring (Camargo et al. 2007; Devereux et al. 2007; Miyake et al. 2011; Erkkola et al. 2009).

Notably, the lung diseases related to vitamin D are all inflammatory in nature (Herr et al. 2011). Epidemiologic studies show a significant correlation between low serum concentrations of 25-hydroxyvitamin D and chronic lung diseases such as asthma (Devereux et al. 2009) and chronic obstructive pulmonary disease (Kunisaki et al. 2011; Janssens et al. 2009), as well as accelerated decline in lung function (Gilbert et al. 2009; Black and Scragg 2005). In prevalent asthma cases, increased vitamin D predicts reduced rates of asthma exacerbations in those both on and off inhaled corticosteroids (Brehm et al. 2010; Weiss and Litonjua 2011; Sutherland et al. 2010; Brehm et al. 2009). Vitamin D has also been shown to be involved in upper respiratory tract infections (Bartley 2010) and antiviral states through the NF-κB pathway (Hansdottir et al. 2010).

Low serum 25-hydroxyvitamin D levels and low cardiorespiratory fitness are both associated with increased cardiovascular and all-cause mortality (Ardestani et al. 2011). A significant correlation is also reported between the 25(OH)D level and self-reported period (in hours) of moderate to vigorous physical activity. Vitamin D level is furthermore related to the maximal oxygen uptake in adults (Ardestani et al. 2011).

A cohort study on chronic kidney disease showed that vitamin D deficiency may contribute to inflammation and subsequent albuminuria (Isakova et al. 2011). This report evaluated the relationship between vitamin D, inflammation, and albuminuria. Low 25D and $1,25(OH)_2D_3$ levels are independently associated with albuminuria. Inflammatory cytokine IL-6 may be an important intermediary between vitamin D deficiency and albuminuria (Isakova et al. 2011).

Recently, local $1,25(OH)_2D_3$ production by immune cells has been shown to exert autocrine or paracrine immunomodulating effects (Adorini and Penna 2008). Immune cells also express vitamin D receptor (VDR) and the enzymes needed to metabolize vitamin D_3 (1α-, 25-, and 24-hydroxylases; von Essen et al. 2010). The effects of $1,25(OH)_2D_3$ on the immune system include modulating the T-cell antigen receptor, decreasing Th1/Th17 CD4+ T cells and cytokines, increasing regulatory T cells, downregulating T cell–driven IgG production, and inhibiting dendritic cell differentiation (von Essen et al. 2010; Jeffery et al. 2009; Kamen and Tangpricha 2010). In addition, $1,25(OH)_2D_3$ helps maintain self-tolerance by dampening overly zealous adaptive immune responses while enhancing protective innate immune responses (Guillot et al. 2010; Kamen and Tangpricha 2010).

In summary, the regulation of endothelial cells, epithelial cells, and immune cells by $1,25(OH)_2D_3$ provides a link between vitamin D and inflammatory diseases, including IBD, juvenile diabetes mellitus, multiple sclerosis, asthma, and rheumatoid arthritis (Maruotti and Cantatore 2010).

12.2.2 ANTI-INFLAMMATORY MECHANISMS OF $1,25(OH)_2D_3$ AND VDR

Immune cells, including macrophages, dendritic cells, and activated T cells, express the intracellular VDR and are responsive to $1,25(OH)_2D_3$. One mechanism for the anti-inflammatory role of $1,25(OH)_2D_3$ is to downregulate the expression of many proinflammatory cytokines, such as IL-1, IL-6, IL-8, and TNF-α, in a variety of cell types (Harant et al. 1998; D'Ambrosio et al. 1998; Yu et al. 1995; Wu et al. 2010). A recent study showed that the stimulation of CD4(+)CD25(–) T cells in the presence of $1,25(OH)_2D_3$ inhibits the production of proinflammatory cytokines, including IL-17, IL-21, and IFN-γ, but does not substantially affect T-cell division (Chang et al. 2010). Furthermore, vitamin D is shown to suppress Th17 by inducing C/EBP homologous protein (Chang et al. 2010). In contrast to its inhibitory effects on inflammatory cytokines, $1,25(OH)_2D_3$ stimulates the expression of CTLA-4 and forkhead box P3 (FOXP3), the latter requiring the presence of IL-2. Thus, $1,25(OH)_2D_3$ and IL-2 have direct synergistic effects on activated T cells, acting as potent anti-inflammatory agents and physiologic inducers of adaptive regulatory T cells (Jeffery et al. 2009).

Active VDR affects the transcription of at least 913 genes and affects processes ranging from calcium metabolism to the expression of key antimicrobial peptides (Albert et al. 2009). Interestingly, vitamin D and VDR are also associated with the key pathways, NF-κB, JNK, and STAT, which regulate inflammatory responses.

The NF-κB pathway plays a central role in activating inflammation. The mechanisms of vitamin D_3 inhibition of the NF-κB pathway include increasing IκBα protein levels, decreasing the DNA binding capacity of NF-κB, suppressing NF-κB transcription, and downregulating the expression of inflammatory cytokines, such as IL-1, IL-6, IL-8, and TNF (Harant et al. 1998). VDR physically interacts with NF-κB subunit p65 in human osteoblasts (Lu et al. 2004), mouse embryonic fibroblast cells, human intestinal epithelial cells, and mouse intestine (Sun et al. 2006; Wu et al. 2010). The absence of VDR in cells leads to a reduction of IκBα protein, the inhibitor of NF-κB (Wu et al. 2009).

Chronic inflammation has been supposed as a key event in chronic fatigue syndrome. Knowledge of the effect of vitamin D on chronic NF-κB activation could open a new disease approach for chronic fatigue syndrome (Hoeck and Pall 2011).

A recent study provides molecular links between vitamin D deficiency and Crohn disease (Wang et al. 2010). Pretreatment with $1,25(OH)_2D_3$ and then NOD2 ligand, muramyl dipeptide, synergistically decreased NF-κB activity, the expression of DEFB2/HBD2, and the antimicrobial peptide cathelicidin (Wang et al. 2010). $1,25(OH)_2D_3$ robustly stimulates the expression of the pattern recognition receptor NOD2/CARD15/IBD1, which is associated with susceptibility to Crohn disease. Importantly, this synergistic response has also been seen in macrophages from a donor with wildtype NOD2 but was absent in macrophages from patients with homozygous alleles for nonfunctional NOD2 variants of Crohn disease (Wang et al. 2010).

There is a significant correlation between low serum concentrations of 25-hydroxyvitamin D and chronic lung diseases such as asthma (Devereux et al. 2009) and chronic obstructive pulmonary disease (Kunisaki et al. 2011; Janssens et al. 2009) and accelerated decline in lung function (Gilbert et al. 2009; Black and Scragg 2005). It may be possible that vitamin D or VDR deficiency would invoke lung inflammation and alteration in lung function by proteinase/antiproteinase imbalance (Sundar et al. 2011). Recently, we have reported that VDR deficiency in mice could lead to abnormal lung phenotypes due to differential modulation of signaling mediators, further leading to spontaneous airspace enlargement and altered mechanical properties in the lung (Sundar et al. 2011).

VDR signaling is known to be associated with the JNK pathway (Hughes and Brown 2006; Vertino et al. 2005; Qi et al. 2002). In human osteosarcoma cells, vitamin D–induced reduction of proliferation occurs through the activation of JNK and MEK1/MEK2 pathways. The direct binding of VDR to c-*Jun* is described *in vitro* (Towers et al. 1999). The transcriptional activity of VDR is increased by the p38/JNK pathway stimulation of AP-1 (Qi et al. 2002). Interactions between VDR and JNK/c-*Jun* pathway contribute to an increased VDR activity and subsequent promotion of vitamin D_3–induced growth inhibition. However, the mechanism by which VDR regulates JNK in inflammation is not well established.

Dysregulation of JAK–STAT functionality can cause a variety of diseases, including inflammation and cancer. Vitamin D_3 also regulates the STAT pathways (Morales et al. 2002; Vidal et al. 2002; Zheng et al. 2011). Vitamin D modulates the JAK–STAT signaling pathway in the IL-12/IFN-γ axis, leading to Th1 differentiation (Muthian et al. 2006). STAT1 in THP1 macrophages was found to be essential for the induction of IL-1β and, thus, for the activation of β-catenin signaling in tumor cells. Simultaneous exposure to $1,25(OH)_2D_3$ and IFN-γ increases nuclear VDR and STAT1 levels. STAT1–VDR interactions increase STAT1-induced transcription in human monocytes and macrophages (Vidal et al. 2002). Furthermore, a recent study showed that vitamin D_3 exerts its chemopreventive activity by interrupting the cross talk between tumor epithelial cells and the tumor microenvironment (Kaler et al. 2009). Vitamin D_3 interrupted this cross talk by blocking the constitutive activation of STAT1 and the production of IL-1β in macrophages and, therefore, in a VDR-dependent manner, inhibited the ability of macrophages to activate Wnt signaling in colon carcinoma cells (Kaler et al. 2009).

In summary, increasing evidence indicates that vitamin D_3 influences the process by which immune cells acquire signaling molecules that enable them to migrate to inflammatory sites and normal extralymphoid tissue sites. Moreover, vitamin D/VDR may exhibit anti-inflammatory effects by inhibiting NF-κβ signaling, inhibiting p38 stress kinase signaling, interacting with STAT, and suppressing prostaglandin action (Krishnan and Feldman 2011; Stoffels et al. 2006), thus decreasing the subsequent production of pro-inflammatory cytokines. Hence, these features of vitamin D can be used for anti-inflammatory applications.

12.3 APPLICATIONS OF VITAMIN D IN DISEASE PREVENTION AND TREATMENT

Studies have shown promise when using vitamin D_3 to ameliorate diseases or when using vitamin D deficiency to exacerbate diseases in various experimental models, including rheumatoid arthritis, brain inflammation, cancer, lung diseases, and IBD. So far, in human studies, diseases showing favorable responses to vitamin D_3 treatment include rheumatoid arthritis, scleroderma, sarcoidosis, Sjögren syndrome, autoimmune thyroid disease, psoriasis, ankylosing spondylitis, Reiter syndrome, and types I and II diabetes (Adorini and Penna 2008; Zwerina et al. 2011; Sun et al. 2011; Stein and Shane 2011; Motiwala and Wang 2011; Martineau et al. 2011; Kimball et al. 2011; Jorgensen et al. 2010).

12.3.1 THERAPEUTIC ROLE OF VITAMIN D₃ IN EXPERIMENTAL MODELS

Vitamin D deficiency is associated with increased disease activity in patients with rheumatoid arthritis. Reduced vitamin D intake has been linked to increased susceptibility to developing rheumatoid arthritis (Zwerina et al. 2011). A recent study using a mouse model of rheumatoid arthritis showed that VDR plays an important role in limiting the inflammatory phenotype. Lacking VDR leads to a proinflammatory monocyte phenotype associated with increased inflammation, cartilage damage, and bone erosion (Zwerina et al. 2011).

An aged brain exhibits signs of oxidative stress and inflammatory stress. Vitamin D₃ acts as an anti-inflammatory agent and reverses the age-related increase in microglial activation and the accompanying increase in IL-1β concentration (Moore et al. 2005).

In an experimental rat model, it was shown that vitamin D deficiency could increase baseline brain inflammation, exacerbate the effects of traumatic brain injuries, and attenuate the benefits of neurosteroid progesterone treatment. In contrast, if the deficiency was corrected, these effects may be reversed (Cekic et al. 2011).

Recently, Zosky et al. have reported a direct role for vitamin D in reducing lung function in a mouse model. Female BALB/c mice were given a standard vitamin D–deficient or vitamin D–sufficient diet, had their ultraviolet B light exposure controlled, and then mated with vitamin D–sufficient males, and their offspring was studied by plethysmography and forced oscillometry. The offspring mice of the vitamin D–deficient mothers had reduced lung volumes and borderline reduced number of alveoli compared with the offspring mice of the vitamin D–sufficient mothers (Zosky et al. 2011).

The use of hypocalcemic vitamin D analogs is a strategy for exploiting the immunomodulatory actions of active vitamin D while circumventing its calcemic side effects. A recent article demonstrated that a vitamin D analog, TX527, directly targeted T cells. TX527 suppressed T-cell proliferation and activation status, accompanied by decreased expression of effector cytokines (IFN-γ, IL-4, and IL-17). Furthermore, TX527 triggered the emergence of CD4(+)CD25(high)CD127(low) regulatory T cells. Moreover, TX527 imprinted T cells with a specific homing signature favoring migration to inflammatory sites (Baeke et al. 2011).

Effective innate immunity against many microbial pathogens requires macrophage responses that upregulate phagocytosis and direct antimicrobial pathways. Vitamin D is known to induce antimicrobial peptides in various cell types, including lung, intestinal, and skin cells (Gombart et al. 2005; Gombart 2009; Wang et al. 2004; Weber et al. 2005). Subsequently, Liu et al. 2006 showed that TLR activation induced cathelicidin antimicrobial peptide (CAMP) by activating the vitamin D pathway. Recent data indicate that IL-15 induces the vitamin D–dependent antimicrobial pathway and CD209 (Montoya et al. 2009). There is a long history of using vitamin D supplementation to treat *Mycobacterium tuberculosis* infection. Both autophagy and vitamin D₃–mediated innate immunity have been shown to confer protection against infection with intracellular *M. tuberculosis* (Liu and Modlin 2008; Hoyer-Hansen et al. 2010). Further studies have shown that human cathelicidin serves as a mediator of vitamin D₃–induced autophagy (Yuk et al. 2009; Fabri and Modlin 2009). In oral epithelial cells, vitamin D was shown to enhance the innate immunity in antibacterial activity (McMahon et al. 2011).

Inflammatory biliary diseases involve bacterial factors. D'Aldebert et al. (2009) reported that biliary epithelial cells in the human liver have intense immunoreactivity for cathelicidin and for the VDR. Bile salts may contribute to biliary tract sterility by controlling epithelial cell innate immunity. Bile acids and vitamin D induce the cathelicidin gene through Farsenoid X receptor and VDR, respectively. This may explain the role of VDR in sterility. These data suggest a strategy that systematically combines therapeutic bile salt, ursodeoxycholic acid, and vitamin D to increase therapeutic efficacy (D'Aldebert et al. 2009).

Many studies have implicated vitamin D and VDR in IBD (Lim et al. 2005). Low vitamin D levels have been reported in patients with IBD. The VDR protein is significantly lower in patients with IBD and patients with colitis-associated colon cancer (Abreu et al. 2004; Wada et al. 2009). In experimental models, both local and endocrine synthesis of $1,25(OH)_2D_3$ affect the development of murine colitis (Froicu and Cantorna 2007). VDR status affects the development of murine colitis and T-cell homing in the intestine (Yu et al. 2008). It has also been shown that VDR stabilizes cell tight-junction structures in the intestinal epithelial cells (Kong et al. 2008; Fujita et al. 2008); hence, proper functioning of VDR is needed to control intestinal homeostasis.

Cells of the gastrointestinal tract, including epithelial cells and lamina propria macrophages, are constantly exposed to lumenal bacteria, which play key roles in normal intestinal development and innate immunity. The intestinal Paneth cells are known to secrete antimicrobial peptides, which are regulated by VDR signaling (Yuk et al. 2009). Paneth cells seem to be key players, highlighting the paramount importance of the antimicrobial host defense in the pathogenesis of Crohn disease (Koslowski et al. 2010; Wehkamp and Stange 2006). Recent studies indicate that VDR$^{-/-}$ mice have increased bacterial loads in the intestine (Lagishetty et al. 2010). Our studies demonstrate that VDR signaling responds to pathogenic *Salmonella* in the intestine *in vivo* (Wu et al. 2010). After *Salmonella* infection, VDR$^{-/-}$ mice had increased bacterial burdens and mortalities. Reduced vitamin D/VDR levels, an altered number and diversity of gut bacteria or both, will promote inflammation, which might disrupt the mucosal barrier and promote food and other allergen sensitization or abnormal tolerization (Weiss 2011). It is also possible that vitamin D status in the intestine determines how the gut flora interacts with the immune system (Weiss 2011).

Taken together, vitamin D/VDR has anti-inflammatory functions: protecting the mucus membrane against infection, preventing the uptake of microorganisms, and moderating the organism's immune responses in various tissues. VDR's ability to exert anti-infection and anti-inflammatory actions sets the stage for therapeutic exploitation of VDR ligands for the treatment of various intestinal inflammatory conditions.

12.3.2 APPLICATIONS AND LIMITATIONS OF VITAMIN D IN ANTI-INFLAMMATION

Because of the broad physiological relevance of vitamin D and VDR, vitamin D, its analogs, or VDR agonists may potentially become anti-inflammatory reagents for human diseases.

In patients with multiple sclerosis, the suppression of T-cell reactivity by vitamin D was very encouraging (Kimball et al. 2011). In a recent open-label randomized controlled trial, treated patients received increasing doses of cholecalciferol (4000–40,000 IU/d) plus calcium (1200 mg/d), followed by equilibration to a moderate, physiological intake (10,000 IU/d). Control patients did not receive supplements. At 12 months, mean serum 25-hydroxyvitamin D concentrations were significantly high in the treated participants, whereas serum $1,25(OH)_2D$ did not differ between the baseline and 1 year. In treated patients, 12-month *peripheral blood mononuclear cell* proliferative responses to neuron antigens myelin basic protein and exon-2 were suppressed. Multiple sclerosis–associated, abnormal T-cell reactivity was suppressed by cholecalciferol (Kimball et al. 2011).

Vitamin D_3 and its analogs are used to treat IBD in clinical trials. A recent study showed the efficacy of a potent VDR agonist for the treatment of IBD (Laverny et al. 2010). In a double-blind randomized controlled trial, daily oral supplementation with vitamin D_3 increased the mean serum vitamin D_3 level and diminished relapse rates over the 1-year follow-up compared with the placebo in Ulcerative colitis (UC) (Jorgensen et al. 2010).

Because of its anti-inflammatory and antiproliferatory actions, vitamin D and its derivatives are extensively explored as chemopreventive and even chemotherapeutic agents (Fedirko et al. 2010). Serum vitamin D levels and colorectal cancer has an inverse association. According to a recent review (Manson et al. 2011), subjects with a serum 25-hydroxyvitamin D level of 33 ng/mL or higher had about half the risk of colorectal cancer of those with levels of 12 ng/mL or lower. The European Prospective Investigation into Cancer and Nutrition study has recently reported a similarly strong inverse association. A prospective study from the Japan Public Health Center did not find an inverse relation between plasma 25-hydroxyvitamin D levels and the occurrence of colon cancer, although an inverse association with rectal cancer was apparent (Manson et al. 2011).

Despite biological plausibility and widespread enthusiasm, the 2010 report from the Institute of Medicine found evidence that vitamin D reduces cancer incidence and related mortality was inconsistent and inconclusive as to causality (Institute of Medicine 2010). New trials assessing moderate-to high-dose vitamin D supplementation for cancer prevention are in progress (Manson et al. 2011).

The biggest obstacle to the clinical use of vitamin D is its potent hypercalcemic effect. Given the potentially toxic effects of vitamin D, the recommendations for the optimal dose of vitamin D are still debated. *CYP24A1* encodes 25-hydroxyvitamin D 24-hydroxylase, the key enzyme of 1,25-dihydroxyvitamin D_3 degradation. A recent study identified *CYP24A1* mutations that resulted in the loss of function. This explains the increased sensitivity to vitamin D in patients with idiopathic infantile hypercalcemia and is a genetic risk factor for the development of symptomatic hypercalcemia that may be triggered by vitamin D prophylaxis in otherwise apparently healthy infants (Schlingmann et al. 2011).

There is considerable discussion of the serum concentrations of 25(OH)D associated with deficiency (e.g., rickets), adequacy for bone health, and optimal overall health. Moreover, the mutated p53 interacts with VDR in cancer cells. These data indicate that extra caution should be taken in using vitamin D or its analogs in individualized cancer therapy (Stambolsky et al. 2010). Researchers continue to develop potent and safe VDR agonists (Laverny et al. 2010). Because of the uncertainty of vitamin D and VDR in the pathogenesis of diseases, additional research is required to determine the proper dosage for optimal health (Sun 2010).

In summary, the immunomodulating effects of vitamin D may explain the reported epidemiological associations between vitamin D status and a large number of autoimmune and inflammatory diseases. Such associations have been suggested by observational studies not only in rheumatoid arthritis, lupus, IBD, and type 1 diabetes but also in infections, malignancies, transplant rejection, and cardiovascular disease. VDR is activated by 1,25(OH)$_2$ vitamin D_3. In the nucleus, VDR binds to the vitamin D–response element (VDRE) in the promoter of target genes to regulate gene transcription. Therefore, one of the effective and efficient ways of taking advantage of the anti-inflammatory action of vitamin D is to enhance the functions of VDR (Figure 12.2). These include

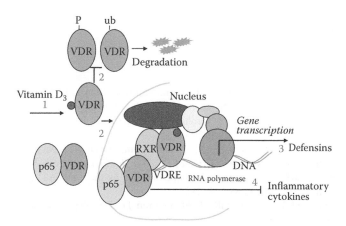

FIGURE 12.2 Molecular mechanism of 1,25(OH)2D₃/VDR action and strategies for enhancing the anti-inflammatory ability of VDR. VDR is activated by 1,25(OH)$_2$D$_3$. In the nucleus, VDR binds to the VDRE in the target gene promoter to regulate gene transcription. (RXR: retinoid X receptor.) VDR negatively regulates the proinflammatory NF-κB activity. VDR/NF-κB p65 interaction exists in cells. The level of VDR is regulated through ubiquitination. Ubiquitinated VDR is degraded by the proteasome. Potential strategies for enhancing the anti-inflammatory ability of vitamin D/VDR can be developed at different levels: (1) activate VDR signaling by vitamin D or vitamin D analogs; (2) increase VDR expression by VDR agonists or stabilize VDR by preventing its degradation; (3) enhance the expression of defensins, VDR target genes; or (4) correct VDR dysfunction by blocking inflammatory cytokines. (From Sun, J., *Functional Food Reviews*, 2012. With permission.)

(1) using vitamin D or its analogs to activate vitamin D/VDR binding, (2) using VDR agonists to enhance VDR expression and activity, (3) blocking inflammatory cytokines, and (4) targeting the microenvironment and correcting the dysfunction of vitamin D/VDR by using alternative methods, such as probiotics, to increase VDR expression in intestine (Yoon et al. 2011). Clinical studies and experimental models show that vitamin D supplementation or VDR agonists produce therapeutic effects. Thus, modulation of vitamin D and VDR signaling may hold promise for the treatment of inflammatory diseases.

12.4 CONCLUSION

There is increasing concern regarding the use of vitamin D as an inexpensive and easy supplement for disease prevention. However, little is known about the molecular mechanisms responsible for the dysfunction of VDR in human diseases. $1,25(OH)_2D_3$ has potent immunomodulatory properties that have promoted its potential use in the prevention and treatment of infectious disease and autoimmune conditions. However, we need to advance our understanding of how VDR is involved in the pathogenesis of human diseases. We need further insights into the balance between good inflammation and bad inflammation from being altered to favor adaptive immunity. Manipulating the level of $1,25(OH)_2D_3$ in the body and restoring the function of VDR may represent a new approach to prevent and treat various illnesses, including chronic inflammation.

ACKNOWLEDGMENTS

This work was supported by the National Institutes of Health under Grants DK075386-0251 and R03DK089010-01, the American Cancer Society under Grant RSG-09-075-01-MBC, and the Empire State Stem Cell Board of the State of New York under Grant N09G-279.

DISCLOSURE

The authors report no conflicts of interest.

REFERENCES

Abreu, M. T., V. Kantorovich, E. A. Vasiliauskas, U. Gruntmanis, R. Matuk, K. Daigle, S. Chen, D. Zehnder, Y. C. Lin, H. Yang, M. Hewison, and J. S. Adams. 2004. Measurement of vitamin D levels in inflammatory bowel disease patients reveals a subset of Crohn's disease patients with elevated 1,25-dihydroxyvitamin D and low bone mineral density. *Gut* 53:1129–36.
Adams, J. S., and M. Hewison. 2008. Unexpected actions of vitamin D: New perspectives on the regulation of innate and adaptive immunity. *Nature Clinical Practice. Endocrinology & Metabolism* 4(2):80–90.
Adorini, L., and G. Penna. 2008. Control of autoimmune diseases by the vitamin D endocrine system. *Nature Clinical Practice. Rheumatology* 4(8):404–412.
Albert, P. J., A. D. Proal, and T. G. Marshall. 2009. Vitamin D: The alternative hypothesis. *Autoimmunity Reviews* 8(8):639–644.
Allavena, P., G. Germano, F. Marchesi, and A. Mantovani. 2011. Chemokines in cancer related inflammation. *Experimental Cell Research* 317(5):664–673.
Ardestani, A., B. Parker, S. Mathur, P. Clarkson, L. S. Pescatello, H. J. Hoffman, D. M. Polk, and P. D. Thompson. 2011. Relation of vitamin D level to maximal oxygen uptake in adults. *American Journal of Cardiology* 107(8):1246–1249.
Baeke, F., H. Korf, L. Overbergh, A. Verstuyf, L. Thorrez, L. Van Lommel, M. Waer, F. Schuit, C. Gysemans, and C. Mathieu. 2011. The vitamin D analog, TX527, promotes a human CD4+CD25 high CD 127 low regulatory T cell profile and induces a migratory signature specific for homing to sites of inflammation. *The Journal of Immunology* 186(1):132–142.
Bartley, J. 2010. Vitamin D, innate immunity and upper respiratory tract infection. *The Journal of Laryngology and Otology* 124(5):465–469.

Ben Dror, I., R. Koren, U. A. Liberman, A. Erman, R. Ziegler, and A. Ravid. 1993. Foreign body granulomatous inflammation increases the sensitivity of splenocytes to immunomodulation by 1,25-dihydroxy vitamin D_3. *International Journal of Immunopharmacology* 15(3):275–280.

Black, P. N., and R. Scragg. 2005. Relationship between serum 25-hydroxyvitamin D and pulmonary function in the Third National Health and Nutrition Examination Survey. *Chest* 128(6):3792–3798.

Bouillon, R., G. Carmeliet, L. Verlinden, E. van Etten, A. Verstuyf, H. F. Luderer, L. Lieben, C. Mathieu, and M. Demay. 2008. Vitamin D and human health: lessons from vitamin D receptor null mice. *Endocr Rev* 29:726–76.

Brehm, J. M., J. C. Celedon, M. E. Soto-Quiros, L. Avila, G. M. Hunninghake, E. Forno, D. Laskey, J. S. Sylvia, B. W. Hollis, S. T. Weiss, and A. A. Litonjua. 2009. Serum vitamin D levels and markers of severity of childhood asthma in Costa Rica. *American Journal of Respiratory and Critical Care Medicine* 179(9):765–771.

Brehm, J. M., B. Schuemann, A. L. Fuhlbrigge, B. W. Hollis, R. C. Strunk, R. S. Zeiger, S. T. Weiss, and A. A. Litonjua. 2010. Serum vitamin D levels and severe asthma exacerbations in the Childhood Asthma Management Program study. *The Journal of Allergy and Clinical Immunology* 126(1):52–58 e5.

Burke, R. R., B. A. Rybicki, and D. S. Rao. 2010. Calcium and vitamin D in sarcoidosis: How to assess and manage. *Seminars in Respiratory and Critical Care Medicine* 31(4):474–484.

Camargo, C. A., Jr., S. L. Rifas-Shiman, A. A. Litonjua, J. W. Rich-Edwards, S. T. Weiss, D. R. Gold, K. Kleinman, and M. W. Gillman. 2007. Maternal intake of vitamin D during pregnancy and risk of recurrent wheeze in children at 3 years of age. *The American Journal of Clinical Nutrition* 85(3):788–795.

Carlberg C., and S. Seuter. 2009. A genomic perspective on vitamin D signaling. *Anticancer Res* 29:3485–93.

Cekic, M., S. M. Cutler, J. W. VanLandingham, and D. G. Stein. 2011. Vitamin D deficiency reduces the benefits of progesterone treatment after brain injury in aged rats. *Neurobiology of Aging* 32(5):864–874.

Chang, S. H., Y. Chung, and C. Dong. 2010. Vitamin D suppresses Th17 cytokine production by inducing C/EBP homologous protein (CHOP) expression. *Journal of Biological Chemistry* 285(50):38751–38755.

D'Ambrosio, D., M. Cippitelli, M. G. Cocciolo, D. Mazzeo, P. Di Lucia, R. Lang, F. Sinigaglia, and P. Panina-Bordignon. 1998. Inhibition of IL-12 production by 1,25-dihydroxyvitamin D_3: Involvement of NF-kappaB downregulation in transcriptional repression of the p40 gene. *Journal of Clinical Investigation* 101(1):252–262.

Devereux, G., A. A. Litonjua, S. W. Turner, L. C. Craig, G. McNeill, S. Martindale, P. J. Helms, A. Seaton, and S. T. Weiss. 2007. Maternal vitamin D intake during pregnancy and early childhood wheezing. *The American Journal of Clinical Nutrition* 85(3):853–859.

Devereux, G., H. Macdonald, and C. Hawrylowicz. 2009. Vitamin D and asthma: Time for intervention? *American Journal of Respiratory and Critical Care Medicine* 17(9):739–740.

Erkkola, M., M. Kaila, B. I. Nwaru, C. Kronberg-Kippila, S. Ahonen, J. Nevalainen, R. Veijola, J. Pekkanen, J. Ilonen, O. Simell, M. Knip, and S. M. Virtanen. 2009. Maternal vitamin D intake during pregnancy is inversely associated with asthma and allergic rhinitis in 5-year-old children. *Clinical and Experimental Allergy* 39(6):875–882.

Fabri, M., and R. L. Modlin. 2009. A vitamin for autophagy. *Cell Host & Microbe* 6(3):201–203.

Fedirko, V., R. M. Bostick, Q. Long, W. D. Flanders, M. L. McCullough, E. Sidelnikov, C. R. Daniel, R. E. Rutherford, and A. Shaukat. 2010. Effects of supplemental vitamin D and calcium on oxidative DNA damage marker in normal colorectal mucosa: A randomized clinical trial. *Cancer Epidemiology, Biomarkers & Prevention* 19(1):280–291.

Franceschi, C., and M. Bonafe. 2003. Centenarians as a model for healthy aging. *Biochemical Society Transactions* 31(2):457–461.

Froicu, M., and M. T. Cantorna. 2007. Vitamin D and the vitamin D receptor are critical for control of the innate immune response to colonic injury. *BMC Immunology* 8:5.

Fujita, H., K. Sugimoto, S. Inatomi, T. Maeda, M. Osanai, Y. Uchiyama, Y. Yamamoto, T. Wada, T. Kojima, H. Yokozaki, T. Yamashita, S. Kato, N. Sawada, and H. Chiba. 2008. Tight junction proteins claudin-2 and -12 are critical for vitamin D–dependent Ca^{2+} absorption between enterocytes. *Molecular Biology of the Cell* 19(5):1912–1921.

Galli, S. J., M. Tsai, and A. M. Piliponsky. 2008. The development of allergic inflammation. *Nature* 454(7203):445–454.

Ghoreishi, M., P. Bach, J. Obst, M. Komba, J. C. Fleet, and J. P. Dutz. 2009. Expansion of antigen-specific regulatory T cells with the topical vitamin D analog calcipotriol. *The Journal of Immunology* 182(10):6071–6078.

Gilbert, C. R., S. M. Arum, and C. M. Smith. 2009. Vitamin D deficiency and chronic lung disease. *Canadian Respiratory Journal* 16(3):75–80.

Gombart, A. F. 2009. The vitamin D–antimicrobial peptide pathway and its role in protection against infection. *Future Microbiol* 4 (9):1151–1165.

Gombart, A. F., N. Borregaard, and H. P. Koeffler. 2005. Human cathelicidin antimicrobial peptide (CAMP) gene is a direct target of the vitamin D receptor and is strongly upregulated in myeloid cells by 1,25-dihydroxyvitamin D3. *FASEB Journal* 19(9):1067–1077.

Grivennikov, S. I., F. R. Greten, and M. Karin. 2010. Immunity, inflammation, and cancer. *Cell* 140(6):883–899.

Guillot, X., L. Semerano, N. Saidenberg-Kermanac'h, G. Falgarone, and M. C. Boissier. 2010. Vitamin D and inflammation. *Joint, Bone, Spine* 77(6):552–557.

Handschin, C., and B. M. Spiegelman. 2008. The role of exercise and PGC1alpha in inflammation and chronic disease. *Nature* 454 (7203):463–469.

Hansdottir, S., M. M. Monick, N. Lovan, L. Powers, A. Gerke, and G. W. Hunninghake. 2010. Vitamin D decreases respiratory syncytial virus induction of NF-kappaB-linked chemokines and cytokines in airway epithelium while maintaining the antiviral state. *The Journal of Immunology* 184(2):965–974.

Harant, H., B. Wolff, and I. J. Lindley. 1998. 1Alpha,25-dihydroxyvitamin D3 decreases DNA binding of nuclear factor-kappaB in human fibroblasts. *FEBS Letters* 436(3):329–334.

Herr, C., T. Greulich, R. A. Koczulla, S. Meyer, T. Zakharkina, M. Branscheidt, R. Eschmann, and R. Bals. 2011. The role of vitamin D in pulmonary disease: COPD, asthma, infection, and cancer. *Respiratory Research* 12:31.

Hoeck, A. D., and M. L. Pall. 2011. Will vitamin D supplementation ameliorate diseases characterized by chronic inflammation and fatigue? *Medical Hypotheses* 76(2):208–213.

Hoyer-Hansen, M., S. P. Nordbrandt, and M. Jaattela. 2010. Autophagy as a basis for the health-promoting effects of vitamin D. *Trends in Molecular Medicine* 16(7):295–302.

Hughes, P. J., and G. Brown. 2006. 1Alpha,25-dihydroxyvitamin D3–mediated stimulation of steroid sulphatase activity in myeloid leukaemic cell lines requires VDRnuc-mediated activation of the RAS/RAF/ERK-MAP kinase signaling pathway. *Journal of Cellular Biochemistry* 98(3):590–617.

Institute of Medicine, Food and Nutrition Board. 2010. *Dietary Reference Intakes for Calcium and Vitamin D*. Washington, DC: National Academy Press (http://www.iom.edu/Reports/2010/Dietary-Reference-Intakes-for-Calcium-and-Vitamin-D.aspx).

Isakova, T., O. M. Gutierrez, N. M. Patel, D. L. Andress, M. Wolf, and A. Levin. 2011. Vitamin D deficiency, inflammation, and albuminuria in chronic kidney disease: Complex interactions. *Journal of Renal Nutrition* 21(4):295–302.

Jablonski, K. L., M. Chonchol, G. L. Pierce, A. E. Walker, and D. R. Seals. 2011. 25-Hydroxyvitamin D deficiency is associated with inflammation-linked vascular endothelial dysfunction in middle-aged and older adults. *Hypertension* 57(1):63–69.

Janssens, W., A. Lehouck, C. Carremans, R. Bouillon, C. Mathieu, and M. Decramer. 2009. Vitamin D beyond bones in chronic obstructive pulmonary disease: Time to act. *American Journal of Respiratory and Critical Care Medicine* 179(8):630–636.

Jeffery, L. E., F. Burke, M. Mura, Y. Zheng, O. S. Qureshi, M. Hewison, L. S. Walker, D. A. Lammas, K. Raza, and D. M. Sansom. 2009. 1,25-Dihydroxyvitamin D_3 and IL-2 combine to inhibit T cell production of inflammatory cytokines and promote development of regulatory T cells expressing CTLA-4 and FoxP3. *The Journal of Immunology* 183(9):5458–5467.

Jorgensen, S. P., J. Agnholt, H. Glerup, S. Lyhne, G. E. Villadsen, C. L. Hvas, L. E. Bartels, J. Kelsen, L. A. Christensen, and J. F. Dahlerup. 2010. Clinical trial: Vitamin D_3 treatment in Crohn's disease: A randomized double-blind placebo-controlled study. *Alimentary Pharmacology & Therapeutics* 32(3):377–383.

Kaler, P., L. Augenlicht, and L. Klampfer. 2009. Macrophage-derived IL-1beta stimulates Wnt signaling and growth of colon cancer cells: A cross talk interrupted by vitamin D_3. *Oncogene* 28(44):3892–3902.

Kamen, D. L., and V. Tangpricha. 2010. Vitamin D and molecular actions on the immune system: Modulation of innate and autoimmunity. *Journal of Molecular Medicine (Berlin, Germany)* 88(5):441–450.

Kimball, S., R. Vieth, H. M. Dosch, A. Bar-Or, R. Cheung, D. Gagne, P. O'Connor, C. D'Souza, M. Ursell, and J. M. Burton. 2011. Cholecalciferol plus calcium suppresses abnormal PBMC reactivity in patients with multiple sclerosis. *The Journal of Clinical Endocrinology and Metabolism* 96(9):2826–2834.

Kong, J., Z. Zhang, M. W. Musch, G. Ning, J. Sun, J. Hart, M. Bissonnette, and Y. C. Li. 2008. Novel role of the vitamin D receptor in maintaining the integrity of the intestinal mucosal barrier. *American Journal of Physiology. Gastrointestinal and Liver Physiology* 294(1):G208–G216.

Koslowski, M. J., J. Beisner, E. F. Stange, and J. Wehkamp. 2010. Innate antimicrobial host defense in small intestinal Crohn's disease. *International Journal of Medical Microbiology* 300(1):34–40.

Krishnan, A. V., and D. Feldman. 2011. Mechanisms of the anti-cancer and anti-inflammatory actions of vitamin D. *Annu Rev Pharmacol Toxicol* 51:311–336.

Kunisaki, K. M., D. E. Niewoehner, R. J. Singh, and J. E. Connett. 2011. Vitamin D status and longitudinal lung function decline in the Lung Health Study. *The European Respiratory Journal* 37(2):238–243.

Lagishetty, V., A. V. Misharin, N. Q. Liu, T. S. Lisse, R. F. Chun, Y. Ouyang, S. M. McLachlan, J. S. Adams, and M. Hewison. 2010. Vitamin D deficiency in mice impairs colonic antibacterial activity and predisposes to colitis. *Endocrinology* 151(6):2423–2432.

Lagishetty, V., N. Q. Liu, and M. Hewison. 2011. Vitamin D metabolism and innate immunity. *Molecular and Cellular Endocrinology* 347(1–2):97–105.

Laing, K. J., and C. J. Secombes. 2004a. Chemokines. *Developmental and Comparative Immunology* 28(5):443–460.

Laing, K. J., and C. J. Secombes. 2004b. Trout CC chemokines: Comparison of their sequences and expression patterns. Molecular Immunology 41(8):793–808.

Laverny, G., G. Penna, S. Vetrano, C. Correale, M. Nebuloni, S. Danese, and L. Adorini. 2010. Efficacy of a potent and safe vitamin D receptor agonist for the treatment of inflammatory bowel disease. *Immunology Letters* 131(1):49–58.

Lim, W. C., S. B. Hanauer, and Y. C. Li. 2005. Mechanisms of disease: Vitamin D and inflammatory bowel disease. *Nature Clinical Practice. Gastroenterology & Hepatology* 2(7):308–315.

Liu, P. T., and R. L. Modlin. 2008. Human macrophage host defense against *Mycobacterium tuberculosis*. *Current Opinion in Immunology* 20(4):371–376.

Liu, N. Q., A. T. Kaplan, V. Lagishetty, Y. B. Ouyang, Y. Ouyang, C. F. Simmons, O. Equils, and M. Hewison. 2011. Vitamin D and the regulation of placental inflammation. *The Journal of Immunology* 186(10):5968–5974

Lu, X., P. Farmer, J. Rubin, and M. S. Nanes. 2004. Integration of the NFkappaB p65 subunit into the vitamin D receptor transcriptional complex: Identification of p65 domains that inhibit 1,25-dihydroxyvitamin D_3–stimulated transcription. *Journal of Cellular Biochemistry* 92(4):833–848.

Manson, J. E., S. T. Mayne, and S. K. Clinton. 2011. Vitamin D and prevention of cancer—Ready for prime time? *New England Journal of Medicine* 364(15):1385–1387.

Mantovani, A., P. Allavena, A. Sica, and F. Balkwill. 2008. Cancer-related inflammation. *Nature* 454(7203): 436–444.

Martineau, A. R., P. M. Timms, G. H. Bothamley, Y. Hanifa, K. Islam, A. P. Claxton, G. E. Packe, J. C. Moore-Gillon, M. Darmalingam, R. N. Davidson, H. J. Milburn, L. V. Baker, R. D. Barker, N. J. Woodward, T. R. Venton, K. E. Barnes, C. J. Mullett, A. K. Coussens, C. M. Rutterford, C. A. Mein, G. R. Davies, R. J. Wilkinson, V. Nikolayevskyy, F. A. Drobniewski, S. M. Eldridge, and C. J. Griffiths. 2011. High-dose vitamin D(3) during intensive-phase antimicrobial treatment of pulmonary tuberculosis: A double-blind randomized controlled trial. *Lancet* 377(9761):242–250.

Maruotti, N., and F. P. Cantatore. 2010. Vitamin D and the immune system. *The Journal of Rheumatology* 37(3):491–495.

McMahon, L., K. Schwartz, O. Yilmaz, E. Brown, L. K. Ryan, and G. Diamond. 2011. Vitamin D–mediated induction of innate immunity in gingival epithelial cells. *Infection and Immunity* 79(6):2250–2256.

Medzhitov, R. 2008. Origin and physiological roles of inflammation. *Nature* 454(7203):428–435.

Miyake, Y., S. Sasaki, K. Tanaka, and Y. Hirota. 2011. Maternal B vitamin intake during pregnancy and wheeze and eczema in Japanese infants aged 16–24 months: The Osaka Maternal and Child Health Study. *Pediatric Allergy and Immunology* 22(1 Pt 1):69–74.

Montoya, D., D. Cruz, R. M. Teles, D. J. Lee, M. T. Ochoa, S. R. Krutzik, R. Chun, M. Schenk, X. Zhang, B. G. Ferguson, A. E. Burdick, E. N. Sarno, T. H. Rea, M. Hewison, J. S. Adams, G. Cheng, and R. L. Modlin. 2009. Divergence of macrophage phagocytic and antimicrobial programs in leprosy. *Cell Host & Microbe* 6 (4):343–353.

Moore, M. E., A. Piazza, Y. McCartney, and M. A. Lynch. 2005. Evidence that vitamin D3 reverses age-related inflammatory changes in the rat hippocampus. *Biochemical Society Transactions* 33(Pt 4):573–577.

Morales, O., M. H. Faulds, U. J. Lindgren, and L. A. Haldosen. 2002. 1Alpha,25-dihydroxyvitamin D3 inhibits GH-induced expression of SOCS-3 and CIS and prolongs growth hormone signaling via the Janus kinase (JAK2)/signal transducers and activators of transcription (STAT5) system in osteoblast-like cells. *Journal of Biological Chemistry* 277 (38):34879–34884.

Motiwala, S. R., and T. J. Wang. 2011. Vitamin D and cardiovascular disease. *Current Opinion in Nephrology and Hypertension* 20(4):345–353.

Muthian, G., H. P. Raikwar, J. Rajasingh, and J. J. Bright. 2006. 1,25 Dihydroxyvitamin-D3 modulates JAK-STAT pathway in IL-12/IFNgamma axis leading to Th1 response in experimental allergic encephalomyelitis. *Journal of Neuroscience Research* 83(7):1299–1309.

Perlstein, T. S., R. Pande, N. Berliner, and G. J. Vanasse. 2011. Prevalence of 25-hydroxyvitamin D deficiency in subgroups of elderly persons with anemia: association with anemia of inflammation. *Blood*
117(10):2800–2806.

Qi, X., R. Pramanik, J. Wang, R. M. Schultz, R. K. Maitra, J. Han, H. F. DeLuca, and G. Chen. 2002. The
p38 and JNK pathways cooperate to trans-activate vitamin D receptor via c-Jun/AP-1 and sensitize
human breast cancer cells to vitamin D(3)–induced growth inhibition. *Journal of Biological Chemistry*
277(29):25884–25892.

Rieber, N., A. Hector, T. Kuijpers, D. Roos, and D. Hartl. 2012. Current concepts of hyperinflammation in
chronic granulomatous disease. *Clinical & Developmental Immunology* 2012:252460.

Schlingmann, K. P., M. Kaufmann, S. Weber, A. Irwin, C. Goos, U. John, J. Misselwitz, G. Klaus, E. Kuwertz-
Broking, H. Fehrenbach, A. M. Wingen, T. Guran, J. G. Hoenderop, R. J. Bindels, D. E. Prosser, G. Jones,
and M. Konrad. 2011. Mutations in CYP24A1 and idiopathic infantile hypercalcemia. *New England
Journal of Medicine* 365(5):410–421.

Stambolsky, P., Y. Tabach, G. Fontemaggi, L. Weisz, R. Maor-Aloni, Z. Sigfried, I. Shiff, I. Kogan, M. Shay,
E. Kalo, G. Blandino, I. Simon, M. Oren, and V. Rotter. 2010. Modulation of the vitamin D_3 response by
cancer-associated mutant p53. *Cancer Cell* 17(3):273–285.

Stein, E. M., and E. Shane. 2011. Vitamin D in organ transplantation. *Osteoporosis International* 22(7):
2107–2118.

Stoffels, K., L. Overbergh, A. Giulietti, L. Verlinden, R. Bouillon, and C. Mathieu. 2006. Immune regulation of
25-hydroxyvitamin-D_3-1alpha-hydroxylase in human monocytes. *Journal of Bone and Mineral Research*
21(1):37–47.

Sun, J. 2010. Vitamin D and mucosal immune function. *Current Opinion in Gastroenterology* 26(6):591–595.

Sun, J. 2012. Vitamin D and inflamation. *Functional Food Reviews* 3(4):34–43.

Sun, J., J. Kong, Y. Duan, F. L. Szeto, A. Liao, J. L. Madara, and Y. C. Li. 2006. Increased NF-kappaB activity in fibroblasts lacking the vitamin D receptor. *American Journal of Physiology. Endocrinology and
Metabolism* 291(2):E315–E322.

Sun, Q., L. Shi, E. B. Rimm, E. L. Giovannucci, F. B. Hu, J. E. Manson, and K. M. Rexrode. 2011. Vitamin
D intake and risk of cardiovascular disease in US men and women. *The American Journal of Clinical
Nutrition* 94(2):534–542.

Sundar, I. K., J. W. Hwang, S. Wu, J. Sun, and I. Rahman. 2011. Deletion of vitamin D receptor leads to premature emphysema/COPD by increased matrix metalloproteinases and lymphoid aggregates formation.
Biochemical and Biophysical Research Communications 406(1):127–133.

Sutherland, E. R., E. Goleva, L. P. Jackson, A. D. Stevens, and D. Y. Leung. 2010. Vitamin D levels, lung function, and steroid response in adult asthma. *American Journal of Respiratory and Critical Care Medicine*
181(7):699–704.

Towers, T. L., T. P. Staeva, and L. P. Freedman. 1999. A two-hit mechanism for vitamin D_3–mediated transcriptional repression of the granulocyte-macrophage colony-stimulating factor gene: Vitamin D receptor competes for DNA binding with NFAT1 and stabilizes c-Jun. *Molecular and Cellular Biology*
19(6):4191–4199.

Vertino, A. M., C. M. Bula, J. R. Chen, M. Almeida, L. Han, T. Bellido, S. Kousteni, A. W. Norman, and S. C.
Manolagas. 2005. Nongenotropic, anti-apoptotic signaling of 1alpha,25(OH)2-vitamin D_3 and analogs
through the ligand binding domain of the vitamin D receptor in osteoblasts and osteocytes: Mediation by
Src, phosphatidylinositol 3-, and JNK kinases. *Journal of Biological Chemistry* 280(14):14130–14137.

Vidal, M., C. V. Ramana, and A. S. Dusso. 2002. Stat1-vitamin D receptor interactions antagonize
1,25-dihydroxyvitamin D transcriptional activity and enhance stat1-mediated transcription. *Molecular
and Cellular Biology* 22(8):2777–2787.

Visser, M., M. Pahor, D. R. Taaffe, B. H. Goodpaster, E. M. Simonsick, A. B. Newman, M. Nevitt, and T. B.
Harris. 2002. Relationship of interleukin-6 and tumor necrosis factor-alpha with muscle mass and muscle
strength in elderly men and women: The Health ABC Study. *The Journals of Gerontology. Series A,
Biological Sciences and Medical Sciences* 57(5):M326–M332.

von Essen, M. R., M. Kongsbak, P. Schjerling, K. Olgaard, N. Odum, and C. Geisler. 2010. Vitamin D controls
T cell antigen receptor signaling and activation of human T cells. *Nature Immunology* 11(4):344–349.

Wada K., H. Tanaka, K. Maeda, T. Inoue, E. Noda, R. Amano, N. Kubo, K. Muguruma, N. Yamada, M. Yashiro,
T. Sawada, B. Nakata, M. Ohira, and K. Hirakawa. 2009. Vitamin D receptor expression is associated
with colon cancer in ulcerative colitis. *Oncol Rep* 22:1021–5.

Wang, T. T., F. P. Nestel, V. Bourdeau, Y. Nagai, Q. Wang, J. Liao, L. Tavera-Mendoza, R. Lin, J. W. Hanrahan,
S. Mader, and J. H. White. 2004. Cutting edge: 1,25-dihydroxyvitamin D_3 is a direct inducer of antimicrobial peptide gene expression. *The Journal of Immunology* 173(5):2909–2912.

Wang, T. T., B. Dabbas, D. Laperriere, A. J. Bitton, H. Soualhine, L. E. Tavera-Mendoza, S. Dionne, M. J. Servant, A. Bitton, E. G. Seidman, S. Mader, M. A. Behr, and J. H. White. 2010. Direct and indirect induction by 1,25-dihydroxyvitamin D_3 of the NOD2/CARD15-defensin beta2 innate immune pathway defective in Crohn disease. *Journal of Biological Chemistry* 285(4):2227–2231.

Weber, G., J. D. Heilborn, C. I. Chamorro Jimenez, A. Hammarsjo, H. Torma, and M. Stahle. 2005. Vitamin D induces the antimicrobial protein hCAP18 in human skin. *The Journal of Investigative Dermatology* 124(5):1080–1082.

Wehkamp, J., and E. F. Stange. 2006. Paneth cells and the innate immune response. *Current Opinion in Gastroenterology* 22(6):644–650.

Weiss, U. 2008. Inflammation. *Nature* 454(7203):427.

Weiss, S. T. 2011. Bacterial components plus vitamin D: the ultimate solution to the asthma (autoimmune disease) epidemic? *The Journal of Allergy and Clinical Immunology* 127(5):1128–1130.

Weiss, S. T., and A. A. Litonjua. 2011. The *in utero* effects of maternal vitamin D deficiency: How it results in asthma and other chronic diseases. *American Journal of Respiratory and Critical Care Medicine* 183(10):1286–1287.

Wu, S., and J. Sun. 2011. Vitamin D, vitamin D receptor, and macroautophagy in inflammation and infection. *Discovery Medicine* 11(59):325–335.

Wu, S., Y. Xia, X. Liu, and J. Sun. 2009. Vitamin D receptor deletion leads to reduced level of IkappaBalpha protein through protein translation, protein–protein interaction, and post-translational modification. *The International Journal of Biochemistry & Cell Biology* 42(2):329–336.

Wu, S., A. P. Liao, Y. Xia, Y. C. Li, J. D. Li, R. B. Sartor, and J. Sun. 2010. Vitamin D receptor negatively regulates bacterial-stimulated NF-{kappa}B activity in intestine. *American Journal of Pathology* 177(2):686–697.

Yoon, S., Wu, S., Zhang, Y., Lu, R, Petrof, E., Yuan, L., Claud, E., Sun, J. 2011. *Probiotic Regulation of Vitamin D Receptor in Intestinal Inflammation*. Oral Presentation at the Annual Meeting of the American Gastroenterological Associatiation, Digestive Disease Week (Chicago).

Yu, X. P., T. Bellido, and S. C. Manolagas. 1995. Down-regulation of NF-kappa B protein levels in activated human lymphocytes by 1,25-dihydroxyvitamin D_3. *Proceedings of the National Academy of Sciences of the United States of America* 92(24):10990–10994.

Yu, S., D. Bruce, M. Froicu, V. Weaver, and M. T. Cantorna. 2008. Failure of T cell homing, reduced CD4/CD8alphaalpha intraepithelial lymphocytes, and inflammation in the gut of vitamin D receptor KO mice. *Proceedings of the National Academy of Sciences of the United States of America* 105(52):20834–20839.

Yuk, J. M., D. M. Shin, H. M. Lee, C. S. Yang, H. S. Jin, K. K. Kim, Z. W. Lee, S. H. Lee, J. M. Kim, and E. K. Jo. 2009. Vitamin D3 induces autophagy in human monocytes/macrophages via cathelicidin. *Cell Host & Microbe* 6(3):231–243.

Zheng, W., K. E. Wong, Z. Zhang, U. Dougherty, R. Mustafi, J. Kong, D. K. Deb, H. Zheng, M. Bissonnette, and Y. C. Li. 2011. Inactivation of the vitamin D receptor in APC($^{-/+}$) mice reveals a critical role for the vitamin D receptor in intestinal tumor growth. *International Journal of Cancer* 130(1):10–19.

Zosky, G. R., L. J. Berry, J. G. Elliot, A. L. James, S. Gorman, and P. H. Hart. 2011. Vitamin D deficiency causes deficits in lung function and alters lung structure. *American Journal of Respiratory and Critical Care Medicine* 183(10):1336–1343.

Zwerina, K., W. Baum, R. Axmann, G. R. Heiland, J. H. Distler, J. Smolen, S. Hayer, J. Zwerina, and G. Schett. 2011a. Vitamin D receptor regulates TNF-mediated arthritis. *Annals of the Rheumatic Diseases* 70(6):1122–1129.

13 Vitamin D and Infection

Jim Bartley and Carlos A. Camargo, Jr.

CONTENTS

13.1 INTRODUCTION

In the preantibiotic era, sun exposure and cod-liver oil were recognized treatments for tuberculosis (TB) (Table 13.1).[1] In the 1920s, for example, pulmonary TB was treated routinely with graded sun exposure. No one knew exactly why it worked, except that many sick patients with TB sent to rest in sunny locales or in the mountains became free of the disease.[2] With the advent of sulfonamides after the First World War, and the subsequent development of more effective antibiotics, these observations were neglected.

Although not known at the time, ultraviolet (UV) light at wavelengths between 270 and 290 nm makes pre–vitamin D_3 from 7-dehydrocholesterol.[3] In addition to vitamin D, the sun also creates other important substances in the skin, such as proopiomelanocortin. This molecule is then broken down into corticotrophin-releasing hormone, adrenocorticotrophic hormone, a range of melanocyte-stimulating hormones (α-MSH, β-MSH, and γ-MSH), and β-endorphins, which act as natural painkillers.[4] MSH has an important anti-inflammatory role in the skin as well as possible antimicrobial actions.[5] Likewise, cod-liver oil contains vitamin A, vitamin D, and omega-3 fatty acids, which all have anti-infective and immune-regulating properties.[6]

Recent basic scientific discoveries have renewed interest in the potential anti-infective actions of vitamin D.[7–10] Currently, these basic scientific advances are being translated into clinical practice. A number of interventional trials have been performed (Table 13.2), and other trials are currently under way (Table 13.3). Nevertheless, more studies are required to establish the role of vitamin D treatment in the prevention and management of infectious diseases; the scientific literature includes opinions ranging from extreme optimism to guarded caution.[11,12] Until more trials are completed

TABLE 13.1
TB Treatments in the Preantibiotic Era

Heliotherapy (Sun Exposure)

1855	Slovenia: Rikli, a Swiss physician, healed people using sun tanning.
1903	Denmark: Finsen was awarded the Nobel Prize for Medicine for the successful treatment of cutaneous TB (lupus vulgaris) using UV radiation.
1903	Switzerland: Rollier opened a hospital to treat TB by using graded sun exposure. In 1914, he published a book based on his experience.
1922	UK: Increasing popularity of sunlight treatment led to the establishment of "Committee on Sunlight" and Light Department at London Hospital; there was widespread adoption of sun-exposure practices.

Cod-Liver Oil

1833	Germany: Henkel reported on the successful use of cod-liver oil in the treatment of TB.
1841	Scotland: Bennett published the medical uses of cod-liver oil, including in TB. Five case reports showed improvement in TB symptoms, but two relapsed after stopping cod-liver oil.
1849	UK: Hospital for Consumption and Diseases of the Chest in London published outcomes of cod-liver oil in TB: 18% disease arrested, 63% improved, and 19% unchanged.
1855	USA: Woods from Philadelphia attributed the 19% decrease in deaths due to TB between 1847 and 1852 to the use of cod-liver oil.

Source: Ralph, A. et al., *Trends in Microbiology*, 16:336–344, 2008.

TABLE 13.2
Controlled Trials Testing the Effect of Vitamin D Supplementation on Infection

Study Authors (*n*)	Type of Study, Vitamin D Dosage, and Duration of Intervention	Results	Comment
Dental Disease			
McBeath and Zucker[62] (*n* = 296; variable)	RCT with four groups: control group, 250, 400, and 800 IU/day of vitamin D_3 for 12 months	Incidence of caries showed an inverse linear dose–relationship with vitamin D_3 dose	800 IU/day of vitamin D_3 reduced the incidence of caries
Krall et al.[71] (*n* = 145)	3-year RCT using 700 IU/day of vitamin D_3 and calcium assessment at 18 months	Tooth loss was 13% in patients taking supplements versus 27% in patients not taking supplements	Vitamin D was not independently related to risk of losing teeth
Respiratory System			
Aloia and Li-Ng[126] (*n* = 204)	3-year RCT using 2000 IU/day of vitamin D_3 in African American women	Number of flu or cold episodes in the treated group were 1/3 of the placebo group (8 vs. 26, respectively)	Significant reduction of reported flu, small sample
Avenell et al.[127] (*n* = 1700)	RCT 800 IU D_3/day 24–62 months	No difference in infection or antibiotic usage in previous week	Small dose
Li-Ng et al.[129] (*n* = 162)	12-week RCT using 2000 IU vitamin D_3/day	No significant difference in incidence of flu or cold symptoms	3-month trial—with this dose regimen, it may take more than 3 months to achieve adequate vitamin D levels

(continued)

TABLE 13.2 (Continued)

Controlled Trials Testing the Effect of Vitamin D Supplementation on Infection

Study Authors (n)	Type of Study, Vitamin D Dosage, and Duration of Intervention	Results	Comment
Urashima et al.[130] (n = 334)	4-month RCT using 1200 IU vitamin D₃/day in schoolchildren	Relative risk of .58 compared with control group (p = .04). Asthma attacks significantly reduced in treatment group (p = .006; secondary outcome)	Significant reduction of influenza A but not influenza B
Laaksi et al.[131] (n = 164)	RCT using 400 IU vitamin D₃/day for 6 months	No statistically significant difference in days off (p = .06); supplemented group reported as healthier	Low vitamin D supplement and supplemented group almost showed a significant result
Manaseki-Holland et al.[132] (n = 453)	RCT with a single dose of 100,000 IU vitamin D₃	No difference in days to recovery, but time to repeat pneumonia reduced	
Majak et al.[140] (n = 48 children)	RCT with of a single dose of 500 IU D₃/day	A significant reduction in asthma exacerbations (p = .029)	Children with a decreased 25(OH)D level eight times more likely to have an asthma exacerbation
TB			
Morcos et al.[231] (n = 24)	RCT 1000 IU vitamin D/day for 8 weeks	A greater clinical improvement was reported in the vitamin D group	Small numbers and methodology poorly reported
Nursyam et al.[86] (n = 67)	RCT using 10,000 IU vitamin D/day	77.1% sputum conversion rate in antibiotic-only group (placebo) compared with 100% in vitamin D group	Highly significant results. Small sample size, however
Wejse et al.[87] (n = 365)	RCT with 100,000 IU vitamin D₃ at 1, 5, and 8 months	No significant difference in groups	Dose may not have been high enough to result in a difference
Martineau et al.[88] (n = 126)	RCT—100,000 IU vitamin D₃ at baseline, 12, 28, and 42 days	No significant difference in time to sputum culture conversion (p = .14)	Sputum culture conversion hastened in tt genotype of the TaqI VDR polymorphism (p = .02)
GI Tract/Liver			
Abu-Mouch[153] (n = 58)	RCT using 1000–4000 IU vitamin D to achieve 25(OH)D levels > 75 nmol/L in patients with hepatitis C	Response rate with usual antiviral treatment: with vitamin D (26/27), without vitamin D (15/31)	96% responded with vitamin D treatment
Postoperative Complications			
Bischoff-Ferrari et al.[204] (n = 173)	RCT using 800 or 2000 IU vitamin D₃/day and intense physiotherapy	90% reduction in infection rate in the group using 2000 IU daily of vitamin D	Reduction in number of a wide range of infections but individual infection condition numbers were small

Abbreviations: RCT, randomized controlled trial; IU, International Units; RR, relative risk; VDR, vitamin D receptor.

TABLE 13.3
Ongoing Clinical Trials Looking at the Effect of Vitamin D Supplementation on Infectious Disease Outcomes (as of July 2011)*

General

Vitamin D Assessment (ViDA) trial (Scragg, R., and Camargo, C.A., Jr., New Zealand)—recruiting

Vitamin D and Omega-3 Trial (VITAL) (Manson, J., USA; PI of ancillary trial on infectious disease is Camargo, C.A., Jr.)—recruiting

Skin

Antimicrobial response to oral vitamin D_3 in patients with psoriasis (Gallo, R., USA)—completed

Vitamin D and psoriasis study (von Hurst, P., New Zealand)—not yet recruiting

Vitamin D deficiency and atopic dermatitis (Chiu, Y., USA)—recruiting

Dental Disease

Dose-dependent anti-inflammatory effects of vitamin D in a human gingivitis model (Garcia, R., USA)—active, not recruiting

TB

Pulmonary TB and vitamin D (Goswami, R., India)—active, not recruiting

Impact of vitamin D supplementation on host immunity to *Mycobacterium tuberculosis* and response to treatment (Ziegler, T., USA)—recruiting

Role of vitamin D in innate immunity to TB (Davaasambuu, G., USA)—active, not recruiting

L-arginine and vitamin D adjunctive therapy in pulmonary TB (Anstey, N., Australia)—recruiting

Replacement of vitamin D in patients with active TB (Sallahudin, S., Afghanistan)—completed

The effect of vitamin D on TB among Koreans (Yim, J.-J., Korea)—recruiting

A clinical trial to study the effect of the addition of vitamin D to conventional treatment in new pulmonary TB patients (Mathai, D., India)—recruiting

Upper Respiratory System

Effect of vitamin D supplementation on upper respiratory infections in adults (Murdoch, D., New Zealand)—recruitment closed

Vitamin D and *S. aureus* in the Diabetes Study (Jorde, R., Norway)—recruiting

Vitamin D supplementation and acute respiratory infection in older nursing home residents (Ginde, A., USA)—recruiting

A trial of vitamin D and health advice for the prevention of upper respiratory tract infections (Smieja, M., Canada)—not yet recruiting

Vitamin D supplementation in chronic rhinosinusitis with nasal polyps (VdinCRS; Schlosser, R., USA)—not yet recruiting

Vitamin D supplementation and upper respiratory tract infections in adolescent swimmers (Dubnov-Raz, G., Israel) —completed

Vitamin D for chronic sinusitis (Pinto, J., USA)—recruiting

Vitamin D and the influenza epidemic (Jorde, R., Norway)—completed

Vitamin D levels in children undergoing adenotonsillectomies and controls (Baroody, F., USA)—recruiting

Trial of vitamin D supplementation for the prevention of influenza and other respiratory infections (ViDiFlu) (Martineau, A., UK)—recruiting

Lower Respiratory System

Kabul vitamin D supplementation trial—pneumonia/diarrhea (Chandramohan, D., Afghanistan)—status not known

The effect of vitamin D replacement on airway reactivity, allergy, and inflammatory mediators in exhaled breath condensate in vitamin D–deficient asthmatic children (Bentur, L., Israel)—not yet recruiting

Randomized trial: maternal vitamin D supplementation to prevent childhood asthma (VDAART) (Weiss, S., USA)—recruiting

Vitamin D prevents elderly pneumonia (Arai, H., Japan)—completed

(continued)

TABLE 13.3 (Continued)
Ongoing Clinical Trials Looking at the Effect of Vitamin D Supplementation on Infectious Disease Outcomes (as of July 2011)

Trial of vitamin D supplementation for the prevention of influenza and other respiratory infections (ViDiFlu) (Martineau, A., UK)—recruiting

Trial of vitamin D supplementation in asthma (ViDiAs) (Martineau, A., UK)—recruiting

Trial of vitamin D supplementation in chronic obstructive pulmonary disease (ViDiCO) (Martineau, A., UK)—recruiting

Vitamin D to treat asthma in children: a randomized, double-blind, placebo-controlled trial (Elremeli M, United Arab Emirates)—not yet recruiting

Vitamin D for the treatment of severe asthma (Breitenbuecher A. Switzerland)—enrollment by invitation

Vitamin D supplementation during pregnancy for prevention of asthma in childhood (ABC vitamin D) (Bisgaard H, Denmark)—ongoing, not recruiting

Vitamin D in bronchiolitis obliterans syndrome (VIT001) (Verleden G, Belgium)—recruiting

Prospective intervention study on vitamin D in patients with cystic fibrosis (D-vitamin) (Pincikova T, Sweden)—recruiting

Randomized trial: maternal vitamin D supplementation to prevent childhood asthma (VDAART) (Weiss S, USA)—recruiting

Study of the effect of vitamin D as an add-on therapy to corticosteroids in asthma (VIDA) (Mauger D, USA)—recruiting

Gastrointestinal Tract (Including Hepatobiliary)

The beneficial effect of vitamin D supplement to peg interferon α2a or to telbivudine monotherapy in patients with chronic hepatitis B virus (HBV) infection (Nimer, A., Israel)—recruiting

Does vitamin D improve sustained virologic response (SVR) in Genotype 2,3 chronic hepatitis C patients? (Nimer, A., Israel)—not yet recruiting

Efficacy of vitamin D on top of pegylated interferon and ribavirin in patients with chronic viral hepatitis C nonresponders to antiviral therapy (Cacoub P, France)—not yet recruiting

Vitamin D treatment in patients with chronic hepatitis C (Abu-Mouch S, Israel)—recruiting

Vitamin D supplementation in Crohn's patients (CTSA) (Cantorna M, USA)—ongoing not recruiting

Vitamin D levels in children with IBD (Pappa H, USA)—recruiting

Vitamin D supplementation as nontoxic immunomodulation in children with Crohn disease (Ziring D, USA)—recruiting

Vitamin D in pediatric Crohn disease (Green T, Canada)—recruiting

Vitamin D supplementation in adult Crohn disease (VITD_CD) (Raftery T, Ireland)—not yet recruiting

HIV

Vitamin D supplementation and CD4 count in HIV-infected children (Bitnun A, Canada)—active, not recruiting

Vitamin D HIV study (HIV Vitamin D) (Yin Y, USA)—recruiting

Vitamin D supplements for HIV-positive patients on cART (Branch A, USA)—not yet recruiting

Study of vitamin D supplementation to male HIV seropositive patients (Bang U, Denmark)—not yet recruiting

Vaccination

Vitamin D supplementation and *Varicella zoster* virus vaccine responsiveness in older nursing home residents (Ginde A, USA)—recruiting

Vitamin D supplementation enhances immune response to Bacille–Calmette–Guerin (BCG) vaccination in infants (BCG-25-D) (Moodley A, USA)—recruiting

Miscellaneous

Study of vitamin D_3 substitution to patients with primary immunodeficiency (VITAPID) (Andersson J, Sweden)—enrolling by invitation

Vitamin D supplementation before surgery for colorectal cancer (McCall, J., New Zealand)—not yet recruiting

Vitamin D status in heart surgery patients and effects on clinical outcomes (Braun, L., Australia)—not yet recruiting

*Trials were identified through searches of www.clinicaltrials.gov and www.anzctr.org.au

and published, clinicians will need to rely on the best information currently available to guide their decisions. This chapter reviews the current evidence on the relationships of vitamin D with a range of infectious diseases.

13.2 OVERVIEW OF POTENTIAL MECHANISMS

Vitamin D has an important role in innate immunity through the production of two antimicrobial peptides (AMPs): cathelicidin and defensin β2. AMPs are synthesized and released largely by epithelial cells and neutrophils. Although they have a broad spectrum of antimicrobial activity against viruses, bacteria, and fungi,[13–15] some bacteria have developed AMP resistance mechanisms.[15] AMPs kill bacteria by inserting themselves into the bacterial cell membrane bilayers to form pores by "barrel-stave," "carpet," or "toroidal-pore" mechanisms. Recent evidence also suggests that AMPs inhibit cell wall synthesis, nucleic acid synthesis, protein synthesis, and enzymatic activity and disrupt mitochondrial membranes.[13,15]

Bacteria can exist in sophisticated communities called biofilms.[16] In a biofilm state, bacteria produce an extracellular matrix that protects its inhabitants against environmental threats, including "biocides, antibiotics, antibodies, surfactants, bacteriophages, and foraging predators such as free-living amoebae and white blood cells."[17] Bacteria in biofilms can make proteinases that can degrade AMPs.[15,18–20] The exopolysaccharide made by *S. epidermidis* and *Staphylococcus aureus* is positively charged; thus, it repels positively charged AMPs.[15,18] AMPs are also inactivated by the products of inflammation such as lipopolysaccharide from Gram-negative bacteria.[21] LL-37, the free C-terminal peptide of hCAP-18, *in vitro* seems able to both prevent and break down *P. aeruginosa* biofilms.[22,23] Innate immune mechanisms through AMP production may prevent infection, but may be less effective once biofilms are established.

Vitamin D's interactions with the adaptive immune system are complex, and mechanisms continue to be elucidated. Mahon and colleagues[24] have identified more than 102 genes that are the targets of 1,25-dihydroxyvitamin D [1,25(OH)$_2$D] in mature T helper (T$_H$) cells; 1,25(OH)$_2$D downregulated 57 genes and upregulated 45 genes. Along with T and B lymphocytes, vitamin D receptors (VDRs) are present in other cells important in adaptive immunity such as neutrophils, macrophages, and dendritic cells. More recently, vitamin D has been shown to have an important role in T-cell receptor maturation and the development of the T-cell response.[25] These issues are reviewed in Chapters 10 and 12. Vitamin D's potential anti-infective mechanism(s) of action will be elaborated below in the context of each organ system or infectious disease grouping.

UNDERSTANDING VDR POLYMORPHISMS

The human VDR is polymorphic.[26] Carriage of the T allele in the *Taq*I VDR receptor polymorphism is associated with increased anti-infective action against TB.[27,28] By contrast, carriage of the F allele in the *Fok*I VDR polymorphism is associated with a reduction in anti-infective action against TB.[27–29] The significance of VDR polymorphisms on vitamin D production and activity is an active area of investigation.[26,30]

13.3 SKIN DISEASE

Low vitamin D status has been linked to a number of skin diseases, including atopic dermatitis (AD), psoriasis, acne, and leprosy. At the beginning of the 20th century, the therapeutic effects of light on skin diseases were controversial and an area of active investigation. In Denmark, smallpox patients were treated routinely by exposure to "red light" only.[31] Dr. Niels Finsen noted that exposure to blue, violet, and UV rays seemed to make smallpox worse.[31] In 1903, Finsen was awarded the second Nobel Prize in Physiology and Medicine for discovering that high-intensity light produced from an electric arc lamp cured people with lupus vulgaris (TB of the skin). Because Finsen's lens systems largely transmitted at wavelengths of more than 340 nm, the beneficial effects may have been mediated by factors beyond vitamin D production.[32] In the same era, the Swiss physician Dr. Oskar Bernhard successfully treated infected wounds with sunlight.[2] During the First World War, heliotherapy was used to heal the wounds from casualties on both sides.[2] No further work has been done looking into the role of vitamin D status or supplementation in wound infection, although UV light is used to prevent wound infection in clean surgical wounds.[33]

Complex interrelationships exist between the bacterial microflora and the immune system of the skin.[34] Keratinocytes produce vitamin D_3 and also contain the enzymes necessary to form $1,25(OH)_2D$. Keratinocyte development and function is dependent on vitamin D.[35] Mice without VDRs have abnormal skin barrier formation, lipid secretion, and composition.[36] Suberythemal UVB in mice increases the AMP levels, the enzymes necessary for epidermal lipid synthesis, and the permeability barrier function of the skin.[37] In inflammatory skin conditions, LL-37 production is increased.[38]

Patients with AD are more susceptible to skin infections from both bacteria (particularly *S. aureus*)[39] and viruses (particularly *Herpes simplex*).[40] AD is associated with low AMP levels, particularly LL-37.[39] Children at higher latitudes (with less UVB exposure and presumably lower vitamin D status) seem to be more susceptible to AD.[41] UVB therapy is an effective treatment for severe AD.[42] Norwegian children randomized to live in a subtropical climate for 4 weeks (compared with staying in Norway) showed an improvement in AD.[42] In Italian children, 25(OH)D levels also correlate with AD severity.[43] These observations are probably mediated by vitamin D, because topical vitamin D has been shown to increase cathelicidin levels in patients with AD.[44] For example, oral vitamin D (4,000 IU/day) increased skin cathelicidin levels by sixfold in patients with AD.[45] Sidbury and colleagues conducted a small pilot randomized controlled trial (RCT) that suggested that vitamin D supplementation benefited children with winter-related AD;[46] these findings were confirmed recently in an RCT of more than 100 Mongolian children.[47] Thus, although further research is needed, vitamin D supplementation seems to be a useful adjunct therapy for at least some cases of AD, and we suspect that antimicrobial effects mediate this therapeutic benefit.

Anecdotally, sun exposure improves acne. Excessive sebum production is implicated in acne pathogenesis, and sebocytes contain all the enzymes necessary for the local synthesis and metabolism of $1,25(OH)_2D$.[48] Proliferation of the SZ95 human sebaceous gland cell line is suppressed by $1,25(OH)_2D$.[48] UVB radiation reduces *Propionibacterium acnes* and *S. aureus* colony counts *in vitro*.[49] Blue light also directly reduces keratinocyte cell inflammatory cytokines (IL-1α and ICAM-1) *in vitro*.[50] These other light influences aside, vitamin D supplementation might have a role in acne treatment. Further research is needed.

Topical vitamin D analogs are useful adjunct treatments for psoriasis.[51] Psoriatic lesions show increased AMP and inflammatory marker (IL-17A, IL-17F, and IL-8) concentrations.[52] In psoriatic skin, topical calcipotriol (a vitamin D analog) inhibits human β-defensin (HBD)-2, HBD-3, IL-17A, IL-17F, and IL-8. Paradoxically, and for unknown reasons, the skin concentration of the cathelicidin precursor protein hCAP18, but not LL-37, is increased.[52] This has been postulated as being due to deficiencies in the factors required for hCAP18 processing.[52] In a mouse model, topical calcipotriol and UV radiation induce regulatory T cells. A selective immunosuppression by the regulatory T cells is then thought to decrease inflammation.[53,54] Although these mechanistic findings support the possible role of vitamin D supplementation in psoriasis, the current evidence in humans is too sparse to draw conclusions.

As noted earlier, VDR polymorphisms seem to influence infectious diseases. For example, they have been linked to leprosy, which is caused by *Mycobacterium leprae*.[55] The tt VDR polymorphism was associated with tuberculoid leprosy in Bengali patients, whereas the TT genotype was associated with lepromatous leprosy.[56] The tt genotype was also associated with susceptibility to leprosy in a case-control study of patients in Malawi.[57] LL-37 levels, but not 25-hydroxyvitamin D

[25(OH)D] levels, are lower in Yemeni patients with leprosy when compared with a control group.[58] In the last decade, relatively little work has been done on this topic.

13.4 DENTAL DISEASE

Low 25(OH)D levels have been associated with dental caries, gingivitis, and periodontal disease. In dental caries, alterations in dental microflora shift the balance of teeth demineralization/remineralization toward an overall mineral loss. The dental microflora metabolize carbohydrates and produce acid, which reduces the environmental pH, leading to tooth surface demineralization. The reasons for the changes in dental microflora and the subsequent production of caries continue to be debated.[59] In the 1930s, East linked the incidence of dental caries in adolescent white boys to sun exposure hours. The group residing in areas having more than 3000 hours of sunshine per year had 291 cavities per 100 boys, whereas the group living in areas with less than 2200 hours of sunshine per year had 486 cavities per 100 boys.[60] As well as hours of sunshine, an increased number of cavities has also been linked to higher latitude (above 40°N)[61] and the winter–spring seasons.[62] Thus, vitamin D supplementation (800 IU/day) and UV light use would seem to prevent dental caries.[62,63] Simplistically, vitamin D plays a critical role in calcium homeostasis, bone modeling, and bone repair; however, vitamin D through AMP production potentially influences the underlying dental microflora. However, much of the work reviewed in the previous sections was done more than 80 years ago and does not seem to have been replicated.

Epithelium attaches to the teeth as it protrudes into the oral cavity. The gingival epithelium that attaches to the tooth is called the junctional epithelium. The junctional epithelium is composed of less differentiated cells with less tight intercellular junctions, making this region more susceptible to infection. Bacterial biofilm (dental plaque) overlying teeth surfaces causes gingivitis (inflammation of the mucosa overlying the alveolar bone). Gingivitis may evolve to periodontitis, which involves periodontal ligament destruction and alveolar bone loss. Patients with morbus Kostmann, which involves a severe congenital neutropenia and deficient cathelicidin levels in saliva and plasma, develop severe gingivitis and periodontitis.[64] Dietrich and colleagues[65] found that chronic gingivitis, as measured by bleeding on probing, had a linear inverse association with 25(OH)D serum concentrations of up to 90–100 nmol/L. This association was independent of age, sex, income, body mass index, vitamin C intake, full crown coverage, calculus, frequency of dental visits, use of oral contraceptives and hormone replacement therapy among women, number of missing teeth, and diabetes.

VDR gene polymorphisms have been linked to periodontal disease and tooth loss.[66,67] Osteoporosis is associated with alveolar bone and tooth loss, suggesting that poor alveolar bone quality may be a risk factor for tooth loss.[68] The prevalence of periodontal disease, as measured by periodontal attachment loss, is inversely associated with serum concentrations of 25(OH)D in people older than 55 years, an association that is independent of bone mineral density.[69] In pregnant women, vitamin D insufficiency [25(OH)D <75 nmol/L] is also associated with periodontal disease.[70] In an interventional study of hip bone loss, Krall and colleagues[71] found that patients supplemented with both calcium and vitamin D for 2 years had a rate of tooth loss that was 60% lower than those who had taken a placebo containing no vitamin D or calcium. After the randomized trial had been completed, long-term follow-up studies suggested that calcium intake may have been a more important factor. Calcium and vitamin D supplementation (up to 1000 IU daily) also has a "modest positive effect" on periodontal health. All parameters of periodontal health, including attachment loss, alveolar crest-height, probing depth, gingival index, and furcation involvement were better in subjects who took vitamin D and oral calcium supplements.[72]

Bisphosphonates are being increasingly used in the treatment of osteoporosis. Most cases of bisphosphonate-related necrosis of the jaw (BRONJ) have occurred after 3 years.[73] Infection is a factor in BRONJ; however, it is not known whether this represents a primary or secondary phenomenon.[74] In a recent animal study, the prevalence of BRONJ was studied among three different animal groups: a vitamin D–deficient group taking zoledronate, a vitamin D–deficient group not using zoledronate, and a zoledronate alone group. This animal study has implicated vitamin D deficiency as a factor in the development of BRONJ.[75] Further research into the correction of vitamin D deficiency in patients receiving a bisphosphonate is needed.

13.5 TUBERCULOSIS

As previously noted, both sun exposure and cod-liver oil have links to vitamin D and were accepted therapies for TB in the preantibiotic era (Table 13.1).[76] In the 1940s, high-dose vitamin D_2 was used in the treatment of lupus vulgaris;[77] however, with the advent of antibiotics, this work was forgotten. Although TB can affect a number of organ systems, most recent research has focused on the lower respiratory tract.

VDR polymorphisms have been investigated as possible genetic determinants of TB risk and treatment outcome. In a case-control study of 2015 African subjects, homozygotes for *Taq*I polymorphism (genotype tt) were significantly underrepresented in patients with TB.[78] In Gujarati Asians, the ff genotype of the *Fok*I RFLP was associated with the extent of pulmonary TB in 25(OH)D-deficient patients.[79] In Peru, sputum mycobacterial culture conversion was significantly faster among patients with the *Fok*I FF genotype (compared with non-FF genotypes) and patients with the *Taq*I Tt genotype (compared with the TT genotype).[28] Decreased serum 25(OH)D concentrations have been linked with the activation of latent TB,[80] a higher risk of active TB,[81] and disease progression.[82] In a prospective study in Pakistan, Talat and colleagues[82] examined 109 household contacts of 20 people with sputum-positive pulmonary TB. Over 4 years of follow-up, contacts with 25(OH)D levels lower than 17.5 nmol/L had a statistically significant increased risk of progressing to active TB compared with contacts with levels higher than 32.5 nmol/L. Thus, vitamin D supplementation could be potentially useful in people with latent TB or in preventing TB.[83]

> ### THE FIRST DOCUMENTED CASE OF TB DUE TO VITAMIN D DEFICIENCY (500,000 BC)
>
> *Homo erectus* is believed to be among the first hominids to leave Africa, and Turkey was probably one of their routes as they left. A 500,000-year-old skull from Turkey has shown evidence of TB. The findings support the theory that dark-skinned people, who migrate out of tropical climates, tend to have lower levels of vitamin D, which places them at increased risk of infection. People with paler skins were better adapted to a colder environment as they migrated north.
>
> From Kappelman J., Alçiçek M. C., Kazanci N., Schultz M., Ozkul M., Sen S. 2008. First Homo erectus from Turkey and implications for migrations into temperate Eurasia. *Am. J. Phys. Anthropol.* 135:110–116.

In recent years, three RCTs and at least 10 prospective case-series have evaluated vitamin D supplementation in patients with TB. Ergocalciferol (vitamin D_2), which may be less potent than cholecalciferol (vitamin D_3), was used in many of these studies.[84,85] For this and other reasons, Martineau and colleagues[76] feel that most of these studies have methodologic limitations. Nevertheless, the studies generally support a role for vitamin D in the prevention and management of TB. In Indonesian patients with pulmonary TB, vitamin D supplementation with 0.25 mg (10,000 IU)/day for 6 weeks led to more rapid sputum clearance. Only 54% of patients had follow-up chest radiographs at 6 weeks; however, in these patients, the radiological improvement was also greater in the vitamin D supplemented group although no statistical analysis was reported.[86] In another recent trial, 100,000 IU of cholecalciferol at inclusion into the trial and again 5 and 8 months after the start of treatment did not improve the clinical outcome or overall mortality among patients with TB. The vitamin D dose used may have been inadequate.[87] Recently, Martineau and colleagues found that the administration of four doses of 100,000 IU of vitamin D_3 did not significantly influence time to sputum culture conversion in the whole pulmonary TB study population, but it did significantly hasten sputum culture conversion in participants with the tt genotype of the *Taq*I VDR polymorphism.[88] Additional appropriately designed trials investigating the influence of adjunctive vitamin D in patients with active TB are under way (Table 13.3).

13.6 RESPIRATORY SYSTEM

Low vitamin D status is associated with an increased risk of upper and lower respiratory tract infection. VDR receptor polymorphisms in patients with type I diabetes are associated with methicillin resistant *Staphylococcus aureus* (MRSA) nasal carriage[89] and 25(OH)D deficiency is also associated with an increased risk of MRSA nasal carriage.[90] Whether vitamin D supplementation reduces the risk of MRSA colonization is not known. Finnish army recruits with 25(OH)D levels that were lower than 40 nmol/L were at an increased risk of developing upper respiratory tract infections.[91] Parents of Dutch children with the least sun exposure were twice as likely to report that their child developed a cough and were three times as likely to report that their child had a runny nose compared with children with the most sun exposure.[92] The seasonal variation in influenza has also been linked to low vitamin D levels.[93] Low 25(OH)D levels are seen in children with otitis media with effusion[94] and tonsil disease;[95] however, neither of these small studies had a control group. A case-control study of 149 Yemeni children has linked the duration of chronic suppurative otitis media with low 25(OH)D levels, but this association was lost with adjustment for age.[96] Pinto and colleagues[97] found low 25(OH)D levels in urban African American, but not white, subjects with chronic rhinosinusitis. In American adults, serum 25(OH)D levels of more than 75 nmol/L are associated with a reduced incidence of upper respiratory tract infection.[98] Sabetta and colleagues[99] found that 25(OH)D concentrations of more than 95 nmol/L are associated with a twofold reduction in the risk of developing acute viral respiratory tract infection as well as a marked reduction in the number of days ill. In a large cross-sectional study of English adults, Berry and colleagues[100] found a 7% reduction in respiratory tract infection for each 10 nmol/L increase in 25(OH)D levels up to 100 nmol/L.

Before the discovery of vitamin D, rickets was thought to have an infectious etiology.[101] Rickets is associated with an increased risk of acute respiratory tract infection, particularly pneumonia.[102–107] Cord plasma 25(OH)D levels predict the development of respiratory syncytial virus with lower respiratory tract infection.[108] In Turkey, newborns with subclinical vitamin D deficiency have an increased risk of acquiring acute lower respiratory tract infections (ALRI).[109] In Canada, significantly more children who were vitamin D–deficient were admitted to a pediatric intensive care unit with ALRI. Vitamin D might influence ALRI disease severity.[110] A case-control study has also reported an association between the serum 25(OH)D levels of more than 50 nmol/L and ALRI in children.[111] This study led to the recognition of a link between subclinical vitamin D deficiency, nonexclusive breastfeeding, and increased risk for severe ALRIs. Similar observations have been made between ALRI severity and vitamin D deficiency in Japanese children.[112] Although one Canadian study found that 25(OH)D status was not associated with an increased risk of hospitalization for ALRI, many of the infants were already taking vitamin D through fortified infant formula or supplements. The average 25(OH)D level of these infants was 77 nmol/L, indicating that their vitamin D status was not going to be an important factor in their risk of developing infection.[113] The *Fok*I ff genotype, which is associated with a reduction in the anti-infective action of vitamin D, has been linked to an increased risk of developing an ALRI in early childhood.[114]

Low vitamin D status has been potentially linked to an increased rate of pneumonia during the 1918 influenza epidemic in the United States[115] as well as pneumonia deaths in England.[116] In New Zealand patients 25(OH)D deficiency is associated with increased mortality from pneumonia during winter.[117] Low serum 25(OH)D concentrations are seen in cystic fibrosis, which is characterized by recurrent infections.[118] In patients with cystic fibrosis, serum IgG levels are negatively associated with serum 25(OH)D.[119] 25(OH)D deficiency is frequent in chronic obstructive pulmonary disease (COPD) and correlates with disease severity.[120] RCT evidence on the effect of vitamin D supplementation in patients with severe COPD has not been reported,[120] but such work is under way (Table 13.3). Surprisingly, no work seems to have been published investigating the role, if any, of vitamin D supplementation in bronchiectasis.[121]

Historically, cod-liver oil supplementation was shown to reduce upper respiratory tract infection frequency. The beneficial effect was attributed to vitamin A.[122,123] A number of interventional studies indicate that vitamin D supplementation probably will protect against respiratory tract infection. For example, suberythemal courses of UV radiation given twice a week for 3 years to 410 teenage athletes compared with 446 nonirradiated athletes resulted in 50% fewer respiratory viral infections and 300% fewer days of absences.[124] In one interventional cohort study in which 60,000 IU of vitamin D was given weekly for 6 weeks to children with recurrent respiratory tract infection, the incidence of disease in the children receiving supplementation was reduced to that of the control group.[125] In some studies in which 25(OH)D was given for skeletal health, a reduction in the risk of infection was also noted.[126,127] In Japanese patients undergoing dialysis, vitamin D supplementation also prevents acute respiratory tract infections.[128]

Three RCTs looking at the role of vitamin D supplementation in the prevention of upper respiratory tract infection,[129–131] and one looking at the role of vitamin D supplementation in pneumonia,[132] have been published recently. In the first study, Li-Ng and colleagues[129] randomized 162 adults to receive 2000 IU of vitamin D_3 daily or a matching placebo for 12 weeks. No difference in the duration or severity of URI symptoms was found. The authors attributed this to a number of reasons. First, the subjects started vitamin D supplementation during winter, rather than at the beginning of winter. Because the half-life of 25(OH)D is at least 2–3 weeks, it is generally accepted that it takes 2–3 months for blood 25(OH)D levels to plateau with vitamin D supplementation if a loading dose is not given.[133] This meant that the subjects were reaching optimum 25(OH)D levels at the end of winter and the end of the trial. Second, the vitamin D dosage may have been inadequate; and third, the baseline 25(OH)D levels were higher than in previous studies, meaning that vitamin D supplementation may have been less effective.

In the second RCT,[130] Urashima and colleagues gave 334 Japanese schoolchildren 1200 IU daily of vitamin D_3 or a placebo. There was a 50% reduction in children who were diagnosed with influenza A (primary outcome). However, if one combines the number of cases of influenza A and influenza B, there was no reduction in total influenza infections between the vitamin D–treated and the control groups. In the third RCT, Laaksi and colleagues[131] supplemented 164 young Finnish men with only 400 IU/day of vitamin D_3. Absence from duty due to respiratory tract infection and number of days absent were lower in the treated group. The proportion of subjects without any days absent was slightly higher in the vitamin D supplementation group. Lastly, in an RCT of 453 Afghani children with pneumonia, aged 1–36 months, subjects were given either a single high dose of oral vitamin D_3 (100,000 IU) or placebo drops and routine pneumonia treatment. Although the recovery times were similar (i.e., no therapeutic benefit), the risk of a future episode of pneumonia was reduced, as was the time to a further episode of pneumonia.[132] In a recent study by Martineau and colleagues,[88] in which high-dose vitamin D_3 was used in the treatment of pulmonary TB, the number of people who had upper respiratory tract infection symptoms was recorded. One of 71 patients receiving at least one dose of vitamin D_3 as compared with 6 of 70 receiving at least one dose of placebo reported symptoms. This secondary outcome was of borderline statistical significance ($p = .06$). Further research studies into the protective actions of vitamin D on respiratory tract infection are under way (Table 13.3).

Childhood wheezing and exacerbation of obstructive lung diseases, such as asthma and COPD, provide insight into the role of vitamin D on infectious diseases because they usually represent uncomplicated acute respiratory infections. Vitamin D also has other potential links with these disorders. For example, low 25(OH)D levels are associated with reduced lung function[134] and asthma prevalence seems higher in industrialized countries that are furthest from the equator[135] and lower at high altitudes.[41] Moreover, the VDR gene maps to the long arm of chromosome 12, a region frequently linked to asthma and allergy-related phenotypes in genomewide linkage analyses. Associations have been shown between VDR polymorphisms and the risk of asthma and atopy.[136,137] Camargo and colleagues[138] found an inverse relationship between cord blood 25(OH)D level and cumulative risk of wheezing at the ages of 15 months, 3 years, and 5 years. Brehm and colleagues[139] examined the association between serum 25(OH)D and risk of a severe asthma exacerbation (defined as an asthma-related emergency department visit or hospitalization) in North America. In 1022 children, subjects with 25(OH)D levels lower than 75 nmol/L were more likely to have a severe exacerbation over the next 4 years. In the RCT where Japanese children were given 1,200 IU daily or placebo, children with a previous diagnosis of asthma, there was also a significant reduction in number of asthma attacks;[130] only 2 asthmatic children taking vitamin D and 12 children taking placebo had asthma attacks. Recently, Majak and colleagues[140] reported an RCT on the role of vitamin D supplementation (500 IU daily of vitamin D_3) versus placebo for 6 months in Polish children with newly diagnosed asthma; the investigators observed a significant reduction in asthma exacerbations due to acute upper respiratory tract infections ($p = .029$). Children whose 25(OH)D levels decreased over the study period were eight times more likely to have an exacerbation of asthma compared with those children with stable or increased 25(OH)D levels. Lastly, self-reported maternal asthma was also linked to low vitamin D levels.[141]

In contrast to these favorable findings, Hyppönen and colleagues[142] have reported that regular vitamin D supplementation ($\geq$2000 IU/day) of Finnish infants increases the risk of developing atopy, allergic rhinitis, and asthma at 31 years of age. In a separate study in the United Kingdom, they also linked deficient (<25 nmol/L) and excessively high (>135 nmol/L) serum 25(OH)D_3 levels with elevated serum IgE levels.[143] Thus, the relationships of vitamin D with asthma and allergy seems complex. Vitamin D might influence asthma in at least two levels: (1) predisposition toward versus protection against asthma, and (2) prevention of infection-related asthma exacerbations.

13.7 CIRCULATORY SYSTEM

In the United States, rheumatic fever rates are higher in the winter than in the summer, higher in the northeast than in the south, and higher in African Americans than in white Americans.[144] Endocarditis shows a similar seasonal variation being more common in the fall/winter.[145] All of these observations suggest that vitamin D deficiency might have a role in these disorders, but little other work seems to have been undertaken.

13.8 GASTROINTESTINAL TRACT (INCLUDING THE HEPATOBILIARY SYSTEM)

Vitamin D seems important in hepatic health. VDRs and AMPs are expressed in human liver biliary epithelial cells.[146] 25(OH)D deficiency is common among patients with chronic liver disease.[147] VDR polymorphisms are linked to occult hepatitis B viral (HBV) infection[148] and HBV load.[149] Low 25(OH)D levels have been linked to genotype 1 chronic hepatitis C virus (HCV), more severe fibrosis, and a poorer response to interferon-based therapy.[150,151] Indeed, serum 25(OH)D levels predict response to antiviral treatment in HCV.[152] In a recent RCT, Dr. Abu-Mouch[153] found that vitamin D supplementation (1000–4000 IU daily) to achieve 25(OH)D levels higher than 80 nmol/L led to 96% of the vitamin D–supplemented group being HCV RNA-negative as opposed to 48% in the control group. Vitamin D may have an important role in hepatitis management and further research is under way (Table 13.3).

The intestinal mucosa is constantly exposed to a vast array of microbes and food antigens. It must "tolerate" the commensal flora and sense potentially harmful pathogens.[154] In *Helicobacter pylori* infection, cathelicidin and defensin production in gastric epithelial cells are upregulated.[155,156] Serum 25(OH)D levels correlate with serum *H. pylori*–specific IgG antibody titers ($r = .36$, $p = .04$).[157] A reduction in *H. pylori* infection in 34 elderly Japanese women taking 40 IU of 1α-hydroxyvitamin

D_3 a day for 20 years has been reported.[158] 1α-Hydroxyvitamin D_3 is a vitamin D analog used to compensate for the reduced renal hydroxylation experienced in renal failure.[159] There are major limitations in this retrospective study, including the small sample and the low dose of vitamin D given.[160] Further studies would seem worthwhile.

In a mouse model, vitamin D deficiency predisposes to colon inflammation, suggesting that vitamin D deficiency might contribute to inflammatory bowel disease.[161] In a cohort of African adults, Kelly and colleagues[162] found that low α-defensin expression was associated with a higher risk of diarrhea. The VDR also functions as a secondary intestinal bile acid receptor and 1-hydroxyvitamin D stimulates bile acid metabolism.[163] Receptors for primary bile salts include the farsenoid X receptor, the pregnane X receptor, and the constitutive androstane receptor.[163,164] Bile flow obstruction leads to bacterial overgrowth in the small intestine, mucosal injury barrier disruption, and systemic infection. The overlap in the ligands and target genes of the VDRs, farsenoid X receptors, pregnane X receptors, and constitutive androstane receptors may be important in gut detoxification and immunity.[165]

In many developing countries, *Shigella* infection is an important cause of childhood morbidity and mortality. *Shigella* infection survives by inhibiting the endogenous expression of AMPs in human colon epithelial cells.[166,167] Butyrate and phenylbutyrate are short-chain fatty acids. AMP production is enhanced by butyrate, phenylbutyrate, and vitamin D. When the colon epithelium is exposed to butyrate, cathelicidin is produced, killing *Shigella*.[168] A depletion in the population of bacteria that produces butyrate in the colon has been implicated in ulcerative colitis and Crohn disease.[169] Increased intake of fermentable dietary fiber leading to the increased production of short-chain fatty acids may be clinically beneficial in the treatment of colitis.[170] Vitamin D has a potential role in gut health through increased AMP production; a number of trials exploring this are under way (Table 13.3).

13.9   HIV

Vitamin D influences HIV disease, but the precise mechanisms remain unclear. Vitamin D could potentially influence HIV infection at a number of levels. Vitamin D could protect against initial HIV infection as well as influence disease progression and survival.[171] *In vitro*, LL-37 inhibits HIV-1 replication by target cells.[172] VDR polymorphisms are associated with susceptibility to HIV infection[173] and seem to influence disease progression.[174]

Most of the studies looking at the relationships between HIV disease and vitamin D status have been case-series or cross sectional. In Danish adult male HIV-positive patients, serum 25(OH)D levels, although low, were not associated with age, years of HIV infection, highly active antiretroviral therapy or CD4 count.[175] Some cross-sectional studies have found positive correlations between 1,25(OH)$_2$D and CD4 cell counts.[171] Several studies suggest that 1,25(OH)$_2$D concentrations are often decreased among HIV-infected persons. One Norwegian study found that HIV-infected patients with low 1,25(OH)$_2$D levels had significantly shorter survival times compared with those with normal concentrations.[176] A European study involving 1985 HIV-positive persons has linked 25(OH)D deficiency with a higher risk of mortality.[177] However, a correlation between mortality from HIV infection and vitamin D deficiency has not been clearly established.[171]

In Tanzania, Mehta and colleagues[178] measured maternal 25(OH)D levels between 12 and 27 weeks of pregnancy. A low maternal 25(OH)D level (<80 nmol/L) was associated with a 50% higher risk of mother-to-child transmission of HIV at 6 weeks, a twofold higher risk of mother-to-child transmission of HIV through breastfeeding among children who were not infected with HIV at 6 weeks, and a 46% higher overall risk of HIV infection. Children born to women with a low 25(OH)D level had a 61% higher risk of dying during follow-up. In HIV-infected women, low initial 25(OH)D levels were also associated with significant HIV disease progression, severe anemia, and hypochromic microcytosis. Women with higher 25(OH)D levels had a significantly lower risk of all-cause mortality although there was no significant association with AIDS-related mortality or

T-cell counts.[179] To our knowledge, RCTs investigating the role of vitamin D supplementation on protection against initial HIV infection as well as HIV disease progression and survival have not been reported.

13.10 GENITOURINARY SYSTEM

Vitamin D may have an important role in the overall health of the urinary tract.[180] AMPs are found throughout the genitourinary system. Although HBD-1 urinary concentrations are lower than the effective antimicrobial concentrations, the fluid layer close to the urinary tract epithelium into which AMPs are initially secreted is assumed to have higher anti-infective AMP concentrations. Kidney infection induces the expression of HBD-2[181] and cathelicidin.[182] $25(OH)D_3$ supplementation increases cathelicidin production in human bladder cells infected with uropathogenic *Escherichia coli*.[183] Genetically engineered mice in which a defensin gene, analogous to the constitutively expressed HBD-1, was deleted do not maintain urine sterility. Approximately 30% of the healthy Defb1$^{-/-}$ mice had *Staphylococcus* species in bladder urine.[184] The possibility has also been raised that the asymptomatic bacteriuria said to occur in 2% to 10% of pregnant women, and in up to 20% of women ages 80 years or older, might be linked to vitamin D deficiency.[180] Although one study in children has suggested that vitamin D supplementation might increase the risk of urinary tract infection in the first 3 months of life,[185] this retrospective study has been criticized because of biological implausibility, inappropriate statistical analysis, and poor recording of vitamin D supplementation.[186]

Studies suggest that vitamin D may also be important in vaginal health. After sexual intercourse, cathelicidin in vaginal secretions is processed into the AMP ALL-38 (LL-37 with an *N*-terminal alanine residue) by gastricsin that is present in the seminal fluid.[187] Bacterial vaginosis is a syndrome characterized by the loss of normal vaginal flora, predominantly hydrogen peroxide–producing *Lactobacillus* species, and an increased prevalence of anaerobic bacteria. 25(OH)D deficiency is associated with bacterial vaginosis during pregnancy.[188,189] A variety of cationic polypeptides, which includes defensins and LL-37, seem to play a role in combating HIV infections.[190] A lack of fungicidal killing of *Candida albicans* has been demonstrated in hereditary resistance to active vitamin D.[191] Vitamin D could reduce *C. albicans* vaginal infections after treatment of urinary tract infections.[192] Low levels of HBD-2 and HBD-3, and increased levels of LL-37 are seen in condylomata acuminata.[193,194] Although existing evidence is strongly suggestive that vitamin D and associated AMP production might have an important role in protecting the genitourinary tract against infection, no interventional studies seem to have been published.

13.11 NERVOUS SYSTEM (INCLUDING EYE)

Neisseria meningitidis is an important cause of bacterial meningitis. In the United States, the occurrence of meningitis, as with many other infectious diseases, is highest in late winter and early spring, when vitamin D levels are lowest. The risk of invasive meningococcal disease correlates with the clear-sky UVB index using a lag of 1–4 days.[195] No other similar studies seem to have been published nor do 25(OH)D levels seem to have been measured in patients affected by meningitis.

Although herpes zoster does not show with any seasonal variation,[196] UVB phototherapy in the acute stage of the zoster rash seems to reduce the incidence and severity of postherpetic neuralgia (PHN).[197] After 3 months, treatment of PHN with UVB radiation is no longer beneficial.[197] These data are consistent with the hypothesis that early treatment with a vitamin D cream could reduce the incidence of PHN.[198]

With regard to ocular health, AMPs are known to play an important role.[199] Defensins and cathelicidin are produced by corneal and conjunctival epithelial cells. On the ocular surface, cathelicidin may prevent bacterial biofilm formation by *P. aeruginosa*[200] and HBD-2 and HBD-3 promote resistance to *P. aeruginosa* infection.[201] Cathelicidin (LL-37) eye drops might have a role in the treatment of ocular infections. The evidence base is currently very limited, and further research is needed.

13.12 MUSCULOSKELETAL SYSTEM AND CONNECTIVE TISSUE

Osteomyelitis often requires multiple surgical interventions and local or systemic antibiotic therapy. In a rabbit model, AMP hLF1-11, when added to the cement used in implants, significantly reduced *S. aureus* infection.[202] The T allele ($p = .007$) and T/T genotype ($p = .028$) are significantly associated with bone loss secondary to deep infection after hip arthroplasty.[203] In a double-blind RCT of 2000 IU versus 800 IU of vitamin D_3 daily after hip fractures in elderly patients, investigators reported a 39% reduction in hospital readmissions in the group using 2000 IU of vitamin D_3 daily. This reduction of readmissions was a result of a 60% reduction in fall-related injuries and a 90% reduction in overall infections. The infections influenced included infection of the hip prosthesis, pneumonia, bronchitis, colitis, sepsis and infection of an indwelling catheter.[204] The numbers involved with each individual infection were small but vitamin D seems to have potential in the prevention of infection related to, and after orthopedic surgery.

13.13 SEPSIS

Sepsis, although commonly caused by bacterial infections, is also associated with both viral and fungal infections. A number of epidemiologic features suggest that low vitamin D levels might be associated with an increased risk of septicemia and sepsis.[205] In the United States, rates of sepsis show a seasonal variation—being highest in winter and lowest in fall, highest in the Northeast and lowest in the West, higher in African Americans than in white Americans, and being higher with age.[206] All of these observations are consistent with a potential role for low vitamin D status in the prevention or management of sepsis.

In patients undergoing renal dialysis, low 25(OH)D levels were linked to fatal infection.[207] In individuals initiating chronic hemodialysis, low baseline cathelicidin levels were also independently associated with an increased risk of death attributable to infection.[208] Lower 25(OH)D levels are seen in critically ill patients compared with healthy controls. In the same study, 25(OH)D status was positively associated with plasma LL-37 levels.[209] 25(OH)D levels have been associated with higher sepsis severity in emergency department patients hospitalized with suspected infection.[210] These studies indicate that vitamin D status may be important in sepsis. To our knowledge, however, interventional studies with infectious disease outcomes have yet to be performed.

13.14 VACCINATION

Although vitamin D may have important direct effects in the prevention of infectious disease, vitamin D could play another more indirect role through improving vaccination efficacy. 1,25(OH)$_2$D administered topically with trivalent influenza vaccine in mice enhanced both mucosal and systemic antibody responses.[211] In mice, the addition of topical 1,25(OH)$_2$D augments induced immunity to hepatitis B surface antigen.[212] However, an RCT of coadministration of 1 mg (40,000 IU) of 1,25(OH)$_2$D$_3$ intramuscularly with influenza vaccine was not found to enhance the humoral immune response as measured by change in H1N1, H3N2, or influenza B antigen titers in healthy, young, mostly white volunteers; the vitamin D status of these volunteers, as measured by serum 25(OH)D levels, was not assessed.[213] In patients with prostate cancer, a higher 25(OH)D level was associated with a better serologic response to trivalent influenza vaccine.[214] All patients in the upper quartile of baseline 25(OH)D concentrations mounted a serological response to the trivalent influenza vaccine (defined as a response to any of the three strains); the apparent beneficial effect of high baseline 25(OH)D concentrations was largest for vaccination against influenza A/H3N2 strain.[214] BCG vaccination has also been shown to increase 25(OH)D concentrations in British children.[215] Future studies are needed to address the question of whether vitamin D supplementation enhances immune response, particularly to influenza vaccination.

13.15 PRACTICAL ADVICE FOR CLINICIANS

Because of the well-known risks of skin damage and skin cancer with exposure to UV radiation,[216] many health professionals prefer a recommendation of oral vitamin D supplementation to maintain a healthy vitamin D status.[217] A loading dose is often useful, as plateau levels are not reached for 2–3 months with regular oral supplementation.[133,218] Humans can obtain vitamin D orally in two forms—vitamin D_3 (cholecalciferol) and vitamin D_2 (ergocalciferol). Ergocalciferol is considered by most experts to be less potent than cholecalciferol[84,85] although some investigators dispute this assertion.[219] Both oral forms of vitamin D are hydroxylated in the liver by the cytochrome P450 enzyme CYP27A1 to 25(OH)D. Hydroxylation of vitamin D to 25(OH)D slows when 25(OH)D levels reach 100 nmol/L. At this level, the enzyme seems to become saturated.[220] In adult men, serum 25(OH)D levels increase by approximately 1.75 nmol/L for each 100 IU/day (0.70 nmol/L for each microgram per day) of vitamin D consumed.[221] In many vitamin D–deficient people, current vitamin D supplementation recommendations,[222] even to reach levels of 50 nmol/L or higher, may be inadequate.[140,223] Vitamin D requirements in various disease states are also not known; however, in patients with gastrointestinal absorption difficulties, vitamin D requirements are higher.[118] A standard oral vitamin D supplementation recommendation in the presence of ongoing infection may also be inadequate, but this remains to be proven. The interaction of active disease and 25(OH)D levels is an understudied area. A recent study of knee operations found that plasma 25(OH)D levels decreased significantly after surgery.[224] The possibility that increased inflammation increases vitamin D requirements is a topic that merits further investigation.

Although many laboratories use 150 or 200 nmol/L as the safe upper limit for serum 25(OH)D, the evidence indicates that 350 nmol/L is also safe for most individuals.[3] Some have even advocated 500 nmol/L as a safe upper limit before hypercalcemia occurs.[225,226] However, a safe 25(OH)D level may not necessarily be the optimal level for all people. For example, serum 25(OH)D levels of more than 135 nmol/L are associated cross sectionally with elevated serum IgE levels.[143] A birth cohort study in Tucson, Arizona (a locale with abundant sunlight and UVB exposure) suggests that higher cord blood 25(OH)D levels are associated with a higher risk of atopic outcomes;[227] higher vitamin D levels could promote T_H2 cell development and thus an increased allergic response. Vitamin D supplementation of people with normal or high levels of 25(OH)D may also increase TB risk.[228] Others have also found that 25(OH)D levels in older men may have a U-shaped relationship with overall mortality,[229] the small numbers raise doubts about the reproducibility of this finding. Further research is necessary to determine the optimal vitamin D levels. Based on currently available evidence, we believe that the optimal level of serum 25(OH)D is probably approximately 100 nmol/L (40 ng/mL). Several observational studies support this target level.[230] Even higher levels may be required in specific disease states, but not in others. It seems unlikely that there will be one optimal level for a myriad of health problems.

13.16 CONCLUSIONS

Vitamin D has important links to both innate and acquired immune systems. Moreover, many clinical studies from around the world suggest that vitamin D may have an important role in the prevention and treatment of a diverse range of infectious diseases (Table 13.2). Nevertheless, appropriate double-blind RCT data are still lacking for most infectious disease outcomes. Accordingly, definitive statements on the optimal 25(OH)D level and vitamin D treatment regimen for the prevention or management of infectious diseases are premature. The completion of many ongoing RCTs worldwide (Table 13.3) will provide critical evidence on the role of vitamin D in infectious diseases.

REFERENCES

1. Ralph, A., P. Kelly, and N. Anstey. 2008. L-arginine and vitamin D: Novel adjunctive immunotherapies in tuberculosis. *Trends in Microbiology* 16:336–344.
2. Hobday, R. 1997. Sunlight therapy and solar architecture. *Medical History* 41:455–472.
3. Holick, M. 2007. Vitamin D deficiency. *New England Journal of Medicine* 357:266–281.
4. Bicknell, A. B. 2008. The tissue-specific processing of pro-opiomelanocortin. *Journal of Neuroendocrinology* 20:692–699.
5. Brzoska, T., T. A. Luger, C. Maaser, C. Abels, and M. Bohm. 2008. Alpha-melanocyte–stimulating hormone and related tripeptides: Biochemistry, antiinflammatory and protective effects *in vitro* and *in vivo*, and future perspectives for the treatment of immune-mediated inflammatory diseases. *Endocrine Reviews* 29:581–602.
6. Selmi, C., and K. Tsuneyama. 2010. Nutrition, geoepidemiology, and autoimmunity. *Autoimmunity Reviews* 9:A267–A270.
7. Wang, T., F. Nestel, V. Bourdeau et al. 2004. Cutting edge: 1,25-dihydroxyvitamin D_3 is a direct inducer of antimicrobial peptide gene expression. *The Journal of Immunology* 173:2909–2912.
8. Liu, P. T., S. Stenger, H. Li et al. 2006. Toll-like receptor triggering of a vitamin D–mediated human antimicrobial response. *Science* 311:1770–1773.
9. Adams, J. S., and M. Hewison. 2008. Unexpected actions of vitamin D: New perspectives on the regulation of innate and adaptive immunity. *Nature Clinical Practice. Endocrinology & Metabolism* 4:80–90.
10. Liu, P., S. Stenger, D. Tang, and R. Modlin. 2007. Cutting edge: Vitamin D–mediated human antimicrobial activity against *Mycobacterium tuberculosis* is dependent on the induction of cathelicidin. *The Journal of Immunology* 179:2060–2063.
11. Bruce, D., J. Ooi, S. Yu, and M. Cantorna. 2010. Vitamin D and host resistance to infection? Putting the cart in front of the horse. *Experimental Biology and Medicine* 235:921–927.
12. Grey, A., and M. Bolland. 2010. Vitamin D a place in the sun? *Archives of Internal Medicine* 170:1099–1020.
13. Gudmundsson, G., and B. Agerberth. 1999. Neutrophil antibacterial peptides, multifunctional effector molecules in the mammalian immune system. *The Journal of Immunology Methods* 232:45–54.
14. Ganz, T. 2004. Antimicrobial polypeptides. *Journal of Leukocyte Biology* 75:34–38.
15. Brogden, K. 2005. Antimicrobial peptides: Pore inhibitors or metabolic inhibitors in bacteria? *Nature Reviews. Microbiology* 3:238–350.
16. Costerton, J. W. 1999. Bacterial biofilms: A common cause of persistent infections. *Science* 284:1318–1322.
17. Dunne, W. M. 2002. Bacterial adhesion: Seen any good biofilms lately? *Clinical Microbiology Reviews*;15:155–166.
18. Otto, M. 2006. Bacterial evasion of antimicrobial peptides by biofilm formation. *Current Topics in Microbiology and Immunology* 306:251–258.
19. Taggart, C., C. Greene, S. Smith et al. 2003. Inactivation of human [beta]-defensins 2 and 3 by elastolytic cathepsins. *The Journal of Immunology* 171:931–937.
20. Schmidtchen, A., I. Frick, E. Andersson, H. Tapper, and L. Björck. 2002. Proteinases of common pathogenic bacteria degrade and inactivate the antibacterial peptide LL-37. *Molecular Microbiology* 46:157–168.
21. Rosenfeld, Y., and Y. Shai. 2006. Lipopolysaccharide (endotoxin)-host defense antibacterial peptides interactions: Role in bacterial resistance and prevention of sepsis. *Biochimica et Biophysica Acta* 1758:1513–1522.
22. Overhage, J., A. Campisano, M. Bains, E. C. W. Torfs, B. H. A. Rehm, and R. E. W. Hancock. 2008. Human host defense peptide LL-37 prevents bacterial biofilm formation. *Infection and Immunity* 76:4176–4182.
23. Chennupati, S. K., A. G. Chiu, E. Tamashiro et al. 2009. Effects of an LL-37-derived antimicrobial peptide in an animal model of biofilm *Pseudomonas* sinusitis. *American Journal of Rhinology & Allergy* 23:46–51.
24. Mahon, B. D., A. Wittke, V. Weaver, and M. T. Cantorna. 2003. The targets of vitamin D depend on the differentiation and activation status of CD4 positive T cells. *Journal of Cellular Biochemistry* 89:922–932.
25. von Essen, M. R., M. Kongsbak, P. Schjerling, K. Olgaard, N. Ødum, and C. Geisler. 2010. Vitamin D controls T cell antigen receptor signaling and activation of human T cells. *Nature Immunology* 11:344–349.

26. Uitterlinden, A., Y. Fang, J. Van Meurs, H. Pols, and J. Van Leeuwen. 2004. Genetics and biology of vitamin D receptor polymorphisms. *Gene* 338:143–156.
27. Selvaraj, P., G. Chandra, M, Jawahar, M. Rani, D. Rajeshwari, and P. Narayanan. 2004. Regulatory role of vitamin D receptor gene variants of Bsm I, Apa I, Taq I, and Fok I polymorphisms on macrophage phagocytosis and lymphoproliferative response to mycobacterium tuberculosis antigen in pulmonary tuberculosis. *Journal of Clinical Immunology* 24:423–432.
28. Roth, D., G. Soto, F. Arenas et al. 2004. Association between vitamin D receptor gene polymorphisms and response to treatment of pulmonary tuberculosis. *The Journal of Infectious Diseases* 190:920–927.
29. Arai, H., K. Miyamoto, Y. Taketani et al. 1997. A vitamin D receptor gene polymorphism in the translation initiation codon: Effect on protein activity and relation to bone mineral density in Japanese women. *Journal of Bone and Mineral Research* 12:915–921.
30. McGrath, J. J., S. Saha, T. H. J. Burne, and D. W. Eyles. 2010. A systematic review of the association between common single nucleotide polymorphisms and 25-hydroxyvitamin D concentrations. *The Journal of Steroid Biochemistry and Molecular Biology* 121:471–477.
31. Finsen, N. 1903. Remarks on the red-light treatment of small-pox. Is the treatment of small-pox patients in broad daylight warrantable? *British Medical Journal* 1(2214).
32. Møller, K., B. Kongshoj, P. Philipsen, V. Thomsen, and H. Wulf. 2005. How Finsen's light cured lupus vulgaris. *Photodermatology, Photoimmunology & Photomedicine* 21:118–124.
33. Ritter, M. A., E. M. Olberding, and R. A. Malinzak. 2007. Ultraviolet lighting during orthopaedic surgery and the rate of infection. *Journal of Bone and Joint Surgery. American Volume* 89:1935–1940.
34. Krutmann, J. 2009. Pre- and probiotics for human skin. *Journal of Dermatological Science* 54:1–5.
35. Bikle, D. D. 2004. Vitamin D regulated keratinocyte differentiation. *Journal of Cellular Biochemistry* 92:436–444.
36. Oda, Y., Y. Uchida, S. Moradian, D. Crumrine, P. M. Elias, and D. D. Bikle. 2008. Vitamin D receptor and coactivators SRC2 and 3 regulate epidermis-specific sphingolipid production and permeability barrier formation. *The Journal of Investigative Dermatology* 129:1367–1378.
37. Hong, S. P., M. J. Kim, M.-Y. Jung et al. 2008. Biopositive effects of low-dose UVB on epidermis: Coordinate upregulation of antimicrobial peptides and permeability barrier reinforcement. *The Journal of Investigative Dermatology* 128:2880–2887.
38. Frohm, M., B. Agerberth, G. Ahangari et al. 1997. The expression of the gene coding for the antibacterial peptide LL-37 is induced in human keratinocytes during inflammatory disorders. *Journal of Biological Chemistry* 272:15258–15263.
39. Ong, P., T. Ohtake, C. Brandt et al. 2002. Endogenous antimicrobial peptides and skin infections in atopic dermatitis. *New England Journal of Medicine* 347:1151–1160.
40. Howell, M., A. Wollenberg, R. Gallo et al. 2006. Cathelicidin deficiency predisposes to eczema herpeticum. *The Journal of Allergy and Clinical Immunology* 117:836–841.
41. Weiland, S. K. 2004. Climate and the prevalence of symptoms of asthma, allergic rhinitis, and atopic eczema in children. *Occupational and Environmental Medicine* 61:609–615.
42. Byremo, G., G. Rød, and K. H. Carlsen. 2006. Effect of climatic change in children with atopic eczema. *Allergy* 61:1403–1410.
43. Peroni, D. G., G. L. Piacentini, E. Cametti, I. Chinellato, and A. L. Boner. 2011. Correlation between serum 25-hydroxyvitamin D levels and severity of atopic dermatitis in children. *The British Journal of Dermatology* 164:1078–1082.
44. Mallbris, L., L. Carlén, T. Wei et al. 2010. Injury downregulates the expression of the human cathelicidin protein hCAP18/LL-37 in atopic dermatitis. *Experimental Dermatology* 19:442–449.
45. Hata, T., P. Kotol, M. Jackson et al. 2008. Administration of oral vitamin D induces cathelicidin production in atopic individuals. *The Journal of Allergy and Clinical Immunology* 122:829–831.
46. Sidbury, R., A. F. Sullivan, R. I. Thadhani, and C. A. Camargo, Jr. 2008. Randomized controlled trial of vitamin D supplementation for winter-related atopic dermatitis in Boston: A pilot study. *The British Journal of Dermatology* 159:245–247.
47. Khandsuren, B., D. Ganmaa, R. Sidbury, K. Erdenedelger, N. Radnaakhand, and C. A. Camargo, Jr. 2009. Randomized controlled trial of vitamin D supplementation for winter-related atopic dermatitis in Mongolian children [abstract] Presented at the World Congress of Pediatric Dermatology (Bangkok, Thailand; November 17–19, 2009).
48. Krämer, C., H. Seltmann, M. Seifert, W. Tilgen, C. C. Zouboulis, and J. Reichrath. 2009. Characterization of the vitamin D endocrine system in human sebocytes *in vitro*. *The Journal of Steroid Biochemistry and Molecular Biology* 113:9–16.

49. Kalayciyan, A., O. Oguz, H. Bahar, M. Torun, and E. Aydemir. 2002. *In vitro* bactericidal effect of low-dose ultraviolet B in patients with acne. *Journal of the European Academy of Dermatology and Venereology* 16:642–643.

50. Shnitkind, E., E. Yaping, S. Geen, A. Shalita, and W. Lee. 2006. Anti-inflammatory properties of narrow-band blue light. *Journal of Drugs in Dermatology* 5:605–610.

51. Sigmon, J. R., B. A. Yentzer, S. R. Feldman. 2009. Calcitriol ointment: A review of a topical vitamin D analog for psoriasis. *The Journal of Dermatological Treatment* 20:208–212.

52. Peric, M., S. Koglin, Y. Dombrowski et al. 2009. Vitamin D analogs differentially control antimicrobial peptide/"alarmin" expression in psoriasis. *PLoS One* 4:e6340.

53. Ghoreishi, M., P. Bach, J. Obst, M. Komba, J. C. Fleet, J. P. Dutz. 2009. Expansion of antigen-specific regulatory T cells with the topical vitamin D analog calcipotriol. *The Journal of Immunology* 182:6071–6078.

54. Loser, K., and S. Beissert. 2009. Regulation of cutaneous immunity by the environment: An important role for UV irradiation and vitamin D. *International Immunopharmacology* 9:587–589.

55. Fitness, J., K. Tosh, and A. Hill. 2002. Genetics of susceptibility to leprosy. *Genes and Immunity* 3:441–453.

56. Roy, S., A. Frodsham, B. Saha, S. Hazra, C. Mascie-Taylor, and A. Hill. 1999. Association of vitamin D receptor genotype with leprosy type. *The Journal of Infectious Diseases* 179:187–191.

57. Fitness, J., S. Floyd, D. Warndorff et al. 2004. Large-scale candidate gene study of leprosy susceptibility in the Karonga district of northern Malawi. *The American Journal of Tropical Medicine and Hygiene* 71:330–340.

58. Matzner, M., A. R. Al Samie, H-M. Winkler et al. 2011. Low serum levels of cathelicidin LL-37 in leprosy. *Acta Tropica* 117:56–59.

59. Takahashi, N., and B. Nyvad. 2011. The role of bacteria in the caries process: ecological perspectives. *Journal of Dental Research* 90:294–303.

60. East, B. 1939. Mean annual hours of sunshine and incidence of dental caries. *American Journal of Public Health* 29:777–780.

61. East, B., and H. Kaiser. 1940. Relation of dental caries in rural children to sex, age and environment. *American Journal of Diseases of Children* 60:1289–1303.

62. McBeath, E., and T. Zucker. 1938. The role of vitamin D in the control of dental caries. *American Journal of Public Health* 15:547–564.

63. Hargreaves, J. A., and G. W. Thompson. 1989. Ultraviolet light and dental caries in children. *Caries Research* 23:389–392.

64. Putsep, K., G. Carlsson, H. Boman, and M. Andersson. 2002. Deficiency of antibacterial peptides in patients with morbus Kostmann: an observation study. *Lancet* 360:1144–1149.

65. Dietrich, T., M. Nunn, B. Dawson-Hughes, and H. Bischoff-Ferrari. 2005. Association between serum concentrations of 25-hydroxyvitamin D and gingival inflammation. *The American Journal of Clinical Nutrition* 82:575–580.

66. Tachi, Y. 2003. Vitamin D receptor gene polymorphism is associated with chronic periodontitis. *Life Sciences* 73:3313–3321.

67. De Brito, R., R. Scarel-Caminiaga, P. Trevilatto, A. de Souza, and S. Barros. 2004. Polymorphisms in the vitamin D receptor gene are associated with periodontal disease. *Journal of Periodontology* 75:1090–1095.

68. Krall, E. 2006. Osteoporosis and the risk of tooth loss. *Clinical Calcium* 16:287–290.

69. Dietrich, T., K. J. Joshipura, B. Dawson-Hughes, and H. A. Bischoff-Ferrari. 2004. Association between serum concentrations of 25-hydroxyvitamin D_3 and periodontal disease in the US population. *The American Journal of Clinical Nutrition* 80:108–113.

70. Boggess, K. A., J. A. Espinola, K. Moss, J. Beck, S. Offenbacher, C. A. Camargo, Jr. 2011. Vitamin D status and periodontal disease among pregnant women. *Journal of Periodontology* 82:195–200.

71. Krall, E., C. Wehler, R. Garcia, S. Harris, B. Dawson-Hughes. 2001. Calcium and vitamin D supplements reduce tooth loss in the elderly. *American Journal of Medicine* 111:452–456.

72. Garcia, M. N., C. F. Hildebolt, D. D. Miley et al. 2011. One-year effects of vitamin D and calcium supplementation on chronic periodontitis. *Journal of Periodontology* 82:25–32.

73. Sawatari, Y., and R. Marx. 2007. Bisphosphonates and bisphosphonate induced osteonecrosis. *Oral and Maxillofacial Surgery Clinics of North America* 19:487–498.

74. Vescovi, P., and S. Nammour. 2010. Bisphosphonate-related osteonecrosis of the jaw (BRONJ) therapy. A critical review. *Minerva Stomatologica* 59:181–213.

75. Hokugo, A., R. Christensen, E. M. Chung et al. 2010. Increased prevalence of bisphosphonate-related osteonecrosis of the jaw with vitamin D deficiency in rats. *Journal of Bone and Mineral Research* 25:1337–1349.
76. Martineau, A., F. Honecker, R. Wilkinson, and C. Griffiths. 2007. Vitamin D in the treatment of pulmonary tuberculosis. *The Journal of Steroid Biochemistry and Molecular Biology* 103:793–798.
77. Gaumond, E. 1948. Lupus vulgaris and vitamin D. *Canadian Medical Association Journal* 59:522–526.
78. Bellamy, R., C. Ruwende, T. Corrah et al. 1999. Tuberculosis and chronic hepatitis B virus infection in Africans and variation in the vitamin D receptor gene. *The Journal of Infectious Diseases* 179:721–724.
79. Wilkinson, R. J., M. Llewelyn, Z. Toosi et al. 2000. Influence of vitamin D deficiency and vitamin D receptor polymorphisms on tuberculosis among Gujarati Asians in west London: A case-control study. *Lancet* 355:618–621.
80. Rook, G. 1988. The role of vitamin D in tuberculosis. *The American Review of Respiratory Disease* 138:768–770.
81. Nnoaham, K. E., and A. Clarke. 2007. Low serum vitamin D levels and tuberculosis: a systematic review and meta-analysis. *International Journal of Epidemiology* 37:113–119.
82. Talat, N., S. Perry, J. Parsonnet, G. Dawood, and R. Hussain. 2010. Vitamin D deficiency and tuberculosis progression. *Emerging Infectious Diseases* 16:853–855.
83. Vieth, R. 2011. Vitamin D nutrient to treat TB begs the prevention question. *Lancet* 377:189–190.
84. Trang, H., D. Cole, L. Rubin, A. Pierratos, S. Siu, and R. Vieth. 1998. Evidence that vitamin D_3 increases serum 25-hydroxyvitamin D more efficiently than does vitamin D_2. *The American Journal of Clinical Nutrition* 68:854–858.
85. Armas, L. A. G. 2004. Vitamin D_2 Is much less effective than vitamin D_3 in humans. *The Journal of Clinical Endocrinology and Metabolism* 89:5387–5391.
86. Nursyam, E., Z. Amin, and C. Rumende. 2006. The effect of vitamin D as supplementary treatment in patients with moderately advanced pulmonary tuberculous lesion. *Acta Medica Indonesiana* 38:3–5.
87. Wejse, C., V. F. Gomes, P. Rabna et al. 2009. Vitamin D as supplementary treatment for tuberculosis: A double-blind, randomized, placebo-controlled trial. *American Journal of Respiratory and Critical Care Medicine* 179:843–850.
88. Martineau, A. R., P. M. Timms, G. H. Bothamley et al. 2011. High-dose vitamin D_3 during intensive-phase antimicrobial treatment of pulmonary tuberculosis: A double-blind randomised controlled trial. *Lancet* 377:242–250.
89. Panierakis, C., G. Goulielmos, D. Mamoulakis, S. Maraki, E. Papavasiliou, and E. Galanakis. 2009. *Staphylococcus aureus* nasal carriage might be associated with vitamin D receptor polymorphisms in type 1 diabetes. *International Journal of Infectious Diseases* 13:e437–e443.
90. Matheson, E. M., A. G. Mainous, W. J. Hueston, V. A. Diaz, and C. J. Everett. 2010. Vitamin D and methicillin-resistant *Staphylococcus aureus* nasal carriage. *Scandinavian Journal of Infectious Diseases* 42:455–460.
91. Laaksi, I., J. Ruohola, P. Tuohimaa et al. 2007. An association of serum vitamin D concentrations < 40 nmol/L with acute respiratory tract infection in young Finnish men. *The American Journal of Clinical Nutrition* 86:714–717.
92. Temorschuizen, F., A. Wijga, J. Gerritsen, H. Neijens, and H. van Loveren. 2004. Exposure to solar ultraviolet radiation and respiratory tract symptoms in 1-year-old children. *Photodermatology, Photoimmunology & Photomedicine* 20:270–271.
93. Cannell, J. J., M. Zasloff, C. F. Garland, R. Scragg, and E. Giovannucci. 2008. On the epidemiology of influenza. *Virology Journal* 5:29.
94. Linday, L., R. Shindledecker, J. Dolitsky, T. Chen, and M. Holick. 2008. Plasma 25-hydroxyvitamin D levels in young children undergoing placement of tympanostomy tubes. *The Annals of Otology, Rhinology, and Laryngology* 117:740–744.
95. Reid, D., R. Morton, L. Salkeld, and J. Bartley. 2011. Vitamin D and tonsil disease—preliminary observations. *International Journal of Pediatric Otorhinolaryngology* 75:261–264.
96. Elemraid, M. A., I. J. Mackenzie, W. D. Fraser et al. 2011. A case-control study of nutritional factors associated with chronic suppurative otitis media in Yemeni children. *European Journal of Clinical Nutrition*.
97. Pinto, J., J. Schneider, R. Perez, M. DeTineo, F. Baroody, R. Naclerio. 2008. Serum 25-hydroxyvitamin D levels are lower in urban African American subjects with chronic rhinosinusitis. *The Journal of Allergy and Clinical Immunology* 122:415–417.
98. Ginde, A., J. Mansbach, and C. A. Camargo, Jr. 2009. Association between serum 25-hydroxyvitamin D level and upper respiratory tract infection in the third National Health and Nutrition Examination Survey. *Archives of Internal Medicine* 169:384–390.

99. Sabetta, J. R., P. DePetrillo, R. J. Cipriani, J. Smardin, L. A., Burns, and M. L. Landry. 2010. Serum 25-hydroxyvitamin D and the incidence of acute viral respiratory tract Infections in healthy adults. *PLoS One* 5:e11088.

100. Berry, D. J., K. Hesketh, C. Power, and E. Hyppönen. 2011. Vitamin D status has a linear association with seasonal infections and lung function in British adults. *The British Journal of Nutrition* 106:1433–1440.

101. Chesney, R. 2010. Vitamin D and the magic mountain: The anti-infectious role of the vitamin. *Journal of Pediatrics* 156:698–703.

102. Mariam, T., and G. Sterky. 1973. Severe rickets in infancy and childhood in Ethiopia. *Journal of Pediatrics* 82:876–878.

103. El-Radhi, A., M. Majeed, N. Mansor, and M. Ibrahim. 1982. High incidence of rickets in children with wheezy bronchitis in a developing country. *Journal of the Royal Society of Medicine* 75:884–887.

104. Muhe, L., S. Lulseged, K. Mason, and E. Simoes. 1997. Case-control study of the role of nutritional rickets in the risk of developing pneumonia in Ethiopian children. *Lancet* 349:1801–1804.

105. Banajeh, S., N. al-Sunbali, and S. al-Sanahan. 1997. Clinical characteristics and outcome of children aged under 5 years hospitalized with severe pneumonia in Yemen. *Annals of Tropical Paediatrics* 17:321–326.

106. Najada, A,, M. Habashneh, and M. Khader. 2004. The frequency of nutritional rickets among hospitalized infants and its relation to respiratory diseases. *Journal of Tropical Pediatrics* 50:364–368.

107. Banajeh, S. M. 2009. Nutritional rickets and vitamin D deficiency association with the outcomes of childhood very severe pneumonia: A prospective cohort study. *Pediatric Pulmonology* 44:1207–1215.

108. Belderbos, M. E., M. L. Houben, B. Wilbrink et al. 2011. Cord blood vitamin D deficiency is associated with respiratory syncytial virus bronchiolitis. *Pediatrics* 127:e1513–e1520.

109. Karatekin, G., A. Kaya, Ö. Salihoğlu, H. Balci, and A. Nuhoğlu. 2007. Association of subclinical vitamin D deficiency in newborns with acute lower respiratory infection and their mothers. *European Journal of Clinical Nutrition* 63:473–477.

110. McNally, J. D., K. Leis, L. A. Matheson, C. Karuananyake, K. Sankaran, and A. M. Rosenberg. 2009. Vitamin D deficiency in young children with severe acute lower respiratory infection. *Pediatric Pulmonology* 44:981–988.

111. Wayse, V., A. Yousafzai, K. Mogale, and S. Filteau. 2004. Association of subclinical vitamin D deficiency with severe acute lower respiratory infection in Indian children under 5 y. *European Journal of Clinical Nutrition* 58:563–567.

112. Inamo, Y., M. Hasegawa, K. Saito et al. 2011. Serum vitamin D concentrations and associated severity of acute lower respiratory tract infections in Japanese hospitalized children. *Pediatrics International* 53:199–201.

113. Roth, D. E., A. B. Jones, C. Prosser, J. L. Robinson, and S. Vohra. 2007. Vitamin D status is not associated with the risk of hospitalization for acute bronchiolitis in early childhood. *European Journal of Clinical Nutrition* 63:297–299.

114. Roth, D. E., A. B. Jones, C. Prosser, J. L. Robinson, and S. Vohra. 2008. Vitamin D receptor polymorphisms and the risk of acute lower respiratory tract infection in early childhood. *The Journal of Infectious Diseases* 197:676–680.

115. Grant, W., and E. Giovannucci. 2009. The possible roles of solar ultraviolet-B radiation and vitamin D in reducing case-fatality rates from the 1918–1919 influenza pandemic in the United States. *Dermatoendocrinology* 1:215–219.

116. Douglas, A., D. Strachan, and J. Maxwell. 1996. Seasonality of tuberculosis: the reverse of other respiratory diseases in the UK. *Thorax* 51:944–946.

117. Leow, L., T. Simpson, R. A. Y. Cursons, N. Karalus, and R. J. Hancox. 2011. Vitamin D, innate immunity and outcomes in community acquired pneumonia. *Respirology* 16:611–616.

118. Green, D., K. Carson, A. Leonard et al. 2008. Current treatment recommendations for correcting vitamin D deficiency in pediatric patients with cystic fibrosis are inadequate. *Journal of Pediatrics* 153:554–559.

119. Pincikova, T., K. Nilsson, I. E. Moen et al. 2010. Inverse relation between vitamin D and serum total immunoglobulin G in the Scandinavian Cystic Fibrosis Nutritional Study. *European Journal of Clinical Nutrition* 65:102–109.

120. Janssens, W., R. Bouillon, B. Claes et al. 2009. Vitamin D deficiency is highly prevalent in COPD and correlates with variants in the vitamin D–binding gene. *Thorax* 65:215–220.

121. Chishimba, L., D. R. Thickett, R. A. Stockley, and A. M. Wood. 2010. The vitamin D axis in the lung: A key role for vitamin D–binding protein. *Thorax* 65:456–462.

122. Holmes, A., M. Pigott, W. Sawyer, and L. Comstock. 1932. Vitamins aid reduction of lost time in industry. *Industrial and Engineering Chemistry* 24:1058–1060.

123. Holmes, A., M. Pigott, W. Sawyer, and L. Comstock. 1936. Cod liver oil—a five year study of its value for reducing industrial absenteeism caused by colds and respiratory diseases. *Industrial Medicine* 5:359–361.
124. Gigineishvili, G., N. Il'in, R. Suzdal'nitskiii, and V. Levando. 1990. The use of irradiation to correct the immune system and decrease morbidity in athletes [in Russian] *Vopr Kurotol Fizioter Lech Fiz Kult* May–Jun, 30-3 cited in Cannell, J. J., R. Vieth, W. Willett, M. Zasloff, J. N. Hathcock, J. H. White et al. 2008. Cod liver oil, vitamin A toxicity, frequent respiratory tract infections and the vitamin D epidemic. *The Annals of Otology, Rhinology, and Laryngology* 117:864–870.
125. Rehman, P. 1994. Sub-clinical rickets and recurrent infection. *Journal of Tropical Pediatrics* 40:58.
126. Aloia, J. F., M. Li-Ng. 2007. Re: Epidemic influenza and vitamin D. *Epidemiology and Infection* 135:1095–1096.
127. Avenell, A., J. Cook, G. MacLennan, and G. MacPherson. 2007. Vitamin D supplementation to prevent infections: A substudy of a randomised placebo-controlled trial in older people. *Age and Ageing* 36:574–577.
128. Tsujimoto, Y., H. Tahara, T. Shoji et al. 2011. Active vitamin D and acute respiratory infections in dialysis patients. *Clinical Journal of the American Society of Nephrology* 6:1361–1367.
129. Li-Ng, M., J. F. Aloia, S. Pollack et al. 2009. A randomized controlled trial of vitamin D$_3$ supplementation for the prevention of symptomatic upper respiratory tract infections. *Epidemiology and Infection* 137:1396–1404.
130. Urashima, M., T. Segawa, M. Okazaki, M. Kurihara, Y. Wada, and H. Ida. 2010. Randomized trial of vitamin D supplementation to prevent seasonal influenza A in schoolchildren. *The American Journal of Clinical Nutrition* 91:1255–1260.
131. Laaksi, I., J. P. Ruohola, V. Mattila, A. Auvinen, T. Ylikomi, and H. Pihlajamäki. 2010. Vitamin D supplementation for the prevention of acute respiratory tract infection: a randomized, double-blinded trial among young Finnish men. *Journal of Infectious Diseases* 202:809–814.
132. Manaseki-Holland, S., G. Qader, M. Isaq Masher et al. 2010. Effects of vitamin D supplementation to children diagnosed with pneumonia in Kabul: a randomized controlled trial. *Tropical Medicine & International Health* 15:1148–1155.
133. Bacon, C. J., G. D. Gamble, A. M. Horne, M. A. Scott, and I. R. Reid. 2008. High-dose oral vitamin D$_3$ supplementation in the elderly. *Osteoporosis International* 20:1407–1415.
134. Black, P., and R. Scragg. 2005. Relationship between serum 25-hydroxyvitamin D and pulmonary function in the third National Health and Nutrition Examination Survey. *Chest* 128:3792–3798.
135. Masoli, M., D. Fabian, S. Holt, and R. Beasley; Global Initiative for Asthma (GINA) Program. 2004. The global burden of asthma: Executive summary of the GINA Dissemination Committee report. *Allergy* 59:469–478.
136. Raby, B., R. Lazarus, E. Silverman et al. 2004. Association of vitamin D receptor gene polymorphisms with childhood and adult asthma. *American Journal of Respiratory and Critical Care Medicine* 170:1057–1065.
137. Saadi, A., G. Gao, H. Li, C. Wei, Y. Gong, and Q. Liu. 2009. Association study between vitamin D receptor gene polymorphisms and asthma in the Chinese Han population: A case-control study. *BMC Medical Genetics* 10:71.
138. Camargo, C. A. Jr., T. Ingham, K. Kristin Wickens et al. 2011. Cord-blood 25-hydroxyvitamin D levels and risk of respiratory infection, wheezing, and asthma. *Pediatrics* 127:e180–e187.
139. Brehm, J., B. Schuemann, A. Fuhlbrigge et al. 2010. Serum vitamin D levels and severe asthma exacerbations in the childhood asthma management program study. *The Journal of Allergy and Clinical Immunology* 126:52–58.
140. Majak, P., M. Olszowiec-Chlebna, K. Smejda, and I. Stelmach. 2011. Vitamin D supplementation in children may prevent asthma exacerbation triggered by acute respiratory infection. *The Journal of Allergy and Clinical Immunology* 127:1294–1296. Feb 9. [Epub ahead of print].
141. Carroll, K. N., T. Gebretsadik, E. K. Larkin et al. 2011. Relationship of maternal vitamin D level with maternal and infant respiratory disease. *American Journal of Obstetrics and Gynecology*.
142. Hyppönen, E., U. Sovio, M. Wjst et al. 2004. Infant vitamin D supplementation and allergic conditions in adulthood: Northern Finland birth cohort 1966. *Annals of the New York Academy of Sciences* 1037:84–95.
143. Hyppönen, E., D. J. Berry, M. Wjst, and C. Power. 2009. Serum 25-hydroxyvitamin D and IgE—a significant but nonlinear relationship. *Allergy* 64:613–620.
144. Miyake, C. Y., K. Gauvreau, L. Y. Tani, R. P. Sundel, and J. W. Newburger. 2007. Characteristics of children discharged from hospitals in the United States in 2000 with the diagnosis of acute rheumatic fever. *Pediatrics* 120:503–508.

145. Finkelhor, R., G. Cater, A. Qureshi, D. Einstadter, M. Hecker, and G. Bosich. 2005. Seasonal diagnosis of echocardiographically demonstrated endocarditis. *Chest* 128:2588–2592.

146. D'Aldebert, E., Bi. Biyeyeme, M. J. Mve, M. Mergey et al. 2009. Bile salts control the antimicrobial peptide cathelicidin through nuclear receptors in the human biliary epithelium. *Gastroenterology* 136:1435–1443.

147. Arteh, J., S. Narra, and S. Nair. 2009. Prevalence of vitamin D deficiency in chronic liver disease. *Digestive Diseases and Sciences* 55:2624–2628.

148. Arababadi, M. K., A. A. Pourfathollah, A. Jafarzadeh, G. Hassanshahi, and M. E. Rezvani. 2010. Association of exon 9 but not intron 8 VDR polymorphisms with occult HBV infection in south-eastern Iranian patients. *Journal of Gastroenterology and Hepatology* 25:90–93.

149. Suneetha, P., S. Sarin, A. Goyal, G. Kumar, D. Shukla, and S. Hissar. 2006. Association between vitamin D receptor, CCR5, TNF-α and TNF-β gene polymorphisms and HBV infection and severity of liver disease. *Journal of Hepatology* 44:856–863.

150. Petta, S., C. Cammà, C. Scazzone et al. 2010. Low vitamin D serum level is related to severe fibrosis and low responsiveness to interferon-based therapy in genotype 1 chronic hepatitis C. *Hepatology* 51:1158–1167.

151. Lange, C. M., J. Bojunga, E. Ramos-Lopez et al. 2011. Vitamin D deficiency and a CYP27B1-1260 promoter polymorphism are associated with chronic hepatitis C and poor response to interferon-alfa based therapy. *Journal of Hepatology* 54:887–893.

152. Bitetto, D., G. Fattovich, C. Fabris et al. 2011. Complementary role of vitamin D deficiency and the interleukin-28B rs12979860 C/T polymorphism in predicting antiviral response in chronic hepatitis C. *Hepatology* 53:1118–1126.

153. Abu-Mouch, S. 2009. Abstract LB20. The Liver Meeting 2009: 60th Annual Meeting of the American Association for the Study of Liver Diseases.

154. McCracken, V., and R. Lorenz. 2001. The gastrointestinal ecosystem: A precarious alliance among epithelium, immunity and microbiota. *Cellular Microbiology* 3:1–11.

155. Hase, K., M. Murakami, M. Iimura et al. 2003. Expression of LL-37 by human gastric epithelial cells as a potential host defense mechanism against *Helicobacter pylori*. *Gastroenterology* 125.1613–1625.

156. George, J. 2003. Host anti-microbial response to *Helicobacter pylori* infection. *Molecular Immunology* 40:451–456.

157. Nasri, H., and A. Baradaran. 2007. The influence of serum 25-hydroxy vitamin D levels on *Helicobacter pylori* infections in patients with end-stage renal failure on regular hemodialysis. *Saudi Journal of Kidney Diseases and Transplantation* 18:215–219.

158. Kawaura, A. 2006. Inhibitory effect of long term 1-α hydroxyvitamin D_3 administration on *Helicobacter pylori* infection. *Journal of Clinical Biochemistry and Nutrition* 38:103–106.

159. Brandi, L. 2008. 1α (OH)D_3 One-α-hydroxy-cholecalciferol—an active vitamin D analog. Clinical studies on prophylaxis and treatment of secondary hyperparathyroidism in uremic patients on chronic dialysis. *Danish Medical Bulletin* 55:186–210.

160. Yamshchikov, A., N. Desai, H. Blumberg, T. Ziegler, and V. Tangpricha. 2009. Vitamin D for treatment and prevention of infectious diseases: A systematic review of randomized controlled trials. *Endocrine Practice* 15:438–449.

161. Lagishetty, V., A. V. Misharin, N. Q. Liu et al. 2010. Vitamin D deficiency in mice impairs colonic antibacterial activity and predisposes to colitis. *Endocrinology* 151:2423–2432.

162. Kelly, P., M. Bajaj-Elliott, M. Katubulushi et al. 2006. Reduced gene expression of intestinal [α]-defensins predicts diarrhea in a cohort of African adults *The Journal of Infectious Diseases* 193:1464–1470.

163. Nishida, S., J. Ozeki, and M. Makishima. 2009. Modulation of bile acid metabolism by 1-hydroxyvitamin D_3 administration in mice. *Drug Metabolism and Disposition* 37:2037–2044.

164. Chawla, A. 2001. Nuclear receptors and lipid physiology: Opening the X-files. *Science* 294:1866–1870.

165. Gombart, A. 2009. The vitamin D–antimicrobial peptide pathway and its role in protection against infection. *Future Microbiology* 4:1151–1165.

166. Islam, D., L. Bandholtz, J. Nilsson et al. 2001. Downregulation of bactericidal peptides in enteric infections: A novel immune escape mechanism with bacterial DNA as a potential regulator. *Nature Medicine* 7:180–185.

167. Gudmundsson, G. H., P. Bergman, J. Andersson, R. Raqib, and B. Agerberth. 2010. Battle and balance at mucosal surfaces—the story of *Shigella* and antimicrobial peptides. *Biochemical and Biophysical Research Communications* 396:116–119.

168. Schwab, M., V. Reynders, Y. Shastri, S. Loitsch, J. Stein, and O. Schroder. 2007. Role of nuclear hormone receptors in butyrate-mediated up-regulation of the antimicrobial peptide cathelicidin in epithelial colorectal cells. *Molecular Immunology* 44:2107–2114.

169. Frank, D. N., A. L. St. Amand, R. A. Feldman, E. C. Boedeker, N. Harpaz, N. R. Pace. 2007. Molecular-phylogenetic characterization of microbial community imbalances in human inflammatory bowel diseases. *Proceedings of the National Academy of Sciences of the United States of America* 104:13780–13785.
170. Maslowski, K. M., A. T. Vieira, A. Ng et al. 2009. Regulation of inflammatory responses by gut micro-biota and chemoattractant receptor GPR43. *Nature* 461:1282–1286.
171. Villamor, E. 2006. A potential role for vitamin D on HIV infection? *Nutrition Review* 64:226–233.
172. Bergman, P., L. Walter-Jallow, K. Broliden, B. Agerberth, and J. Södelund. 2007. The antimicrobial peptide LL-37 inhibits HIV-1 replication. *Current HIV Research* 5:410–415.
173. de la Torre Manuel, S., C. Torres, G. Nieto et al. 2008. Vitamin D receptor gene haplotypes and suscep-tibility to HIV-1 infection in injection drug users. *The Journal of Infectious Diseases* 197:405–410.
174. Nieto, G., Y. Barber, M. Rubio, M. Rubio, and J. Fibla. 2004. Association between AIDS disease progres-sion rates and the Fok-I polymorphism of the VDR gene in a cohort of HIV-1 seropositive patients. *The Journal of Steroid Biochemistry and Molecular Biology* 89–90:199–207.
175. Bang, U. C., S. A. Shakar, M. F. Hitz et al. 2010. Deficiency of 25-hydroxyvitamin D in male HIV-positive patients: A descriptive cross-sectional study. *Scandinavian Journal of Infectious Diseases* 42:306–310.
176. Haug, C., F. Muller, P. Aukrust, and S. Froland. 1994. Subnormal serum concentration of 1,25-vitamin D in human immunodeficiency virus infection: correlation with degree of immune deficiency and survival. *The Journal of Infectious Diseases* 169:889–893.
177. Viard, J-P., J-C. Souberbielle, O. Kirk et al. 2011. Vitamin D and clinical disease progression in HIV infection: results from the EuroSIDA study. *AIDS* 25:1305–1315.
178. Mehta, S., D. J. Hunter, F. M. Mugusi et al. 2009. Perinatal outcomes, including mother-to-child trans-mission of HIV, and child mortality and their association with maternal vitamin D status in Tanzania. *The Journal of Infectious Diseases* 200:1022–1030.
179. Mehta, S. E. Giovannucci, F. M. Mugusi et al. 2010. Vitamin D status of HIV-infected women and its association with HIV disease progression, anemia, and mortality. *PLoS One* 5:e8770.
180. Zasloff, M. 2007. Antimicrobial peptides, innate Immunity, and the normally sterile urinary tract. *Journal of the American Society of Nephrology* 18:2810–2816.
181. Lehmann, J., M. Retz, J. Harder et al. 2002. Expression of human beta-defensins 1 and 2 in kidneys with chronic bacterial infection. *BMC Infectious Diseases* 2:20.
182. Chromek, M., Z. Slamova, P. Bergman et al. 2006. The antimicrobial peptide cathelicidin protects the urinary tract against invasive bacterial infection. *Nature Medicine* 12:636–641.
183. Hertting, O., A. Holm, P. Luthje et al. 2010. Vitamin D induction of the human antimicrobial peptide cathelicidin in the urinary bladder. *PLoS One* 5:e15580.
184. Morrison, G. 2002. Characterization of the mouse beta defensin 1, Defb1, mutant mouse model. *Infection and Immunity* 70:3053–3060.
185. Katikaneni, R., T. Ponnapakkam, A. Ponnapakkam, and R. Gensure. 2009. Breastfeeding does not pro-tect against urinary tract infection in the first 3 months of life, but vitamin D supplementation increases the risk by 76%. *Clinical Pediatrics* 48:750–755.
186. Linday, L. A., R. D. Shindledecker, F. R. Greer, and M. F. Holick. 2009. Commentary on "Breastfeeding does not protect against urinary tract infection in the first 3 months of life, but vitamin D supplementation increases the risk by 76%." *Clinical Pediatrics* 49:93–94.
187. Sorensen, O. 2003. Processing of seminal plasma hCAP-18 to ALL-38 by gastricsin: a novel mechanism of generating antimicrobial peptides in the vagina. *Journal of Biological Chemistry* 278:28540–28546.
188. Bodnar, L. M., M. A. Krohn, and H. N. Simhan. 2009. Maternal vitamin D deficiency is associated with bacterial vaginosis in the first trimester of pregnancy. *Journal of Nutrition* 139:1157–1161.
189. Hensel, K., T. Randis, S. Gelber, and A. Ratner. 2011. Pregnancy-specific association of vitamin D defi-ciency and bacterial vaginosis. *American Journal of Obstetrics and Gynecology* 204:41e1–e9.
190. Venkataraman, N., A. Cole, P. Svoboda, J. Pohl, and A. Cole. 2005. Cationic polypeptides are required for anti–HIV-1 activity of human vaginal fluid. *The Journal of Immunology* 175:7560–7567.
191. Etzioni, A., Z. Hochberg, and S. Pollak et al. 1989. Defective leukocyte fungicidal activity in end-organ resistance to 1,25-dihydroxyvitamin D. *Pediatric Research* 25:276–279.
192. Lopez-Garcia, B., P. Lee, K. Yamasaki, and R. Gallo. 2005. Antifungal activity of cathelicidins and their potential role in *Candida albicans* skin infection. *The Journal of Investigative Dermatology* 125:108–115.
193. Conner, K., K. Nern, J. Rudisill, T. Ogrady, and R. Gallo. 2002. The antimicrobial peptide LL-37 is expressed by keratinocytes in condyloma acuminatum and verruca vulgaris. *Journal of the American Academy of Dermatology* 47:347–350.

194. Meyer-Hoffert, U., T. Schwarz, J. M. Schröder, and R. Gläser. 2008. Expression of human beta-defensin-2 and -3 in verrucae vulgares and condylomata acuminata. *Journal of the European Academy of Dermatology and Venereology* 22:1050–1054.

195. Kinlin, L. M., C. V. Spain, V. Ng, C. C. Johnson, A. N. J. White, and D. N. Fisman. 2008. Environmental exposures and invasive meningococcal disease: an evaluation of effects on varying time scales. *American Journal of Epidemiology* 169:588–595.

196. Hope-Simpson, R. 1965. The nature of herpes zoster: A long-term study and a new hypothesis. *Proceedings of the Royal Society of Medicine* 58:9–20.

197. Jalali, M. H. A., H. Ansarin, and R. Soltani-Arabshahi. 2006. Broad-band ultraviolet B phototherapy in zoster patients may reduce the incidence and severity of postherpetic neuralgia. *Photodermatology, Photoimmunology & Photomedicine* 22:232–237.

198. Bartley, J. 2009. Post herpetic neuralgia, Schwann cell activation and vitamin D. *Medical Hypotheses* 73:927–929.

199. McDermott, A. M. 2009. The role of antimicrobial peptides at the ocular surface. *Ophthalmic Research* 41:60–75.

200. Wu, M., S. A. McClellan, R. P. Barrett, Y. Zhang, and L. D. Hazlett. 2009. Beta-defensins 2 and 3 together promote resistance to *Pseudomonas aeruginosa* keratitis. *The Journal of Immunology* 183:8054–8060.

201. Li, Q., A. Kumar, J.-F. Gui, and F-SX. Yu. 2008. *Staphylococcus aureus* lipoproteins trigger human corneal epithelial innate response through toll-like receptor-2. *Microbial Pathogenesis* 44:426–434.

202. Stallmann, H., C. Faber, A. Nieuw Amerongen, and P. Wuisman. 2006. Antimicrobial peptides: review of their application in musculoskeletal infections. *Injury* 37:S34–40.

203. Malik, M. H. A., F. Jury, A. Bayat, W. E. R. Ollier, and P. R. Kay. 2007. Genetic susceptibility to total hip arthroplasty failure: a preliminary study on the influence of matrix metalloproteinase 1, interleukin 6 polymorphisms and vitamin D receptor. *Annals of the Rheumatic Diseases* 66:1116–1120.

204. Bischoff-Ferrari, H. A., B. Dawson-Hughes, A. Platz et al. 2010. Effect of high-dosage cholecalciferol and extended physiotherapy on complications after hip fracture: a randomized controlled trial. *Archives of Internal Medicine* 170:813–820.

205. Grant, W. 2009. Solar ultraviolet-B irradiance and vitamin D may reduce the risk of septicemia. *Dermatoendocrinology* 1:37–42.

206. Danai, P. A., S. Sinha, M. Moss, M. J. Haber, and G. S. Martin. 2007. Seasonal variation in the epidemiology of sepsis. *Critical Care Medicine* 35:410–415.

207. Drechsler, C., S. Pilz, B. Obermayer-Pietsch et al. 2010. Vitamin D deficiency is associated with sudden cardiac death, combined cardiovascular events, and mortality in haemodialysis patients. *European Heart Journal* 31:2253–2261.

208. Gombart, A., I. Bhan, N. Borregaard et al. 2009. Low plasma level of cathelicidin antimicrobial peptide (hCAP18) predicts increased infectious disease mortality in patients undergoing hemodialysis. *Clinical Infectious Diseases* 48:418–424.

209. Jeng, L., A. V. Yamshchikov, S. E. Judd et al. 2009. Alterations in vitamin D status and anti-microbial peptide levels in patients in the intensive care unit with sepsis. *Journal of Translational Medicine* 7:28.

210. Ginde, A. A., C. A. Camargo, Jr., and N. I. Shapiro. 2011. Vitamin D insufficiency and sepsis severity in emergency department patients with suspected infection. *Academic Emergency Medicine* 18:551–554.

211. Daynes, R. A., and B. A. Araneo. 1994. The development of effective vaccine adjuvants employing natural regulators of T-cell lymphokine production *in vivo*. *Annals of the New York Academy of Sciences* 730:144–161.

212. Daynes, R. A., E. Y. Enioutina, S. Butler, H.-H. Mu, Z. A. McGee, B. A. Araneo. 1996. Induction of common mucosal immunity by hormonally immunomodulated peripheral immunization. *Infection and Immunity* 64:1100–1109.

213. Kriesel, J., and J. Spruance. 1999. Calcitriol (1,25-dihydroxy-vitamin D3) coadministered with the influenza vaccine does not enhance humoral immunity in human volunteers. *Vaccine* 17:1883–1888.

214. Chadha, M. K., M. Fakih, J. Muindi et al. 2011. Effect of 25-hydroxyvitamin D status on serological response to influenza vaccine in prostate cancer patients. *Prostate* 71:368–372.

215. Lalor, M., S. Floyd, P. Gorak-Stolinska et al. 2011. BCG vaccination: A role for vitamin D? *PLoS One* 6:e16709.

216. Narayanan, D., R. Saladi, and J. Fox. 2010. Ultraviolet radiation and skin cancer. *International Journal of Dermatology* 49:978–986.

217. LoPiccolo, M., and H. Lim. 2010. Vitamin D in health and disease. *Photodermatology, Photoimmunology & Photomedicine* 26:224–229.

218. Pearce, S. H., and T. D. Cheetham. 2010. Diagnosis and management of vitamin D deficiency. *British Medical Journal* 340:b5664–b.
219. Biancuzzo, R., A. Young, D. Bibuld et al. 2010. Fortification of orange juice with vitamin D_2 or vitamin D_3 is as effective as an oral supplement in maintaining vitamin D status in adults. *The American Journal of Clinical Nutrition* 91:1621–1626.
220. Heaney, R. P., L. A. Armas, J. R. Shary, N. H. Bell, N. Binkley, and B. W. Hollis. 2008. 25-Hydroxylation of vitamin D_3: Relation to circulating vitamin D_3 under various input conditions. *The American Journal of Clinical Nutrition* 87:1738–1742.
221. Heaney, R., K. Davies, T. Chen, M. Holick, and M. Barger-Lux. 2003. Human serum 25-hydroxycholecalciferol response to extended oral dosing with cholecalciferol *The American Journal of Clinical Nutrition* 77:204–210 [Erratum, *Am. J. Clin. Nutr.* 78:1047].
222. Institute of Medicine (IOM). 2011. *Dietary Reference Intakes for Calcium and Vitamin D*. Washington, DC: The National Academies Press.
223. Aloia, J., M. Patel, R. Dimaano et al. 2008. Vitamin D intake to attain a desired serum 25-hydroxyvitamin D concentration. *The American Journal of Clinical Nutrition* 87:1952–1958.
224. Reid, D., B. Toole, S. Knox et al. 2011. The relation between acute changes in the systemic inflammatory response and plasma 25-hydroxyvitamin D concentrations after elective knee arthroplasty. *The American Journal of Clinical Nutrition* 93:1006–1011.
225. Hathcock, J. N., A. Shao, R. Vieth, and R. Heaney. 2007. Risk assessment for vitamin D. *The American Journal of Clinical Nutrition* 85:6–18.
226. Heaney, R. P. 2008. Vitamin D: criteria for safety and efficacy. *Nutrition Review* 66:S178–S181.
227. Rothers, J., A. Wright, M. Halonen, C. A. Camargo, Jr. 2011. Cord blood 25-hydroxyvitamin D levels are associated with aeroallergen sensitization in children from Tucson, Arizona. *The Journal of Allergy and Clinical Immunology* 128:1093–1099.
228. Nielsen, N. O., T. Skifte, M. Andersson et al. 2010. Both high and low serum vitamin D concentrations are associated with tuberculosis: a case–control study in Greenland. *The British Journal of Nutrition* 104:1487–1491.
229. Michaelsson, K., J. A. Baron, G. Snellman et al. 2010. Plasma vitamin D and mortality in older men: a community-based prospective cohort study. *The American Journal of Clinical Nutrition* 92:841–848.
230. Bischoff-Ferrari, H. A., A. Shao, B. Dawson-Hughes, J. Hathcock, E. Giovannucci, and W. C. Willett. 2010. Benefit–risk assessment of vitamin D supplementation. *Osteoporosis International* 21:1121–1132.
231. Morcos, M., A. Gabr, S. Samuel et al. 1998. Vitamin D administration to tuberculous children and its value. *Bollettino Chimico Farmaceutico* 137:157–164.

Section IV

Aging

14 Potential Role of Vitamin D and Fibroblast Growth Factor 23–Klotho System in Aging

Nasimul Ahsan, Syed K. Rafi, Beate Lanske,
and Mohammed S. Razzaque

CONTENTS

14.1 VITAMIN D SYNTHESIS

Although vitamin D is commonly perceived as a bone-forming vitamin, it in fact exerts numerous essential physiological functions ranging from the maintenance of mineral ion homeostasis to the regulation of the immune system [1–5]. The synthesis of vitamin D is a complex and multiorgan process that begins in the skin and continues further in the liver, and is then converted into biologically active 1,25 dihydroxyvitamin D_3 [$1,25(OH)_2D_3$] in the kidney. Two sequential hydroxylations are involved in the production of the bioactive metabolite $1,25(OH)_2D_3$; first, by the enzyme 25 hydroxylase (CYP27A1) in the liver to 25 hydroxyvitamin D [$25(OH)_2D$], and then by 1α hydroxylase [$1\alpha(OH)$ase; CYP27B1] in the kidney to $1,25(OH)_2D_3$. To keep the homeostatic balance of $1,25(OH)_2D_3$, the enzyme 24-hydroxylase (CYP24) generates $24,25(OH)_2D$ in the liver or kidney for eventual inactivation. The bioactive form, $1,25(OH)_2D_3$, can interact with the high-affinity vitamin D receptor (VDR), which is a ligand-dependent transcription factor [6]. Besides classic genomic functions through VDR, $1,25(OH)_2D_3$ could also exert its bioactivities through a VDR-independent rapid response system that is presumed to be mediated either by separate sets of cell membrane receptors or its caveolae components [6]. In addition to intestine, bone, kidney, and the parathyroid gland, the VDR receptors are also widely present in tissues and organs that are not considered to be involved in calcium metabolism, implicating a wide range of functions for vitamin D; indeed, such involvement is proposed in hypertension, immunoregulation, embryogenesis, and tumorigenesis [5].

14.2 HYPOVITAMINOSIS D

Studies have shown a strong association between hypovitaminosis D and various aging-related conditions such as osteoporosis, cancer, diabetes, autoimmune disorders, hypertension, atherosclerosis, and muscle weakness. Several central nervous system disorders, including multiple sclerosis, Alzheimer's disease, Parkinson's disease, and schizophrenia, have also been found to be associated with altered serum vitamin D concentrations [7]. Studies have also shown that low serum vitamin D levels are associated with inflammation, and oxidative load with increased risk of mortality; a recent meta-analysis has demonstrated that intake of vitamin D supplement at normal doses was associated with an improvement in the mortality risk factors [9]. Similarly, the National Health and Nutrition Examination Survey III showed that people with vitamin D levels in the lowest quartile had a mortality rate ratio of 1.26 [95% confidence interval (CI); 1.08–1.46] [9]. A retrospective study has also shown that vitamin D supplementation can reduce mortality in patients with end stage renal disease and low vitamin D levels [10].

Calcium supplements along with vitamin D are commonly prescribed to treat or prevent the progression of osteoporosis, particularly among older women. The beneficial effects of such a combination of supplements, however, are minimized by the potential risk of harmful cardiovascular events [11–13]. In fact, it has been reported that calcium supplements can increase the risk of myocardial infarction and stroke by up to 30% [14,15]. In contrast, there are studies that claim that daily intake of calcium (1000 mg) with vitamin D (400 IU) supplementation neither increased nor decreased the risk of coronary heart disease or cerebrovascular disorders in healthy postmenopausal women [16]. Although the debate about the beneficial effects of calcium and vitamin D supplementation for osteoporotic patients is yet to be settled, we believe that harmful cardiovascular effects following calcium and vitamin D supplementation may be partly related to phosphate balance [17–20].

The recommendations on how to prevent hypovitaminosis D in various age groups need careful evaluation and estimation because the published methods are mostly targeted toward subsets of populations, and therefore, may not be relevant at the individual level. In general, hypovitaminosis D is indicated when $25(OH)_2D$ levels range from 50 to 100 nmol/L; levels between 50 and 25 nmol/L are considered as vitamin D insufficiency, whereas levels lower than 25 nmol/L indicate vitamin D deficiency [21]. It is, however, worthwhile to mention that precisely defining hypovitaminosis, insufficiency, and deficiency groups are not always straightforward, as the values of $25(OH)_2D$ could be influenced by the secondary effects of serum levels of parathyroid hormone (PTH) or the levels of calcium in a particular individual. Moreover, most of the peripheral tissues are able to convert circulating $25(OH)_2D$ to the active metabolite $1,25(OH)_2D_3$ to cover their local requirements, and might not reflect it in their serum levels [22]. The list of the factors that can influence vitamin D synthesis is growing, and PTH is an important factor that can stimulate the activity of the $1\alpha(OH)$ase to facilitate the biosynthesis of $1,25(OH)_2D_3$ [23]. Furthermore, hypocalcemia or hypophosphatemia (or both) can also exert stimulatory effects on $1\alpha(OH)$ase [24], whereas fibroblast growth factor 23 (FGF23) can suppress $1\alpha(OH)$ase activity, and thereby can also suppress $1,25(OH)_2D_3$ synthesis [25,26].

14.3 FIBROBLAST GROWTH FACTOR 23

FGF23 is an approximately 30 kDa protein, and is proteolytically processed to generate a NH_2-terminal (~18 kDa) fragment and a COOH-terminal (~12 kDa) fragment. The NH_2-terminal fragment of FGF23 contains the FGF receptor (FGFR)–binding domain. Genetically manipulated *Fgf23* mouse models have convincingly shown the phosphaturic activities of FGF23 [27–33], and also gave an experimental clarification of human diseases that are associated with altered FGF23 function [34–38]. For instance, elevated FGF23 levels are also observed in various genetic bone disorders due to excessive phosphate wasting, resulting in rickets/osteomalacia (Table 14.1) [36]; in autosomal dominant hypophosphatemic rickets, gain of function mutations were identified in the *Fgf23* gene [39]. In X-linked hypophosphatemia, decreased clearance of FGF23 due to mutation in PHEX (a phosphate-regulating gene with homologies to endopeptidases on the X-chromosome) result in vitamin D–resistant rickets/osteomalacia [40,41]. Recently, mutations in the dentin matrix protein-1 (DMP-1) gene have been shown to be associated with autosomal recessive hypophosphatemic

TABLE 14.1

Serum Profiling of Human Diseases Associated with Increased FGF23 Activities

	XLH	ADHR	TIO	ARHR
Phosphorus	Low	Low	Low	Low
Calcium	Normal	Normal	Normal	Normal
PTH	Normal	Normal	Normal	Normal
$1,25(OH)_2D_3$	Low/normal	Low	Low/normal	Normal
FGF23	High	High	High	High

Source: Adapted with modification from Razzaque, M.S. and Lanske, B., *J. Endocrinol.*, 194, 1–10, 2007.

Note: XLH, X-linked hypophosphatemia; ADHR, autosomal dominant hypophosphatemic rickets; TIO, tumor-induced osteomalacia; ARHR: autosomal recessive hypophosphatemic rickets/osteomalacia.

rickets/osteomalacia [42]. Consistent with human disease, transgenic mice overexpressing *Fgf23* develop hypophosphatemia and rickets/osteomalacia due to increased urinary phosphate excretion [29,31,33].

In contrast to phosphate wasting diseases due to gain-of-function of FGF23, in familial tumoral calcinosis, loss-of-function of FGF23 is associated with hyperphosphatemia and ectopic calcification [43]. Likewise, *Fgf23* knockout mice develop severe hyperphosphatemia due to increased renal phosphate uptake, causing skeletal mineralization anomalies and extensive soft tissue calcification [27,28]; *Fgf23* knockout mice also have extremely high serum levels of $1,25(OH)_2D_3$ [44,45], whereas transgenic mice producing high amounts of FGF23 have reduced serum levels of $1,25(OH)_2D_3$ [29,31,33]. Of relevance, FGF23 can suppress the expression of $1\alpha(OH)$ase to reduce the production of the active vitamin D metabolite $1,25(OH)_2D_3$ [46], whereas vitamin D can exert stimulating effects on FGF23 and Klotho [46,47]; Klotho allows FGF23 to bind to its receptor with much higher affinity to induce downstream signaling events.

14.4 KLOTHO

Klotho is a type 1 membrane protein; the human *KLOTHO* gene has five exons and can generate two transcripts, and has approximately 86% homology to mouse *Klotho*. Klotho expression is mostly detected in the distal convoluted tubules of the kidney, the parathyroid gland, and the epithelium of the choroid plexus in the brain [48,49]. The genetic inactivation of *Klotho* results in hyperphosphatemia and hypervitaminosis D in mice; such biochemical changes in the serum are associated with premature aging–like features, including infertility and a shortened life span in the *klotho*-ablated mice [50]. Interestingly, the phenotype of *Klotho* knockout mice resembles that of *Fgf23* knockout mice, which led to the identification of FGF23 and Klotho in similar signaling pathways (Table 14.2) [51].

FGFs are secreted factors that are able to bind to cell surface FGFRs to activate downstream signaling events. Studies have shown that FGF23 requires Klotho as a cofactor to interact with the FGFR. FGF23 has been shown to bind to multiple FGFRs, including FGFR1c, FGFR3c, and FGFR4 [52,53]; however, some *in vivo* studies have shown that neither FGFR3 nor FGFR4 is the principal mediator of FGF23 effects [54]. Klotho seems to allow FGF23 to bind to its receptor complex with much higher affinity compared with FGFR alone. Moreover, FGF23, in the presence of Klotho, can activate downstream signaling events, as indicated by the activation of early growth response element-1 and the phosphorylation of FGFR substrate-2a, and extracellular signal-regulated kinase proteins [52,53,55]. A point mutation in the human *KLOTHO* gene was found to be associated with hyperphosphatemia, despite significantly elevated serum levels of FGF23 [56]; FGF23 became nonfunctional in this patient, due to lack of KLOTHO function, and thereby could

TABLE 14.2

Similarities of Premature Aging–Like Phenotype of *Fgf23* and *Klotho* Knockout Mice

	Fgf23 Knockout Mouse	*Klotho* Knockout Mouse	*Fgf23*/Klotho Double-Knockout Mouse
Gross Appearance			
Body weight	Reduced	Reduced	Reduced
Growth retardation	Present	Present	Present
Kyphosis	Present	Present	Present
Gait walk	Present	Present	Present
Body hair	Sparse	Sparse	Sparse
Generalized Atrophy			
Muscle wasting	Present	Present	Present
Skin atrophy	Present	Present	Present
Thymus/spleen atrophy	Present	Present	Present
Morphological Changes			
Atherosclerosis	Present	Present	Present
Ectopic calcification	Present	Present	Present
Altered skeletal mineral	Present	Present	Present
Osteopenia/rickets	Present	Present	Present
Emphysema	Present	Present	Present
Renal NaPi-2a expression	Increased	Increased	Increased
Biochemical Changes			
Serum $1,25(OH)_2D_3$	High	High	High
Scrum phosphate	High	High	High
Serum calcium	High	High	High
Serum PTH	Low	Low	Low
Serum FGF23	Absent	High	Absent
Overall Effect			
Physical activity	Sluggish	Sluggish	Sluggish
Infertility	Present	Present	Present
Life span	Short	Short	Short

Source: Adapted with modification from Razzaque, M.S. and Lanske, B., *J. Endocrinol.*, 194, 1–10, 2007; Nakatani, T. et al., *FASEB J.*, 23, 433–441, 2009.

not exert its phosphate-lowering effects, which eventually lead to severe vascular and soft tissue calcifications in the patient [56]. Of relevance, *Klotho* knockout mice also have extremely high serum levels of $1,25(OH)_2D_3$ with vascular calcification [57].

Klotho knockout mice develop a syndrome resembling chronic kidney disease (CKD) in patients undergoing dialysis, including hypoactivity, muscle atrophy, skeletal anomalies, and widespread calcifications. Circulatory FGF23 levels are found to be extremely high in patients with CKD [58]. Although the exact role of elevated circulatory levels of FGF23 in patients with CKD is unclear, it is plausible that low levels of vitamin D and secondary hyperparathyroidism in patients with CKD may be influenced by increased circulatory FGF23 levels. FGF23 is a counter-regulatory hormone for vitamin D; therefore, it is likely that elevated circulatory levels of FGF23 might contribute to the hypovitaminosis D in patients with CKD. In patients with CKD, long-term use of active vitamin D metabolites is one of the common treatment options, and in these populations, some studies have shown that vitamin D supplementation may be independently associated with cardiovascular anomalies, including coronary artery calcification [59].

14.5   VITAMIN D AND VASCULAR CHANGES

Experimental studies indicate that $1,25(OH)_2D_3$ and its analogues can inhibit angiogenesis [60]. In fact, through its inhibitory effects on vascular endothelial growth factor–induced proliferation of retinal blood vessels, vitamin D is effective in controlling macular degeneration. A close association between vitamin D and vascular calcification has been reported in various human diseases, including atherosclerosis, osteoporosis, and CKD. As mentioned previously, vitamin D and its metabolites exert biological activities mainly through interacting with VDR, which is present in most human tissues, including endothelial and vascular smooth muscle cells [61]. In addition, vascular smooth muscle cells also possess the $1\alpha(OH)$ase system and thus can locally synthesize vitamin D metabolites [62]. In patients with CKD undergoing dialysis treatment, vitamin D therapy has been reported to be associated with a higher incidence of vascular calcification [63]. Uremic tumoral calcinosis is another severe complication of such therapy in patients undergoing dialysis [64]. The most commonly affected sites are the elbow, wrist, shoulder, and hip. Use of a high calcium dialyzer bath and prolonged administration of vitamin D are believed to be the main underlying cause of tumoral calcinosis [65], and the use of a low calcium bath and withdrawal of vitamin D therapy markedly improved this condition [66].

Vitamin D–associated vascular calcification was also noted in various genetically modified mouse models. Genetic inactivation of either *Fgf23* or *Klotho* leads to abnormal mineral ion metabolism and increased vitamin D activities, resulting in widespread soft tissue and vascular calcification [45,51,67]. More importantly, soft tissue and vascular calcification can be eliminated from both *Fgf23* and *Klotho* knockout mice by suppressing vitamin D activities [44,45,68]. It should, however, be noted that reducing vitamin D activity in *Fgf23* and *Klotho* knockout mice also reduced their serum phosphate levels, and the observed elimination of vascular calcification is likely to be associated with reduced phosphate burden in these mutant mice [44,45,57,68].

In addition to the effects of vitamin D on various vascular components, it also exerts a major role in skeletogenesis. It is well documented that PTH enhances the tubular absorption of calcium and stimulates $1\alpha(OH)$ase to produce $1,25(OH)_2D_3$, which is recognized by its receptor in osteoblasts, causing an increase in the expression of the RANK ligand. RANK, the receptor of RANK ligand on preosteoclasts, binds its ligand to facilitate the maturation of preosteoclasts to osteoclasts. Mature osteoclasts remove calcium and phosphorus from the bone, which is required to promote bone mineralization. Approximately 33% of women 60–70 years of age and 66% of those 80 years of age or older have osteoporosis and are likely to sustain an osteoporotic fracture in their lifetime [69,70]. A World Health Organization committee has defined osteoporosis based on bone density. Standardized bone density measurements of the total hip that are higher than 833 mg/cm^2 are considered as normal bone, between 833 and 648 mg/cm^2 is considered as osteopenia, and lower than 648 mg/cm^2 is considered as osteoporosis, and when there has been a fragility fracture, the condition is referred to as severe or established osteoporosis. Subsequently, further modifications have been suggested and adopted to clinically classify various stages of osteoporosis. In a study comparing calcium or placebo and 700–800 IU of vitamin D_3 per day, the relative risk of hip fracture was found to be reduced by 26% (pooled risk ratio, 0.77; 95% CI, 0.68–0.87) with vitamin D_3 [71]. Similar beneficial effects of vitamin D treatment have been documented in various human tumors.

14.6   VITAMIN D AND FGF23–KLOTHO SYSTEM IN TUMORS

Observational studies in both human and animal models support that vitamin D has a beneficial role in tumor prevention, progression, and survival. Epidemiologic studies indicate that low vitamin D levels are associated with a 30% to 50% increased risk of colon, prostate, and breast tumors

[72–74]. A population-based randomized control trial found that postmenopausal women who were receiving calcium and vitamin D supplements had a reduced risk of tumor after the first year of treatment (risk ratio, 0.232; 95% CI, 0.09–0.60) [75]. The likely explanation is that colon, prostate, and breast tissues express 1α(OH)ase, and local production of $1,25(OH)_2D_3$ might control the induction of genes that help suppress tumor growth by inhibiting cellular proliferation and differentiation [76–78]. It has also been suggested that if a cell becomes malignant, $1,25(OH)_2D_3$ can induce apoptosis and thus prevent angiogenesis, thereby creating a microenvironment that is less helpful for the malignant cell to survive [60,79].

A significant correlation between serum FGF23 concentration and various stages of cancer is also reported elsewhere; for instance, in women with ovarian cancer, serum or plasma FGF23 concentrations were found to be elevated in patients with advanced-stage epithelial ovarian cancer [80]. Several case reports also described increased serum FGF23 concentrations in patients with skeletal metastasis [81]. There are also reports that evaluate the FGF system as a potential target for therapy in prostate cancer [82]. In contrast to FGF23, expression of Klotho was reduced in breast cancer tissue [83], whereas high Klotho expression was associated with reduced tumor size. In a similar line of observation, forced expression of the *Klotho* gene, or treatment with soluble Klotho, inhibited the activation of insulin-like growth factor (IGF-1) and insulin signaling pathways, and thereby reduced breast tumors [83]. Klotho can also inhibit the proliferation and increase apoptosis of lung cancer A549 cells, and such inhibition is also partly achieved by the inhibition of IGF-1/insulin pathways and through the expression of apoptosis-promoting genes, including bax/bcl-2 [84]. More recently, investigators have shown that secreted Klotho was effective in blocking three major signaling pathways—TGF-β1, Wnt, and IGF-1—that cause tissue fibrosis and metastatic spread of tumor cells [85].

14.7   VITAMIN D AND FGF23–KLOTHO SYSTEM IN AGING

Aging is a progressive biological process that is regulated by complex genetic interactions and is influenced by several environmental factors. The aging process is related to an overall decline in the functionality of the vital systems, decreased cognition and reproductive capacity, and increased mortality. Human and experimental studies suggest that oxidative stress, DNA damage, and mitochondrial dysfunction contribute to the mammalian aging process, possibly by altering normal cellular and subcellular activities [86,87]. Also, numerous endocrine and humoral factors, including insulin, IGF-1, Klotho, and FGF23 might influence mammalian aging and survival by abnormally regulating metabolic balance or by exerting effects on mineral ion homeostasis (Table 14.3) [45,50,67,88]. The relationship between many of these diseases and aging-related changes in physiology show a U-shaped response curve to serum vitamin D concentrations, suggesting that both

---

TABLE 14.3
Potential Factors Involved in Mammalian Aging

Factors that regulate:
1. DNA damage, repair, and nuclear function
2. Genetic maintenance of telomere
3. Oxidative stress and mitochondrial dysfunction
4. Effects of caloric restriction
5. Insulin and IGF-1 signaling
6. Mineral ion homeostasis

Source: Adapted from Razzaque, M.S. and Lanske, B., *Trends Mol. Med.* 12, 298–305, 2006.

---

high and low levels of vitamin D levels are associated with an increased risk of chronic diseases and the aging process [89].

FGF23 has recently been shown to affect aging by influencing vitamin D and phosphate homeostasis [29,44,68]. For instance, *Fgf23* knockout mice show the early aging phenotypes, which include thin skin, atrophy of the intestine, spleen, and muscle, weight loss, short life span, osteoporosis, and atherosclerosis due to abnormal vitamin D and phosphate homeostasis. Similarly, mice lacking *Klotho* activity also show phenotypes resembling human aging [50]. The strikingly similar phenotypes of *Fgf23* and *Klotho* knockout mice imply that the premature aging process is a consequence of the disruption of a common humoral signaling pathway, affecting vitamin D and phosphate metabolism. Interestingly, reducing either vitamin D or serum phosphate levels from *Fgf23* and *Klotho* knockout mice can rescue most of the premature aging–like phenotypes, implicating their role in mammalian aging [50,51].

14.8 CONCLUDING REMARKS

Mild vitamin D deficiency is more prevalent in elderly individuals, and vitamin D supplementation is, therefore, a reasonable approach to minimize the risk for osteoporosis and metabolic bone disease. However, uncontrolled use of vitamin D, without specific objectives, is an unrealistic approach and may induce unexpected complications. For instance, treatment with vitamin D and calcium supplements might not be equally advisable to an elderly individual with the risk of developing cardiovascular calcification, as opposed to an elderly individual who has a healthy heart with unhealthy bone [90–92]. Finally, in this brief chapter, we wanted to highlight our findings of how the recently identified FGF23–Klotho axis might affect aging and other diseases, by influencing vitamin D and phosphate homeostasis. Further studies will determine if therapeutic manipulation of the FGF23–Klotho system might be able to fine-tune vitamin D homeostasis to avert the undesired side effects (Figure 14.1).

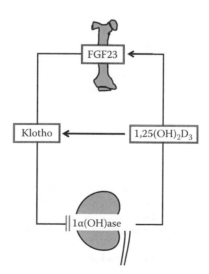

FIGURE 14.1 Simplified diagram showing osteorenal communication of FGF23 and vitamin D. The osteocyte-derived FGF23, in the presence of Klotho, can act on the kidney to suppress the expression of the 1α(OH)ase gene, and it can thereby reduce the production of $1,25(OH)_2D_3$. Because $1,25(OH)_2D_3$ can induce both FGF23 and Klotho, reduced levels of $1,25(OH)_2D_3$ may have inhibitory effects on FGF23 production in bone. In physiological conditions, osteorenal communication through $1,25(OH)_2D_3$ and FGF23 may fine-tune their levels to maintain normal mineral ion balance. (Adapted with modification from Razzaque, M.S., *IUBMB Life* 63, 240–247, 2011.)

REFERENCES

1. Binderup, L., S. Latini, E. Binderup, C. Bretting M. Calverley and K. Hansen. 1991. 20-Epi-vitamin D3 analogues: A novel class of potent regulators of cell growth and immune responses. *Biochemical Pharmacology* 42:1569–1575.
2. Baeke, F., C. Gysemans, H. Korf, and C. Mathieu. 2010. Vitamin D insufficiency: Implications for the immune system. *Pediatric Nephrology* 25:1597–1606.
3. Daniel, C., H. H. Radeke, N. A. Sartory et al. 2006. The new low calcemic vitamin D analog 22-ene-25-oxa-vitamin D prominently ameliorates T helper cell type 1–mediated colitis in mice. *Journal of Pharmacology and Experimental Therapeutics* 319:622–631.
4. Martineau, A. R., K. A. Wilkinson, S. M. Newton et al. 2007. IFN-gamma- and TNF-independent vitamin D–inducible human suppression of mycobacteria: The role of cathelicidin LL-37. *The Journal of Immunology* 178:7190–7198.
5. Veldman, C. M., M. T. Cantorna, and H. F. DeLuca. 2000. Expression of 1,25-dihydroxyvitamin D(3) receptor in the immune system. *Archives of Biochemistry and Biophysics* 374:334–338.
6. Dusso, A. S., A. J. Brown, and E. Slatopolsky. 2005. Vitamin D. *American Journal of Physiology. Renal Physiology* 289:F8–28.
7. Annweiler, C., A. M. Schott, G. Berrut et al. 2010. Vitamin D and ageing: Neurological issues. *Neuropsychobiology* 62:139–150.
8. Autier, P., and S. Gandini. 2007. Vitamin D supplementation and total mortality: A meta-analysis of randomized controlled trials. *Archives of Internal Medicine* 167:1730–1737.
9. Melamed, M. L., E. D. Michos, W. Post, and B. Astor. 2008. 25-Hydroxyvitamin D levels and the risk of mortality in the general population. *Archives of Internal Medicine* 168:1629–1637.
10. Bhan, I., and R. Thadhani. 2009. Vitamin D therapy for chronic kidney disease. *Seminars in Nephrology* 29:85–93.
11. Reid, I. R., and M. J. Bolland. 2008. Calcium supplementation and vascular disease. *Climacteric* 11:280–286.
12. Reid, I. R., M. J. Bolland, and A. Grey. 2008. Effect of calcium supplementation on hip fractures. *Osteoporosis International* 19:1119–1123.
13. Reid, I. R., M. J. Bolland, A. Avenell, and A. Grey. 2011. Cardiovascular effects of calcium supplementation. *Osteoporosis International* 22:1649–1658.
14. Bolland, M. J., P. A. Barber, R. N. Doughty et al. 2008. Vascular events in healthy older women receiving calcium supplementation: Randomized controlled trial. *British Medical Journal* 336:262–266.
15. Bolland, M. J., A. Grey, A. Avenell, G. D. Gamble, and I. R. Reid. 2011. Calcium supplements with or without vitamin D and risk of cardiovascular events: Reanalysis of the Women's Health Initiative limited access dataset and meta-analysis. *British Medical Journal* 342:d2040.
16. Hsia, J., G. Heiss, H. Ren et al. 2007. Calcium/vitamin D supplementation and cardiovascular events. *Circulation* 115:846–854.
17. Razzaque, M. S. 2011. Phosphate toxicity: New insights into an old problem. *Clinical Science* 120:91–97.
18. Razzaque, M. S. 2011. The dualistic role of vitamin D in vascular calcifications. *Kidney International* 79:708–714.
19. Razzaque, M. S. 2011. Osteo-renal regulation of systemic phosphate metabolism. *IUBMB Life* 63:240–247.
20. Razzaque, M. S. 2009. FGF23-mediated regulation of systemic phosphate homeostasis: Is Klotho an essential player? *American Journal of Physiology. Renal Physiology* 296:F470–476.
21. Lips, P. 2001. Vitamin D deficiency and secondary hyperparathyroidism in the elderly: Consequences for bone loss and fractures and therapeutic implications. *Endocrine Reviews* 22:477–501.
22. Segersten, U., P. Correa, M. Hewison et al. 2002. 25-Hydroxyvitamin D(3)-1alpha-hydroxylase expression in normal and pathological parathyroid glands. *The Journal of Clinical Endocrinology and Metabolism* 87:2967–2972.
23. Brenza, H. L., and H. F. DeLuca. 2000. Regulation of 25-hydroxyvitamin D3 1alpha-hydroxylase gene expression by parathyroid hormone and 1,25-dihydroxyvitamin D3. *Archives of Biochemistry and Biophysics* 381:143–152.
24. Brown, A. J., A. Dusso, and E. Slatopolsky. 1999. Vitamin D. *American Journal of Physiology* 277:F157–F175.
25. Horst, R. L., J. P. Goff, and T. A. Reinhardt. 1997. Calcium and vitamin D metabolism during lactation. *Journal of Mammary Gland Biology and Neoplasia* 2:253–263.

26. Tsujikawa, H., Y. Kurotaki, T. Fujimori, K. Fukuda, and Y. Nabeshima. 2003. Klotho, a gene related to a syndrome resembling human premature aging, functions in a negative regulatory circuit of vitamin D endocrine system. *Molecular Endocrinology* 17:2393–2403.

27. Shimada, T., M. Kakitani, Y. Yamazaki et al. 2004. Targeted ablation of Fgf23 demonstrates an essential physiological role of FGF23 in phosphate and vitamin D metabolism. *Journal of Clinical Investigation* 113:561–568.

28. Sitara, D., M. S. Razzaque, M. Hesse et al. 2004. Homozygous ablation of fibroblast growth factor-23 results in hyperphosphatemia and impaired skeletogenesis, and reverses hypophosphatemia in Phex-deficient mice. *Matrix Biology* 23:421–432.

29. DeLuca, S., D. Sitara, K. Kang et al. 2008. Amelioration of the premature ageing–like features of Fgf-23 knockout mice by genetically restoring the systemic actions of FGF-23. *Journal of Pathology* 216:345–355.

30. Liu, S., and L. D. Quarles. 2007. How fibroblast growth factor 23 works. *Journal of the American Society of Nephrology* 18:1637–1647.

31. Shimada, T., I. Urakawa, Y. Yamazaki et al. 2004. FGF-23 transgenic mice demonstrate hypophospha-temic rickets with reduced expression of sodium phosphate cotransporter type IIa. *Biochemical and Biophysical Research Communications* 314:409–414.

32. Liu, S., J. Zhou, W. Tang, X. Jiang, D. W. Rowe, and L. D. Quarles. 2006. Pathogenic role of Fgf23 in Hyp mice. *American Journal of Physiology. Endocrinology and Metabolism* 291:E38–49.

33. Bai, X., D. Miao, J. Li, D. Goltzman, and A. C. Karaplis. 2004. Transgenic mice overexpressing human fibroblast growth factor 23 (R176Q) delineate a putative role for parathyroid hormone in renal phosphate wasting disorders. *Endocrinology* 145:5269–5279.

34. Berndt, T., and R. Kumar. 2007. Phosphatonins and the regulation of phosphate homeostasis. *Annual Review of Physiology* 69:341–359.

35. Razzaque, M. S. 2007. Can fibroblast growth factor 23 fine-tune therapies for diseases of abnormal min-eral ion metabolism? *Nature Clinical Practice. Endocrinology & Metabolism* 3:788–789.

36. Razzaque, M. S., B. Lanske. 2007. The emerging role of the fibroblast growth factor-23–klotho axis in renal regulation of phosphate homeostasis. *Journal of Endocrinology* 194:1–10.

37. Bergwitz, C., and H. Juppner. 2010. Regulation of phosphate homeostasis by PTH, vitamin D, and FGF23. *Annual Review of Medicine* 61:91–104.

38. Juppner, H. 2011. Phosphate and FGF-23. *Kidney International. Supplement* S24–27.

39. White, K. E., G. Carn, B. Lorenz-Depiereux, A. Benet-Pages, T. M. Strom, and M. J. Econs. 2000. Autosomal dominant hypophosphataemic rickets is associated with mutations in FGF23. *Nature Genetics* 26:345–348.

40. Tenenhouse, H. S. 1999. X-linked hypophosphataemia: A homologous disorder in humans and mice. *Nephrology, Dialysis, Transplantation* 14:333–341.

41. Jap, T. S., C. Y. Chiu, D. M. Niu, and M. A. Levine. 2011. Three novel mutations in the PHEX gene in Chinese subjects with hypophosphatemic rickets extends genotypic variability. *Calcified Tissue International* 88:370–377.

42. Lorenz-Depiereux, B., M. Bastepe, A. Benet-Pages et al. 2006. DMP1 mutations in autosomal recessive hypophosphatemia implicate a bone matrix protein in the regulation of phosphate homeostasis. *Nature Genetics* 38:1248–1250.

43. Benet-Pages, A., P. Orlik, T. M. Strom, and B. Lorenz-Depiereux. 2005. An FGF23 missense mutation causes familial tumoral calcinosis with hyperphosphatemia. *Human Molecular Genetics* 14:385–390.

44. Sitara, D., M. S. Razzaque, R. St-Arnaud et al. 2006. Genetic ablation of vitamin D activation pathway reverses biochemical and skeletal anomalies in Fgf-23–null animals. *The American Journal of Pathology* 169:2161–2170.

45. Razzaque, M. S., D. Sitara, T. Taguchi, R. St-Arnaud, and B. Lanske. 2006. Premature aging–like phenotype in fibroblast growth factor 23 null mice is a vitamin D–mediated process. *FASEB Journal* 20:720–722.

46. Shimada, T., H. Hasegawa, Y. Yamazaki et al. 2004. FGF-23 is a potent regulator of vitamin D metabo-lism and phosphate homeostasis. *Journal of Bone and Mineral Research* 19:429–435.

47. Shimada, T., Y. Yamazaki, M, Takahashi et al. 2005. Vitamin D receptor–independent FGF23 actions in regulating phosphate and vitamin D metabolism. *American Journal of Physiology. Renal Physiology* 289:F1088–1095.

48. Nabeshima, Y. 2009. Discovery of alpha-Klotho unveiled new insights into calcium and phosphate homeostasis. *Proceedings of the Japan Academy. Series B, Physical and Biological Sciences* 85:125–141.

49. Nabeshima, Y. 2002. Klotho: A fundamental regulator of aging. *Ageing Research Reviews* 1:627–638.

50. Kuro-o, M., Y. Matsumura, H. Aizawa et al. 1997. Mutation of the mouse klotho gene leads to a syndrome resembling aging. *Nature* 390:45–51.
51. Nakatani, T., B. Sarraj, M. Ohnishi et al. 2009. *In vivo* genetic evidence for klotho-dependent, fibroblast growth factor 23 (Fgf23)–mediated regulation of systemic phosphate homeostasis. *FASEB Journal* 23:433–441.
52. Kurosu, H., Y. Ogawa, M. Miyoshi et al. 2006. Regulation of fibroblast growth factor-23 signaling by klotho. *Journal of Biological Chemistry* 281:6120–6123.
53. Urakawa, I., Y. Yamazaki, T. Shimada et al. 2006. Klotho converts canonical FGF receptor into a specific receptor for FGF23. *Nature* 444:770–774.
54. Liu, S., L. Vierthaler, W. Tang, J. Zhou, L. D. Quarles. 2008. FGFR3 and FGFR4 do not mediate renal effects of FGF23. *Journal of the American Society of Nephrology* 19:2342–2350.
55. Medici, D., M. S. Razzaque, S. Deluca et al. 2008. FGF-23–Klotho signaling stimulates proliferation and prevents vitamin D–induced apoptosis. *Journal of Cell Biology* 182:459–465.
56. Ichikawa, S., E. A. Imel, M. L. Kreiter et al. 2007. A homozygous missense mutation in human klotho causes severe tumoral calcinosis. *Journal of Clinical Investigation* 117:2684–2691.
57. Ohnishi, M., T. Nakatani, B. Lanske, and M. S. Razzaque. 2009. *In vivo* genetic evidence for suppressing vascular and soft-tissue calcification through the reduction of serum phosphate levels, even in the presence of high serum calcium and 1,25-dihydroxyvitamin d levels. *Circulation. Cardiovascular Genetics* 2:583–590.
58. Gutierrez, O. M., M. Mannstadt, T. Isakova et al. 2008. Fibroblast growth factor 23 and mortality among patients undergoing hemodialysis. *New England Journal of Medicine* 359:584–592.
59. Briese, S., S. Wiesner, J. C. Will et al. 2006. Arterial and cardiac disease in young adults with childhood-onset end-stage renal disease—impact of calcium and vitamin D therapy. *Nephrology, Dialysis, Transplantation* 21:1906–1914.
60. Bernardi, R. J., C. S. Johnson, R. A. Modzelewski, and D. L. Trump. 2002. Antiproliferative effects of 1alpha,25-dihydroxyvitamin D(3) and vitamin D analogs on tumor-derived endothelial cells. *Endocrinology* 143:2508–2514.
61. Merke, J., W. Hofmann, D. Goldschmidt, and E. Ritz. 1987. Demonstration of 1,25(OH)2 vitamin D3 receptors and actions in vascular smooth muscle cells *in vitro*. *Calcified Tissue International* 41: 112–114.
62. Somjen, D., Y. Weisman, F. Kohen et al. 2005. 25-Hydroxyvitamin D3-1alpha-hydroxylase is expressed in human vascular smooth muscle cells and is upregulated by parathyroid hormone and estrogenic compounds. *Circulation* 111:1666–1671.
63. Goldsmith, D. J., A. Covic, P. A. Sambrook, and P. Ackrill. 1997. Vascular calcification in long-term haemodialysis patients in a single unit: A retrospective analysis. *Nephron* 77:37–43.
64. Quarles, L. D., G. Murphy, M. J. Econs, S. Martinez, B. Lobaugh, and K. W. Lyles. 1991. Uremic tumoral calcinosis: Preliminary observations suggesting an association with aberrant vitamin D homeostasis. *American Journal of Kidney Diseases* 18:706–710.
65. Cofan, F., S. Garcia, A. Combalia, J. M. Campistol, F. Oppenheimer, and R. Ramon. 1999. Uremic tumoral calcinosis in patients receiving longterm hemodialysis therapy. *The Journal of Rheumatology* 26:379–385.
66. Franco, M., L. Albano, H. Gaid, M. Gigante, D. Barrillon, and P. Jaeger. 2005. Resolution of tumoral calcinosis in a hemodialysis patient using low calcium dialysate. *Joint Bone Spine* 72:95–97.
67. Razzaque, M. S., and B. Lanske. 2006. Hypervitaminosis D and premature aging: Lessons learned from Fgf23 and Klotho mutant mice. *Trends in Molecular Medicine* 12:298–305.
68. Ohnishi, M., T. Nakatani, B. Lanske, and M. S. Razzaque. 2009. Reversal of mineral ion homeostasis and soft-tissue calcification of klotho knockout mice by deletion of vitamin D 1alpha-hydroxylase. *Kidney International* 75:1166–1172.
69. Boonen, S., H. A. Bischoff-Ferrari, C. Cooper et al. 2006. Addressing the musculoskeletal components of fracture risk with calcium and vitamin D: A review of the evidence. *Calcified Tissue International* 78:257–270.
70. Larsen, E. R., L. Mosekilde, and A. Foldspang. 2004. Vitamin D and calcium supplementation prevents osteoporotic fractures in elderly community dwelling residents: A pragmatic population-based 3-year intervention study. *Journal of Bone and Mineral Research* 19:370–378.
71. Bischoff-Ferrari, H. A., E. Giovannucci, W. C. Willett, T. Dietrich, and B. Dawson-Hughes. 2006. Estimation of optimal serum concentrations of 25-hydroxyvitamin D for multiple health outcomes. *The American Journal of Clinical Nutrition* 84:18–28.

72. Giovannucci, E., Y. Liu, M. J. Stampfer, and W. C. Willett. 2006. A prospective study of calcium intake and incident and fatal prostate cancer. *Cancer Epidemiology, Biomarkers & Prevention* 15:203–210.
73. Ahonen, M. H., L. Tenkanen, L. Teppo, M. Hakama, and P. Tuohimaa. 2000. Prostate cancer risk and prediagnostic serum 25-hydroxyvitamin D levels (Finland). *Cancer Causes & Control* 11:847–852.
74. Feskanich, D., J. Ma, C. S. Fuchs et al. 2004. Plasma vitamin D metabolites and risk of colorectal cancer in women. *Cancer Epidemiology, Biomarkers & Prevention* 13:1502–1508.
75. Lappe, J. M., D. Travers-Gustafson, K. M. Davies, R. R. Recker, and R. P. Heaney. 2007. Vitamin D and calcium supplementation reduces cancer risk: Results of a randomized trial. *The American Journal of Clinical Nutrition* 85:1586–1591.
76. Nagpal, S., S. Na, and R. Rathnachalam. 2005. Noncalcemic actions of vitamin D receptor ligands. *Endocrine Reviews* 26:662–687.
77. Gorham, E. D., C. F. Garland, F. C. Garland et al. 2005. Vitamin D and prevention of colorectal cancer. *Journal of Steroid Biochemistry and Molecular Biology* 97:179–194.
78. Grant, W. B. 2002. An estimate of premature cancer mortality in the U.S. due to inadequate doses of solar ultraviolet-B radiation. *Cancer* 94:1867–1875.
79. Mantell, D. J., P. E. Owens, N. J. Bundred, E. B. Mawer, and A. E. Canfield. 2000. 1 alpha,25-Dihydroxyvitamin D(3) inhibits angiogenesis *in vitro* and *in vivo*. *Circulation Research* 87:214–220.
80. Tebben, P. J., K. R. Kalli, W. A. Cliby et al. 2005. Elevated fibroblast growth factor 23 in women with malignant ovarian tumors. *Mayo Clinic Proceedings* 80:745–751.
81. Fournier, P., E. Imel, and J. Chirgwin. 2010. Osteocyte-derived FGF23 acts on bone-metastatic breast cancer cells to increase resistance to 1,25-dihydroxy-vitamin D. *ASBMR Annual Meeting*.
82. Gowardhan, B., D. A. Douglas, M. E. Mathers et al. 2005. Evaluation of the fibroblast growth factor system as a potential target for therapy in human prostate cancer. *British Journal of Cancer* 92:320–327.
83. Wolf, I., S. Levanon-Cohen, S. Bose et al. 2008. Klotho: A tumor suppressor and a modulator of the IGF-1 and FGF pathways in human breast cancer. *Oncogene* 27:7094–7105.
84. Chen, B., X. Wang, W. Zhao, and J. Wu. 2010. Klotho inhibits growth and promotes apoptosis in human lung cancer cell line A549. *Journal of Experimental & Clinical Cancer Research* 29:99.
85. Doi, S., Y. Zou, O. Togao et al. 2011. Klotho inhibits transforming growth factor-β1 (TGF-β1) signaling and suppresses renal fibrosis and cancer metastasis in mice. *Journal of Biological Chemistry* 286:8655–8665.
86. Marciniak, R., and L. Guarente. 2001. Human genetics. Testing telomerase. *Nature* 413:370–371, 373.
87. Rudolph, K. L., S. Chang, H. W. Lee et al. 1999. Longevity, stress response, and cancer in aging telomerase–deficient mice. *Cell* 96:701–712.
88. Holzenberger, M., L. Kappeler, C. De Magalhaes Filho. 2004. IGF-1 signaling and aging. *Experimental Gerontology* 39:1761–1764.
89. Tuohimaa, P., T. Keisala, A. Minasyan, J. Cachat, and A. Kalueff. 2009. Vitamin D, nervous system and aging. *Psychoneuroendocrinology* 34 Suppl 1:S278–286.
90. Trivedi, D. P., R. Doll, and K. T. Khaw. 2003. Effect of four monthly oral vitamin D3 (cholecalciferol) supplementation on fractures and mortality in men and women living in the community: Randomized double blind controlled trial. *British Medical Journal* 326:469.
91. Heikinheimo, R. J., J. A. Inkovaara, E. J. Harju et al. 1992. Annual injection of vitamin D and fractures of aged bones. *Calcified Tissue International* 51:105–110.
92. Jackson, R. D., A. Z. LaCroix, M. Gass et al. 2006. Calcium plus vitamin D supplementation and the risk of fractures. *New England Journal of Medicine* 354:669–683.

15 Vitamin D and Cardiovascular Disease

Jared P. Reis and Pamela L. Lutsey

CONTENTS

15.1 INTRODUCTION

Vitamin D is widely known for its important role in bone health, preventing the development of rickets, osteoporosis, and fractures. However, there is a growing body of compelling evidence from several lines of scientific inquiry suggesting that vitamin D may also play an important role in the pathogenesis of cardiovascular disease (CVD). CVD is the leading cause of death worldwide, accounting for approximately 22% of deaths globally (World Health Organization 2008). Vitamin D deficiency has been linked with several cardiovascular disorders, including congestive heart failure, left ventricular hypertrophy, peripheral arterial disease, subclinical vascular disease, myocardial infarction, stroke, and related mortalities—associations that remain even after careful consideration for traditional CVD risk factors and lifestyle behaviors, including differences in dietary intakes and physical activity levels.

Serum 25-hydroxyvitamin D [25(OH)D] levels, the best indicator of circulating vitamin D status, vary with geography, seasonality, latitude, and altitude, presumably as a result of differences in duration or intensity of sunlight. Interestingly, the risk of CVD is noted to be highest in areas of increased geographic latitude and during winter months—findings which parallel trends in low 25(OH)D levels (Zittermann et al. 2005). However, to date, a direct causal relationship between 25(OH)D deficiency and the risk of CVD has not been completely established.

In this chapter, we explore the evidence regarding the role that vitamin D may play in the development of CVD, including cardiovascular risk factors, subclinical disease, and clinical cardiovascular outcomes. Later sections of the chapter will review evidence pertaining to clinical trials of vitamin D supplementation, the potential for differences in vitamin D levels to explain certain disparities in cardiovascular outcomes, the potential cardiovascular effects of vitamin D toxicity, and the clinical implications of the totality of the research evidence. The research presented in this chapter is mostly acquired from the observational and experimental population-based and clinical patient–based literature; however, when appropriate, a brief discussion of the biological mechanisms will make use of research emanating from laboratory investigations.

15.2 CVD RISK FACTORS

15.2.1 HYPERTENSION

Several plausible biological mechanisms have been proposed linking vitamin D in the regulation of blood pressure. Vitamin D has been implicated in the proximal regulation of the renin–angiotensin system, interacting with the renin–angiotensin system and sodium to influence vascular smooth muscle tone, and indirectly influencing the endothelium (Vaidya and Forman 2010). Mice lacking the vitamin D receptor experience increased renin activity, develop hypertension, and are more susceptible to obstructive renal injury (Li et al. 2002). This unfavorable phenotype can be reversed with an infusion of 1,25 dihydroxyvitamin D [1,25(OH)$_2$D] (active vitamin D; Li et al. 2002).

There are now more than 20 cross-sectional studies that have examined the association of vitamin D with blood pressure or prevalent hypertension (Vaidya and Forman 2010). The majority of these studies have demonstrated an association of lower vitamin D intake or 25(OH)D levels with higher blood pressure or increased rates of hypertension. In the Third National Health and Nutrition Examination Survey, a nationally representative sample of adults from the United States, those with serum 25(OH)D concentrations in the lowest quartile had 30% higher odds of prevalent hypertension compared with similar adults with the highest vitamin D levels (Martins et al. 2007). The inverse association between vitamin D and prevalent hypertension has also been confirmed in younger populations. For example, as shown in Figure 15.1, among adolescents from the United States, aged 12–19 years, participating in the National Health and Nutrition Examination Survey 2001–2004, the odds ratio for hypertension was more than two times higher for those with vitamin D levels lower than 15.0 ng/mL compared with those with levels higher than 26.0 ng/mL, even after adjustment for age, gender, race/ethnicity, socioeconomic status, physical activity, and body mass index (Reis et al. 2009). However, in the Rancho Bernardo Study of older adults living in Southern California, no evidence of a cross-sectional association was observed between serum 25(OH)D concentrations and prevalent hypertension among men or women; although it is possible that the low prevalence of vitamin D deficiency contributed to the lack of an association (Reis et al. 2007).

Prospective observational studies have also documented an inverse relation between vitamin D and incident hypertension. In a subsample of men from the Health Professionals Follow-up Study, and in women from the Nurses' Health Study, the multivariable-adjusted relative risks of hypertension for those with vitamin D levels lower than 15 ng/mL compared with those with levels of

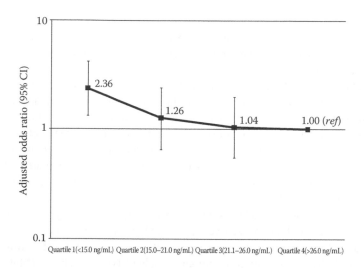

FIGURE 15.1 Adjusted odds ratios (95% CIs) for prevalent hypertension according to quartiles of serum vitamin D levels among adolescents in the United States aged 12–19 years, National Health and Nutrition Examination Survey 2001–2004. *Ref* indicates referent group.

30 ng/mL or higher was 6.13 [95% confidence intervals (CI), 1.00–37.8] and 2.67 (95% CI, 1.05–6.79), respectively (Forman et al. 2007). A nested case-control study of women from the Nurses' Health Study also confirmed this inverse association (Forman et al. 2008). However, in the Tromso Study of adults from Norway, baseline vitamin D levels did not predict future hypertension or change in blood pressure over time, and there was no association between change in vitamin D over time and change in blood pressure (Jorde et al. 2010). Thus, accumulating evidence suggests that vitamin D deficiency may increase the risk of future hypertension, although the findings have been inconsistent.

15.2.2 Obesity, Metabolic Syndrome, and Diabetes

Several studies have reported a cross-sectional association between vitamin D deficiency and obesity, the metabolic syndrome (a constellation of cardiovascular risk factors, including obesity, hypertension, elevated glucose levels, high triglycerides, and low high-density lipoprotein cholesterol concentrations), glucose intolerance, hyperinsulinemia, and diabetes (Ford et al. 2005; Martins et al. 2007; Reis et al. 2008a; Baynes et al. 1997). Although circulating vitamin D is fat-soluble and is therefore sequestered in adipose tissue, the association with metabolic syndrome and diabetes persists after simultaneous adjustment for adiposity, physical activity, and other potential confounding factors. In the National Health and Nutrition Examination Survey, lower vitamin D levels were strongly associated with prevalent metabolic syndrome after multivariable adjustment (Ford et al. 2005; Reis et al. 2008a).

As mentioned previously, adipose cells sequester vitamin D, making stores less available to become biologically activated, leading to an increased risk of deficiency among overweight or obese individuals (Wortsman et al. 2000). Those who are overweight or obese also may be less likely to be physically active outdoors or may have clothing habits that limit skin production of vitamin D from sunlight. Thus, it remains to be determined whether vitamin D deficiency precedes the development of obesity or is a consequence of increased adiposity. A recent study from the Insulin Resistance Atherosclerosis Family Study of Hispanic and African American Adults confirmed that higher 25(OH)D levels were inversely associated with body mass index and measures of visceral and subcutaneous adiposity in cross-sectional analyses (Young et al. 2009). However, baseline 25(OH)D levels were not predictive of these markers of adiposity over the next 5 years (Young et al. 2009).

Vitamin D deficiency has long been suspected of being a risk factor for type 1 diabetes on the basis of indirect evidence that countries at higher latitudes report higher incidence and prevalence rates of type 1 diabetes compared with countries at lower latitudes (Mohr et al. 2008). Type 1 diabetes is also more commonly diagnosed during the winter months (Christau et al. 1977; Fishbein et al. 1982). These ecological studies hypothesize that geography and season are indicators for reduced sunlight and vitamin D deficiency; however, there are other potential explanations that may support these findings (e.g., viral infections and reduced activity levels, which are also more commonly observed in the winter months).

The association of low vitamin D levels with diabetes does have biological basis, particularly for type 1 diabetes (Mathieu et al. 2005). Active vitamin D binding to its receptor located in pancreatic islet cells is thought to promote insulin secretion. Active vitamin D also has immunomodulatory effects that lead to diabetes prevention in animal models of type 1 diabetes (Mathieu et al. 2005). Vitamin D may also prevent insulin resistance, although the mechanism is less clear than for insulin secretion (Chiu et al. 2004). Certain vitamin D receptor polymorphisms are associated with increased risk of diabetes (Forouhi et al. 2008). Smaller interventional trials in humans suggest the potential benefit of vitamin D supplementation on parameters of glucose/insulin homeostasis (Chiu et al. 2004; Mathieu et al. 2005).

A meta-analysis of case-control studies reported a lower risk for developing type 1 diabetes with self-reported vitamin D supplementation in early childhood (odds ratio, 0.71; 95% CI, 0.60–0.84; Zipitis and Akobeng 2008). Results of a rare prospective study also found a reduced risk of type 1 diabetes among children in Finland who were given vitamin D supplements early in life (Hypponen et al. 2001). Interestingly, children who were suspected of having vitamin D–deficient rickets during the first year of life were three times more likely to develop diabetes (relative risk, 3.0; 95% CI, 1.0–9.0) than children who were not diagnosed with rickets (Hypponen et al. 2001).

Several cross-sectional studies have examined the association between vitamin D status and the prevalence of glucose intolerance or type 2 diabetes (Pittas et al. 2007). Although most have reported an inverse association between vitamin D status and glucose intolerance, others have failed to show an association (Pittas et al. 2007).

Three prospective observational studies from the United States and one from Finland have reported an association between vitamin D status or intake and risk of incident type 2 diabetes (Knekt et al. 2008; Liu et al. 2005, 2010; Pittas et al. 2006). In the Women's Health Study, vitamin D intake from diet and supplements was inversely associated with incident type 2 diabetes; however, this analysis was not adjusted for many potential confounding factors (Liu et al. 2005). In a subsample of women from the Nurses' Health Study, there was a lower risk of diabetes among women who reported the highest intake of both vitamin D and calcium, but not vitamin D alone (Pittas et al. 2006). In the Framingham Offspring Study, which included children from the original Framingham Heart Study population, Liu et al. (2010) used measured 25(OH)D levels and potential determinants of vitamin D in a subsample of the population to predict vitamin D levels in the larger cohort. The researchers then examined the relation between this predicted vitamin D score and incident type 2 diabetes. As shown in Figure 15.2, a significant inverse association was observed between tertiles of predicted vitamin D and diabetes (Liu et al. 2010). In a nested case-control study from the Mini-Finland Health Survey cohort, higher vitamin D status was associated with a lower risk of incident type 2 diabetes among men, but not among women (Knekt et al. 2008). Thus, although cross-sectional studies have reported relatively consistent associations between low vitamin D and prevalent type 1 or type 2 diabetes, the evidence from prospective studies is sparse and inconsistent.

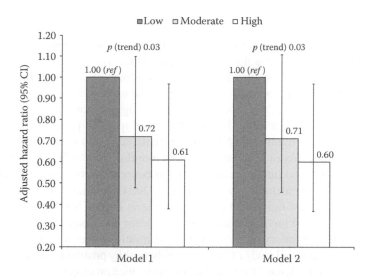

FIGURE 15.2 Adjusted hazard ratios (95% CIs) for type 2 diabetes according to tertiles of predicted 25(OH)D scores, Framingham Offspring Cohort. Model 1 is adjusted for waist circumference, parental history of diabetes, hypertension, low high-density lipoprotein concentrations, elevated triglycerides, impaired fasting glucose, and Dietary Guidelines for Americans Adherence Index. Model 2 is additionally adjusted for age and sex. *Ref* indicates referent group.

15.2.3 LIPIDS

A number of cross-sectional studies have shown an association between low vitamin D levels and hypertriglyceridemia, but not high-density lipoprotein or low-density lipoprotein cholesterol concentrations (Martins et al. 2007; Ford et al. 2005). The biological mechanism linking vitamin D with triglycerides has not been completely elucidated. Interestingly, cholesterol and vitamin D share the 7-dehydrocholesterol pathway. Cholesterol is synthesized from 7-dehydrocholesterol, which is also a precursor of 25(OH)D. Several recent studies have shown that statin medications, used primarily to lower cholesterol, also increase vitamin D levels. In a study of 83 patients with acute coronary artery disease, use of a statin post-diagnosis was associated with lower cholesterol and triglyceride concentrations and higher serum vitamin D levels after 12 months (Perez-Castrillon et al. 2007). The percentage of patients who were vitamin D–deficient decreased from 75% to 57% (Perez-Castrillon et al. 2007). Similarly, in another small study of 91 hyperlipidemic patients not previously on lipid-lowering medications, statin treatment for 8 weeks significantly increased 25(OH)D and 1,25(OH)$_2$D levels (Yavuz et al. 2009).

Randomized trials have shown significant cardiovascular benefit from statins (Scandinavian Simvastatin Survival Study (4S) 1994; Shepherd et al. 1995). In the West of Scotland Coronary Prevention Study Group of men with hyperlipidemia and no previous history of CVD, treatment with a statin lowered plasma cholesterol and low-density lipoprotein levels as well as the risk of coronary heart disease by more than 30% (Shepherd et al. 1995). However, it has been asserted that the effects of statins on CVD risk reduction may be due to factors other than cholesterol lowering, including beneficial effects on inflammation and endothelial function. Statins have been shown to increase bone mass and lower the risk of fractures in the elderly (Garrett et al. 2001; Wang et al. 2000). Vitamin D supplementation also increases bone mass, reduces the risk of fractures, and has anti-inflammatory properties (Dawson-Hughes et al. 1997; Timms et al. 2002). Thus, it has been hypothesized that statins may be analogs of vitamin D (Grimes 2006). Although intriguing and potentially biologically plausible, randomized placebo-controlled trials are needed to confirm the influence statin medications may have on vitamin D.

15.2.4 INFLAMMATION

There is increasing evidence that vitamin D acts as a modulator of the immune system. One of the first indications for this role was the finding of vitamin D receptor expression in a wide range of immune cells (e.g., monocytes, activated lymphocytes; Veldman et al. 2000). *In vitro* data suggest that active vitamin D suppresses proinflammatory cytokine expression and regulates immune cell activity (van Etten and Mathieu 2005). *In vivo* data also suggest that vitamin D supplementation may reduce proinflammatory cytokines. In a small clinical trial, 40 postmenopausal women with osteoporosis received active vitamin D (0.5 μg/day) and calcium (1000 mg/day), whereas 30 received calcium only (1000 mg/day; Inanir et al. 2004). After 6 months, vitamin D treatment was associated with reduced interleukin-1 and tumor necrosis factor-α concentrations compared with no changes in the women who received calcium only (Inanir et al. 2004). Similarly, in another small trial of patients with congestive heart failure, 50 μg/day of vitamin D plus 500 mg/day of calcium compared with calcium only (500 mg/day) resulted in significant decreases in tumor necrosis factor-α and increases in anti-inflammatory interleukin-10 concentrations after 9 months (Schleithoff et al. 2006). However, in another small study of 47 presumably healthy postmenopausal women, 12 weeks of vitamin D (800 IU) and calcium (1000 mg) had no influence on circulating interleukin-6, tumor necrosis factor-α, or C-reactive protein levels (Gannage-Yared et al. 2003).

Observational studies have provided little evidence for the role of vitamin D in immune system functioning. In a study of 1381 participants from the Framingham Offspring Study, no association was observed between 25(OH)D concentrations and a panel of more than 10 circulating proinflammatory biomarkers, including C-reactive protein, tumor necrosis factor receptor-2, and fibrinogen (Shea et al. 2008). In another large cross-sectional study of 6538 participants from the 1958 British Birth Cohort, 25(OH)D levels were inversely associated with several inflammatory and hemostatic biomarkers, including C-reactive protein (Hypponen et al. 2010). However, these associations were largely explained by differences in adiposity. Among the Old Order Amish, a founder population in which confounding influences such as geography and lifestyle would be expected to be minimized, no association was observed between 25(OH)D levels and C-reactive protein concentrations (Michos et al. 2009). There was also no association between 25(OH)D and C-reactive protein levels in the Third National Health and Nutrition Examination Survey (Melamed et al. 2008a).

In summary, much of the evidence supporting a role for vitamin D in immune system functioning has come from *in vitro* studies of the active vitamin D metabolite. Although *in vivo* studies have been mostly unable to document an association between 25(OH)D concentrations and biomarkers of inflammation, the reported effects of vitamin D may not have been detected because active vitamin D was not measured. Additional research is needed to further elucidate whether vitamin D may modify CVD risk through any effect on the immune system.

15.2.5 RENAL DISEASE

The kidney is the primary organ responsible for the conversion of 25(OH)D to active vitamin D. To maintain circulating calcium levels within a tight physiologic range, parathyroid hormone levels rise, increasing the risk of secondary hyperparathyroidism. Patients with chronic kidney disease generally have more severe vitamin D deficiencies than the general population (Gonzalez et al. 2004). Vitamin D deficiency and insufficiency are also more common among predialysis patients (Jean et al. 2008). Predialysis chronic kidney disease patients who live in countries exposed to greater amounts of sunlight are less likely to be vitamin D deficient or insufficient than those who live in less sunny countries (Cuppari et al. 2008). Among 825 dialysis patients in the United States, only 22% had 25(OH)D levels higher than 30 ng/mL, whereas 60% had levels between 10 and 30 ng/mL (Wolf et al. 2007). In a study of dialysis patients in France, 42% had severe vitamin D deficiency (Jean et al. 2008). Thus, in patients with end-stage renal disease, vitamin D deficiency is an even more prevalent and severe problem than among those with mild-to-moderate kidney disease.

In patients with chronic kidney disease, numerous epidemiologic studies have shown that vitamin D levels are inversely associated with the rate of renal function loss and mortality. In a study of 140 patients in France with stage II to V chronic kidney disease, 25(OH)D levels were significantly associated with all-cause mortality independent of several potential confounding factors (Barreto et al. 2009). In a study of 168 patients with stage I or II chronic kidney disease, 25(OH)D levels were significantly associated with estimated glomerular filtration rate, and independently predictive of end-stage renal disease and all-cause mortality (Ravani et al. 2009). In the general population, vitamin D deficiency has also been associated with an increased risk of albuminuria, a risk factor for chronic kidney disease progression, end-stage renal disease, cardiovascular events, and mortality (de Boer et al. 2007).

Compared with the absence of any form of vitamin D therapy, treatment of chronic kidney patients with active vitamin D has been associated with decreased mortality in observational studies, with the newer vitamin D analogs seeming to have a greater effect than the native hormone (Teng et al. 2003, 2005; Tentori et al. 2006). The active vitamin D analogs reportedly decrease parathyroid hormone levels while moderating the increase in calcium and phosphate experienced with the naturally occurring hormone treatment. Calcium and phosphate are important intermediate factors in the development of vascular calcification in chronic kidney disease. Use of any form of vitamin D treatment in renal disease has become controversial, however, since a recent meta-analysis of 76 trials testing the effects of vitamin D on mineral metabolism and cardiovascular and mortality outcomes, including 3667 patients with chronic kidney disease, suggested that treatment did not consistently reduce parathyroid hormone levels (Palmer et al. 2007). Furthermore, when compared with placebo, active vitamin D was associated with hypercalcemia and hyperphosphatemia, whereas the vitamin D analogs were also associated with hypercalcemia, but not increased phosphate levels. Vitamin D treatment did not show a significant reduction in risk of death, bone pain, vascular calcification, or parathyroidectomy. The researchers concluded that vitamin D treatment for chronic kidney disease remains uncertain. Nevertheless, large randomized placebo controlled trials testing the effects of vitamin D in patients with chronic kidney disease with hard clinical end points have not yet been conducted. Until the results of these studies are available, vitamin D treatment will likely continue to be an important adjuvant therapy to lessen the effects of secondary hyperparathyroidism in chronic kidney disease.

15.3 SUBCLINICAL CVD

15.3.1 VASCULAR CALCIFICATION

Vascular calcification, a strong risk factor for CVD morbidity and mortality, accumulates when there is a net calcium efflux from bone. The inverse association frequently observed between vascular and skeletal calcium may help explain why vascular calcification is often associated with osteoporosis. Given the essential role of vitamin D in bone formation, it can be hypothesized that vitamin D may play a role in the development of vascular calcification. Vitamin D receptors have been identified in almost all tissues, including vascular smooth muscle cells, cardiomyocytes, and endothelial cells (Towler and Clemens 2005). Animal models given supraphysiologic doses of vitamin D develop vascular calcification (Fleckenstein-Grun et al. 1995), although vitamin D toxicity is rare in humans. Laboratory studies suggest that vitamin D deficiency may also lead to biological processes that facilitate vascular calcification, suggesting a biphasic dose–response curve whereby both low and excessive vitamin D levels may be harmful (Zittermann et al. 2007).

In one of the first clinical studies of vitamin D and subclinical atherosclerosis, Watson et al. (1997) found serum $1,25(OH)_2D$ levels were inversely correlated with the extent of coronary artery calcification in a sample of 153 patients at moderately high risk for coronary heart disease who underwent electron beam computed tomography scanning. These findings were confirmed by Doherty et al. (1997) who also found that $1,25(OH)_2D$ levels were inversely correlated with calcium mass in a sample of 283 white and African American adults. Levels of $1,25(OH)_2D$ did not, however, explain racial differences in coronary artery calcium mass between African Americans and whites. In another study, Arad et al. (1998) found no correlation between serum $1,25(OH)_2D$ levels and coronary calcification in 50 patients undergoing coronary angiography.

Among the Old Order Amish population described earlier, no association was observed between 25(OH)D levels and the prevalence or severity of coronary artery calcification (Michos et al. 2009). In a subset of participants in the Multi-Ethnic Study of Atherosclerosis who were free of clinical CVD at baseline, de Boer et al. (2009) found no evidence of a cross-sectional association of 25(OH)D levels with prevalent coronary calcification after adjustment for CVD risk factors, including body mass index and kidney function. However, there was a significant inverse association of 25(OH)D with risk of developing incident coronary calcification during a 3-year follow-up, an association that seemed strongest among participants with reduced kidney function.

In a prospective study of 374 non-Hispanic white subjects with type 1 diabetes, a cross-sectional association between vitamin D deficiency and coronary artery calcification was observed (Young et al. 2011). In addition, among subjects who were initially free of coronary calcification, lower vitamin D levels were associated with an increased likelihood of developing incident calcification over the next 3 years among those with the vitamin D receptor M1T CC genotype (Young et al. 2011). In another study, a common single-nucleotide polymorphism in the CYP24A1 gene was associated with coronary artery calcification in three independent populations, suggesting a role for vitamin D metabolism in the development of vascular calcification (Shen et al. 2010).

Most studies of vitamin D and vascular calcification have been performed in samples comprised primarily of white participants. However, in a study of 340 middle-aged African American adults with type 2 diabetes, Freedman et al. (2010) observed a *positive* cross-sectional association between 25(OH)D levels and carotid artery and aortic calcification that was independent of age, gender, body mass index, kidney function, and glycosylated hemoglobin concentration. No association was observed between $1,25(OH)_2D$ levels and coronary, carotid, or aortic calcification, or thoracic or lumbar bone mineral density. The positive association observed between 25(OH)D and vascular calcification in this sample of African Americans is in contrast with many of the previously described findings of an inverse association. These results suggest that there may be biologically mediated racial differences in the regulation of bone and vascular calcification. Interestingly, relative to whites, African Americans have less vascular calcification despite a greater CVD risk factor burden, lower rates of osteoporosis in the face of a lower calcium intake, form fewer calcium-containing kidney stones, and manifest skeletal resistance to the effects of parathyroid hormone (Acheson 2005; Luckey et al. 1996; Stamatelou et al. 2003).

Thus, accumulating research suggests that vitamin D may play a role in the regulation of vascular calcification; however, additional studies are necessary to confirm these findings. Further work is also required to determine whether race or ethnicity modifies these effects.

15.3.2 Intima–Media Thickness

Intima–media thickness is a measurement of the thickness of artery walls, typically performed using external high-resolution ultrasound. The intima–media thickness of the carotid arteries (cIMT) has been the most commonly measured site in cardiovascular research and is widely recognized as a surrogate measure of atherosclerosis. The measurement is made at the carotid bifurcation, where the common carotid artery divides into the internal and external carotid arteries. Increased cIMT values indicate a higher likelihood of having stroke or myocardial infarction. In epidemiologic studies, higher cIMT values in asymptomatic individuals also predict future cardiovascular events.

In a sample of 390 consecutively enrolled patients with type 2 diabetes, those with low 25(OH)D levels had a marked increase in common cIMT when compared with their vitamin D–sufficient counterparts (Targher et al. 2006). In the Rancho Bernardo Study of older adults, Reis et al. (2007) found that 25(OH)D levels were inversely associated with internal cIMT, but not with common cIMT. The authors hypothesized that internal cIMT may be a better marker of atherosclerosis whereas common cIMT may reflect more vascular changes from sheer stress. However, among the previously described population of Old Order Amish, no evidence of a cross-sectional association of 25(OH) D levels with cIMT was observed (Michos et al. 2009). In a population-based sample of 542 adults from Belgium, Richart et al. (2011) observed a significant positive association between the ratio of parathyroid hormone to 25(OH)D levels and common cIMT. Thus, preliminary data suggest that vitamin D may be inversely associated with cIMT; however, this evidence has been generated primarily from cross-sectional investigations.

15.3.3 PERIPHERAL ARTERIAL DISEASE

Peripheral arterial disease is a manifestation of atherosclerosis resulting in the obstruction of blood flow to the lower extremities, significant morbidity, and an elevated risk of all-cause and cardiovascular mortality (Golomb et al. 2006; Ouriel 2001). Fahrleitner et al. (2002) reported that patients with advanced peripheral arterial disease had significantly lower vitamin D levels than patients with less severe disease. Patients with symptomatic peripheral arterial disease typically experience muscle weakness and pain in the lower extremities during exertion, leading to restricted mobility and potentially limited outdoor exposure to sunlight. This reverse causation bias may help explain why patients with more severe disease have lower vitamin D levels than those with less severe disease. However, low vitamin D levels have also been associated with asymptomatic peripheral arterial disease, suggesting that vitamin D deficiency may also precede the development of clinical disease. As displayed in Figure 15.3, among ambulatory participants of the National Health and Nutrition Examination Survey 2001–2004, low serum 25(OH)D levels were associated with an increased risk for prevalent peripheral arterial disease determined via ankle–brachial blood pressure measurement after demographics, known cardiovascular risk factors, and physical activity were taken into account, although there was no association with calcium, phosphate, and parathyroid hormone

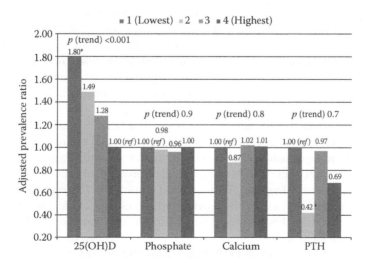

FIGURE 15.3 Adjusted prevalence ratios for peripheral arterial disease according to quartiles of serum 25(OH)D, phosphate, calcium, and parathyroid hormone (PTH). (From National Health and Nutrition Examination Survey 2001–2004.) Ratios were adjusted for age, sex, race/ethnicity, education, smoking, physical activity, diabetes, total to HDL–cholesterol ratio, body mass index, systolic blood pressure, glycosylated hemoglobin, statin use, antihypertensive medication use, vitamin D supplement use, history of myocardial infarction, log homocysteine, elevated C-reactive protein, and chronic kidney disease (*, $p < .001$). *Ref* indicates referent group.

(Melamed et al. 2008b). Hence, preliminary data suggest that vitamin D may be involved in the development of peripheral arterial disease; however, prospective studies are warranted.

15.3.4 LEFT VENTRICULAR HYPERTROPHY

Left ventricular hypertrophy is a pathological feature of CVD involving the thickening of the myocardium of the left ventricle of the heart, typically in response to high blood pressure. To compensate for an increased systemic pressure, the left ventricle must work harder to pump the same amount of blood. As the workload increases, the walls of the chamber grow thicker, lose elasticity, and eventually may fail to pump with as much force as a healthy heart.

Most of the scientific evidence linking vitamin D with myocardial hypertrophy have come from small clinical trials of 1,25(OH)$_2$D infusion in patients with chronic kidney disease. In two small studies of hemodialysis patients with secondary hyperparathyroidism, intravenous 1,25(OH)$_2$D for 15 weeks resulted in significant reductions in intraventricular wall thickness, left ventricular posterior wall thickness, and left ventricle mass index compared with control patients with secondary hyperparathyroidism who did not receive vitamin D (Park et al. 1999; Kim et al. 2006). Thus, preliminary evidence suggests that vitamin D may favorably influence myocardial hypertrophy; however, studies in presumably healthy population–based samples are necessary to confirm these findings.

15.4 CVD EVENTS

15.4.1 MYOCARDIAL INFARCTION

Major risk factors for myocardial infarction or heart attack include abnormal lipid levels, diabetes, hypertension, and the presence of subclinical CVD. The association between vitamin D levels and myocardial infarction has been explored in data from the Health Professional's Follow-up Study, the Intermountain Healthcare System, the Third National Health and Nutrition Examination Survey, and numerous small case-control studies. In a nested case-control study of 18,225 men from the Health Professional's Follow-up Study, there was a dose-dependent relationship between vitamin D and risk of myocardial infarction (Giovannucci et al. 2008). As compared with men with serum vitamin D levels higher than 30 ng/mL, men with levels lower than 15 ng/mL were more than two times more likely to experience a myocardial infarction, whereas men with vitamin D levels of 15 to 30 ng/mL were 50% more likely to experience a myocardial infarction, after adjusting for matching characteristics and numerous CVD risk factors.

Using medical record data from the Intermountain Healthcare System, investigators identified 41,504 individuals with at least one vitamin D measurement, and evaluated the relation between vitamin D and incident myocardial infarction (Anderson et al. 2010). Participants with vitamin D levels lower than 15 ng/mL were at 45% greater risk of myocardial infarction than their counterparts with levels higher than 30 ng/mL. A limitation of this study is that only participants for whom serum vitamin D levels were drawn at the providers' discretion for clinical indications (e.g., osteoporosis risk) were included in this analysis. Thus, the results may have been influenced by selection bias or other unknown factors.

Data from the third National Health and Nutrition Examination Survey showed that the prevalence of coronary heart disease was higher among those with vitamin D levels lower than 20 ng/mL, as compared with those with values 30 ng/mL or higher (Kim et al. 2008). This association did, however, become nonsignificant after adjustment for traditional CVD risk factors (e.g., hypertension, diabetes, and chronic kidney disease), suggesting that the relation of vitamin D to coronary heart disease risk may be mediated by traditional CVD risk factors (Kim et al. 2008). Small case-control studies from New Zealand, Denmark, and Norway have also reported an inverse association between vitamin D levels and myocardial infarction (Lund et al. 1978; Scragg et al. 1990; Vik et al. 1979).

Thus, preliminary data suggest that vitamin D may be inversely associated with the development of myocardial infarction.

15.4.2 Cerebrovascular Disease

Cerebrovascular disease, also known as stroke, is a leading cause of death and disability (Anderson et al. 2004; World Health Organization 2008). Risk factors for stroke are generally similar to those for myocardial infarction, although hypertension is believed to play a more important etiologic role in stroke risk. Evidence relating vitamin D levels to cerebrovascular disease is relatively sparse. Low vitamin D levels were found to predict fatal stroke in the Mini-Finland Health Survey, which included 6219 individuals who were followed for more than 27 years (Kilkkinen et al. 2009). The hazard ratio for stroke among those in the highest quintile of vitamin D levels versus the lowest was 0.48 (95% CI, 0.31–0.75). Similarly, there was an inverse association between vitamin D levels and fatal stroke in the Ludwigshafen Risk and Cardiovascular Health study, which included 3316 participants who were referred for coronary angiography between 1997 and 2000 (Pilz et al. 2008a). The association remained significant even after accounting for CVD risk factors as well as calcium and parathyroid hormone levels. Likewise, members of the Intermountain Healthcare System with vitamin D levels lower than 15 ng/mL were at 78% greater risk of stroke than members with levels higher than 30 ng/mL (Anderson et al. 2010). Two small case-control studies of patients who experienced acute strokes also reported a relation with serum vitamin D (Marniemi et al. 2005; Poole et al. 2006).

In sum, although sparse, the observational evidence shows an inverse relation between vitamin D levels and stroke risk. Additional studies are needed which stratify by stroke subtype, as it is presently unclear whether vitamin D is related to both ischemic and hemorrhagic stroke.

15.4.3 Congestive Heart Failure

In the general population, congestive heart failure frequently occurs as a consequence of hypertension, coronary, and valvular CVD. In addition to acting through hypertension and other traditional CVD risk factors, low levels of serum vitamin D may also affect congestive heart failure risk by directly influencing cardiac morphology. Low vitamin D, and the accompanying elevated levels of parathyroid hormone, are known to increase cardiac contractility, lead to cardiomyocyte hypertrophy, and interstitial fibrosis of the heart (Rostand and Drueke 1999; Weishaar et al. 1990).

The Ludwigshafen Risk and Cardiovascular Health study evaluated the prospective relation between low vitamin D levels and incidence of coronary heart failure mortality. In this population of 3299 Germans referred for coronary angiography, the hazard ratio (95% CI) for heart failure mortality was 2.84 (1.20–6.74) when comparing patients with low vitamin D ($\approx$10 ng/mL) to those in the optimal range ($\approx$25 ng/mL; Pilz et al. 2008b). Vitamin D levels were also linked to a greater incidence of heart failure in the Intermountain Healthcare System (Anderson et al. 2010). Furthermore, the prevalence of congestive heart failure was higher among participants of the National Health and Nutrition Examination Survey 2001–2004 with vitamin D levels lower than 20 ng/mL as compared with those with levels of 30 ng/mL or higher (Kim et al. 2008). Low vitamin D levels have also been linked to congestive heart failure in a number of small clinical studies (Arroyo et al. 2006; Laguardia et al. 2006; Shane et al. 1997; Zittermann et al. 2003). Additionally, among patients with end-stage heart failure, lower circulating levels of $1,25(OH)_2D$ were related to poorer patient outcomes at 1 year (Zittermann et al. 2008).

Like the other CVD phenotypes discussed previously, the majority of evidence suggests a relation between low vitamin D levels and heart failure risk. However, as mentioned previously with peripheral arterial disease, one must be wary of reverse causality bias when considering a relation between vitamin D and heart failure. The symptoms of heart failure may lead to participation in

fewer outdoor activities compared with healthy adults, which may in turn lead to less sunlight exposure and a greater propensity for suboptimal vitamin D levels. Prospective studies are needed to
measure vitamin D levels among individuals without clinical evidence of heart failure at baseline
and follow them over time to determine whether low vitamin D predicts the future occurrence of
incident congestive heart failure.

15.4.4 Composite End Points and Mortality

The relation of vitamin D to combined CVD outcomes was explored in a nested case-control sample
of 1739 Framingham Offspring Study participants (Wang et al. 2008b). There was a graded increase
in CVD risk across categories of 25(OH)D. Relative to their counterparts with vitamin D levels
15 ng/mL or higher, those with vitamin D levels of 10–15 ng/mL were at 53% greater risk, whereas
those with vitamin D levels lower than 10 ng/mL were at 80% greater risk of a CVD event, after
multivariate adjustment. Vitamin D was also inversely related to the incidence of combined CVD
outcomes in data from the Intermountain Healthcare System (Anderson et al. 2010).

Numerous studies have evaluated the relation between vitamin D and mortality. As shown in
Figure 15.4, lower baseline vitamin D levels among members of the Intermountain Healthcare
System were significantly associated with lower survival (increased mortality) during the follow-
up period (Anderson et al. 2010). Low vitamin D levels were also associated with a greater risk
of CVD mortality in the population-based Mini-Finland Health Survey (Kilkkinen et al. 2009),
Kuopio Ischaemic Heart Disease Risk Factor Study (Finland; Virtanen et al. 2011), Hoorn Study
(Netherlands; Pilz et al. 2009), InCHIANTI Study (Italy; Semba et al. 2010), and the third National
Health and Nutrition Examination Survey (United States; Fiscella and Franks 2010; Ginde et al.
2009b). Vitamin D levels were also inversely related to CVD mortality in patients undergoing dialysis (Wang et al. 2008a) and those scheduled for coronary angiography (Dobnig et al. 2008). Thus,
the preponderance of evidence suggests that lower vitamin D levels are associated with greater
CVD mortality.

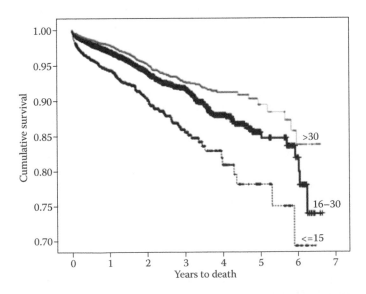

FIGURE 15.4 Kaplan–Meier survival curves according to baseline serum vitamin D levels ≤ 15, 16–30, and
>30 ng/mL. Survival differed significantly by initial vitamin D level (log rank $p < .0001$).

15.5 VITAMIN D SUPPLEMENTATION AND CARDIOVASCULAR OUTCOMES

It is tempting to assume that suboptimal vitamin D levels are causally related to CVD risk, given the biological plausibility and the observational epidemiologic evidence. History, however, instructs us to use caution when making inferences regarding causality because several micronutrients have seemed to be associated with lower CVD risk in observational epidemiologic studies, but have failed to show an effect when tested in supplementation trials. The most recent example of this phenomenon involves the B vitamins, which include folate, vitamin B6, and vitamin B12. Both experimental and observational studies have shown inverse relations between vitamin B intake and homocysteine concentrations (Homocysteine Lowering Trialists' Collaboration 1998), and homocysteine has been related to CVD risk in numerous observational studies (Splaver et al. 2004). In the clinical trial setting, B vitamin supplementation successfully reduced homocysteine concentrations; however, there was no evidence of reductions in the risk of CVD (Bonaa et al. 2006; Lonn et al. 2006; Toole et al. 2004).

Experimental evidence relating vitamin D supplementation to CVD events is extremely sparse. To date, the Women's Health Initiative is the only large randomized clinical trial that has explicitly evaluated the relation between vitamin D supplementation and CVD outcomes. In the Women's Health Initiative, 36,282 postmenopausal women aged 50–79 years were randomized to 400 IU of vitamin D and 1000 mg of calcium carbonate daily, and were followed for an average of 7 years. For all CVD outcomes (i.e., myocardial infarction or coronary heart disease, death, and stroke) the findings were null (Hsia et al. 2007). Although these results were disappointing, it is widely believed that the vitamin D supplement dose was insufficient to alter serum vitamin D levels enough to see a difference in CVD rates between the treatment groups (Hsia et al. 2007; Michos and Blumenthal 2007). As of the 2010 Institute of Medicine report, the recommended daily allowance of vitamin D for adults younger than 70 years is 600 IU/day, and for those older than 70 years it is 800 IU/day (Committee to Review Dietary Reference Intakes for Vitamin D and Calcium 2010). In instances of vitamin D deficiency, individuals are typically prescribed 50,000 IU/week for 8 weeks to achieve repletion, and 2000 IU per day is then used as a maintenance dose. Every 100 IU of vitamin D ingested daily increases the serum 25(OH)D levels by approximately 1 ng/mL (Heaney et al. 2003; Lee et al. 2008; Vieth 1999). Thus, despite the large sample size and long follow-up period of the Women's Health Initiative, the low vitamin D supplement dose prohibits drawing strong conclusions from this study.

In a recent meta-analysis, investigators were interested in determining the effects of supplementation on the risk of death by combining the results of several vitamin D supplementation trials on any health condition (Autier and Gandini 2007). A total of 18 independent randomized clinical trials were included. These trials included 57,311 participants and 4777 deaths. Doses of vitamin D in the trials varied from 300 to 2000 IU daily. The meta-analysis summary relative risk for mortality from any cause was 0.93 (95% CI, 0.87–0.99) for those randomized to supplements as compared with placebo. The lower risk of mortality among supplement users is intriguing, despite being unclear which conditions contributed to the lower mortality experienced by the vitamin D supplement users.

To establish whether vitamin D is causally related to CVD risk, large-scale randomized clinical trials are needed. The Vitamin D and Omega-3 Trial (VITAL) was designed to address this research gap. VITAL began recruitment in 2010 and will follow participants for 5 years for CVD and cancer outcomes (Brigham and Women's Hospital Public Affairs). It is a 2×2 factorial, double-blind, placebo-controlled clinical trial that will randomize 20,000 participants to 2000 IU/day of vitamin D and fish oil supplements (~1 g/day ω-3). It is hoped that VITAL will provide additional clues about the effects of vitamin D supplementation on the risk of CVD.

15.6 VITAMIN D AND DISPARITIES IN CVD

CVD rates vary greatly by country and race/ethnicity. In the United States, data have consistently shown that coronary heart disease rates are highest among African Americans, followed by

Caucasians, Hispanics, and Asians, respectively (Mensah et al. 2005). Racial/ethnic differences in CVD rates may result from a myriad of factors, including differences in behaviors, genetics, exposure to traditional CVD risk factors, and access to health care. It has also been suggested that differential exposure to low vitamin D may explain some of the racial/ethnic differences in CVD outcomes.

Vitamin D levels vary substantially by race/ethnicity. In the National Health and Nutrition Examination Survey 2001–2004, mean serum vitamin D levels were significantly higher ($p < .001$) in whites (26 ng/mL) than in Mexican Americans (22 ng/mL) and African Americans (16 ng/mL; Ginde et al. 2009a). Figure 15.5 shows that approximately 3% of whites, 28% of African Americans, and 8% of Mexican Americans had deficient vitamin D levels (<10 ng/mL), whereas only 30% of whites, 3% of African Americans, and 11% of Mexican Americans had optimal levels (≥30 ng/mL; Ginde et al. 2009a).

There are several possible explanations for the lower serum vitamin D levels among non-whites, and particularly African Americans. African Americans have a greater cutaneous melanin content, which inhibits the initial conversion of 7-dihydrocholesterol to previtamin D_3 in the skin (Clemens et al. 1982; Loomis 1967). As a result, African Americans require proportionally more sun exposure to synthesize equivalent amounts of vitamin D compared with individuals with lighter skin coloration (Armas et al. 2007; Clemens et al. 1982). This is the most salient explanation for the strikingly low 25(OH)D levels observed in African Americans. African Americans also have lower than usual vitamin D intake from supplements and food, which may also contribute to their lower vitamin D levels (Calvo et al. 2004; Moore et al. 2005). Additionally, racial differences in vitamin D metabolism may provide some explanation (Awumey et al. 1998; Cosman et al. 1997; Dawson-Hughes et al. 1995; Reasner et al. 1990).

Using data from the third National Health and Nutrition Examination Survey, investigators recently evaluated whether differences in serum vitamin D levels explained racial/ethnic differences in CVD risk (Fiscella and Franks 2010). As expected, the age- and sex-adjusted cardiovascular mortality was higher among African Americans than whites. However, once the authors adjusted for 25(OH)D levels, the difference in CVD risk between African Americans and whites was no longer statistically significant (Fiscella and Franks 2010). These findings suggest that black/white differences in 25(OH)D levels may contribute to the excess CVD mortality among blacks.

Similar analyses have explored whether 25(OH)D levels explain racial/ethnic differences in peripheral arterial disease and end-stage renal disease. After controlling for established and novel

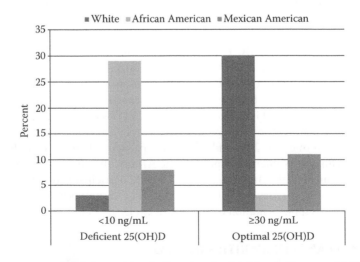

FIGURE 15.5 Prevalence of deficient (<10 ng/mL) and optimal (≥30 ng/mL) serum 25(OH)D levels among white, African American, and Mexican American adults; National Health and Nutrition Examination Survey 2001–2004.

CVD risk factors, racial differences in vitamin D concentrations explained nearly one-third of the excess risk of peripheral arterial disease in blacks as compared with whites (Reis et al. 2008b). Recent research has also estimated that up to 60% of the excess risk of end-stage renal disease in blacks was conferred by vitamin D deficiency (Melamed et al. 2009).

15.7 VITAMIN D INTOXICATION AND CVD

Vitamin D intoxication (which usually occurs when serum levels are >70 ng/mL) is associated with hypercalcemia, vascular calcification, and nephrocalcinosis. A number of studies have confirmed that vitamin D analogs promote vascular calcification (Mathew et al. 2008). A higher risk or a trend for higher risk for cardiovascular events has been observed among those in the treatment arms of recent randomized clinical trials of calcium supplementation (without vitamin D) in the maintenance of bone mineral density among older adults (Bolland et al. 2008, 2010; Reid et al. 2008). Although vitamin D intoxication is a rare occurrence in humans, animal models given supraphysiologic vitamin D doses have developed vascular calcification (Fleckenstein-Grun et al. 1995). Zittermann et al. (2007) proposed a biphasic dose–response curve for vitamin D in vascular calcification, wherein both lower and higher levels may be stimulatory. Observational studies have not demonstrated a dose-dependent survival benefit of higher dosages over lower dosages (Teng et al. 2003, 2005; Tentori et al. 2006). Calcium-containing phosphate binders and vitamin D are involved in calcification pathogenesis. However, no association between the vitamin D dose administered and the progression of cardiac vascular calcification was observed when a normal calcium × phosphate product was maintained in patients undergoing chronic hemodialysis (Morosetti et al. 2008). Mathew et al. (2008) also found that clinically relevant doses of vitamin D agonists, calcitriol and paricalcitol, were protective against vascular calcification in a mouse model of chronic kidney disease (CKD)-stimulated atherosclerosis, whereas higher dosages of both agonists were stimulatory. It is important to note that, in animal studies, the vitamin D doses used to produce vascular calcification are much greater than those used for human therapeutic purposes.

15.8 PUBLIC HEALTH AND CLINICAL IMPLICATIONS

CVD is the leading cause of death worldwide (World Health Organization 2008). It is therefore important to identify novel risk factors which affect a large proportion of the population and can be inexpensively intervened upon to limit CVD morbidity and mortality. If low vitamin D levels are, in fact, causally related to CVD risk, then vitamin D supplementation might provide a new and inexpensive means of reducing CVD morbidity and mortality.

From a public health perspective, low vitamin D is an ideal exposure upon which to intervene as a large proportion of the population has suboptimal levels (40%–60%; Ginde et al. 2009a; Looker et al. 2008; Martins et al. 2007); low levels are highly amenable to correction through supplementation or increased sun exposure, and treatment options are inexpensive and easy to implement. Both the number-needed-to-treat and the cost-to-prevent one CVD event would be low. Population-based strategies to elevate vitamin D levels could include increased fortification of foods, greater concentrations of vitamin D from supplements, and public health messages to encourage reasonable sun exposure and intake of oily fish. Clinically, if it was shown in trials that replacement was effective at reducing CVD risk in populations at high risk, then screening for, and treatment of, suboptimal vitamin D could be a primary prevention priority.

15.9 CONCLUSION

Low vitamin D levels may influence CVD risk by acting on numerous CVD risk factors (i.e., hypertension, diabetes, lipids, inflammation, and renal disease), and promoting subclinical CVD (i.e., vascular calcification, greater carotid intima–medial thickness, peripheral arterial disease, and left ventricular hypertrophy). Furthermore, epidemiologic evidence has related low vitamin D levels to risk of myocardial infarction, stroke, heart failure, and CVD mortality. Given the high prevalence of vitamin D deficiency and insufficiency among people of African ancestry, suboptimal vitamin D may also help explain racial disparities in CVD risk. The possibilities of supplementing vitamin D for CVD prevention and treatment are exciting; however, results from large clinical trials with hard clinical end points are needed to confirm that vitamin D is efficacious in preventing CVD.

REFERENCES

Acheson, L. S. 2005. Bone density and the risk of fractures: Should treatment thresholds vary by race? *JAMA* 293(17):2151–2154.

Anderson, C. S., K. N. Carter, W. J. Brownlee, M. L. Hackett, J. B. Broad, and R. Bonita. 2004. Very long-term outcome after stroke in Auckland, New Zealand. *Stroke* 35(8):1920–1924.

Anderson, J. L., H. T. May, B. D. Horne, T. L. Bair, N. L. Hall, J. F. Carlquist, D. L. Lappe, and J. B. Muhlestein. 2010. Relation of vitamin D deficiency to cardiovascular risk factors, disease status, and incident events in a general healthcare population. *American Journal of Cardiology* 106(7):963–968.

Arad, Y., L. A. Spadaro, M. Roth, J. Scordo, K. Goodman, S. Sherman, G. Lerner, D. Newstein, and A. D. Guerci. 1998. Serum concentration of calcium, 1,25 vitamin D and parathyroid hormone are not correlated with coronary calcifications. An electron beam computed tomography study. *Coronary Artery Disease* 9(8):513–518.

Armas, L. A., S. Dowell, M. Akhter, S. Duthuluru, C. Huerter, B. W. Hollis, R. Lund, and R. P. Heaney. 2007. Ultraviolet-B radiation increases serum 25-hydroxyvitamin D levels: The effect of UVB dose and skin color. *Journal of the American Academy of Dermatology* 57(4):588–593.

Arroyo, M., S. P. Laguardia, S. K. Bhattacharya, M. D. Nelson, P. L. Johnson, L. D. Carbone, K. P. Newman, and K. T. Weber. 2006. Micronutrients in African-Americans with decompensated and compensated heart failure. *Translational Research* 148(6):301–308.

Autier, P., and S. Gandini. 2007. Vitamin D supplementation and total mortality: A meta-analysis of randomized controlled trials. *Archives of Internal Medicine* 167(16):1730–1737.

Awumey, E. M., D. A. Mitra, B. W. Hollis, R. Kumar, and N. H. Bell. 1998. Vitamin D metabolism is altered in Asian Indians in the southern United States: A clinical research center study. *The Journal of Clinical Endocrinology and Metabolism* 83(1):169–173.

Barreto, D. V., F. C. Barreto, S. Liabeuf, M. Temmar, F. Boitte, G. Choukroun, A. Fournier, and Z. A. Massy. 2009. Vitamin D affects survival independently of vascular calcification in chronic kidney disease. *Clinical Journal of the American Society of Nephrology* 4(6):1128–1135.

Baynes, K. C., B. J. Boucher, E. J. Feskens, and D. Kromhout. 1997. Vitamin D, glucose tolerance and insulinaemia in elderly men. *Diabetologia* 40(3):344–347.

Bolland, M. J., P. A. Barber, R. N. Doughty, B. Mason, A. Horne, R. Ames, G. D. Gamble, A. Grey, and I. R. Reid. 2008. Vascular events in healthy older women receiving calcium supplementation: Randomized controlled trial. *BMJ* 336(7638):262–266.

Bolland, M. J., A. Avenell, J. A. Baron, A. Grey, G. S. MacLennan, G. D. Gamble, and I. R. Reid. 2010. Effect of calcium supplements on risk of myocardial infarction and cardiovascular events: Meta-analysis. *BMJ* 341:c3691.

Bonaa, K. H., I. Njolstad, P. M. Ueland, H. Schirmer, A. Tverdal, T. Steigen, H. Wang, J. E. Nordrehaug, E. Arnesen, and K. Rasmussen. 2006. Homocysteine lowering and cardiovascular events after acute myocardial infarction. *New England Journal of Medicine* 354(15):1578–1588.

Brigham and Women's Hospital Public Affairs. Largest study of vitamin D and omega-3s set to begin soon at Brigham and Women's Hospital.

Calvo, M. S., S. J. Whiting, and C. N. Barton. 2004. Vitamin D fortification in the United States and Canada: Current status and data needs. *The American Journal of Clinical Nutrition* 80(6 Suppl):1710S–1716S.

Chiu, K. C., A. Chu, V. L. Go, and M. F. Saad. 2004. Hypovitaminosis D is associated with insulin resistance and beta cell dysfunction. *The American Journal of Clinical Nutrition* 79(5):820–825.

Christau, B., H. Kromann, O. O. Andersen, M. Christy, K. Buschard, K. Arnung, I. H. Kristensen, B. Peitersen, J. Steinrud, and J. Nerup. 1977. Incidence, seasonal and geographical patterns of juvenile-onset insulin-dependent diabetes mellitus in Denmark. *Diabetologia* 13(4):281–284.

Clemens, T. L., J. S. Adams, S. L. Henderson, and M. F. Holick. 1982. Increased skin pigment reduces the capacity of skin to synthesise vitamin D3. *Lancet* 1(8263):74–76.

Committee to Review Dietary Reference Intakes for Vitamin D and Calcium. 2010. Dietary reference intakes for calcium and vitamin D: Institute of Medicine.

Cosman, F., D. C. Morgan, J. W. Nieves, V. Shen, M. M. Luckey, D. W. Dempster, R. Lindsay, and M. Parisien. 1997. Resistance to bone resorbing effects of PTH in black women. *Journal of Bone and Mineral Research* 12(6):958–966.

Cuppari, L., A. B. Carvalho, and S. A. Draibe. 2008. Vitamin D status of chronic kidney disease patients living in a sunny country. *Journal of Renal Nutrition* 18(5):408–414.

Dawson-Hughes, B., S. S. Harris, S. Finneran, and H. M. Rasmussen. 1995. Calcium absorption responses to calcitriol in black and white premenopausal women. *The Journal of Clinical Endocrinology and Metabolism* 80(10):3068–3072.

Dawson-Hughes, B., S. S. Harris, E. A. Krall, and G. E. Dallal. 1997. Effect of calcium and vitamin D supplementation on bone density in men and women 65 years of age or older. *New England Journal of Medicine* 337(10):670–676.

de Boer, I. H., G. N. Ioannou, B. Kestenbaum, J. D. Brunzell, and N. S. Weiss. 2007. 25-Hydroxyvitamin D levels and albuminuria in the Third National Health and Nutrition Examination Survey (NHANES III). *American Journal of Kidney Diseases* 50(1):69–77.

de Boer, I. H., B. Kestenbaum, A. B. Shoben, E. D. Michos, M. J. Sarnak, and D. S. Siscovick. 2009. 25-Hydroxyvitamin D levels inversely associate with risk for developing coronary artery calcification. *Journal of the American Society of Nephrology* 20(8):1805–1812.

Dobnig, H., S. Pilz, H. Scharnagl, W. Renner, U. Seelhorst, B. Wellnitz, J. Kinkeldei, B. O. Boehm, G. Weihrauch, and W. Maerz. 2008. Independent association of low serum 25-hydroxyvitamin D and 1,25-dihydroxyvitamin D levels with all-cause and cardiovascular mortality. *Archives of Internal Medicine* 168(12):1340–1349.

Doherty, T. M., W. Tang, S. Dascalos, K. E. Watson, L. L. Demer, R. M. Shavelle, and R. C. Detrano. 1997. Ethnic origin and serum levels of 1alpha,25-dihydroxyvitamin D3 are independent predictors of coronary calcium mass measured by electron-beam computed tomography. *Circulation* 96(5):1477–1481.

Fahrleitner, A., H. Dobnig, A. Obernosterer, E. Pilger, G. Leb, K. Weber, S. Kudlacek, and B. M. Obermayer-Pietsch. 2002. Vitamin D deficiency and secondary hyperparathyroidism are common complications in patients with peripheral arterial disease. *Journal of General Internal Medicine* 17(9):663–669.

Fiscella, K., and P. Franks. 2010. Vitamin D, race, and cardiovascular mortality: Findings from a national US sample. *Annals of Family Medicine* 8(1):11–18.

Fishbein, H. A., R. E. LaPorte, T. J. Orchard, A. L. Drash, L. H. Kuller, and D. K. Wagener. 1982. The Pittsburgh insulin-dependent diabetes mellitus registry: Seasonal incidence. *Diabetologia* 23(2):83–85.

Fleckenstein-Grun, G., F. Thimm, M. Frey, and S. Matyas. 1995. Progression and regression by verapamil of vitamin D3–induced calcific medial degeneration in coronary arteries of rats. *Journal of Cardiovascular Pharmacology* 26(2):207–213.

Ford, E. S., U. A. Ajani, L. C. McGuire, and S. Liu. 2005. Concentrations of serum vitamin D and the metabolic syndrome among U.S. adults. *Diabetes Care* 28(5):1228–1230.

Forman, J. P., E. Giovannucci, M. D. Holmes, H. A. Bischoff-Ferrari, S. S. Tworoger, W. C. Willett, and G. C. Curhan. 2007. Plasma 25-hydroxyvitamin D levels and risk of incident hypertension. *Hypertension* 49(5):1063–1069.

Forman, J. P., G. C. Curhan, and E. N. Taylor. 2008. Plasma 25-hydroxyvitamin D levels and risk of incident hypertension among young women. *Hypertension* 52(5):828–832.

Forouhi, N. G., J. Luan, A. Cooper, B. J. Boucher, and N. J. Wareham. 2008. Baseline serum 25-hydroxy vitamin D is predictive of future glycemic status and insulin resistance: The Medical Research Council Ely Prospective Study 1990–2000. *Diabetes* 57(10):2619–2625.

Freedman, B. I., L. E. Wagenknecht, K. G. Hairston, D. W. Bowden, J. J. Carr, R. C. Hightower, E. J. Gordon, J. Xu, C. D. Langefeld, and J. Divers. 2010. Vitamin D, adiposity, and calcified atherosclerotic plaque in African-Americans. *The Journal of Clinical Endocrinology and Metabolism* 95(3):1076–1083.

Gannage-Yared, M. H., M. Azoury, I. Mansour, R. Baddoura, G. Halaby, and R. Naaman. 2003. Effects of a short-term calcium and vitamin D treatment on serum cytokines, bone markers, insulin and lipid concentrations in healthy post-menopausal women. *Journal of Endocrinological Investigation* 26(8): 748–753.

Garrett, I. R., G. Gutierrez, and G. R. Mundy. 2001. Statins and bone formation. *Current Pharmaceutical Design* 7(8):715–736.

Ginde, A. A., M. C. Liu, and C. A. Camargo, Jr. 2009a. Demographic differences and trends of vitamin D insufficiency in the US population, 1988–2004. *Archives of Internal Medicine* 169(6):626–632.

Ginde, A. A., R. Scragg, R. S. Schwartz, and C. A. Camargo, Jr. 2009b. Prospective study of serum 25-hydroxy-vitamin D level, cardiovascular disease mortality, and all-cause mortality in older U.S. adults. *Journal of the American Geriatrics Society* 57(9):1595–1603.

Giovannucci, E., Y. Liu, B. W. Hollis, and E. B. Rimm. 2008. 25-Hydroxyvitamin D and risk of myocardial infarction in men: A prospective study. *Archives of Internal Medicine* 168(11):1174–1180.

Golomb, B. A., T. T. Dang, and M. H. Criqui. 2006. Peripheral arterial disease: Morbidity and mortality implications. *Circulation* 114(7):688–699.

Gonzalez, E. A., A. Sachdeva, D. A. Oliver, and K. J. Martin. 2004. Vitamin D insufficiency and deficiency in chronic kidney disease. A single center observational study. *American Journal of Nephrology* 24(5):503–510.

Grimes, D. S. 2006. Are statins analogues of vitamin D? *Lancet* 368(9529):83–86.

Heaney, R. P., K. M. Davies, T. C. Chen, M. F. Holick, and M. J. Barger-Lux. 2003. Human serum 25-hydroxy-cholecalciferol response to extended oral dosing with cholecalciferol. *The American Journal of Clinical Nutrition* 77(1):204–210.

Homocysteine Lowering Trialists' Collaboration. 1998. Lowering blood homocysteine with folic acid based supplements: Meta-analysis of randomised trials. *BMJ* 316(7135):894–898.

Hsia, J., G. Heiss, H. Ren, M. Allison, N. C. Dolan, P. Greenland, S. R. Heckbert, K. C. Johnson, J. E. Manson, S. Sidney, and M. Trevisan. 2007. Calcium/vitamin D supplementation and cardiovascular events. *Circulation* 115(7):846–854.

Hypponen, E., E. Laara, A. Reunanen, M. R. Jarvelin, and S. M. Virtanen. 2001. Intake of vitamin D and risk of type 1 diabetes: A birth-cohort study. *Lancet* 358(9292):1500–1503.

Hypponen, E., D. Berry, M. Cortina-Borja, and C. Power. 2010. 25-Hydroxyvitamin D and pre-clinical alterations in inflammatory and hemostatic markers: A cross-sectional analysis in the 1958 British Birth Cohort. *PLoS One* 5(5):e10801.

Inanir, A., K. Ozoran, H. Tutkak, and B. Mermerci. 2004. The effects of calcitriol therapy on serum interleukin-1, interleukin-6 and tumour necrosis factor-alpha concentrations in post-menopausal patients with osteoporosis. *The Journal of International Medical Research* 32(6):570–582.

Jean, G., B. Charra, and C. Chazot. 2008. Vitamin D deficiency and associated factors in hemodialysis patients. *Journal of Renal Nutrition* 18(5):395–399.

Jorde, R., Y. Figenschau, N. Emaus, M. Hutchinson, and G. Grimnes. 2010. Serum 25-hydroxyvitamin D levels are strongly related to systolic blood pressure but do not predict future hypertension. *Hypertension* 55(3):792–798.

Kilkkinen, A., P. Knekt, A. Aro, H. Rissanen, J. Marniemi, M. Heliovaara, O. Impivaara, and A. Reunanen. 2009. Vitamin D status and the risk of cardiovascular disease death. *American Journal of Epidemiology* 170(8):1032–1039.

Kim, H. W., C. W. Park, Y. S. Shin, Y. S. Kim, S. J. Shin, Y. S. Kim, E. J. Choi, Y. S. Chang, and B. K. Bang. 2006. Calcitriol regresses cardiac hypertrophy and QT dispersion in secondary hyperparathyroidism on hemodialysis. *Nephron. Clinical Practice* 102(1):c21–29.

Kim, D. H., S. Sabour, U. N. Sagar, S. Adams, and D. J. Whellan. 2008. Prevalence of hypovitaminosis D in cardiovascular diseases (from the National Health and Nutrition Examination Survey 2001 to 2004). *American Journal of Cardiology* 102(11):1540–1544.

Knekt, P., M. Laaksonen, C. Mattila, T. Harkanen, J. Marniemi, M. Heliovaara, H. Rissanen, J. Montonen, and A. Reunanen. 2008. Serum vitamin D and subsequent occurrence of type 2 diabetes. *Epidemiology* 19(5):666–671.

Laguardia, S. P., B. K. Dockery, S. K. Bhattacharya, M. D. Nelson, L. D. Carbone, and K. T. Weber. 2006. Secondary hyperparathyroidism and hypovitaminosis D in African-Americans with decompensated heart failure. *American Journal of the Medical Sciences* 332(3):112–118.

Lee, J. H., J. H. O'Keefe, D. Bell, D. D. Hensrud, and M. F. Holick. 2008. Vitamin D deficiency an important, common, and easily treatable cardiovascular risk factor? *Journal of the American College of Cardiology* 52(24):1949–1956.

Li, Y. C., J. Kong, M. Wei, Z. F. Chen, S. Q. Liu, and L. P. Cao. 2002. 1,25-Dihydroxyvitamin D(3) is a negative endocrine regulator of the renin–angiotensin system. *Journal of Clinical Investigation* 110(2):229–238.

Liu, S., Y. Song, E. S. Ford, J. E. Manson, J. E. Buring, and P. M. Ridker. 2005. Dietary calcium, vitamin D, and the prevalence of metabolic syndrome in middle-aged and older U.S. women. *Diabetes Care* 28(12):2926–2932.

Liu, E., J. B. Meigs, A. G. Pittas, C. D. Economos, N. M. McKeown, S. L. Booth, and P. F. Jacques. 2010. Predicted 25-hydroxyvitamin D score and incident type 2 diabetes in the Framingham Offspring Study. *The American Journal of Clinical Nutrition* 91(6):1627–1633.

Lonn, E., S. Yusuf, M. J. Arnold, P. Sheridan, J. Pogue, M. Micks, M. J. McQueen, J. Probstfield, G. Fodor, C. Held, and J. Genest, Jr. 2006. Homocysteine lowering with folic acid and B vitamins in vascular disease. *New England Journal of Medicine* 354(15):1567–1577.

Looker, A. C., C. M. Pfeiffer, D. A. Lacher, R. L. Schleicher, M. F. Picciano, and E. A. Yetley. 2008. Serum 25-hydroxyvitamin D status of the US population: 1988–1994 compared with 2000–2004. *The American Journal of Clinical Nutrition* 88(6):1519–1527.

Loomis, W. F. 1967. Skin-pigment regulation of vitamin-D biosynthesis in man. *Science* 157(788):501–506.

Luckey, M. M., S. Wallenstein, R. Lapinski, and D. E. Meier. 1996. A prospective study of bone loss in African-American and white women—a clinical research center study. *The Journal of Clinical Endocrinology and Metabolism* 81(8):2948–2956.

Lund, B., J. Badskjaer, B. Lund, and O. H. Soerensen. 1978. Vitamin D and ischaemic heart disease. *Hormone and Metabolic Research* 10(6):553–556.

Marniemi, J., E. Alanen, O. Impivaara, R. Seppanen, P. Hakala, T. Rajala, and T. Ronnemaa. 2005. Dietary and serum vitamins and minerals as predictors of myocardial infarction and stroke in elderly subjects. *Nutrition, Metabolism, and Cardiovascular Diseases* 15(3):188–197.

Martins, D., M. Wolf, D. Pan, A. Zadshir, N. Tareen, R. Thadhani, A. Felsenfeld, B. Levine, R. Mehrotra, and K. Norris. 2007. Prevalence of cardiovascular risk factors and the serum levels of 25-hydroxyvitamin D in the United States: Data from the Third National Health and Nutrition Examination Survey. *Archives of Internal Medicine* 167(11):1159–1165.

Mathew, S., R. J. Lund, L. R. Chaudhary, T. Geurs, and K. A. Hruska. 2008. Vitamin D receptor activators can protect against vascular calcification. *Journal of the American Society of Nephrology*.

Mathieu, C., C. Gysemans, A. Giulietti, and R. Bouillon. 2005. Vitamin D and diabetes. *Diabetologia* 48(7):1247–1257.

Melamed, M. L., E. D. Michos, W. Post, and B. Astor. 2008a. 25-hydroxyvitamin D levels and the risk of mortality in the general population. *Archives of Internal Medicine* 168(15):1629–1637.

Melamed, M. L., P. Muntner, E. D. Michos, J. Uribarri, C. Weber, J. Sharma, and P. Raggi. 2008b. Serum 25-hydroxyvitamin D levels and the prevalence of peripheral arterial disease: Results from NHANES 2001 to 2004. *Arteriosclerosis, Thrombosis, and Vascular Biology* 28(6):1179–85.

Melamed, M. L., B. Astor, E. D. Michos, T. H. Hostetter, N. R. Powe, and P. Muntner. 2009. 25-Hydroxyvitamin D levels, race, and the progression of kidney disease. *Journal of the American Society of Nephrology* 20(12):2631–2639.

Mensah, G. A., A. H. Mokdad, E. S. Ford, K. J. Greenlund, and J. B. Croft. 2005. State of disparities in cardiovascular health in the United States. *Circulation* 111(10):1233–1241.

Michos, E. D., and R. S. Blumenthal. 2007. Vitamin D supplementation and cardiovascular disease risk. *Circulation* 115(7):827–828.

Michos, E. D., E. A. Streeten, K. A. Ryan, E. Rampersaud, P. A. Peyser, L. F. Bielak, A. R. Shuldiner, B. D. Mitchell, and W. Post. 2009. Serum 25-hydroxyvitamin D levels are not associated with subclinical vascular disease or C-reactive protein in the Old Order Amish. *Calcified Tissue International* 84(3):195–202.

Mohr, S. B., C. F. Garland, E. D. Gorham, and F. C. Garland. 2008. The association between ultraviolet B irradiance, vitamin D status and incidence rates of type 1 diabetes in 51 regions worldwide. *Diabetologia* 51(8):1391–1398.

Moore, C. E., M. M. Murphy, and M. F. Holick. 2005. Vitamin D intakes by children and adults in the United States differ among ethnic groups. *Journal of Nutrition* 135(10):2478–85.

Morosetti, M., L. Jankovic, G. Palombo, S. Cipriani, S. Dominijanni, A. Balducci, G. Splendiani, G. Simonetti, A. Romagnoli, and G. Coen. 2008. High-dose calcitriol therapy and progression of cardiac vascular calcifications. *Journal of Nephrology* 21(4):603–608.

Ouriel, K. 2001. Peripheral arterial disease. *Lancet* 358(9289):1257–1264.

Palmer, S. C., D. O. McGregor, P. Macaskill, J. C. Craig, G. J. Elder, and G. F. Strippoli. 2007. Meta-analysis: Vitamin D compounds in chronic kidney disease. *Annals of Internal Medicine* 147(12):840–853.

Park, C. W., Y. S. Oh, Y. S. Shin, C. M. Kim, Y. S. Kim, S. Y. Kim, E. J. Choi, Y. S. Chang, and B. K. Bang. 1999. Intravenous calcitriol regresses myocardial hypertrophy in hemodialysis patients with secondary hyperparathyroidism. *American Journal of Kidney Diseases* 33(1):73–81.

Perez-Castrillon, J. L., G. Vega, L. Abad, A. Sanz, J. Chaves, G. Hernandez, and A. Duenas. 2007. Effects of atorvastatin on vitamin D levels in patients with acute ischemic heart disease. *American Journal of Cardiology* 99(7):903–905.

Pilz, S., H. Dobnig, J. E. Fischer, B. Wellnitz, U. Seelhorst, B. O. Boehm, and W. Marz. 2008a. Low vitamin D levels predict stroke in patients referred to coronary angiography. *Stroke* 39(9):2611–2613.

Pilz, S., W. Marz, B. Wellnitz, U. Seelhorst, A. Fahrleitner-Pammer, H. P. Dimai, B. O. Boehm, and H. Dobnig. 2008b. Association of vitamin D deficiency with heart failure and sudden cardiac death in a large cross-sectional study of patients referred for coronary angiography. *The Journal of Clinical Endocrinology and Metabolism* 93(10):3927–3935.

Pilz, S., H. Dobnig, G. Nijpels, R. J. Heine, C. D. Stehouwer, M. B. Snijder, R. M. van Dam, and J. M. Dekker. 2009. Vitamin D and mortality in older men and women. *Clinical Endocrinology* 71(5):666–672.

Pittas, A. G., B. Dawson-Hughes, T. Li, R. M. Van Dam, W. C. Willett, J. E. Manson, and F. B. Hu. 2006. Vitamin D and calcium intake in relation to type 2 diabetes in women. *Diabetes Care* 29(3):650–656.

Pittas, A. G., J. Lau, F. B. Hu, and B. Dawson-Hughes. 2007. The role of vitamin D and calcium in type 2 diabetes. A systematic review and meta-analysis. *The Journal of Clinical Endocrinology and Metabolism* 92(6):2017–2029.

Poole, K. E., N. Loveridge, P. J. Barker, D. J. Halsall, C. Rose, J. Reeve, and E. A. Warburton. 2006. Reduced vitamin D in acute stroke. *Stroke* 37(1):243–245.

Ravani, P., F. Malberti, G. Tripepi, P. Pecchini, S. Cutrupi, P. Pizzini, F. Mallamaci, and C. Zoccali. 2009. Vitamin D levels and patient outcome in chronic kidney disease. *Kidney International* 75(1):88–95.

Reasner, C. A., 2nd, J. F. Dunn, D. A. Fetchick, Y. Liel, B. W. Hollis, S. Epstein, J. Shary, G. R. Mundy, and N. H. Bell. 1990. Alteration of vitamin D metabolism in Mexican–Americans. *Journal of Bone and Mineral Research* 5(1):13–17.

Reid, I. R., R. Ames, B. Mason, H. E. Reid, C. J. Bacon, M. J. Bolland, G. D. Gamble, A. Grey, and A. Horne. 2008. Randomized controlled trial of calcium supplementation in healthy, nonosteoporotic, older men. *Archives of Internal Medicine* 168(20):2276–2282.

Reis, J. P., D. von Muhlen, D. Kritz-Silverstein, D. L. Wingard, and E. Barrett-Connor. 2007. Vitamin D, parathyroid hormone levels, and the prevalence of metabolic syndrome in community-dwelling older adults. *Diabetes Care* 30(6):1549–1555.

Reis, J. P., E. D. Michos, D. von Muhlen, and E. R. Miller, 3rd. 2008a. Differences in vitamin D status as a possible contributor to the racial disparity in peripheral arterial disease. *The American Journal of Clinical Nutrition* 88(6):1469–1477.

Reis, J., D. von Muhlen, and E. Miller. 2008b. Relation of 25-hydroxyvitamin D and parathyroid hormone levels with metabolic syndrome among US adults. *Eur J Endocrinol*.

Reis, J. P., D. von Muhlen, E. R. Miller, 3rd, E. D. Michos, and L. J. Appel. 2009. Vitamin D status and cardiometabolic risk factors in the United States adolescent population. *Pediatrics* 124(3):e371–e379.

Richart, T., L. Thijs, T. Nawrot, J. Yu, T. Kuznetsova, E. J. Balkestein, H. A. Struijker-Boudier, and J. A. Staessen. 2011. The metabolic syndrome and carotid intima–media thickness in relation to the parathyroid hormone to 25-OH-D(3) ratio in a general population. *American Journal of Hypertension* 24(1):102–109.

Rostand, S. G., and T. B. Drueke. 1999. Parathyroid hormone, vitamin D, and cardiovascular disease in chronic renal failure. *Kidney International* 56(2):383–392.

Scandinavian Simvastatin Survival Study (4S). 1994. Randomised trial of cholesterol lowering in 4444 patients with coronary heart disease. *Lancet* 344(8934):1383–1389.

Schleithoff, S. S., A. Zittermann, G. Tenderich, H. K. Berthold, P. Stehle, and R. Koerfer. 2006. Vitamin D supplementation improves cytokine profiles in patients with congestive heart failure: A double-blind, randomized, placebo-controlled trial. *The American Journal of Clinical Nutrition* 83(4):754–759.

Scragg, R., R. Jackson, I. M. Holdaway, T. Lim, and R. Beaglehole. 1990. Myocardial infarction is inversely associated with plasma 25-hydroxyvitamin D3 levels: A community-based study. *International Journal of Epidemiology* 19(3):559–563.

Semba, R. D., D. K. Houston, S. Bandinelli, K. Sun, A. Cherubini, A. R. Cappola, J. M. Guralnik, and L. Ferrucci. 2010. Relationship of 25-hydroxyvitamin D with all-cause and cardiovascular disease mortality in older community-dwelling adults. *European Journal of Clinical Nutrition* 64(2):203–209.

Shane, E., D. Mancini, K. Aaronson, S. J. Silverberg, M. J. Seibel, V. Addesso, and D. J. McMahon. 1997. Bone mass, vitamin D deficiency, and hyperparathyroidism in congestive heart failure. *The American Journal of Medicine* 103(3):197–207.

Shea, M. K., S. L. Booth, J. M. Massaro, P. F. Jacques, R. B. D'Agostino, Sr., B. Dawson-Hughes, J. M. Ordovas, C. J. O'Donnell, S. Kathiresan, J. F. Keaney, Jr., R. S. Vasan, and E. J. Benjamin. 2008. Vitamin K and vitamin D status: Associations with inflammatory markers in the Framingham Offspring Study. *American Journal of Epidemiology* 167(3):313–320.

Shen, H., L. F. Bielak, J. F. Ferguson, E. A. Streeten, L. M. Yerges-Armstrong, J. Liu, W. Post, J. R. O'Connell, J. E. Hixson, S. L. Kardia, Y. V. Sun, M. A. Jhun, X. Wang, N. N. Mehta, M. Li, D. L. Koller, H. Hakonarson, B. J. Keating, D. J. Rader, A. R. Shuldiner, P. A. Peyser, M. P. Reilly, and B. D. Mitchell.

2010. Association of the vitamin D metabolism gene CYP24A1 with coronary artery calcification. *Arteriosclerosis, Thrombosis, and Vascular Biology* 30(12):2648–2654.

Shepherd, J., S. M. Cobbe, I. Ford, C. G. Isles, A. R. Lorimer, P. W. MacFarlane, J. H. McKillop, and C. J. Packard. 1995. Prevention of coronary heart disease with pravastatin in men with hypercholesterolemia. West of Scotland Coronary Prevention Study Group. *New England Journal of Medicine* 333(20):1301–1307.

Splaver, A., G. A. Lamas, and C. H. Hennekens. 2004. Homocysteine and cardiovascular disease: Biological mechanisms, observational epidemiology, and the need for randomized trials. *American Heart Journal* 148(1):34–40.

Stamatelou, K. K., M. E. Francis, C. A. Jones, L. M. Nyberg, and G. C. Curhan. 2003. Time trends in reported prevalence of kidney stones in the United States: 1976–1994. *Kidney International* 63(5):1817–1823.

Targher, G., L. Bertolini, R. Padovani, L. Zenari, L. Scala, M. Cigolini, and G. Arcaro. 2006. Serum 25-hydroxyvitamin D3 concentrations and carotid artery intima–media thickness among type 2 diabetic patients. *Clinical Endocrinology* 65(5):593–597.

Teng, M., M. Wolf, E. Lowrie, N. Ofsthun, J. M. Lazarus, and R. Thadhani. 2003. Survival of patients undergoing hemodialysis with paricalcitol or calcitriol therapy. *New England Journal of Medicine* 349(5):446–456.

Teng, M., M. Wolf, M. N. Ofsthun, J. M. Lazarus, M. A. Hernan, C. A. Camargo, Jr., and R. Thadhani. 2005. Activated injectable vitamin D and hemodialysis survival: A historical cohort study. *Journal of the American Society of Nephrology* 16(4):1115–1125.

Tentori, F., W. C. Hunt, C. A. Stidley, M. R. Rohrscheib, E. J. Bedrick, K. B. Meyer, H. K. Johnson, and P. G. Zager. 2006. Mortality risk among hemodialysis patients receiving different vitamin D analogs. *Kidney International* 70(10):1858–1865.

Timms, P. M., N. Mannan, G. A. Hitman, K. Noonan, P. G. Mills, D. Syndercombe-Court, E. Aganna, C. P. Price, and B. J. Boucher. 2002. Circulating MMP9, vitamin D and variation in the TIMP-1 response with VDR genotype: Mechanisms for inflammatory damage in chronic disorders? *QJM* 95(12):787–796.

Toole, J. F., M. R. Malinow, L. E. Chambless, J. D. Spence, L. C. Pettigrew, V. J. Howard, E. G. Sides, C. H. Wang, and M. Stampfer. 2004. Lowering homocysteine in patients with ischemic stroke to prevent recurrent stroke, myocardial infarction, and death: The Vitamin Intervention for Stroke Prevention (VISP) randomized controlled trial. *JAMA* 291(5):565–575.

Towler, D.A., and T. L. Clemens. 2005. Vitamin D and cardiovascular medicine. In *Vitamin D*, 2nd edition, edited by D. Feldman, J. W. Pike and F. H. Gloriuex. Amsterdam: Elsevier Academic Press.

Vaidya, A., and J. P. Forman. 2010. Vitamin D and hypertension: Current evidence and future directions. *Hypertension* 56(5):774–779.

van Etten, E., and C. Mathieu. 2005. Immunoregulation by 1,25-dihydroxyvitamin D3: Basic concepts. *The Journal of Steroid Biochemistry and Molecular Biology* 97(1–2):93–101.

Veldman, C. M., M. T. Cantorna, and H. F. DeLuca. 2000. Expression of 1,25-dihydroxyvitamin D(3) receptor in the immune system. *Archives of Biochemistry and Biophysics* 374(2):334–338.

Vieth, R. 1999. Vitamin D supplementation, 25-hydroxyvitamin D concentrations, and safety. *The American Journal of Clinical Nutrition* 69(5):842–856.

Vik, B., K. Try, D. S. Thelle, and O. H. Forde. 1979. Tromso Heart Study: Vitamin D metabolism and myocardial infarction. *British Medical Journal* 2(6183):176.

Virtanen, J. K., T. Nurmi, S. Voutilainen, J. Mursu, and T. P. Tuomainen. 2011. Association of serum 25-hydroxyvitamin D with the risk of death in a general older population in Finland. *European Journal of Nutrition*.

Wang, P. S., D. H. Solomon, H. Mogun, and J. Avorn. 2000. HMG-CoA reductase inhibitors and the risk of hip fractures in elderly patients. *JAMA* 283(24):3211–3216.

Wang, A. Y., C. W. Lam, J. E. Sanderson, M. Wang, I. H. Chan, S. F. Lui, M. M. Sea, and J. Woo. 2008a. Serum 25-hydroxyvitamin D status and cardiovascular outcomes in chronic peritoneal dialysis patients: A 3-y prospective cohort study. *The American Journal of Clinical Nutrition* 87(6):1631–1638.

Wang, T. J., M. J. Pencina, S. L. Booth, P. F. Jacques, E. Ingelsson, K. Lanier, E. J. Benjamin, R. B. D'Agostino, M. Wolf, and R. S. Vasan. 2008b. Vitamin D deficiency and risk of cardiovascular disease. *Circulation* 117(4):503–511.

Watson, K. E., M. L. Abrolat, L. L. Malone, J. M. Hoeg, T. Doherty, R. Detrano, and L. L. Demer. 1997. Active serum vitamin D levels are inversely correlated with coronary calcification. *Circulation* 96(6):1755–1760.

Weishaar, R. E., S. N. Kim, D. E. Saunders, and R. U. Simpson. 1990. Involvement of vitamin D3 with cardiovascular function. III. Effects on physical and morphological properties. *American Journal of Physiology* 258(1 Pt 1):E134–E142.

Wolf, M., A. Shah, O. Gutierrez, E. Ankers, M. Monroy, H. Tamez, D. Steele, Y. Chang, C. A. Camargo, Jr., M. Tonelli, and R. Thadhani. 2007. Vitamin D levels and early mortality among incident hemodialysis patients. *Kidney International* 72(8):1004–1013.

World Health Organization. The top 10 causes of death.

Wortsman, J., L. Y. Matsuoka, T. C. Chen, Z. Lu, and M. F. Holick. 2000. Decreased bioavailability of vitamin D in obesity. *The American Journal of Clinical Nutrition* 72(3):690–693.

Yavuz, B., D. T. Ertugrul, H. Cil, N. Ata, K. O. Akin, A. A. Yalcin, M. Kucukazman, K. Dal, M. S. Hokkaomeroglu, B. B. Yavuz, and E. Tutal. 2009. Increased levels of 25 hydroxyvitamin D and 1,25-dihydroxyvitamin D after rosuvastatin treatment: A novel pleiotropic effect of statins? *Cardiovascular Drugs and Therapy* 23(4):295–299.

Young, K. A., C. D. Engelman, C. D. Langefeld, K. G. Hairston, S. M. Haffner, M. Bryer-Ash, and J. M. Norris. 2009. Association of plasma vitamin D levels with adiposity in Hispanic and African Americans. *The Journal of Clinical Endocrinology and Metabolism* 94(9):3306–3313.

Young, K. A., J. K. Snell-Bergeon, R. G. Naik, J. E. Hokanson, D. Tarullo, P. A. Gottlieb, S. K. Garg, and M. Rewers. 2011. Vitamin D deficiency and coronary artery calcification in subjects with type 1 diabetes. *Diabetes Care* 34(2):454–458.

Zipitis, C. S., and A. K. Akobeng. 2008. Vitamin D supplementation in early childhood and risk of type 1 diabetes: A systematic review and meta-analysis. *Archives of Disease in Childhood* 93(6):512–517.

Zittermann, A., S. S. Schleithoff, G. Tenderich, H. K. Berthold, R. Korfer, and P. Stehle. 2003. Low vitamin D status: A contributing factor in the pathogenesis of congestive heart failure? *Journal of the American College of Cardiology* 41(1):105–112.

Zittermann, A., S. S. Schleithoff, and R. Koerfer. 2005. Putting cardiovascular disease and vitamin D insufficiency into perspective. *British Journal of Nutrition* 94(4):483–492.

Zittermann, A., S. S. Schleithoff, and R. Koerfer. 2007. Vitamin D and vascular calcification. *Curr Opin Lipidol* 18(1):41–46.

Zittermann, A., S. S. Schleithoff, C. Gotting, O. Dronow, U. Fuchs, J. Kuhn, K. Kleesiek, G. Tenderich, and R. Koerfer. 2008. Poor outcome in end-stage heart failure patients with low circulating calcitriol levels. *European Journal of Heart Failure* 10(3):321–327.

16 Vitamin D, Aging, and Chronic Diseases

Pentti Tuohimaa

CONTENTS

16.1 INTRODUCTION

Aging is a complex biological process at the molecular, cellular, and organismal levels and is influenced by genetic and environmental factors. It is generally characterized by a declining ability to respond to stress, an increasing homeostatic imbalance, and an increased risk of aging-related diseases. Although an accumulation of cellular and DNA damages mainly caused by oxidative stress (wear and tear) might be the cause of aging, the weakening repair mechanisms also play an important role [1].

Around midlife, even in the absence of chronic diseases, vitamin D metabolism begins to change toward an insufficiency [2]. Serum concentrations of $25OHD_3$ (calcidiol) decrease, calcium absorption in the intestine diminishes, and calcium bioavailability declines. This in turn stimulates parathyroid hormone (PTH) secretion [3,4]. The mild hyperparathyroidism associated with aging stimulates bone turnover leading to accelerated osteoporosis. In the past decades, our knowledge of vitamin D_3 and its biological activities has significantly developed: its hormonally active forms are not only bone hormones but also have a wide spectrum of actions in health and diseases [5]. During the last decade, there has been accumulating evidence that an excessive production of $1\alpha,25$ $(OH)_2D_3$ (calcitriol) plays a crucial role in the premature aging caused by fibroblast growth factor-23 (FGF-23)/klotho mutations [6]. On the other hand, a vitamin D_3 insufficiency also seems to enhance premature aging [7]. The negative vitamin D_3 balance with aging in turn accelerates premature aging phenomena as well as the expression of chronic diseases. Therefore, a vitamin D_3 substitution in the elderly is important, but it is not an easy task, because the most optimal substitution is difficult to define. This chapter deals with the complexity of vitamin D_3 substitution with a special reference to the unique endocrinology of cholecalciferol hormones and their role in aging and aging-related diseases.

16.2 FUNCTIONS OF CHOLECALCIPHEROL HORMONES: DUAL HORMONE THEORY

Our understanding of how vitamin D mediates biological responses has entered a new era. Calcium homeostasis and bone are no longer the only targets of vitamin D, but practically all cells in the body contain vitamin D receptor (VDR). Therefore, vitamin D is involved in a wide range of diseases [7,8]. The main functions of hormonal forms of vitamin D could be classified as follows: (1) antiproliferation action; (2) differentiation [9]; (3) antimicrobial innate immunity [10]; (4) immunomodulation [5]; (5) apoptosis [11]; (6) genomic stability [12]; (7) calcium homeostasis; and (8) detoxification [13–15]. All of these functions are important in aging.

Since 1968, there has been a general agreement that the active form of vitamin D_3 is $25OHD_3$ (calcidiol) [16]; however, this dogma was rejected in 1971 when $1\alpha,25\ (OH)_2D_3$ (calcitriol) was isolated and found to be almost 1000-fold more active when assayed on the molar basis [17]. The latter view was strengthened once VDR binding assays with radioactive ligands were developed and VDR was cloned [18]. As a consequence, the basic dogma in vitamin D endocrinology was that calcidiol is an inactive metabolite hydroxylated by CYP27A1 mainly in the liver. Calcidiol is thereafter activated in the kidney through 1α-hydroxylation (CYP27B1) to the biologically active $1\alpha,25\ (OH)_2D_3$ (calcitriol). All forms are carried in the blood bound to vitamin D binding protein, with calcidiol, having the highest affinity and the highest proportion, being protein bound. The prevailing hypothesis was that the free calcitriol was the biologically active hormone. Today, almost every detail of the classical dogma has changed, or there are proposals for alternative explanations. 25-Hydroxylation does not only occur in the liver, but also in many other organs, including the skin, which may contribute significantly to the serum calcidiol concentration [19]. Even though the kidney seems to be the major source of serum calcitriol, extrarenal calcitriol production seems to be physiologically important as an autocrine and paracrine factor [20–24]. These findings led to the conclusion that the local production of calcitriol regulates cell growth and differentiation. However, the low concentrations of the locally produced calcitriol [25], and normally in the serum, make it unlikely that calcitriol could elicit these effects. The half-life of calcidiol in the circulation is approximately 2 weeks, whereas that of calcitriol is less than 4 h [26]. *In vitro*, calcitriol can produce biological responses in serum-free medium being in favor of the "free hormone" hypothesis. However, the concentrations of calcitriol needed for these responses are usually 100- to 1000-fold higher than the physiological ones. Moreover, the serum level of calcidiol is approximately 1000 times higher than that of calcitriol. Only 0.04% of calcidiol and 0.4% of calcitriol are free in plasma and the rest are tightly bound to either a vitamin D binding protein or serum albumin [27]. This means that the concentration of free calcidiol is 100 times higher than that of the free calcitriol. Based on the free hormone hypothesis [28], calcidiol is 100 times more accessible to the target cells compared with calcitriol, but the absolute concentrations of free hormone in serum are far lower than the concentrations known to give any biological response. On the other hand, based on the "bound hormone" hypothesis [29–31], the bound calcidiol that can be taken up by the target cells has approximately 1000 times higher serum concentrations than that of calcitriol. If calcidiol were an inactive metabolite, it would competitively inhibit the action of calcitriol. The binding affinities of the metabolites to the VDR have been measured in different cell types with varying results, but the binding affinity of calcitriol is only approximately 50-fold higher than that of calcidiol [32], suggesting that in physiological situations, most of the VDR molecules are occupied by calcidiol.

To solve the calcidiol enigma, we planned a set of experiments with calcidiol, in which the role of calcitriol was eliminated. Once calcidiol and calcitriol enter the target cell, they bind to the VDR [33]. Calcidiol may also bind to 1α-hydroxylase and be converted to calcitriol. We eliminated 1α-hydroxylase pharmacologically and genetically. By blocking 1α-hydroxylase activity with a specific enzyme inhibitor, we have demonstrated that calcidiol itself can regulate gene expression in human primary prostate stromal cells [34] and mouse primary prostate cells [35]. In addition, we demonstrated that calcidiol can promote the activity of 24-hydroxylase gene promoter in MCF-7 human breast cancer cells and inhibit the growth of LNCaP human prostate cancer cells [34]. We also isolated primary cells from the kidney and skin of 1α-hydroxylase knockout mice. In the 1α-hydroxylase knockout kidney and skin cells, we verified the gene-regulatory action of calcidiol [36]. In addition, calcidiol seems to inhibit precancerous and alveolar lesions in mouse mammary organ culture systems derived from 1α-hydroxylase knockout mice [37]. All the data demonstrate clearly that calcidiol has an inherent hormonal activity regulating genes and cell proliferation.

All the products of 24-hydroxylation (CYP24) were previously presumed to be inactive degradation products. However, there is accumulating evidence suggesting that at least $24,25(OH)_2D_3$ (24,25-cholecalciferol or 24-calcitriol) might be biologically active in chondrocytes [38]. This action seems to be mediated by membrane VDR (mVDR) because the effect can also be detected in nuclear VDR$^{-/-}$ cells. 24-Calcitriol also seems to act through the nuclear VDR, stimulating differentiation of human osteoblasts as well as bone mineralization [39]. Furthermore, 24-calcitriol seems to be necessary for bone fracture healing because the callus formation is impaired in CYP24$^{-/-}$ mice [40,41]. It can be concluded that there are at least three hormonally active forms of vitamin D_3: calcidiol, calcitriol, and 24-calcitriol (Figure 16.1). We recently proposed that these hormones should be called calcipherol or cholecalcipherol hormones (CHs) [42]. There is a good argument for each term: The prefix "calci" is already used in the names of individual hormones (i.e., calcidiol). On the other hand, the basic structure is known as cholecalcipherol and it favors the latter name. It is important to avoid the use of the term "vitamin D" because there is no biologically active vitamin D, but only CHs are biologically functional.

It is possible that new active cholecalcipherol hormones will be found in the future [43,44]. Hydroxylations occur at all positions from 20 to 28, except for positions 21 and 27. Therefore, there are several hormone candidates to be tested for their activity in physiological situations. 1α,20 $(OH)_2D_3$ metabolized by CYP11A1 is a hormonally active metabolite in skin keratinocytes [45] and its activity is VDR dependent.

A dilemma in cholecalcipherol hormone endocrinology has been the reciprocal feedback between serum calcidiol and calcitriol. Chronic administration of calcitriol increases the metabolic clearance of calcidiol [46,47] and decreases its production [48], with consequent depletion of calcidiol stores.

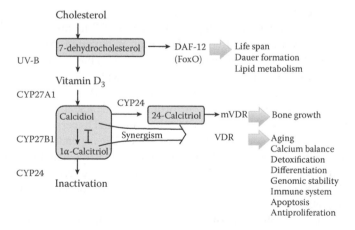

FIGURE 16.1 Simplified model of the endocrine system of cholecalcipherol hormones. There are three active hormonal forms: calcidiol [25(OH)D$_3$, 25-hydroxy-cholecalcipherol], calcitriol [1,25(OH)$_2$D$_3$, 1α-calcitriol, 1α,25-dihydroxy-cholecalcipherol], and 24-calcitriol [24,25(OH)$_2$D$_3$, 24,25-dihydroxy-cholecalcipherol]. Calcidiol and calcitriol act synergistically through the VDR, and they have a negative mutual feedback regulation (I–I). 24-Calcitriol probably acts through the mVDR. DAF-12 (dauer formation protein-12) regulates the life span of *C. elegans*, and it is a homolog of mammalian VDR and FoxO (forkhead box protein). The putative ligand of DAF-12 is 7-dehydrocholesterol. CYP27A1, 25-hydroxylase; CYP27B1, 1α-hydroxylase; CYP24, 24-hydroxylase. For details and references, see text.

On the other hand, administration of elevated doses of vitamin D_3 or calcidiol increases the metabolic rate of calcitriol [49]. The "hormonal calcidiol" concept helps in understanding the feedback system because both hormones seem to be active and are able to regulate the mutual expression. The physiological significance of the negative feedback (Figure 16.1) could be that the hormonal sum effect of serum/intracellular calcidiol and calcitriol is under control. As a consequence, 24-hydroxylated products in the serum always increase when either calcidiol or calcitriol increases in the serum. Therefore, an increase of serum 24-calcitriol might be the best indicator of an overdose of oral vitamin D_3.

A strong argument against the hormonal calcidiol concept is the phenotype of the 1α-hydroxylase knockout mice (CYP27B1$^{-/-}$), which is quite similar, although not identical, to that of VDR$^{-/-}$ mice [7]. The CYP27B1$^{-/-}$ mice have an elevated serum calcidiol concentration, but it cannot prevent the phenotype. This discrepancy can be explained by our recent findings on the mutual synergistic action of calcidiol and calcitriol [35]. In CYP27B1$^{-/-}$ cells, calcidiol and calcitriol show a synergistic action on CYP24 expression. We treated 1α-hydroxylase knockout kidney and skin cells with both calcidiol and calcitriol at different concentrations, and found that the combined effects of both metabolites were significantly higher than the sum of the effects of either alone. We do not know the mechanism of the synergism, but a possible explanation could be homodimerization of VDR occupied with the two ligands (calcitriol and calcidiol). Calcidiol could bind to the AP site and calcitriol to the GP site, causing different conformational changes allowing apoVDR homodimer stabilization [50–56]. It is apparent that the coactivator complex associated with homodimer is different from that bound to VDR–RXR heterodimers, suggesting that the genes regulated are also different. The hormonal synergism suggests that both calcidiol and calcitriol, in fact, are important hormones and that they act in concert. Therefore, calcidiol, with its variable serum concentrations, has always been clinically and epidemiologically more important than calcitriol with its rather stable serum concentrations. Within the cell, neither the concentration of calcitriol nor calcidiol alone is sufficient for a significant physiological response but, in combination, they can cause a perfect response [36]. This we call "dual hormone theory." The dual cholecalcipherol hormone synergism is schematically presented in Figure 16.1.

A dilemma in the VDR action is the regulation of the hair cycle. Alopecia is a part of the phenotype of many patients with hereditary vitamin D–resistant rickets caused by VDR mutations [57]. VDR-KO mice develop their first coat of hair normally, but begin to lose their hair rostrally at the age of 6 months, and hair loss is complete at the age of 8 months [42]. The reinitiation of anagen occurs after the first cycle or after the depilation is impaired [58]. Reconstitution of the VDR in keratinocytes to the VDR-KO mice reverses the defect in the hair cycle and the mice do not develop alopecia [59]. Correction of metabolic disturbances with a high calcium diet prevents rickets and hyperparathyroidism but does not prevent the alopecia [60], which suggests that alopecia is not calcium-dependent. Because CYP27B1$^{-/-}$ (1α-hydroxylase deficient) mice show rickets, growth retardation, osteomalacia, hypocalcemia, hypophosphatemia, and hyperparathyroidism, but no alopecia [61], it seemed that the development of alopecia was VDR-dependent, but was not dependent on calcitriol. This raised the question of whether the hair cycle is regulated by apoVDR or if there is an unknown ligand of VDR for hair follicle development. Because serum calcidiol concentrations increase in CYP27B1$^{-/-}$ mice, it is possible that this high calcidiol concentration would be sufficient to maintain the hair cycle.

A new aspect in cholecalcipherol hormone endocrinology is the U-shaped dose–response to serum calcidiol concentrations of some diseases and aging phenomena. This is also known as hormesis [62]: both insufficient and excess concentrations of CHs are harmful to health. Hormesis is the term for the usually favorable biological responses to low exposures to toxins, toxic substances, and other stressors. The concept has been explored extensively with respect to aging [63,64]. According to the hormesis paradigm, agents induce dose–response relationships having two or three distinct phases (biphasic, U-, J-, or reversed U-shaped = bell-shaped). A mild stress exposure such as sun radiation usually has antiaging effects [65]. A U-shaped hormesis typically occurs in all natural ligands of nuclear receptors such as retinoic acid, thyroid hormones, and steroids. A hormetic response to serum calcidiol concentration is reported in some epidemiologic studies: risk of prostate cancer [66,67], all-cause deaths, and cardiovascular deaths [68–71]. Also, the aging process in general shows a U-shaped response to cholecalciferol hormones [7]: both low and high action of CHs enhance aging phenomena such as skin aging, osteoporosis, ectopic calcification, immunodeficiency, muscle atrophy, and loss of hearing. Another type of vitamin D hormesis is a reversed response. Usually, higher serum calcidiol concentrations seem to protect against cancer as mentioned previously but sometimes they increase the risk of cancer such as esophageal cancer in men [72] and pancreatic cancer [73]. In skin, low doses of calcitriol increase epidermal cell proliferation, whereas high doses inhibit it [74]. Recently, several U-shaped vitamin D responses have been reported such as the risk of falls and fractures in elderly women [75], schizophrenia [76], breast cancer [77], cancer mortality [78], small-for-gestational age births [79], common single nucleotide polymorphism [80], cardiovascular complications in chronic kidney disease [81], cardiovascular mortality [68,69,71], and all-cause mortality [69,70,78]. Thus, it seems that there is an optimal serum concentration of calcidiol, which may delay aging and prevent aging-related diseases. Based on the epidemiologic studies mentioned previously, the optimal concentration is approximately 40–80 nmol/L, but it may be different in different diseases and among different populations; therefore, more studies are needed.

16.3 CHOLECALCIPHEROLPAUSE AND SUBSTITUTION

The insufficiency of CHs during aging (cholecalciperholpause) is common in the elderly. The main cause for the common vitamin D insufficiency in the elderly [82–84] seems to be a decreased skin production of vitamin D_3 due to lack of 7-dehydrocholesterol [85,86]. The diminished skin production in elderly is often aggravated by changes in lifestyle. Many older people are homebound and, when exposed to the sun, often wear protective clothing. The cholecalciferol hormone balance develops more and more negatively with aging, and this hormone insufficiency in turn may facilitate the aging processes.

Intestinal absorption of vitamin D_3 is not significantly affected in the elderly people in the absence of diseases associated with intestinal malabsorption [87]. Synthesis of calcidiol does not seem to be influenced by aging [2]. The effect of aging on serum concentration of calcitriol is controversial. A number of studies suggest either a decrease, no change, or an increase (see the review in [2]). However, bone responsiveness to CHs seems to decrease with aging [88]. This is most likely due to a decrease of VDR concentration in target tissues. Also, intestinal calcium absorption is affected in the elderly [89]. As a consequence of the hypovitaminosis D, a mild hyperparathyroidism is often seen in the elderly, which in turn stimulates bone turnover accelerating osteoporosis.

The main reason for vitamin D substitution in the elderly is to reduce the incidence of bone fractures [90]. There are three factors playing role in the risk of bone fracture: (1) muscle strength; (2) balance; and (3) bone mineral density. Several studies suggest that cholecalciferol hormones regulate muscle strength [91–93]. This is apparently due to the regulation of calcium balance within the muscle cell. It has been proposed that vitamin D–deficient elderly people are at a higher risk of falling because of impaired muscle tonus. Much less attention has been paid to the role of balance in the risk of falling during aging. The balance deficit is one of the first signs of aging. Using aging VDR-KO mice, we found that their balance deficit developed faster than in normal mice [94]. Because otolites in semicircular canals contain calcium, it is possible that their development and turnover needs CHs and VDR, which we demonstrated in the epithelium of crista ampullaris. It is interesting that the sensorineural hearing loss is also associated with hypoparathyroidism [95,96]. The vitamin D substitution may not effectively increase the bone density, but it is extremely important in preventing the further development of osteoporosis in the elderly. Because the mild PTH elevation in the elderly is the main reason for osteoporosis, the substitution needed should be effective to suppress PTH. For fracture risk reduction in the elderly, the weight of evidence indicates that the minimum amount of serum calcidiol needed is 75 nmol/L or more [90].

In the elderly, the prevention of chronic diseases with vitamin D substitution (see below) may not be as important as earlier in life, because even with the most optimal substitution, only a few healthy months (not years) can be achieved. Certainly, bone fracture may be an important invalidity limiting the life span and the quality of life, but the most interesting new areas of preventive medicine in the elderly are the mood, senses, and neurological degeneration.

The degeneration of senses occurring with aging might be faster in the presence of vitamin D insufficiency than in normovitaminosis. A common reason for blindness in the elderly is age-related macular degeneration (AMD). There is preliminary evidence suggesting that cholecalciferol hormones may protect against AMD [97]. Vitamin D–dependent deafness, and its reversal with vitamin D substitution, has been known for almost three decades [98–101]. In the VDR-KO mice, we found unilateral loss of hearing as a sign of premature aging [102]. These animals also showed signs of mild balance deficits [94], but not any loss of gustatory nor olfactory senses [103]. A degeneration of senses is clearly associated with aging. The apparent cause of deafness might be disturbances in calcium metabolism of the inner ear and, therefore, it is also associated with hypoparathyroidism [95,96]. It is interesting that hypervitaminosis D is also associated with loss of hearing [104]. In conclusion, the senses, some of which are calcium dependent, are more likely affected in cholecalciferol hormone disturbances.

As to the quality of life of the elderly, putative neuroprotective properties of vitamin D are of interest in the future [42]. Calcitriol seems to enhance the glutathione content of neurons and protects them from reactive oxygen species [105]. A vitamin D insufficiency seems to be associated with the cognitive impairment in the elderly and in patients with Alzheimer disease [106–109]. One of the most important reasons for poor quality of life in the elderly is depression; therefore, relevant studies on the role of vitamin D are needed. Preliminary studies suggest that sufficient administration of vitamin D would improve seasonal affective disorder as well as other types of depression [110–113]. Vitamin D sufficiency may delay the appearance of neurodegenerative diseases [42,114,115].

In conclusion, the evident indications for vitamin D supplementation to the elderly include (1) for the prevention of osteoporosis [241], (2) for the prevention of hip fractures [242], (3) for enhancing muscle strength [91], and (4) for the home- or hospital-bound elderly [243]. Further indications, which need more studies, might be as follows: (1) for vitamin D–dependent hearing loss [98,101,102]; (2) for AMD [97]; (3) for seasonal affective disorder [110,113]; (4) for depression in the elderly [111,112]; (5) for Alzheimer's disease [107]; (6) for Parkinson's disease [162]; and (7) for cognitive ability [106,108,109].

The vitamin D_3 photosynthesis during sun exposure is enormous [116]. It is obvious that the high photosynthesis rate became a problem for living organisms. However, part of the problem was solved by photodegradation of the precursors of vitamin D_3, but all living organisms have also developed an effective detoxification system based on CYP24, 24-hydroxylase, which initiates a series of hydroxylations finally leading to the excretion in the kidney as calcitroic acid [117]. There is a basic difference in the physiology of vitamin D transport in the skin versus the intestine. Absorption of oral vitamin D in the ileum is uncontrolled. The intestinal absorption is fast and the 25-hydroxylation in the liver is not rate limited if the body stores are within the normal range [118]. This in turn means that the serum calcidiol concentration shows a peak value shortly after oral administration of vitamin D. This kind of daily variation is smaller when vitamin D is produced in skin exposed to sun. The sun exposure slowly increases serum calcidiol, reaching a maximum 1 or 2 weeks later [119]. This suggests that the negative feedback systems of CHs (discussed previously) are probably less activated if the vitamin D is obtained from the skin. Therefore, it seems that sun exposure or narrow band UV-B irradiation could be the method of choice for elderly people. This choice is supported by the fact that the potential risk of skin cancer in elderly people need not be considered. However, the significant reduction (up to 70%) of vitamin D production in the aging skin may limit the efficiency of UV irradiation.

Oral vitamin D substitution in the elderly is compromised because of the increased incidence of intestinal malabsorption diseases with aging. However, a similar response in serum calcidiol

(22 nmol/L) was found in both young and old men at a daily dose of 20 µg of vitamin D_3 [87]. A very similar result was found in a study on Finnish elderly women (65–85 years old) [120]. A daily dose of 5, 10, or 20 µg of vitamin D_3 increases serum calcidiol by 11, 14, and 24 nmol/L, respectively, within 6 weeks. The more vitamin D is given, the smaller the relative net increase in serum calcidiol. This is apparently due to the negative feedback system caused by the daily peak values of calcidiol. Furthermore, vitamin D–deficient people respond more efficiently to supplementation than healthy subjects with an adequate vitamin D status [121]. This suggests that at the population level, the general recommendations for substitution should be made with caution. The upper limit for the general recommendation for the elderly could be higher than for the younger population, but not higher than 20 µg (800 IU). Higher recommendations are common nowadays [122,123], but more clinical studies are needed before these recommendations are practiced. A good indicator of the too high dosage would be serum 24OH metabolites, because the induction of CYP24 can be regarded as a detoxification defense system as discussed previously. A solution for vitamin D administration for the elderly (as well as for younger people) would be a topical administration of vitamin D using skin patches similar to HRT, which theoretically should give an even calcidiol serum concentration (plateau) without peak values. This idea would be worth studying in more detail.

16.4 AGE-RELATED DEGENERATIVE DISEASES

Epidemiological and clinical studies show that the serum 25-hydroxyvitamin D (calcidiol) levels are inversely associated with osteoporosis [124,125], muscle weakness [126], common solid cancers [127], including breast cancer [128,129], prostate cancer [130], ovarian cancer [131], and colorectal cancer [132,133], and many other chronic diseases such as cardiovascular diseases [71], hypertension [134,135], psoriasis [136], rheumatoid arthritis [137], diabetes [138–140], respiratory infections [141], multiple sclerosis [142,143], and several neurodegenerative diseases [42]. The potential of cholecalciferol hormones in the prevention and treatment of degenerative diseases has been widely reviewed by Armin Zittermann [144,145]. In some degenerative diseases, CHs are directly involved in the basic mechanism of the disease, such as negative regulation of the renin–angiotensin system in hypertension [142,146,147], insulin regulation in diabetes [139,148,149], several osteotrophic genes (osteoporosis) [150], antimicrobial peptides (infections) [10], and mitotic control in cancer and psoriasis [9]. However, in most diseases, the mechanism of action of CHs is unknown. It is possible that these diseases reflect the cholecalciferol hormone-dependent aging process as described later. Thus, an increased risk of chronic diseases means that the degenerative disease genetically disposed appears earlier because of premature aging in the presence of cholecalciferol hormone imbalance. In conclusion, several chronic degenerative diseases are at least partially vitamin D–dependent and their early appearance is a sign of premature aging as well as often being a sign of vitamin D imbalance. In the prevention of these chronic diseases, the administration of a high monthly or daily dose of vitamin D might not be useful because it might lead to resistance as discussed previously, whereas a topical vitamin D administration or sun exposure might be more physiological for the endocrine feedback systems. Therefore, the method and dose of vitamin D administration should be completely reevaluated in preventive medicine.

16.5 ROLE OF CHOLECALCIPHEROL HORMONES IN AGING

The main models for premature aging come from inherited progeroid syndromes. Segmental progerias, such as dyskeratosis congenita, Werner disease, Bloom syndrome, and ataxia telangiectasia display symptoms of accelerated aging, mainly due to reduced DNA repair and increased genetic instability [151]. Some progerias display symptoms—such as alopecia, osteoporosis, and fingernail atrophy—associated with shortened telomeres. It is interesting that the latter symptoms are also typical of vitamin D_3 deficiency.

Within the last decade, calcitriol has become a key mediator of aging. Klotho and FGF-23 mutations in human and experimental animals caused premature aging [152–154]. The overexpression of Klotho could extend the life span of mice [155]. Klotho protein can repress insulin/insulin-like growth factor-1 (IGF-I) signaling and induce insulin resistance. Recently, Maekawa et al. [156] demonstrated that the Klotho protein is able to inhibit the tumor necrosis factor-α (TNFα)–induced activation of nuclear factor κB (NFκB) signaling and subsequently reduce the inflammatory reaction. It seems that the aging process caused by klotho or FGF-23 mutations is calcitriol dependent [6]. Although dysregulation of cholecalciferol hormones, especially the upregulation of 1α-hydroxylation, can explain all the aging phenomena caused by FGF-23 mutations, the role of calcium signaling cannot be totally excluded [157] because almost all the aging phenotype changes could be corrected by reducing serum calcium as well as by deletion of 1α-hydroxylase gene. A dysregulation of calcium signaling seems to play an important role in the development of aging-related diseases [158] and neurodegeneration [159]. It is important to note that FGF-23/klotho seems to be the main negative regulator of 1α-hydroxylation. In conclusion, increased calcitriol or calcium levels (or both) seem to mediate all aging phenomena associated with impaired FGF-23/klotho signaling (Figure 16.2).

Aging can be understood as a process of homeostatic imbalance. Homeostasis means the ability or tendency of an organism or cell to maintain internal equilibrium by adjusting its physiological processes. An imbalance will lead to aging, diseases, and death. According to the hormonal theories of aging, the gradual or rapid decline of hormones (*hormonal imbalance*), such as androgens and estrogens, may initiate aging [160]. Like estrogen decline (menopause) or androgen decline (andropause) CHs are also known to decrease significantly with aging (cholecalcipherolpause) [2,7]. Another possibility for homeostatic disturbance is the *immunological imbalance*. CHs are known to regulate acquired immune systems at several levels [150]. Calcitriol is known to suppress, as well as stimulate, immune responses [161]. The consequence of imbalance of defense mechanisms is immunosenescence. Also, *neurohumoral imbalance* leads to the aging of the nervous system. It is possible that CHs exert a combination of neurohumoral and immunological action in the central nervous system [162]. CHs are often classified as neurosteroids [163], a more appropriate term could be neuroactive secosteroid [42]. In studies of aging, the *hormesis* concept has often been applied. A mild stress to cells and organisms induces adaptive stress resistance, which is called hormesis. Hormesis suggests that at low doses, its effects may be beneficial, although at higher doses, it may be pleiotropically harmful. A mild irradiation can delay aging and, in fact, have some health-promoting effects in contrast to heavy irradiation [65]. The U-shaped hormesis to varying serum

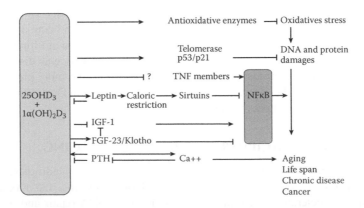

FIGURE 16.2 Schematic presentation of the aging mechanisms regulated by calcipherol hormones, 25OHD$_3$ (calcidiol) and 1α,25(OH)$_2$D$_3$ (calcitriol). →, stimulatory action; –|, inhibitory action; ?, regulation of the different members of the TNF family is variable. For details and references, see text.

concentrations of calcidiol has been discussed previously. Hormesis may explain why both very low and high serum CHs enhance aging.

The antiaging effect of a low caloric diet has been known for about a century. Low caloric diet and fasting delays aging and the effect is thought to be mediated by sirtuins [164]. Sirtuins are homologs of yeast Sir2 (silent information regulator 2 protein) [165]. They are histone deacetylases or ADP-ribosyl transferases located in the nucleus, cytoplasm, or mitochondria. The mammalian seven sirtuins have important functions in the regulation of metabolism, growth, differentiation, cell survival, aging, and life span [166]. The antiaging effect of caloric restriction seems to be mediated especially by Sir1 [167,168]. Sirtuins are known to enhance autophagy [169] and suppress NFκB activity [170], which are strongly involved in aging processes (see "disposable soma"). Thus far, there has been no evidence that CHs could directly regulate sirtuins, but there is an indirect hormonal connection. Serum calcidiol correlates inversely with serum leptin [171] and may regulate leptin secretion. Because leptin controls appetite, it will lead to caloric restriction and in turn to an induction of sirtuins. Leptin is a negative regulator 1α-hydroxylase [172], suggesting that there is a feedback loop between calcitriol and leptin. Leptin is involved in brain aging and Alzheimer's disease [173]. There is another interesting hormonal connection: leptin also seems to regulate PTH [171]. On the other hand, PTH seems to be involved in the regulation of aging [174–176], and serum calcium might be the key mediator of aging. Both primary and secondary hyperparathyroidism increase death rate independently of vitamin D status or general health factors [177]. These hormonal factors of aging are schematically shown in Figure 16.2.

Vitamin D has played a significant role in evolution [178]. During human evolution, selection has been effective, for example, with regard to skin color because vitamin D deficiency will lead to pelvic deformation and difficulties in childbirth. The lighter skin types are able to respond to UV-B by up to 50-fold more effectively than darker skin types. Aging of organisms is a precondition for evolution because without death no evolution is possible. There are three mainstream evolutionary hypotheses of aging: (1) the *accumulation of deleterious somatic mutations* and reduced ability to repair DNA [179]; (2) *antagonistic pleiotropy*, referring to genes that enhance reproductive success early in life, the by-product of which is later decline and death; and (3) a *disposable soma* says that finite food energy is preferentially used for reproduction but compromises repair [180].

16.5.1 Accumulation of Deleterious Mutations

In 1956, Harman [181] formulated his free radical theory of aging and later identified mitochondria as the major endogenous source of oxidative stress [182]. Aging mechanisms include oxidative stress, DNA mutations, and shorter telomeres [183]. Antioxidant defense, DNA repair, and telomerase temper the effects of these deleterious events. Active forms of vitamin D seem to regulate some antioxidant mechanisms [12,184–187]. Calcidiol increases the activity of superoxide dismutase, glutathione peroxidase, as well as that of catalase (Figure 16.2). Thus, cholecalciferol hormones may, to a certain extent, directly oppose the oxidative stress, decrease the risk of DNA mutations, and increase genomic stability [12].

The length of telomeres is dependent on the activity of telomerase reverse transcriptase (TERT). Along with cell division and aging, telomeres gradually shorten due to insufficient reverse transcription. A combination treatment with $1\alpha,25$ $(OH)_2D_3$ and $9\text{-}cis$-retinoic acid inhibits human TERT [188]. It was proposed that this could be a mechanism to inhibit cell proliferation and as a tool against carcinogenesis therefore extending the life span. Cancer cells, to become immortal, acquire the capacity of expressing telomerase [189,190]. Short telomeres lead to senescence [191] and an efficient telomerase activity is necessary for longevity. Therefore, the inhibitory effect of calcitriol on TERT raises the question of whether it inhibits or accelerates aging.

The tumor suppressor protein, p53, is essential for the detection of DNA damage and repair. Its inactivation enhances cancer development, whereas its overexpression leads to premature aging [192,193]. The phosphorylation of retinoblastoma protein (Rb) does not occur in the cells with

DNA damage and they cannot enter the cell cycle, in which p21 regulated by p53 plays a central role. CHs seem to upregulate the expression of both p21 and p53 [194] and enhance apoptosis. Accordingly, we found that the expression of p53 was significantly decreased in mature VDR-KO mice [195]. It is interesting that VDR is induced by DNA damage as well as by p53, which may mediate the effect of DNA damage on VDR expression [196]. The lower level of p53 expression in VDR-KO mice might explain the higher susceptibility of their skin to UV-induced cancer [197]. Collectively, these data suggest that p53 and p21 might be key mediators of antiaging effects of CHs (Figure 16.2).

It has been demonstrated that IGF-I deficiency increases life span [198,199]. CHs seem to regulate the expression of IGF-I and IGF-binding proteins [200,201]. The basal and stimulated IGF-I and IGF-II production is inhibited by calcitriol [202]. We found that IGF-I receptor was downregulated in VDR-KO mice, suggesting a lower IGF activity [195]; however, the aging of these animals was enhanced. The activity of growth hormone/insulin/IGF signaling is at least partially mediated by catalase and superoxide dismutase and therefore oxidative stress is antagonized [1]. Insulin/IGF-I signaling can also enhance NFκB signaling (see below) [199] and subsequently potentiate the aging process and aggravate age-related degenerative diseases. However, the final output of cholecalcipherol hormones on the IGF-I signaling system remains unknown because there are opposite regulations.

16.5.2 ANTAGONISTIC PLEIOTROPY

Antagonistic pleiotropy means that some genes are important for reproductive success early in life, but they enhance aging later in life [203]. A similar hypothesis was later defined as an evolutionary compromise by Darwinian medicine [204–209]. In species with long life spans, the repair mechanisms need to be less active than in species with shorter life spans because the same genes are involved in aging-related diseases such as cancer, and therefore, more active repair mechanisms would cause cancer later in life. On the contrary, effective repair mechanisms are very useful for species with shorter life spans because there is not enough lifetime for cancer development. An example of this is the repair of lost body parts: in lizards, a tail lost after autotomy can regenerate completely [210], whereas even a small damaged part of a human finger does not regenerate. Lizard tail regeneration is dependent on growth factors [211]. Growth factors such as IGF-I are proposed to be involved in carcinogenesis [212]. Several repair genes are clearly involved in and are associated with inherited predisposition to cancer and aging such as APC, MLH, and MSH genes for colon cancer or BRCA1, BRCA2, and BARD1 genes for breast and ovarian cancer [1]. Only BRCA1 seems to be vitamin D–regulated in prostate cancer.

The pleiotropic shift of the action of the growth hormone/insulin/IGF-I signaling system during aging is a good example of the antagonistic pleiotropy hypothesis. A high activity of this signaling is needed for growth in early life, whereas dwarf individuals live longer than their normal counterparts. This evolutionarily conserved pleiotropy is called the insulin/IGF paradox [213]. Laron syndrome is a human dwarf disease that is caused by inactive growth hormone receptor associated with IGF-I deficiency [214]. Despite their obesity, patients with Laron syndrome are long-lived due to their resistance against cancer. The role of CHs in the regulation of IGF-I signaling has been discussed previously.

16.5.3 DISPOSABLE SOMA

The disposable soma theory on aging states that each individual has to allocate limited energy resources to either the maintenance of soma or reproduction [180]. This trade-off between reproduction and longevity was shown to also exist in humans [215]. The original theory was genetic. A compatible nongenetic theory called *entropic host defense catastrophe* has been recently proposed [216].

In *Caenorhabditis elegans*, DAF-12 seems to be a decisive checkpoint for many life history traits, including longevity [217]. DAF-12 is a member of the nuclear hormone receptor family and is a close homolog of liver X receptor, pregnane X receptor, VDR, constitutive androsterone receptor, and farnesoid X receptor—all belonging to subfamily 1. DAF-12 regulates lipid metabolism, stimulates dauer formation and therefore increases the life span of *C. elegans*. It is interesting that 7-dehydrocholesterol is a putative ligand of the DAF-12 (Figure 16.1) [218]. DAF-12 is downstream of the signaling systems that translate environmental cues as well as germ line signals [219,220]. This positions DAF-12 at the crossing point of maintenance of soma and germ line signaling. Also, DAF-16 influences the rate of aging of *C. elegans* in response to insulin/IGF-I signaling [221]. Both DAF-12 and DAF-16 belong to the mammalian FoxO family of transcription factors (forkhead box proteins), which is composed of more than 100 members (Figure 16.1). FoxO members have a critical role in the regulation of the immune system, immunosenescence, and general aging as antiaging factors [222]. FoxO factors (similar with VDR) can induce the expression of several antioxidative enzymes such as catalase and SOD2 as well as stress resistance inducers [223]. FoxO factors play a key role in the maintenance of the energy metabolism, especially glucose balance [224]. The activity of FoxO factors is upregulated by a deficiency of IGF-I signaling [223,225], suggesting a negative regulation by IGF-I.

Similar with DAF-12, VDR mediates environmental stimulus (sun irradiation) to the soma and regulates reproduction. VDR and vitamin D–metabolizing enzymes expressed in human testis [23]. Studies on VDR-KO mice demonstrate that a complete lack of cholecalciferol hormone action impairs fertility. Female mice show uterine hypoplasia and an impaired ovarian folliculogenesis [226]. The fertility of vitamin D–deficient female rats decreases by 75% [227]. This infertility cannot be reversed by calcium repletion, suggesting that the effect was directly mediated by VDR [228]. Aromatase is one of the targets of VDR [229]. Also, male fertility is reduced, the sperm count is decreased, and testicular morphology is changed [229]. As described previously, CHs are important for cell cycle control, differentiation, calcium balance, and detoxification early in life; thus, they control homeostatic balance. However, an excess as well as a lack of CHs seems to increase the risk of some degenerative diseases and aging (see hormesis). It seems that VDR acts at the crossing point of reproduction and soma.

NFκB is known to be strongly associated with aging [230], cancer [231], and skin aging [232]. The mammalian Rel/NFκB family includes three Rel proteins and two NFκB components, p50 and p52. These components form dimeric complexes with each other, which are trapped in the cytoplasm bound to several inhibitory proteins (IκBα, IκBβ, IκBε, IκBζ, and Bcl-3). Several protein kinases can phosphorylate IκB proteins. Subsequently, the NFκB complex is translocated into the nucleus, where it can transactivate several genes, especially inflammatory and immunomodulatory genes [233]. Several pro-aging factors (oxidative stress, innate immunity, TNF family, and insulin/IGF-I signaling) can activate NFκB and several antiaging factors (SIRT1, SIRT6, FoxOs, p53, Klotho, and hormetic phytochemicals) can inhibit NFκB activation [216]. Members of the TNF family, interleukin-6 and TNF-α can regulate adaptive immunity and are upregulated with aging [234]. The effects of CHs are somewhat complex: interleukin-6 is either downregulated or not regulated [235,236], whereas TNF-α seems to be upregulated by calcitriol [236]. Another member of the TNF family, RANK (receptor activator of NFκB) has a key role in osteoporosis. The RANK ligand is regulated by CHs [237–239].

There is no substantial evidence of whether CHs can directly regulate the activity of NFκB. A study using mouse embryonic fibroblasts derived from VDR[+/−] and VDR[−/−] mice suggests that VDR may act directly on IκBα expression and thus prevent NFκB activation [240]. We found a decreased expression of NFκB in aging VDR[−/−] mice [195]. It is evident that CHs are involved in the regulation of NFκB activity, but the regulation is complex and occurs at different upstream levels. With aging, the imbalance between the negative and positive regulation of NFκB activity may lead to NFκB-driven entropic senescence and decreased autophagy as proposed by Salminen and Kaarniranta [216]. The central role of NFκB and CHs in aging is depicted in Figure 16.2.

16.6 CONCLUSIONS

There seems to be three cholecalciferol hormones (CHs): calcitriol [$1\alpha,25(OH)_2D_3$], calcidiol ($25OHD_3$), and 24-calcitriol [$24,25(OH)_2D_3$]. A combination of calcitriol and calcidiol seems to act synergistically in the target cells, regulating several functions, including aging. The physiological circulating levels of calcidiol or calcitriol alone are not sufficient for biological responses. There is a reciprocal relationship between these metabolites (a negative feedback control), which seems to regulate the endocrine balance of CHs.

Aging is a complex event and several separate molecular mechanisms lead to the same end point—normal or premature senescence. The interrelationships between different signaling systems involved in aging remain speculative. Aging is characterized by gradual loss of stress tolerance due to an accumulation of DNA and protein damage caused by oxidative stress. The repair mechanisms, antioxidative enzymes, IGF signaling, sirtuins, and NFκB play an important role in premature aging. CHs have been demonstrated to be key factors in the regulation of most of these mediators and therefore they play a central role in aging. It seems that an optimal calcidiol serum concentration might delay aging. On the contrary, calcidiol imbalance may lead to premature aging and the earlier appearance of chronic diseases as a sign of aging.

Epidemiological studies on the role of hormonal forms of vitamin D_3 in chronic diseases are inconsistent. There are several reasons for this inconsistency: one of the reasons might be a non-linear dependency on calcidiol serum concentrations. A U- or J-shaped risk curve is typical of hormones such as steroids, thyroid hormones, and retinoids as well as cholecalcipherol hormones. This phenomenon is known as hormesis. CHs seem to have harmful effects on health and accelerate aging at both low and high serum concentrations. This suggests that there is an optimal serum concentration of calcidiol that delays aging. The elderly are at a high risk of vitamin D insufficiency with aging (cholecalcipherolpause) because they are not exposed to enough sun and their skin has a lower capacity to produce vitamin D_3.

Studies based on bone health of the elderly suggest that serum calcidiol concentrations should be higher than 70 nmol/L. The selected studies based on several health outcomes suggest that the optimum is between 40 and 80 nmol/L (16–32 ng/mL). This level can be reached if the daily vitamin D dose in the elderly is 10–20 µg. However, more studies are needed on several common degenerative diseases before the final vitamin D recommendations for the elderly can be given.

REFERENCES

1. Irminger-Finger, I. 2007. Science of cancer and aging. *Journal of Clinical Oncology* 25:1844–1851.
2. Halloran, B. P., and A. A. Portale. 2005. Vitamin D Metabolism and Aging. In *Vitamin D*, edited by D. Feldman, J. W. Pike, and F. H. Glorieux, 2nd ed., 823–838. New York: Academic Press.
3. Dawson-Hughes, B., S. S. Harris, and G. E. Dallal. 1997. Plasma calcidiol, season, and serum parathyroid hormone concentrations in healthy elderly men and women. *The American Journal of Clinical Nutrition* 65:67–71.
4. Lips, P., T. Duong, A. Oleksik, D. Black, S. Cummings, D. Cox, and T. Nickelsen. 2001. A global study of vitamin D status and parathyroid function in postmenopausal women with osteoporosis: baseline data from the multiple outcomes of raloxifene evaluation clinical trial. *The Journal of Clinical Endocrinology and Metabolism* 86:1212–1221.
5. Bouillon, R., G. Carmeliet, L. Verlinden, E. van Etten, A. Verstuyf, H. F. Luderer, L. Lieben, C. Mathieu, and M. Demay. 2008. Vitamin D and human health: lessons from vitamin D receptor null mice. *Endocrine Reviews* 29:726–776.
6. Razzaque, M. S., D. Sitara, T. Taguchi, R. St-Arnaud, and B. Lanske. 2006. Premature aging–like phenotype in fibroblast growth factor 23 null mice is a vitamin D–mediated process. *FASEB J* 20:720–722.
7. Tuohimaa, P. 2009. Vitamin D and aging. *The Journal of Steroid Biochemistry and Molecular Biology* 114:78–84.
8. Norman, A. W. 2008. From vitamin D to hormone D: fundamentals of the vitamin D endocrine system essential for good health. *The American Journal of Clinical Nutrition* 88:491S–499S.

9. Ylikomi, T,, I. Laaksi, Y. R. Lou, P. Martikainen, S. Miettinen, P. Pennanen, S. Purmonen, H. Syvala, A. Vienonen, and P. Tuohimaa. 2002. Antiproliferative action of vitamin D. *Vitamins and Hormones* 64:357–406.

10. Schauber, J., and R. L. Gallo. 2008. The vitamin D pathway: a new target for control of the skin's immune response? *Experimental Dermatology* 17:633–639.

11. Guzey, M., S. Kitada, and J. C. Reed. 2002. Apoptosis induction by 1alpha,25-dihydroxyvitamin D3 in prostate cancer. *Molecular Cancer Therapeutics* 1:667–677.

12. Chatterjee, M. 2001. Vitamin D and genomic stability. *Mutation Research* 475:69–87.

13. Makishima, M., T. T. Lu, W. Xie, G. K. Whitfield, H. Domoto, R. M. Evans, M. R. Haussler, D. J. Mangelsdorf. 2002. Vitamin D receptor as an intestinal bile acid sensor. *Science* 296:1313–1316.

14. Thompson, P. D., P. W. Jurutka, G. K. Whitfield, S. M. Myskowski, K. R. Eichhorst, C. E. Dominguez, C. A. Haussler, and M. R. Haussler. 2002. Liganded VDR induces CYP3A4 in small intestinal and colon cancer cells via DR3 and ER6 vitamin D responsive elements. *Biochemical and Biophysical Research Communications* 299:730–738.

15. Lin, R., and J. H. White. 2004. The pleiotropic actions of vitamin D. *Bioessays* 26:21–28.

16. Blunt, J. W., H. F. DeLuca, and H. K. Schnoes. 1968. 25-Hydroxycholecalciferol. A biologically active metabolite of vitamin D3. *Biochemistry* 7:3317–3322.

17. Holick, M. F., H. K. Schnoes, H. F. DeLuca, T. Suda, and R. J. Cousins. 1971. Isolation and identification of 1,25-dihydroxycholecalciferol. A metabolite of vitamin D active in intestine. *Biochemistry* 10:2799–2804.

18. McDonnell, D. P., D. J. Mangelsdorf, J. W. Pike, M. R. Haussler, and B. W. O'Malley. 1987. Molecular cloning of complementary DNA encoding the avian receptor for vitamin D. *Science* 235:1214–1217.

19. Seifert, M., W. Tilgen, and J. Reichrath. 2009. Expression of 25-hydroxyvitamin D-1alpha-hydroxylase (1alphaOHase, CYP27B1) splice variants in HaCaT keratinocytes and other skin cells: modulation by culture conditions and UV-B treatment *in vitro*. *Anticancer Research* 29:3659–3667.

20. Hewison, M., F. Burke, K. N. Evans, D. A. Lammas, D. M. Sansom, P. Liu, R. L. Modlin, and J. S. Adams. 2007. Extra-renal 25-hydroxyvitamin D3-1alpha-hydroxylase in human health and disease. *The Journal of Steroid Biochemistry and Molecular Biology* 103:316–321.

21. Anderson, P. H., and G. J. Atkins. 2008 The skeleton as an intracrine organ for vitamin D metabolism. *Molecular Aspects of Medicine* 29:397–406.

22. McCarthy, K., C. Laban, S. A. Bustin, W. Ogunkolade, S. Khalaf, R. Carpenter, and P. J. Jenkins. 2009. Expression of 25-hydroxyvitamin D-1-alpha-hydroxylase, and vitamin D receptor mRNA in normal and malignant breast tissue. *Anticancer Research* 29:155–157.

23. Blomberg Jensen, M., J. E. Nielsen, A. Jorgensen, E. Rajpert-De Meyts, D. M. Kristensen, N. Jorgensen, N. E. Skakkebaek, A. Juul, and H. Leffers. 2010. Vitamin D receptor and vitamin D metabolizing enzymes are expressed in the human male reproductive tract. *Human Reproduction* 25:1303–1311.

24. Li, J., M. E. Byrne, E. Chang, Y. Jiang, S. S. Donkin, K. K. Buhman, J. R. Burgess, and D. Teegarden. 2008. 1alpha,25-Dihydroxyvitamin D hydroxylase in adipocytes. *The Journal of Steroid Biochemistry and Molecular Biology* 112:122–126.

25. van Driel, M., M. Koedam, C. J. Buurman, M. Hewison, H. Chiba, A. G. Uitterlinden, H. A. Pols, and J. P. van Leeuwen. 2010. Evidence for auto/paracrine actions of vitamin D in bone: 1alpha-hydroxylase expression and activity in human bone cells. *FASEB J* 20:2417–2419. *J. Steroid Biochem Mol Biol* 118:162–170.

26. Holick, M. F. 2004. Sunlight and vitamin D for bone health and prevention of autoimmune diseases, cancers, and cardiovascular disease. *The American Journal of Clinical Nutrition* 80:1678S–1688S.

27. Campbell, F. C., H. Xu, M. El-Tanani, P. Crowe, and V. Bingham. 2010. The yin and yang of vitamin D receptor (VDR) signaling in neoplastic progression: operational networks and tissue-specific growth control. *Biochemical Pharmacology* 79:1–9.

28. Bikle, D. D., and E. Gee. 1989. Free, and not total, 1,25-dihydroxyvitamin D regulates 25-hydroxyvitamin D metabolism by keratinocytes. *Endocrinology* 124:649–654.

29. Nykjaer, A., D. Dragun, D. Walther, H. Vorum, C. Jacobsen, J. Herz, F. Melsen, E. I. Christensen, and T. E. Willnow. 1999. An endocytic pathway essential for renal uptake and activation of the steroid 25-(OH) vitamin D3. *Cell* 96:507–515.

30. Adams, J. S. 2005. "Bound" to work: the free hormone hypothesis revisited. *Cell* 122:647–649.

31. Rowling, M. J., C. M. Kemmis, D. A. Taffany, and J. Welsh. 2010. Megalin-mediated endocytosis of vitamin D binding protein correlates with 25-hydroxycholecalciferol actions in human mammary cells. *Journal of Nutrition* 136:2754–2759. *J. Steroid Biochem Mol Biol* 121:254–256.

32. Bouillon, R., W. H. Okamura, and A. W. Norman. 1995 Structure–function relationships in the vitamin D endocrine system. *Endocrine Reviews* 16:200–257.

33. Haussler, M. R., C. A. Haussler, L. Bartik, G. K. Whitfield, J. C. Hsieh, S. Slater, and P. W. Jurutka. 2008. Vitamin D receptor: molecular signaling and actions of nutritional ligands in disease prevention. *Nutrition Review* 66:S98–112.

34. Lou, Y. R., I. Laaksi, H. Syvälä, M. Bläuer, T. L. Tammela, T. Ylikomi, and P. Tuohimaa. 2004. 25-Hydroxyvitamin D3 is an active hormone in human primary prostatic stromal cells. *FASEB J* 18:332–334.

35. Lou, Y-R., F. Molnár, S. Qiao, A. V. Kalueff, R. St-Arnaud, C. Carlberg, and P. Tuohimaa. 2009. 25-Hydroxyvitamin D3 is an agonistic vitamin D receptor ligand. *The Journal of Steroid Biochemistry and Molecular Biology*.

36. Lou, Y. R., F. Molnar, M. Perakyla, S. Qiao, A. V. Kalueff, R. St-Arnaud, C. Carlberg, and P. Tuohimaa. 2010. 25-Hydroxyvitamin D(3) is an agonistic vitamin D receptor ligand. *The Journal of Steroid Biochemistry and Molecular Biology* 118:162–170.

37. Peng, X., M. Hawthorne, A. Vaishnav, R. St-Arnaud, and R. G. Mehta. 2009. 25-Hydroxyvitamin D3 is a natural chemopreventive agent against carcinogen induced precancerous lesions in mouse mammary gland organ culture. *Breast Cancer Research and Treatment* 113:31–41.

38. Boyan, B. D., V. L. Sylvia, N. McKinney, and Z. Schwartz. 2003. Membrane actions of vitamin D metabolites 1alpha,25(OH)2D3 and 24R,25(OH)2D3 are retained in growth plate cartilage cells from vitamin D receptor knockout mice. *Journal of Cellular Biochemistry* 90:1207–1223.

39. van Driel, M., M. Koedam, C. J. Buurman, M. Roelse, F. Weyts, H. Chiba, A. G. Uitterlinden, H. A. Pols, and J. P. van Leeuwen. 2006. Evidence that both 1alpha,25-dihydroxyvitamin D3 and 24-hydroxylated D3 enhance human osteoblast differentiation and mineralization. *Journal of Cellular Biochemistry* 99:922–935.

40. St-Arnaud, R., A. Arabian, and R.-P. Naja. 2009. CYP24A1, the 25-hydroxyvitamin D-24-hydroxylase, is necessary for optimal fracture repair. 14th Workshop on Vitamin D, Book of Abstracts 54.

41. St-Arnaud, R. 2010. CYP24A1-deficient mice as a tool to uncover a biological activity for vitamin D metabolites hydroxylated at position 24. *The Journal of Steroid Biochemistry and Molecular Biology*.

42. Tuohimaa, P., T. Keisala, A. Minasyan, J. Cachat, and A. Kalueff. 2009. Vitamin D, nervous system and aging. *Psychoneuroendocrinology* 34 Suppl 1:S278–286.

43. Zbytek, B., Z. Janjetovic, R. C. Tuckey, M. A. Zmijewski, T. W. Sweatman, E. Jones, M. N. Nguyen, and A. T. Slominski. 2008. 20-Hydroxyvitamin D3, a product of vitamin D3 hydroxylation by cytochrome P450scc, stimulates keratinocyte differentiation. *The Journal of Investigative Dermatology* 128:2271–2280.

44. Molnar, F., R. Sigueiro, Y. Sato, C. Araujo, I. Schuster, P. Antony, J. Peluso, C. Muller, A. Mourino, D. Moras, and N. Rochel. 1alpha,25(OH)2-3-epi-vitamin D3, a natural physiological metabolite of vitamin D3: its synthesis, biological activity and crystal structure with its receptor. *PLoS One* 6:e18124.

45. Tuckey, R. C., Z. Janjetovic, W. Li, M. N. Nguyen, M. A. Zmijewski, J. Zjawiony, A. Slominski. 2008. Metabolism of 1alpha-hydroxyvitamin D3 by cytochrome P450scc to biologically active 1alpha,20-dihydroxyvitamin D3. *The Journal of Steroid Biochemistry and Molecular Biology* 112:213–219.

46. Halloran, B. P., D. D. Bikle, M. J. Levens, M. E. Castro, R. K. Globus, and E. Holton. 1986. Chronic 1,25-dihydroxyvitamin D3 administration in the rat reduces the serum concentration of 25-hydroxyvitamin D by increasing metabolic clearance rate. *Journal of Clinical Investigation* 78:622–628.

47. Clements, M. R., M. Davies, M. E. Hayes, C. D. Hickey, G. A. Lumb, E. B. Mawer, and P. H. Adams. 1992. The role of 1,25-dihydroxyvitamin D in the mechanism of acquired vitamin D deficiency. *Clinical Endocrinology* 37:17–27.

48. Reinholz, G. G., and H. F. DeLuca. 1998. Inhibition of 25-hydroxyvitamin D3 production by 1, 25-dihydroxyvitamin D3 in rats. *Archives of Biochemistry and Biophysics* 355:77–83.

49. Vieth, R., K. McCarten, and K. H. Norwich. 1990. Role of 25-hydroxyvitamin D3 dose in determining rat 1,25-dihydroxyvitamin D3 production. *American Journal of Physiology* 258:E780–789.

50. Mizwicki, M. T., and A. W. Norman. 2009. The vitamin D sterol-vitamin D receptor ensemble model offers unique insights into both genomic and rapid-response signaling. *Science Signaling* 2:re4.

51. Shaffer, P. L., and D. T. Gewirth. 2004. Structural analysis of RXR–VDR interactions on DR3 DNA. *The Journal of Steroid Biochemistry and Molecular Biology* 89–90:215–219.

52. Seoane, S., and R. Perez-Fernandez. 2006. The vitamin D receptor represses transcription of the pituitary transcription factor Pit-1 gene without involvement of the retinoid X receptor. *Molecular Endocrinology* 20:735–748.

53. Shaffer, P. L., and D. T. Gewirth. 2002. Structural basis of VDR–DNA interactions on direct repeat response elements. *EMBO J* 21:2242–2252.

54. Jurutka, P. W., P. N. MacDonald, S. Nakajima, J. C. Hsieh, P. D. Thompson, G. K. Whitfield, M. A. Galligan, C. A. Haussler, and M. R. Haussler. 2002. Isolation of baculovirus-expressed human vitamin D

receptor: DNA responsive element interactions and phosphorylation of the purified receptor. *Journal of Cellular Biochemistry* 85:435–457.

55. Koszewski, N. J., T. A. Reinhardt, and R. L. Horst. 1999. Differential effects of 20-epi vitamin D analogs on the vitamin D receptor homodimer. *Journal of Bone and Mineral Research* 14:509–517.

56. Nishikawa, J., M. Kitaura, M. Matsumoto, M. Imagawa, and T. Nishihara. 1994. Difference and similarity of DNA sequence recognized by VDR homodimer and VDR/RXR heterodimer. *Nucleic Acids Research* 22:2902–2907.

57. Marx, S. J., M. M. Bliziotes, and M. Nanes. 1986. Analysis of the relation between alopecia and resistance to 1,25-dihydroxyvitamin D. *Clinical Endocrinology* 25:373–381.

58. Sakai, Y., and M. B. Demay. 2000. Evaluation of keratinocyte proliferation and differentiation in vitamin D receptor knockout mice. *Endocrinology* 141:2043–2049.

59. Chen, C. H., Y. Sakai, and M. B. Demay. 2001. Targeting expression of the human vitamin D receptor to the keratinocytes of vitamin D receptor null mice prevents alopecia. *Endocrinology* 142:5386–5389.

60. Li, Y. C., M. Amling, A. E. Pirro, M. Priemel, J. Meuse, R. Baron, G. Delling, and M. B. Demay. 1998. Normalization of mineral ion homeostasis by dietary means prevents hyperparathyroidism, rickets, and osteomalacia, but not alopecia in vitamin D receptor–ablated mice. *Endocrinology* 139:4391–4396.

61. Dardenne, O., J. Prud'homme, A. Arabian, F. H. Glorieux, and R. St-Arnaud. 2001. Targeted inactivation of the 25-hydroxyvitamin D(3)-1(alpha)-hydroxylase gene (CYP27B1) creates an animal model of pseudovitamin D-deficiency rickets. *Endocrinology* 142:3135–3141.

62. Hayes, D. P. 2010. Vitamin D and ageing. *Biogerontology* 11:1–16.

63. Rattan, S. I. 2008. Principles and practice of hormetic treatment of aging and age-related diseases. *Human & Experimental Toxicology* 27:151–154.

64. Rattan, S. I. 2008. Hormesis in aging. *Ageing Research Reviews* 7:63–78.

65. Richardson, R. B. 2009. Ionizing radiation and aging: rejuvenating an old idea. *Aging* 1:887–902.

66. Tuohimaa, P., L. Tenkanen, M. Ahonen, S. Lumme, E. Jellum, G. Hallmans, P. Stattin, S. Harvei, T. Hakulinen, T. Luostarinen, J. Dillner, M. Lehtinen, and M. Hakama. 2004. Both high and low levels of blood vitamin D are associated with a higher prostate cancer risk: a longitudinal, nested case-control study in the Nordic countries. *International Journal of Cancer* 108:104–108.

67. Faupel-Badger, J. M., L. Diaw, D. Albanes, J. Virtamo, K. Woodson, and J. A. Tangrea. 2007. Lack of association between serum levels of 25-hydroxyvitamin D and the subsequent risk of prostate cancer in Finnish men. *Cancer Epidemiology, Biomarkers & Prevention* 16:2784–2786.

68. Visser, M., D. J. Deeg, M. T. Puts, J. C. Seidell, and P. Lips. 2006. Low serum concentrations of 25-hydroxyvitamin D in older persons and the risk of nursing home admission. *The American Journal of Clinical Nutrition* 84:616–622; quiz 671–612.

69. IARC. 2008. Vitamin D and cancer. IARC Working Group Reports 5.

70. Melamed, M. L., E. D. Michos, W. Post, and B. Astor. 2008. 25-Hydroxyvitamin D levels and the risk of mortality in the general population. *Archives of Internal Medicine* 168:1629–1637.

71. Wang, T. J., M. J. Pencina, S. L. Booth, P. F. Jacques, E. Ingelsson, K. Lanier, E. J. Benjamin, R. B. D'Agostino, M. Wolf, and R. S. Vasan. 2008. Vitamin D deficiency and risk of cardiovascular disease. *Circulation* 117:503–511.

72. Chen, W., S. M. Dawsey, Y. L. Qiao, S. D. Mark, Z. W. Dong, P. R. Taylor, P. Zhao, and C. C. Abnet. 2007. Prospective study of serum 25(OH)-vitamin D concentration and risk of oesophageal and gastric cancers. *British Journal of Cancer* 97:123–128.

73. Stolzenberg-Solomon, R. Z., R. Vieth, A. Azad, P. Pietinen, P. R. Taylor, J. Virtamo, and D. Albanes. 2006. A prospective nested case-control study of vitamin D status and pancreatic cancer risk in male smokers. *Cancer Research* 66:10213–10219.

74. Stumpf, W. E. 2006. The dose makes the medicine. *Drug Discovery Today* 11:550–555.

75. Sanders, K. M., A. L. Stuart, E. J. Williamson, J. A. Simpson, M. A. Kotowicz, D. Young, and G. C. Nicholson. 2010. Annual high-dose oral vitamin D and falls and fractures in older women: a randomized controlled trial. *JAMA* 303:1815–1822.

76. McGrath, J. J., D. W. Eyles, C. B. Pedersen, C. Anderson, P. Ko, T. H. Burne, B. Norgaard-Pedersen, D. M. Hougaard, and P. B. Mortensen. 2010. Neonatal vitamin D status and risk of schizophrenia: a population-based case-control study. *Archives of General Psychiatry* 67:889–894.

77. Muenstedt, K., and S. El-Safadi. 2010. Nutritive supplements—help or harm for breast cancer patients? *Breast Care (Basel, Switzerland)* 5:383–387.

78. Michaelsson, K., J. A. Baron, G. Snellman, R. Gedeborg, L. Byberg, J. Sundstrom, L. Berglund, J. Arnlov, P. Hellman, R. Blomhoff, A. Wolk, H. Garmo, L. Holmberg, and H. Melhus. 2010. Plasma vitamin D and

mortality in older men: a community-based prospective cohort study. *The American Journal of Clinical Nutrition* 92:841–848.

79. Bodnar, L. M., J. M. Catov, J. M. Zmuda, M. E. Cooper, M. S. Parrott, J. M. Roberts, M. L. Marazita, and H. N. Simhan. 2010. Maternal serum 25-hydroxyvitamin D concentrations are associated with small-for-gestational age births in white women. *Journal of Nutrition* 140:999–1006.

80. McGrath, J. J., S. Saha, T. H. Burne, and D. W. Eyles. 2010. A systematic review of the association between common single nucleotide polymorphisms and 25-hydroxyvitamin D concentrations. *The Journal of Steroid Biochemistry and Molecular Biology* 121:471–477.

81. Querfeld, U., and R. H. Mak. 2010. Vitamin D deficiency and toxicity in chronic kidney disease: in search of the therapeutic window. *Pediatric Nephrology* 25:2413–2430.

82. Gloth, F. M., 3rd, C. M. Gundberg, B. W. Hollis, J. G. Haddad, Jr., and J. D. Tobin. 1995. Vitamin D deficiency in homebound elderly persons. *JAMA* 274:1683–1686.

83. van der Wielen, R. P., M. R. Lowik, H. van den Berg, L. C. de Groot, J. Haller, O. Moreiras, and W. A. van Staveren. 1995. Serum vitamin D concentrations among elderly people in Europe. *Lancet* 346:207–210.

84. McMurtry, C. T., S. E. Young, R. W. Downs, and R. A. Adler. 1992. Mild vitamin D deficiency and secondary hyperparathyroidism in nursing home patients receiving adequate dietary vitamin D. *Journal of the American Geriatrics Society* 40:343–347.

85. Holick, M. F., L. Y. Matsuoka, and J. Wortsman. 1989. Age, vitamin D, and solar ultraviolet. *Lancet* 2:1104–1105.

86. MacLaughlin, J., and M. F. Holick. 1985. Aging decreases the capacity of human skin to produce vitamin D3. *Journal of Clinical Investigation* 76:1536–1538.

87. Harris, S. S., and B. Dawson-Hughes. 2002. Plasma vitamin D and 25OHD responses of young and old men to supplementation with vitamin D3. *Journal of the American College of Nutrition* 21:357–362.

88. Martinez P., I. Moreno, F. De Miguel, V. Vila, P. Esbrit, and M. E. Martinez. 2001. Changes in osteocalcin response to 1,25-dihydroxyvitamin D(3) stimulation and basal vitamin D receptor expression in human osteoblastic cells according to donor age and skeletal origin. *Bone* 29:35–41.

89. Ebeling, P. R., M. E. Sandgren, E. P. DiMagno, A. W. Lane, H. F. DeLuca, and B. L. Riggs. 1992. Evidence of an age-related decrease in intestinal responsiveness to vitamin D: relationship between serum 1,25-dihydroxyvitamin D3 and intestinal vitamin D receptor concentrations in normal women. *The Journal of Clinical Endocrinology and Metabolism* 75:176–182.

90. Dawson-Hughes, B. 2008. Serum 25-hydroxyvitamin D and functional outcomes in the elderly. *The American Journal of Clinical Nutrition* 88:537S–540S.

91. Boland, R. 1986. Role of vitamin D in skeletal muscle function. *Endocrine Reviews* 7:434–448.

92. Bischoff-Ferrari, H. A., M. Borchers, F. Gudat, U. Durmuller, H. B. Stahelin, and W. Dick. 2004. Vitamin D receptor expression in human muscle tissue decreases with age. *Journal of Bone and Mineral Research* 19:265–269.

93. Sato, Y., J. Iwamoto, T. Kanoko, and K. Satoh. 2005. Low-dose vitamin D prevents muscular atrophy and reduces falls and hip fractures in women after stroke: a randomized controlled trial. *Cerebrovascular Diseases* 20:187–192.

94. Minasyan, A., T. Keisala, A. Zou, Y. Zhang, E. Toppila, H. Syvälä, Y. R. Lou, A. Kalueff, I. Pyykkö, and P. Tuohimaa. 2009. Balance deficits in vitamin D receptor mutant mice. *The Journal of Steroid Biochemistry and Molecular Biology*. Vestibular dysfunction in vitamin D receptor mutant mice. *J. Steroid Biochem Mol Biol* 114:161–166.

95. Ikeda, K., T. Kobayashi, J. Kusakari, T. Takasaka, S. Yumita, and Y. Furukawa. 1987. Sensorineural hearing loss associated with hypoparathyroidism. *Laryngoscope* 97:1075–1079.

96. Ishida, S., H. Isotani, K. Kameoka, and T. Kishi. 2001. Familial idiopathic hypoparathyroidism, sensorineural deafness and renal dysplasia. *Internal Medicine* 40:110–113.

97. Parekh, N., R. J. Chappell, A. E. Millen, D. M. Albert, and J. A. Mares. 2007. Association between vitamin D and age-related macular degeneration in the Third National Health and Nutrition Examination Survey, 1988 through 1994. *Archives of Ophthalmology* 125:661–669.

98. Brookes, E. B., and A. W. Morrison. 1981. Vitamin D deficiency and deafness. *British Medical Journal (Clinical Research Ed.)* 283:273–274.

99. Brookes, G. B. 1983. Vitamin D deficiency—a new cause of cochlear deafness. *The Journal of Laryngology and Otology* 97:405–420.

100. Brookes, G. B. 1985. Vitamin D deficiency and deafness: 1984 update. *The American Journal of Otology* 6:102–107.

101. Ziporyn, T. 1983. Possible link probed: deafness and vitamin D. *JAMA* 250:1951–1952.

102. Zou, J., A. Minasyan, T. Keisala, Y. Zhang, J. H. Wang, Y. R. Lou, A. Kalueff, I. Pyykko, and P. Tuohimaa. 2008. Progressive hearing loss in mice with a mutated vitamin D receptor gene. *Audiology & Neuro-Otology* 13:219–230.

103. Minasyan, A., T. Keisala, Y. R. Lou, A. V. Kalueff, and P. Tuohimaa. 2007. Neophobia, sensory and cognitive functions, and hedonic responses in vitamin D receptor mutant mice. *The Journal of Steroid Biochemistry and Molecular Biology* 104:274–280.

104. Cohen, H. N., I. Fogelman, I. T. Boyle, and J. A. Doig. 1979. Deafness due to hypervitaminosis D. *Lancet* 1:985.

105. Garcion, E., N. Wion-Barbot, C. N. Montero-Menei, F. Berger, and D. Wion. 2002. New clues about vitamin D functions in the nervous system. *Trends in Endocrinology and Metabolism* 13:100–105.

106. Llewellyn, D. J., K. M. Langa, and I. A. Lang. 2009. Serum 25-hydroxyvitamin D concentration and cognitive impairment. *Journal of Geriatric Psychiatry and Neurology* 22:188–195.

107. Oudshoorn, C., F. U. Mattace-Raso, N. van der Velde, E. M. Colin, and T. J. van der Cammen. 2008. Higher serum vitamin D3 levels are associated with better cognitive test performance in patients with Alzheimer's disease. *Dementia and Geriatric Cognitive Disorders* 25:539–543.

108. Lee, D. M., A. Tajar, A. Ulubaev, N. Pendleton, T. W. O'Neill, D. B. O'Connor, G. Bartfai, S. Boonen, R. Bouillon, F. F. Casanueva, J. D. Finn, G. Forti, A. Giwercman, T. S. Han, I. T. Huhtaniemi, K. Kula, M. E. Lean, M. Punab, A. J. Silman, D. Vanderschueren, and F. C. Wu. 2009. Association between 25-hydroxyvitamin D levels and cognitive performance in middle-aged and older European men. *Journal of Neurology, Neurosurgery and Psychiatry* 80:722–729.

109. Annweiler, C., A. M. Schott, G. Allali, S. A. Bridenbaugh, R. W. Kressig, P. Allain, F. R. Herrmann, O. Beauchet. 2010. Association of vitamin D deficiency with cognitive impairment in older women: cross-sectional study. *Neurology* 74:27–32.

110. Gloth, F. M., 3rd, W. Alam, and B. Hollis. 1999. Vitamin D vs broad spectrum phototherapy in the treatment of seasonal affective disorder. *The Journal of Nutrition, Health & Aging* 3:5–7.

111. Nanri, A., T. Mizoue, Y. Matsushita, K. Poudel-Tandukar, M. Sato, M. Ohta, and N. Mishima. 2009. Association between serum 25-hydroxyvitamin D and depressive symptoms in Japanese: analysis by survey season. *European Journal of Clinical Nutrition* 63:1444–1447.

112. Bertone-Johnson, E. R. 2009. Vitamin D and the occurrence of depression: causal association or circumstantial evidence? *Nutrition Review* 67:481–492.

113. Shipowick, C. D., C. B. Moore, C. Corbett, and R. Bindler. 2009. Vitamin D and depressive symptoms in women during the winter: a pilot study. *Applied Nursing Research* 22:221–225.

114. Evatt, M. L., M. R. Delong, N. Khazai, A. Rosen, S. Triche, and V. Tangpricha. 2008. Prevalence of vitamin D insufficiency in patients with Parkinson disease and Alzheimer disease. *Archives of Neurology* 65:1348–1352.

115. Masoumi, A., B. Goldenson, S. Ghirmai, H. Avagyan, J. Zaghi, K. Abel, X. Zheng, A. Espinosa-Jeffrey, M. Mahanian, P. T. Liu, M. Hewison, M. Mizwickie, J. Cashman, and M. Fiala. 2009. 1alpha,25-Dihydroxyvitamin D3 interacts with curcuminoids to stimulate amyloid-beta clearance by macrophages of Alzheimer's disease patients. *Journal of Alzheimer's Disease* 17:703–717.

116. MacLaughlin, J. A., R. R. Anderson, and M. F. Holick. 1982. Spectral character of sunlight modulates photosynthesis of previtamin D3 and its photoisomers in human skin. *Science* 216:1001–1003.

117. Beckman, M. J., P. Tadikonda, E. Werner, J. Prahl, S. Yamada, and H. F. DeLuca. 1996. Human 25-hydroxyvitamin D3-24-hydroxylase, a multicatalytic enzyme. *Biochemistry* 35:8465–8472.

118. Brown, A. J., A. Dusso, and E. Slatopolsky. 1999. Vitamin D. *The American Journal of Physiology* 277:F157–175.

119. Vahavihu, K., L. Ylianttila, R. Salmelin, C. Lamberg-Allardt, H. Viljakainen, P. Tuohimaa, T. Reunala, and E. Snellman. 2008. Heliotherapy improves vitamin D balance and atopic dermatitis. *The British Journal of Dermatology* 158:1323–1328.

120. Viljakainen, H. T., A., Palssa, M. Karkkainen, J. Jakobsen, and C. Lamberg-Allardt. 2006. How much vitamin D3 do the elderly need? *Journal of the American College of Nutrition* 25:429–435.

121. Heaney, R. P., K. M. Davies, T. C. Chen, M. F. Holick, and M. J. Barger-Lux. 2003. Human serum 25-hydroxycholecalciferol response to extended oral dosing with cholecalciferol. *The American Journal of Clinical Nutrition* 77:204–210.

122. Heaney, R. P. 2008. Vitamin D in health and disease. *Clinical Journal of the American Society of Nephrology* 3:1535–1541.

123. Pearce, S. H., and T. D. Cheetham. 2010. Diagnosis and management of vitamin D deficiency. *BMJ* 340:b5664.

124. Cranney, A., H. A. Weiler, S. O'Donnell, and L. Puil. 2008. Summary of evidence-based review on vitamin D efficacy and safety in relation to bone health. *The American Journal of Clinical Nutrition* 88:513S–519S.

125. Ruohola, J. P., I. Laaksi, T. Ylikomi, R. Haataja, V. M. Mattila, T. Sahi, P, Tuohimaa, and H. Pihlajamaki. 2006. Association between serum 25(OH)D concentrations and bone stress fractures in Finnish young men. *Journal of Bone and Mineral Research* 21:1483–1488.

126. Montero-Odasso, M., and G. Duque. 2005. Vitamin D in the aging musculoskeletal system: an authentic strength preserving hormone. *Molecular Aspects of Medicine* 26:203–219.

127. Tuohimaa, P. 2008. Vitamin D, aging, and cancer. *Nutrition Review* 66:S147–152.

128. Garland, C. F., E. D. Gorham, S. B. Mohr, W. B. Grant, E. L. Giovannucci, M. Lipkin, H. Newmark, M. F. Holick, and F. C. Garland. 2007. Vitamin D and prevention of breast cancer: pooled analysis. *The Journal of Steroid Biochemistry and Molecular Biology* 103:708–711.

129. Abbas, S., J. Chang-Claude, and J. Linseisen. 2009. Plasma 25-hydroxyvitamin D and premenopausal breast cancer risk in a German case-control study. *International Journal of Cancer* 124:250–255.

130. Ahonen, M. H., L. Tenkanen, L. Teppo, M. Hakama, and P. Tuohimaa. 2000. Prostate cancer risk and prediagnostic serum 25-hydroxyvitamin D levels (Finland). *Cancer Causes & Control* 11:847–852.

131. Tworoger, S. S., I. M. Lee, J. E. Buring, B. Rosner, B. W. Hollis, and S. E. Hankinson. 2007. Plasma 25-hydroxyvitamin D and 1,25-dihydroxyvitamin D and risk of incident ovarian cancer. *Cancer Epidemiology, Biomarkers & Prevention* 16:783–788.

132. Feskanich, D., J. Ma, C. S. Fuchs, G. J. Kirkner, S. E. Hankinson, B. W. Hollis, and E. L. Giovannucci. 2004. Plasma vitamin D metabolites and risk of colorectal cancer in women. *Cancer Epidemiology, Biomarkers & Prevention* 13:1502–1508.

133. Tangrea, J., K. Helzlsouer, P. Pietinen, P. Taylor, B. Hollis, J. Virtamo, and D. Albanes. 1997. Serum levels of vitamin D metabolites and the subsequent risk of colon and rectal cancer in Finnish men. *Cancer Causes & Control* 8:615–625.

134. Forman, J. P., G. C. Curhan, and E. N. Taylor. 2008. Plasma 25-hydroxyvitamin D levels and risk of incident hypertension among young women. *Hypertension* 52:828–832.

135. Margolis, K. L., R. M. Ray, L. Van Horn, J. E. Manson, M. A. Allison, H. R. Black, S. A. Beresford, S. A. Connelly, J. D. Curb, and R. H. Grimm, Jr., T. A. Kotchen, L. H. Kuller, S. Wassertheil-Smoller, C. A. Thomson, and J. C. Torner. 2008. Effect of calcium and vitamin D supplementation on blood pressure: the Women's Health Initiative Randomized Trial. *Hypertension* 52:847–855.

136. Fogh, K., and K. Kragballe. 2004. New vitamin D analogs in psoriasis. *Current Drug Targets. Inflammation and Allergy* 3:199–204.

137. Ranganathan, P. 2009. Genetics of bone loss in rheumatoid arthritis—role of vitamin D receptor polymorphisms. *Rheumatology (Oxford)* 48:342–346.

138. Hypponen, E., E. Laara, A. Reunanen, M. R. Jarvelin, and S. M. Virtanen. 2001. Intake of vitamin D and risk of type 1 diabetes: a birth-cohort study. *Lancet* 358:1500–1503.

139. Mathieu, C., C. Gysemans, A. Giulietti, and R. Bouillon. 2005. Vitamin D and diabetes. *Diabetologia* 48:1247–1257.

140. Sloka, S., M. Grant, and L. A. Newhook. 2009. The geospatial relation between UV solar radiation and type 1 diabetes in Newfoundland. *Acta Diabetologica* 47:73–78.

141. Laaksi, I., J. P. Ruohola, P. Tuohimaa, A. Auvinen, R. Haataja, H. Pihlajamaki, and T. Ylikomi. 2007. An association of serum vitamin D concentrations < 40 nmol/L with acute respiratory tract infection in young Finnish men. *The American Journal of Clinical Nutrition* 86:714–717.

142. Munger, K. L., S. M. Zhang, E. O'Reilly, M. A. Hernan, M. J. Olek, W. C. Willett, and A. Ascherio. 2004. Vitamin D intake and incidence of multiple sclerosis. *Neurology* 62:60–65.

143. Munger, K. L., L. I. Levin, B. W. Hollis, N. S. Howard, and A. Ascherio. 2006. Serum 25-hydroxyvitamin D levels and risk of multiple sclerosis. *JAMA* 296:2832–2838.

144. Zittermann, A. 2003. Vitamin D in preventive medicine: are we ignoring the evidence? *The British Journal of Nutrition* 89:552–572.

145. Zittermann, A. 2006. Vitamin D and disease prevention with special reference to cardiovascular disease. *Progress in Biophysics and Molecular Biology* 92:39–48.

146. Li, Y. C., G. Qiao, M. Uskokovic, W. Xiang, W. Zheng, and J. Kong. 2004. Vitamin D: a negative endocrine regulator of the renin–angiotensin system and blood pressure. *The Journal of Steroid Biochemistry and Molecular Biology* 89–90:387–392.

147. Freundlich, M., Y. Quiroz, Z. Zhang, Y. Zhang, Y. Bravo, J. R. Weisinger, Y. C. Li, and B. Rodriguez-Iturbe. 2008. Suppression of renin–angiotensin gene expression in the kidney by paricalcitol. *Kidney International* 74:1394–1402.

148. Norman, A. W., J. B. Frankel, A. M. Heldt, and G. M. Grodsky. 1980. Vitamin D deficiency inhibits pancreatic secretion of insulin. *Science* 209:823–825.

149. Billaudel, B. J., P. M. Bourlon, B. C. Sutter, and A. G. Faure-Dussert. 1995. Regulatory effect of 1,25-dihydroxyvitamin D3 on insulin release and calcium handling via the phospholipid pathway in islets from vitamin D–deficient rats. *Journal of Endocrinological Investigation* 18:673–682.

150. Bouillon, R., H. Bischoff-Ferrari, and W. Willett. 2008. Vitamin D and health: perspectives from mice and man. *Journal of Bone and Mineral Research* 23:974–979.

151. Hofer, A. C., R. T. Tran, O. Z. Aziz, W. Wright, G. Novelli, J. Shay, and M. Lewis. 2005. Shared phenotypes among segmental progeroid syndromes suggest underlying pathways of aging. *The Journals of Gerontology. Series A, Biological Sciences and Medical Sciences* 60:10–20.

152. Kuro-o, M., Y. Matsumura, H. Aizawa, H. Kawaguchi, T. Suga, T. Utsugi, Y. Ohyama, M. Kurabayashi, T. Kaname, E. Kume, H. Iwasaki, A. Iida, T. Shiraki-Iida, S. Nishikawa, R. Nagai, and Y. I. Nabeshima. 1997. Mutation of the mouse klotho gene leads to a syndrome resembling ageing. *Nature* 390:45–51.

153. Arking, D. E., A. Krebsova, M. Macek, Sr., M. Macek, Jr., A. Arking, I. S. Mian, L. Fried, A. Hamosh, S. Dey, I. McIntosh, and H. C. Dietz. 2002. Association of human aging with a functional variant of klotho. *Proceedings of the National Academy of Sciences of the United States of America* 99:856–861.

154. Shimada, T., H. Hasegawa, Y. Yamazaki, T. Muto, R. Hino, Y. Takeuchi, T. Fujita, K. Nakahara, S. Fukumoto, and T. Yamashita. 2004. FGF-23 is a potent regulator of vitamin D metabolism and phosphate homeostasis. *Journal of Bone and Mineral Research* 19:429–435.

155. Kurosu, H., M. Yamamoto, J. D. Clark, J. V. Pastor, A. Nandi, P. Gurnani, O. P. McGuinness, H. Chikuda, M. Yamaguchi, H. Kawaguchi, I. Shimomura, Y. Takayama, J. Herz, C. R. Kahn, K. P. Rosenblatt, and M. Kuro-o. 2005. Suppression of aging in mice by the hormone Klotho. *Science* 309:1829–1833.

156. Maekawa, Y., K. Ishikawa, O. Yasuda, R. Oguro, H. Hanasaki, I. Kida, Y. Takemura, M. Ohishi, T. Katsuya, and H. Rakugi. 2009. Klotho suppresses TNF-alpha–induced expression of adhesion molecules in the endothelium and attenuates NF-kappaB activation. *Endocrine* 35:341–346.

157. Lanske, B., and M. S. Razzaque. 2007. Mineral metabolism and aging: the fibroblast growth factor 23 enigma. *Current Opinion in Nephrology and Hypertension* 16:311–318.

158. Peacock, M. 2010. Calcium metabolism in health and disease. *Clinical Journal of the American Society of Nephrology* 5 Suppl 1:S23–30.

159. Yu, J. T., R. C. Chang, and L. Tan. 2009. Calcium dysregulation in Alzheimer's disease: from mechanisms to therapeutic opportunities. *Progress in Neurobiology* 89:240–255.

160. Russell, S. J., and C. R. Kahn. 2007. Endocrine regulation of ageing. *Nature Reviews. Molecular Cell Biology* 8:681–691.

161. May, E., K. Asadullah, and U. Zugel. 2004. Immunoregulation through 1,25-dihydroxyvitamin D3 and its analogs. *Current Drug Targets. Inflammation and Allergy* 3:377–393.

162. Fernandes de Abreu, D. A., D. Eyles, and F. Feron. 2009. Vitamin D, a neuro-immunomodulator: implications for neurodegenerative and autoimmune diseases. *Psychoneuroendocrinology* 34 Suppl 1: S265–277.

163. Kalueff, A. V., and P. Tuohimaa. 2007. Neurosteroid hormone vitamin D and its utility in clinical nutrition. *Current Opinion in Clinical Nutrition and Metabolic Care* 10:12–19.

164. Finkel, T., C. X. Deng, and R. Mostoslavsky. 2009. Recent progress in the biology and physiology of sirtuins. *Nature* 460:587–591.

165. Sinclair, D., K. Mills, and L. Guarente. 1998. Aging in *Saccharomyces cerevisiae*. *Annual Review of Microbiology* 52:533–560.

166. Longo, V. D., and B. K. Kennedy. 2006. Sirtuins in aging and age-related disease. *Cell* 126:257–268.

167. Li, Y., W. Xu, M. W. McBurney, and V. D. Longo. 2008. SirT1 inhibition reduces IGF-I/IRS-2/Ras/ERK1/2 signaling and protects neurons. *Cell Metabolism* 8:38–48.

168. Boily, G., E. L. Seifert, L. Bevilacqua, X. H. He, G. Sabourin, C. Estey, C. Moffat, S. Crawford, S. Saliba, K. Jardine, J. Xuan, M. Evans, M. E. Harper, and M. W. McBurney. 2008. SirT1 regulates energy metabolism and response to caloric restriction in mice. *PLoS One* 3:e1759.

169. Salminen, A., and K. Kaarniranta. 2009. SIRT1: regulation of longevity via autophagy. *Cellular Signalling* 21:1356–1360.

170. Salminen, A., A. Kauppinen, T. Suuronen, and K. Kaarniranta. 2008. SIRT1 longevity factor suppresses NF-kappaB–driven immune responses: regulation of aging via NF-kappaB acetylation? *Bioessays* 30:939–942.

171. Maetani, M., G. Maskarinec, A. A. Franke, and R. V. Cooney. 2009. Association of leptin, 25-hydroxyvitamin D, and parathyroid hormone in women. *Nutrition and Cancer* 61:225–231.

172. Matsunuma, A., T. Kawane, T. Maeda, S. Hamada, and N. Horiuchi. 2004. Leptin corrects increased gene expression of renal 25-hydroxyvitamin D3-1 alpha-hydroxylase and -24-hydroxylase in leptin-deficient, ob/ob mice. *Endocrinology* 145:1367–1375.

173. Lieb, W., A. S. Beiser, R. S. Vasan, Z. S. Tan, R. Au, T. B. Harris, R. Roubenoff, S. Auerbach, C. DeCarli, P. A. Wolf, and S. Seshadri. 2009. Association of plasma leptin levels with incident Alzheimer's disease and MRI measures of brain aging. *JAMA* 302:2565–2572.

174. Imura, A., Y. Tsuji, M. Murata, R. Maeda, K. Kubota, A. Iwano, C. Obuse, K. Togashi, M. Tominaga, N. Kita, K. Tomiyama, J. Iijima, Y. Nabeshima, M. Fujioka, R. Asato, S. Tanaka, K. Kojima, J. Ito, K. Nozaki, N. Hashimoto, T. Ito, T. Nishio, T. Uchiyama, and T. Fujimori. 2007. Alpha-Klotho as a regulator of calcium homeostasis. *Science* 316:1615–1618.

175. Green, K. N., and F. M. LaFerla. 2008 Linking calcium to Abeta and Alzheimer's disease. *Neuron* 59:190–194.

176. Gentili, C., S. Morelli, and A. R. de Boland. 2004. PTH and phospholipase A2 in the aging process of intestinal cells. *Journal of Cellular Biochemistry* 93:312–326.

177. Sambrook, P. N., J. S. Chen, L. M. March, I. D. Cameron, R. G. Cumming, S. R. Lord, J. Schwarz, and M. J. Seibel. 2004. Serum parathyroid hormone is associated with increased mortality independent of 25-hydroxy vitamin D status, bone mass, and renal function in the frail and very old: a cohort study. *The Journal of Clinical Endocrinology and Metabolism* 89:5477–5481.

178. Holick, M. F., X. Q. Tian, and M. Allen. 1995. Evolutionary importance for the membrane enhancement of the production of vitamin D3 in the skin of poikilothermic animals. *Proceedings of the National Academy of Sciences of the United States of America* 92:3124–3126.

179. Failla, G. 1958. The aging process and cancerogenesis. *Annals of the New York Academy of Sciences* 71:1124–1140.

180. Kirkwood, T. B. 1977. Evolution of ageing. *Nature* 270:301–304.

181. Harman, D. 1956. Aging: a theory based on free radical and radiation chemistry. *Journal of Gerontology* 11:298–300.

182. Harman, D. 1972. The biologic clock: the mitochondria? *Journal of the American Geriatrics Society* 20:145–147.

183. Cawthon, R. M., K. R. Smith, E. O'Brien, A. Sivatchenko, and R. A. Kerber. 2003 Association between telomere length in blood and mortality in people aged 60 years or older. *Lancet* 361:393–395.

184. Hamden, K., S. Carreau, K. Jamoussi, S. Miladi, S. Lajmi, D. Aloulou, F. Ayadi, and A. Elfeki. 2009. 1Alpha,25 dihydroxyvitamin D3: therapeutic and preventive effects against oxidative stress, hepatic, pancreatic and renal injury in alloxan-induced diabetes in rats. *Journal of Nutritional Science and Vitaminology* 55:215–222.

185. Willson, R. L. 1992. Free radical–induced biological damage and the critical roles of vitamin A, vitamin C, vitamin D and vitamin E and of copper, iron, selenium and zinc. *Journal of Nutritional Science and Vitaminology* Spec No:541–544.

186. Bao, B. Y., H. J. Ting, J. W. Hsu, and Y. F. Lee. 2008. Protective role of 1 alpha, 25-dihydroxyvitamin D3 against oxidative stress in nonmalignant human prostate epithelial cells. *International Journal of Cancer* 122:2699–2706.

187. Chen, K. B., A. M. Lin, and T. H. Chiu. 2003. Systemic vitamin D3 attenuated oxidative injuries in the locus coeruleus of rat brain. *Annals of the New York Academy of Sciences* 993:313–324; discussion 345–319.

188. Ikeda, N., H. Uemura, H. Ishiguro, M. Hori, M. Hosaka, S. Kyo, K. Miyamoto, E. Takeda, and Y. Kubota. 2003. Combination treatment with 1alpha,25-dihydroxyvitamin D3 and 9-cis-retinoic acid directly inhibits human telomerase reverse transcriptase transcription in prostate cancer cells. *Molecular Cancer Therapeutics* 2:739–746.

189. de Lange, T., and R. A. DePinho. 1999. Unlimited mileage from telomerase? *Science* 283:947–949.

190. Artandi, S. E., S. Alson, M. K. Tietze, N. E. Sharpless, S. Ye, R. A. Greenberg, D. H. Castrillon, J. W. Horner, S. R. Weiler, R. D. Carrasco, and R. A. DePinho. 2002. Constitutive telomerase expression promotes mammary carcinomas in aging mice. *Proceedings of the National Academy of Sciences of the United States of America* 99:8191–8196.

191. Sharpless, N. E., and R. A. DePinho. 2004. Telomeres, stem cells, senescence, and cancer. *Journal of Clinical Investigation* 113:160–168.

192. Itahana, K., G. Dimri, and J. Campisi. 2001 Regulation of cellular senescence by p53. *European Journal of Biochemistry* 268:2784–2791.

193. Matheu, A., A. Maraver, and M. Serrano. 2008 The Arf/p53 pathway in cancer and aging. *Cancer Research* 68:6031–6034.

194. Audo, I., S. R. Darjatmoko, C. L. Schlamp, J. M. Lokken, M. J. Lindstrom, D. M. Albert, and R. W. Nickells. 2003. Vitamin D analogues increase p53, p21, and apoptosis in a xenograft model of human retinoblastoma. *Investigative Ophthalmology & Visual Science* 44:4192–4199.

195. Keisala, T., A. Minasyan, A. Kalueff, I. Pyykko, and P. Tuohimaa. 2009. Premature aging of vitamin D mutant mice. *The Journal of Steroid Biochemistry and Molecular Biology* 115:91–97.

196. Kommagani, R., V. Payal, and M. P. Kadakia. 2007. Differential regulation of vitamin D receptor (VDR) by the p53 family: p73-dependent induction of VDR upon DNA damage. *Journal of Biological Chemistry* 282:29847–29854.

197. Ellison, T. I., M. K. Smith, A. C. Gilliam, and P. N. MacDonald. 2008. Inactivation of the vitamin D receptor enhances susceptibility of murine skin to UV-induced tumorigenesis. *The Journal of Investigative Dermatology* 128:2508–2517.

198. Berryman, D. E., J. S. Christiansen, G. Johannsson, M. O. Thorner, and J. J. Kopchick. 2008. Role of the GH/IGF-1 axis in lifespan and healthspan: lessons from animal models. *Growth Hormone & IGF Research* 18:455–471.

199. Salminen, A., and K. Kaarniranta. 2010. Insulin/IGF-1 paradox of aging: regulation via AKT/IKK/NF-kappaB signaling. *Cellular Signalling* 22:573–577.

200. Gomez, J. M. 2006. The role of insulin-like growth factor I components in the regulation of vitamin D. *Current Pharmaceutical Biotechnology* 7:125–132.

201. Matilainen, M., M. Malinen, K. Saavalainen, and C. Carlberg. 2005. Regulation of multiple insulin-like growth factor binding protein genes by 1alpha,25-dihydroxyvitamin D3. *Nucleic Acids Research* 33:5521–5532.

202. Linkhart, T. A., and M. J. Keffer. 1991. Differential regulation of insulin-like growth factor-I (IGF-I) and IGF-II release from cultured neonatal mouse calvaria by parathyroid hormone, transforming growth factor-beta, and 1,25-dihydroxyvitamin D3. *Endocrinology* 128:1511–1518.

203. Williams, G. C. 1957. Pleiotrophy, natural selection and evolution of senescence. *Evolution* 11:398–411.

204. Nesse, R. M. 2001. On the difficulty of defining disease: a Darwinian perspective. *Medicine, Health Care, and Philosophy* 4:37–46.

205. Greaves, M. 2007. Darwinian medicine: a case for cancer. *Nature Reviews. Cancer* 7:213–221.

206. Greaves, M. 2002. Cancer causation: the Darwinian downside of past success? *The Lancet Oncology* 3:244–251.

207. Wick, G., P. Berger, P. Jansen-Durr, and B. Grubeck-Loebenstein. 2003. A Darwinian-evolutionary concept of age-related diseases. *Experimental Gerontology* 38:13–25.

208. Kirkwood, T. B. 1997. The origins of human ageing. *Philosophical Transactions of the Royal Society of London. Series B, Biological Sciences* 352:1765–1772.

209. Stearns, S. C., and D. Ebert. 2001. Evolution in health and disease: work in progress. *The Quarterly Review of Biology* 76:417–432.

210. Clause, A. R., and E. A. Capaldi. 2006 Caudal autotomy and regeneration in lizards. *Journal of Experimental Zoology. Part A, Comparative Experimental Biology* 305:965–973

211. Liu, Y., Z. Fan, Y. Zhou, M. Liu, F. Ding, and X. Gu. 2009. The molecular cloning of platelet-derived growth factor-C (PDGF-C) gene of *Gekko japonicus* and its expression change in the spinal cord after tail amputation. *Cell Mol Neurobiol* 29:263–271.

212. Papatsoris, A. G., M. V. Karamouzis, and A. G. Papavassiliou. 2005. Novel insights into the implication of the IGF-1 network in prostate cancer. *Trends in Molecular Medicine* 11:52–55.

213. Rincon, M., R. Muzumdar, G. Atzmon, and N. Barzilai. 2004. The paradox of the insulin/IGF-1 signaling pathway in longevity. *Mechanisms of Ageing and Development* 125:397–403.

214. Laron, Z. 2008. The GH–IGF1 axis and longevity. The paradigm of IGF1 deficiency. *Hormones (Athens)* 7:24–27.

215. Westendorp, R. G., and T. B. Kirkwood. 1998. Human longevity at the cost of reproductive success. *Nature* 396:743–746.

216. Salminen, A., and K. Kaarniranta. 2010. Genetics vs. entropy: Longevity factors suppress the NF-kappaB–driven entropic aging process. *Ageing Research Reviews* 9:298–314.

217. Mooijaart, S. P., B. W. Brandt, E. A. Baldal, J. Pijpe, M. Kuningas, M. Beekman, B. J. Zwaan, P. E. Slagboom, R. G. Westendorp, and D. van Heemst. 2005. *C. elegans* DAF-12, nuclear hormone receptors and human longevity and disease at old age. *Ageing Research Reviews* 4:351–371.

218. Rottiers, V., and A. Antebi. 2006. Control of *Caenorhabditis elegans* life history by nuclear receptor signal transduction. *Experimental Gerontology* 41:904–909.

219. Tatar, M., A. Bartke, and A. Antebi. 2003. The endocrine regulation of aging by insulin-like signals. *Science* 299:1346–1351.

220. Hsin, H., and C. Kenyon. 1999. Signals from the reproductive system regulate the lifespan of *C. elegans*. *Nature* 399:362–366.
221. Murphy, C. T., S. A. McCarroll, C. I. Bargmann, A. Fraser, R. S. Kamath, J. Ahringer, H. Li, and C. Kenyon. 2003. Genes that act downstream of DAF-16 to influence the lifespan of *Caenorhabditis elegans*. *Nature* 424:277–283.
222. Peng, S. L. 2008. Foxo in the immune system. *Oncogene* 27:2337–2344.
223. Calnan, D. R., and A. Brunet. 2008. The FoxO code. *Oncogene* 27:2276–2288.
224. Gross, D. N., A. P. van den Heuvel, and M. J. Birnbaum. 2008. The role of FoxO in the regulation of metabolism. *Oncogene* 27:2320–2336.
225. Huang, H., and D. J. Tindall. 2007. Dynamic FoxO transcription factors. *Journal of Cell Science* 120:2479–2487.
226. Yoshizawa, T., Y. Handa, Y. Uematsu, S. Takeda, K. Sekine, Y. Yoshihara, T. Kawakami, K. Arioka, H. Sato, Y. Uchiyama, S. Masushige, A. Fukamizu, T. Matsumoto, and S. Kato. 1997. Mice lacking the vitamin D receptor exhibit impaired bone formation, uterine hypoplasia and growth retardation after weaning. *Nature Genetics* 16:391–396.
227. Halloran, B. P., and H. F. DeLuca. 1980. Effect of vitamin D deficiency on fertility and reproductive capacity in the female rat. *Journal of Nutrition* 110:1573–1580.
228. Kwiecinksi, G. G., G. I. Petrie, and H. F. DeLuca. 1989. 1,25-Dihydroxyvitamin D3 restores fertility of vitamin D–deficient female rats. *American Journal of Physiology* 256:E483–487.
229. Kinuta, K., H. Tanaka, T. Moriwake, K. Aya, S. Kato, and Y. Seino. 2000. Vitamin D is an important factor in estrogen biosynthesis of both female and male gonads. *Endocrinology* 141:1317–1324.
230. Adler, A. S., S. Sinha, T. L. Kawahara, J. Y. Zhang, E. Segal, and H. Y. Chang. 2007. Motif module map reveals enforcement of aging by continual NF-kappaB activity. *Genes & Development* 21:3244–3257.
231. Sarkar, F. H., Y. Li, Z. Wang, and D. Kong. 2008. NF-kappaB signaling pathway and its therapeutic implications in human diseases. *International Reviews of Immunology* 27:293–319.
232. Bernard, D., K. Gosselin, D. Monte, C. Vercamer, F. Bouali, A. Pourtier, B. Vandenbunder, and C. Abbadie. 2004. Involvement of Rel/nuclear factor-kappaB transcription factors in keratinocyte senescence. *Cancer Research* 64:472–481.
233. Vallabhapurapu, S., and M. Karin. 2009. Regulation and function of NF-kappaB transcription factors in the immune system. *Annual Review of Immunology* 27:693–733.
234. Krabbe, K. S., M. Pedersen, and H. Bruunsgaard. 2004. Inflammatory mediators in the elderly. *Experimental Gerontology* 39:687–699.
235. Nonn, L., L. Peng, D. Feldman, and D. M. Peehl. 2006. Inhibition of p38 by vitamin D reduces interleukin-6 production in normal prostate cells via mitogen-activated protein kinase phosphatase 5: implications for prostate cancer prevention by vitamin D. *Cancer Research* 66:4516–4524.
236. Golovko, O., N. Nazarova, and P. Tuohimaa. 2005. Vitamin D–induced up-regulation of tumour necrosis factor alpha (TNF-alpha) in prostate cancer cells. *Life Sciences* 77:562–577.
237. Anderson, P. H., R. K. Sawyer, A. J. Moore, B. K. May, P. D. O'Loughlin, and H. A. Morris. 2008. Vitamin D depletion induces RANKL-mediated osteoclastogenesis and bone loss in a rodent model. *Journal of Bone and Mineral Research* 23:1789–1797.
238. Kitazawa, R., K. Mori, A. Yamaguchi, T. Kondo, and S. Kitazawa. 2008. Modulation of mouse RANKL gene expression by Runx2 and vitamin D3. *Journal of Cellular Biochemistry* 105:1289–1297.
239. Tang, X., and H. Meng. 2009. Osteogenic induction and 1,25-dihydroxyvitamin D3 oppositely regulate the proliferation and expression of RANKL and the vitamin D receptor of human periodontal ligament cells. *Archives of Oral Biology* 54:625–633.
240. Szeto, F. L., J. Sun, J. Kong, Y. Duan, A. Liao, J. L. Madara, and Y. C. Li. 2007. Involvement of the vitamin D receptor in the regulation of NF-kappaB activity in fibroblasts. *The Journal of Steroid Biochemistry and Molecular Biology* 103:563–566.
241. Ooms, M. E., J. C. Roos, P. D. Bezemer, W. J. van der Vijgh, L. M. Bouter, and P. Lips. 1995. Prevention of bone loss by vitamin D supplementation in elderly women: a randomized double-blind trial. *The Journal of Clinical Endocrinology and Metabolism* 80:1052–1058.
242. Bergman, G. J., T. Fan, J. T. McFetridge, and S. S. Sen. 2010. Efficacy of vitamin D3 supplementation in preventing fractures in elderly women: a meta-analysis. *Current Medical Research and Opinion* 26:1193–1201.
243. Prentice, A. 2008. Vitamin D deficiency: a global perspective. *Nutrition Review* 66:S153–164.

17 Vitamin D: Defending the Aging Nervous System

Cédric Annweiler

CONTENTS

Vitamin D is a secosteroid hormone classically involved in the regulation of phosphocalcic homeo-stasis and osteogenesis (Annweiler et al. 2010e; Kalueff and Tuohimaa 2007). The past decade was characterized by an increased number of publications highlighting the adverse effects of hypovita-minosis D not only on bone diseases but also on non–bone health outcomes (Zittermann 2003; Sutra Del Galy et al. 2009; Bouvard et al. 2011), including immunosuppression, cancer, infections, or car-diovascular diseases, as described in previous chapters. These observations could be explained by the fact that vitamin D exhibits multiple biological targets mediated by the vitamin D receptor (VDR), which is present in numerous cells and tissues (Annweiler et al. 2010e; Kalueff and Tuohimaa 2007).

The action of vitamin D on the central nervous system (CNS) has been less frequently described compared with other target organs (Annweiler and Beauchet 2010; Annweiler et al. 2010e; Grant 2009; Kalueff and Tuohimaa 2007; McCann and Ames 2008; Garcion et al. 2002). Nevertheless, it seems that hypovitaminosis D—including age-related hypovitaminosis D—is closely associ-ated with neurological dysfunction (Kalueff and Tuohimaa 2007; Zittermann 2003; Annweiler et al. 2010e,f; Beauchet et al. 2011a,b). The objective of this chapter is to describe the relationships between vitamin D and the CNS while aging.

17.1 VITAMIN D: THE NEUROSTEROID FUNCTION

17.1.1 NEUROSTEROID PROPERTIES

Most of the rationale supporting a neurosteroid involvement of vitamin D comes from experimental research on rodents. To the best of our knowledge, no argument precludes the generalization of these findings to humans.

17.1.1.1 Metabolism

Serum vitamin D reaches the CNS by crossing the blood–brain barrier (Garcion et al. 2002; Langub et al. 2001). At this level, 1,25-dihydroxyvitamin D (1,25OH$_2$D) binds to the VDRs expressed in neuronal and glial cells in the temporal, cingulated and orbital cortex, thalamus, nucleus accum-bens, stria terminalis, amygdala, olfactory network, and spinal cord (Baas et al. 2000; Cornet et al. 1998; Langub et al. 2001; Prufer et al. 1999; Walbert et al. 2001; Jia and Nemere 1999).

It is interesting to note that the enzymes required to synthesize the active form 1,25OH$_2$D (i.e., the vitamin D 25-hydroxylase and 25-hydroxyvitamin D-1α-hydroxylase), historically identified in the liver and kidneys, respectively, were also found in the brain (Garcion et al. 2002; Miller and Portale 2000). The CNS can therefore synthesize its own active form of vitamin D, which has an intracrine (i.e., autocrine or paracrine) neurosteroid action by binding to brain VDRs (Garcion et al. 2002; McGrath et al. 2001; Miller and Portale 2000).

The existence of such metabolic pathways supports the hypothesis of an involvement of vitamin D in brain homeostasis. Numerous effects have been described.

17.1.1.2 Neurotrophic Function

Vitamin D has a trophic function in neuronal differentiation and maturation by controlling the rate of neurotrophins and the number of mitoses (Brown et al. 2003). *In vitro*, vitamin D increases the synthesis of neurotrophic agents such as the nerve growth factor (Brown et al. 2003; Naveilhan et al. 1996; Neveu et al. 1994), the glial cell line–derived neurotrophic factor (Naveilhan et al. 1996), or neurotrophin 3 (Saporito et al. 1993), as well as the synthesis of low-affinity neurotrophin receptor p75NTR (Naveilhan et al. 1996). Vitamin D thus accelerates neuronal growth in a dose-dependent manner in cultured rat hippocampal neurons (Brown et al. 2003).

17.1.1.3 Neurotransmission

Vitamin D is involved in the regulation of several neurotransmitters, including acetylcholine, dopa-mine, serotonin, and γ-aminobutyric acid (Annweiler et al. 2010e). For example, high levels of serotonin were reported in the hypothalamus of rats receiving vitamin D–rich foods (Stumpf et al. 1991). Similarly, vitamin D supplementation in vitamin D–deficient rats resulted in increased activity of choline acetyltransferase (i.e., more acetylcholine bioavailable) in several specific brain regions, including the hypothalamus (Sonnenberg et al. 1986). Also, vitamin D deficiency in young rats was associated with increased cortical dopamine levels compared with control rats (Baksi and Hughes 1982). Finally, a link between vitamin D and γ-aminobutyric acid was highlighted in transgenic mice lacking effective brain VDRs [or VDR-knockout mice (VDR-KO mice); Kalueff et al. 2005b, 2006c].

17.1.1.4 Neuroimmunomodulation

VDR-dependent immunosuppressive effects have been described in the CNS, including the decrease in pro-inflammatory cytokines (i.e., tumor necrosis factor-α or interleukin 6), or the increase in anti-inflammatory cytokines (transforming growth factor or interleukin 4) and in the number and sensitization of macrophages (Cantorna et al. 1996, 1998; Garcion et al. 2003). For instance, it has been suggested that vitamin D could reduce the accumulation of amyloid-β42 peptide in stimulating innate immunity and accelerating the clearance of amyloid-β peptide by macrophage phagocytosis (Masoumi et al. 2009).

Vitamin D–related expression of major histocompatibility complex class II and cofactor 4 also plays an important role in autoimmunity (Bemiss et al. 2002; Garcion et al. 2002). Vitamin D deficiency in mice was accompanied by autoimmune damage to the CNS (Bemiss et al. 2002; Garcion et al. 2003). In contrast, supplementation with 1,25OH$_2$D inhibited autoimmune processes in a mouse model of experimental allergic encephalitis (Kalueff and Tuohimaa 2007). However, this effect disappeared in VDR-KO mice (Meehan and DeLuca 2002), confirming the involvement of vitamin D in the neuroimmunomodulation process and the need for VDRs.

17.1.1.5 Neuroprotection

First, the trophic induction of neuronal cells by vitamin D could be neuroprotective in cerebral ischemia (Wang et al. 2000), as well as in Parkinson's disease (Ibi et al. 2001; Sanchez et al. 2002; Wang et al. 2001), and could also prevent neurotrophic deficits in diabetic rats (Riaz et al. 1999).

Second, the neuroprotective effect of vitamin D results from the maintenance of intraneuronal calcium homeostasis through the regulation of voltage-gated calcium channels and through the synthesis of parvalbumin or calbindin (de Viragh et al. 1989). Increased intracellular calcium levels result in neuronal death either by immediate necrosis (due to loss of energy charge and collapse of internal homeostasis) or by delayed apoptosis (due to oxidative stress and irreversible cell damage; Ankarcrona et al. 1995; Bonfoco et al. 1995).

Third, vitamin D is also involved in brain detoxification by interacting with reactive oxygen and nitrogen species (Garcion et al. 1996, 1999; Kalueff et al. 2004) and by regulating γ-glutamyl transpeptidase activity (Garcion et al. 2002, 2003), a key enzyme in glutathione metabolism (Dringen et al. 2000; Shinpo et al. 2000). *In vitro*, 1,25OH$_2$D concentrations from 0.1 to 100 nmol provide effective protection of neurons against superoxide and hydrogen peroxide (Ibi et al. 2001; Lin et al. 2003). It has also been shown that vitamin D inhibited the synthesis of inducible nitric oxide synthase (Garcion et al. 1998), an enzyme produced in CNS cells in response to stress, with high doses altering neurons (Dawson and Dawson 1996; Mitrovic et al. 1994).

Fourth, the anticalcification effect of vitamin D should be noted (Kalueff et al. 2006a; Kalueff and Tuohimaa 2007). Calcium deposits have been described in the basal ganglia of subjects with normocalcemia and normophosphoremia, but hypovitaminosis D (Kalueff et al. 2006a; Kalueff and Tuohimaa 2007).

These observations highlight the central role of vitamin D in controlling CNS homeostasis. Serum vitamin D concentration decreases with aging. Maintaining a concentration high enough throughout life may therefore be an effective solution to prevent the onset of age-related neurological disorders, including behavioral disorders and cognitive decline.

17.1.2 HYPOVITAMINOSIS D AND AGING: EFFECT ON BEHAVIOR

Behavior is an objectively observable set of reactions that an organism with a nervous system typically performs in response to environmental stimuli (Bloch 1994).

The origin, development, and complexity of behavior of an animal organism is closely related to the complexity of its nervous system (Math 2008). The more complex the brain is, the more developed and adapted the behaviors are. For example, the innate fundamental behaviors, which include reactions of defense and survival, and social or maternal behaviors (Math 2008), include a complex stereotyped sequence of movements triggered by hypothalamic stimulation. Thus, stimulation of the ventral medial hypothalamus triggers the fight, and that of its dorsal part, the flight. Cholinergic stimulation of the lateral hypothalamus triggers aggression (Bandler 1969), whereas its destruction suppresses maternal behaviors (Bandler 1969).

In other words, behavior is directly related to brain function. This could explain the onset of behavioral disorders in the case of hypovitaminosis D.

17.1.2.1 Animal Testing

Experimental data are mainly based on the study of rodents. Note that the distribution of VDR is the same in rodents and humans (Eyles et al. 2005). Animal experimentation has shown that VDRs appear early in embryonic life, from the 12th embryonic day (Veenstra et al. 1998). They are initially in the neuroepithelium, then in the subventricular zone of the lateral ventricles (Cui et al. 2007; Veenstra et al. 1998). The subventricular zone, a germinal zone identified in the mammalian brain, is rich in neural stem cells (Cui et al. 2007).

The presence of VDRs within the embryonic nervous tissue, at this level, implies that vitamin D plays a role in the embryonic stage. Pathology, that is, prenatal vitamin D deficiency, is an excellent model to better understand the nature of this effect. McGrath et al. (1999) therefore created a developmental vitamin D model of rats that are deficient in vitamin D during the *in utero* period. This experimental model confirmed the existence of the clinical effect of hypovitaminosis D in the embryonic brain. Indeed, the brains of rats exposed to vitamin D deficiency during pregnancy were morphologically altered compared with newborn controls (Eyles et al. 2003). In addition, the developmental vitamin D model in rats exhibited behavioral changes, including a spontaneous hyperlocomotion (Burne et al. 2004) or disruption of latent inhibition (Becker et al. 2005).

In adults, hypovitaminosis D also results in behavioral disorders. More specifically, because the effect of $1,25OH_2D$ is mediated by VDRs, the genetic ablation of functional VDRs provided a powerful model of resistance to vitamin D (similar to avitaminosis D; Li et al. 1998; Yoshizawa et al. 1997; Burne et al. 2005; Zou et al. 2008) and showed an effect on instinctual behaviors in VDR-KO mice (Burne et al. 2006; Kalueff al. 2005, 2006b). For instance, the VDR-KO mouse had altered behavioral sequencing of grooming with a haphazardly atypical distribution focusing on the legs, with frequent interruptions and extended duration of toilet, instead of following the typical cephalocaudal progression of wild-type mice without any genetic engineering (Kalueff et al. 2005a). Also, VDR-KO mice were unable to swim because of excessive stress and motor disorders, and drowned (Burne et al. 2006). Similarly, VDR-KO mice exhibited a change in social behaviors, as illustrated by aggressivity or aberrant maternal behaviors, including neglect and cannibalism (Kalueff et al. 2006b).

17.1.2.2 Generalization to Humans

In humans, experimental data also argue in favor of a link between vitamin D and behavioral regulation. In particular, the VDRs have been described in brain regions, including the cortex, the cerebellum, or the limbic system that are key areas for behavior (Stumpf et al. 1982; Walbert et al. 2001). At this level, vitamin D may act as a central regulator of brain processes that result in appropriate motor, psychological, and social behaviors. Conversely, hypovitaminosis D observed in elderly populations may participate in the aging brain producing alterations in neuronal processes and related functions.

To the best of our knowledge, no study has yet focused on the effect of hypovitaminosis D on behavior in older adults. Nevertheless, it is interesting to draw parallels between the fundamental behavioral disorders described in VDR-KO mice, and those expressed by older humans with severe cognitive impairments. The question is whether hypovitaminosis D could be the link between these behavioral disorders and, in particular, whether age-related hypovitaminosis D could explain age-related cognitive decline. The current literature provides arguments supporting such a relationship.

17.2 AGE-RELATED HYPOVITAMINOSIS D: COGNITIVE ISSUES

17.2.1 COGNITIVE IMPAIRMENT: DEFINITION AND EPIDEMIOLOGY IN OLDER ADULTS

Cognitive impairment means a deterioration of the mental processes such as memory, decision, understanding, and reasoning. Whatever its nature and severity, cognitive impairment may be associated with behavioral disorders. In general, an overall slowing down of information processing speed accompanies advancing age (Plassman et al. 2008). All other cognitive impairments encountered while aging are pathologic.

The vast majority of cognitive impairments occur after 70 years. In the United States, about a quarter of adults aged 70 and older present with cognitive impairment (Plassman et al. 2008). The progression to dementia is 10% a year (Plassman et al. 2008), except in subjects with mild cognitive impairment, in whom there is a higher risk of dementia at 10 years (Mitchell and Shiri-Feshki 2009; Petersen et al. 2001).

17.2.2 VITAMIN D AND COGNITION: CLINICAL EVIDENCE

It has long been recognized that the consumption of fish is inversely associated with the onset of dementia (Annweiler et al. 2010a; Morris 2009). For example, in 815 older adults from the Chicago Health and Aging Project cohort (mean age, 73.1 years; 61.9% female), eating one fish meal per week was associated with a 60% lower risk of developing Alzheimer's disease (AD) over 4 years compared with rarely or never eating fish (Morris et al. 2003). This effect was initially attributed to the consumption of omega-3 polyunsaturated fatty acids contained in fish, because they are involved in 30% to 40% of the phospholipid membrane of cortical neurons as well as in the maintenance of brain function (Morris 2009). However, recent evidence does not support this hypothesis. Indeed, the Rotterdam Study, which reported a 70% lower risk of developing AD after 2 years of regular consumption of fish (Kalmijn et al. 1997), failed to find an association between the consumption of omega-3 specifically and the risk of AD at 6 years (Engelhart et al. 2002). Similarly, in the Canadian Study of Health and Aging, initial serum omega-3 concentrations could not predict the onset of dementia (Laurin et al. 2003). Finally, no clinical trial has yet shown a protective effect of omega-3 fatty acid supplements against cognitive decline (Morris 2009).

These results indicate that the neuroprotective effect of the regular consumption of fish may be explained by another mechanism. For instance, it could be explained by a healthy lifestyle combining healthy diet and regular physical activity (Scarmeas et al. 2009), or by another constituent of fish such as vitamin D (Annweiler et al. 2010a). Epidemiologic studies provide arguments supporting this hypothesis.

17.2.3 EPIDEMIOLOGICAL EVIDENCE

Clinical data support the hypothesis of a link between vitamin D and cognitive function. For instance, lower serum 25OHD concentrations were found in demented subjects compared with nondemented subjects (Sutherland et al. 1992; Sato et al. 1998). Similarly, a link was reported between cognitive decline and bone mineral density (BMD) in hypovitaminosis D–related senile osteoporosis (Lui et al. 2003). This link was independent of baseline BMD and could not be explained by functional status, or by the use of estrogen or ApoE supplements. Even if a temporal relationship was possible in this study, the action of vitamin D on cognitive function was also likely.

17.2.4 VITAMIN D AND GLOBAL COGNITIVE PERFORMANCE

Global cognitive performance means all thought processes. This concept includes all functions of the mind by which we construct an operative representation of the reality from our perception, in particular, to feed our thinking and guide our actions (Matlin 2008). In other words, cognition includes not only higher level processes such as reasoning, memory, decision making, or executive functions, but also more basic processes such as perception, motor skills, and emotions or affectivity (Matlin 2008; Singer 1980).

Global cognitive function depends on the efficiency of all targeted cognitive processes (Matlin 2008; Singer 1980) and can be assessed using composite psychometric tests, including several specific cognitive scales. These tests have been developed to compare patients between them and to classify them either quantitatively or typologically, or to follow their evolution separately. The choice of the test used is based on patient characteristics (i.e., age and education level).

A systematic literature review published in 2009 found that only three studies have addressed the issue of the association between serum 25OHD and global cognitive performance in adults (Annweiler et al. 2009a). Using several composite cognitive tests to assess global cognitive function, including the Short Blessed Test, the Clinical Dementia Rating—Sum of Boxes, and the Mini-Mental State Examination (MMSE), these three studies consistently showed a significant positive association between vitamin D concentration and global cognitive function, even after adjustment for a number of clinical confounders (Wilkins et al. 2006; Przybelski and Binkley 2007; Oudshoorn et al. 2008). However, this association could not be considered as definitively established because none of these studies took into account the serum parathyroid hormone (PTH) and calcium concentrations, although these molecules are recognized confounding factors when considering the association between vitamin D and cognition (Annweiler et al. 2010b).

This issue justified another study in 2010, in which we examined the association between vitamin D concentrations and the Short Portable Mental State Questionnaire (SPMSQ) score (Pfeiffer 1975) among a randomized sample from the EPIDOS (EPIDémiologie de l'OStéoporose) cohort study (Annweiler et al. 2010d). In this study, we found, among 752 community-dwelling women aged 80.2 ± 3.5 years on average, a significant association between vitamin D deficiency and cognitive impairment [adjusted odds ratios (OR) = 1.99 with $p = .017$] after adjustment for a list of confounders, including serum PTH and calcium.

Since this publication, two new studies have further confirmed this association. The first one showed that among 1766 subjects without dementia or with minor neurocognitive disorder (mean age, 78.2 years; 59.9% female) lower serum 25OHD concentrations were directly associated with a higher risk of a pathological abbreviated mental test score (Hodkinson 1972), with a trend to higher ORs for each quartile of vitamin D ($p < .001$; Llewellyn et al. 2009). The most recent one confirmed a significant positive linear association between serum vitamin D concentration and the Short Blessed Test score among 60 elderly patients without dementia or with minor neurocognitive disorders (mean age, 75 years; 50% Caucasian and 50% African American; Wilkins et al. 2009).

Thus, the association between hypovitaminosis D and global cognitive impairment has been established by previously published studies. The absence of negative results could be explained by a publication bias, that is to say the tendency of researchers and publishers to communicate mainly "positive" results (i.e., showing a significant difference) rather than inconclusive or "negative" results (i.e., supporting the null hypothesis; Sackett 1979). Nevertheless, the neurosteroid properties of vitamin D cited above leads us to suggest that the absence of negative results could thus far be explained by the fact that this association is real.

All quoted studies were cross-sectional (Annweiler et al. 2010d; Llewellyn et al. 2009; Oudshoorn et al. 2008; Przybelski and Binkley 2007; Wilkins et al. 2006, 2009), that is, they provided a snapshot of a given population regardless of the time. If this study design has the advantage of being easy to implement, it does not establish the temporal sequence of events (disease and exposition are determined at the same time), and prevents the finding of a causal relationship. Longitudinal cohort studies are required to test the hypothesis that hypovitaminosis D precedes the incident decline of global cognitive performance.

17.2.4.1 Hypovitaminosis D and Global Cognitive Decline

To clarify this issue, two recent studies tested the ability of hypovitaminosis D to predict the incident onset of cognitive decline. The first one explored incident cognitive decline with the Modified Mini-Mental State Examination (3MS) score as a function of initial serum 25OHD (Slinin et al. 2010). The 3MS, a composite test with a score ranging from 0 (worst) to 100 (best cognitive performance), was performed among 1138 men from the Osteoporotic Fractures in Men Study cohort aged 65 years and older, free of initial cognitive impairment, and were followed for an average of 4.6 years. The cognitive decline was defined by the loss of at least five points in the 3MS score during the follow-up. The authors found that the incidence of cognitive decline increased with decreasing serum 25OHD concentrations ($p = .04$ between each 25OHD quartile) after adjustment for age, study center, and season tested (Slinin et al. 2010). It should be noted that this trend was not significant ($p = .10$) after adjustment for ethnicity, educational level, comorbidities, autonomy, alcohol and tobacco consumption, body mass index (BMI), and physical activity (Slinin et al. 2010).

More recently, baseline serum 25OHD concentrations were also associated with a significant decline in MMSE score (defined by the loss of at least three points in MMSE) among 858 adults ages 65 and older, followed for 6 years in the InCHIANTI cohort (Llewellyn et al. 2010). The relative risk of significant decline in MMSE score was 1.60 (95% CI, 1.19–2.00) in subjects with vitamin D deficiency compared with subjects with normal vitamin D status, even after adjustment for age, gender, education level, baseline MMSE score, season, alcohol and tobacco consumption, depression, energy intake, intake of vitamin E, and degree of mobility (Llewellyn et al. 2010). Moreover, participants with the lowest 25OHD concentrations lost 0.3 MMSE points per year higher than those with normal vitamin D status (Llewellyn et al. 2010).

Both of these longitudinal studies were interesting in that they tested the assumption that vitamin D status could predict the risk of cognitive decline in older adults, thereby confirming the effect of hypovitaminosis D on global cognitive performance.

It remains difficult to determine which specific cognitive function is altered in hypovitaminosis D, that is, which domain-specific cognitive function explains the association between hypovitaminosis D and the deterioration of global cognitive performance.

17.2.5 WHICH SPECIFIC COGNITIVE FUNCTION IS ASSOCIATED WITH VITAMIN D?

17.2.5.1 It's not...

17.2.5.1.1 Language

Language is the instrument used for the expression of thought. It is the human ability to communicate using vocal signs that can also be transcribed as writing (Matlin 2008). Language disorders (aphasia) are disorders of the expression or comprehension of spoken or written language due to cortical lesions (Jakobson 1960; Sabouraud 1995). Anatomically, language is underpinned by the frontal (phonemic arrangement) and temporal lobes (semantic). Language can be assessed using verbal fluency tests in which the participant lists as many words as possible within a specified class in a given time. The class may be semantic (e.g., animal names) or phonemic (e.g., words beginning with a letter designated by the examiner; Lezak 1995). For instance, the Controlled Oral Word Association Test (COWAT) is a common phonemic verbal fluency test (Benton and Hamsher 1976; Loonstra et al. 2001). It consists of listing words beginning with the letters "f," "a, " or "s" (60 s for each letter). The score is the total number of listed words (Benton and Hamsher 1976; Loonstra et al. 2001).

Based on the neurosteroid properties of vitamin D, it has been hypothesized that vitamin D may be involved in language. The association between serum 25OHD (63.6 ± 18.1 nmol/L on average) and COWAT score was examined among 148 participants from the fifth Tromsø cohort study (mean age, 62 years; 46% female; Jorde et al. 2006). In this cross-sectional study, the authors failed to find a significant linear association between serum 25OHD and COWAT score ($\beta = 0.13$ with $t = 1.08$; $|t| > 1.96$ corresponds to $p < .05$; Jorde et al. 2006). The link between vitamin D and COWAT score was also examined unsuccessfully in 1080 adults aged 75 years on average (76% female) participating in the Nutrition and Memory in Elders Study ($\beta = 0.05$, $p = .20$; Buell et al. 2009b).

17.2.5.1.2 Episodic Memory

Memory cannot be reduced to a single memory. There are several types of memory, such as procedural (or implicit) and declarative (or explicit) memories. Specifically, within declarative memories, episodic memory refers to the processes of encoding, storage, and retrieval of personally experienced information (Matlin 2008), and is different from the decontextualized semantic memory (concepts, meanings of words and symbols). Anatomically, memory depends on the hippocampus and adjacent structures (i.e., the parahippocampal gyrus, the subiculum, the cortex overlying the amygdala, the nucleus dorsomedial, and the anterior nuclear group). Episodic memory is usually tested by learning and recall of a word list (Grober et al. 1988; Wechsler 1997; Cowppli-Bony et al. 2005). These tests are based on the achievement of three successive stages: first, an encoding step with an immediate recall to control the encoding; second, an information storage step coupled to a distractive test; and third, a delayed refund word list step (free recall ± help).

Three studies have explored the link between vitamin D and episodic memory. The first one found no significant linear association among 148 older adults aged 62 years on average (46% female), between the serum 25OHD concentration and the number of words returned successfully from a 12-word list ($\beta = -0.03$, $t = -0.26$; Jorde et al. 2006). The second study examined the learning and memory abilities of 4809 individuals aged 60–90 years from the National Health and Nutrition Examination Survey cohort (McGrath et al. 2007). The episodic memory score was derived from two different tasks based on the recall of a list of words and a story. Unlike the hypothesis, the authors found that subjects in the highest 25OHD quintile were those with poorer memory performance ($p = .02$). This cross-sectional analysis remained purely descriptive, and no association was sought. Finally, a third study found no association between serum 25OHD concentration and the recovery of a 12-word list (Wechsler 1997) among 1080 adults aged 75 years (76% female; $\beta = 0.01$ with $p = .54$; Buell et al. 2009b).

17.2.5.1.3 Visual Memory

Visual memory is a sensory memory related to visual perceptions. It helps to remember things, places, or people as a mental image. Several psychometric tests explore visual memory, including the visual memory score of the revised Wechsler memory scale (Wechsler 1987). This score is composed of visual memory subtests of figurative memory, image matching, and visual reproduction (Wechsler 1987). Similarly, the Rey Complex Figure (RCF) is a test of not only perceptual organization but also of visual memory (Osterrieth 1944). It consists of copying a complex geometric figure and then, after a period not exceeding 3 min, of reproducing it without having been advised in advance (Osterrieth 1944). Finally, the Camden Topographical Recognition Memory Test (CTRM) is based on the presentation of 30 color photographs of outdoor scenes, each displayed for 3 s, followed by a recognition step, with a maximum score of 30 corresponding to the absence of errors (Warrington 1966).

Two studies have thus far explored the link between vitamin D and visual memory. The first one found no significant association between serum 25OHD and the number of recalls on the task of image matching from the revised Wechsler memory scale with a 30-min delayed recall ($\beta = 0.13$ with $t = 1.21$; Jorde et al. 2006). The second one assessed the visual memory with two tests (RCF and CTRM) among 3369 men (mean age, 60 years) from the European Male Ageing Study cohort

(Lee et al. 2009). No linear association was found between serum 25OHD concentrations and RCF score [$\beta = -0.021$ for 10 nmol/L of 25OHD (95% CI, -0.163 to 0.121)], or CTRM score [$\beta = -0.001$ for 10 nmol/L of 25OHD (95% CI, -0.146 to 0.144)].

In summary, vitamin D has been associated with neither language nor episodic and verbal memories in the elderly (Jorde et al. 2006; Lee et al. 2009; McGrath et al. 2007; Buell et al. 2009b). Consequently, none of these specific cognitive functions seems to explain the association found between hypovitaminosis D and global cognitive impairment. Recent studies have provided new answers, raising the idea that executive functions may be the missing link.

17.2.5.2 Vitamin D and Executive Functions

17.2.5.2.1 Definition

Executive functions refer to a heterogeneous set of high-level processes required to have a flexible and adapted behavior (Bérubé 1991; Godefroy et al. 2008; Luria 1969; Miyake et al. 2000; Rabbit 1997; Matlin 2008). They include the capacities for anticipation, planning, organization, problem solving, logical reasoning, working memory, cognitive control, abstract thinking, selective attention, motor response, motivation, and initiative (Godefroy et al. 2008). All these functions are inseparable from attention (Bérubé 1991; Deutsch and Deutsch 1963). Attentional and executive functions are called "high level" because they regulate other cognitive activities. Their alteration thus results in slower processing speed and impaired mental processes (also called executive dysfunction). Recently, Miyake et al. (2000) summarized the executive functions into three specific areas: mental shifting (i.e., the ability to move from one cognitive operation to another), cognitive inhibition (i.e., the ability to inhibit an automatic response), and information updating (i.e., updating information in working memory).

Anatomically, executive functions are mainly related to the frontal lobes (prefrontal cortex; Godefroy et al. 2008). Subcortical structures, including caudate nucleus, putamen, pallidum, nucleus accumbens, and thalamus are also involved through frontal–subcortical networks. Damage to these networks, whatever the location, results in executive dysfunction.

Specific psychometric tools have been developed to explore executive functions.

For example, the Digit Symbol Substitution Test (DSST) is a subtest of the Wechsler Adult Intelligence Scale–Third Edition (WAIS-III) that measures information processing speed and visuospatial attention (Wechsler 1997). It consists of learning a list of numbers paired with symbols, and then designating the symbols corresponding to a new list of numbers as quickly as possible. The DSST score is the number of symbols correctly completed in the allotted time.

The Trail Making Test (TMT) measures mental shifting and is divided into two parts (Reitan 1958). The first part (TMT A) is for connecting numbers in ascending order as quickly as possible (1–2–3–4 …). This section assesses visual perceptual speed and psychomotor speed. The second part (TMT B) assesses mental shifting per se, and also divided attention because the participant is subjected to a dual cognitive task: the participant must proceed in the same manner as for the TMT A, but with alternating numbers and letters (1–A–2–B–3–C …; Reitan 1958; Gaudino et al. 1995).

The Stroop test measures cognitive inhibition (Stroop 1935). Initially, the subject names the color of colored rectangles. The participant must then read the names of the colors printed in black. Finally, an interference is introduced: the participant must name the ink color in which words are written and inhibit the reading of each word (Stroop 1935).

Finally, information updating can be approximated by carrying out a task of counting backward (working memory; Baddeley and Della Sala 1996; Marsh and Hicks 1998; Oberauer 2002).

17.2.5.2.2 Vitamin D and Executive Functions

Several studies have recently examined the association of vitamin D with executive functions. In particular, strong associations have been found between vitamin D status and information processing speed. For instance, significant linear associations were reported between serum 25OHD concentration and DSST score among 3369 men aged 60 years on average [β = 0.152 for 10 nmol/L of 25OHD (95% CI, 0.051–0.253); (Lee et al. 2009)], as well as among 1080 community-dwellers aged 75 on average (β = 0.19 with $p < .001$; Buell et al. 2009b).

Mental shifting is one of the main executive subfunctions. It refers to the ability to alter the course of thinking or behavior to adapt to the changing environmental needs (Bérubé 1991). It can be explored with the second part of the TMT (TMT B; Reitan 1958; Gaudino et al. 1995). Importantly, serum 25OHD concentrations were associated with better performance on TMT B (β = −0.73 with $p = .02$) among 1080 people (75 years; 76% female; Buell et al. 2009b). A longitudinal cohort study also showed that among 858 adults (aged 65 and older and followed for 6 years), the relative risk for significant decline in TMT B score was 1.31 (95% CI, 1.03–1.51) in subjects initially deficient in vitamin D compared with subjects with a normal vitamin D status (Llewellyn et al. 2010), even after adjustment for age, gender, educational level, initial TMT B score, alcohol and tobacco consumption, depression, energy intake, intake of vitamin E, degree of mobility, and season tested (Llewellyn et al. 2010).

Another executive subfunction is based on cognitive inhibition. Only one study has thus far explored the potential link between vitamin D and cognitive inhibition using the Stroop test in 148 people with a mean age of 62 years (46% female; Jorde et al. 2006). No significant association was found between serum 25OHD concentrations and the Stroop test score parts 1 and 2 (β = 0.12 with $t = 1.05$) or part 3 (β = −0.07; $t = −0.68$; Jorde et al. 2006).

Information updating is the third executive subfunction according to Miyake's model (Miyake et al. 2000). It is the ability to replace and update information stored a moment before. It is therefore inseparable from the working memory, that is, the mental manipulation of information (Baddeley and Della Sala 1996; Oberauer 2002; Marsh and Hicks 1998). Working memory is used as a mental workspace to store and manipulate information for short periods or when performing a task (Baddeley 1993; Baddeley and Wilson 1985; Oberauer 2002; Matlin 2008). It uses two different storage systems: the phonological loop (for retention of verbal information heard or read) and the visuospatial notebook (for retention of visuospatial information and mental imagery; Baddeley and Wilson 1985). The central administrator coordinates the activities of the phonological loop and visuospatial notebook, and allows the performance of two tasks simultaneously (i.e., storage and processing; Baddeley and Wilson 1985). Anatomically, working memory involves several cortical regions: prefrontal cortex for attentional control and more posterior areas, including the parietal cortex for information updating (Bledowski et al. 2009; Curtis and D'Esposito 2003; Postle 2006). The role of the hippocampus remains unclear. From a psychometric point of view, working memory is the short-term memory required for the temporary retention of information and its cognitive manipulation. It is explored with a counting backward task that assesses more specifically the integrity of the phonological loop and the central administrator (Hittmair-Delazer et al. 1994). The visuospatial component is apprehended through tests such as the task of spatial working memory from the computerized Cambridge Neuropsychological Testing Automated Battery (Luciana and Nelson 2002). This test consists of a series of white squares that appear on the screen. Some squares change color according to a variable order. At the end of each sequence, the subject has to touch each colored box in the order in which the squares appeared. The maximum score is 9, corresponding to the longest possible sequence successfully reproduced by the subject.

To test the association between vitamin D deficiency and impaired working memory in the elderly, we re-analyzed the data from a randomized sample of 752 women from the EPIDOS cohort, this time, not using the total SPMSQ score, but a subtest of counting aloud backward by 3 from 20 (Pfeiffer 1975). Working memory was considered normal in the absence of counting errors.

Among 129 women with serum 25OHD levels lower than 10 ng/mL, the mean 25OHD concentration was 7.21 ± 2.09 ng/mL. As indicated in Table 17.1, these women presented more often with impaired performance on the counting backward task than women without vitamin D deficiencies

TABLE 17.1

Characteristics and Comparison of the Randomized Sample Subjects (*n* = 752) Separated into Two Groups Based on Serum 25OHD Concentrations

	Serum 25OHD Concentration (ng/mL)		
	<10 (*n* = 129)	≥10 (*n* = 623)	*p**
Clinical Measures			
Age, mean ± SD (years)	80.71 ± 3.84	80.05 ± 3.40	0.129
Use of psychoactive drugs,[†] *n* (%)	65 (50.39)	290 (46.55)	0.440
Number of chronic diseases,[‡] mean ± SD	3.14 ± 1.84	2.97 ± 1.81	0.340
Hypertension,[‡] *n* (%)	63 (48.84)	303 (49.03)	1.000
Low education level,[§] *n* (%)	107 (83.59)	498 (80.19)	0.460
Neuropsychological Measures			
Pfeiffer's SPMSQ score (/10), mean ± SD	8.56 ± 1.67	9.05 ± 1.34	**0.0001**
Working memory impairment,[**] *n* (%)	49 (38.58)	179 (28.87)	**0.034**
Depression,[‡] *n* (%)	14 (10.85)	85 (13.67)	0.475
Serum Measures			
25OHD concentration, mean ± SD (ng/mL)	7.21 ± 2.09	20.13 ± 10.81	**<0.0001**
iPTH concentration, mean ± SD (pg/mL)	70.70 ± 35.17	57.70 ± 25.15	**<0.0001**
Calcium concentration,[††] mean ± SD (mmol/L)	2.24 ± 0.095	2.24 ± 0.12	0.695

Notes: SD, standard deviation; 25OHD, 25-hydroxyvitamin D; iPTH, intact PTH *p* significant (i.e., <0.05) indicated in boldface.

*Comparisons based on one-way analysis of variance or chi-square test, as appropriate.

[†]Use of benzodiazepines or antidepressants or neuroleptics.

[‡]Obtained from physical examination and a health status questionnaire to target comorbid diseases.

[§]Obtained from a structured questionnaire, considered when Elementary School Recognition Certificate not passed.

[**]Subtest of the SPMSQ; working memory was considered as impaired in case of backward counting error.

[††]Corrected value based on the formula (Ca + 0.02 [46-albumin]).

(*p* = .034). There was no significant difference for the other characteristics, except with regard to secondary hyperparathyroidism within the vitamin D–deficient group (*p* < .001; Table 17.1).

Table 17.2 reports the univariate and multiple logistic regressions between working memory impairment, 25OHD deficiency, and subjects' characteristics. Vitamin D deficiency was significantly associated with impairment of working memory (unadjusted OR = 1.54 with *p* = .031; adjusted OR = 1.57 with *p* = .033), even after adjustment for age, number of chronic diseases, use of psychoactive drugs, current hypertension or depression, educational level, and serum iPTH and calcium concentrations. In addition, vitamin D deficiency remained associated with working memory impairment in the stepwise backward logistic regression model (adjusted OR = 1.52 with *p* = .041; Table 17.2).

This work underlined the existence of an association between hypovitaminosis D and working memory impairment. It remained yet limited by the use of a subtest of the SPMSQ that was not specifically developed to explore working memory.

Recently, these findings were confirmed using a task of spatial working memory from the Cambridge Neuropsychological Testing Automated Battery (Seamans et al. 2010). In this study, the authors examined 387 Europeans aged 55–87 years (49.4% female) from the Zinc Effects in Nutrient/Nutrient Interactions and Trends in Health and Ageing cohort study. The results showed that serum 25OHD concentrations inversely correlated with the total number of errors in the task

TABLE 17.2

Univariate and Multiple Logistic Regressions Showing the Cross-Sectional Association between Impaired Working Memory (Dependent Variable) and 25OHD Deficiency (Independent Variable) Adjusted for Subjects' Characteristics ($n = 752$)

	Working Memory Impairment*		
	Model 1 (95% CI) [p]	Model 2 (95% CI) [p]	Model 3 (95% CI) [p]
25OHD < 10 ng/mL	**1.54** (1.04–2.30) [0.031]	**1.57** (1.04–2.37) [0.033]	**1.52** (1.02–2.28) [0.041]
Age	**0.94** (0.90–0.98) [0.005]	**0.93** (0.89–0.98) [0.004]	**0.94** (0.90–0.98) [0.004]
Use of psychoactive drugs[†]	0.80 (0.59–1.10) [0.169]	0.81 (0.59–1.13) [0.217]	
Number of chronic diseases[‡]	0.98 (0.90–1.07) [0.695]	0.97 (0.88–1.06) [0.491]	
Hypertension[‡]	0.98 (0.72–1.34) [0.902]	1.03 (0.74–1.44) [0.867]	
Depression[‡]	1.18 (0.73–1.89) [0.494]	1.17 (0.71–1.96) [0.534]	
Low education level[§]	1.35 (0.89–2.03) [0.156]	1.40 (0.92–2.15) [0.120]	1.42 (0.93–2.18) [0.103]
iPTH (pg/mL)	1.00 (0.99–1.01) [0.777]	1.01 (0.99–1.01) [0.215]	
Calcium** (mmol/L)	3.18 (0.77–13.11) [0.109]	4.15 (0.96–17.92) [0.057]	3.73 (0.88–15.87) [0.075]

Notes: Model 1, unadjusted logistic regression. Model 2, fully adjusted logistic regression. Model 3, stepwise backward logistic regression. SPMSQ, Short Portable Mental State Questionnaire; iPTH, intact PTH. OR significant (i.e., $p < 0.05$) indicated in boldface.

*Subtest of the SPMSQ. Working memory was considered as impaired in case of backward counting error.

[†]Use of benzodiazepines, antidepressants or neuroleptics.

[‡]Obtained from physical examination and a health status questionnaire to target comorbid diseases (hypertension, diabetes, dyslipidemia, coronary heart disease, chronic obstructive pulmonary disease, peripheral vascular disease, cancer, stroke, Parkinson's disease, and depression).

[§]Obtained from a structured questionnaire, considered when Elementary School Recognition Certificate not passed.

**Corrected value based on the formula (Ca + 0.02 [46-albumin]).

($r = -.174$, $p < .003$; Seamans et al. 2010). In addition, subjects belonging to the highest tertile of 25OHD made fewer errors on the spatial working memory task than those in the lowest tertile of 25OHD ($p = .04$; Seamans et al. 2010). Thus, vitamin D seems to be associated with both verbal and spatial working memories.

17.2.5.3 Vitamin D and Specific Cognitive Performance: Summary

In summary, these cross-sectional and longitudinal studies support the association of vitamin D with processing speed (DSST), mental shifting (TMT B), and information updating (working memory). These vitamin D–related cognitive functions are essential to the real-time operations of all cognitive processes, and their clinical manifestations as appropriate behaviors (Miyake et al. 2000; Salthouse 2001). Therefore, it seems that these three vitamin D–related specific cognitive functions could shape the effect of hypovitaminosis D on global cognitive performance.

Older adults with hypovitaminosis D therefore exhibit cognitive slowing, impaired attention, and difficulties in adapting to different situations of everyday life. These difficulties can affect their ability to correctly perform activities of daily living because executive dysfunction is directly correlated with functional disabilities (Hanninen et al. 1997). It is of particular importance and marks a major turning point in the history of aging subjects with a minor neurocognitive disorder because this induced loss of autonomy defines the onset of dementia (American Psychiatric Association 1994).

17.3 VITAMIN D AND DEMENTIA

17.3.1 EPIDEMIOLOGICAL EVIDENCE

The question of the relationship between hypovitaminosis D and dementia was recently addressed in 288 inpatients hospitalized in a geriatric acute care unit (mean age, 86 years; 66.1% female; Annweiler et al. 2011). We hypothesized an association between severe hypovitaminosis D and moderately-severe-to-severe dementia, irrespective of the nature of dementia (defined according to the DSM-IV criteria, and based on a MMSE score of <15). The results showed that a concentration of 25OHD lower than 10 ng/mL was the only variable significantly associated with the existence of advanced stage dementia, that is, dementia with a negative effect on autonomy.

These data confirmed the results of the study by Buell et al. (2009a), which showed that among 318 community-dwellers aged 73.5 years on average (72.6% female), that subjects with serum 25OHD levels of 20 ng/mL or lower had significantly more dementia from any cause than those with normal vitamin D concentrations (30.5% versus 14.5%, respectively, $p < .01$; Buell et al. 2009a). Moreover, in this study, the risk of dementia was multiplied by 2.21 (95% CI, 1.13–4.32) for 25OHD levels of 20 ng/mL or lower compared with 25OHD levels of more than 20 ng/mL ($p = .02$; Buell et al. 2009a). Specifically, hypovitaminosis D increased the risk of AD (OR = 2.51; 95% CI, 1.04–6.09; $p = .04$), as well as of stroke, a major cause of vascular dementia (OR = 2.29; 95% CI, 1.09–4.83; $p = .03$).

These results supported the assumption that vitamin D deficiency could play a role in the onset of dementia. Nevertheless, their cross-sectional designs precluded the finding of a causal link. In particular, the loss of autonomy in the context of severe dementia may explain the existence of vitamin D deficiency due to difficulties of accessing sunlight or vitamin D–rich foods.

To clarify this issue, a prospective 7-year longitudinal study was recently published to determine whether hypovitaminosis D at baseline could predict the onset of non-Alzheimer dementias within 7 years among 40 initially nondemented older women from the EPIDOS Toulouse study (Annweiler et al. 2012a). The results showed an association between vitamin D deficiency of less than 10 ng/mL at baseline and the onset of non-Alzheimer dementias (OR = 19.57, $p = .042$). Conversely, vitamin D deficiency was not associated with AD ($p = .222$). Despite a small sample size, this study suggested that vitamin D deficiency may precipitate the onset of dementia.

17.3.2 UNDERLYING MECHANISMS

Cognitive disorders are mainly explained using two mechanisms: neurodegenerative processes and vascular mechanisms (Perl 2010; Shagam 2009). The finding of an association of hypovitaminosis D with both AD (Buell et al. 2009a) and non-Alzheimer dementias (Annweiler et al. 2012a) suggests that subjects with hypovitaminosis D are exposed to both these risks due to the loss of the neuroprotective and vasculoprotective effects of vitamin D.

17.3.2.1 Neuroprotection

AD is the most common degenerative dementia. AD is due to a threefold process of inflammation secondary to deposits of Aβ42 peptide, oxidative stress, and calcic excitotoxicity, responsible for the death of cholinergic and glutamatergic neurons with subsequent cerebral atrophy.

It is interesting to note that the experimentally described neurosteroid properties of vitamin D could fight point by point against each of these degenerative processes. In particular, it has been suggested that vitamin D may attenuate Aβ42 accumulation by stimulating the phagocytosis of the Aβ peptide (Masoumi et al. 2009), together with enhancing brain-to-blood Aβ efflux transport at the blood–brain barrier (Ito et al. 2011), with a decreased number of amyloid plaques as a result (Yu et al. 2011). Vitamin D also exhibits anti-inflammatory and antioxidant properties in the CNS (Cantorna et al. 1996, 1998; Garcion et al. 1996, 2002, 2003; Kalueff et al. 2004). In addition, vitamin D regulates the concentration of intracellular calcium in hippocampal neurons and modulates neuronal excitability through the regulation of voltage-gated calcium channels in the hippocampus (Annweiler et al. 2010e; Brewer et al. 2001; Garcion et al. 2002). Furthermore, vitamin D supplementation in rats results in the increased hypothalamic bioavailability of acetylcholine (Sonnenberg et al. 1986). Finally, the neurotrophic effects of vitamin D, including the stimulation of dendritic growth, could help fight against brain atrophy encountered in AD (Brown et al. 2003).

The experimentally demonstrated neuroprotective properties of vitamin D seem to stabilize neurophysiological functioning, and may protect brain neurons against the degenerative processes involved in AD.

17.3.2.2 Vasculoprotection

Vascular dementia occurs in case of repeated ischemic strokes resulting in neuronal death by ischemia. In addition to the emboligenic mechanisms, strokes are mainly explained by vascular risk factors such as hypertension or diabetes (Bérubé 1991). Current protection strategies are mainly based on prevention (World Health Organization 2000), that is, primary prevention (before the disease) by acting on the risk factors to reduce the incidence of the disease, and secondary prevention (during the disease) to reduce the duration and prevalence of the disease (Wilson and Jungner 1968). Vitamin D seems to play a role in both steps of vascular prevention (Grant 2011).

17.3.2.2.1 Primary Prevention: Vitamin D and Vascular Risk

Correcting hypovitaminosis D could prevent the onset of vascular-induced cognitive impairments because hypovitaminosis D seems to promote several cardiovascular risk factors (Kendrick et al. 2009). First, atherosclerosis is a systemic inflammatory disease (Zittermann 2003) that seems worse in the case of hypovitaminosis D. For instance, C-reactive protein—a marker of inflammation and atherosclerosis (Van Lente 2000)—is regulated by interleukin 6 and tumor necrosis factor-α (Mendall et al. 1997), which secretion decreases in a dose-dependent way in the presence vitamin D (Muller et al. 1992).

Second, demographic studies (Scragg 1981), longitudinal cohort studies (Forman et al. 2007), and clinical trials (Lind et al. 1987, 1988; Pfeifer et al. 2001) suggest that hypovitaminosis D may be a risk factor for hypertension in adults by suppressing the expression of the renin–angiotensin system in the juxtaglomerular apparatus (Li et al. 2002; Qiao et al. 2005) and through an effect on arterial compliance (Carthy et al. 1989; Merke et al. 1989; Somjen et al. 2005; Zehnder et al. 2002). For instance, it has been reported that the relative risk of developing hypertension during a 4-year follow-up was increased by 6.13 in men deficient in vitamin D (25OHD < 15 ng/mL) compared with nondeficient men (25OHD ≥ 30 ng/mL; Forman et al. 2007). The risk was also multiplied by 2.67 among vitamin D–deficient women compared with nondeficient women. Similarly, indirectly, skin exposure to UVB has been associated with low blood pressure (Cooper and Rotimi 1994; Krause et al. 1998; Kunes et al. 1991; Rostand 1997; Woodhouse et al. 1993) and the Intersalt Study, which gathered more than 10,000 people from around the world, found a positive association between blood pressures and distance to the equator (Intersalt Cooperative Research Group 1988; Rostand 1997). Finally, a trial showed that the daily intake of 800 IU vitamin D (combined with 1200 mg of calcium) decreased systolic blood pressures by more than 9% (Pfeifer et al. 2001). Other trials found a decrease in blood pressures in hypertensive patients receiving 0.75 µg (Lind et al. 1988) or 1 µg of vitamin D (Lind et al. 1987).

Third, hypovitaminosis D negatively influences glycemia, and vitamin D supplementation improves glucose metabolism (Pittas et al. 2007). A literature review has highlighted a significant association between vitamin D deficiency and type 2 diabetes (OR = 0.36; 95% CI, 0.16–1.80) for normal vitamin D status compared with hypovitaminosis D (Pittas et al. 2007). Similarly, vitamin D supplementation protects against type 2 diabetes (OR = 0.86; 95% CI, 0.79–0.93; Pittas et al. 2007). This effect could be explained by an action of vitamin D specifically on pancreatic β cells (which express both the VDRs and 1-α-hydroxylase; Bland et al. 2004). Within these cells, insulin secretion is calcium-dependent (Milner and Hales 1967). Vitamin D, which regulates the membrane flow of calcium, could therefore influence insulin resistance. The second explanation is based on the fact that type 2 diabetes is an inflammatory disease (Duncan et al. 2003; Hu et al. 2004; Pradhan et al. 2001). Vitamin D could therefore improve insulin sensitivity and β-cell survival by modulating the genetic expression, synthesis and effects of inflammatory cytokines (Pittas et al. 2007).

Finally, it should be noted that hypovitaminosis D may itself be a cardiovascular risk factor, as the National Health and Nutrition Examination Survey study, involving 4839 patients followed from 2001 to 2004, showed that low serum 25OHD was associated with a higher prevalence of peripheral arterial disease defined by an ankle brachial pressure index lower than 0.9 (Melamed et al. 2008). Similarly, a prospective study of 3408 patients showed an independent inverse association between serum vitamin D concentrations and cardiovascular mortality (Ginde et al. 2009).

17.3.2.2.2 *Secondary Prevention: Vitamin D and Defense Reaction in Cerebral Ischemia*

Correcting hypovitaminosis D in older patients could also provide a measure of protection against cerebral ischemia (and indirectly against vascular dementia). Indeed, Wang et al. (2000) showed that ligation of the middle cerebral artery caused a cerebral infarct size reduction in rats pretreated with vitamin D compared with control rats without vitamin D supplementation. This phenomenon could be explained by the neurotrophic action of vitamin D. Extrapolating these observations to humans reinforces the need to maintain a normal vitamin D status in older patients with a vascular risk.

In summary, hypovitaminosis D in the elderly could be an explanatory factor for dementia because of its involvement in both degenerative and vascular brain damage. The question is then, how to determine whether correction of hypovitaminosis D could prevent the onset of these lesions and of dementia. The multi-target therapies are primarily designed to fight against neuronal damage. The neuroprotective and vasculoprotective properties of vitamin D are therefore potentially attractive to prevent the onset of dementia, or at least to prevent the decline of cognitive performance in demented people.

17.4 EFFECTS OF VITAMIN D INTAKES ON COGNITION

No placebo-controlled randomized trial has yet explored the benefits of vitamin D supplementation to prevent or treat cognitive cognitive decline or dementia in older adults (Annweiler et al. 2009a,b,c, 2010c,e). For the moment, it seems worthwhile to determine whether high vitamin D intakes are associated with better cognitive performance in the elderly.

17.4.1 OBSERVATIONAL STUDIES

Vitamin D is a fat-soluble vitamin that exists in two forms: vitamin D_2 produced by irradiation of ergosterol and vitamin D_3 produced by the action of ultraviolet (UV) radiation in the skin/vitamin D_3 provided directly by foods. Three sources of vitamin D can, therefore, be distinguished: dietary intakes, sun exposure, and supplements.

17.4.1.1 Dietary Intakes of Vitamin D and Cognition

An association was recently highlighted between the weekly dietary intakes of vitamin D and global cognitive performance ($\beta = 0.002$; 95% CI, 0.001–0.003; $p < .001$; Annweiler et al. 2010g). The findings showed among 5596 community-dwelling older women (mean age, 80.4 years) from the EPIDOS study, that inadequate vitamin D dietary intakes were associated with cognitive impairment (OR = 1.30; 95% CI, 1.04–1.63; $p = .024$). Although this association could be explained in a more general way by a healthy lifestyle illustrated by a rich and varied diet and by regular physical activity (Annweiler et al. 2010a; Scarmeas et al. 2009), it was consistent with previous literature. In particular, in 2007, a pilot study found a correlation between the 3-day vitamin D dietary intakes and cognitive impairment on MMSE score ($r = .35$, $p < .01$) among 69 community-dwellers aged 84 years on average (59.4% women; Rondanelli et al. 2007). This work proposed a precise quantification of the dietary intakes of vitamin D, but was restricted to a limited size. Despite these methodologic divergences, both studies found that high dietary intakes of vitamin D were associated with high cognitive performance.

17.4.1.2 Sun-Related Intakes of Vitamin D and Cognition

Sun exposure of skin is the second natural source of vitamin D. To the best of our knowledge, no study has yet explored the effect of solar radiation on cognition. Only the study cited above assessed the link between sun exposure and cognitive status in older women (Annweiler et al. 2010g). In this study, we found an unadjusted association between sun exposure of hands and face at least 15 min per day between 11:00 a.m. and 3:00 p.m., and global cognitive performance assessed with Pfeiffer's SPMSQ ($\beta = 0.248$; 95% CI, 0.183–0.313; $p < .001$; Annweiler et al. 2010g). In addition, sun exposure protected against cognitive impairment (OR = 0.641; 95% CI, 0.542–0.758; $p < .001$; Annweiler et al. 2010g). Nevertheless, both associations were not significant after adjustment for age, disability level, and other covariates ($\beta = 0.037$ with $p = .282$; and OR = 0.944 with $p = .552$, respectively; Annweiler et al. 2010g), which was consistent with the observations that the skin synthesis of vitamin D from UVB decreases while aging (MacLaughlin and Holick 1985; Holick et al. 1989).

17.4.1.3 Vitamin D Supplements and Cognition

The third source of vitamin D for the elderly relies on supplements. To date, only three pre/post studies have examined the effect of the use of vitamin D supplements on cognition.

The first one assessed the change in scores on semantic fluency task, clock-drawing test, and neuropsychiatric inventory after the administration of vitamin D_2 (50,000 IU per os three times a week for 4 weeks) among 25 older residents with hypovitaminosis D (mean age, 86.2 years; 68% female), compared with 38 residents with no hypovitaminosis D who received no vitamin D supplementation (mean age, 87.4 years; 78.9% female; mean baseline 25OHD, 34.8 ± 1.8 ng/mL; Przybelski et al. 2008). No significant between-group difference was observed after 4 weeks of treatment. Nevertheless, among the treatment groups, the increase in serum 25OHD concentrations (from 17.3 to 63.8 ng/mL, $p < .0001$) was coupled with an improved neuropsychiatric inventory score [from 7.3 ± 2.0 before vitamin D supplementation to 5.5 ± 1.9 after treatment (lower scores indicate better performance)] and an improved clock-drawing test (5.7 ± 0.5 after treatment versus 5.1 ± 0.5 before treatment).

In line with this, it has been recently found that vitamin D supplementation of up to 3000 IU per day for 8 weeks in 13 patients with mild to moderate AD was associated with a 6-point improvement of the Alzheimer's Disease Assessment Scale-cognition (ADAS-cog) score (range, 4.5–8.5, $p < .001$; Stein et al. 2011).

Finally, it was shown in a cohort of older patients, followed for 15 months and taking no specific antidementia drugs, that at equal initial cognitive performance and at equal initial vitamin D status, those who received vitamin D supplementation exhibited a significant increase in cognitive scores at the end of the follow-up period compared with those who did not receive any vitamin D supplementation (Annweiler et al. 2012b). This effect was found for the MMSE score ($p = .013$), the Cognitive Assessment Battery score ($p = .020$), and specifically, for the Frontal Assessment Battery score ($p = .014$), which is an aggregate measure of executive functions.

In summary, the intakes of vitamin D—whether from the diet, sun exposure, or supplements—seem to be positively associated with cognitive performance in elderly subjects. The designs of most previous studies likely prevented finding a cause-and-effect link. At this stage, clinical trials are needed to determine the effect of vitamin D administration on cognition and AD with higher levels of evidence.

17.4.2 Interventional Studies

Only three randomized double-blind trials have focused on this issue.

The objective of the first one was to determine the effect of 3 years of estroprogestagen replacement versus placebo on cognitive function in older women (Greenspan et al. 2005). Both intervention and placebo arms received vitamin D supplementation at the same time. Results showed a modest but significant improvement of the MMSE score in the placebo group, which received both placebo and vitamin D supplementation ($n = 186$; mean age, 71.3 ± 4.8 years, 100% female; Greenspan et al. 2005). Furthermore, there were no significant between-group differences with the intervention group that received both estroprogestagen and vitamin D supplements, reinforcing the hypothesis of a vitamin D–related cognitive improvement (Greenspan et al. 2005).

The second randomized control trial was designed to correct hypovitaminosis D among 32 participants with AD using open label 1000 IU of vitamin D_2 per day for 8 weeks, and then to examine in a double-blinded manner the cognitive effectiveness of supraphysiological doses of vitamin D_2 in 16 patients with AD (i.e., 7000 IU/day) compared with 16 other patients with AD receiving physiological doses of vitamin D_2 (i.e., 1000 IU/day; Stein et al. 2011). The authors failed to find a significant between-group difference regarding the change in ADAS-cog and Wechsler Memory Scale-Revised Logical memory scores in the whole cohort, highlighting the fact that supraphysiological doses of vitamin D were not more efficient in improving cognitive abilities than physiological doses.

Finally, the third randomized double-blind trial focused on the effect of vitamin D supplementation on neuromuscular function in older adults during a 6-month follow-up (Dhesi et al. 2004). In this trial, 139 community-dwelling subjects aged 65 and older with a history of falls and serum 25OHD levels of 12 ng/mL or lower received a single intramuscular injection of 600,000 IU of ergocalciferol ($n = 70$; mean age, 76.6 years; 79.7% female) or placebo ($n = 69$; mean age, 77.0 years; 76.8% female) at the beginning of the study. The outcome was the choice reaction time measured with a computerized program (Kalra et al. 1993). Each participant performed the test three times at baseline, and then three times after 6 months of follow-up. The results showed that the intervention group accelerated the mean choice reaction time of 0.41 s, which was not the case in the placebo group (-0.41 vs. 0.06 s, respectively; $t = 2.52$, $p = .01$). This result was consistent with Chapter 17.2.5.2, in which we highlighted an association between serum vitamin D concentration and information processing speed.

All these findings suggest an additional rationale supporting the efficacy of vitamin D supplementation on cognitive performance and progression of dementia in vitamin D–deficient older adults.

17.5 CONCLUSIONS

Vitamin D is a steroid hormone involved in neurophysiology. Age-related hypovitaminosis D results in CNS malfunction, particularly in terms of cognition. Observational studies have highlighted an association between low vitamin D status and global cognitive decline. This association could be specifically explained by hypovitaminosis D–induced executive dysfunction, with adverse consequences on processing speed and autonomy. Correcting hypovitaminosis D in the elderly through an exogenous supply seems to be an effective neuroprotective strategy, both easily applicable and inexpensive, for the prevention of cognitive disorders and dementia. The therapeutic benefit remains to be defined by future placebo-controlled clinical trials.

REFERENCES

American Psychiatric Association. 1994. Diagnostic and statistical manual of mental disorders. In *DSM-IV*, edited by A. P. P., Inc. Washington, DC.

Ankarcrona, M., J. M. Dypbukt, E. Bonfoco, B. Zhivotovsky, S. Orrenius, S. A. Lipton, and P. Nicotera. 1995. Glutamate-induced neuronal death: A succession of necrosis or apoptosis depending on mitochondrial function. *Neuron* 15 (4):961–973.

Annweiler, C., G. Allali, P. Allain, S. Bridenbaugh, A. M. Schott, R. W. Kressig, and O. Beauchet. 2009a. Vitamin D and cognitive performance in adults: A systematic review. *European Journal of Neurology* 16 (10):1083–1089.

Annweiler, C., A. M. Schott, G. Berrut, and O. Beauchet. 2009b. Vitamine D et cognition, du soleil pour le crépuscule de l'esprit. *Annales de Gérontologie* 2 (4):239–242.

Annweiler, C., A. M. Schott, G. Berrut, B. Fantino, and O. Beauchet. 2009c. Vitamin D–related changes in physical performance: A systematic review. *The Journal of Nutrition, Health & Aging* 13 (10):893–898.

Annweiler, C., and O. Beauchet. 2010. Vitamin D and bone fracture mechanisms: What about the nonbone 'D'efense? *International Journal of Clinical Practice* 64 (5):541–543.

Annweiler, C., D. Le Gall, B. Fantino, and O. Beauchet. 2010a. Fish consumption and dementia: Keep the vitamin D in memory. *European Journal of Neurology* e40.

Annweiler, C., D. Le Gall, B. Fantino, O. Beauchet, K. L. Tucker, and J. S. Buell. 2010b. 25-Hydroxyvitamin D, dementia, and cerebrovascular pathology in elders receiving home services. *Neurology* 75 (1):95; author reply 95–96.

Annweiler, C., M. Montero-Odasso, A. M. Schott, G. Berrut, B. Fantino, and O. Beauchet. 2010c. Fall prevention and vitamin D in the elderly: An overview of the key role of the non-bone effects. *Journal of Neuroengineering and Rehabilitation* 7:50.

Annweiler, C., A. M. Schott, G. Allali, S. A. Bridenbaugh, R. W. Kressig, P. Allain, F. R. Herrmann, and O. Beauchet. 2010d. Association of vitamin D deficiency with cognitive impairment in older women: Cross-sectional study. *Neurology* 74 (1):27–32.

Annweiler, C., A. M. Schott, G. Berrut, V. Chauvire, D. Le Gall, M. Inzitari, and O. Beauchet. 2010e. Vitamin D and ageing: Neurological issues. *Neuropsychobiology* 62 (3):139–150.

Annweiler, C., A. M. Schott, M. Montero-Odasso, G. Berrut, B. Fantino, F. R. Herrmann, and O. Beauchet. 2010f. Cross-sectional association between serum vitamin D concentration and walking speed measured at usual and fast pace among older women: The EPIDOS Study. *Journal of Bone and Mineral Research* 25 (8):1858–1866.

Annweiler, C., A. M. Schott, Y. Rolland, H. Blain, F. R. Herrmann, and O. Beauchet. 2010g. Dietary intakes of vitamin D and cognition in older women: A large population-based study. *Neurology* 75 (20):1810–1816.

Annweiler, C., B. Fantino, D. Le Gall, A. M. Schott, G. Berrut, and O. Beauchet. 2011. Severe vitamin D deficiency is associated with advanced-stage dementia amongst geriatric inpatients. *Journal of the American Geriatrics Society* 59 (1):169–171.

Annweiler, C., Y. Rolland, A. M. Schott, H. Blain, B. Vellas, and O. Beauchet. 2012a. Serum vitamin D deficiency as a predictor of incident non-Alzheimer dementias: A 7-year longitudinal follow-up. *Dementia and Geriatric Cognitive Disorders* 32 (4):273–278.

Annweiler, C., B. Fantino, J. Gautier, M. Beaudenon, S. Thiery, and O. Beauchet. 2012b. Cognitive effects of vitamin D supplementation amongst older outpatients visiting memory clinic: A pre–post study. *Journal of the American Geriatrics Society* 60 (4):793–795.

Baas, D., K. Prufer, M. E. Ittel, S. Kuchler-Bopp, G. Labourdette, L. L. Sarlieve, and P. Brachet. 2000. Rat oligodendrocytes express the vitamin D(3) receptor and respond to 1,25-dihydroxyvitamin D(3). *Glia* 31 (1):59–68.

Baddeley, A. 1993. *La mémoire humaine*. Grenoble: Presses Universitaires de Grenoble

Baddeley, A., and S. Della Sala. 1996. Working memory and executive control. *Philosophical Transactions of the Royal Society of London. Series B, Biological Sciences* 351 (1346):1397–1403; discussion 1403-4.

Baddeley, A. D., and B. A. Wilson. 1985. Phonological coding and short-term memory in patients without speech. *Journal of Memory and Language* 24:490–502.

Baksi, S. N., and M. J. Hughes. 1982. Chronic vitamin D deficiency in the weanling rat alters catecholamine metabolism in the cortex. *Brain Research* 242 (2):387–390.

Bandler, R. J., Jr. 1969. Facilitation of aggressive behaviour in rat by direct cholinergic stimulation of the hypothalamus. *Nature* 224 (5223):1035–1036.

Beauchet, O., C. Annweiler, J. Verghese, B. Fantino, F. R. Herrmann, and G. Allali. 2011a. Biology of gait control: Vitamin D involvement. *Neurology* 76 (19):1617–1622.

Beauchet, O., D. Milea, A. Graffe, B. Fantino, and C. Annweiler. 2011b. Association between serum 25-hydroxyvitamin D concentrations and vision: A cross-sectional population-based study of older adults. *Journal of the American Geriatrics Society* 59 (3):568–570.

Becker, A., D. W. Eyles, J. J. McGrath, and G. Grecksch. 2005. Transient prenatal vitamin D deficiency is associated with subtle alterations in learning and memory functions in adult rats. *Behavioural Brain Research* 161 (2):306–312.

Bemiss, C. J., B. D. Mahon, A. Henry, V. Weaver, and M. T. Cantorna. 2002. Interleukin-2 is one of the targets of 1,25-dihydroxyvitamin D3 in the immune system. *Archives of Biochemistry and Biophysics* 402 (2):249–254.

Benton, A. L., and K. Hamsher. 1976. *Multilingual Aphasia Examination*. Iowa City: University of Iowa.

Bérubé, L. 1991. *Terminologie de neuropsychologie et de neurologie du comportement*. Montréal: Les Éditions de la Chenelière, Inc.

Bland, R., D. Markovic, C. E. Hills, S. V. Hughes, S. L. Chan, P. E. Squires, and M. Hewison. 2004. Expression of 25-hydroxyvitamin D_3-1alpha-hydroxylase in pancreatic islets. *The Journal of Steroid Biochemistry and Molecular Biology* 89–90 (1–5):121–125.

Bledowski, C., B. Rahm, and J. B. Rowe. 2009. What "works" in working memory? Separate systems for selection and updating of critical information. *Journal of Neuroscience* 29 (43):13735–13741.

Bloch, H. 1994. In *Grand dictionnaire de la psychologie*, edited by Larousse, Paris.

Bonfoco, E., D. Krainc, M. Ankarcrona, P. Nicotera, and S. A. Lipton. 1995. Apoptosis and necrosis: Two distinct events induced, respectively, by mild and intense insults with N-methyl-D-aspartate or nitric oxide/superoxide in cortical cell cultures. *Proceedings of the National Academy of Sciences of the United States of America* 92 (16):7162–7166.

Bouvard, B., C. Annweiler, A. Salle, O. Beauchet, D. Chappard, M. Audran, and E. Legrand. 2011. Extraskeletal effects of vitamin D: Facts, uncertainties, and controversies. *Joint Bone Spine* 78 (1):10–16.

Brewer, L. D., V. Thibault, K. C. Chen, M. C. Langub, P. W. Landfield, and N. M. Porter. 2001. Vitamin D hormone confers neuroprotection in parallel with downregulation of L-type calcium channel expression in hippocampal neurons. *Journal of Neuroscience* 21 (1):98–108.

Brown, J., J. I. Bianco, J. J. McGrath, and D. W. Eyles. 2003. 1,25-dihydroxyvitamin D_3 induces nerve growth factor, promotes neurite outgrowth and inhibits mitosis in embryonic rat hippocampal neurons. *Neuroscience Letters* 343 (2):139–143.

Buell, J. S., B. Dawson-Hughes, T. M. Scott, D. E. Weiner, G. E. Dallal, W. Q. Qui, P. Bergethon, I. H. Rosenberg, M. F. Folstein, S. Patz, R. A. Bhadelia, and K. L. Tucker. 2009. 25-Hydroxyvitamin D, dementia, and cerebrovascular pathology in elders receiving home services. *Neurology* 74 (1):18–26.

Buell, J. S., T. M. Scott, B. Dawson-Hughes, G. E. Dallal, I. H. Rosenberg, M. F. Folstein, and K. L. Tucker. 2009. Vitamin D is associated with cognitive function in elders receiving home health services. *The Journals of Gerontology. Series A, Biological Sciences and Medical Sciences* 64 (8):888–895.

Burne, T. H., A. Becker, J. Brown, D. W. Eyles, A. Mackay-Sim, and J. J. McGrath. 2004. Transient prenatal Vitamin D deficiency is associated with hyperlocomotion in adult rats. *Behavioural Brain Research* 154 (2):549–555.

Burne, T. H., J. J. McGrath, D. W. Eyles, and A. Mackay-Sim. 2005. Behavioural characterization of vitamin D receptor knockout mice. *Behavioural Brain Research* 157 (2):299–308.

Burne, T. H., A. N. Johnston, J. J. McGrath, and A. Mackay-Sim. 2006. Swimming behaviour and post-swimming activity in vitamin D receptor knockout mice. *Brain Research Bulletin* 69 (1):74–78.

Cantorna, M. T., C. E. Hayes, and H. F. DeLuca. 1996. 1,25-Dihydroxyvitamin D_3 reversibly blocks the progression of relapsing encephalomyelitis, a model of multiple sclerosis. *Proceedings of the National Academy of Sciences of the United States of America* 93 (15):7861–7864.

Cantorna, M. T., W. D. Woodward, C. E. Hayes, and H. F. DeLuca. 1998. 1,25-dihydroxyvitamin D_3 is a positive regulator for the two anti-encephalitogenic cytokines TGF-beta 1 and IL-4. *The Journal of Immunology* 160 (11):5314–5319.

Carthy, E. P., W. Yamashita, A. Hsu, and B. S. Ooi. 1989. 1,25-Dihydroxyvitamin D3 and rat vascular smooth muscle cell growth. *Hypertension* 13 (6 Pt 2):954–959.

Cooper, R., and C. Rotimi. 1994. Hypertension in populations of West African origin: Is there a genetic predisposition? *Journal of Hypertension* 12 (3):215–227.

Cornet, A., C. Baudet, I. Neveu, A. Baron-Van Evercooren, P. Brachet, and P. Naveilhan. 1998. 1,25-Dihydroxyvitamin D_3 regulates the expression of VDR and NGF gene in Schwann cells *in vitro*. *Journal of Neuroscience Research* 53 (6):742–746.

Cowppli-Bony, P., C. Fabrigoule, L. Letenneur, K. Ritchie, A. Alperovitch, J. F. Dartigues, and B. Dubois. 2005. [Validity of the five-word screening test for Alzheimer's disease in a population based study]. *Revue Neurologique* 161 (12 Pt 1):1205–1212.

Cui, X., J. J. McGrath, T. H. Burne, A. Mackay-Sim, and D. W. Eyles. 2007. Maternal vitamin D depletion alters neurogenesis in the developing rat brain. *International Journal of Developmental Neuroscience* 25 (4):227–232.

Curtis, C. E., and M. D'Esposito. 2003. Persistent activity in the prefrontal cortex during working memory. *Trends in Cognitive Sciences* 7 (9):415–423.

Dawson, V. L., and T. M. Dawson. 1996. Nitric oxide actions in neurochemistry. *Neurochemistry International* 29 (2):97–110.

de Viragh, P. A., K. G. Haglid, and M. R. Celio. 1989. Parvalbumin increases in the caudate putamen of rats with vitamin D hypervitaminosis. *Proceedings of the National Academy of Sciences of the United States of America* 86 (10):3887–3890.

Deutsch, J. A., and D. Deutsch. 1963. Some theoretical considerations. *Psychological Review* 70:80–90.

Dhesi, J. K., S. H. Jackson, L. M. Bearne, C. Moniz, M. V. Hurley, C. G. Swift, and T. J. Allain. 2004. Vitamin D supplementation improves neuromuscular function in older people who fall. *Age and Ageing* 33 (6):589–595.

Dringen, R., J. M. Gutterer, and J. Hirrlinger. 2000. Glutathione metabolism in brain metabolic interaction between astrocytes and neurons in the defense against reactive oxygen species. *European Journal of Biochemistry* 267 (16):4912–4916.

Duncan, B. B., M. I. Schmidt, J. S. Pankow, C. M. Ballantyne, D. Couper, A. Vigo, R. Hoogeveen, A. R. Folsom, and G. Heiss. 2003. Low-grade systemic inflammation and the development of type 2 diabetes: The atherosclerosis risk in communities study. *Diabetes* 52 (7):1799–1805.

Engelhart, M. J., M. I. Geerlings, A. Ruitenberg, J. C. Van Swieten, A. Hofman, J. C. Witteman, and M. M. Breteler. 2002. Diet and risk of dementia: Does fat matter? The Rotterdam Study. *Neurology* 59 (12):1915–1921.

Eyles, D., J. Brown, A. Mackay-Sim, J. McGrath, and F. Feron. 2003. Vitamin D3 and brain development. *Neuroscience* 118 (3):641–653.

Eyles, D. W., S. Smith, R. Kinobe, M. Hewison, and J. J. McGrath. 2005. Distribution of the vitamin D receptor and 1 alpha-hydroxylase in human brain. *Journal of Chemical Neuroanatomy* 29 (1):21–30.

Forman, J. P., E. Giovannucci, M. D. Holmes, H. A. Bischoff-Ferrari, S. S. Tworoger, W. C. Willett, and G. C. Curhan. 2007. Plasma 25-hydroxyvitamin D levels and risk of incident hypertension. *Hypertension* 49 (5):1063–1069.

Garcion, E., X. D. Thanh, F. Bled, E. Teissier, M. P. Dehouck, F. Rigault, P. Brachet, A. Girault, G. Torpier, and F. Darcy. 1996. 1,25-Dihydroxyvitamin D3 regulates gamma 1 transpeptidase activity in rat brain. *Neuroscience Letters* 216 (3):183–186.

Garcion, E., L. Sindji, C. Montero-Menei, C. Andre, P. Brachet, and F. Darcy. 1998. Expression of inducible nitric oxide synthase during rat brain inflammation: Regulation by 1,25-dihydroxyvitamin D3. *Glia* 22 (3):282–294.

Garcion, E., L. Sindji, G. Leblondel, P. Brachet, and F. Darcy. 1999. 1,25-dihydroxyvitamin D3 regulates the synthesis of gamma-glutamyl transpeptidase and glutathione levels in rat primary astrocytes. *Journal of Neurochemistry* 73 (2):859–866.

Garcion, E., N. Wion-Barbot, C. N. Montero-Menei, F. Berger, and D. Wion. 2002. New clues about vitamin D functions in the nervous system. *Trends in Endocrinology and Metabolism* 13 (3):100–105.

Garcion, E., L. Sindji, S. Nataf, P. Brachet, F. Darcy, and C. N. Montero-Menei. 2003. Treatment of experimental autoimmune encephalomyelitis in rat by 1,25-dihydroxyvitamin D3 leads to early effects within the central nervous system. *Acta Neuropathologica* 105 (5):438–448.

Gaudino, E. A., M. W. Geisler, and N. K. Squires. 1995. Construct validity in the Trail Making Test: What makes Part B harder? *Journal of Clinical and Experimental Neuropsychology* 17 (4):529–535.

Ginde, A. A., R. Scragg, R. S. Schwartz, and C. A. Camargo, Jr. 2009. Prospective study of serum 25-hydroxyvitamin D level, cardiovascular disease mortality, and all-cause mortality in older U.S. adults. *Journal of the American Geriatrics Society* 57 (9):1595–1603.

Godefroy, O., M. Jeannerod, P. Allain, and D. Le Gall. 2008. [Frontal lobe, executive functions and cognitive control]. *Revue Neurologique* 164 Suppl 3:S119–127.

Grant, W. B. 2009. Does vitamin D reduce the risk of dementia? *Journal of Alzheimer's Disease* 17 (1):151–159.

Grant, W. B. 2011. Vitamin D might reduce some vascular risk factors and, consequently, risk of dementia. *The Netherlands Journal of Medicine* 69 (1):51.

Greenspan, S. L., N. M. Resnick, and R. A. Parker. 2005. The effect of hormone replacement on physical performance in community-dwelling elderly women. *The American Journal of Medicine* 118 (11):1232–1239.

Grober, E., H. Buschke, H. Crystal, S. Bang, and R. Dresner. 1988. Screening for dementia by memory testing. *Neurology* 38 (6):900–903.

Hanninen, T., M. Hallikainen, K. Koivisto, K. Partanen, M. P. Laakso, P. J. Riekkinen, Sr., and H. Soininen. 1997. Decline of frontal lobe functions in subjects with age-associated memory impairment. *Neurology* 48 (1):148–153.

Hittmair-Delazer, M., C. Semenza, and G. Denes. 1994. Concepts and facts in calculation. *Brain* 117 (Pt 4):715–728.

Hodkinson, H. M. 1972. Evaluation of a mental test score for assessment of mental impairment in the elderly. *Age and Ageing* 1 (4):233–238.

Holick, M. F., L. Y. Matsuoka, and J. Wortsman. 1989. Age, vitamin D, and solar ultraviolet. *Lancet* 2 (8671):1104–1105.

Hu, F. B., J. B. Meigs, T. Y. Li, N. Rifai, and J. E. Manson. 2004. Inflammatory markers and risk of developing type 2 diabetes in women. *Diabetes* 53 (3):693–700.

Ibi, M., H. Sawada, M. Nakanishi, T. Kume, H. Katsuki, S. Kaneko, S. Shimohama, and A. Akaike. 2001. Protective effects of 1 alpha,25-(OH)(2)D($_3$) against the neurotoxicity of glutamate and reactive oxygen species in mesencephalic culture. *Neuropharmacology* 40 (6):761–771.

Ito, S., S. Ohtsuki, Y. Nezu, Y. Koitabashi, S. Murata, and T. Terasaki. 2011. 1α,25-Dihydroxyvitamin D$_3$ enhances cerebral clearance of human amyloid-b peptide(1–40) from mouse brain across the blood–brain barrier. *Fluids and Barriers of the CNS* 8:20.

Intersalt Cooperative Research Group. 1988. Intersalt: An international study of electrolyte excretion and blood pressure. Results for 24 hour urinary sodium and potassium excretion. *BMJ* 297 (6644):319–328.

Jakobson, R. 1960. Closing statements: Linguistics and poetics. In *Style in Langage*. New York: T.A. Sebeok.

Jia, Z., and I. Nemere. 1999. Immunochemical studies on the putative plasmalemmal receptor for 1,25-dihydroxyvitamin D$_3$. II. Chick kidney and brain. *Steroids* 64 (8):541–550.

Jorde, R., K. Waterloo, F. Saleh, E. Haug, and J. Svartberg. 2006. Neuropsychological function in relation to serum parathyroid hormone and serum 25-hydroxyvitamin D levels. The Tromso Study. *Journal of Neurology* 253 (4):464–470.

Kalmijn, S., L. J. Launer, A. Ott, J. C. Witteman, A. Hofman, and M. M. Breteler. 1997. Dietary fat intake and the risk of incident dementia in the Rotterdam Study. *Annals of Neurology* 42 (5):776–782.

Kalra, L., S. H. Jackson, and C. G. Swift. 1993. Effect of antihypertensive treatment on psychomotor performance in the elderly. *Journal of Human Hypertension* 7 (3):285–290.

Kalueff, A. V., K. O. Eremin, and P. Tuohimaa. 2004. Mechanisms of neuroprotective action of vitamin D($_3$). *Biochemistry (Mosc)* 69 (7):738–741.

Kalueff, A. V., Y. R. Lou, I. Laaksi, and P. Tuohimaa. 2005a. Abnormal behavioral organization of grooming in mice lacking the vitamin D receptor gene. *Journal of Neurogenetics* 19 (1):1–24.

Kalueff, A. V., A. Minasyan, and P. Tuohimaa. 2005b. Anticonvulsant effects of 1,25-dihydroxyvitamin D in chemically induced seizures in mice. *Brain Research Bulletin* 67 (1–2):156–160.

Kalueff, A., E. Loseva, H. Haapasalo, I. Rantala, J. Keranen, Y. R. Lou, A. Minasyan, T. Keisala, S. Miettinen, M. Kuuslahti, and P. Tuohimaa. 2006a. Thalamic calcification in vitamin D receptor knockout mice. *Neuroreport* 17 (7):717–721.

Kalueff, A. V., T. Keisala, A. Minasyan, M. Kuuslahti, S. Miettinen, and P. Tuohimaa. 2006b. Behavioural anomalies in mice evoked by "Tokyo" disruption of the vitamin D receptor gene. *Neuroscience Research* 54 (4):254–260.

Kalueff, A. V., A. Minasyan, T. Keisala, M. Kuuslahti, S. Miettinen, and P. Tuohimaa. 2006c. Increased severity of chemically induced seizures in mice with partially deleted vitamin D receptor gene. *Neuroscience Letters* 394 (1):69–73.

Kalueff, A. V., and P. Tuohimaa. 2007. Neurosteroid hormone vitamin D and its utility in clinical nutrition. *Current Opinion in Clinical Nutrition and Metabolic Care* 10 (1):12–19.

Kendrick, J., G. Targher, G. Smits, and M. Chonchol. 2009. 25-Hydroxyvitamin D deficiency is independently associated with cardiovascular disease in the Third National Health and Nutrition Examination Survey. *Atherosclerosis* 205 (1):255–260.

Krause, R., M. Buhring, W. Hopfenmuller, M. F. Holick, and A. M. Sharma. 1998. Ultraviolet B and blood pressure. *Lancet* 352 (9129):709–710.

Kunes, J., J. Tremblay, F. Bellavance, and P. Hamet. 1991. Influence of environmental temperature on the blood pressure of hypertensive patients in Montreal. *American Journal of Hypertension* 4 (5 Pt 1): 422–426.

Langub, M. C., J. P. Herman, H. H. Malluche, and N. J. Koszewski. 2001. Evidence of functional vitamin D receptors in rat hippocampus. *Neuroscience* 104 (1):49–56.

Laurin, D., R. Verreault, J. Lindsay, E. Dewailly, and B. J. Holub. 2003. Omega-3 fatty acids and risk of cognitive impairment and dementia. *Journal of Alzheimer's Disease* 5 (4):315–322.

Lee, D. M., A. Tajar, A. Ulubaev, N. Pendleton, T. W. O'Neill, D. B. O'Connor, G. Bartfai, S. Boonen, R. Bouillon, F. F. Casanueva, J. D. Finn, G. Forti, A. Giwercman, T. S. Han, I. T. Huhtaniemi, K. Kula, M. E. Lean, M. Punab, A. J. Silman, D. Vanderschueren, and F. C. Wu. 2009. Association between 25-hydroxyvitamin D levels and cognitive performance in middle-aged and older European men. *Journal of Neurology, Neurosurgery and Psychiatry* 80 (7):722–729.

Lezak, M. D. 1995. *Neuropsychological Assessment*. Oxford, UK: Oxford University Press.

Li, Y. C., A. E. Pirro, and M. B. Demay. 1998. Analysis of vitamin D–dependent calcium-binding protein messenger ribonucleic acid expression in mice lacking the vitamin D receptor. *Endocrinology* 139 (3):847–851.

Li, Y. C., J. Kong, M. Wei, Z. F. Chen, S. Q. Liu, and L. P. Cao. 2002. 1,25-Dihydroxyvitamin D($_3$) is a negative endocrine regulator of the renin–angiotensin system. *Journal of Clinical Investigation* 110 (2):229–238.

Lin, A. M., S. F. Fan, D. M. Yang, L. L. Hsu, and C. H. Yang. 2003. Zinc-induced apoptosis in substantia nigra of rat brain: Neuroprotection by vitamin D_3. *Free Radical Biology & Medicine* 34 (11):1416–1425.

Lind, L., B. Wengle, and S. Ljunghall. 1987. Blood pressure is lowered by vitamin D (alphacalcidol) during long-term treatment of patients with intermittent hypercalcaemia. A double-blind, placebo-controlled study. *Acta Medica Scandinavica* 222 (5):423–427.

Lind, L., H. Lithell, E. Skarfors, L. Wide, and S. Ljunghall. 1988. Reduction of blood pressure by treatment with alphacalcidol. A double-blind, placebo-controlled study in subjects with impaired glucose tolerance. *Acta Medica Scandinavica* 223 (3):211–217.

Llewellyn, D. J., K. M. Langa, and I. A. Lang. 2009. Serum 25-hydroxyvitamin D concentration and cognitive impairment. *Journal of Geriatric Psychiatry and Neurology* 22 (3):188–195.

Llewellyn, D. J., I. A. Lang, K. M. Langa, G. Muniz-Terrera, C. L. Phillips, A. Cherubini, L. Ferrucci, and D. Melzer. 2010. Vitamin D and risk of cognitive decline in elderly persons. *Archives of Internal Medicine* 170 (13):1135–1141.

Loonstra, A. S., A. R. Tarlow, and A. H. Sellers. 2001. COWAT metanorms across age, education, and gender. *Applied Neuropsychology* 8 (3):161–166.

Luciana, M., and C. A. Nelson. 2002. Assessment of neuropsychological function through use of the Cambridge Neuropsychological Testing Automated Battery: Performance in 4- to 12-year-old children. *Developmental Neuropsychology* 22 (3):595–624.

Lui, L. Y., K. Stone, J. A. Cauley, T. Hillier, and K. Yaffe. 2003. Bone loss predicts subsequent cognitive decline in older women: The study of osteoporotic fractures. *Journal of the American Geriatrics Society* 51 (1):38–43.

Luria, A. 1969. *The Working Brain: An Introduction to Neuropsychology*. New York: Basic Books.

MacLaughlin, J., and M. F. Holick. 1985. Aging decreases the capacity of human skin to produce vitamin D_3. *Journal of Clinical Investigation* 76 (4):1536–1538.

Marsh, R. L., and J. L. Hicks. 1998. Event-based prospective memory and executive control of working memory. *Journal of Experimental Psychology. Learning, Memory, and Cognition* 24 (2):336–349.

Masoumi, A., B. Goldenson, S. Ghirmai, H. Avagyan, J. Zaghi, K. Abel, X. Zheng, A. Espinosa-Jeffrey, M. Mahanian, P. T. Liu, M. Hewison, M. Mizwickie, J. Cashman, and M. Fiala. 2009. 1alpha,25-Dihydroxyvitamin D_3 interacts with curcuminoids to stimulate amyloid-beta clearance by macrophages of Alzheimer's disease patients. *Journal of Alzheimer's Disease* 17 (3):703–717.

Math, F. 2008. *Neurosciences cliniques: De la perception aux troubles du comportement*. Paris: De Boeck.

Matlin, M. W. 2008. *Cognition*. 7th revised ed. New York: Wiley-Blackwell.

McCann, J. C., and B. N. Ames. 2008. Is there convincing biological or behavioral evidence linking vitamin D deficiency to brain dysfunction? *FASEB J* 22 (4):982–1001.

McGrath, J. 1999. Hypothesis: Is low prenatal vitamin D a risk-modifying factor for schizophrenia? *Schizophrenia Research* 40 (3):173–177.

McGrath, J., F. Feron, D. Eyles, and A. Mackay-Sim. 2001. Vitamin D: The neglected neurosteroid? *Trends in Neurosciences* 24 (10):570–572.

McGrath, J., R. Scragg, D. Chant, D. Eyles, T. Burne, and D. Obradovic. 2007. No association between serum 25-hydroxyvitamin D_3 level and performance on psychometric tests in NHANES III. *Neuroepidemiology* 29 (1–2):49–54.

Meehan, T. F., and H. F. DeLuca. 2002. The vitamin D receptor is necessary for 1alpha,25-dihydroxyvitamin $D_{(3)}$ to suppress experimental autoimmune encephalomyelitis in mice. *Archives of Biochemistry and Biophysics* 408 (2):200–204.

Melamed, M. L., P. Muntner, E. D. Michos, J. Uribarri, C. Weber, J. Sharma, and P. Raggi. 2008. Serum 25-hydroxyvitamin D levels and the prevalence of peripheral arterial disease: Results from NHANES 2001 to 2004. *Arteriosclerosis, Thrombosis, and Vascular Biology* 28 (6):1179–1185.

Mendall, M. A., P. Patel, M. Asante, L. Ballam, J. Morris, D. P. Strachan, A. J. Camm, and T. C. Northfield. 1997. Relation of serum cytokine concentrations to cardiovascular risk factors and coronary heart disease. *Heart* 78 (3):273–277.

Merke, J., P. Milde, S. Lewicka, U. Hugel, G. Klaus, D. J. Mangelsdorf, M. R. Haussler, E. W. Rauterberg, and E. Ritz. 1989. Identification and regulation of 1,25-dihydroxyvitamin D_3 receptor activity and biosynthesis of 1,25-dihydroxyvitamin D_3. Studies in cultured bovine aortic endothelial cells and human dermal capillaries. *Journal of Clinical Investigation* 83 (6):1903–1915.

Miller, W. L., and A. A. Portale. 2000. Vitamin D1 alpha-hydroxylase. *Trends in Endocrinology and Metabolism* 11 (8):315–319.

Milner, R. D., and C. N. Hales. 1967. The role of calcium and magnesium in insulin secretion from rabbit pancreas studied *in vitro*. *Diabetologia* 3 (1):47–49.

Mitchell, A. J., and M. Shiri-Feshki. 2009. Rate of progression of mild cognitive impairment to dementia—meta-analysis of 41 robust inception cohort studies. *Acta Psychiatrica Scandinavica* 119 (4):252–265.

Mitrovic, B., B. A. St Pierre, A. J. Mackenzie-Graham, and J. E. Merrill. 1994. The role of nitric oxide in glial pathology. *Annals of the New York Academy of Sciences* 738:436–446.

Miyake, A., N. P. Friedman, M. J. Emerson, A. H. Witzki, A. Howerter, and T. D. Wager. 2000. The unity and diversity of executive functions and their contributions to complex "frontal lobe" tasks: A latent variable analysis. *Cognitive Psychology* 41 (1):49–100.

Morris, M. C. 2009. The role of nutrition in Alzheimer's disease: Epidemiological evidence. *European Journal of Neurology* 16 Suppl 1:1–7.

Morris, M. C., D. A. Evans, J. L. Bienias, C. C. Tangney, D. A. Bennett, R. S. Wilson, N. Aggarwal, and J. Schneider. 2003. Consumption of fish and *n*-3 fatty acids and risk of incident Alzheimer disease. *Archives of Neurology* 60 (7):940–946.

Muller, K., P. M. Haahr, M. Diamant, K. Rieneck, A. Kharazmi, and K. Bendtzen. 1992. 1,25-Dihydroxyvitamin D_3 inhibits cytokine production by human blood monocytes at the post-transcriptional level. *Cytokine* 4 (6):506–512.

Naveilhan, P., I. Neveu, D. Wion, and P. Brachet. 1996. 1,25-Dihydroxyvitamin D3, an inducer of glial cell line–derived neurotrophic factor. *Neuroreport* 7 (13):2171–2175.

Neveu, I., P. Naveilhan, C. Baudet, P. Brachet, and M. Metsis. 1994. 1,25-dihydroxyvitamin D3 regulates NT-3, NT-4 but not BDNF mRNA in astrocytes. *Neuroreport* 6 (1):124–126.

Oberauer, K. 2002. Access to information in working memory: Exploring the focus of attention. *Journal of Experimental Psychology. Learning, Memory, and Cognition* 28 (3):411–421.

Osterrieth, PA. 1944. Le test de copie d'une figure complexe. *Archives de Psychologie* 30:206–356.

Oudshoorn, C., F. U. Mattace-Raso, N. van der Velde, E. M. Colin, and T. J. van der Cammen. 2008. Higher serum vitamin D_3 levels are associated with better cognitive test performance in patients with Alzheimer's disease. *Dementia and Geriatric Cognitive Disorders* 25 (6):539–543.

Perl, D. P. 2010. Neuropathology of Alzheimer's disease. *The Mount Sinai Journal of Medicine, New York* 77 (1):32–42.

Petersen, R. C., J. C. Stevens, M. Ganguli, E. G. Tangalos, J. L. Cummings, and S. T. DeKosky. 2001. Practice parameter—Early detection of dementia: Mild cognitive impairment (An evidence-based review). Report of the Quality Standards Subcommittee of the American Academy of Neurology. *Neurology* 56 (9):1133–1142.

Pfeifer, M., B. Begerow, H. W. Minne, D. Nachtigall, and C. Hansen. 2001. Effects of a short-term vitamin $D_{(3)}$ and calcium supplementation on blood pressure and parathyroid hormone levels in elderly women. *The Journal of Clinical Endocrinology and Metabolism* 86 (4):1633–1637.

Pfeiffer, E. 1975. A short portable mental status questionnaire for the assessment of organic brain deficit in elderly patients. *Journal of the American Geriatrics Society* 23 (10):433–441.

Pittas, A. G., J. Lau, F. B. Hu, and B. Dawson-Hughes. 2007. The role of vitamin D and calcium in type 2 diabetes. A systematic review and meta-analysis. *The Journal of Clinical Endocrinology and Metabolism* 92 (6):2017–2029.

Plassman, B. L., K. M. Langa, G. G. Fisher, S. G. Heeringa, D. R. Weir, M. B. Ofstedal, J. R. Burke, M. D. Hurd, G. G. Potter, W. L. Rodgers, D. C. Steffens, J. J. McArdle, R. J. Willis, and R. B. Wallace. 2008. Prevalence of cognitive impairment without dementia in the United States. *Annals of Internal Medicine* 148 (6):427–434.

Postle, B. R. 2006. Working memory as an emergent property of the mind and brain. *Neuroscience* 139 (1): 23–38.

Pradhan, A. D., J. E. Manson, N. Rifai, J. E. Buring, and P. M. Ridker. 2001. C-reactive protein, interleukin 6, and risk of developing type 2 diabetes mellitus. *JAMA* 286 (3):327–334.

Prufer, K., T. D. Veenstra, G. F. Jirikowski, and R. Kumar. 1999. Distribution of 1,25-dihydroxyvitamin D_3 receptor immunoreactivity in the rat brain and spinal cord. *Journal of Chemical Neuroanatomy* 16 (2):135–145.

Przybelski, R. J., and N. C. Binkley. 2007. Is vitamin D important for preserving cognition? A positive correlation of serum 25-hydroxyvitamin D concentration with cognitive function. *Archives of Biochemistry and Biophysics* 460 (2):202–205.

Przybelski, R., S. Agrawal, D. Krueger, J. A. Engelke, F. Walbrun, and N. Binkley. 2008. Rapid correction of low vitamin D status in nursing home residents. *Osteoporosis International* 19 (11):1621–1628.

Qiao, G., J. Kong, M. Uskokovic, and Y. C. Li. 2005. Analogs of 1alpha,25-dihydroxyvitamin $D_{(3)}$ as novel inhibitors of renin biosynthesis. *The Journal of Steroid Biochemistry and Molecular Biology* 96 (1):59–66.

Rabbit, P. 1997. *Methodology of Frontal and Executive Functions*. Hove, UK: Psychology Press.

Reitan, R. M. 1958. Validity of the Trail Making Test as an indicator of organic brain damage. *Perceptual and Motor Skills* 8:271–276.

Riaz, S., M. Malcangio, M. Miller, and D. R. Tomlinson. 1999. A vitamin D($_3$) derivative (CB1093) induces nerve growth factor and prevents neurotrophic deficits in streptozotocin-diabetic rats. *Diabetologia* 42 (11):1308–1313.

Rondanelli, M., R. Trotti, A. Opizzi, and S. B. Solerte. 2007. Relationship among nutritional status, pro/anti-oxidant balance and cognitive performance in a group of free-living healthy elderly. *Minerva Medica* 98 (6):639–645.

Rostand, S. G. 1997. Ultraviolet light may contribute to geographic and racial blood pressure differences. *Hypertension* 30 (2 Pt 1):150–156.

Sabouraud, O. 1995. *Le langage et ses maux*. Paris: Ed. Odile Jacob.

Sackett, D. L. 1979. Bias in analytic research. *Journal of Chronic Diseases* 32 (1–2):51–63.

Salthouse, T. A. 2001. Attempted decomposition of age-related influences on two tests of reasoning. *Psychology and Aging* 16 (2):251–263.

Sanchez, B., E. Lopez-Martin, C. Segura, J. L. Labandeira-Garcia, and R. Perez-Fernandez. 2002. 1,25-Dihydroxyvitamin D($_3$) increases striatal GDNF mRNA and protein expression in adult rats. *Brain Research. Molecular Brain Research* 108 (1–2):143–146.

Saporito, M. S., H. M. Wilcox, K. C. Hartpence, M. E. Lewis, J. L. Vaught, and S. Carswell. 1993. Pharmacological induction of nerve growth factor mRNA in adult rat brain. *Experimental Neurology* 123 (2):295–302.

Sato, Y., T. Asoh, and K. Oizumi. 1998. High prevalence of vitamin D deficiency and reduced bone mass in elderly women with Alzheimer's disease. *Bone* 23 (6):555–557.

Scarmeas, N., J. A. Luchsinger, N. Schupf, A. M. Brickman, S. Cosentino, M. X. Tang, and Y. Stern. 2009. Physical activity, diet, and risk of Alzheimer disease. *JAMA* 302 (6):627–637.

Scragg, R. 1981. Seasonality of cardiovascular disease mortality and the possible protective effect of ultra-violet radiation. *International Journal of Epidemiology* 10 (4):337–341.

Seamans, K. M., T. R. Hill, L. Scully, N. Meunier, M. Andrillo-Sanchez, A. Polito, I. Hininger-Favier, D. Ciarapica, E. E. Simpson, B. J. Stewart-Knox, J. M. O'Connor, C. Coudray, and K. D. Cashman. 2010. Vitamin D status and measures of cognitive function in healthy older European adults. *European Journal of Clinical Nutrition* 64 (10):1172–1178.

Shagam, J. Y. 2009. The many faces of dementia. *Radiologic Technology* 81 (2):153–168.

Shinpo, K., S. Kikuchi, H. Sasaki, F. Moriwaka, and K. Tashiro. 2000. Effect of 1,25-dihydroxyvitamin D($_3$) on cultured mesencephalic dopaminergic neurons to the combined toxicity caused by L-buthionine sulfoximine and 1-methyl-4-phenylpyridine. *Journal of Neuroscience Research* 62 (3):374–382.

Singer, R. N. 1980. Cognitive processes, learner strategies, and skilled motor behaviors. *Canadian Journal of Applied Sport Sciences* 5 (1):25–32.

Slinin, Y., M. L. Paudel, B. C. Taylor, H. A. Fink, A. Ishani, M. T. Canales, K. Yaffe, E. Barrett-Connor, E. S. Orwoll, J. M. Shikany, E. S. Leblanc, J. A. Cauley, and K. E. Ensrud. 2010. 25-Hydroxyvitamin D levels and cognitive performance and decline in elderly men. *Neurology* 74 (1):33–41.

Somjen, D., Y. Weisman, F. Kohen, B. Gayer, R. Limor, O. Sharon, N. Jaccard, E. Knoll, and N. Stern. 2005. 25-Hydroxyvitamin D$_3$-1alpha-hydroxylase is expressed in human vascular smooth muscle cells and is upregulated by parathyroid hormone and estrogenic compounds. *Circulation* 111 (13):1666–1671.

Sonnenberg, J., V. N. Luine, L. C. Krey, and S. Christakos. 1986. 1,25-Dihydroxyvitamin D3 treatment results in increased choline acetyltransferase activity in specific brain nuclei. *Endocrinology* 118 (4):1433–1439.

Stroop, J. R. 1935. Studies of interference in serial verbal reactions. *Journal of Experimental Psychology* 18:643–662.

Stumpf, W. E., M. Sar, S. A. Clark, and H. F. DeLuca. 1982. Brain target sites for 1,25-dihydroxyvitamin D$_3$. *Science* 215 (4538):1403–1405.

Stumpf, W. E., R. A. Mueler, and B. W. Hollis. 1991. Serum 1,25 dihydroxyvitamin D$_3$ (soltriol) levels influence serotonin levels in the hypothalamus of the rat. *Abstracts—Society for Neuroscience* 17:498.

Stein, M.S., S. C. Scherer, K. S. Ladd, and L. C. Harrisson. 2011. A randomized controlled trial of high-dose vitamin D2 followed by intranasal insulin in Alzheimer's disease. *Journal of Alzheimer's Disease* 26 (3):477–484.

Sutherland, M. K., M. J. Somerville, L. K. Yoong, C. Bergeron, M. R. Haussler, and D. R. McLachlan. 1992. Reduction of vitamin D hormone receptor mRNA levels in Alzheimer as compared to Huntington hippocampus: Correlation with calbindin-28k mRNA levels. *Brain Research. Molecular Brain Research* 13 (3):239–250.

Sutra Del Galy, A., M. Bertrand, F. Bigot, P. Abraham, R. Thomlinson, M. Paccalin, O. Beauchet, and C. Annweiler. 2009. Vitamin D insufficiency and acute care in geriatric inpatients. *Journal of the American Geriatrics Society* 57 (9):1721–1723.

Van Lente, F. 2000. Markers of inflammation as predictors in cardiovascular disease. *Clinica Chimica Acta* 293 (1–2):31–52.

Veenstra, T. D., K. Prufer, C. Koenigsberger, S. W. Brimijoin, J. P. Grande, and R. Kumar. 1998. 1,25-Dihydroxyvitamin D_3 receptors in the central nervous system of the rat embryo. *Brain Research* 804 (2):193–205.

Walbert, T., G. F. Jirikowski, and K. Prufer. 2001. Distribution of 1,25-dihydroxyvitamin D_3 receptor immuno-reactivity in the limbic system of the rat. *Hormone and Metabolic Research* 33 (9):525–531.

Wang, Y., Y. H. Chiang, T. P. Su, T. Hayashi, M. Morales, B. J. Hoffer, and S. Z. Lin. 2000. Vitamin $D_{(3)}$ attenuates cortical infarction induced by middle cerebral arterial ligation in rats. *Neuropharmacology* 39 (5):873–880.

Wang, J. Y., J. N. Wu, T. L. Cherng, B. J. Hoffer, H. H. Chen, C. V. Borlongan, and Y. Wang. 2001. Vitamin $D_{(3)}$ attenuates 6-hydroxydopamine–induced neurotoxicity in rats. *Brain Research* 904 (1):67–75.

Warrington, E. K. 1966. *The Camden Memory Tests Manual.* Hove: UK: Psychology Press.

Wechsler, D. 1987. *Wechsler Memory Scale-Revised. Manual.* New York: The Psychological Corporation.

Wechsler, D. 1997. *Manual for the Wechsler Memory Scale—Third Edition (WMS-III)*, 3rd ed. San Antonio, TX: The Psychological Corporation.

Wilkins, C. H., Y. I. Sheline, C. M. Roe, S. J. Birge, and J. C. Morris. 2006. Vitamin D deficiency is associated with low mood and worse cognitive performance in older adults. *The American Journal of Geriatric Psychiatry* 14 (12):1032–1040.

Wilkins, C. H., S. J. Birge, Y. I. Sheline, and J. C. Morris. 2009. Vitamin D deficiency is associated with worse cognitive performance and lower bone density in older African Americans. *Journal of the National Medical Association* 101 (4):349–354.

Wilson, J. M., and Y. G. Jungner. 1968. [Principles and practice of mass screening for disease]. *Boletín de la Oficina Sanitaria Panamericana* 65 (4):281–393

Woodhouse, P. R., K. T. Khaw, and M. Plummer. 1993. Seasonal variation of blood pressure and its relationship to ambient temperature in an elderly population. *Journal of Hypertension* 11 (11):1267–1274.

World Health Organization. 2000. The World Health Report 2000. Health systems: Improving performance. Geneva, Switzerland.

Yoshizawa, T., Y. Handa, Y. Uematsu, S. Takeda, K. Sekine, Y. Yoshihara, T. Kawakami, K. Arioka, H. Sato, Y. Uchiyama, S. Masushige, A. Fukamizu, T. Matsumoto, and S. Kato. 1997. Mice lacking the vitamin D receptor exhibit impaired bone formation, uterine hypoplasia and growth retardation after weaning. *Nature Genetics* 16 (4):391–396.

Yu, J., M. Gattoni-Celli, H. Zhu, N. R. Bat, K. Sambamurti, S. Gattoni-Celli, M. S. Kindy. 2011. Vitamin D3–enriched diet correlates with a decrease of amyloid plaques in the brain of AbetaPP transgenic mice. *Journal of Alzheimer's Disease* 25 (2):295–307.

Zehnder, D., R. Bland, R. S. Chana, D. C. Wheeler, A. J. Howie, M. C. Williams, P. M. Stewart, and M. Hewison. 2002. Synthesis of 1,25-dihydroxyvitamin $D_{(3)}$ by human endothelial cells is regulated by inflammatory cytokines: A novel autocrine determinant of vascular cell adhesion. *Journal of the American Society of Nephrology* 13 (3):621–629.

Zittermann, A. 2003. Vitamin D in preventive medicine: Are we ignoring the evidence? *The British Journal of Nutrition* 89 (5):552–572.

Zou, J., A. Minasyan, T. Keisala, Y. Zhang, J. H. Wang, Y. R. Lou, A. Kalueff, I. Pyykko, and P. Tuohimaa. 2008. Progressive hearing loss in mice with a mutated vitamin D receptor gene. *Audiology & Neuro-Otology* 13 (4):219–230.

Index

Note: Page numbers followed by "*f*" indicate figure; those followed by "*t*" indicate table.

Printed and bound by CPI Group (UK) Ltd, Croydon, CR0 4YY

21/10/2024

01777040-0013